AF358827

Abiotic Stresses in Plants: From Molecules to Environment

Abiotic Stresses in Plants: From Molecules to Environment

Editor

Martin Bartas

Basel • Beijing • Wuhan • Barcelona • Belgrade • Novi Sad • Cluj • Manchester

Editor
Martin Bartas
Department of Biology
University of Ostrava
Ostrava
Czech Republic

Editorial Office
MDPI AG
Grosspeteranlage 5
4052 Basel, Switzerland

This is a reprint of articles from the Special Issue published online in the open access journal *International Journal of Molecular Sciences* (ISSN 1422-0067) (available at: https://www.mdpi.com/journal/ijms/special_issues/0GR24E915V).

For citation purposes, cite each article independently as indicated on the article page online and as indicated below:

Lastname, A.A.; Lastname, B.B. Article Title. *Journal Name* **Year**, *Volume Number*, Page Range.

ISBN 978-3-7258-1948-5 (Hbk)
ISBN 978-3-7258-1947-8 (PDF)
doi.org/10.3390/books978-3-7258-1947-8

Cover image courtesy of Nela Klusová

Contents

About the Editor

Martin Bartas

Martin Bartas (born 1992 in Karviná, Czech Republic) is a postdoctoral researcher affiliated with the University of Ostrava, Czech Republic. He received his Ph.D. degree in 2020 and currently is working at the Department of Biology, Faculty of Science, with a focus on various questions of molecular biology. His passion for science and discoveries stems from his curious and playful mind, often covering a range of relatively diverse fields, from plants to viruses to animal longevity. In his research, he uses mainly methods of bioinformatics and computational biology, but nonetheless sometimes also likes to pipette a little bit. In his free time, he likes to do sports (running, cycling, hiking, and swimming), learn new skills, and travel. Martin is an open-minded person and always welcomes new research cooperations and exchange of ideas.

Preface

Thank you for opening this reprint, which is of a Special Issue entitled "Abiotic Stresses in Plants: From Molecules to Environment". Plants must contend with a diverse range of abiotic stresses, including drought, extreme temperatures, high salinity, waterlogging, nutrient imbalances, and numerous other factors. If we want to understand these processes, we shall dive into the molecular level and start our journey right here, going from nucleic acids to proteins, from cells to tissues, and from "trees to the forest", illuminating fantastic and still undescribed natural laws and interplays. Research focused on abiotic stresses also provides valuable insights into how plants respond to environmental changes, offering solutions to mitigate the effects of climate change on agriculture and ecosystems. As the focus of this Special Issue is quite complex and exceeds the restricted space of this Preface, a brief introduction to the problematics can be found a few pages down in my Editorial article. Whether you are a specialist in plant research or a random guest in this exciting field of science, allow me to wish you a pleasant experience.

In this place, I would like to express my deepest gratitude to all contributors, which comprise both renowned authors and young scientists at the beginning of their careers. When I browsed through their affiliations covering extensive and diverse sets of countries and institutions, I more than ever realized that a passion for making discoveries and treading new avenues is something we have in common, not only scientists but all curious people in general, and that the passion for science is in fact a passion for peace and love, from people for people, bridging gaps, and helping others prosper.

My sincere thanks and respect deserve also all my past teachers, current colleagues, and friends, not only at my alma mater (University of Ostrava), but also at the Institute of Biophysics in Brno, Max Planck Institute for Multidisciplinary Sciences in Göttingen, University of Cambridge, and all others.

I would like to especially acknowledge Ms. Nela Klusová, who is affiliated with the University of Ostrava, Department of Physics, for providing microphotography depicting a cross-section of the petiole from *Ficus lyrata*.

Last but not least, my grateful thanks go to my mom Andrea, dad Robert, and the rest of my family, including dwarf dachshund Rony, as without their continual support only a little of what I am doing would become possible.

Martin Bartas
Editor

International Journal of
Molecular Sciences

MDPI

Editorial

Abiotic Stresses in Plants: From Molecules to Environment

Martin Bartas

Department of Biology and Ecology, Faculty of Science, University of Ostrava, 710 00 Ostrava, Czech Republic; martin.bartas@osu.cz

Citation: Bartas, M. Abiotic Stresses in Plants: From Molecules to Environment. *Int. J. Mol. Sci.* **2024**, *25*, 8072. https://doi.org/10.3390/ijms25158072

Received: 19 July 2024
Accepted: 23 July 2024
Published: 24 July 2024

1. Foreword

Plants face several challenges during their growth and development, including environmental factors (mainly abiotic ones), that can lead to/induce oxidative stress—specifically, adverse temperatures (both hot and cold), drought, salinity, radiation, nutrient deficiency (or excess), toxic metals, waterlogging, air pollution, and mechanical stimuli. With the increasing frequency of extreme weather events due to ongoing climate change, understanding how plants sense and adapt to these stresses will become more and more crucial for ensuring food security and sustainable agriculture. In this Special Issue, a diverse array of original articles was released, encompassing multidisciplinary approaches and methods of bioinformatics, in vitro and in vivo experiments, and innovative field trials, some of which are poised for practical implementation. Additionally, a collection of comprehensive reviews dedicated to various aspects of the issue was also published. This editorial aims to provide a concise entry point into the problems in this area and introduce the reader to the newest advances in this exciting field, where basic scientific discoveries meet future applications.

2. Abiotic Stresses in Plants

All living organisms must cope with various stresses during their lifecycle. In contrast to most animals, plants must withstand adverse conditions in the place where they were "born"; they cannot quickly escape from sun to shadow, from nutrient-poor spot to nutrient-rich venue, or from flooded space to safety. Therefore, during evolution, plants acquired what is likely one of the most sophisticated mechanisms for dealing with a variety of abiotic stresses [1], which are schematically depicted in Figure 1. It is worth mentioning that an extreme condition for one plant species might be an optimal condition for another.

Plants exhibit an ability to adapt to a wide range of temperatures by reprogramming their transcriptome, proteome, and metabolome [2,3]. *Temperature stress* covers both cold and heat stress; interestingly, the involved molecular pathways are distinct to some extent, leading to the expression of either cold-responsive or heat-responsive genes [2].

Drought stress in plants significantly disrupts the balance between water uptake and loss. Subsequent physiological and molecular responses aim to conserve water and tolerate stress via stomatal closure, reduced cell expansion, altered photosynthetic activity, or synthesis of osmoprotectants [4] to mitigate the damaging effects of drought stress and enhance plants' survival in water-deficient environments.

Salinity stress in plants arises from excessive salt accumulation in the soil. When exposed to salinity, plants activate homeostatic mechanisms to counter the imbalance caused by salt. This activation involves triggering various signaling components, including the Salt Overly Sensitive (SOS) pathway, the abscisic acid (ABA) pathway, reactive oxygen species (ROS), and osmotic stress signaling, to mitigate the effects of salinity [5].

Radiation stress in plants usually refers to exposure to various excessive sources of radiation (including gamma rays and ultraviolet (UV) radiation) that can induce DNA damage, hinder photosynthesis, and impair overall plant growth and development [6]. Photoreceptor protein UVR8 is considered a main sensing component for both UV-B and

UV-A [7]. Subsequently, plants respond to radiation stress by activating various protective mechanisms, including the synthesis of UV-absorbing compounds and the enhancement of antioxidant defense systems [8]. In addition, the spectral quality and quantity of visible light affect plant growth and stress reactions [9,10].

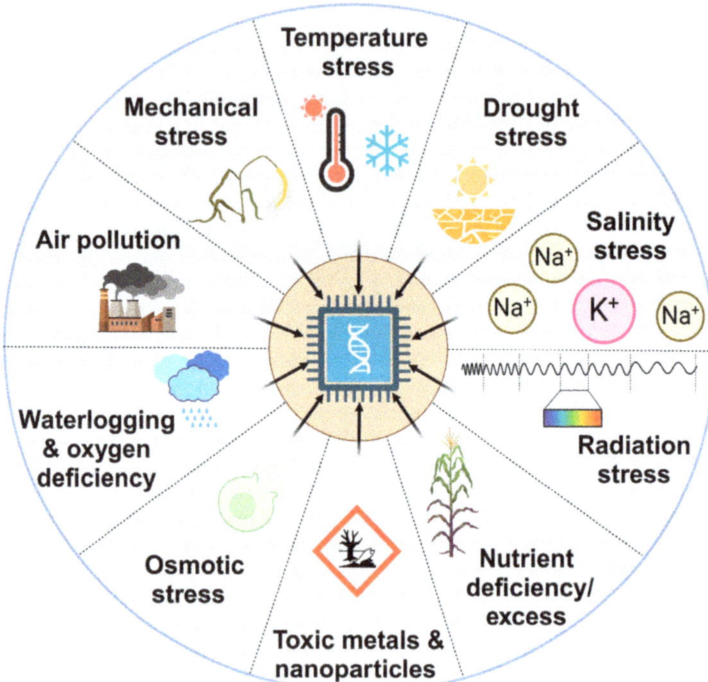

Figure 1. Abiotic stresses in plants. Diagram of most important abiotic stresses in plants (outer circle) which, sooner or later, elicit an integrative response, often involving complex signaling pathways and proteins, interacting with DNA, and inducing the expression of target "stress-responsive genes" (inner circle). This figure was created in BioRender.

Nutrient deficiency stress refers to plants lacking a sufficient quantity and quality (e.g., redox state and form of compound/salt) of essential nutrients required for their growth and development. Plants respond to this through a range of adaptive mechanisms, including altered root architecture [11], increased nutrient uptake efficiency, and the reprogramming of metabolic pathways [12]. Additionally, nutrient deficiency stress can impact the synthesis of important biomolecules, such as chlorophyll and proteins, further influencing plant growth. Generally, nutrients in plants exist in three states: deficiency, optimality, and excess [13]. The latter can be damaging not only to the plant but also to the local soil microenvironment and plant symbionts.

Stress induced by toxic metals in plants occurs when exposure to excessive levels of metals such as cadmium, lead, or arsenic disrupts essential physiological processes, leading to oxidative stress, growth inhibition, and cellular damage [14]. Plants respond to this by activating detoxification mechanisms, sequestering toxic metals in vacuoles, and producing chelating compounds [15]. *Nanoparticles* are also capable of inducing oxidative stress and adverse effects in plants [16]; however, some of them can also have a beneficial effect on plant growth, such as in cross-tolerance (when plants are exposed to a particular type of stress, it can trigger shared signals and pathways, leading to increased resilience against diverse forms of stress [17]) and related mitigation of stress induced by high salinity in the environment [18].

Osmotic stress occurs when plants are subjected to conditions that disrupt the balance of water and solutes within their cells, leading to significant changes in water potential (ψw) [19]. It can occur at the initial stage of plant exposure to other abiotic stresses and induces specific kinases, such as the SNF1-related protein kinase (SnRK) 2 family, which is essential for osmotic stress signaling [20]. Osmotic stress often leads to physiological damage, impairs ABA-mediated stomatal closing, and decreases survival in drought and salinity stresses [20].

The main cause of *oxygen deficiency stress* in plants is environmental factors such as excessive rainfall leading to soil waterlogging or complete submergence of plants, which can result in a shortage of oxygen (hypoxia) or a total absence of oxygen (anoxia) [21,22]. Under these conditions, a series of rapid molecular and metabolic responses are activated to endure such stress, including the induction of specific kinases and the activation of various cellular signaling networks to cope with the low-oxygen environment [23].

Air pollution stress (represented mainly by ozone, nitrogen oxides, and suspended particulate pollutants) increases the production of reactive oxygen species (ROS) in plants, leading to oxidative stress and potential damage at various levels (including the disruption of essential functions like transpiration, reduced resistance to heat stress, nutrient deficiencies, and subsequent illnesses) [24]. Furthermore, air pollution can shift the competitive balance among various plant species and may lead to changes in the composition of the plant community, altering the structure and function of the ecosystem [25].

Last but not least, *mechanical stress* encompasses a range of physical forces such as wind, touch, heavy rain, or vibrations, which can significantly influence plant growth and development [26]. These stresses trigger a variety of responses, including changes in gene expression, hormone signaling, and cell wall composition, to help the plant adapt and withstand these mechanical forces. Additionally, mechanical stress can lead to alterations in plant architecture, including stem thickening, root system modification, and changes in leaf orientation, as the plant adapts to its environment and attempts to minimize damage from external forces [27]. Interestingly, herbivores can also induce significant mechanical stress in plants, e.g., through feeding or trampling, prompting the production of "bitter" secondary metabolites that serve as a defense mechanism against further herbivory [28].

It is worth mentioning that not all stresses are equal. Wild plants in their natural habitats are usually *adapted* to the specific extreme conditions of a particular environment and have developed very effective protective mechanisms. On the other hand, cultural plants/crops often have to deal with severe abiotic stresses which are partly a consequence of unsustainable agriculture and accompanying phenomena, such as soil erosion and loss of groundwater [29,30].

From a basic research perspective, there is still a limited understanding of the combined effects of two or more abiotic stresses on plant physiology and development, as well as their synergistic or antagonistic impacts [31,32]. Moreover, the combination of abiotic and biotic stresses can lead to even more intricate interactions, affecting the plant's molecular and metabolic responses [33].

Abiotic stresses in plants represent one of the main challenges in current agriculture. Due to ongoing climate change, there is a need for a deeper understanding of how plants deal with various environmental cues, especially on a molecular level. The recent integration of multidisciplinary approaches, such as omics technologies, bioinformatics, and computational modeling, is facilitating a comprehensive understanding of the complex interactions between plants and their environment [34], pointing to the original name of this Special Issue, "Abiotic Stresses in Plants: From Molecules to Environment".

3. Short Overview and Summary of Published Articles

Let us delve into specific articles featured in this Special Issue and highlight their key focus and main message. I encourage readers to peruse this list and explore the full papers that pique their interest.

A short communication by Bartosz J. Płachno et al. highlights the effectiveness of the continuous and well-developed cuticle in *Drosophyllum lusitanicum* for water conservation in arid environments and provides insights into the challenges and techniques associated with studying the composition of cell walls in glands of this carnivorous plant.

Yucun Yang et al. developed a rapid and nondestructive method based on hyperspectral imaging for the evaluation of wheat (*Triticum aestivum* L.) chlorophyll under drought stress, which may be particularly useful for high-throughput phenotypic analysis and the genetic breeding of crops in general.

The article by Hongying Chen, Anne M. Visscher et al. investigates the intra-speci variation in desiccation tolerance of *Citrus sinensis* 'bingtangcheng' (L.) seeds, highlighting the influence of maternal environmental conditions, such as annual sunshine hours and temperature, on seed survival during dehydration, as well as the stable expression levels of stress-responsive genes in modulating desiccation tolerance.

Research conducted by Lixian Wei, Xin Zhao et al. identified 17 Caffeoyl-coenzyme A-O-methyltransferase encoding genes (*CCoAOMTs*) in *Dendrocalamus farinosus*, revealing that they are significantly associated with lignin content, xylem thickness, and drought resistance in transgenic plants. This posits them as potential candidate genes involved in the drought response and lignin synthesis pathway in plants, with potential implications for the genetic improvement of important traits in *Dendrocalamus farinosus* and other species.

A study by Jinghan Peng, Siyu Liu et al., based mostly on bioinformatics, facilitated a comprehensive identification of twenty *HSP90* genes in oats (*Avena sativa*), elucidating their evolutionary pathways and responses to various abiotic stresses and demonstrating the significant up-regulation of specific *HSP90* genes under heat stress.

Dariel López et al. demonstrate that nocturnal warming has a more pronounced impact than diurnal warming on the cold deacclimation process in *Deschampsia antarctica*, influencing the expression of freezing tolerance-related genes and the plant's response to cold stress. These findings provide valuable insights into the effects of asymmetric warming on plants' adaptive responses and survival in a changing environment.

The research by Mason T. MacDonald et al. uncovers the relationship between lipid changes during cold acclimation and postharvest needle abscission in balsam fir (*Abies balsamea*), highlighting the practical significance of identifying peak needle retention and suggesting the potential use of the monogalactosyldiacylglycerol (MGDG) to digalactosyldiacylglycerol (DGDG) ratio as a screening tool for balsam fir genotypes with stronger needle retention characteristics.

The genome-wide analysis conducted by Jiafang Shen et al. identified seven genes encoding sucrose phosphate synthase (*SPS*) in soybean (*Glycine max*). The authors then investigated their tissue expression patterns and cold stress response. The study highlights the potential roles of *SPS* genes in soybean's response to cold stress and their implications for future research in this area.

Valentin Ambroise et al. investigated the impact of long-term mild metal exposure on the cold acclimation of *Salix viminalis* roots and revealed that simultaneous exposure to both stressors results in unique effects. The study also highlights the under-studied role of lignans and the ROS damage repair and removal system in plants facing multiple stressors.

The study by Xuan Ma, Qiang Zhang et al. identified the molecular basis of salt resistance in two poplar species (*Populus alba* and *Populus russkii*), revealing that *Populus alba* exhibited enhanced energy metabolism, superior Na$^+$ transportation, different regulation of stress-related genes, and an overall improved ability to thrive under salt stress.

Yong Chen, Wanling Yang et al. focused their efforts on the miRNA-mediated salt stress response in Dongxiang wild rice (*Oryza rufipogon* Griff.), identifying 874 known and 476 novel miRNAs. This sheds light on the regulatory mechanisms of salt stress tolerance in this wild rice, offering insights that could potentially improve salt tolerance in cultivated rice in the future.

The study by Ziwei Li et al. sheds light on the regulation mechanism of a polymer soil amendment (PPM) on rapeseed (*Brassica napus* L.) photosynthesis under drip irrigation with brackish water. The research highlights the differential responses of rapeseed phenotype, photosynthetic physiology, transcriptomics, and metabolomics to PPM application with different types of brackish water, providing insights for addressing salinity stress in crops caused by drip irrigation with brackish water.

The study by Mingjie Ren, Jingjing Ma et al. unveils the molecular mechanisms behind Magnolia's (*Magnolia sinostellata*) sensitivity to shade, highlighting the critical role of the *STAY-GREEN* gene in regulating chlorophyll degradation under light deficiency stress.

Miho Ohnishi, Shu Maekawa et al. author two articles in this issue. The first investigates the activity of ferredoxin (Fd)-dependent cyclic electron flow (Fd-CEF) around photosystem I (PSI) in intact leaves of *Arabidopsis thaliana*, highlighting the relationship between the oxidation rate of Fd reduced by PSI (vFd) and photosynthetic linear electron flow activity and discussing the physiological significance of the excessive vFd observed under higher photosynthesis conditions. The second (follow-up) study is devoted to the enhanced reduction of Fd in the PGR5-deficient mutant of *Arabidopsis thaliana*, which is stimulated by Fd-CEF around PSI. The authors found that the protein PROTON GRADIENT REGULATION 5 (PGR5) is probably not catalyze Fd-CEF, which challenges previous assumptions about its role in this process.

The study by Karolina Stałanowska et al. reveals the size-dependent harmful effects of zinc oxide (ZnO) nanoparticles on the germination and seedling growth of garden pea (*Pisum sativum* L.) and wheat (*Triticum aestivum* L.), emphasizing the potential phytotoxic risks associated with soil contamination by ZnO nanoparticles.

The study by Sandra S. Scholz et al. demonstrates that the root-colonizing endophytic fungus *Piriformospora indica* facilitates the metabolomic adaptation of its host plant (*Arabidopsis thaliana*) against nitrogen limitation by delivering reduced nitrogen metabolites, thereby alleviating nitrogen starvation responses and reprogramming the expression of nitrogen metabolism-related genes.

Vida Nasrollahi et al. investigated the role of Squamosa-Promoter Binding Protein-Like 9 (SPL9) in nodulation in the important crop alfalfa (*Medicago sativa*), revealing that MsSPL9 regulates nodulation under high-nitrate conditions by modulating the expression of nitrate-responsive genes.

The study by Ying Guo, Yongli Qi et al. focuses on grafted ginkgo trees (*Ginkgo biloba* L.) in different regions of China and involves RNA-seq, small RNA-seq, and metabolomics data to elucidate the regulatory mechanism of terpene trilactone (TTL) biosynthesis in ginkgo leaves. The authors identified differentially expressed miRNAs and transcription factors that play a role in the environmental response and provide insights for improving the medicinal value of ginkgo leaves under global climate change.

The research by Xi-Min Zhang et al., which focused on *Rhododendron delavayi* under waterlogging stress, revealed significant physiological and molecular changes, including significantly reduced CO_2 assimilation and chlorophyll content, excessive H_2O_2 accumulation, and altered metabolic pathways.

The review by Bowen Tan and Sixue Chen focuses on the mechanisms of the C3-to-CAM photosynthesis transition, an adaptive form of photosynthesis in hot and arid regions, which provides promising solutions for creating stress-resilient crops in the face of global climate challenges.

The review by Shuya Tan, Yueqi Sha et al. provides an overview of the molecular mechanisms triggering leaf senescence under abiotic stresses, emphasizing the significance of understanding these processes to enhance crop resilience and productivity in unfavorable environments.

The review by Li Yang et al. highlights the impact of salinity on cotton growth and fiber yield, focusing on the roles of S-adenosylmethionine in ethylene (ET) and polyamine

(PA) biosynthesis and signal transduction pathways. Improved regulatory pathways of ET and PAs under salt stress in cotton could be potentially used for the breeding of salt-tolerant varieties.

The last published review is authored by Adriana Volná et al. and provides insights into the regulation of genes related to phenolic compound production concerning the full route from photoperception to transcription control and gene expression. Such knowledge enhances our understanding of plant responses to UV and visible light and offers potential avenues for manipulating the content and profile of phenolic compounds, with implications for horticulture and food production.

4. Conclusions

In this Special Issue, a total of 25 articles were published, completely outperforming my modest expectations (List of Contributions). Out of this number, there were 20 original research articles, 4 comprehensive reviews, and this editorial, which is slowly concluding. I hope that the readers of this Special Issue find the articles published herein useful and inspirational; the field of *Abiotic stresses in plants* is highly dynamic and rapidly evolving, which provides many opportunities for young scientists to find their research niche.

Funding: This work was supported by the European Union under the LERCO project number CZ.10.03.01/00/22_003/0000003 via the Operational Programme Just Transition.

Acknowledgments: I would like to express sincere gratitude to all contributors who decided to send their excellent research work to this Special Issue. My cordial thanks are devoted to Adriana Volná and Jiří Červeň for their kind help with manuscript proofreading.

Conflicts of Interest: The author declares no conflicts of interest.

List of Contributions:

1. Ambroise, V.; Legay, S.; Jozefczak, M.; Leclercq, C.C.; Planchon, S.; Hausman, J.-F.; Renaut, J.; Cuypers, A.; Sergeant, K. Impact of Heavy Metals on Cold Acclimation of Salix Viminalis Roots. *Int. J. Mol. Sci.* **2024**, *25*, 1545. https://doi.org/10.3390/ijms25031545.
2. Guo, Y.; Qi, Y.; Feng, Y.; Yang, Y.; Xue, L.; El-Kassaby, Y.A.; Wang, G.; Fu, F. Inferring the Regulatory Network of miRNAs on Terpene Trilactone Biosynthesis Affected by Environmental Conditions. *Int. J. Mol. Sci.* **2023**, *24*, 17002. https://doi.org/10.3390/ijms242317002.
3. Chen, H.; Visscher, A.M.; Ai, Q.; Yang, L.; Pritchard, H.W.; Li, W. Intra-Specific Variation in Desiccation Tolerance of *Citrus sinensis* 'bingtangcheng' (L.) Seeds under Different Environmental Conditions in China. *Int. J. Mol. Sci.* **2023**, *24*, 7393. https://doi.org/10.3390/ijms24087393.
4. Chen, Y.; Yang, W.; Gao, R.; Chen, Y.; Zhou, Y.; Xie, J.; Zhang, F. Genome-Wide Analysis of microRNAs and Their Target Genes in Dongxiang Wild Rice (*Oryza rufipogon* Griff.) Responding to Salt Stress. *Int. J. Mol. Sci.* **2023**, *24*, 4069. https://doi.org/10.3390/ijms24044069.
5. Li, Z.; Fan, H.; Yang, L.; Wang, S.; Hong, D.; Cui, W.; Wang, T.; Wei, C.; Sun, Y.; Wang, K.; et al. Multi-Omics Analysis of the Effects of Soil Amendment on Rapeseed (*Brassica napus* L.) Photosynthesis under Drip Irrigation with Brackish Water. *Int. J. Mol. Sci.* **2024**, *25*, 2521. https://doi.org/10.3390/ijms25052521.
6. López, D.; Larama, G.; Sáez, P.L.; Bravo, L.A. Transcriptome Analysis of Diurnal and Nocturnal-Warmed Plants, the Molecular Mechanism Underlying Cold Deacclimation Response in *Deschampsia antarctica*. *Int. J. Mol. Sci.* **2023**, *24*, 11211. https://doi.org/10.3390/ijms241311211.
7. Ma, X.; Zhang, Q.; Ou, Y.; Wang, L.; Gao, Y.; Lucas, G.R.; Resco de Dios, V.; Yao, Y. Transcriptome and Low-Affinity Sodium Transport Analysis Reveals Salt Tolerance Variations between Two Poplar Trees. *Int. J. Mol. Sci.* **2023**, *24*, 5732. https://doi.org/10.3390/ijms24065732.
8. MacDonald, M.T.; Lada, R.R.; MacDonald, G.E.; Caldwell, C.D.; Udenigwe, C.C. Changes in Polar Lipid Composition in Balsam Fir during Seasonal Cold Acclimation and Relationship to Needle Abscission. *Int. J. Mol. Sci.* **2023**, *24*, 15702. https://doi.org/10.3390/ijms242115702.
9. Maekawa, S.; Ohnishi, M.; Wada, S.; Ifuku, K.; Miyake, C. Enhanced Reduction of Ferredoxin in PGR5-Deficient Mutant of *Arabidopsis thaliana* Stimulated Ferredoxin-Dependent Cyclic Electron Flow around Photosystem I. *Int. J. Mol. Sci.* **2024**, *25*, 2677. https://doi.org/10.3390/ijms25052677.

10. Nasrollahi, V.; Allam, G.; Kohalmi, S.E.; Hannoufa, A. MsSPL9 Modulates Nodulation under Nitrate Sufficiency Condition in Medicago Sativa. *Int. J. Mol. Sci.* **2023**, *24*, 9615. https://doi.org/10.3390/ijms24119615.

11. Ohnishi, M.; Maekawa, S.; Wada, S.; Ifuku, K.; Miyake, C. Evaluating the Oxidation Rate of Reduced Ferredoxin in *Arabidopsis thaliana* Independent of Photosynthetic Linear Electron Flow: Plausible Activity of Ferredoxin-Dependent Cyclic Electron Flow around Photosystem I. *Int. J. Mol. Sci.* **2023**, *24*, 12145. https://doi.org/10.3390/ijms241512145.

12. Peng, J.; Liu, S.; Wu, J.; Liu, T.; Liu, B.; Xiong, Y.; Zhao, J.; You, M.; Lei, X.; Ma, X. Genome-Wide Analysis of the Oat (*Avena sativa*) HSP90 Gene Family Reveals Its Identification, Evolution, and Response to Abiotic Stress. *Int. J. Mol. Sci.* **2024**, *25*, 2305. https://doi.org/10.3390/ijms25042305.

13. Płachno, B.J.; Kapusta, M.; Stolarczyk, P.; Świątek, P. Do Cuticular Gaps Make It Possible to Study the Composition of the Cell Walls in the Glands of *Drosophyllum lusitanicum*? *Int. J. Mol. Sci.* **2024**, *25*, 1320. https://doi.org/10.3390/ijms25021320.

14. Ren, M.; Ma, J.; Lu, D.; Wu, C.; Zhu, S.; Chen, X.; Wu, Y.; Shen, Y. STAY-GREEN Accelerates Chlorophyll Degradation in *Magnolia sinostellata* under the Condition of Light Deficiency. *Int. J. Mol. Sci.* **2023**, *24*, 8510. https://doi.org/10.3390/ijms24108510.

15. Shen, J.; Xu, Y.; Yuan, S.; Jin, F.; Huang, Y.; Chen, H.; Shan, Z.; Yang, Z.; Chen, S.; Zhou, X.; et al. Genome-Wide Identification of GmSPS Gene Family in Soybean and Expression Analysis in Response to Cold Stress. *Int. J. Mol. Sci.* **2023**, *24*, 12878. https://doi.org/10.3390/ijms241612878.

16. Scholz, S.S.; Barth, E.; Clément, G.; Marmagne, A.; Ludwig-Müller, J.; Sakakibara, H.; Kiba, T.; Vicente-Carbajosa, J.; Pollmann, S.; Krapp, A.; et al. The Root-Colonizing Endophyte *Piriformospora indica* Supports Nitrogen-Starved *Arabidopsis thaliana* Seedlings with Nitrogen Metabolites. *Int. J. Mol. Sci.* **2023**, *24*, 15372. https://doi.org/10.3390/ijms242015372.

17. Stałanowska, K.; Szablińska-Piernik, J.; Okorski, A.; Lahuta, L.B. Zinc Oxide Nanoparticles Affect Early Seedlings' Growth and Polar Metabolite Profiles of Pea (*Pisum sativum L.*) and Wheat (*Triticum aestivum L.*). *Int. J. Mol. Sci.* **2023**, *24*, 14992. https://doi.org/10.3390/ijms241914992.

18. Tan, B.; Chen, S. Defining Mechanisms of C_3 to CAM Photosynthesis Transition toward Enhancing Crop Stress Resilience. *Int. J. Mol. Sci.* **2023**, *24*, 13072. https://doi.org/10.3390/ijms241713072.

19. Tan, S.; Sha, Y.; Sun, L.; Li, Z. Abiotic Stress-Induced Leaf Senescence: Regulatory Mechanisms and Application. *Int. J. Mol. Sci.* **2023**, *24*, 11996. https://doi.org/10.3390/ijms241511996.

20. Volná, A.; Červeň, J.; Nezval, J.; Pech, R.; Špunda, V. Bridging the Gap: From Photoperception to the Transcription Control of Genes Related to the Production of Phenolic Compounds. *Int. J. Mol. Sci.* **2024**, *25*, 7066. https://doi.org/10.3390/ijms25137066.

21. Wei, L.; Zhao, X.; Gu, X.; Peng, J.; Song, W.; Deng, B.; Cao, Y.; Hu, S. Genome-Wide Identification and Expression Analysis of Dendrocalamus Farinosus CCoAOMT Gene Family and the Role of DfCCoAOMT14 Involved in Lignin Synthesis. *Int. J. Mol. Sci.* **2023**, *24*, 8965. https://doi.org/10.3390/ijms24108965.

22. Yang, L.; Wang, X.; Zhao, F.; Zhang, X.; Li, W.; Huang, J.; Pei, X.; Ren, X.; Liu, Y.; He, K.; et al. Roles of S-Adenosylmethionine and Its Derivatives in Salt Tolerance of Cotton. *Int. J. Mol. Sci.* **2023**, *24*, 9517. https://doi.org/10.3390/ijms24119517.

23. Yang, Y.; Nan, R.; Mi, T.; Song, Y.; Shi, F.; Liu, X.; Wang, Y.; Sun, F.; Xi, Y.; Zhang, C. Rapid and Nondestructive Evaluation of Wheat Chlorophyll under Drought Stress Using Hyperspectral Imaging. *Int. J. Mol. Sci.* **2023**, *24*, 5825. https://doi.org/10.3390/ijms24065825.

24. Zhang, X.-M.; Duan, S.-G.; Xia, Y.; Li, J.-T.; Liu, L.-X.; Tang, M.; Tang, J.; Sun, W.; Yi, Y. Transcriptomic, Physiological, and Metabolomic Response of an Alpine Plant, *Rhododendron delavayi*, to Waterlogging Stress and Post-Waterlogging Recovery. *Int. J. Mol. Sci.* **2023**, *24*, 10509. https://doi.org/10.3390/ijms241310509.

References

1. Des Marais, D.L.; Juenger, T.E. Pleiotropy, Plasticity, and the Evolution of Plant Abiotic Stress Tolerance. *Ann. N. Y. Acad. Sci.* **2010**, *1206*, 56–79. [CrossRef] [PubMed]

2. Ding, Y.; Yang, S. Surviving and Thriving: How Plants Perceive and Respond to Temperature Stress. *Dev. Cell* **2022**, *57*, 947–958. [CrossRef]

3. Guy, C.; Kaplan, F.; Kopka, J.; Selbig, J.; Hincha, D.K. Metabolomics of Temperature Stress. *Physiol. Plant.* **2008**, *132*, 220–235. [CrossRef] [PubMed]

4. Ozturk, M.; Turkyilmaz Unal, B.; García-Caparrós, P.; Khursheed, A.; Gul, A.; Hasanuzzaman, M. Osmoregulation and Its Actions during the Drought Stress in Plants. *Physiol. Plant.* **2021**, *172*, 1321–1335. [CrossRef]

5. Rai, G.K.; Mishra, S.; Chouhan, R.; Mushtaq, M.; Chowdhary, A.A.; Rai, P.K.; Kumar, R.R.; Kumar, P.; Perez-Alfocea, F.; Colla, G. Plant Salinity Stress, Sensing, and Its Mitigation through WRKY. *Front. Plant Sci.* **2023**, *14*, 1238507. [CrossRef]
6. Kovács, E.; Keresztes, Á. Effect of Gamma and UV-B/C Radiation on Plant Cells. *Micron* **2002**, *33*, 199–210. [CrossRef] [PubMed]
7. Rai, N.; O'Hara, A.; Farkas, D.; Safronov, O.; Ratanasopa, K.; Wang, F.; Lindfors, A.V.; Jenkins, G.I.; Lehto, T.; Salojärvi, J.; et al. The Photoreceptor UVR8 Mediates the Perception of Both UV-B and UV-A Wavelengths up to 350 Nm of Sunlight with Responsivity Moderated by Cryptochromes. *Plant Cell Environ.* **2020**, *43*, 1513–1527. [CrossRef] [PubMed]
8. Shi, C.; Liu, H. How Plants Protect Themselves from Ultraviolet-B Radiation Stress. *Plant Physiol.* **2021**, *187*, 1096–1103. [CrossRef] [PubMed]
9. Yavari, N.; Tripathi, R.; Wu, B.-S.; MacPherson, S.; Singh, J.; Lefsrud, M. The Effect of Light Quality on Plant Physiology, Photosynthetic, and Stress Response in Arabidopsis Thaliana Leaves. *PLoS ONE* **2021**, *16*, e0247380. [CrossRef]
10. Volná, A.; Červeň, J.; Nezval, J.; Pech, R.; Špunda, V. Bridging the Gap: From Photoperception to the Transcription Control of Genes Related to the Production of Phenolic Compounds. *Int. J. Mol. Sci.* **2024**, *25*, 7066. [CrossRef]
11. Lopez, G.; Ahmadi, S.H.; Amelung, W.; Athmann, M.; Ewert, F.; Gaiser, T.; Gocke, M.I.; Kautz, T.; Postma, J.; Rachmilevitch, S.; et al. Nutrient Deficiency Effects on Root Architecture and Root-to-Shoot Ratio in Arable Crops. *Front. Plant Sci.* **2023**, *13*, 1067498. [CrossRef] [PubMed]
12. Morcuende, R.; Bari, R.; Gibon, Y.; Zheng, W.; Pant, B.D.; Bläsing, O.; Usadel, B.; Czechowski, T.; Udvardi, M.K.; Stitt, M.; et al. Genome-Wide Reprogramming of Metabolism and Regulatory Networks of Arabidopsis in Response to Phosphorus. *Plant Cell Environ.* **2007**, *30*, 85–112. [CrossRef] [PubMed]
13. Kamali Aliabad, K.; Zamani, E.; Shahbazi Manshadi, S. Investigation of Deficiency and Excess Nutrients Under in Vitro Culture Conditions in Petunia Hybrida. *Proc. Natl. Acad. Sci. India Sect. B Biol. Sci.* **2024**. [CrossRef]
14. Feng, Z.; Ji, S.; Ping, J.; Cui, D. Recent Advances in Metabolomics for Studying Heavy Metal Stress in Plants. *TrAC Trends Anal. Chem.* **2021**, *143*, 116402. [CrossRef]
15. Ghuge, S.A.; Nikalje, G.C.; Kadam, U.S.; Suprasanna, P.; Hong, J.C. Comprehensive Mechanisms of Heavy Metal Toxicity in Plants, Detoxification, and Remediation. *J. Hazard. Mater.* **2023**, *450*, 131039. [CrossRef] [PubMed]
16. Ranjan, A.; Rajput, V.D.; Minkina, T.; Bauer, T.; Chauhan, A.; Jindal, T. Nanoparticles Induced Stress and Toxicity in Plants. *Environ. Nanotechnol. Monit. Manag.* **2021**, *15*, 100457. [CrossRef]
17. Ramegowda, V.; Da Costa, M.V.J.; Harihar, S.; Karaba, N.N.; Sreeman, S.M. Chapter 17—Abiotic and Biotic Stress Interactions in Plants: A Cross-Tolerance Perspective. In *Priming-Mediated Stress and Cross-Stress Tolerance in Crop Plants*; Hossain, M.A., Liu, F., Burritt, D.J., Fujita, M., Huang, B., Eds.; Academic Press: Cambridge, MA, USA, 2020; pp. 267–302. ISBN 978-0-12-817892-8.
18. Zulfiqar, F.; Ashraf, M. Nanoparticles Potentially Mediate Salt Stress Tolerance in Plants. *Plant Physiol. Biochem.* **2021**, *160*, 257–268. [CrossRef] [PubMed]
19. Xiong, L.; Zhu, J.-K. Molecular and Genetic Aspects of Plant Responses to Osmotic Stress. *Plant Cell Environ.* **2002**, *25*, 131–139. [CrossRef] [PubMed]
20. Fàbregas, N.; Yoshida, T.; Fernie, A.R. Role of Raf-like Kinases in SnRK2 Activation and Osmotic Stress Response in Plants. *Nat. Commun.* **2020**, *11*, 6184. [CrossRef]
21. Zahra, N.; Hafeez, M.B.; Shaukat, K.; Wahid, A.; Hussain, S.; Naseer, R.; Raza, A.; Iqbal, S.; Farooq, M. Hypoxia and Anoxia Stress: Plant Responses and Tolerance Mechanisms. *J. Agron. Crop Sci.* **2021**, *207*, 249–284. [CrossRef]
22. Parent, C.; Capelli, N.; Berger, A.; Crèvecoeur, M.; Dat, J.F. An Overview of Plant Responses to Soil Waterlogging. *Plant Stress* **2008**, *2*, 20–27.
23. León, J.; Castillo, M.C.; Gayubas, B. The Hypoxia–Reoxygenation Stress in Plants. *J. Exp. Bot.* **2021**, *72*, 5841–5856. [CrossRef] [PubMed]
24. Saxena, P.; Kulshrestha, U. Biochemical Effects of Air Pollutants on Plants. In *Plant Responses to Air Pollution*; Kulshrestha, U., Saxena, P., Eds.; Springer: Singapore, 2016; pp. 59–70. ISBN 978-981-10-1201-3.
25. Lovett, G.M.; Tear, T.H.; Evers, D.C.; Findlay, S.E.G.; Cosby, B.J.; Dunscomb, J.K.; Driscoll, C.T.; Weathers, K.C. Effects of Air Pollution on Ecosystems and Biological Diversity in the Eastern United States. *Ann. N. Y. Acad. Sci.* **2009**, *1162*, 99–135. [CrossRef] [PubMed]
26. Brenya, E.; Pervin, M.; Chen, Z.-H.; Tissue, D.T.; Johnson, S.; Braam, J.; Cazzonelli, C.I. Mechanical Stress Acclimation in Plants: Linking Hormones and Somatic Memory to Thigmomorphogenesis. *Plant Cell Environ.* **2022**, *45*, 989–1010. [CrossRef] [PubMed]
27. Hamant, O.; Haswell, E.S. Life behind the Wall: Sensing Mechanical Cues in Plants. *BMC Biol.* **2017**, *15*, 59. [CrossRef] [PubMed]
28. Harborne, J.B. Role of Secondary Metabolites in Chemical Defence Mechanisms in Plants. In *Ciba Foundation Symposium 154—Bioactive Compounds from Plants*; John Wiley & Sons, Ltd.: Hoboken, NJ, USA, 2007; pp. 126–139. ISBN 978-0-470-51400-9.
29. Suprasanna, P. Plant Abiotic Stress Tolerance: Insights into Resilience Build-Up. *J. Biosci.* **2020**, *45*, 120. [CrossRef]
30. Rhodes, C.J. Soil Erosion, Climate Change and Global Food Security: Challenges and Strategies. *Sci. Prog.* **2014**, *97*, 97–153. [CrossRef] [PubMed]
31. Sewelam, N.; Brilhaus, D.; Bräutigam, A.; Alseekh, S.; Fernie, A.R.; Maurino, V.G. Molecular Plant Responses to Combined Abiotic Stresses Put a Spotlight on Unknown and Abundant Genes. *J. Exp. Bot.* **2020**, *71*, 5098–5112. [CrossRef] [PubMed]
32. Zandalinas, S.I.; Mittler, R. Plant Responses to Multifactorial Stress Combination. *New Phytol.* **2022**, *234*, 1161–1167. [CrossRef]

33. Leisner, C.P.; Potnis, N.; Sanz-Saez, A. Crosstalk and Trade-Offs: Plant Responses to Climate Change-Associated Abiotic and Biotic Stresses. *Plant Cell Environ.* **2023**, *46*, 2946–2963. [CrossRef]
34. Ambrosino, L.; Colantuono, C.; Diretto, G.; Fiore, A.; Chiusano, M.L. Bioinformatics Resources for Plant Abiotic Stress Responses: State of the Art and Opportunities in the Fast Evolving-Omics Era. *Plants* **2020**, *9*, 591. [CrossRef] [PubMed]

International Journal of
Molecular Sciences

Communication

Do Cuticular Gaps Make It Possible to Study the Composition of the Cell Walls in the Glands of *Drosophyllum lusitanicum*?

Bartosz J. Płachno [1,*], Małgorzata Kapusta [2], Piotr Stolarczyk [3] and Piotr Świątek [4]

[1] Department of Plant Cytology and Embryology, Institute of Botany, Faculty of Biology, Jagiellonian University in Kraków, 9 Gronostajowa St., 30-387 Kraków, Poland
[2] Bioimaging Laboratory, Faculty of Biology, University of Gdańsk, 59 Wita Stwosza St., 80-308 Gdańsk, Poland; malgorzata.kapusta@ug.edu.pl
[3] Department of Botany, Physiology and Plant Protection, Faculty of Biotechnology and Horticulture, University of Agriculture in Kraków, 29 Listopada 54 Ave., 31-425 Kraków, Poland; piotr.stolarczyk@urk.edu.pl
[4] Institute of Biology, Biotechnology and Environmental Protection, Faculty of Natural Sciences, University of Silesia in Katowice, 9 Bankowa St., 40-007 Katowice, Poland; piotr.swiatek@us.edu.pl
* Correspondence: bartosz.plachno@uj.edu.pl; Tel.: +48-12-664-60-39

Abstract: Carnivorous plants can survive in poor habitats because they have the ability to attract, capture, and digest prey and absorb animal nutrients using modified organs that are equipped with glands. These glands have terminal cells with permeable cuticles. Cuticular discontinuities allow both secretion and endocytosis. In *Drosophyllum lusitanicum*, these emergences have glandular cells with cuticular discontinuities in the form of cuticular gaps. In this study, we determined whether these specific cuticular discontinuities were permeable enough to antibodies to show the occurrence of the cell wall polymers in the glands. Scanning transmission electron microscopy was used to show the structure of the cuticle. Fluorescence microscopy revealed the localization of the carbohydrate epitopes that are associated with the major cell wall polysaccharides and glycoproteins. We showed that *Drosophyllum* leaf epidermal cells have a continuous and well-developed cuticle, which helps the plant inhibit water loss and live in a dry environment. The cuticular gaps only partially allow us to study the composition of cell walls in the glands of *Drosophyllum*. We recoded arabinogalactan proteins, some homogalacturonans, and hemicelluloses. However, antibody penetration was only limited to the cell wall surface. The localization of the wall components in the cell wall ingrowths was missing. The use of enzymatic digestion improves the labeling of hemicelluloses in *Drosophyllum* glands.

Keywords: arabinogalactan proteins; carbohydrate epitopes; carnivorous plants; cell wall; cuticle; cuticular gaps; Drosophyllaceae; mucilage glands; transfer cells; scanning transmission electron microscopy

Citation: Płachno, B.J.; Kapusta, M.; Stolarczyk, P.; Świątek, P. Do Cuticular Gaps Make It Possible to Study the Composition of the Cell Walls in the Glands of *Drosophyllum lusitanicum*? *Int. J. Mol. Sci.* **2024**, 25, 1320. https://doi.org/10.3390/ijms25021320

Academic Editor: Martin Bartas

Received: 28 December 2023
Revised: 18 January 2024
Accepted: 19 January 2024
Published: 21 January 2024

1. Introduction

Carnivorous plants have the ability to attract, capture, and digest prey using modified organs—traps—and then absorb the nutrients that are obtained from the bodies of the victims [1–3]. Typical carnivorous plants occur in swamps, bog habitats, or humid tropical rainforests [4–9]. There are also completely aquatic carnivorous plant species [10–13]. In these habitats, there is a wide diversity of carnivorous plant species. This is because carnivory is most likely to evolve and be favored ecologically (optimal investment in carnivory) in these sunny, nutrient-poor habitats that are rich in water [14,15]. However, in contrast to most carnivorous plants, *Drosophyllum lusitanicum* (L.) Link inhabits heathlands and ruderal sites with both very tough climatic and pedological conditions. In these places, *Drosophyllum* is exposed to an extreme climate with very high air and soil temperatures and very low air humidity throughout the day [16–18]. According to Adamec [19], *Drosophyllum* tissues do not store water. *Drosophyllum* has a xeromorphic root system [19–21] that is not able to supply the plant with sufficient water to cover the transpiration rates when the plant is exposed to an extreme temperature and the soil is dry [19]. The key to their survival are the stalked, glandular emergences (glands) that occur on the *Drosophyllum* leaves. These produce a polysaccharide mucilage that is hygroscopic [4,22,23], and water then condenses from the oceanic fog as dew onto these mucilage droplets. Adlassnig et al. [16] and Adamec [19] proposed that this water can then be absorbed by the glands and transported to the leaves; thus, plants can survive harsh conditions. This mucilage is also used to trap prey [24,25] and contains volatile organic compounds, so it has a stake in attracting victims [26]. Experimentally, it was shown that *Drosophyllum* has been found to take up nutrients (N, P, K, and Mg) from insects [27].

Due to their peripheral location, the gland cells of carnivorous plants are easily accessible to fixatives, which is extremely important for keeping the cell structure intact. They are also rich in cytoplasm and display a wide range of cellular activities, and therefore all of these characteristics make the cells of the carnivorous plant glands an interesting research model [28]. Rottloff et al. [29,30] showed the carnivorous plant glands could be easily isolated and directly used for further gene expression analysis using PCR techniques after preparing the RNA. Carnivorous plants have glands with permeable cuticles because of the occurrence of cuticular discontinuities that allow secretion and endocytosis to occur [4]. In some carnivorous plants, endocytosis has been confirmed as one way that they absorb the nutrients from the bodies of captured prey [31–33]. There are several types of these discontinuities [4,34–37]. For example, in the glandular cells of the tentacles of *Drosera*, there are cuticular pores (about 30 nm in diameter), which can be visualized using SEM [38,39]. Similar pores were also observed in the glands of *Genlisea* [40,41] and in the epidermal cells of *Roridula* [42]. In the mucilage glands of *Drosophyllum lusitanicum*, cuticular discontinuities were found by Schnepf [43], who later described them as cutin-free wall regions that are invisible in the SEM. The term 'cuticular gaps' was proposed for such discontinuities [36,37]. Cuticle discontinuities in carnivorous plant glands are easily detected using vital dyes, e.g., an aqueous solution of neutral red or methylene blue [24,31,36,37,40]. Moreover, these cuticle discontinuities in carnivorous plant glands allow intact cells to be penetrated by dyes such as $DiOC_6$ and styryl dye FM4-64, thereby making it possible to visualize the organelles and membranes, as was shown by Płachno et al. [41], Lichtscheidl et al. [44], and Li et al. [45]. Thus, these cuticle discontinuities make the cells of the carnivorous plant glands an excellent research model.

Cell wall polymers play an important role in the functionality of plant cell walls. For example, esterified homogalacturonans are involved in the porosity, elasticity, expansibility, and hydration of the cell wall [46–50]. De-esterified homogalacturonans increase the rigidity and resistance to mechanical stress of the cell walls [47,50,51]. Glactan also plays a role in cell wall rigidity [50,52]. Hemicelluloses contribute to strengthening the cell wall by interacting with cellulose [53]. Arabinogalactans have many important functions in plant cells that range from participating in the formation of wall outgrowths to playing the role of signal molecules [54–58].

Recently, using sectioned material, we showed the occurrence of the major cell wall polysaccharides and glycoproteins in the mucilage glands of *Drosophyllum lusitanicum* [59]. In this study, we wanted to assess whole-mount immunolabeled glands using the same species. We wanted to determine whether the specific discontinuous cuticle of the secretory cells of glands is permeable enough to antibodies. Knowing the occurrence of wall components in the sliced material where the walls are accessible to antibodies, we compared them with the whole-mount immunolabeling method, which was mainly used for root hairs and pollen tubes [60–64]. Recently, we used this method in another carnivorous plant in order to localize the cell wall components in the trichomes from *Utricularia dichotoma* traps [65]; however, we had only partial success as the labeling was limited to the glandular cell fragments in which the cuticle was discontinuous.

2. Results

2.1. Cuticle and Cuticular Discontinuities

The epidermal cells of mature leaves have a well-developed cuticle with a continuous cuticularized wall layer and a thick cutinized wall. The cuticularized wall layer contains electron-translucent lamellae, which are parallel to the surface. The cutinized wall also contained some electron-translucent lamella. There was also cuticular wax (Figure 1A,B). In the mucilage gland head cells, there is a well-developed cutinized wall, however, with some parts that are cutin-free. These regions without cutin (cuticular gaps) were connected to the cell surface (Figure 1C,D). The cuticular gaps were up to 450 nm (130–450 nm) in diameter. The cuticular material/wax, or other external detritus were observed on the cell surface (Figure 2C). In some glands, a thin, cuticularized wall layer also occurred (Figure 2D). In the digestive gland head cells, there was a cuticle with a thin, cuticularized wall layer and a thick, cutinized wall. Cuticular gaps also occurred, and they were up to 450 nm (60–450 nm) in diameter (Figure 2E). Some cuticular gaps were covered by a cuticularized wall layer (Figure 2F).

2.2. Distribution of Arabinogalactan Proteins (AGPs)

We used the JIM8, JIM13, and JIM14 antibodies in order to localize the AGPs. JIM8 reacted with the fluorescence in the cell walls of the outer glandular cell of the stalked mucilage glands (Figure 2A,B). JIM8 reacted with a strong fluorescence in the cell walls of the sectioned cells of the leaf and digestive glands (Figure 2C). JIM8 yielded fluorescence signals in the debris on the cell surfaces (Figure 2D). JIM13 yielded fluorescence signals in the debris or secretions on the epidermal cell surfaces and also gland cells (Figure 2E). JIM13 reacted with a strong fluorescence in the cell walls of the sectioned cells of the leaf and digestive glands (Figure 2G). JIM14 yielded fluorescence signals in the debris on the cell surfaces (Figure 2H) and in the cell walls of the sectioned cells of the leaf (Figure 2I).

2.3. Distribution of Homogalacturonan

Low-methyl-esterified Homogalacturonans (HG) were detected by JIM5 and LM19 antibodies. JIM5 yielded fluorescence signals in the debris on the cell surfaces (Figure 3A) and in the cell walls of the sectioned cells of the leaf (Figure 3B). The fluorescence signal that was detected by LM19 occurred in the external cell walls of the outer glandular cells of the mucilage glands (Figure 3C). It also occurred in the debris on the cell surfaces and in the cell walls of the sectioned cells of a leaf (Figure 3D).

Highly esterified HGs were detected by JIM7 antibodies. A fluorescence signal was observed in the external cell walls of the outer glandular cells of the mucilage glands (Figure 3E) and in the cell walls of the sectioned cells of the leaf (Figure 3F).

The pectic polysaccharide (1–4)-β-D-galactan was detected by LM5. A weak signal (spot signal) occurred in the external cell walls of the outer glandular cells of both gland types (Figure 3G). In the areas where the cuticle was damaged, the signal was stronger (Figure 3H).

2.4. Distribution of Hemicellulose

Xyloglucan was detected by LM15 and LM25 antibodies. The LM15 antibody gave a fluorescence signal in the cell walls of the sectioned cells of the leaf and gland. When the sections were pre-treated with pectate lyase (during which the pectins were removed), the LM15 antibody gave a fluorescence signal in the external cell walls of the outer glandular cells of the mucilage glands (Figure 4A). LM25 gave a weak fluorescence signal from xyloglucan in the walls of the outer glandular cells of the mucilage glands (Figure 4B). When the sections were pre-treated with pectate lyase (during which the pectins were removed), the LM25 antibody gave a stronger fluorescence signal in the external cell walls of the outer glandular cells of the mucilage glands (Figure 4C).

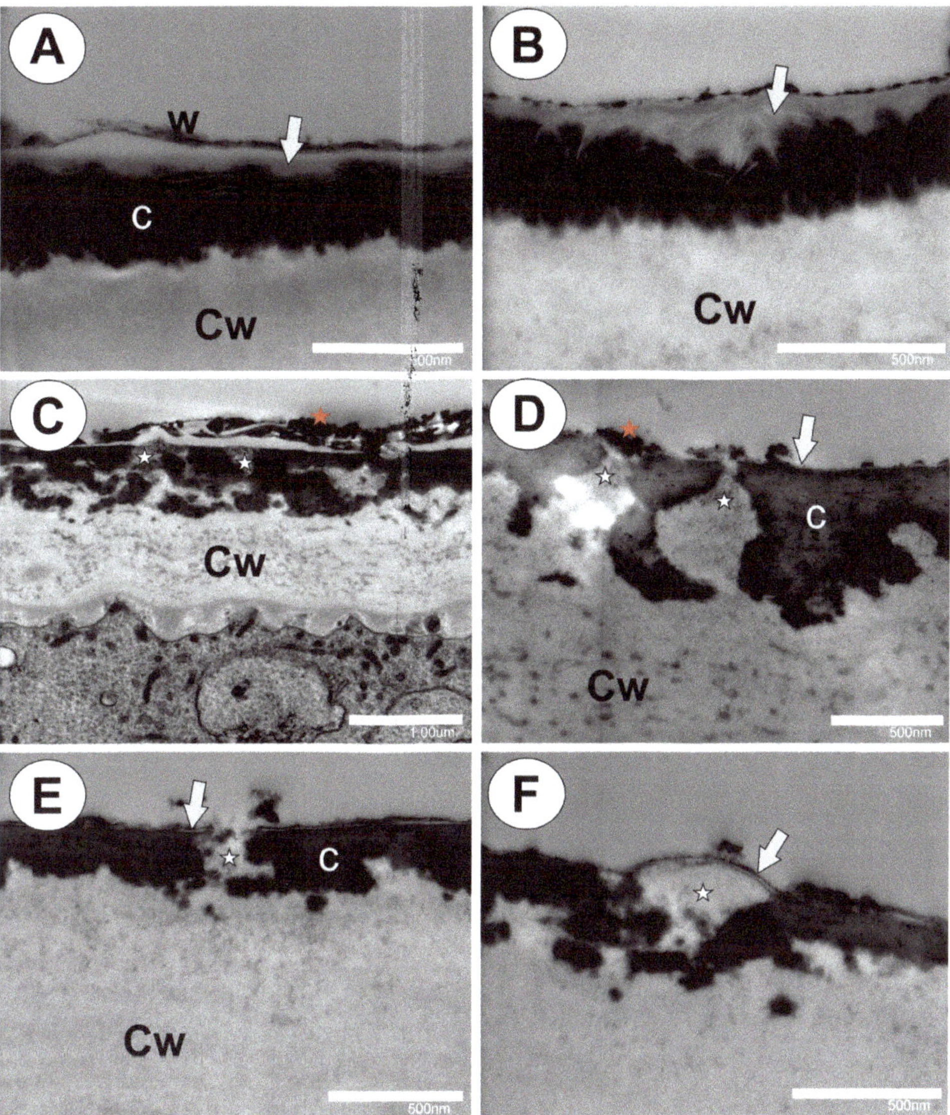

Figure 1. The structure of the cuticle and cuticular discontinuities of the epidermal cells and glands of *Drosophyllum lusitanicum*. (**A**,**B**) STEM of a mature *Drosophyllum lusitanicum* epidermal cell showing the cell wall and cuticle, bars 500 nm and 500 nm. (**C**) STEM of the outer glandular cell of a stalked mucilage gland showing the cuticular gaps with wall elements that extend to the cell surface, bar 1 μm. (**D**) STEM of the outer glandular cell of the mucilage gland showing cuticular gaps (white star) with wall elements that extend to the cell surface, bars 500 nm and 500 nm. (**E**) STEM of the outer glandular cell of the sessile digestive gland showing the cuticular gap (white star); note the cuticularized wall layer (arrow); bar 500 nm. (**F**) STEM of the outer glandular cell of the sessile digestive gland showing a large cuticular gap (white star) covered by a cuticularized wall layer (arrow), bar 500 nm. Abbreviations: w—wax; Cw—cell wall; c—cutinized wall; arrow—cuticularized wall layer; red star—wax or other external detritus; white star—cuticular gap.

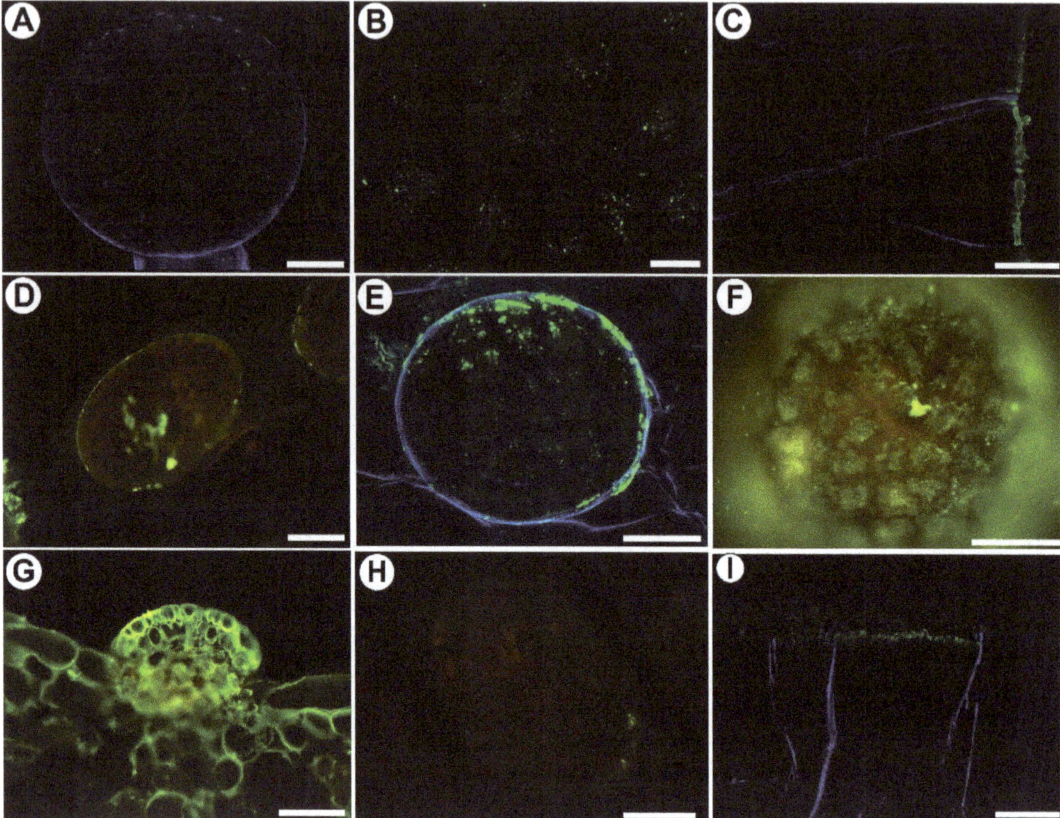

Figure 2. The arabinogalactan proteins that were detected in the glands of *Drosophyllum lusitanicum*. Green fluorescence is a sign of antibodies. (**A**) Arabinogalactan proteins (labeled with JIM8, green fluorescence) were detected in the mucilage gland, bar 50 μm. (**B**) Arabinogalactan proteins (labeled with JIM8) were detected in the outer glandular cell of the mucilage gland, bar 10 μm. (**C**) JIM8 fluorescence signals in the cell walls of the sectioned cells of the leaf, bar 50 μm. (**D**) JIM8 fluorescence signals in the debris on the digestive gland surface, bar 100 μm. (**E**) Arabinogalactan proteins (labeled with JIM13, green fluorescence) were detected on the digestive gland surface, bar 50 μm. (**F**) Arabinogalactan proteins (labeled with JIM13) were detected on the mucilage gland surface, bar 100 μm. (**G**) JIM13 fluorescence signals in the cell walls of the sectioned cells of the leaf and digestive gland, bar 100 μm. (**H**) JIM14 fluorescence signals in the debris on the digestive gland surface, bar 100 μm. (**I**) JIM14 fluorescence signals in the cell walls of the sectioned cells of the leaf, bar 50 μm.

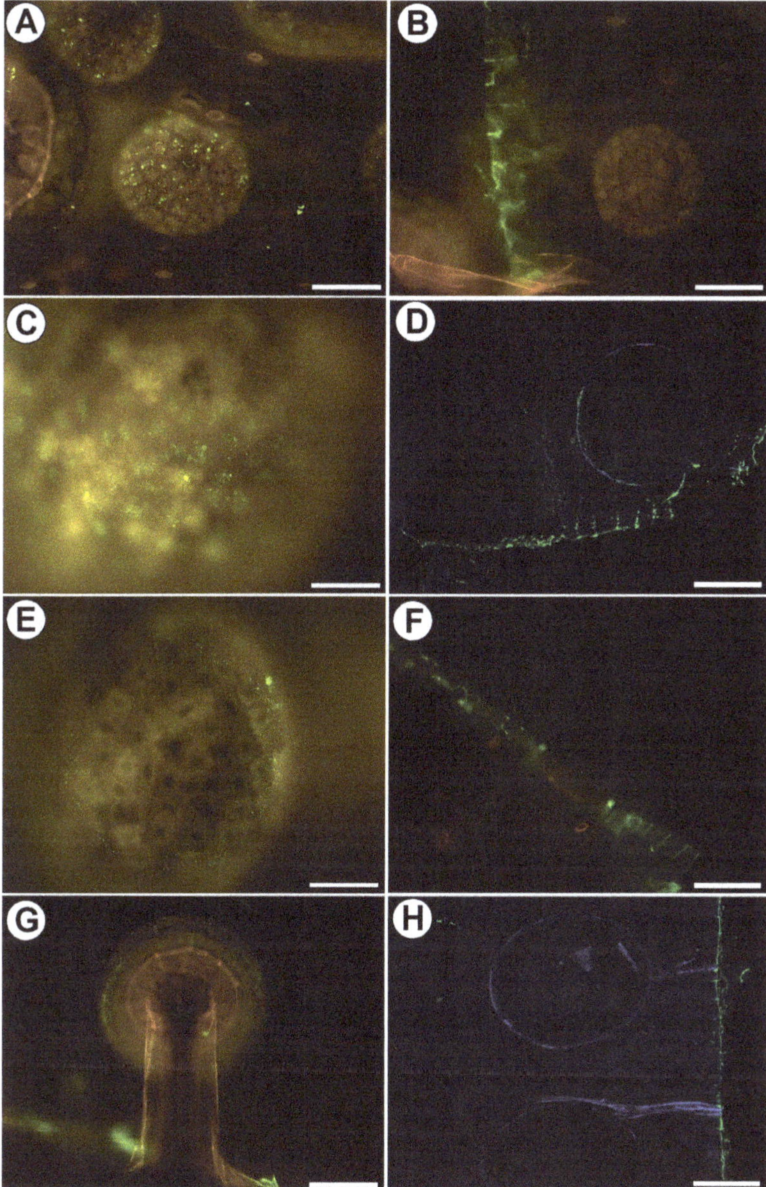

Figure 3. Homogalacturonan (HG) was detected in the glands of *Drosophyllum lusitanicum*. Green fluorescence is a sign of antibodies. (**A**) JIM5 green fluorescence signals in the debris on the cell surfaces, bar 100 μm. (**B**) JIM5 fluorescence signals in the cell walls of the sectioned cells of the leaf, bar 100 μm. (**C**) Homogalacturonan (labeled with LM19) was detected in the mucilage gland cells, bar 50 μm. (**D**) LM19 fluorescence signals in the cell walls of the sectioned cells of the leaf and cell surfaces, bar 100 μm. (**E**) A weak signal (spot signal) of JIM7 in the external cell walls of the outer glandular cells of mucilage gland, bar 50 μm. (**F**) Signal of JIM7 in cell walls in the sectioned cells of the leaf, bar 100 μm. (**G**) A weak signal (spot signal) of LM5 in the external cell walls of the outer glandular cells of mucilage gland, bar 100 μm. (**H**) Signal of LM5 in cell walls in places where cuticles were damaged in digestive gland and in the sectioned cells of the leaf, bar 100 μm.

Figure 4. Xyloglucan was detected in the glands of *Drosophyllum lusitanicum*. Green fluorescence is a sign of antibodies. (**A**) Xyloglucan (labeled with LM15) detected in the mucilage gland cells after they had been pre-treated with pectate lyase, bar 20 μm. (**B**) Xyloglucan (labeled with LM25) detected in the mucilage gland cells bar 20 μm. (**C**) Xyloglucan (labeled with LM25) detected in the mucilage gland cells after they had been pre-treated with pectate lyase, bar 20 μm.

3. Discussion

3.1. Cuticle Structure

Adamec [19] suggested that *Drosophyllum* leaves have a thick cuticle that efficiently prevents water losses via transpiration. In this study, we proved this because we found a thick, continuous, and well-developed cuticle (with a continuous cuticularized wall layer and a thick cutinized wall) in the epidermal cells, which acts as a barrier to excessive water loss. Similar to the observations of Joel and Juniper [36], Juniper et al. [4], and Vassilyev [66], we found cuticular gaps in the cuticle of mature *Drosophyllum* gland cells. However, we observed that some of these cuticular gaps were covered by a cuticularized wall layer, which may call into question whether these gaps are functional. Perhaps such discontinuities are still emerging, and those that are covered by a cuticularized wall layer are in a state of immaturity.

3.2. Pros and Cons of the Whole-Mount Immunolabeled Gland Technique

A cuticularized wall layer is probably a barrier to antibodies. Anderson [42] observed that in *Roridula*, some gaps have a plug of either cuticular wax or other external detritus. He also wondered whether such gaps would be permeable to liquid compounds. We also observed external detritus on the gland cell surface. Such material can block the access of antibodies to the cell walls. In addition, as we also showed, such material reacts with some of the antibodies, which may be due to the fact that it is partly derived from the plant's secretions. In the trichomes from *Utricularia dichotoma* traps, Płachno and Kapusta [65] showed that the success of the labeling was connected to the permeability of the cuticle (the occurrence of discontinuities). In the case of *Drosophyllum*, some of the results are ambiguous, and it is sometimes difficult to distinguish whether the signal came from the peripheral part of the wall or from the secretion or the material on the cell surface. Therefore, it is interesting to compare the occurrence of the cell wall components in the cells of the outer mucilage glands, which was our method, with the traditional method used by Płachno et al. [59] (Table 1). As for the AGPs, their detection was possible using both methods. However, in the sliced material, the AGPs were also found in the cell wall ingrowths. We know from our previous analyses of the glands of carnivorous plants that cell wall growths contain arabinogalactans. This has been found in the glands of *Dionaea muscipula* [67,68] and *Aldrovanda vesiculosa* [69,70]. AGPs have also been reported in cell wall ingrowths in various plant species, both in lower plants and angiosperms [71–75]. In the whole-mount immunolabeled glands, there was a lack of detection of the AGPs in the deeper parts of the cell walls; thus, the detection of the AGPs in the cell wall ingrowths is ineffective. Therefore, we do not recommend this technique for the analysis of cell wall ingrowth components.

Table 1. Comparison of the occurrence of cell wall components in the cells of the outer mucilage glands using our method with traditional method used by Płachno et al. [59].

Antibody	Whole-Mount Immunolabeled Glands	Sectioned Glands
	AGPs	
JIM8	Occurred	Occurred
JIM13	Lack of or connection with secretion	Occurred
JIM14	Lack	Weak signal
	Homogalacturonan	
JIM5	Lack	Lack
LM19	Occurred	Lack or weak signal
JIM7	Occurred	Lack
LM5	Weak signal	Weak signal
	Hemicelluloses	
LM15	Occurred after being pre-treated with pectate lyase	Occurred
LM25	Occurred	Occurred

As for the occurrence of homogalacturonans, the results from both methods were similar in peripheral parts of outer cell walls, and the only difference was that in the whole-mount immunolabeled mucilage glands, signals of LM19 and JIM7 antibodies were detected. However, it should be remembered that we analyzed the surface signals, and therefore, secretion contamination cannot be ruled out. We found that using enzymatic digestion improved the labeling of hemicelluloses in the *Drosophyllum* glands. Pectins mask the presence of the hemicelluloses; thus, the digestion of pectin makes the antibodies available to the hemicelluloses [76].

We think that to better understand both the cell wall polymers and cuticle structure in *Drosophyllum* glands, immunogold techniques are required.

4. Materials and Methods

4.1. Plant Material

Drosophyllum lusitanicum (L.) Link plants were grown in the greenhouses of the Botanical Garden of the Jagiellonian University. The plants were kept under high sunlight exposure.

4.2. Histological and Immunochemical Analysis

Leaf fragments (from three plants) were fixed in 8% (*w/v*) paraformaldehyde (PFA, Sigma-Aldrich, Sigma-Aldrich Sp. z o.o., Poznań, Poland) mixed with 0.25% (*v/v*) glutaraldehyde (GA, Sigma-Aldrich, Sigma-Aldrich Sp. z o.o., Poznań, Poland) in a PIPES buffer overnight at 4 °C. The PIPES buffer contained 50 mM PIPES (piperazine-N,N'-bis [2-ethanesulfonic acid], Sigma-Aldrich, Sigma-Aldrich Sp. z o.o., Poznań, Poland), 10 mM EGTA (ethylene gly-col-bis[β-aminoethyl ether]N,N,N',N'-tetraacetic acid, Sigma Aldrich Sp. z o.o., Poznań, Poland), and 1 mM MgCl$_2$ (Sigma-Aldrich, Sigma-Aldrich Sp. z o.o., Poznań, Poland), pH 6.8. The plant material was washed in a PBS buffer and later blocked with 1% bovine serum albumin (BSA, Sigma-Aldrich) in a PBS buffer and incubated with the following primary antibodies (Table 2) [47,76–82]—anti-AGP: JIM8, JIM13; JIM14; anti-pectin: JIM5, JIM7, LM19, LM5, and anti-hemicelluloses: LM25, LM15 overnight at 4 °C. All of the primary antibodies were used in a 1:20 dilution. They were purchased from Plant Probes, Leeds, UK, and the goat anti-rat secondary antibody conjugated with FITC was purchased from Abcam (Abcam plc, Cambridge, UK). The samples were then cover-slipped using a Mowiol mounting medium: a mixture of Mowiol® 4-88 (Sigma-Aldrich, Sigma-Aldrich Sp. z o.o., Poznań, Poland) and glycerol for fluorescence microscopy (Merck, Poland) with the addition of 2.5% DABCO (The Carl Roth GmbH + Co. KG, Germany). The material was viewed using a Nikon Eclipse E800 microscope (Precoptic, Warsaw, Poland) and a Leica DM6000 B (KAWA.SKA Sp. z o.o., Piaseczno, Poland, and Leica Microsystems GmbH, Wetzlar, Germany). The signal of the antibodies (green fluorescence) was visualized using a Nikon B-2A filter (Longpass Emission; excitation 450–490 nm, DM. 500 nm, emission 515 nm cut-on) and a Leica GFP filter emission 525/50. The autofluorescence of cell walls and cutin was visualized with a Nikon B-2A filter (yellow fluorescence) and Leica DAPI filter emission 460/50 (blue fluorescence). Part of the photos were acquired as Z stacks and deconvolved using 5 iterations of a 3D nonblind algorithm (AutoQuant™, Media Cybernetics Inc., Rockville, Maryland, USA). In order to maximize the spatial resolution, images are presented as the maximum projections. At least two different replications were performed for each of the analyzed traps, and about five to ten glands were analyzed for each antibody that was used. Negative controls were created by omitting the primary antibody step, which caused no fluorescence signal in any of the control frames for any of the stained slides (Figure S1). To remove the HG from the cell walls, the material was pretreated with 0.1 M sodium carbonate pH = 11.4 for 2 h at room temperature. This was followed by digestion with a pectate lyase 10 A (Nzytech) at 10 µg/mL for 2 h at room temperature in 50 mM N-cyclohexyl-3-aminopropane sulfonic acid (CAPS), to which 2 mM of a $CaCl_2$ buffer at pH 10 was added [76], and then incubation with the LM15 and LM25 antibodies as described above.

Table 2. List of the monoclonal antibodies that were used in the current study and the epitopes they recognize [77].

Antibody	Epitope
	AGPs
JIM8	Arabinogalactan
JIM13	Arabinogalactan/arabinogalactan protein
JIM14	Arabinogalactan/arabinogalactan protein
	Homogalacturonan
JIM5	Homogalacturonan (HG) domain of c pectic polysaccharides recognizes partially methyl-esterified epitopes of HG and can also bind to unesterified HG
JIM7	HG domain of the pectic polysaccharides recognizes partially methyl-esterified epitopes of HG but does not bind to unesterified HG
LM5	Linear tetrasaccharide in (1–4)-β-D-galactans (RGI side chain)
LM19	HG domain in pectic polysaccharides recognizes a range of HG with a preference to bind strongly to unesterified HG
	Hemicelluloses
LM15	XXXG motif of xyloglucan
LM25	XLLG, XXLG, and XXXG motifs of xyloglucan

4.3. Light Microscopy (LM)

The cuticle was stained using Auramine O (Sigma-Aldrich Sp. z o.o., Poznań, Poland), and later, the traps were examined using a Leica DM6000 B (Leica Microsystems GmbH, Wetzlar, Germany) and also a Nikon Eclipse E400 light microscope (Nikon, Tokyo, Japan) with a UV-2A filter (Ex. 330–380 nm, DM. 400 nm, Em. 420-α nm).

4.4. Scanning Transmission Electron Microscopy

The glands were also examined using electron microscopy, as follows: Fragments of the traps were fixed in a mixture of 2.5% glutaraldehyde with 2.5% formaldehyde in a 0.05 M cacodylate buffer (Sigma-Aldrich, Sigma-Aldrich Sp. z o.o., Poznań, Poland; pH 7.2) overnight or for several days, and later, the material was processed as in Płachno et al. [83]. The material was dehydrated with acetone and embedded in an Epoxy Embedding Medium Kit (Fluka). Ultrathin sections were cut on a Leica ultracut UCT ultramicrotome. The sections were examined using a Hitachi UHR FE-SEM SU 8010 microscope, which is housed at the University of Silesia in Katowice.

5. Conclusions

Here we show that *Drosophyllum* leaf epidermal cells have a continuous and well-developed cuticle, which helps the plant inhibit water loss and live in a dry environment. The cuticular gaps only partially allow us to study the composition of cell walls in the glands of *Drosophyllum*. We recorded arabinogalactan proteins, some homogalacturonans, and hemicelluloses; however, antibody penetration was only on the cell wall surface. The location of the wall components in the cell wall ingrowths was missing; thus, we do not recommend the whole-mount immunolabelling organ technique for the analysis of cell wall ingrowth components. The use of enzymatic digestion improves the labeling of hemicelluloses in *Drosophyllum* glands.

Supplementary Materials: The following supporting information can be downloaded at: https://www.mdpi.com/article/10.3390/ijms25021320/s1.

Author Contributions: Conceptualization, B.J.P.; methodology, B.J.P., M.K. and P.Ś.; investigation, B.J.P., M.K., P.S. and P.Ś.; resources, B.J.P.; data curation, B.J.P.; writing–preparing the original draft, B.J.P.; writing–review and editing, B.J.P., M.K., P.Ś. and P.S.; visualization, B.J.P. and M.K.; supervision, B.J.P.; project administration, B.J.P.; funding acquisition, B.J.P. All authors have read and agreed to the published version of the manuscript.

Funding: This research was financially supported by the program "Excellence Initiative-Research University" at the Jagiellonian University in Kraków, Poland.

Institutional Review Board Statement: Not applicable.

Informed Consent Statement: Not applicable.

Data Availability Statement: Data are contained within the article and supplementary materials.

Acknowledgments: We would like to thank the Botanical Garden of Jagiellonian University for the opportunity to use plants from their garden collection.

Conflicts of Interest: The authors declare no conflicts of interest.

Abbreviations

AGPs—arabinogalactan proteins; HG—homogalacturonans; SEM—scanning electron microscope; STEM—Scanning transmission electron microscopy; LM—light microscope.

References

1. Król, E.; Płachno, B.J.; Adamec, L.; Stolarz, M.; Dziubińska, H.; Trębacz, K. Quite a few reasons for calling carnivores 'the most wonderful plants in the world'. *Ann. Bot.* **2012**, *109*, 47–64. [CrossRef] [PubMed]
2. Ellison, A.M.; Adamec, L. *Carnivorous Plants: Physiology, Ecology and Evolution*; Oxford University Press: Oxford, UK, 2018; p. 510.
3. Hedrich, R.; Fukushima, K. On the Origin of Carnivory: Molecular physiology and evolution of plants on an animal diet. *Annu. Rev. Plant Biol.* **2021**, *72*, 133–153. [CrossRef] [PubMed]
4. Juniper, B.E.; Robbins, R.J.; Joel, D.M. *The Carnivorous Plants*; Academic Press: London, UK, 1989.
5. Barthlott, W.; Porembski, S.; Seine, R.; Theisen, I. *Karnivoren: Biologie Und Kultur Fleischfressender Pflanzen*; Ulmer: Stuttgart, Germany, 2004.
6. McPherson, S. *Pitcher Plants of the Americas*; Redfern Natural History Productions: Poole, UK, 2006.
7. McPherson, S. *Glistening Carnivores—The Sticky-Leaved Insect-Eating Plants*; Redfern Natural History Productions: Poole, UK, 2008.
8. McPherson, S. *Pitcher Plants of the Old World*; Redfern Natural History Productions: Poole, UK, 2009.
9. McPherson, S. *Carnivorous Plants and Their Habitats*; Fleischmann, A., Robinson, A., Eds.; Redfern Natural History Productions: Poole, UK, 2010; Volume 2, p. 1442.
10. Taylor, P. The genus *Utricularia*: A taxonomic monograph. *Kew Bull.* **1989**, *4*, 1–724.
11. Cross, A. *Aldrovanda: The Waterwheel Plant*; Redfern Natural History Productions Ltd.: Poole, UK, 2012.
12. Adamec, L. Ecophysiology of aquatic carnivorous plants. In *Carnivorous Plants: Physiology, Ecology and Evolution*; Ellison, A., Adamec, L., Eds.; Oxford Academic: Oxford, UK, 2017.
13. Adamec, L. Biological flora of Central Europe: *Aldrovanda vesiculosa* L. *Perspect. Plant Ecol. Evol. Syst.* **2018**, *35*, 8–21. [CrossRef]

14. Givnish, T.J.; Burkhardt, E.L.; Happel, R.E.; Weintraub, J.D. Carnivory in the bromeliad *Brocchinia reducta*, with a cost/benefit model for the general restriction of carnivorous plants to sunny, moist, nutrient-poor habitats. *Am. Nat.* **1984**, *124*, 479–497. [CrossRef]

15. Givnish, T.J.; Sparks, K.W.; Hunter, S.J.; Pavlovič, A. Why are plants carnivorous? Cost/benefit analysis, whole-plant growth and the context-specific advantages of botanical carnivory. In *Carnivorous Plants: Physiology, Ecology and Evolution*; Ellison, A., Adamec, L., Eds.; Oxford University Press: Oxford, UK, 2018; pp. 232–255.

16. Adlassnig, W.; Peroutka, M.; Eder, G.; Pois, W.; Lichtscheidl, I.K. Ecophysiological observations on *Drosophyllum lusitanicum*. *Ecol. Res.* **2006**, *21*, 255–262. [CrossRef]

17. Paniw, M.; Gil-Cabeza, E.; Ojeda, F. Plant carnivory beyond bogs: Reliance on prey feeding in *Drosophyllum lusitanicum* (Drosophyllaceae) in dry Mediterranean heathland habitats. *Ann. Bot.* **2017**, *119*, 1035–1041. [CrossRef]

18. Skates, L.M.; Paniw, M.; Cross, A.T.; Ojeda, F.; Dixon, K.W.; Stevens, J.C.; Gebauer, G. An ecological perspective on 'plant carnivory beyond bogs': Nutritional benefits of prey capture for the Mediterranean carnivorous plant *Drosophyllum lusitanicum*. *Ann. Bot.* **2019**, *124*, 65–76. [CrossRef]

19. Adamec, L. Ecophysiological investigation on *Drosophyllum lusitanicum*: Why doesn't the plant dry out? *Carniv. Plant Newsl.* **2009**, *38*, 71–74. [CrossRef]

20. Carlquist, S.; Wilson, E.J. Wood anatomy of *Drosophyllum* (Droseraceae): Ecological and phylogenetic considerations. *Bull. Torrey Bot. Club* **1995**, *122*, 185–189. [CrossRef]

21. Adlassnig, W.; Peroutka, M.; Lambers, H.; Lichtscheidl, I.K. The roots of carnivorous plants. *Plant Soil* **2005**, *274*, 127–140. [CrossRef]

22. Heslop-Harrison, Y. Enzyme secretion and digest uptake in carnivorous plants. In *Perspectives in Experimental Biology, Proceedings of the 50th Anniversary Meeting, Sydney, Australia, 6–8 March 2024*; SEB symposium; Pergamon: Oxford, UK, 1976; Volume 2.

23. Adlassnig, W.; Lendl, T.; Peroutka, M.; Lang, I. Deadly glue—Adhesive traps of carnivorous plants. In *Biological Adhesive Systems*; von Byern, J., Grunwald, I., Eds.; Springer: Vienna, Austria; New York, NY, USA, 2010; pp. 15–28.

24. Lloyd, F.E. *The Carnivorous Plants*; Chronica Botanica Company: Waltham, MA, USA, 1942.

25. Bertol, N.; Paniw, M.; Ojeda, F. Effective prey attraction in the rare *Drosophyllum lusitanicum*, a flypaper-trap carnivorous plant. *Am. J. Bot.* **2015**, *102*, 689–694. [CrossRef] [PubMed]

26. Ojeda, F.; Carrera, C.; Paniw, M.; García-Moreno, L.; Barbero, G.F.; Palma, M. Volatile and semi-volatile organic compounds may help reduce pollinator-prey overlap in the carnivorous plant *Drosophyllum lucitanicum* (Drosophyllaceae). *J. Chem. Ecol.* **2021**, *47*, 73–86. [CrossRef]

27. Płachno, B.J.; Adamec, L.; Huet, H. Mineral nutrient uptake from prey and glandular phosphatase activity as a dual test of carnivory in semi-desert plants with glandular leaves suspected of carnivory. *Ann. Bot.* **2009**, *104*, 649–654. [CrossRef]

28. Adlassnig, W.; Peroutka, M.; Lang, I.; Lichtscheidl, I.K. Glands of carnivorous plants as a model system in cell biological research. *Acta Bot. Gall.* **2005**, *152*, 111–124. [CrossRef]

29. Rottloff, S.; Müller, U.; Kilper, R.; Mithöfer, A. Micropreparation of single secretory glands from the carnivorous plant *Nepenthes*. *Anal. Biochem.* **2009**, *394*, 135–137. [CrossRef] [PubMed]

30. Rottloff, S.; Mithöfer, A.; Müller, U.; Kilper, R. Isolation of viable multicellular glands from tissue of the carnivorous plant, *Nepenthes*. *J. Vis. Exp.* **2013**, *82*, e50993.

31. Adlassnig, W.; Koller-Peroutka, M.; Bauer, S.; Koshkin, E.; Lendl, T.; Lichtscheidl, I.K. Endocytotic uptake of nutrients in carnivorous plants. *Plant J.* **2012**, *71*, 303–313. [CrossRef]

32. Koller-Peroutka, M.; Krammer, S.; Pavlik, A.; Edlinger, M.; Lang, I.; Adlassnig, W. Endocytosis and Digestion in Carnivorous Pitcher Plants of the Family Sarraceniaceae. *Plants* **2019**, *8*, 367. [CrossRef]

33. Ivesic, C.; Krammer, S.; Koller-Peroutka, M.; Laarouchi, A.; Gruber, D.; Lang, I.; Lichtscheidl, I.K.; Adlassnig, W. Quantification of Protein Uptake by Endocytosis in Carnivorous Nepenthales. *Plants* **2023**, *12*, 341. [CrossRef]

34. Fineran, B.A.; Lee, M.S.L. Organization of quadrifid and bifid hairs in the trap of *Utricularia monanthos*. *Protoplasma* **1975**, *84*, 43–70. [CrossRef]

35. Fineran, B.A. Glandular trichomes in *Utricularia*: A review of their structure and function. *Isr. J. Bot.* **1985**, *34*, 295–330.

36. Joel, D.M.; Juniper, B.E. Cuticular gaps in carnivorous plant glands. In *The Plant Cuticle*; Cutler, D.F., Alvin, K.L., Price, C.E., Eds.; Academic Press: London, UK, 1982; pp. 121–130.

37. Joel, D.M.; Rea, P.A.; Juniper, B.E. The cuticle of *Dionaea muscipula* Ellis (Venus's Flytrap) in relation to stimulation, secretion and absorption. *Protoplasma* **1983**, *114*, 44–51. [CrossRef]

38. Williams, S.E.; Pickard, B.G. Secretion, absorption and cuticular structure in *Drosera* tentacles. *Plant Physiol.* **1969**, *44*, 5.

39. Williams, S.E.; Pickard, B.G. Connections and barriers between cells of *Drosera* tentacles in relation to their electrophysiology. *Planta* **1974**, *116*, 1–16. [CrossRef] [PubMed]

40. Płachno, B.J.; Faber, J.; Jankun, A. Cuticular discontinuities in glandular hairs of *Genlisea* St.-Hil. in relation to their functions. *Acta Bot. Gall.* **2005**, *152*, 125–130. [CrossRef]

41. Płachno, B.J.; Kozieradzka-Kiszkurno, M.; Świątek, P. Functional Ultrastructure of *Genlisea* (Lentibulariaceae) Digestive Hairs. *Ann. Bot.* **2007**, *100*, 195–203. [CrossRef] [PubMed]

42. Anderson, B. Adaptations to foliar absorption of faeces: A pathway in plant carnivory. *Ann. Bot.* **2005**, *95*, 757–761. [CrossRef]

43. Schnepf, E. Zur feinstruktur der drüsen von *Drosophyllum lusitanicum*. *Planta* **1960**, *54*, 641–674. [CrossRef]

44. Lichtscheidl, I.; Lancelle, S.; Weidinger, M.; Adlassnig, W.; Koller-Peroutka, M.; Bauer, S.; Krammer, S.; Hepler, P.K. Gland cell responses to feeding in *Drosera capensis*, a carnivorous plant. *Protoplasma* **2021**, *258*, 1291–1306. [CrossRef]

45. Li, Y.-X.; Chen, A.; Leu, W.-M. Sessile Trichomes Play Major Roles in Prey Digestion and Absorption, While Stalked Trichomes Function in Prey Predation in *Byblis guehoi*. *Int. J. Mol. Sci.* **2023**, *24*, 5305. [CrossRef] [PubMed]

46. Ridley, B.L.; O'Neill, M.A.; Mohnen, D. Pectins: Structure, biosynthesis and oligogalacturonide-related signaling. *Phytochemistry* **2001**, *57*, 929–967. [CrossRef] [PubMed]

47. Verhertbruggen, Y.; Marcus, S.E.; Haeger, A.; Verhoef, R.; Schols, H.A.; McCleary, B.V.; McKee, L.; Gilbert, H.J.; Knox, P.J. Developmental complexity of arabinan polysaccharides and their processing in plant cell walls. *Plant J.* **2009**, *59*, 413–425. [CrossRef] [PubMed]

48. Peaucelle, A.; Braybrook, S.; Höfte, H. Cell wall mechanics and growth control in plants: The role of pectins revisited. *Front. Plant Sci.* **2012**, *3*, 121. [CrossRef] [PubMed]

49. Braybrook, S.A.; Jönsson, H. Shifting foundations: The mechanical cell wall and development. *Curr. Opin. Plant Biol.* **2016**, *29*, 115–120. [CrossRef] [PubMed]

50. Cornuault, V.; Buffetto, F.; Marcus, S.E.; Crépeau, M.J.; Guillon, F.; Ralet, M.C.; Knox, P. LM6-M: A high avidity rat monoclonal antibody to pectic α-1, 5-L-arabinan. *BioRxiv* **2017**. [CrossRef]

51. Verhertbruggen, Y.; Marcus, S.E.; Chen, J.; Knox, J.P. Cell wall pectic arabinans influence the mechanical properties of *Arabidopsis thaliana* inflorescence stems and their response to mechanical stress. *Plant Cell Physiol.* **2013**, *54*, 1278–1288. [CrossRef]

52. McCartney, L.; Steele-King, C.G.; Jordan, E.; Knox, J.P. Cell wall pectic (1→4)-β-d-galactan marks the acceleration of cell elongation in the *Arabidopsis* seedling root meristem. *Plant J.* **2003**, *33*, 447–454. [CrossRef]

53. Scheller, H.V.; Ulvskov, P. Hemicelluloses. *Annu. Rev. Plant Biol.* **2010**, *61*, 263–289. [CrossRef]

54. Showalter, A.M. Arabinogalactan-proteins: Structure, expression and function. *Cell. Mol. Life Sci.* **2001**, *58*, 1399–1417. [CrossRef]

55. McCurdy, D.W.; Patrick, J.W.; Offler, C.E. Wall ingrowth formation in transfer cells: Novel examples of localized wall deposition in plant cells. *Curr. Opin. Plant Biol.* **2008**, *11*, 653–661. [CrossRef] [PubMed]

56. Nguema-Ona, E.; Coimbra, S.; Vicré-Gibouin, M.; Mollet, J.C.; Driouich, A. Arabinogalactan proteins in root and pollen-tube cells: Distribution and functional aspects. *Ann. Bot.* **2012**, *110*, 383–404. [CrossRef] [PubMed]

57. Lin, S.; Miao, Y.; Huang, H.; Zhang, Y.; Huang, L.; Cao, J. Arabinogalactan Proteins: Focus on the Role in Cellulose Synthesis and Deposition during Plant Cell Wall Biogenesis. *Int. J. Mol. Sci.* **2022**, *23*, 6578. [CrossRef] [PubMed]

58. Leszczuk, A.; Kalaitzis, P.; Kulik, J.; Zdunek, A. Review: Structure and modifications of arabinogalactan proteins (AGPs). *BMC Plant Biol.* **2023**, *23*, 45. [CrossRef] [PubMed]

59. Płachno, B.J.; Kapusta, M.; Stolarczyk, P.; Świątek, P.; Lichtscheidl, I. Differences in the Occurrence of Cell Wall Components between Distinct Cell Types in Glands of *Drosophyllum lusitanicum*. *Int. J. Mol. Sci.* **2023**, *24*, 15045. [CrossRef] [PubMed]

60. Li, Y.Q.; Bruun, L.; Pierson, E.S.; Cresti, M. Periodic deposition of arabinogalactan epitopes in the cell wall of pollen tubes of *Nicotiana tabacum* L. *Planta* **1992**, *188*, 532–538. [CrossRef]

61. Willats, W.G.; McCartney, L.; Knox, J.P. In-situ analysis of pectic polysaccharides in seed mucilage and at the root surface of *Arabidopsis thaliana*. *Planta* **2001**, *213*, 37–44. [CrossRef]

62. Chebli, Y.; Kaneda, M.; Zerzour, R.; Geitmann, A. The cell wall of the *Arabidopsis* pollen tube--spatial distribution, recycling and network formation of polysaccharides. *Plant Physiol.* **2012**, *160*, 1940–1955. [CrossRef]

63. Larson, E.R.; Tierney, M.L.; Tinaz, B.; Domozych, D.S. Using monoclonal antibodies to label living root hairs: A novel tool for studying cell wall microarchitecture and dynamics in *Arabidopsis*. *Plant Methods* **2014**, *10*, 30. [CrossRef]

64. Marzec, M.; Szarejko, I.; Melzer, M. Arabinogalactan proteins are involved in root hair development in barley. *J. Exp. Bot.* **2015**, *5*, 1245–1257. [CrossRef]

65. Płachno, B.J.; Kapusta, M. The Localization of Cell Wall Components in the Quadrifids of Whole-Mount Immunolabeled *Utricularia dichotoma* Traps. *Int. J. Mol. Sci.* **2024**, *25*, 56. [CrossRef]

66. Vassilyev, A.E. Dynamics of ultrastructural characters of *Drosophyllum lusitanicum* Link (Droseraceae) digestive glands during maturation and after stimulation. *Taiwania* **2005**, *50*, 167–182.

67. Płachno, B.J.; Kapusta, M.; Stolarczyk, P.; Świątek, P. Arabinogalactan proteins in the digestive glands of *Dionaea muscipula* J. Ellis Traps. *Cells* **2022**, *11*, 586. [CrossRef]

68. Płachno, B.J.; Kapusta, M.; Stolarczyk, P.; Świątek, P. Stellate trichomes in *Dionaea muscipula* Ellis (Venus Flytrap) Traps, Structure and Functions. *Int. J. Mol. Sci.* **2023**, *24*, 553. [CrossRef]

69. Płachno, B.J.; Kapusta, M.; Stolarczyk, P.; Świątek, P.; Strzemski, M.; Miranda, V.F.O. Immunocytochemical analysis of the wall ingrowths in the digestive gland transfer cells in *Aldrovanda vesiculosa* L. (Droseraceae). *Cells* **2022**, *11*, 2218. [CrossRef]

70. Płachno, B.J.; Kapusta, M.; Stolarczyk, P.; Wójciak, M.; Świątek, P. Immunocytochemical analysis of bifid trichomes in *Aldrovanda vesiculosa* L. Traps. *Int. J. Mol. Sci.* **2023**, *24*, 3358. [CrossRef]

71. Dahiya, P.; Brewin, N.J. Immunogold localization of callose and other cell wall components in pea nodule transfer cells. *Protoplasma* **2000**, *214*, 210–218. [CrossRef]

72. Ligrone, R.; Vaughn, K.C.; Rascio, N. A cytochemical and immunocytochemical analysis of the wall labyrinth apparatus in leaf transfer cells in *Elodea canadensis*. *Ann. Bot.* **2011**, *107*, 717–722. [CrossRef]

73. Henry, J.S.; Lopez, R.A.; Renzaglia, K.S. Differential localization of cell wall polymers across generations in the placenta of *Marchantia polymorpha*. *J. Plant Res.* **2020**, *133*, 911–924. [CrossRef]

74. Henry, J.S.; Renzaglia, K.S. The placenta of *Physcomitrium patens*: Transfer cell wall polymers compared across the three bryophyte groups. *Diversity* **2021**, *13*, 378. [CrossRef]
75. Henry, J.S.; Ligrone, R.; Vaughn, K.C.; Lopez, R.A.; Renzaglia, K.S. Cell wall polymers in the *Phaeoceros* placenta reflect developmental and functional differences across generations. *Bryophyt. Divers. Evol.* **2021**, *43*, 265–283. [CrossRef] [PubMed]
76. Marcus, S.E.; Verhertbruggen, Y.; Hervé, C.; Ordaz-Ortiz, J.J.; Farkas, V.; Pedersen, H.L.; Willats, W.G.; Knox, J.P. Pectic homogalacturonan masks abundant sets of xyloglucan epitopes in plant cell walls. *BMC Plant Biol.* **2008**, *8*, 60. [CrossRef]
77. Available online: https://www.kerafast.com/cat/799/paul-knox-phd (accessed on 13 November 2023).
78. Knox, J.P.; Day, S.; Roberts, K. A set of cell surface glycoproteins forms an early position, but not cell type, in the root apical meristem of *Daucus carota* L. *Development* **1989**, *106*, 47–56. [CrossRef]
79. Pennell, R.I.; Knox, J.P.; Scofield, G.N.; Selvendran, R.R.; Roberts, K. A family of abundant plasma membrane-associated glycoproteins related to the arabinogalactan proteins is unique to flowering plants. *J. Cell Biol.* **1989**, *108*, 1967–1977. [CrossRef] [PubMed]
80. Pennell, R.I.; Janniche, L.; Kjellbom, P.; Scofield, G.N.; Peart, J.M.; Roberts, K. Developmental regulation of a plasma membrane arabinogalactan protein epitope in oilseed rape flowers. *Plant Cell* **1991**, *3*, 1317–1326. [CrossRef] [PubMed]
81. Knox, J.P.; Linstead, P.J.; Cooper, J.P.C.; Roberts, K. Developmentally regulated epitopes of cell surface arabinogalactan proteins and their relation to root tissue pattern formation. *Plant J.* **1991**, *1*, 317–326. [CrossRef]
82. McCartney, L.; Marcus, S.E.; Knox, J.P. Monoclonal antibodies to plant cell wall xylans and arabinoxylans. *J. Histochem. Cytochem.* **2005**, *53*, 543–546. [CrossRef]
83. Płachno, B.J.; Swiatek, P.; Jobson, R.W.; Malota, K.; Brutkowski, W. Serial block face SEM visualization of unusual plant nuclear tubular extensions in a carnivorous plant (*Utricularia*, Lentibulariaceae). *Ann. Bot.* **2017**, *120*, 673–680. [CrossRef]

International Journal of
Molecular Sciences

Rapid and Nondestructive Evaluation of Wheat Chlorophyll under Drought Stress Using Hyperspectral Imaging

Yucun Yang [1], Rui Nan [1], Tongxi Mi [1], Yingxin Song [1], Fanghui Shi [1], Xinran Liu [1], Yunqi Wang [1], Fengli Sun [1,2], Yajun Xi [1,2] and Chao Zhang [1,2,*]

[1] State Key Laboratory of Crop Stress Biology for Arid Areas, College of Agronomy, Northwest A&F University, Xianyang 712100, China
[2] Key Laboratory of Wheat Biology and Genetic Improvement on Northwestern China, Ministry of Agriculture, Xianyang 712100, China
* Correspondence: ahzc2015@nwafu.edu.cn

Abstract: Chlorophyll drives plant photosynthesis. Under stress conditions, leaf chlorophyll content changes dramatically, which could provide insight into plant photosynthesis and drought resistance. Compared to traditional methods of evaluating chlorophyll content, hyperspectral imaging is more efficient and accurate and benefits from being a nondestructive technique. However, the relationships between chlorophyll content and hyperspectral characteristics of wheat leaves with wide genetic diversity and different treatments have rarely been reported. In this study, using 335 wheat varieties, we analyzed the hyperspectral characteristics of flag leaves and the relationships thereof with SPAD values at the grain-filling stage under control and drought stress. The hyperspectral information of wheat flag leaves significantly differed between control and drought stress conditions in the 550–700 nm region. Hyperspectral reflectance at 549 nm (r = −0.64) and the first derivative at 735 nm (r = 0.68) exhibited the strongest correlations with SPAD values. Hyperspectral reflectance at 536, 596, and 674 nm, and the first derivatives bands at 756 and 778 nm, were useful for estimating SPAD values. The combination of spectrum and image characteristics (L*, a*, and b*) can improve the estimation accuracy of SPAD values (optimal performance of RFR, relative error, 7.35%; root mean square error, 4.439; R^2, 0.61). The models established in this study are efficient for evaluating chlorophyll content and provide insight into photosynthesis and drought resistance. This study can provide a reference for high-throughput phenotypic analysis and genetic breeding of wheat and other crops.

Keywords: wheat leaves chlorophyll; drought stress; machine learning; regression model; high-resolution spectral imaging; high-throughput phenotypic identification

Citation: Yang, Y.; Nan, R.; Mi, T.; Song, Y.; Shi, F.; Liu, X.; Wang, Y.; Sun, F.; Xi, Y.; Zhang, C. Rapid and Nondestructive Evaluation of Wheat Chlorophyll under Drought Stress Using Hyperspectral Imaging. *Int. J. Mol. Sci.* **2023**, *24*, 5825. https://doi.org/10.3390/ijms24065825

Academic Editor: Martın Bartas

Received: 21 February 2023
Revised: 11 March 2023
Accepted: 17 March 2023
Published: 18 March 2023

1. Introduction

Wheat (*Triticum aestivum* L.) is a staple food crop crucial for global food security. However, wheat crops experience many abiotic stresses, including low temperatures, drought, high temperatures, and dry and hot winds, which can strongly affect growth, development, and productivity [1,2]. Drought is one of the most severe abiotic stresses worldwide and can significantly reduce the number of tillers, grains per spike, and 1000-grain weight of wheat [3,4]. In 2021, both the United States and Brazil suffered historic severe droughts, which increased global food prices to the highest cost in a recent decade. Therefore, effective monitoring of the impact of drought stress during wheat growth is essential for improving yields, varieties, and food security.

Drought stress decreases crop chlorophyll content, destroys photosynthetic machinery, inhibits growth, and ultimately reduces yield [5,6]. In addition, flag leaf chlorophyll content can reflect the growth status of plants [7,8]. Chlorophyll is involved in capturing light energy and converting it into chemical energy during photosynthesis. It is fundamental for plant photosynthesis and directly determines capacity, net primary productivity, and carbon budget (Figure S1A) [9–11]. The synthesis of chlorophyll in plants is affected by a variety of factors, including temperature, water, disease, and other stresses. Water shortage of plant leaves will affect the synthesis of chlorophyll, accelerate the decomposition of chlorophyll, reduce the absorption capacity of light energy, and inhibit photosynthesis [12,13]. Meanwhile, plants also regulate carotenoid synthesis in response to stress and play a role in light energy capture and light protection during photosynthesis [14,15]. Crop chlorophyll content may be increased or decreased under drought stress [16–18], and the degree of change is closely related to drought resistance [19–22]. Therefore, monitoring flag leaf chlorophyll content can provide key information regarding wheat photosynthesis and drought resistance. Traditional methods for determining chlorophyll content include spectrophotometry and the use of hand-held chlorophyll content meters [23]. However, such methods have several disadvantages, such as leaf destruction and low-throughput, which are not conducive to the measurement of wheat chlorophyll content on a large scale. Compared to traditional methods, hyperspectral imaging (HSI) can determine plant chlorophyll content rapidly, nondestructively, and efficiently [24,25]. In addition, hyperspectral images contain rich spectral information that can be used for precision agriculture research and the establishment of complex mathematical models [26,27]. In recent years, the application of (HSI) technology to plant monitoring has developed rapidly. Such methods are based on the principle that different substances exhibit different absorption and reflection characteristics in distinct spectral bands [28]. The optical characteristics of the leaves include light reflected from the leaf's external surface and internal leaf structures, light absorbed by internal leaf chemicals (e.g., various leaf pigments), and light transmission (Figure S1B) [29]. The visible band (400–750 nm) reflects the synthesis and decomposition of chemical pigments critical for plant photosynthetic activity [30]; the near-infrared band (750–1000 nm) characteristics depend on leaf internal structures, absorption by water, and scattering processes [31]. Stress changes the pigments, water, and structures of plant leaves, and shows different spectral characteristics [32]. Certain hyperspectral curve characteristics, such as chlorophyll absorption troughs at 450 and 650 nm, a reflection peak near 540 nm, and typical red-edge effects in the range of 700–760 nm, reflect plant chlorophyll content and growth status [33–36]. Furthermore, drought stress causes leaf spectral reflectance to increase in the visible region, while chlorophyll content tends to decrease [37,38]. These aspects of HSI make it a feasible approach for estimating plant chlorophyll content. Extensive research has focused on developing models based on the spectral index for estimating chlorophyll content [24,39–43]. However, several problems have been identified in terms of general applicability and stability, which limits the application of the estimation models to other systems [44–46]. Compared to the spectral index, full-band hyperspectral information represents various aspects of physiology more comprehensively; combined with machine learning methods, such information can substantially improve model accuracy [47–49]. Several studies have monitored the physiological characteristics of plants based on information from sensitive bands [50,51]. However, the small number of sensitive bands does not adequately represent all hyperspectral information. Moreover, most studies used few wheat varieties and ignored the heterogeneous inter-multiple variety. Therefore, the applicability of previous models to other systems is limited [44], and the models would be ineffective for large-scale chlorophyll content and drought resistance assessments.

In the present study, we collected 335 wheat varieties (total 2010 leaf samples) under drought stress and control conditions at the filling stage to explore the relationship between hyperspectral image characteristics and chlorophyll content. A machine learning method was used to construct models to establish a rapid, nondestructive, accurate, and widely applicable method for assessing wheat chlorophyll content, photosynthesis, and drought resistance.

2. Results

2.1. Chlorophyll Changes in Wheat Leaves under Different Soil Moisture Conditions

SPAD values reflect changes in crop chlorophyll content and can thus be used to closely track photosynthesis. In this study, the median and upper quartile (Q3) SPAD values indicated that the chlorophyll content of several varieties tended to increase under drought stress, while lower quartile (Q1), lower whisker, and small outlier values suggested that the chlorophyll content of other varieties significantly declined after drought stress (Figure 1). These results indicate that the chlorophyll content of wheat changed significantly during drought stress and that the extent of the change varied among wheat varieties. Thus, we can infer that different wheat varieties exhibit differences in photosynthetic capacity under drought stress, which is closely related to drought resistance.

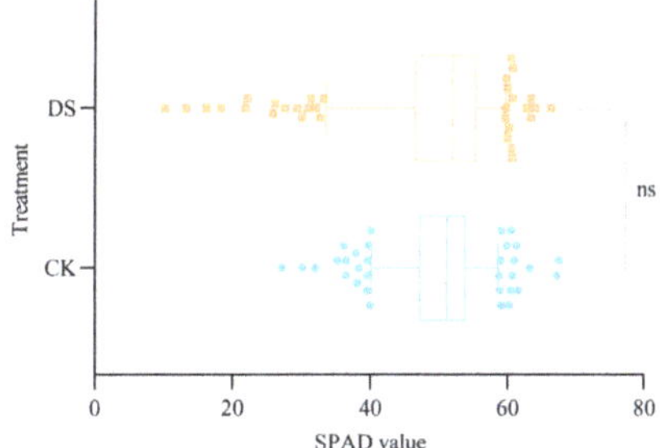

Figure 1. SPAD values of 335 varieties of wheat leaves under different soil moisture conditions. CK, control; DS, drought stress; whisker, 5–95%; ns, no statistically significant difference in SPAD values between the CK and DS treatments ($p > 0.05$).

2.2. Hyperspectral Characteristics of Wheat Leaves under Different Soil Moisture Conditions

2.2.1. Spectral Reflectance Characteristics under Different Soil Moisture Conditions

The leaf is the most important organ for photosynthesis in wheat, and the spectral reflection curves of leaves reflect their growth status. After drought stress, the spectral reflectance of several wheat varieties increased significantly in the visible region (400–700 nm) (Figure 2A). The mean spectral reflectance of wheat leaves markedly increased between 600–700 nm, which represents an absorption trough for chlorophyll (Figure 2C).

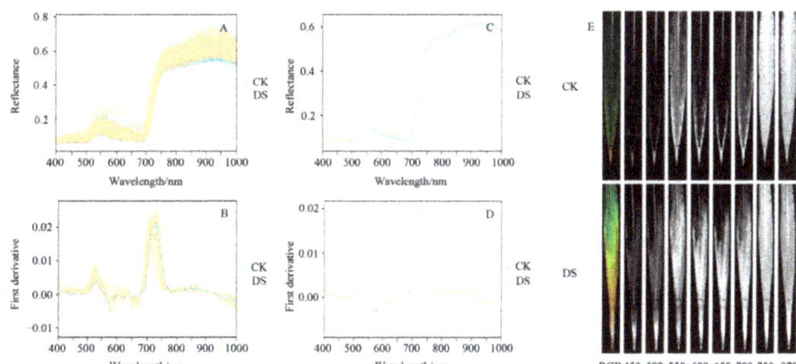

Figure 2. Hyperspectral curves and single-band hyperspectral images of wheat leaves under different soil moisture conditions. Hyperspectral reflectance and first derivative curves of leaves from 335 wheat varieties under control (CK) and drought stress (DS) conditions; (**A,B**) hyperspectral reflectance curves of leaves and first derivative values; (**C,D**) mean hyperspectral reflectance curves of leaves and first derivative values; (**E**) single-band hyperspectral images; hyperspectral band region (400–1000 nm); hyperspectral images obtained using the Pika L hyperspectral imaging system; RGB images (red = 640 nm, green = 550 nm, blue = 460 nm).

2.2.2. Spectral First Derivative Values under Different Soil Moisture Conditions

The first derivative can reduce the influence of noise in hyperspectral information. In the present study, spectral first derivative values of wheat leaves significantly differed between control and drought stress conditions, with increases and decreases seen in the regions between 550–650 and 700–750 nm, respectively, under drought stress (Figure 2B,D). Thus, we can infer that differences in the first derivative values were related to differences in the chlorophyll content and growth status of wheat leaves.

2.2.3. Spectral Images under Different Soil Moisture Conditions

Single-band hyperspectral images are often used to observe sensitive areas of the plant spectral response. Single-band hyperspectral images corresponding to the characteristics of the leaf hyperspectral curve were collected (Figure 2E). The hyperspectral images obtained under drought stress conditions at 450 and 500 nm (in the blue band), and at 970 nm (water absorption zone), did not significantly differ from the images of the controls. In contrast, significant differences were observed for the green and red bands (550, 600, 650, and 700 nm). These results indicate that the green and red bands represent sensitive areas for monitoring wheat chlorophyll content and growth status under drought stress.

2.3. Correlation Analysis between Hyperspectral Characteristics and SPAD Values

Correlation analysis revealed strong negative correlations between spectral reflectance and SPAD values in the visible region (400–735 nm), particularly at 549 and 708 nm (correlation coefficients of −0.64 and −0.61, respectively). However, no significant correlations were detected in the near-infrared region (750–1000 nm) (Figure 3A). The correlation coefficient between the first derivative of the spectrum and SPAD values was −0.61 at 541 nm, 0.68 at 735 nm, and −0.44 near 946 nm. Meanwhile, the first derivative and SPAD values were positively correlated near 600 nm under control conditions but were negatively correlated under drought stress (Figure 3B). These results indicate that the chlorophyll content was strongly correlated with hyperspectral features in the visible region but only weakly correlated in the near-infrared region. In addition, the spectral first derivative had a stronger correlation with SPAD values compared to reflectance.

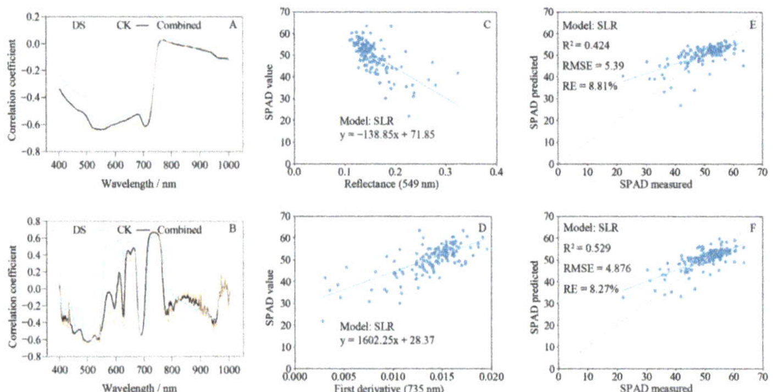

Figure 3. Correlation analysis and fitting results between leaf hyperspectral and SPAD values. (**A,B**) Correlations of spectral reflectance and the first derivative with SPAD values; (**C,D**) simple linear regression (SLR) analysis based on spectral reflectance at 549 nm and the spectral first derivative at 735 nm; (**E,F**) fitting results of predicted and measured values of SPAD based on reflectance at 549 nm and the first derivative at 735 nm. The gray dotted line is the 1:1 fit line between the predicted and measured values. Pearson correlation analysis of hyperspectral reflectance and first derivatives with SPAD values in CK, DS, and combined data sets was performed using Prism 9. Python 3.6 was used for SLR analysis of the data. CK, control; DS, drought stress; combined, all data.

2.4. The Characteristic Bands Identified with the Successive Projections Algorithm for Estimating SPAD Values

The successive projections algorithm (SPA) can eliminate redundant hyperspectral information and improve modeling speed and efficiency, and it is widely used for extracting characteristic bands. The hyperspectral reflectance bands at 536, 596, and 674 nm (Figure 4A), and the first derivative bands at 756 and 778 nm (Figure 4B), were the optimal bands for estimating SPAD values. These results indicate that hyperspectral information at the green, red, and red-edge bands is closely related to wheat SPAD values, which can therefore be used to monitor wheat leaf chlorophyll content.

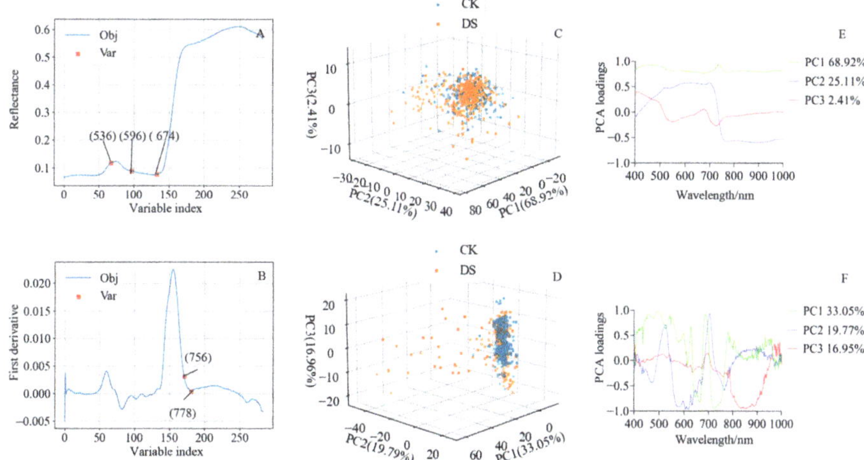

Figure 4. SPA feature band selection and PCA dimension reduction. (**A,B**) Spectral reflectance and first derivative bands extracted by SPA; (**C,D**) 3D spatial distribution of the first three principal components of spectral reflectance and the first derivative; (**E,F**) the first three principal components' loadings of spectral reflectance and the first derivative. Python 3.6 was used to extract SPAD characteristic bands. Prism 9 was used to calculate PCA loadings. The percentage is the proportion of the variance explained by each principal component. SPA, successive projections algorithm; PCA, principal component analysis. Obj, first calibration object; Var, selected variables. CK, control; DS, drought stress.

2.5. Principal Component Analysis of Hyperspectral Information

Principal component analysis (PCA) is a dimensionality reduction technique widely used for hyperspectral data analysis that can reduce dimensionality with minimal information loss. Principal component loading represents the correlation coefficient between a principal component and the original variable. In the present study, PCA was used to reduce the dimensionality of the hyperspectral reflectance data and first derivatives, and the loadings of the first three principal components were calculated. Figure 4C,D present the 3D spatial distribution of leaf spectral reflectance and the first derivative, respectively. The first three principal components were dispersed under drought stress but more aggregated under control conditions (Figure 4C,D). The variance explained by the first principal component of spectral reflectance (68.92%) was higher than that explained by the first derivative (33.05%). These results indicate that drought stress significantly affected leaf blade hyperspectral reflectance, and the collinearity of the hyperspectral first derivative was lower than that of reflectance. Together, the first two principal components of spectral reflectance explained more than 94% of the variance in the data; principal component 1 had high loadings at 446 and 695 nm, while principal component 2 had high loadings at 624, 695, and 770 nm (Figure 4E). The first three principal components of the first derivative of the spectrum explained nearly 70% of the variance; principal component 1 had high loadings at 494, 684, and 722 nm; principal component 2 had high loadings at 523, 616, and 703 nm; and principal component 3 had a high loading at 852 nm (Figure 4F). These results indicate that the drought stress spectra were strongly correlated with the principal components at these wavebands.

2.6. Estimation of SPAD Values Based on Regression Analysis

2.6.1. Estimation of SPAD Values Based on Spectral Characteristics

Henceforth, CA-R and CA-FD are used to denote the highest correlations of spectral reflectance and the first derivative, with SPAD values, while (SPA-R) and (SPA-FD) are used to indicate characteristic band reflectance and the first derivative, as extracted by SPA. Full-R and Full-FD are used to denote full-band reflectance and the first derivative, respectively, whereas PCA-R and PCA-FD are used to indicate spectral reflectance and the first derivative principal components, respectively, derived from the PCA.

The regression model results indicated that the RFR model using Full-R as the independent variable had the best optimal fit (R^2 = 0.60, RMSE = 4.495, RE = 7.35%) among all models. The SLR model based on the CA-R had the worst fit (training set, R^2 = 0.401, RMSE = 5.549, RE = 9.17%) due to the underfitting of the data (Table 1). The models based on Full-R were better than those based on Full-FD (Table 1, Figure 5E,F); although, the models built with CA-FD and SPA-FD provided better fits than CA-R and SPA-R (Figure 5A–D). The models built with SPA-R, SPA-FD, PCA-R, and PCA-FD were superior to the SLR models but inferior to the Full-R and Full-FD models (Table 1). The LASSO and RR models based on SPA-R, SPA-FD, PCA-R, and PCA-FD exhibited underfitting (Table 1).

Table 1. The different models fitting results between SPAD and various spectral characteristics.

Data Set	Model	Training Set	(n = 536)		Testing Set	(n = 134)	
		R^2	RMSE	RE	R^2	RMSE	RE
CA-R (549 nm)	SLR	0.401	5.549	9.17%	0.424	5.390	8.81%
CA-FD (735 nm)	SLR	0.426	5.434	9.01%	0.529	4.876	8.27%
	LASSO	0.405	5.532	9.12%	0.417	5.430	8.86%
SPA-R	RR	0.405	5.532	9.12%	0.417	5.426	8.85%
	RFR	0.478	5.183	8.60%	0.478	5.134	8.40%
	LASSO	0.407	5.523	9.25%	0.518	4.934	8.49%
SPA-FD	RR	0.407	5.523	9.26%	0.517	4.939	8.51%
	RFR	0.510	5.023	8.57%	0.510	4.974	8.47%
	LASSO	0.484	5.150	8.39%	0.571	4.655	7.64%
PCA-R	RR	0.488	5.130	8.20%	0.580	4.607	7.47%
	RFR	0.555	4.788	7.87%	0.555	4.739	7.89%
	LASSO	0.454	5.298	8.62%	0.496	5.045	7.90%
PCA-FD	RR	0.454	5.298	8.64%	0.497	5.039	7.99%
	RFR	0.560	4.755	7.74%	0.560	4.714	7.62%
	LASSO	0.587	4.609	7.43%	0.585	4.578	7.51%
Full-R	RR	0.586	4.617	7.45%	0.585	4.575	7.46%
	RFR	0.600	4.535	7.40%	0.600	4.495	7.35%
	LASSO	0.528	4.929	8.12%	0.528	4.880	7.96%
Full-FD	RR	0.548	4.824	7.81%	0.547	4.784	7.71%
	RFR	0.579	4.653	7.63%	0.577	4.623	7.64%

n, number of samples; R^2, coefficient of determination; RMSE, root mean square error; RE, relative error (percentage); SLR, simple linear regression; LASSO, least absolute shrinkage and selection operator; RR, ridge regression; RFR, random forest regression. CA-R/FD, reflectance/first derivative with the highest correlation with SPAD values; SPA-R/FD, reflectance/first derivative of a characteristic band extracted through SPA; PCA-R/FD, principal components of reflectance/first derivative; Full-R/FD, full-band reflectance/first derivative; SPA, successive projections algorithm; PCA, principal component analysis.

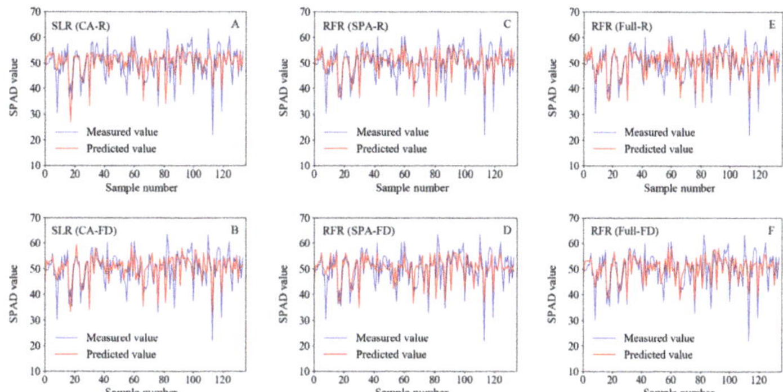

Figure 5. Comparison of predicted and measured values SPAD based on different data sets and models. (**A**,**B**) Simple linear regression (SLR) model based on CA-R and CA-FD; (**C**,**D**) random forest regression (RFR) model based on SPA-R and SPA-FD; (**E**,**F**) RFR model based on Full-R and Full-FD. CA-R/FD, reflectance/first derivative with the highest correlation with SPAD values; SPA-R/FD, reflectance/first derivative of a characteristic band extracted through SPA; Full-R/FD, full-band reflectance/first derivative. CK, control; DS, drought stress.

The fit of the SLR model established with the first derivative at 735 nm (CA-FD) was better than that of the model based on spectral reflectance at 549 nm (CA-R), the data in the former case exhibited a stronger linear relationship with the SPAD values (Figure 3C–F). The dotted lines in Figure 3E,F, Figure 6, and Figure 7 are the 1:1 fit line between the predicted and measured SPAD values, which indicate that the RFR model based on Full-R provided the best fit (Figure 6C). The deviation between the actual and 1:1 fit lines of the models based on Full-R (Figure 6A–C) was smaller compared with the models based on Full-FD (Figure 6D–F). These results indicate that the first derivative transformation can enhance the signal-to-noise ratio of the characteristic bands and improve model accuracy, which may result in the loss of some full-band hyperspectral information. At the same time, the feature band spectrum did not adequately represent whole-leaf hyperspectral information, while the full-band hyperspectral image appears to have more potential for chlorophyll content evaluation.

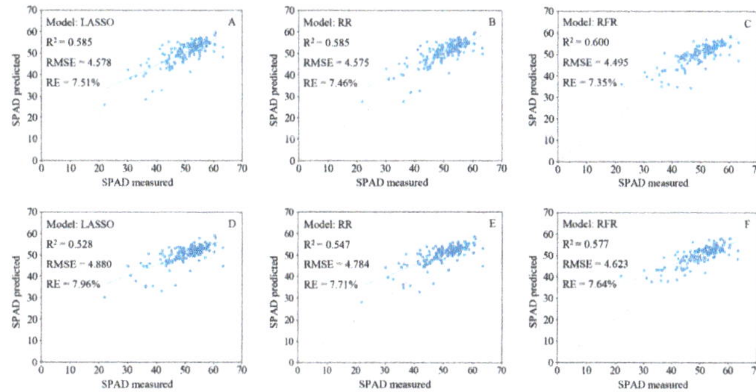

Figure 6. Fitting results of predicted and measured SPAD values of wheat leaves based on a full-band hyperspectral reflectance and first derivative model. (**A–C**) Least absolute shrinkage and selection operator (LASSO) regression, ridge regression (RR), and random forest regression (RFR) models built with full-band hyperspectral reflectance; (**D–F**) LASSO regression, RR, and RFR models built with full-band hyperspectral first derivative. The gray dotted line is the 1:1 fit between the predicted and measured SPAD values, and the blue solid line is the actual fit between predicted and measured values.

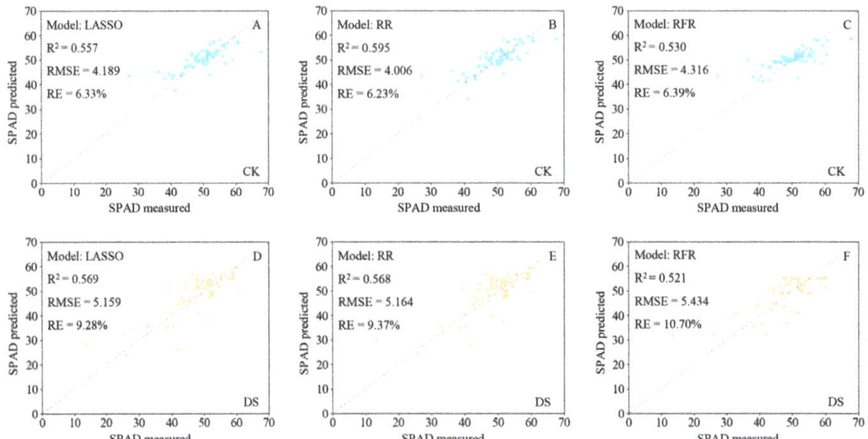

Figure 7. Fitting results of predicted and measured SPAD values of wheat leaves based on a full-band hyperspectral reflectance model under different soil moisture conditions. (**A–C**) Least absolute shrinkage and selection operator (LASSO) regression, ridge regression (RR), and random forest regression (RFR) models under the control condition; (**D–F**) LASSO regression, RR, and RFR models under the drought stress condition. The blue and yellow solid lines are the actual fit between predicted and measured values, and the gray dotted line is the 1:1 fit between the predicted and measured SPAD values. CK, control; DS, drought stress.

The LASSO, RR, and RFR models built with the Full-R demonstrate outstanding stability and accuracy in predicting the SPAD values under different soil moisture conditions (RFR models $R^2 > 0.5$; Figure 7). The LASSO model prediction performance of SPAD values under drought stress is the best with R^2 0.569, RMSE 5.159, and RE 9.28%. As a result, the hyperspectral reflectance in predicting the SPAD values of the wheat leaves under drought stress can provide a robust result and has the potential to be used in drought resistance identification in the future.

To further verify the reliability of models, the SLR models constructed by 549 nm reflectance and 735 nm first derivative were used for SPAD visual mapping (Figure 8). These results are consistent with leaf RGB images (Figure 2E), which indicates that the HSI monitoring SPAD values and drought stress are reliable.

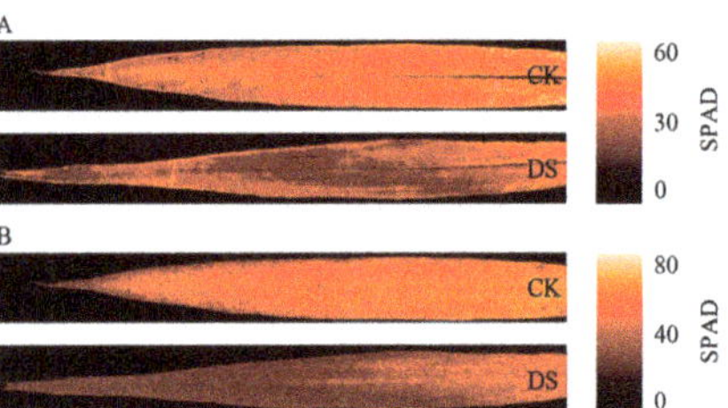

Figure 8. SPAD values map at the leaf level estimated by the characteristic reflectance and first derivative. (**A,B**) SPAD values map at the leaf level estimated by the 549 nm reflectance and 735 nm first derivative. CK, control; DS, drought stress. ENVI 5.3 (ITT, Visual Information Solutions, Boulder, CO, USA) was used to obtain the SPAD values map.

2.6.2. Estimation of SPAD Values Based on Spectral and Image Characteristics

CIEL*a*b* is a visually consistent color model in which the color difference is closer to the actual perceived color difference and is the best way to express the range of perceived colors by the human eye. The L*, a*, and b* values have significant correlations with SPAD values (r = −0.591, −0.164, −0.600; Table 2). Furthermore, we combined L*a*b* features and spectral characteristics to construct RFR models, which indicates that the Full-R combined with L*, a*, and b* has the optimal SPAD values prediction effect (R^2 = 0.61, RMSE = 4.439, RE = 7.35%). Overall, the combination of spectral characteristics and L*a*b* features improve the estimation accuracy for SPAD values (Table 3).

Table 2. Relationships among SPAD, L*, a*, and b* of wheat leaves.

Variable	SPAD	L*	a*	b*
SPAD	1			
L*	−0.591 **			
a*	−0.164 **	0.438 **		
b*	−0.600 **	0.912 **	0.378 **	1

** Significant at the 0.01 probability level.

Table 3. Results of the RFR models based on spectrum and L*a*b* characteristics data set.

Data Set	Training Set (*n* = 536)			Testing Set (*n* = 134)		
	R^2	RMSE	RE	R^2	RMSE	RE
(CA-R) + L*a*b*	0.486	5.14	8.52%	0.486	5.095	8.23%
(CA-FD) + L*a*b*	0.519	4.973	8.19%	0.519	4.925	8.18%
(SPA-R) + L*a*b*	0.502	5.061	8.38%	0.501	5.021	8.27%
(SPA-FD) + L*a*b*	0.51	5.019	8.37%	0.51	4.972	8.13%
(PCA-R) + L*a*b*	0.46	5.271	8.76%	0.46	5.218	8.29%
(PCA-FD) + L*a*b*	0.584	4.626	7.47%	0.584	4.579	7.48%
(Full-R) + L*a*b*	0.61	4.478	7.30%	0.61	4.439	7.35%
(Full-FD) + L*a*b*	0.578	4.661	7.64%	0.578	4.617	7.58%
L*a*b*	0.435	5.39	9.06%	0.434	5.346	8.38%

L*a*b* represent L*, a*, and b* color component values.

3. Discussion

3.1. Feasibility of Estimating Chlorophyll Content of Wheat Leaves Using Hyperspectral Information

Previous studies have demonstrated that plant leaf hyperspectral curves exhibit significant differences in the visible region under stress conditions. For example, Gang et al. (2010) reported a significant correlation between chlorophyll content and spectral characteristics in the visible region (400–680 nm) in castor bean seedlings under salt stress [52]. In the present study, the hyperspectral reflectance curves of wheat leaves under different soil moisture conditions varied remarkably in the visible region (500–700 nm; Figure 2A,C) and were strongly correlated with SPAD values (Figure 3A). These findings indicate that the hyperspectral reflectance curves of wheat leaves under stress conditions were closely related to the chlorophyll content in the visible region (400–700 nm), and, therefore, adequately represent leaf chlorophyll content.

Studies of tea leaves have shown that the hyperspectral reflectance between 552–555 nm and 707–735 nm is strongly correlated with the chlorophyll content [53]. Under drought stress, the spectral first derivative of apple leaves in the visible region (513–539, 564–585, 694, and 699 nm) was also closely related to the chlorophyll content [54]. In the present study, leaf spectral reflectance between 512–603 and 700–723 nm, and the spectral first derivative between 485–542 and 719–760 nm, were strongly correlated with SPAD values (Figure 3A,B). These results highlight the feasibility of using hyperspectral characteristics to estimate chlorophyll content accurately and efficiently under drought stress.

3.2. Models for Estimating SPAD Values in Wheat Leaves

Our SPAD values estimation model based on full-band hyperspectral information exhibited good performance. In a previous study using characteristic bands and full-band hyperspectral reflectance to estimate wheat yield under low-temperature stress, a full-band principal component regression (PCR) model (R^2 = 0.854, root mean square error of prediction 625.7) had the best stability [55]. In the present study, the LASSO, RR, and RFR models based on Full-R were superior to all other models and had R^2 and RMSE values of 0.585 and 4.578, 0.585 and 4.575, and 0.60 and 4.495, respectively (Figure 6A–C, Table 1). These findings demonstrate that models based on Full-R had better stability and accuracy.

The first derivative can reduce the influence of noise and improve the signal-to-noise ratio of the data. Previous chlorophyll content evaluations of pitaya stems revealed that first derivative transformation can enhance the sensitivity of feature bands to chlorophyll content, and a model based on the first derivative of characteristic bands was optimal (R^2 = 0.625, RMSE = 0.048) [56]. A study estimating maize yield found that full-band RR (R^2 = 0.54, RMSE = 2.58) and support vector regression (SVR; R^2 = 0.53, RMSE = 2.69) models performed better than a full-band first derivative model (R^2 = 0.41, RMSE = 3.51 and R^2 = 0.49, RMSE = 2.95, respectively); however, RFR models showed the opposite result [57].

The present study demonstrated that the combining of spectral characteristics with image features (L* a* b*) can improve the estimation accuracy of SPAD values (Tables 1 and 3). These results showed that the fit for diverse data sets varied among the models. Thus, it is necessary to determine which models are most suitable for particular data sets to achieve the best prediction performance. In future studies, we can utilize multiplicative scatter correction (MSC), standard normal variate transformation (SNV), and other algorithms to preprocess hyperspectral data and apply deep learning to enhance stability and accuracy.

3.3. Utility of Hyperspectral Reflectance for Monitoring Wheat Growth and Evaluating Drought Resistance under Drought Stress

The hyperspectral characteristics of wheat leaves were closely related to wheat growth status under different soil moisture conditions. The hyperspectral information of soybean leaves can achieve the accurate estimation for chlorophyll content (R^2 = 0.94, RMSE = 0.201) [49]. Similarly, previous research showed that 617, 675, and 818 nm are the optimal bands for estimating the chlorophyll content of diseased peach fruit [58]. Using the continuous wavelet transform (CWT) to estimate wheat SPAD values under low-temperature stress, a previous study found that 553, 727, 728, 729, and 734 nm are SPAD-sensitive bands, and that spectral reflectance at 553 nm can accurately estimate SPAD values (R^2 = 0.7444, RMSE = 7.359) [59]. One study using multiple feature selection methods to determine SPAD values in pepper leaves reported that the characteristic bands were concentrated within the regions of 415.4–431.5, 526.7–676.2, and 839.3–979.2 nm [60]. Furthermore, 548, 718, and 727 nm were the best wave bands for estimating chlorophyll content in grafted cucumber seedling leaves [61]. Taken together, these studies clearly indicate that hyperspectral information is closely related to chlorophyll content; however, it was still necessary to establish a stable, reliable, and universally applicable model for evaluating diverse species, and for within-species analyses, under different growth conditions. In the present study, full-band hyperspectral reflectance of combined L*, a*, and b* accurately estimated SPAD values (R^2 = 0.61, RMSE = 4.439, RE = 7.35%; Table 3). The spectral reflectance bands at 536, 549, 596, 674, and 708 nm, and the first derivative bands at 735, 756, and 778 nm, were SPAD characteristic bands (Figure 3A,B and Figure 4A,B). Nonetheless, many bands within the 446–770 nm region were closely related to drought stress (Figure 4E,F). Using SLR based on 549 nm reflectance and 735 nm first derivative to assess SPAD values in the leaf level found that the HSI can monitor leaf chlorophyll content and drought stress reliably (Figure 8). In conclusion, plant HSI has great potential for evaluating leaf chlorophyll content under stress conditions; however, studies including large-scale varieties experiencing different stress conditions are necessary to establish stable and reliable models.

In most cases, drought stress reduces leaf chlorophyll content, although several studies have shown that some plants exhibit increased chlorophyll content under drought stress [62,63]. We found that the leaf SPAD values of several wheat varieties increased under drought stress (Table S2), which may be related to the variation in drought resistance seen among different wheat varieties. The characteristics of the hyperspectral curves of different wheat varieties under drought stress also reflected differences in drought resistance among varieties. Therefore, future evaluation models of wheat drought resistance could be applied to aid genetic breeding of crop resistance.

4. Materials and Methods

4.1. Plant Material and Growth Conditions

The study area was located in the National Dryland Plant Variety Rights Trading Center in Yang Ling District (108°4′E, 34°16′N), China, in which new crop varieties are tested. In total, 335 wheat varieties (Table S1) were planted in a steel frame shed on 21 October 2021. The shed had good ventilation and fine sandy loam soils, to which we applied a compound fertilizer (N-P$_2$O5-K$_2$O) at 750 kg/hm^2 prior to sowing. The soil drilling method was used to measure soil water content at a depth of 0.5 m. Beginning at the jointing stage of wheat growth, the relative soil water contents of the control and drought stress treatments were maintained at 75 ± 5% and 50 ± 5%, respectively. The calculation method of relative soil water content is consistent with previous research [64], and uses the formula below:

$$\text{Soil water content }(\%) = \frac{\text{weight of moist soil } - \text{ weight of dried soil}}{\text{weight of dried soil}} \times 100$$

$$\text{Soil water holding capacity }(\%) = \frac{\text{weight of water in saturated soil}}{\text{weight of dried soil}} \times 100$$

$$\text{Relative soil water content }(\%) = \frac{\text{soil water content}}{\text{soil water holding capacity}} \times 100$$

The filling stage is the most rapid stage of wheat grain development, and photosynthetic capacity during this period strongly affects yield. Meanwhile, drought stress at the wheat filling stage has a significant impact on photosynthesis and wheat yield [65,66]. Therefore, it is important to estimate the chlorophyll content of wheat under drought stress in the filling stage for screening the drought-resistant wheat varieties. The experimental data were collected after flowering for 14 days during April and May 2022.

4.2. Hyperspectral Image Acquisition

Leaf hyperspectral images were acquired using the Pika L system (Resonon Inc., Bozeman, MT, USA), which provides 281 spectral bands (2.1-nm spectral resolution) in the visible near-infrared region (400–1000 nm) and has 900 spatial channels. We collected six flag leaves samples from each variety, which were kept fresh [67] and transported to the laboratory for hyperspectral image collection under controlled lighting conditions. Before image acquisition, the dark current was removed and whiteboard calibration was conducted. During the collection stage, the leaf blades were wiped clean and kept flat. The average of three biological replicates was used as the hyperspectral data for subsequent analysis.

4.3. SPAD Values Measurement

A SPAD-502 Plus chlorophyll meter (Minolta, Tokyo, Japan) was used to obtain soil plant analysis development (SPAD; i.e., chlorophyll) values for wheat flag leaves, and this meter has been shown to correlate well with actual chlorophyll levels and to be a reliable method for nondestructive chlorophyll detection [68–70]. SPAD values of the tip, middle, and base of the leaf were obtained (five replicates each), and the average values were used in the analysis.

4.4. Hyperspectral Image Preprocessing

Hyperspectral images were obtained at the same time as SPAD values acquisition, which was preprocessed using the Spectronon Pro software (version 2.116)bundled with the Resonon Pika L acquisition system. A Savitzky–Golay filter was used to smooth the hyperspectral images. The first derivative was calculated using Spectronon Pro. In this study, Spectronon Pro was used to acquire CIEL*a*b* color space images and calculate L*, a*, and b* values.

4.5. Data Processing

Prism 9 (GraphPad Software, San Diego, CA, USA) and Python 3.6 (Python Software Foundation, Beaverton, OR, USA) were used for correlation and regression analyses. The training and testing sets included 80% and 20% of the data ($n = 670$), respectively. Values were calculated for model evaluation, coefficient of determination (R^2), root mean square error (RMSE), and relative error (RE). Four regression models were tested: simple linear regression (SLR) [71], least absolute shrinkage and selection operator (LASSO) regression [72], ridge regression (RR) [73], and random forest regression (RFR) models [74].

5. Conclusions

In this study, the HSI of large-scale wheat varieties under different soil moisture conditions was used to determine the accuracy of rapid leaf chlorophyll content estimation models. The most sensitive bands with respect to chlorophyll content estimation were in the visible band (400–780 nm), and correlation analysis revealed that the optimal bands were located near 541, 549, 708, and 735 nm. The SPA indicated that the reflectance bands at 536, 596, and 674 nm were the optimal bands for estimating SPAD values. The first derivative bands at 756 and 778 nm were the most useful for estimating the relative chlorophyll content. Combining spectral characteristics and L*a*b* features can improve the accuracy of estimating the SPAD values of drought-stressed wheat (RFR model optimal performance: $R^2 = 0.61$, RMSE = 4.439, RE = 7.35%). The technical method established in this study has great potential for evaluating chlorophyll content and stress resistance.

Supplementary Materials: The following supporting information can be downloaded at: https://www.mdpi.com/article/10.3390/ijms24065825/s1.

Author Contributions: C.Z. and Y.Y. conceived and designed the study; Y.Y., R.N., T.M., Y.S. and F.S. (Fanghui Shi) collected the wheat samples; Y.Y., Y.W. and C.Z. analyzed the data; Y.Y. and X.L. wrote the manuscript; F.S. (Fengli Sun), Y.X. and C.Z. revised the manuscript. All authors have read and agreed to the published version of the manuscript.

Funding: This research was funded by the National Natural Science Foundation of China (31970351); the Key Research and Development Program of Shanxi Province, China (2022NY-193); and the Xi'an Science and Technology Planning Project (22NYYF001).

Institutional Review Board Statement: Not applicable.

Informed Consent Statement: Not applicable.

Data Availability Statement: All data are contained within the article.

Conflicts of Interest: The authors declare that the research was conducted in the absence of any commercial or financial relationships that could be construed as a potential conflict of interest.

References

1. Fernie, E.; Tan, D.K.Y.; Liu, S.Y.; Ullah, N.; Khoddami, A. Post-Anthesis Heat Influences Grain Yield, Physical and Nutritional Quality in Wheat: A Review. *Agriculture* **2022**, *12*, 886. [CrossRef]
2. Rane, J.; Singh, A.K.; Kumar, M.; Boraiah, K.M.; Meena, K.K.; Pradhan, A.; Prasad, P.V.V. The Adaptation and Tolerance of Major Cereals and Legumes to Important Abiotic Stresses. *Int. J. Mol. Sci.* **2021**, *22*, 12970. [CrossRef] [PubMed]
3. Daryanto, S.; Wang, L.; Jacinthe, P.-A. Global synthesis of drought effects on cereal, legume, tuber and root crops production: A review. *Agric. Water Manag.* **2017**, *179*, 18–33. [CrossRef]
4. Zhang, J.; Zhang, S.; Cheng, M.; Jiang, H.; Zhang, X.; Peng, C.; Lu, X.; Zhang, M.; Jin, J. Effect of Drought on Agronomic Traits of Rice and Wheat: A Meta-Analysis. *Int. J. Environ. Res. Public Health* **2018**, *15*, 839. [CrossRef]
5. Ali, H.E.; Saad, Z.H.; Elsayed, H. The effect of drought on chlorophyll, proline and chemical composition of three varieties of egyptian rice. *J. Biol. Chem.* **2020**, *15*, 21–30.
6. Sharifi, P.; Mohammadkhani, N. Effects of Drought Stress on Photosynthesis Factors in Wheat Genotypes during Anthesis. *Cereal Res. Commun.* **2016**, *44*, 229–239. [CrossRef]

7. Qian, X.; Liu, L.; Croft, H.; Chen, J. Relationship between leaf maximum carboxylation rate and chlorophyll content preserved across 13 species. *J. Geophys. Res.-Biogeosciences* **2021**, *126*, e2020JG006076. [CrossRef]
8. Pandey, A.; Masthigowda, M.H.; Kumar, R.; Pandey, G.C.; Awaji, S.M.; Singh, G.; Singh, G.P. Physio-biochemical characterization of wheat genotypes under temperature stress. *Physiol. Mol. Biol. Plants* **2023**, *29*, 131–143. [CrossRef]
9. Makino, A. Photosynthesis, Grain Yield, and Nitrogen Utilization in Rice and Wheat. *Plant Physiol.* **2010**, *155*, 125–129. [CrossRef]
10. Tanaka, A.; Tanaka, R. Chlorophyll metabolism. *Curr. Opin. Plant Biol.* **2006**, *9*, 248–255. [CrossRef]
11. Wang, G.; Zeng, F.; Song, P.; Sun, B.; Wang, Q.; Wang, J. Effects of reduced chlorophyll content on photosystem functions and photosynthetic electron transport rate in rice leaves. *J. Plant Physiol.* **2022**, *272*, 153669. [CrossRef] [PubMed]
12. Li, Y.; He, N.; Hou, J.; Xu, L.; Liu, C.; Zhang, J.; Wang, Q.; Zhang, X.; Wu, X. Factors Influencing Leaf Chlorophyll Content in Natural Forests at the Biome Scale. *Front. Ecol. Evol.* **2018**, *6*, 64. [CrossRef]
13. Hörtensteiner, S.; Kräutler, B. Chlorophyll breakdown in higher plants. *Biochim. Et Biophys. Acta (BBA)-Bioenerg.* **2011**, *1807*, 977–988. [CrossRef]
14. Hashimoto, H.; Uragami, C.; Cogdell, R.J. Carotenoids and Photosynthesis. In *Carotenoids in Nature: Biosynthesis, Regulation and Function*; Stange, C., Ed.; Springer International Publishing: Cham, Switzerland, 2016; pp. 111–139.
15. Yang, Y.-Z.; Li, T.; Teng, R.-M.; Han, M.-H.; Zhuang, J. Low temperature effects on carotenoids biosynthesis in the leaves of green and albino tea plant (*Camellia sinensis* (L.) O. Kuntze). *Sci. Hortic.* **2021**, *285*, 110164. [CrossRef]
16. Nikolaeva, M.K.; Maevskaya, S.N.; Shugaev, A.G.; Bukhov, N.G. Effect of drought on chlorophyll content and antioxidant enzyme activities in leaves of three wheat cultivars varying in productivity. *Russ. J. Plant Physiol.* **2010**, *57*, 87–95. [CrossRef]
17. Chowdhury, M.K.; Hasan, M.A.; Bahadur, M.M.; Islam, M.R.; Hakim, M.A.; Iqbal, M.A.; Javed, T.; Raza, A.; Shabbir, R.; Sorour, S.; et al. Evaluation of drought tolerance of some wheat (*Triticum aestivum* L.) genotypes through phenology, growth, and physiological Indices. *Agron.-Basel* **2021**, *11*, 1792. [CrossRef]
18. Chen, J.L.; Zhao, X.Y.; Zhang, Y.Q.; Li, Y.Q.; Luo, Y.Q.; Ning, Z.Y.; Wang, R.X.; Wang, P.Y.; Cong, A.Q. Effects of drought and rehydration on the physiological responses of artemisia halodendron. *Water* **2019**, *11*, 793. [CrossRef]
19. Xie, H.Y.; Li, M.R.; Chen, Y.J.; Zhou, Q.P.; Liu, W.H.; Liang, G.L.; Jia, Z.F. Important physiological changes due to drought stress on oat. *Front. Ecol. Evol.* **2021**, *9*, 644726. [CrossRef]
20. Wasaya, A.; Manzoor, S.; Yasir, T.A.; Sarwar, N.; Mubeen, K.; Ismail, I.A.; Raza, A.; Rehman, A.; Hossain, A.; El Sabagh, A. Evaluation of fourteen bread wheat (*Triticum aestivum* L.) genotypes by observing gas exchange parameters, relative water and chlorophyll content, and yield attributes under drought stress. *Sustainability* **2021**, *13*, 4799. [CrossRef]
21. Saha, S.; Begum, H.H.; Nasrin, S.; Samad, R. Effects of drought stress on pigment and protein contents and antioxidant enzyme activities in five varieties of rice (*Oryza sativa* L.). *Bangladesh J. Bot.* **2020**, *49*, 997–1002. [CrossRef]
22. Hinge, P.; Kale, A.; Pawar, B.; Jadhav, A.; Chimote, V.; Gadakh, S. Effect of PEG induced water stress on chlorophyll content, membrane injury index, osmoprotectants and antioxidant enzymes activities in sorghum (*Sorghum bicolor* (L) Moench). *Maydica* **2015**, *60*, 1–10.
23. Liu, S.; Li, S.; Fan, X.Y.; Yuan, G.D.; Hu, T.; Shi, X.M.; Huang, J.B.; Pu, X.Y.; Wu, C.S. Comparison of two noninvasive methods for measuring the pigment content in foliose macrolichens. *Photosynth. Res.* **2019**, *141*, 245–257. [CrossRef]
24. Zhao, Y.; Yan, C.; Lu, S.; Wang, P.; Li, R. Estimation of chlorophyll content in intertidal mangrove leaves with different thicknesses using hyperspectral data. *Ecol. Indic.* **2019**, *106*, 105511. [CrossRef]
25. Asaari, M.S.M.; Mertens, S.; Verbraeken, L.; Dhondt, S.; Inze, D.; Bikram, K.; Scheunders, P. Non-destructive analysis of plant physiological traits using hyperspectral imaging: A case study on drought stress. *Comput. Electron. Agric.* **2022**, *195*, 106806. [CrossRef]
26. Xia, J.; Cao, H.; Yang, Y.; Zhang, W.; Huang, B. Detection of waterlogging stress based on hyperspectral images of oilseed rape leaves (*Brassica napus* L.). *Comput. Electron. Agric.* **2019**, *159*, 59–68. [CrossRef]
27. Huang, Y.B.; Chen, Z.X.; Yu, T.; Huang, X.Z.; Gu, X.F. Agricultural remote sensing big data: Management and applications. *J. Integr. Agric.* **2018**, *17*, 1915–1931. [CrossRef]
28. Tan, J.Y.; Ker, P.J.; Lau, K.Y.; Hannan, M.A.; Tang, S.G.H. Applications of Photonics in Agriculture Sector: A Review. *Molecules* **2019**, *24*, 2025. [CrossRef]
29. Mahlein, A.K. Plant Disease Detection by Imaging Sensors—Parallels and Specific Demands for Precision Agriculture and Plant Phenotyping. *Plant Dis.* **2016**, *100*, 241–251. [CrossRef]
30. Stewart, E.L.; Lucas, G.B.; Campbell, C.L.; Lucas, L.T. Introduction to plant diseases: Identification and management. *Mycologia* **1991**, *83*, 243. [CrossRef]
31. Mishra, P.; Asaari, M.S.M.; Herrero-Langreo, A.; Lohumi, S.; Diezma, B.; Scheunders, P. Close range hyperspectral imaging of plants: A review. *Biosyst. Eng.* **2017**, *164*, 49–67. [CrossRef]
32. Carter, G.A.; Knapp, A.K. Leaf optical properties in higher plants: Linking spectral characteristics to stress and chlorophyll concentration. *Am. J. Bot.* **2001**, *88*, 677–684. [CrossRef]
33. Cui, B.; Zhao, Q.J.; Huang, W.J.; Song, X.Y.; Ye, H.C.; Zhou, X.F. Leaf chlorophyll content retrieval of wheat by simulated RapidEye, Sentinel-2 and EnMAP data. *J. Integr. Agric.* **2019**, *18*, 1230–1245. [CrossRef]
34. Xie, Q.Y.; Dash, J.; Huang, W.J.; Peng, D.L.; Qin, Q.M.; Mortimer, H.; Casa, R.; Pignatti, S.; Laneve, G.; Pascucci, S.; et al. Vegetation Indices combining the Red and Red-Edge spectral information for leaf area index retrieval. *IEEE J. Sel. Top. Appl. Earth Obs. Remote Sens.* **2018**, *11*, 1482–1493. [CrossRef]

35. Zhao, D.; Raja Reddy, K.; Kakani, V.G.; Read, J.J.; Carter, G.A. Corn (*Zea mays* L.) growth, leaf pigment concentration, photosynthesis and leaf hyperspectral reflectance properties as affected by nitrogen supply. *Plant Soil* **2003**, *257*, 205–218. [CrossRef]

36. Carter, G.A.; Spiering, B.A. Optical properties of intact leaves for estimating chlorophyll concentration. *J. Environ. Qual.* **2002**, *31*, 1424–1432. [CrossRef]

37. Watt, M.S.; Leonardo, E.M.C.; Estarija, H.J.C.; Massam, P.; de Silva, D.; O'Neill, R.; Lane, D.; McDougal, R.; Buddenbaum, H.; Zarco-Tejada, P.J. Long-term effects of water stress on hyperspectral remote sensing indicators in young radiata pine. *For. Ecol. Manage.* **2021**, *502*, 119707. [CrossRef]

38. Xie, Y.K.; Feng, M.C.; Wang, C.; Yang, W.D.; Sun, H.; Yang, C.B.; Jing, B.H.; Qiao, X.X.; Kubar, M.S.; Song, J.Y. Hyperspectral monitor on chlorophyll density in winter wheat under water stress. *Agron. J.* **2020**, *112*, 3667–3676. [CrossRef]

39. Zarco-Tejada, P.J.; Hornero, A.; Beck, P.; Kattenborn, T.; Kempeneers, P.; Hernández-Clemente, R. Chlorophyll content estimation in an open-canopy conifer forest with Sentinel-2A and hyperspectral imagery in the context of forest decline. *Remote Sens. Environ.* **2019**, *223*, 320–335. [CrossRef] [PubMed]

40. Sonobe, R.; Sano, T.; Horie, H. Using spectral reflectance to estimate leaf chlorophyll content of tea with shading treatments. *Biosyst. Eng.* **2018**, *175*, 168–182. [CrossRef]

41. Zhang, H.C.; Ge, Y.F.; Xie, X.Y.; Atefi, A.; Wijewardane, N.K.; Thapa, S. High throughput analysis of leaf chlorophyll content in sorghum using RGB, hyperspectral, and fluorescence imaging and sensor fusion. *Plant Methods* **2022**, *18*, 1–17. [CrossRef]

42. Peng, Y.; Fan, M.; Wang, Q.H.; Lan, W.J.; Long, Y.T. Best hyperspectral indices for assessing leaf chlorophyll content in a degraded temperate vegetation. *Ecol. Evol.* **2018**, *8*, 7068–7078. [CrossRef] [PubMed]

43. Qi, H.X.; Zhu, B.Y.; Kong, L.X.; Yang, W.G.; Zou, J.; Lan, Y.B.; Zhang, L. Hyperspectral inversion model of chlorophyll content in peanut leaves. *Appl. Sci.* **2020**, *10*, 2259. [CrossRef]

44. Zhang, Y.; Hui, J.; Qin, Q.; Sun, Y.; Zhang, T.; Sun, H.; Li, M. Transfer-learning-based approach for leaf chlorophyll content estimation of winter wheat from hyperspectral data. *Remote Sens. Environ.* **2021**, *267*, 112724. [CrossRef]

45. Colombo, R.; Meroni, M.; Marchesi, A.; Busetto, L.; Rossini, M.; Giardino, C.; Panigada, C. Estimation of leaf and canopy water content in poplar plantations by means of hyperspectral indices and inverse modeling. *Remote Sens. Environ.* **2008**, *112*, 1820–1834. [CrossRef]

46. Eitel, J.U.H.; Gessler, P.E.; Smith, A.M.S.; Robberecht, R. Suitability of existing and novel spectral indices to remotely detect water stress in *Populus* spp. *For. Ecol. Manag.* **2006**, *229*, 170–182. [CrossRef]

47. Sytar, O.; Brestic, M.; Zivcak, M.; Olsovska, K.; He, X. Applying hyperspectral imaging to explore natural plant diversity towards improving salt stress tolerance. *Sci. Total Environ.* **2017**, *578*, 90–99. [CrossRef] [PubMed]

48. Wahabzada, M.; Mahlein, A.-K.; Bauckhage, C.; Steiner, U.; Oerke, E.C.; Kersting, K. Plant phenotyping using probabilistic topic models: Uncovering the hyperspectral language of plants. *Sci. Rep.* **2016**, *6*, 22482. [CrossRef]

49. Liu, B.; Yue, Y.M.; Li, R.; Shen, W.J.; Wang, K.L. Plant leaf chlorophyll content retrieval based on a field imaging spectroscopy system. *Sensors* **2014**, *14*, 19910–19925. [CrossRef]

50. George, R.; Padalia, H.; Sinha, S.K.; Kumar, A.S. Evaluating sensitivity of hyperspectral indices for estimating mangrove chlorophyll in Middle Andaman Island, India. *Environ. Monit. Assess.* **2020**, *191*, 785. [CrossRef]

51. Niu, L.; Gao, C.; Sun, J.; Liu, Y.; Zhang, X.; Wang, F. Study on hyperspectral estimation model of chlorophyll content in grape leaves. *Agric. Blotechnol.* **2018**, *7*, 5.

52. Gang, L.; Wan, S.; Jian, Z.; Yang, Z.; Qin, P. Leaf chlorophyll fluorescence, hyperspectral reflectance, pigments content, malondialdehyde and proline accumulation responses of castor bean (*Ricinus communis* L.) seedlings to salt stress levels. *Ind. Crops Prod.* **2010**, *31*, 13–19.

53. Sonobe, R.; Hirono, Y.; Oi, A. Non-destructive detection of tea leaf chlorophyll content using hyperspectral reflectance and machine learning algorithms. *Plants* **2020**, *9*, 368. [CrossRef]

54. Niu, L.; Zhang, X.; Sun, J.; Zheng, J.; Wang, F. Research on estimation models of chlorophyll content in apple leaves based on imaging hyperspectral data. *Agric. Biotechnol.* **2018**, *7*, 220–223+236.

55. Xie, Y.K.; Wang, C.; Yang, W.D.; Feng, M.C.; Qiao, X.X.; Song, J.Y. Canopy hyperspectral characteristics and yield estimation of winter wheat (*Triticum aestivum*) under low temperature injury. *Sci. Rep.* **2020**, *10*, 244. [CrossRef]

56. Li, C.C.; Cui, Y.Q.; Ma, C.Y.; Niu, Q.L.; Li, J.B. Hyperspectral inversion of maize biomass coupled with plant height data. *Crop Sci.* **2021**, *61*, 2067–2079. [CrossRef]

57. Fan, J.H.; Zhou, J.; Wang, B.W.; de Leon, N.; Kaeppler, S.M.; Lima, D.C.; Zhang, Z. Estimation of maize yield and flowering time using multi-temporal UAV-based hyperspectral data. *Remote Sens.* **2022**, *14*, 3052. [CrossRef]

58. Sun, Y.; Wang, Y.H.; Xiao, H.; Gu, X.Z.; Pan, L.Q.; Tu, K. Hyperspectral imaging detection of decayed honey peaches based on their chlorophyll content. *Food Chem.* **2017**, *235*, 194–202. [CrossRef]

59. Wang, H.F.; Huo, Z.G.; Zhou, G.S.; Liao, Q.H.; Feng, H.K.; Wu, L. Estimating leaf SPAD values of freeze-damaged winter wheat using continuous wavelet analysis. *Plant Physiol. Biochem.* **2016**, *98*, 39–45. [CrossRef]

60. Yuan, Z.R.; Ye, Y.; Wei, L.F.; Yang, X.; Huang, C. Study on the optimization of hyperspectral characteristic bands combined with monitoring and visualization of pepper leaf SPAD value. *Sensors* **2022**, *22*, 183. [CrossRef]

61. Jang, S.H.; Hwang, Y.K.; Lee, H.J.; Lee, J.S.; Kim, Y.H. Selecting significant wavelengths to predict chlorophyll content of grafted cucumber seedlings using hyperspectral images. *Korean J. Remote Sens.* **2018**, *34*, 681–692.

62. Rehman, S.U.; Bilal, M.; Rana, R.M.; Tahir, M.N.; Shah, M.K.N.; Ayalew, H.; Yan, G.J. Cell membrane stability and chlorophyll content variation in wheat (*Triticum aestivum*) genotypes under conditions of heat and drought. *Crop Pasture Sci.* **2016**, *67*, 712–718. [CrossRef]
63. Rolando, J.L.; Ramirez, D.A.; Yactayo, W.; Monneveux, P.; Quiroz, R. Leaf greenness as a drought tolerance related trait in potato (*Solanum tuberosum* L.). *Environ. Exp. Bot.* **2015**, *110*, 27–35. [CrossRef]
64. Hou, D.; Bi, J.; Ma, L.; Zhang, K.; Li, D.; Rehmani, M.I.A.; Tan, J.; Bi, Q.; Wei, Y.; Liu, G.; et al. Effects of Soil Moisture Content on Germination and Physiological Characteristics of Rice Seeds with Different Specific Gravity. *Agronomy* **2022**, *12*, 500. [CrossRef]
65. Chen, X.Y.; Zhu, Y.; Ding, Y.; Pan, R.M.; Shen, W.Y.; Yu, X.R.; Xiong, F. The relationship between characteristics of root morphology and grain filling in wheat under drought stress. *PeerJ* **2021**, *9*, e12015. [CrossRef] [PubMed]
66. Farooq, M.; Hussain, M.; Siddique, K.H.M. Drought Stress in Wheat during Flowering and Grain-filling Periods. *Crit. Rev. Plant Sci.* **2014**, *33*, 331–349. [CrossRef]
67. Du, T.; Meng, P.; Huang, J.; Peng, S.; Xiong, D. Fast photosynthesis measurements for phenotyping photosynthetic capacity of rice. *Plant Methods* **2020**, *16*, 6. [CrossRef]
68. Liu, Z.A.; Yang, J.P.; Yang, Z.C. Using a chlorophyll meter to estimate tea leaf chlorophyll and nitrogen contents. *J. Soil Sci. Plant Nutr.* **2012**, *12*, 339–348. [CrossRef]
69. Wakiyama, Y. The Relationship between SPAD Values and Leaf Blade Chlorophyll Content throughout the Rice Development Cycle. *Jarq-Jpn. Agric. Res. Q.* **2016**, *50*, 329–334. [CrossRef]
70. Shibaeva, T.G.; Mamaev, A.V.; Sherudilo, E.G. Evaluation of a SPAD-502 Plus Chlorophyll Meter to Estimate Chlorophyll Content in Leaves with Interveinal Chlorosis. *Russ. J. Plant Physiol.* **2020**, *67*, 690–696. [CrossRef]
71. Eberly, L.E. Correlation and simple linear regression. *Methods Mol. Biol. (Clifton N.J.)* **2007**, *404*, 143–164.
72. Tibshirani, R. Regression Shrinkage and Selection Via the Lasso. *J. R. Stat. Soc. Ser. B (Methodol.)* **1996**, *58*, 267–288. [CrossRef]
73. Hoerl, A.E.; Kennard, R.W. Ridge regression: Biased estimation for nonorthogonal problems. *Technometrics* **2000**, *42*, 80–86. [CrossRef]
74. Breiman, L. Random Forests. *Mach. Learn.* **2001**, *45*, 5–32. [CrossRef]

International Journal of
Molecular Sciences

Article

Intra-Specific Variation in Desiccation Tolerance of *Citrus sinensis 'bingtangcheng'* (L.) Seeds under Different Environmental Conditions in China

Hongying Chen [1,*,†], Anne M. Visscher [2,†], Qin Ai [1], Lan Yang [1], Hugh W. Pritchard [1,2] and Weiqi Li [1,*]

[1] Germplasm Bank of Wild Species, Kunming Institute of Botany, Chinese Academy of Sciences, Kunming 650201, China

[2] Trait Diversity and Function Department, Royal Botanic Gardens, Kew, Wakehurst, Ardingly, West Sussex RH17 6TN, UK

* Correspondence: h.chen@mail.kib.ac.cn (H.C.); weiqili@mail.kib.ac.cn (W.L.); Tel.: +86-871-65223306 (H.C.); +86-871-65223018 (W.L.)

† These authors contributed equally to this work.

Abstract: Intra-specific variation in seed storage behaviour observed in several species has been related to different maternal environments. However, the particular environmental conditions and molecular processes involved in intra-specific variation of desiccation tolerance remain unclear. We chose *Citrus sinensis 'bingtangcheng'* for the present study due to its known variability in desiccation tolerance amongst seed lots. Six seed lots of mature fruits were harvested across China and systematically compared for drying sensitivity. Annual sunshine hours and average temperature from December to May showed positive correlations with the level of seed survival of dehydration. Transcriptional analysis indicated significant variation in gene expression between relatively desiccation-tolerant (DT) and -sensitive (DS) seed lots after harvest. The major genes involved in late seed maturation, such as heat shock proteins, showed higher expression in the DT seed lot. Following the imposition of drying, 80% of stress-responsive genes in the DS seed lot changed to the stable levels seen in the DT seed lot prior to and post-desiccation. However, the changes in expression of stress-responsive genes in DS seeds did not improve their tolerance to desiccation. Thus, higher desiccation tolerance of *Citrus sinensis 'bingtangcheng'* seeds is modulated by the maternal environment (e.g., higher annual sunshine hours and seasonal temperature) during seed development and involves stable expression levels of stress-responsive genes.

Keywords: fruit crop; maternal environment; plant genetic resources; seed storage; sweet orange

Citation: Chen, H.; Visscher, A.M.; Ai, Q.; Yang, L.; Pritchard, H.W.; Li, W. Intra-Specific Variation in Desiccation Tolerance of *Citrus sinensis 'bingtangcheng'* (L.) Seeds under Different Environmental Conditions in China. *Int. J. Mol. Sci.* **2023**, *24*, 7393. https://doi.org/10.3390/ijms24087393

Academic Editor: Martin Bartas

Received: 1 March 2023
Revised: 6 April 2023
Accepted: 7 April 2023
Published: 17 April 2023

1. Introduction

Intra-specific variation in desiccation tolerance (DT) in non-orthodox seeds has been observed in several plant species. Previous studies have suggested that adaptation to different environmental conditions (i.e., the maternal environment) and genetic background possibly influence such variability in desiccation tolerance [1–4]. For example, two subspecies of *Quercus ithaburensis* produced seeds that exhibited different responses to drying [4] and two varieties of *Camellia sinensis* from Kunming (*Camellia sinensis* var. *sinensis*) were less desiccation sensitive than two other collections from Puer and Lincang (*Camellia sinenesis* var. *assamica*) even though all seed lots appeared to be at a similar well-developed stage of maturity. In *Camellia sinensis*, tolerance of dry seasons and genetic background may have contributed to the difference observed in seed DT [3]. In *Acer pseudoplatanus*, seeds from populations of southern Europe showed higher DT than those of northern populations [2]. Other trait parameters of the northern populations, such as lighter seed weight and higher embryo water content, suggested the differences were related to heat sum (determined by temperature and sunshine hours) during seed development [2]. These findings corroborated earlier studies on recalcitrant seeds of *Aesculus hippocastanum*, which showed heat sum-dependent developmental status predicted differential seed quality traits at the point of dispersal across Europe [5]. Thus, intra-specific variation in seed DT can be affected by environmental conditions during seed development.

Orthodox seed development can be divided into three phases: embryogenesis, seed filling and late seed maturation drying [6]. Such seeds generally gain their full DT during the last stage, while non-orthodox seeds tend not to reach this developmental stage and thus have varying levels of desiccation sensitivity (DS). During the late seed maturation stage, key biochemical changes occur such as photosynthetic pigments degradation, and the accumulation of heat shock proteins and soluble sugars, including sucrose, arabinose, galactose and raffinose family oligosaccharides (RFO) [7,8]. Consequently, the particular stage of seed maturity largely influences DT [6]. In the model species *Arabidopsis*, a major metabolic switch to the accumulation of distinct sugars, organic acids, nitrogen-rich amino acids and shikimate-derived metabolisms happens during the transition from the reserve accumulation to seed maturation-desiccation stages [7]. Particular gene families have also been implicated in the development of seed DT. For example, WRKY family genes are positive regulators involved in biotic defence. Defence process-related GO categories were found to be highly enriched during seed maturation [8]. How sensitive this metabolic shift and associated (gene expression) molecular processes are to particular maternal environmental conditions is still unclear. As for the germination trait, intra-specific variation in DT could also be significantly impacted by environmental conditions during development [9]. In this regard, *Citrus sinensis 'bingtangcheng'* is an ideal study material to explore a modulating role of the maternal environment on seed DT as the species has been widely planted across China and genomic resources are available [10].

Citrus sinensis is a major cultivated species of the Rutaceae family [11]. *Citrus* is believed to have originated from the southeast foothills of the Himalayan region that includes the eastern area of Assam, northern Myanmar and western Yunnan [10,12]. The cultivation of this fruit crop started at least 4000 years ago [13,14], and *Citrus* species are a prime source of vitamin C for human nutrition. Conserving the genetic resources of the *Citrus* genus is important to plant breeding, due to the presence of desirable traits in its crop wild relatives. However, there is considerable inter-specific variation in seed storage behaviour (particularly DT). Of the twenty-six species listed on the Seed Information Database (Seed Information Database [SID], RBG, Kew, 2022), only three are listed as having orthodox seeds, while the rest have varying levels of DS. Even for the same species (e.g., *Citrus histrix* DC.), findings have indicated orthodox and recalcitrant storage behaviour in separate studies (SID, RBG, Kew, 2022). The seeds of sweet orange (*C. sinensis 'bingtangcheng'*) are reputed to be oily and partially DT.

In this study, we investigated the differences in DT of *C. sinensis 'bingtangcheng'* from six locations in China across an environmental gradient. Then we examined the transcriptome of seed lots from the two ends of the DT range to dissect the role of seed maturation and stress related genes in DT. Finally, we discussed whether simply switching on key genes is enough to confer DT.

2. Results

2.1. Desiccation Tolerance (DT) of Citrus sinensis 'bingtangcheng' Seed Lots from Different Locations

To gain a better comparison of DT among *C. sinensis 'bingtangcheng'* seeds, six seed lots were collected from latitudes from $24°12'$ N to $29°45'$ N and altitudes varying from 164 to 1133 m a.s.l. (Table 1). The total germination (TG) percentage before desiccation was above 90% for all seed lots and it declined during desiccation (Figure 1). Seeds from Chongqing (Figure 1A) and Shaoguan (Figure 1B) showed the lowest and highest relative DT, compared with seed lots from Huaning, Huaihua, Guilin and Ganzhou (Figure 1C). After 6 d of drying to 2.7% MC, TG in Shaoguan seed lot was still 67% (Figure 1B). Seed lots from Huaning, Guilin and Ganzhou also showed relatively high DT, with 50%, 40%

and 40% TG, respectively, when dried to <5% MC. However, seeds from Huaihua and Chongqing showed low relative desiccation tolerance, with 20% and 29% TG when dried down to 4.9% and 7.7% MC, respectively. A co-plot of TG (probit) against seed MC revealed that when germination was 5.5 probit (i.e., 69% TG at 2.7% MC for the Shaoguan seed lot), the interpolated MCs (i.e., a measure of DT) were 4.5%, 7.9%, 9.2%, 13.6%, 22.3% and 25.5% MC for Shaoguan, Guilin, Huaning, Ganzhou, Huaihua, and Chongqing seed lots, respectively (Figure 1C). These results confirm that Shaoguan was the seed lot with the highest DT and Chongqing the one with the lowest. These findings then framed the molecular studies.

Table 1. Details of the six seed lots of *Citrus sinensis* 'bingtangcheng' studied. MC (%) at TG probit 5.5 is a measure of seed DT (the lower this MC, the higher the DT).

Seed Lot	Location	Altitude (m)	Collection Date	MC (%) at TG Probit 5.5	Whole Seed Dry Mass (mg)	Initial Moisture Content (%)
Huaning	24°12′ N; 103°07′ E	1133	5 December 2014	9.2	77.5 ± 2.5 [d]	48.7 ± 1.4 [b]
Huaihua	27°47′ N; 109°48′ E	244	9 December 2014	22.3	91.7 ± 3.5 [b,c]	49.4 ± 2.6 [b]
Shaoguan	24°41′ N; 113°49′ E	373	26 December 2014	4.5	100.2 ± 2.8 [b]	45.4 ± 1.4 [b,c]
Guilin	25°24′ N; 110°19′ E	164	31 December 2014	7.9	87.7 ± 3.1 [c]	48.4 ± 2.1 [b]
Ganzhou	25°70′ N; 115°20′ E	319	15 December 2014	13.6	120 ± 3.5 [a]	41.8 ± 1.4 [c]
Chongqing	29°45′ N; 106°22′ E	245	28 December 2014	25.5	63.6 ± 2.6 [e]	57.0 ± 1.4 [a]

Values are mean ± SE for 50 individual seeds. Values in the same column with a different letter are significantly different ($p < 0.05$).

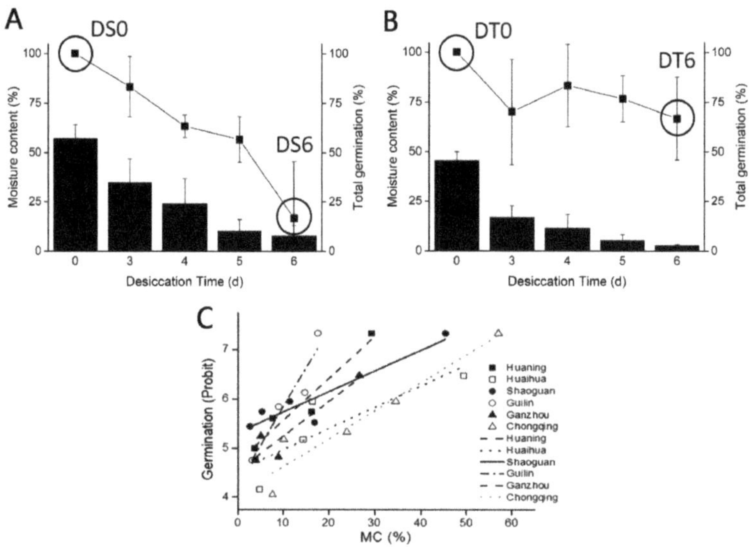

Figure 1. Effect of drying on seed parameters of *Citrus sinensis* 'bingtangcheng' seed lots. Total germination (TG, filled squares) response of (**A**) Chongqing (desiccation sensitive, DS) and (**B**) Shaoguan (desiccation tolerant, DT) seed lots (*n* = 3 biological replicates of 10 seeds each) after drying for up to 6 days (DS6, DT6) to different moisture contents (filled bars). Whole seed moisture contents (MC) were determined for three biological replicates of 10 seeds each. Data represent means ± SD. (**C**) Plot of the relationship between MC and TG (probit scale) for Huaning (■), Huaihua (□), Shaoguan (●), Guilin (○), Ganzhou (▲), Chongqing (△). Data represent mean TG at each MC.

2.2. Correlation Studies between DT of Citrus sinensis 'bingtangcheng' Seed Lots and Other Seed Traits

Seed traits like seed mass (mg), seed dry mass (mg) and initial MC (%) were tested as co-correlants of DT (indicated as the MC at which germination was interpolated to be probit 5.5: the lower this MC, the higher the DT). Variation in seed mass (mg) of the seed lots was normally distributed (Figure S1). Mean seed mass of *C. sinensis 'bingtangcheng'* from Chongqing was the smallest (63.6 mg), while Ganzhou had the largest seed mass at 120 mg ($p < 0.05$) (Table 1). Seed dry mass did not correlate with DT ($r = -0.4$) (Figure 2K). The initial MCs of the seeds ranged from 41.8% for Ganzhou to 57% for Chongqing (Table 1), and this trait showed strong negative correlation with seed DT ($r = 0.64$) (Figure 2L). However, although seed lots from Huaning and Huaihua contained similar MCs of 48.7% and 49.4% respectively, their DT differed (Table 1).

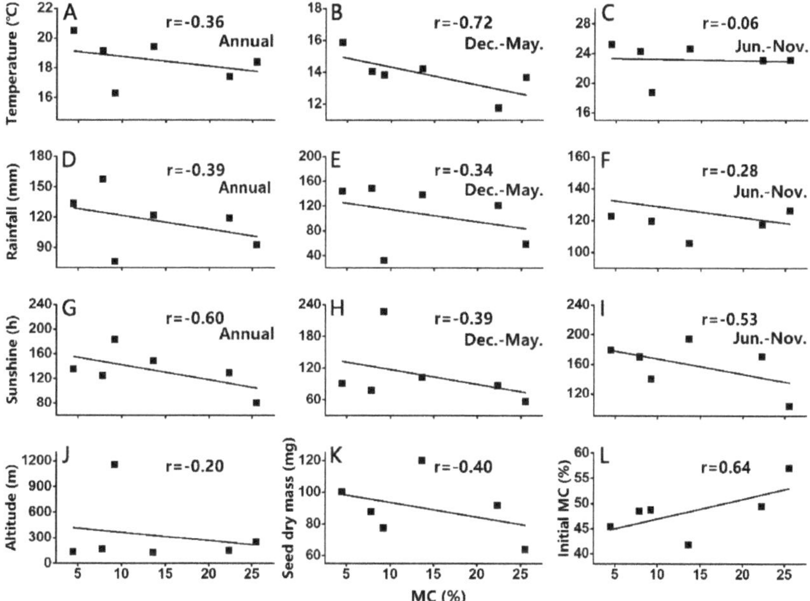

Figure 2. Correlation between the seed MC at which germination was interpolated to be probit 5.5 and *Citrus sinensis 'bingtangcheng'* seed lot characteristics as well as environmental conditions of the collecting sites. (**A–C**) Average temperature (°C), (**D–F**) average rainfall (mm), (**G–I**) sunshine hours (**H**), (**J**) altitude, (**K**) seed dry mass and (**L**) initial moisture content. Climate conditions were averaged for Dec.–May (after fruit dispersal and before flowering) and Jun.–Nov. (after flowering).

2.3. Correlation Studies between Seed DT and Meteorological Data

The environmental conditions of the six locations from which *Citrus sinensis 'bingtangcheng'* fruits were harvested were slightly different (Figure 3, Supplemental Tables S1–S3). Altitude did not correlate with seed DT ($r = -0.20$) (Figure 2J). The average temperature in Shaoguan was highest from December to May, while that in Chongqing was one of the lowest during this time (Figure 3B). Seed DT showed a strong positive correlation with the average temperature over this period ($r = -0.72$) (Figure 2B). The average precipitation was greatest in Guilin (157.3 mm), while Chongqing and Huaning received precipitation below 100 mm: 75.7 mm and 92.3 mm, respectively. Overall, the average annual rainfall ($r = -0.39$), the average rainfall from December to May ($r = -0.34$) and that from June to November ($r = -0.28$) showed only weak positive correlations with seed DT (Figure 2D–F). However, annual average sunshine hours did correlate positively with seed DT ($r = -0.60$; Figure 2G). Annual average sunshine hours from December to May and from June to November were also positively correlated with seed DT, with $r = -0.39$ (Figure 2H) and $r = -0.53$ (Figure 2I) respectively. Besides, analysis of temperature and sunshine hours during the last month before fruit harvesting (November) was analyzed and showed strong correlation ($r = -0.59$ and $r = -0.80$ respectively) (Supplemental Figure S3). Sunshine hours and temperature are two important factors for developmental heat sum, which impacts seed development and DT [2]. Thus, our results showed that variation in sunshine hours and temperature between Chongqing and Shaoguan resulted in different levels of seed DT at the point of fruit maturity.

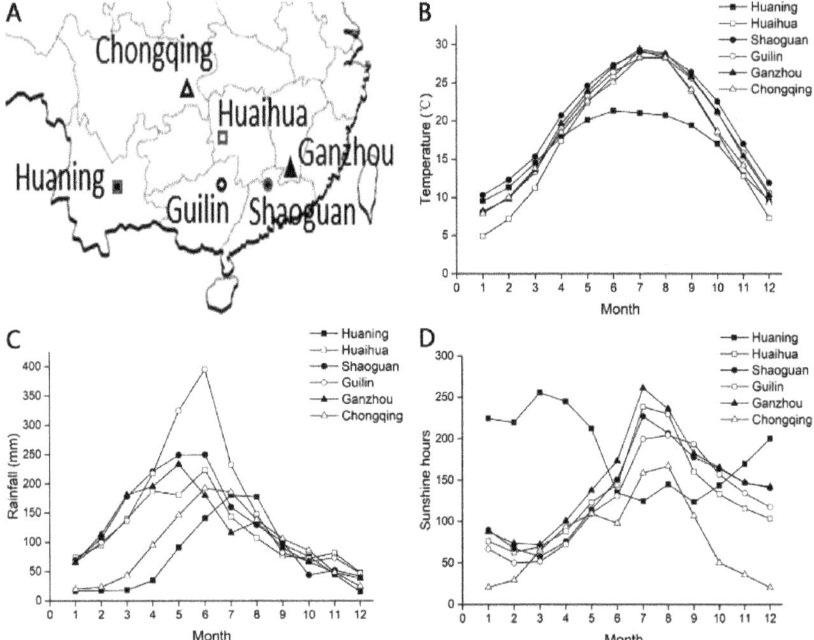

Figure 3. *Citrus sinensis 'bingtangcheng'* collection site environment details. (**A**) map of locations of the six collection sitesacross China, (**B**) Average temperature (°C), (**C**) Average rainfall (mm) and (**D**) Average sunshine duration hours recorded from 1981 to 2010 for Huaning (■), Huaihua (□), Shaoguan (•), Guilin (○), Ganzhou (▲) and Chongqing (△).

2.4. Transcriptome Analysis of DS and DT Seed Lots from Different Locations

Because of the known ecological correlates for varying DT between Chongqing (DS) and Shaoguan (DT), we then explored the molecular factors involved in this variation in stress tolerance. We compared gene expression during desiccation of these two seed lots (Figure 4). We used "FDR < 0.05 and $|\log_2 FC| \geq 1$" as the criteria for significant difference in gene expression (DEGs). The contrast between the two different seed lots before desiccation (DS0 vs. DT0) showed the largest number of DEGs across all comparisons, with 2590 DEGs up-regulated and 1009 DEGs down-regulated (Supplemental Data S1, Figure 4A,B). During the drying process of the DS seed lot (DS0 vs. DS6), 2779 DEGs were identified, of which 1145 were up-regulated and 1634 down-regulated (Supplemental Data S2, Figure 4A,B). In contrast, only 444 DEGs were identified when a relatively DT seed lot was compared before and after the desiccation process (DT0 vs. DT6), with 26 up-regulated genes and 418 down-regulated genes (Supplemental Data S3, Figure 4A,B). Principal component analysis (PCA) was used to reveal the overall variance among DEGs (Figure 4C). Principal component 1 (PC1) explained the separation of the different seed lots. These results suggest that large transcriptomic differences already existed between the two different seed lots before drying. In addition, the dehydration process triggered considerable gene expression change in the DS seed lot, while not much affecting the DT seed lot.

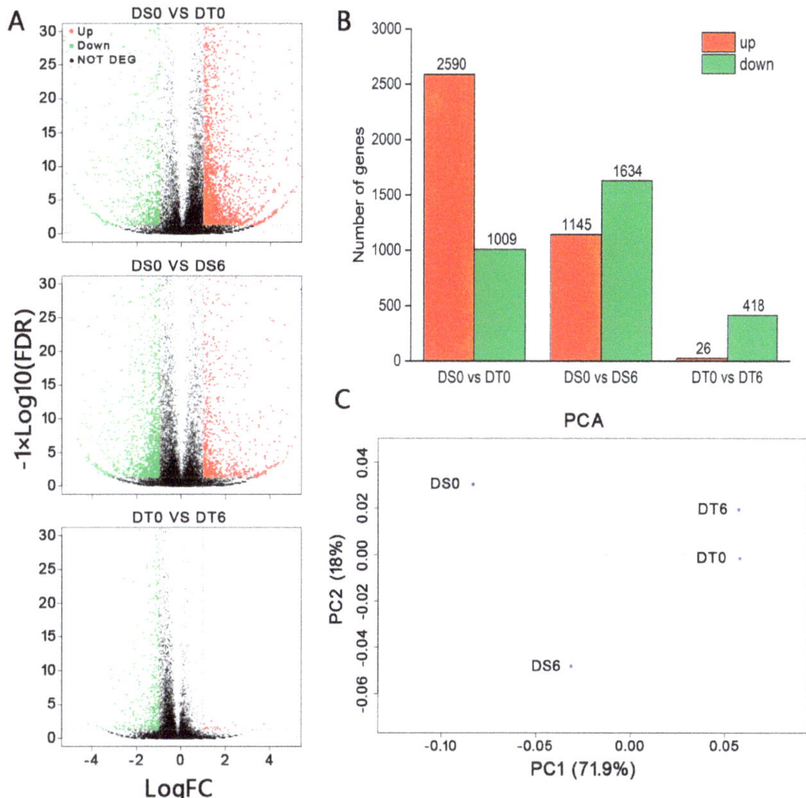

Figure 4. Differentially expressed genes (DEGs: FDR < 0.05 and absolute $\log_2$ fold change $\geq$ 1) were identified for three different comparisons. (**A**) Volcano plots depicting DEGs between two seed lots before desiccation (DS0 vs. DT0) and within seed lots exposed to desiccation (DS0 vs. DS6, DT0 vs. DT6). Red dots indicate significantly up-regulated genes, green dots indicate significantly down-regulated genes, and black dots represent non-DEGs. (**B**) Up-(red) and down-regulated (green) DEGs for each comparison. (**C**) Principal component analysis for the four molecular analysis: DS0, desiccation sensitive seeds dried for 0 d, DS6, desiccation sensitive seeds dehydrated for 6 d, DT0, desiccation tolerant seeds dried for 0 d and DT6, desiccation tolerant seeds dehydrated for 6 d.

2.5. Gene Expression Patterns of DT Citrus sinensis 'bingtangcheng' Seeds Are Characteristic of a Late Seed Maturation Stage

With respect to the genes involved in DT in *Citrus sinensis 'bingtangcheng'* seeds, we first analysed the expression of transcription factors (TFs), which play critical roles in regulating this trait. In the present study, 238 unigenes encoding TFs belonging to 43 families showed significant differences in expression when comparing two seed lots before desiccation (DS0 vs. DT0; Figure 5A, Supplemental Data S4). Focusing on DS seeds during dehydration (DS0 vs. DS6; Figure 5B, Supplemental Data S5), there were 161 unigenes encoding TFs for which gene expression was significantly changed. In contrast, only nine unigenes encoding TFs showed significant changes following the drying of the DT seed lot (i.e., DT0 vs. DT6; Figure 5C, Supplemental Data S6). The four TF families with the highest number of DEGs between DS0 vs. DT0 and DS0 vs. DS6 were ERF, NAC, MYB and WRKY (Figure 5).

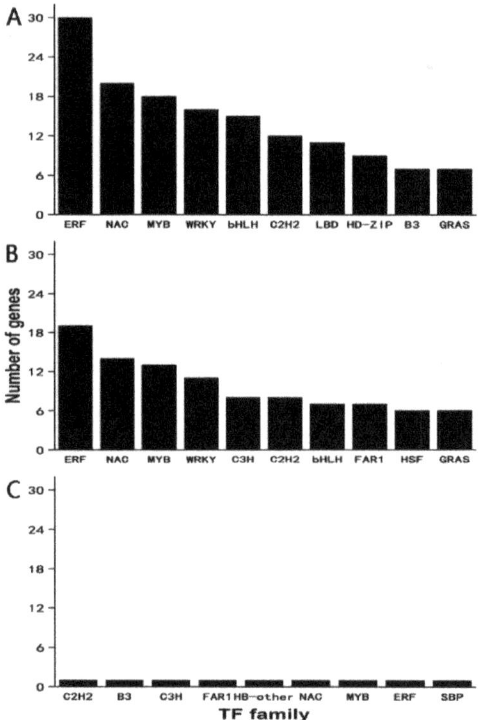

Figure 5. The number distribution of DEGs (FDR < 0.05 and absolute $\log_2$ fold change ≥ 1) that belong to the top 10 transcription factor families identified from three comparisons. (**A**) DS0 vs. DT0; (**B**) DS0 vs. DS6; (**C**) DT0 vs. DT6. DS0 signifies desiccation-sensitive seeds dried for 0 d; DS6: desiccation-sensitive seeds dehydrated for 6 d; DT0, desiccation-tolerant seeds dried for 0 d; DT6, desiccation-tolerant seeds dehydrated for 6 d.

In order to analyse the role of specific late seed maturation genes, we compared our results to a previous report on late maturation of orthodox (fully DT) dicot seeds [6] and identified 82 unigenes involved in desiccation tolerance in *C. sinensis 'bingtangcheng'* (Figure 6). These included genes encoding HSPs (41 unigenes), late embryogenesis abundant (LEA) proteins (3 unigenes), WRKYs (27 unigenes), as well as genes involved in the degradation of photosynthetic pigments (6 unigenes), or non-reducing sugar metabolism (5 unigenes). The expression levels of these unigenes in the two different seed lots (DS0 vs. DT0) are summarized in Supplemental Data S7. Among them were 55 unigenes with up-regulated and 27 unigenes with down-regulated expression in DT vs. DS seeds.

Regarding heat shock protein families, one unigene was predicted to encode for a heat shock factor 24-like protein (HSF24-like), seven for small heat shock proteins (sHSPs), three for HSP60 family proteins, seven for HSP70 family proteins, and one each for HSP90 and HSP100 family proteins. There were 30 HSP unigenes showing high expression levels in the DT seed lot (DT0), with 11 significantly up-regulated compared to the DS seed lot (DS0) (Figure 6B). Eleven HSP unigenes showed lower expression in the DT seed lot, but not significantly so (Figure 6B). Only three LEA-related unigenes were detected, with similar (high) expression in both seed lots (Figure 6B).

With respect to soluble sugars, four unigenes encoding galactinol synthase (two significantly) and raffinose synthase proteins showed higher expression in the DT seed lot (compared to DS) before drying (Figure 6B, Supplemental Data S7). The degradation of photosynthetic pigments is the most visible change associated with late seed maturation and six related genes were detected in the present study. Four unigenes encoding carotenoid cleavage dioxygenases showed higher transcript levels in the DT seed lot (DT0), with two significantly different from the DS seed lot (DS0) (Figure 6B, Supplemental Data S7). In addition, one gene encoding chlorophyll b reductase NYC1 showed significantly lower expression in the DT seed lot (Figure 6B).

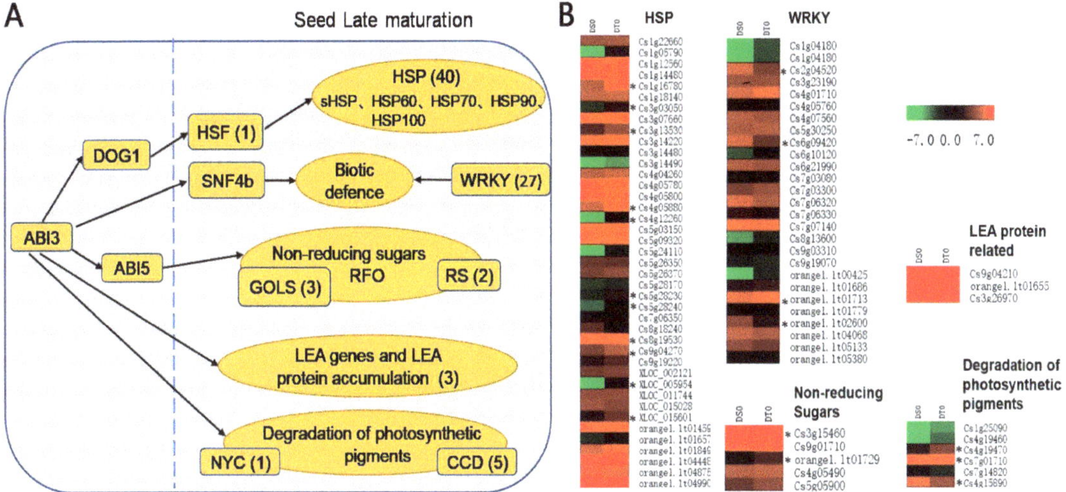

Figure 6. Late seed maturation related unigenes and their expression in *Citrus sinensis 'bingtangcheng'*. (**A**) Putative unigenes involved in late seed maturation. The value in brackets indicates the number of unigenes annotated. (**B**) The heatmap shows the transcript level of each late seed maturation related unigene in the desiccation sensitive (DS0) and the desiccation tolerant (DT0) seed lots. The asterisks indicate the differentially expressed unigenes (FDR < 0.05 and absolute $\log_2$ fold change $\geq$ 1) between the two seed lots before desiccation (DS0 vs. DT0).

In addition, our study identified 27 differentially expressed WRKY unigenes for the DS0 vs. DT0 contrast, among which four WRKY unigenes had significantly higher expression in DT seed lots. Overall, two thirds of WRKY unigenes showed higher expression in the DT seed lot, while one third showed lower expression (Figure 6B, Supplemental Data S6).

Together, the data above indicate that the seed lots from Shaoguan (DT0) and Chongqing (DS0) differed in developmental stage, with the seeds from Shaoguan showing gene expression characteristics associated with later seed maturation.

2.6. DT Citrus sinensis 'bingtangcheng' Seeds Show Stable Levels of Stress Responsive Genes during Drying

To further understand the types and function of genes involved in desiccation tolerance, DEGs of the two *Citrus sinensis 'bingtangcheng'* seed lots, Shaoguan (DT) and Chongqing (DS), were analysed for enrichment of GO terms (Figure 7). For the biological process category, the top 10 significantly enriched GO terms of the DS seed lot during desiccation are listed in Figure 7C, with the top one being 'response to stress'. There were 165 unigenes associated with this response-to-stress category, and the expression trends of those genes were compared in detail (Figure 7A,B, Supplemental Data S8). Of the 165 unigenes, 87 were significantly up-regulated and 78 down-regulated following drying in the DS seed lot (DS0 vs. DS6). In contrast, none showed significant changes following desiccation in the DT seed lot (DT0 vs. DT6). However, among those 165 in the response-to-stress category, the trend for 135 unigenes that changed in the DS seed lot was towards the level of the DT seeds. For example, for an HSP20-like chaperone superfamily protein (Unigene Cs8g19490), the value of FPKM increased in the DS seed from 852 to 2640 during the dehydration process, while the value in the DT seed lot was already at 1759 before drying (Supplemental Data S8, Supplemental Figure S2). Another example is Cytochrome P450 (orange1.1t05138), for which the value of FPKM decreased from 72.23 to 3.36 in the dried DS seed, while the value in the DT seed lot was already at 3.71 before drying. These results show that DT *Citrus sinensis 'bingtangcheng'* seeds have stable levels of stress-responsive genes. Moreover, the expression of 80% of stress-responsive genes in the DS seed lot during drying changed to the levels seen in the DT seed lot prior to dehydration.

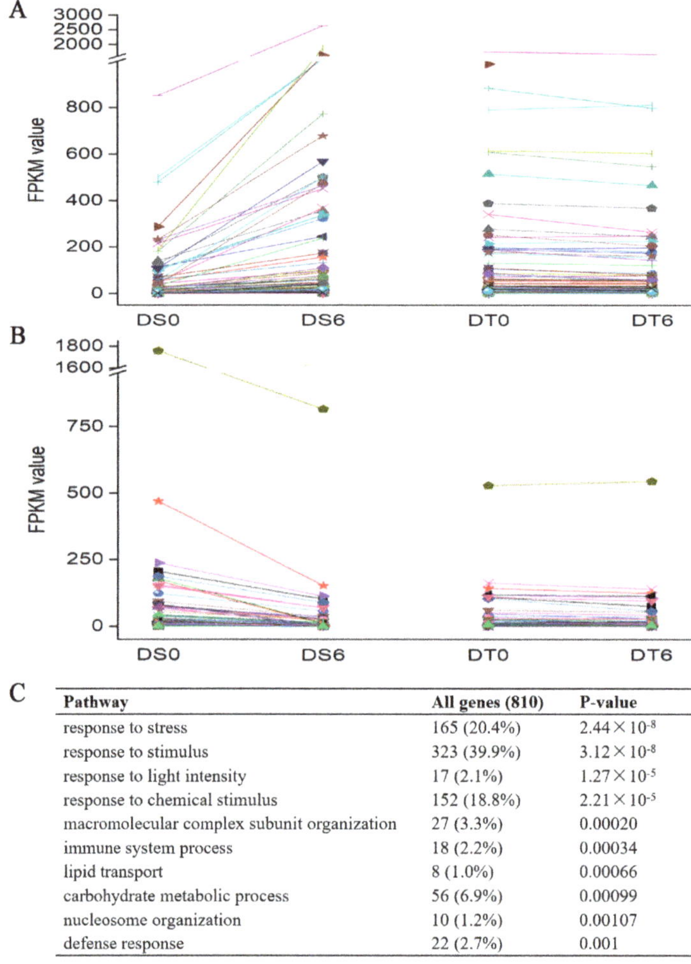

Pathway	All genes (810)	P-value
response to stress	165 (20.4%)	2.44×10^{-8}
response to stimulus	323 (39.9%)	3.12×10^{-8}
response to light intensity	17 (2.1%)	1.27×10^{-5}
response to chemical stimulus	152 (18.8%)	2.21×10^{-5}
macromolecular complex subunit organization	27 (3.3%)	0.00020
immune system process	18 (2.2%)	0.00034
lipid transport	8 (1.0%)	0.00066
carbohydrate metabolic process	56 (6.9%)	0.00099
nucleosome organization	10 (1.2%)	0.00107
defense response	22 (2.7%)	0.001

Figure 7. Analysis of stress-responsive genes that were differentially expressed (DEGs: FDR < 0.05 and absolute $\log^2$ fold change $\geq$ 1) in DS seeds following drying. (**A**) Expression levels of stress-responsive genes that were significantly up-regulated, and (**B**) significantly down-regulated in DS seeds following drying. (**C**) The top 10 enriched GO categories of DEGs (DS0 vs. DS6). DS0 signifies desiccation-sensitive seeds dried for 0 d; DS6, desiccation sensitive seeds dehydrated for 6 d; DT0, desiccation tolerant seeds dried for 0 d; and DT6, desiccation tolerant seeds dehydrated for 6d. The color lines represents the trend of gene changes from 0 d to 6 d.

3. Discussion

3.1. Desiccation Tolerance of Seeds from Mature Citrus sinensis 'bingtangcheng' Fruits and Relationships with Other Seed Traits and Environmental Conditions

It is extremely challenging to judge seed maturity in fleshy fruits. Thus, the relationship between seed and fruit maturation in fleshy fruits has rarely been studied, some exceptions being tomato and cucumber [15]. In contrast, the metabolic, hormonal and biochemical changes occurring during fruit development have been extensively studied [16]. For example, fruit tissue development seems regulated by SBP-box genes and their putative MADS-box promoter targets. If the gene at the ripening inhibitor (rin) locus is mutated, the fruit remains firm and unripe for extended periods [17]. A recent study on *Citrus* reported that seeds can show high viability before fruit maturation is reached [18]. However, whether the acquisition of high germination in seeds correlated with DT was not explored. In our case, although *Citrus sinensis 'bingtangcheng'* fruits achieved maturation as defined within a commercial production setting and total germination before desiccation was high for all seed lots (>90%), seeds exhibited a large variation in DT, indicating potential differences in seed maturation between fruits of apparently similar maturity (Figure 1, Table 1).

Seed mass (or embryo mass) has been used as a surrogate measure of seed maturation and seed DT, with larger seeds indicating greater maturity [2,3]. For example, *Acer pseudoplatanus* fruits/seeds harvested across Europe with greater developmental heat sum, i.e., from southern rather than northern locations, were heavier and more mature at the point of natural dispersal [2]. Although the average dry seed mass of the most DS seed lot (Chongqing) and DT seed lot (Shaoguan) were 63.6 mg and 100.2 mg, respectively (Table 1), we found no overall correlation between dry seed mass and DT in the present study (Figure 2). A dry mass of 137 mg for *C. sinensis 'bingtangcheng'* seeds reported by Hong et al. (2001) was higher than all seed lots analyzed in our study, and drying to 4% MC decreased their viability to 10% [19]. In contrast, the Shaoguan (DT) seed lot in our study retained 67% germination when dried to c. 3% MC (Figure 1), thereby showing higher DT irrespective of the seed mass. Similarly, seed initial MC has also been used as a surrogate measure of seed DT, with lower MCs indicating greater maturation [3]. For example, *Camellia sinensis* seeds with lower initial MC collected from three locations in China were more DT [2]. Although initial MC showed strong negative correlation with DT (Figure 2L), prediction of the precise level of DT is not certain. For example, Huaihua seeds (initial MC 49.4%) had 20% germination when dried to 5% MC while seeds from Huaning with the same initial MC (48.7%) had 50% germination after drying to <5% MC. However, the environmental conditions in those two places showed great differences (Supplemental Tables S1 and S3), suggesting that a number of factors are at play during the development of DT, not just the seed harvest MC.

We then considered how much the environmental conditions at the six locations from which 'mature' *C. sinensis 'bingtangcheng'* fruits were harvested for this study might co-correlate with seed DT [20]. We found that average annual sunshine hours and temperature from December to May showed strong correlations with seed DT (Figure 2). *C. sinensis 'bingtangcheng'* trees from Chongqing (DS) received the least sunshine hours throughout the year and experienced one of the coolest temperatures from December to May compared with the other locations (Figure 3). Such limiting environmental conditions may have resulted in the least seed development despite the fruits being ready to eat. As the development and ripening of fruit may share similar genes with the development of floral organs [21], it is possible that temperatures from December to May might have (negatively) affected flowering time, and therefore the length of post-flowering development achieved at the *C. sinensis 'bingtangcheng'* population sites. In addition, it appears that the temperature and sunshine hours during the month before fruit harvest (November) are important for seed maturation (Supplemental Figure S3). Future studies could address whether or not seed lots identified here as DS are indeed developmentally immature, or genetically less DT.

Precipitation level and timing has also been a co-correlant of seed DT, in coffee [1], tea [3] and African trees [20]. Less precipitation after seed dispersal tends to be associated with more DT. However, the present study on intra-specific differences in *C. sinensis 'bingtangcheng'* seeds prior to dispersal did not find a relationship between precipitation and relative DT spanning a broad range of moisture levels.

3.2. Transcriptome Analysis before and after Drying of Two C. sinensis 'bingtangcheng' Seed Lots Differing in DT

Although *C. sinensis 'bingtangcheng'* is considered to have partial DT, we showed relatively high DT in several of the six seed lots studied (Figure 1, Table 1). Transcriptional analysis revealed that considerable differences in gene expression already existed between the most DS (Chongqing) and DT (Shaoguan) seed lots following harvest, including a large number of transcription factors from ERF, NAC and MYB families, which are implicated in late seed maturation of orthodox seeds [22]. When comparing our results to the late maturation stage in *Arabidopsis* seeds, the top five TF families also included MYB, NAC and ERF [22], which indicates that these TFs are putative transcriptional regulators of DT in seeds. In particular, these TF families are documented to play vital roles in the regulation of ABA mediated and independent drought-signalling pathways [23]. Other important late seed maturation genes in orthodox seeds include those related to degradation of photosynthetic pigments and heat shock protein family genes (*HSPs*) [6]. For example, the presence of chlorophyll molecules in dry non-chlorophyllous seeds is considered a characteristic of immaturity and photosynthetic pigment degradation is a visible change during seed development [24–26]. Moreover, active chlorophyll has recently been identified as a key contributor to seed aging because the molecule is prone to oxidative stress in the absence of metabolic water [27]. In this study, *CCD1* and *CCD8* genes showed higher transcript levels in the DT compared to the DS seed lot prior to drying (Figure 6B). Carotenoid cleavage dioxygenases (CCD) regulate degradation of carotenoids and four CCDs (CCD1, CCD4, CCD7 and CCD8) are active during seed development in *Arabidopsis*. The *CCD1* gene is induced during the onset of the desiccation phase [28], and an increase in *CCD8* transcript levels resulted in a decrease in carotenoid contents in seeds [29].

The function of HSPs and sHSPs as molecular chaperons is to protect proteins from denaturation damage under both stressed and non-stressed conditions [6,30,31]. In this regard, we found that 11 genes encoding HSPs and sHSPs (e.g., *HSP70, HSP22.7, HSP18.2, HSP17.4*) showed significantly higher expression in the DT seeds compared to the DS seeds prior to desiccation (Figure 6B, Supplemental Data S7). HSP families are suggested to have a protective function in seed longevity, and the expression of *HSPs* and *sHSPs* is up-regulated during late seed maturation [32,33]. Longer-term survival in cold, dry storage is known for seeds of some *Citrus* species [34], and the expression of HSP might suggest intra-specific variation in *C. sinensis 'bingtangcheng'* seed lot storage in the dry state. But this remains to be explored.

Improvement of desiccation resistance with high level of HSPs is a universal phenomenon, existing in microorganisms, and cells of plants and animals. For example, the resistance to desiccation of *Azotobacter vinelandii* largely depends on HSP20 which might prevent the aggregation of proteins caused by the lack of water [28]. Moreover, higher constitutive levels of HSP74 in the foot tissue of land snail were correlated with higher desiccation resistance [35]. We discovered that the transcript levels of many HSP genes (including *HSP70, HSP26.5, HSP20, HSP18.2, HSP17.4*) were significantly increased during the desiccation process in the DS seed lot. In contrast, in the DT seed lot HSP gene transcript levels were already high and did not change in response to drying (Supplemental Figure S2). Another group of genes that significantly increased in the DS seed lot during the desiccation process was peroxidase genes (e.g., fold change > 9, Supplemental Data S8, Supplemental Figure S2). Peroxidase, such as ascorbate peroxidase, is considered one of the central ROS detoxification enzymes that function during the development of thermotolerance in plants [36]. Enrichment in stress-responsive genes, such as *HSP* and oxidative stress-related genes, was also observed during acquisition of desiccation tolerance in *Medicago truncatula* during seed development [37]. However, whilst drying activated changes in expression of stress-responsive genes in DS seeds, there was no concomitant improvement in their DT, as assessed by germination tests (Figure 1, Table 1). This suggests that the transient increase in transcripts of potential importance for DT is insufficient to ensure DT in DS seeds. One potential explanation could be that the imposed drying regime was too extreme (RH too low and time period too short), stimulating the high expression of genes that respond to stress, without this translating into sufficient protein synthesis. In this context, different drying regimes or desiccation tolerance re-induction methods, e.g., under higher RH or water potentials [38,39] could be tested in future studies. Now that the citrus genome has been annotated [9,11] and new plant breeding techniques, such as genome editing, are available, many further mechanistic studies can be pursued. For example, Ruby (a MYB-like TF) and its promoter 3'LTR which are one of the main genes responsible for the control of purple pigmentation in citrus fruits [40] could be used to compare fruit maturation across the six sites used in this study. Furthermore, the CRISPR/Cas12a system has already been examined for editing in grapefruit [41]. Future studies on seed DT should embrace these, and other, new plant breeding techniques.

4. Materials and Methods

4.1. Seed Material

Fruits of *Citrus sinensis* 'bingtangcheng' from Huaning, Huaihua, Shaoguan, Guilin, Ganzhou and Chongqing were collected at the point of natural maturity (yellowing and softening) during the winter of 2014 (end November–end December) and used within 6 weeks of harvest.

Fruits of *Citrus sinensis* 'bingtangcheng' were collected from approximately 50 trees at each location and air freighted to the Germplasm Bank of Wild Species of China in Kunming. On arrival, fruits were kept in a cool place (about 10 °C, 8/16 h light/dark) for a maximum of 3 days before all the seeds were removed from the fruit. Cleaned seeds were stored in plastic bags at 5 °C for no more than 4 weeks as the experiments were started. The bags were opened at least once a week to allow for exchange of air. Only seeds with a healthy appearance (i.e., no apparent fungal or bacterial infection and mechanical damage) were pooled and randomly selected for the experiments.

4.2. Desiccation Treatments and Germination Test

Seeds were dried over silica gel (5:1 silica gel to seed weight) in sealed plastic bags at 15 °C for up to 6 days. The silica gel was replenished every 24 h and the bags ventilated. At each sampling time (desiccated for 0, 3, 4, 5, 6 d), four replicates of 10 seeds were withdrawn, three of which were used for the germination test. The remaining 10 seeds were used for moisture content (MC) determination. Seed mass values for the seed lots were calculated from MC determinations (n = 50 seeds).

For the germination test, each replicate of 10 seeds was placed in a 90 mm petri dish containing 1% water agar. Seed coats were removed before germination. The test was run at 30 °C under a day/night cycle of 8/16 h. Germination was defined as radicle emergence of at least 2 mm and it was scored regularly. Total germination (TG) was monitored for 35 d from the onset of imbibition. As no further germination occurring during the final 7 d period, TGs at this time were therefore considered final. Cut-tests were performed on the remaining non-germinated seeds and the vast majority was found to be soft and assessed as being inviable. Seed germination was expressed as a percentage of the total number of seeds sown. The MCs (10 randomly selected seeds/test) were determined gravimetrically by drying the seeds at 103 °C for 17 h (ISTA 2022) (https://www.seedtest.org (accessed on 6 Jun 2022)).

4.3. Environmental Conditions at the Sampling Sites

Meteorological data (Supplemental Tables S1–S3) were obtained from the National Meteorological Information Centre (data.cma.cn (accessed on 12 May 2019)) which provides monthly average temperature, precipitation data and sunshine duration hours recorded from 1981 to 2010 for each location.

4.4. Isolation of RNA, Library Construction and Sequencing

Paired-end transcriptome sequencing (RNA-Seq) was employed to analyse gene expression in seeds from Chongqing (DS) and Shaoguan (DT) at two desiccation stages: dried for 0 d (DS0 and DT0) and for 6 d (DS6 and DT6). Total RNA was isolated using the Omega Plant RNA Kit (Omega Bio-Tek, Norcross, GA, USA) according to the manufacturer's protocol. Three total RNA samples were isolated for each desiccation stage (0 and 6 d) and each sample was isolated from at least 10 ground seeds. The RNA quality was verified by an Agilent 2100 Bioanalyzer (Agilent Technologies, Palo Alto, CA, USA) and checked using RNase free agarose gel electrophoresis. One sample from each time point with the best quality of RNA was selected for RNA-Seq. High quality RNA was enriched by Sera-Mag oligo (dT) beads (Thermo Scientific, Indianapolis, IN, USA). The enriched mRNA was fragmented into small pieces by a fragmentation buffer and reverse-transcribed to cDNA using NEBNext Ultra RNA Library Prep Kit for Illumina (NEB 7530, New EnglandBiolabs, Ipswich, MA, USA). The cDNA was end-repaired with an adapter primer attached and subsequently adaptor-ligated by the addition of a specific adapter, following the manufacturer's protocol (Illumina Inc. San Diego, CA, USA). The cDNA fragments were sequenced using Illumina HiSeq2500 by Gene Denovo Biotechnology Co. (Guangzhou, China).

4.5. Gene Expression and GO Enrichment Analyses

Expression abundance was quantified using 'Fragment Per Kilobase of transcript per Million mapped reads' (FPKM) (Cufflinks software, version 2.2.1) [42]. Differential expression analysis was performed by DESeq2 software (version 1.4.5) for three different contrasts (DS0-vs-DT0, DS0-vs-DS6 and DT0-vs-DT6) [43]. The edgeR software (version 3.6.8) was used to test significant differences in the expression of count data as described previously [44,45] and the selected dispersion value was 0.01. When compared with a reference sequence (CsiDB2011_11), 74.2% of known genes were detected in the four samples, and the number of DEGs between different samples were analysed (Supplemental Datas S1–S3). Differentially expressed genes/transcripts (DEGs) were counted as the genes/transcripts with a false discovery rate (FDR) below 0.05 and absolute $\log_2$ fold change ≥ 1. All DEGs were mapped to Gene Ontology (GO) terms through the Gene Ontology database (https://www.geneontology.org/ (accessed on 6 May 2016)) for GO enrichment and pathway analysis. Gene numbers were calculated for every term, with significantly enriched GO terms in the DEG list (compared to the genome) defined by hypergeometric test. The calculated p-values were subjected to FDR correction, utilizing FDR ≤ 0.05 as a threshold.

4.6. Data Analysis and Presentation

Seed dry mass and initial moisture content were statistically analysed using IBM SPSS Statistics 25.0 software, version SPSS 25.0. Principal component analysis was also conducted with SPSS 25.0 software, version Origin 7.0 (SPSS, Chicago, IL, USA). The data conforming to normal distribution and homogeneity of variance was tested using one-way ANOVA followed by pairwise comparisons of LSD to identify difference. In the results, significant differences ($p < 0.05$) are indicated by different letters. The Origin 7.0 software (Origin 7.0, OriginLab, Northampton, MA, USA) was used to draw figures and for linear fitting (e.g., to analyse correlation in Figure 2. The germination percentage values were converted to probit germination (Figure 1C) use online sources, e.g., http://data.kew.org/sid/viability/convert.jsp (accessed on 3 June 2021).

5. Conclusions

While differences in DT have been reported for the same species from a wide range of locations [3,5] little is known still about the translation of maternal environmental conditions through molecular processes into the acquisition of physiological DT. To gain insight, we systematically compared DT of seed lots harvested from mature *Citrus sinensis 'bingtangcheng'* fruits from six locations across China (Figures 1 and 2, Table 1). We found that these 'mature' fruits yielded seed lots that all had acquired germination competence but varied widely in their DT. The annual average sunshine hours and average temperature from December to May showed strong correlations with seed DT (Table 1, Figure 2). In the future, we could use of real-time data loggers placed at each site to detect higher co-correlants environmental factors of seed DT.

Transcriptional analysis revealed that large differences in gene expression already existed between the most DS and DT seed lots following harvest. Differentially expressed genes implicated in late seed maturation of orthodox seeds showed a large number of transcription factors (e.g., ERF, NAC, MYB and WRKY families), and a higher expression of carotenoid cleavage dioxygenases and HSPs in the more tolerant seed lot. Following the imposition of drying, about 80% of stress-responsive genes in the DS seed lot changed to the stable levels seen in the DT seed lot prior to and post-desiccation (Figure 7). However, the changes in expression of stress-responsive genes in DS seeds did not improve their DT. Future studies could explore whether DS seeds harvested from 'mature' fruits are developmentally immature, and/or genetically less DT.

Supplementary Materials: The following supporting information can be downloaded at: https://www.mdpi.com/article/10.3390/ijms24087393/s1.

Author Contributions: Conceptualization, H.C. and W.L.; methodology, H.C., Q.A. and L.Y.; validation, H.C., A.M.V. and H.W.P.; formal analysis, H.C., W.L. and A.M.V.; writing—original draft preparation, H.C. and A.M.V.; writing—review and editing, H.W.P.; project administration, H.C. and W.L.; funding acquisition, W.L. All authors have read and agreed to the published version of the manuscript.

Funding: This research was funded by The National Natural Science Foundation of China (NSFC 31770375), Science and Technology Basic Resources Investigation Program of China (2021FY100200) and The Royal Botanic Gardens, Kew received grant-in-aid from Defra. This work was facilitated by the Germplasm Bank of Wild Species, Kunming Institute of Botany, Chinese Academy of Sciences.

Institutional Review Board Statement: Not applicable.

Informed Consent Statement: Not applicable.

Data Availability Statement: The authors confirm that the data supporting the findings of this study are available within the article and its Supplementary Materials.

Acknowledgments: We thank Dechang Cao and Xiaojian Hu for comments on the manuscript. We also greatly thank Dong Jiang for kindly providing the Chongqing *Citrus sinensis 'bingtangcheng'* seed lot from the Chinese Academy of Agriculture Science, Chongqing for the seed lot of Chongqing.

Conflicts of Interest: The authors declare no conflict of interest.

References

1. Dussert, S.; Chabrillange, N.; Engelmann, F.; Anthony, F.; Louarn, J.; Hamon, S. Relationship between seed desiccation sensitivity, seed water content at maturity and climatic characteristics of native environments of nine *Coffea* L. species. *Seed Sci. Res.* **2000**, *10*, 293–300. [CrossRef]
2. Daws, M.I.; Cleland, H.; Chmielarz, P.; Gorian, F.; Leprince, O.; Mullins, C.E.; Thanos, C.A.; Vandvik, V.; Pritchard, H.W. Variable desiccation tolerance in *Acer pseudoplatanus* seeds in relation to developmental conditions: A case of phenotypic recalcitrance? *Funct. Plant Biol.* **2006**, *33*, 59–66. [CrossRef] [PubMed]
3. Chen, H.; Pritchard, H.W.; Seal, C.E.; Nadarajan, J.; Li, W.; Yang, S.; Kranner, I. Post desiccation germination of mature seeds of tea (*Camellia sinensis* L.) can be enhanced by pro-oxidant treatment, but partial desiccation tolerance does not ensure survival at −20 °C. *Plant Sci.* **2012**, *184*, 36–44. [CrossRef] [PubMed]

4.   Ganatsas, P.; Tsakaldimi, M.; Zarkadi, P.; Stergiou, D. Intraspecific differences in the response to drying of *Quercus ithaburensis* acorns. *Plant Biosyst.* **2017**, *151*, 878–886. [CrossRef]

5.   Daws, M.I.; Lydall, E.; Chmielarz, P.; Leprince, O.; Matthews, S.; Thanos, C.A.; Pritchard, H.W. Developmental heat sum influences recalcitrant seed traits in *Aesculus hippocastanum* across Europe. *New Phytol.* **2004**, *162*, 157–166. [CrossRef]

6.   Leprince, O.; Pellizzaro, A.; Berriri, S.; Buitink, J. Late seed maturation: Drying without dying. *J. Exp. Bot.* **2017**, *68*, 827–841. [CrossRef]

7.   Fait, A.; Angelovici, R.; Less, H.; Ohad, I.; Urbanczyk-Wochniak, E.; Fernie, A.R.; Galili, G. Arabidopsis seed development and germination is associated with temporally distinct metabolic switches. *Plant Phys.* **2006**, *142*, 839–854. [CrossRef]

8.   Righetti, K.; Vu, J.L.; Pelletier, S.; Vu, B.L.; Glaab, E.; Lalanne, D.; Pasha, A.; Patel, R.V.; Provart, N.J.; Verdier, J.; et al. Inference of longevity-related genes from a robust coexpression network of seed maturation identifies regulators linking seed storability to biotic defense-related pathways. *Plant Cell* **2015**, *27*, 2692–2708. [CrossRef]

9.   Pritchard, H.W.; Sershen Tsan, F.Y.; Wen, B.; Jaganathan, G.K.; Calvi, G.; Pence, V.C.; Mattana, E.; Ferraz, I.D.K.; Seal, C.E. Chapter 19—Regeneration in recalcitrant-seeded species and risks from climate change. In *Plant Regeneration from Seeds: A Global Warming Perspective*; Baskin, C.C., Baskin, J.M., Eds.; Academic Press: Cambridge, MA, USA, 2022; pp. 259–273.

10. Xu, Q.; Chen, L.L.; Ruan, X.; Chen, D.; Zhu, A.; Chen, C.; Bertrand, D.; Jiao, W.B.; Hao, B.H.; Lyon, M.P.; et al. The draft genome of sweet orange (*Citrus sinensis*). *Nat. Genet.* **2013**, *45*, 59–66. [CrossRef]

11. Swingle, W.T.; Reece, P.C. The botany of Citrus and its wild relatives. In *The Citrus Industry*; Reuther, W., Ed.; University of California Press: Berkeley, CA, USA, 1967.

12. Wu, G.A.; Terol, J.; Ibanez, V.; Lopez-Garcia, A.; Perez-Roman, E.; Borreda, C.; Domingo, C.; Tadeo, F.R.; Carbonell-Caballero, J.; Alonso, R.; et al. Genomics of the origin and evolution of *Citrus*. *Nature* **2018**, *554*, 311–316. [CrossRef]

13. Webber, H.J. History and development of the *Citrus* industry. In *The Citrus Industry*; Reuther, W., Ed.; University of California Press: Berkeley, CA, USA, 1967; pp. 1–39.

14. Gmitter, F.G.; Hu, X. The possible role of Yunnan province, China, in the origin of contemporary *Citrus* species (Rutaceae). *Econ. Bot.* **1990**, *44*, 267–277. [CrossRef]

15. Lytovchenko, A.; Eickmeier, I.; Pons, C.; Osorio, S.; Szecowka, M.; Lehmberg, K.; Arrivault, S.; Tohge, T.; Pineda, B.; Anton, M.T.; et al. Tomato fruit photosynthesis is seemingly unimportant in primary metabolism and ripening but plays a considerable role in seed development. *Plant Physiol.* **2011**, *157*, 1650–1663. [CrossRef] [PubMed]

16. Giovannoni, J.J. Genetic regulation of fruit development and ripening. *Plant Cell* **2004**, *16*, 170–180. [CrossRef] [PubMed]

17. Vrebalov, J.; Ruezinsky, D.; Padmanabhan, V.; White, R.; Medrano, D.; Drake, R.; Schuch, W.; Giovannoni, J. A MADS-box gene necessary for fruit ripening at the tomato ripening-inhibitor (Rin) locus. *Science* **2002**, *296*, 343–346. [CrossRef]

18. de Carvalho, D.U.; Boakye, D.A.; Gast, T.; Leite Junior, R.P.; Alferez, F. Determining seed viability during fruit maturation to improve seed production and availability of new Citrus rootstocks. *Front. Plant Sci.* **2021**, *12*, 777078. [CrossRef]

19. Hong, T.D.; Ahmad, N.B.; Murdoch, A.J. Optimum air-dry storage conditions for sweet orange (*Citrus sinensis* (L.) Osbeck) and lemon (*Citrus limon* (L.) Burm. f.) seeds. *Seed Sci. Technol.* **2001**, *29*, 183–192.

20. Pritchard, H.W.; Daws, M.I.; Fletcher, B.J.; Gamene, C.S.; Msanga, H.P.; Omondi, W. Ecological correlates of seed desiccation tolerance in tropical African dryland trees. *Am. J. Bot.* **2004**, *91*, 863–870. [CrossRef]

21. Seymour, G.; Poole, M.; Manning, K.; King, G.J. Genetics and epigenetics of fruit development and ripening. *Curr. Opin. Plant Biol.* **2008**, *11*, 58–63. [CrossRef]

22. Le, B.H.; Cheng, C.; Bui, A.Q.; Wagmaister, J.A.; Henry, K.F.; Pelletier, J.; Kwong, L.; Belmonte, M.; Kirkbride, R.; Horvath, S.; et al. Global analysis of gene activity during Arabidopsis seed development and identification of seed-specific transcription factors. *Proc. Natl. Acad. Sci. USA* **2010**, *107*, 8063–8070. [CrossRef]

23. Aslam, M.M.; Waseem, M.; Jakada, B.H.; Okal, E.J.; Lei, Z.L.; Saqib, H.S.; Yuan, W.; Xu, W.F.; Zhang, Q. Mechanisms of abscisic acid-mediated drought stress responses in plants. *Int. J. Mol. Sci.* **2022**, *23*, 1084. [CrossRef]

24. Nakajima, S.; Ito, H.; Tanaka, R.; Tanaka, A. Chlorophyll b reductase plays an essential role in maturation and storability of Arabidopsis seeds. *Plant Physiol.* **2012**, *160*, 261–273. [CrossRef] [PubMed]

25. Delmas, F.; Sankaranarayanan, S.; Deb, S.; Widdup, E.; Bournonville, C.; Bollier, N.; Northey, J.G.; McCourt, P.; Samuel, M.A. ABI3 controls embryo degreening through Mendel's I locus. *Proc. Natl. Acad. Sci. USA* **2013**, *110*, E3888–E3894. [CrossRef] [PubMed]

26. Teixeira, R.N.; Ligterink, W.; Franca-Neto Jde, B.; Hilhorst, H.W.; da Silva, E.A. Gene expression profiling of the green seed problem in soybean. *BMC Plant Biol.* **2016**, *16*, 37. [CrossRef]

27. Ballesteros, D.; Pritchard, H.W.; Walters, C. Dry architecture-towards an understanding of the variation of longevity in desiccation-tolerance germplasm. *Seed Sci. Res.* **2020**, *30*, 142–155. [CrossRef]

28. Cocotl-Yanez, M.; Moreno, S.; Encarnacion, S.; Lopez-Pliego, L.; Castaneda, M.; Espin, G. A small heat-shock protein (Hsp20) regulated by RpoS is essential for cyst desiccation resistance in *Azotobacter vinelandii*. *Microbiology* **2014**, *160*, 479–487. [CrossRef]

29. Shu, K.; Zhang, H.; Wang, S.; Chen, M.; Wu, Y.; Tang, S.; Liu, C.; Feng, Y.; Cao, X.; Xie, Q. ABI4 regulates primary seed dormancy by regulating the biogenesis of abscisic acid and gibberellins in arabidopsis. *PLoS Genet.* **2013**, *9*, e1003577. [CrossRef]

30. Wang, W.; Vinocur, B.; Shoseyov, O.; Altman, A. Role of plant heat-shock proteins and molecular chaperones in the abiotic stress response. *Trends Plant Sci.* **2004**, *9*, 244–252. [CrossRef]

31. Qin, F.; Yu, B.Z.; Li, W.Q. Heat shock protein 101 (HSP101) promotes flowering under nonstress conditions. *Plant Physiol.* **2021**, *186*, 407–419. [CrossRef]

32. Wehmeyer, N.; Vierling, E. The expression of small heat shock proteins in seeds responds to discrete developmental signals and suggests a general protective role in desiccation tolerance. *Plant Physiol.* **2000**, *122*, 1099–1108.33. [CrossRef]
33. Kotak, S.; Vierling, E.; Baumlein, H.; von Koskull-Doring, P. A novel transcriptional cascade regulating expression of heat stress proteins during seed development of Arabidopsis. *Plant Cell* **2007**, *19*, 182–195. [CrossRef]
34. Malik, S.K.; Chaudhury, R.; Pritchard, H.W. Long-term, large scale banking of citrus species embryos: Comparisons between cryopreservation and other seed banking temperatures. *CryoLetters* **2012**, *33*, 453–464. [PubMed]
35. Mizrahi, T.; Goldenberg, S.; Heller, J.; Arad, Z. Natural variation in resistance to desiccation and heat shock protein expression in the land snail *Theba pisana* along a climatic gradient. *Physiol. Biochem. Zool.* **2015**, *88*, 66–80. [CrossRef] [PubMed]
36. Khanna-Chopra, R. Leaf senescence and abiotic stresses share reactive oxygen species-mediated chloroplast degradation. *Protoplasma* **2012**, *249*, 469–481. [CrossRef]
37. Verdier, J.; Lalanne, D.; Pelletier, S.; Torres-Jerez, I.; Righetti, K.; Bandyopadhyay, K.; Leprince, O.; Chatelain, E.; Vu, B.L.; Gouzy, J.; et al. A regulatory network-based approach dissects late maturation processes related to the acquisition of desiccation tolerance and longevity of *Medicago truncatula* seeds. *Plant Physiol.* **2013**, *163*, 757–774. [CrossRef]
38. Tsou, P.L.; Zhu, H.J.; Godfrey, T.; Blackman, S. Post-excision drying of immature *Phalaenopsis* seeds improves germination and desiccation tolerance. *S. Afr. J. Bot.* **2022**, *150*, 1184–1191. [CrossRef]
39. Peng, L.; Huang, X.; Qi, M.; Prtichard, H.W.; Xue, H. Mechanistic insights derived from re-establishment of desiccation tolerance in germination xerophytic seeds: *Caragana korshinskii* as an example. *Front. Plant Sci.* **2022**, *13*, 1029997. [CrossRef] [PubMed]
40. Salonia, F.; Ciacciulli, A.; Poles, A.; Pappalardo, D.H.; La Malfa, S.; Licciardello, C. New plant breeding techniques in citrus for the improvement of important agronomic traits. A Review. *Front. Plant Sci.* **2020**, *11*, 1234. [CrossRef]
41. Jia, H.; Orbović, V.; Wang, N. CRISPR-LbCas12a-mediated modification of citrus. *Plant Biotechnol. J.* **2019**, *17*, 1928–1937. [CrossRef]
42. Griffith, M.; Walker, K.R.; Spies, N.C.; Ainscough, B.J.; Griffith, O.L. Informatics for RNA sequencing, a web resource for analysis on the cloud. *PLoS Comput. Biol.* **2015**, *11*, e1004393. [CrossRef]
43. Love, M.I.; Huber, W.; Anders, S. Moderated estimation of fold change and dispersion for RNA-seq data with DESeq2. *Genome Biol.* **2014**, *15*, 550. [CrossRef]
44. Robinson, M.D.; McCarthy, D.J.; Smyth, G.K. edgeR: A Bioconductor package for differential expression analysis of digital gene expression data. *Bioinformatics* **2010**, *26*, 139–140. [CrossRef] [PubMed]
45. Xie, Z.; Lin, W.; Luo, J. Promotion of microalgal growth by co-culturing with *Cellvibrio pealriver* using xylan as feedstock. *Bioresour. Technol.* **2016**, *200*, 1050–1054. [CrossRef] [PubMed]

International Journal of *Molecular Sciences*

Article

Genome-Wide Identification and Expression Analysis of *Dendrocalamus farinosus CCoAOMT* Gene Family and the Role of *DfCCoAOMT14* Involved in Lignin Synthesis

Lixian Wei [1,2,†], Xin Zhao [1,2,†], Xiaoyan Gu [1,2], Jiahui Peng [1,2], Wenjuan Song [1,2], Bin Deng [1,2], Ying Cao [1,2,*] and Shanglian Hu [1,2,*]

[1]   Lab of Plant Cell Engineering, Southwest University of Science and Technology, Mianyang 621010, China
[2]   Engineering Research Center for Biomass Resource Utilization and Modification of Sichuan Province, Mianyang 621010, China
*   Correspondence: caoying@swust.edu.cn (Y.C.); hushanglian@swust.edu.cn (S.H.)
†   These authors contributed equally to this work.

Abstract: As the main component of plant cell walls, lignin can not only provide mechanical strength and physical defense for plants, but can also be an important indicator affecting the properties and quality of wood and bamboo. *Dendrocalamus farinosus* is an important economic bamboo species for both shoots and timber in southwest China, with the advantages of fast growth, high yield and slender fiber. Caffeoyl-coenzyme A-O-methyltransferase (CCoAOMT) is a key rate-limiting enzyme in the lignin biosynthesis pathway, but little is known about it in *D. farinosus*. Here, a total of 17 *DfCCoAOMT* genes were identified based on the *D. farinosus* whole genome. *DfCCoAOMT1/14/15/16* were homologs of *AtCCoAOMT1*. *DfCCoAOMT6/9/14/15/16* were highly expressed in stems of *D. farinosus*; this is consistent with the trend of lignin accumulation during bamboo shoot elongation, especially *DfCCoAOMT14*. The analysis of promoter cis-acting elements suggested that *DfCCoAOMTs* might be important for photosynthesis, ABA/MeJA responses, drought stress and lignin synthesis. We then confirmed that the expression levels of *DfCCoAOMT2/5/6/8/9/14/15* were regulated by ABA/MeJA signaling. In addition, overexpression of *DfCCoAOMT14* in transgenic plants significantly increased the lignin content, xylem thickness and drought resistance of plants. Our findings revealed that *DfCCoAOMT14* can be a candidate gene that is involved in the drought response and lignin synthesis pathway in plants, which could contribute to the genetic improvement of many important traits in *D. farinosus* and other species.

Keywords: *Dendrocalamus farinosus*; *CCoAOMT* gene; genome-wide identification; drought response; lignin synthesis

Citation: Wei, L.; Zhao, X.; Gu, X.; Peng, J.; Song, W.; Deng, B.; Cao, Y.; Hu, S. Genome-Wide Identification and Expression Analysis of *Dendrocalamus farinosus CCoAOMT* Gene Family and the Role of *DfCCoAOMT14* Involved in Lignin Synthesis. *Int. J. Mol. Sci.* **2023**, 24, 8965. https://doi.org/10.3390/ijms24108965

Academic Editors: Frank M. You and Martin Bartas

Received: 6 March 2023
Revised: 7 May 2023
Accepted: 12 May 2023
Published: 18 May 2023

1. Introduction

Bamboo forests, with about 1500 species worldwide, cover approximately 31.5 million hectares in tropical and subtropical regions of the world [1]. They help address the issue of global warming by fixing carbon dioxide in the atmosphere [2,3]. Moreover, bamboo resembles wood in appearance, weight, hardness, texture and strength, making bamboo an ideal substitute for many traditional wood products. *D. farinosus* is an important sympodial bamboo species for both bamboo shoots and materials; it is mainly distributed in southwest China, with the advantages of high fiber content, a large biomass, cold tolerance, rapid growth and high yield, which are also important for paper making and ecological greening [4].

Lignin is a complex polyphasic aromatic polymer that is not only a key product of the phenylpropane pathway, but also an important component of plant cell walls, playing an important role in enhancing plant mechanical strength and contributing to the defense against pests and diseases [5]. However, lignin is not conducive to paper making and lignocellulosic biomass fuel production. Lignin is mainly deposited in thick-walled tissues such as vessels and fibers of plants, and by firmly fixing cellulose and hemicellulose components in cell walls, it increases the difficulty to separate and utilize polysaccharide components such as cellulose [6,7]. Overall, lignin is essential for the growth and development of bamboo, resistance to pathogenic bacteria, timber properties of bamboo and quality of bamboo shoots.

The three main lignin monomers are p-hydroxyphenyl (H), guaiacyl (G) and syringyl (S), which are formed from phenylalanine by a multi-step redox catalysis through the phenyl propane metabolic pathway. The enzymes catalyzing lignin synthesis have been clearly identified [8]. CCoAOMT is a key methyltransferase in the lignin synthesis pathway and belongs to the methyltransferase family; it is closely related to plant lignin biosynthesis [9]. The alteration of the *CCoAOMT* gene causes the change of the ratio of S/G and affects wood material and forage digestibility. *CCoAOMT1*, the first identified true *CCoAOMT* family gene involved in lignin synthesis [10], affects plant resistance to *Pseudomonas syringae* DC3000 in *Arabidopsis* [11]; *AtCCoAOMT1* can also improve drought tolerance via ROS- and ABA-dependent signaling pathways [12]. Meanwhile, the overexpression of the *CCoAOMT* gene improved the resistance of plants to different pathogens [13–15]. For example, overexpression of the *Paeonia ostii CCoAOMT* gene in tobacco increased drought tolerance by enhancing lignin content and reactive oxygen species clearing capacity [16]. ZmCCoAOMT2 interacts with NLR Rp1 to regulate plant defense responses in maize [13]. More and more research suggests that CCoAOMT is important for the plant biomass and defense response. At present, the *CCoAOMT* gene family has been identified in *Arabidopsis thaliana*, poplar, tea tree, wheat, grape and other species [17–20], but little is known in *D. farinosus*.

Here, we analyzed the phylogeny, chromosome distribution, gene structure and motif composition of *DfCCoAOMT* family genes by using bioinformatics methods. We identified 17 *DfCCoAOMT* genes based on *D. farinosus* whole-genome data. Meanwhile, RNA-seq and qRT-PCR were used to analyze the expression pattern of *DfCCoAOMT* genes, combined with tissue sections to screen for candidate genes involved in lignin synthesis. *DfCCoAOMT* genes were involved in the ABA/MeJA signaling pathway in response to abiotic and biotic stresses. Overexpression of *DfCCoAOMT14* in tobacco promoted lignin biosynthesis and drought resistance of transgenic plants. Overall, this study laid a foundation for investigating *DfCCoAOMT* family genes in lignin biosynthesis and the drought response during plant growth and development.

2. Results

2.1. Genome-Wide Identification of CCoAOMT Genes in D. farinosus

Based on the reported protein sequence of the *Arabidopsis CCoAOMT* gene family and the methyltransferase structural domain from Pfam (PF01596), a total of 17 CCoAOMT genes were identified based on the *D. farinosus* genome and named as DfCCoAOMT1 to DfCCoAOMT17 (Table 1). Detailed information about these DfCCOAOMT gene sequences is presented in Supplemental Table S1 and Supplementary Document S1. These lengths of DfCCoAOMT proteins ranged from 220 aa (DfCCoAOMT13) to 311 aa (DfCCoAOMT17), the molecular weight ranged from 24.30 KDa (DfCCoAOMT13) to 34.00 KDa (DfCCoAOMT17) and the isoelectric point (pI) ranged from 4.88 (DfCCoAOM6) to 9.44 (DfCCoAOMT3). A subcellular localization prediction indicated that all CCoAOMT proteins were localized in the cytoplasm and chloroplasts.

Table 1. Characteristics of the *CCoAOMT* genes identified in *D. farinosus*.

Gene ID	Gene Name	Protein Length (aa)	MW (KDa)	pI	Subcellular Localization
DfaB03G023550	*DfCCoAOMT1*	240	28.33	5.41	Cytoplasmic
DfaA06G007370	*DfCCoAOMT2*	249	26.75	5.33	Cytoplasmic Chloroplast
DfaB06G012070	*DfCCoAOMT3*	293	32.08	9.44	Chloroplast
DfaA07G002980	*DfCCoAOMT4*	296	32.24	8.35	Chloroplast
DfaA07G014340	*DfCCoAOMT5*	252	27.58	4.95	Cytoplasmic
DfaA07G014350	*DfCCoAOMT6*	248	27.03	4.88	Cytoplasmic
DfaA07G014530	*DfCCoAOMT7*	251	27.85	6.45	Cytoplasmic
DfaA07G014550	*DfCCoAOMT8*	222	24.33	5.16	Cytoplasmic
DfaB07G011900	*DfCCoAOMT9*	250	27.31	5.09	Cytoplasmic
DfaB07G011910	*DfCCoAOMT10*	240	26.75	5.56	Cytoplasmic Chloroplast
DfaB07G011930	*DfCCoAOMT11*	240	26.63	5.78	Cytoplasmic Chloroplast
DfaB07G012160	*DfCCoAOMT12*	250	27.52	5.83	Cytoplasmic Chloroplast
DfaC07G006610	*DfCCoAOMT13*	220	24.30	4.88	Cytoplasmic
DfaA11G016070	*DfCCoAOMT14*	262	28.95	5.22	Cytoplasmic
DfaB11G015250	*DfCCoAOMT15*	260	28.93	5.33	Cytoplasmic
DfaC11G007040	*DfCCoAOMT16*	259	28.86	5.33	Cytoplasmic
Dfa0G042980	*DfCCoAOMT17*	311	34.00	9.09	Chloroplast

2.2. Phylogenetic Relationship, Gene Structure and Conserved Motif Analysis of the CCoAOMT Gene Family in D. farinosus

To clarify the evolutionary relationships and classification of the *CCoAOMT* gene family in *D. farinosus*, a phylogenetic tree was constructed using the full-length protein sequences of CCoAOMT from *D. farinosus*, *Arabidopsis*, rice and *Populus* with IQ-TREE v1.6.12 and the Model Finder algorithm (Figure 1A). The results showed that the *DfCCoAOMT* genes were divided into four groups (Classes I~IV); Classes I, II and IV contained 4, 10 and 3 *DfCCoAOMT* genes, respectively, and Class III had no *DfCCoAOMT* genes. However, in *Arabidopsis*, *AtCCoAOMT* genes were mainly clustered in Class III, only two genes (*AtCCoAOMT3* and *AtCCoAOMT4*) were clustered in Class IV and one (*AtCCoAOMT1*) was clustered in Class I. *DfCCoAOMT1*, *DfCCoAOMT14*, *DfCCoAOMT15* and *DfCCoAOMT16* were homologs of *AtCCoAOMT1*, *OsCCoAOMT6*, *PtCCoAOMT1* and *PtCCoAOMT4*, respectively. *DfCCoAOMT6* and *DfCCoAOMT9* were homologs of *OsCCoAOMT1* and *OsCCoAOMT3*, respectively, suggesting that they might have a similar function.

Next, the gene structure and conserved motif composition of *DfCCoAOMT* were analyzed (Figure 1B). The MEME online website was used to investigate the conservation and diversity of motifs in different DfCCoAOMT proteins. The results showed that all the proteins were found to be related to O-methyltransferase; motifs 2, 3, and 6 were common motifs in all proteins. Motif 1 was present in 16 DfCCoAOMT proteins except DfCCoAOMT13, and motif 9 appeared in 15 DfCCoAOMT proteins apart from DfCCoAOMT4/8. Motif 9 was only observed in Class IV, which might be associated with the functional diversity of *DfCCoAOMT* genes. Then, the exon–intron gene structure of *DfCCoAOMTs* were analyzed. The exon number of *DfCCoAOMTs* ranged from 3 to 9 and the intron number ranged from 2 to 8. Six genes (*DfCCoAOMT6/7/8/10/11/13*) contained four exons and three introns, and *DfCCoAOMT1/5/9/12/14/15* contained three exons and two introns; this difference in gene structure may be an evolutionary genetic difference. Meanwhile, the members of the same group had similar exon–intron structures and motif compositions.

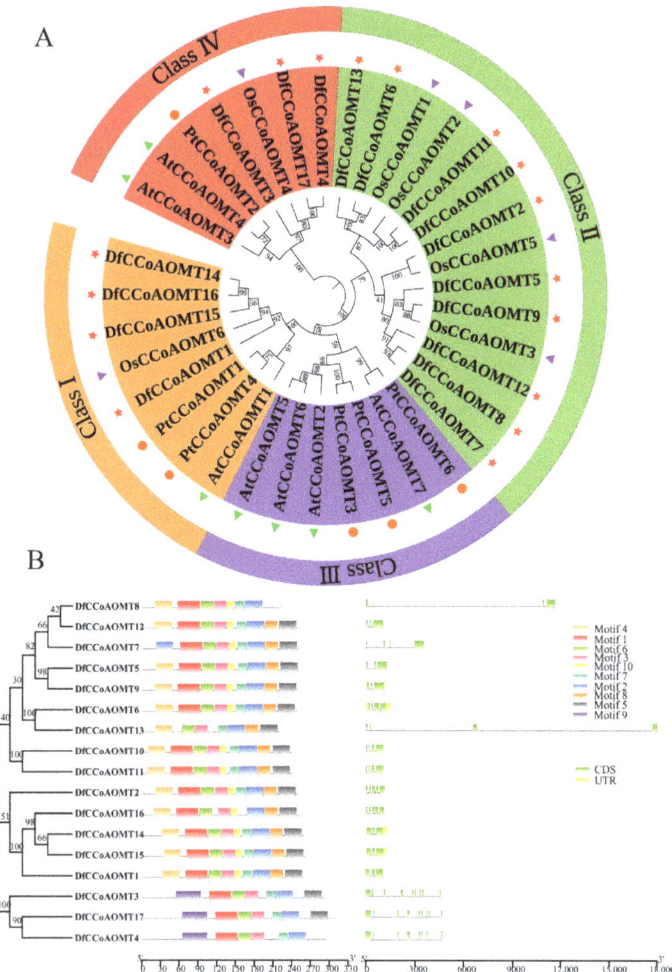

Figure 1. Phylogenetic relationship, conserved motif and gene structure of *D. farinosus CCoAOMTs*. (**A**) Phylogenetic tree construction based on amino acid sequences of the *CCoAOMT* family from *Arabidopsis thaliana* (At), rice (Os), *Populus trichocarpa* (Pt) and *D. farinosus*. The species are labeled with shapes of different colors: Df, (red pentagram); Os, (purple triangle); At, (green triangle); and Pt, (orange circle). (**B**) From left to right, phylogenetic tree constructed by MEGA analysis. The numbers beside the branches represent bootstrap support values from 1000 replications. Conserved motifs identified by MEME analysis. Each motif is represented with a different color. The protein length can be estimated using the scale at the bottom. Gene structures identified by GSDS analysis. Exons and introns are represented by boxes and lines, respectively. The sequence length of exons can be estimated using the scale at the bottom. The intron sizes are not to scale. CDS, coding sequence; UTR, untranslated region.

2.3. Chromosome Distribution and Collinearity Analysis of DfCCoAOMT Genes

Gene shrinkage and expansion are important factors for plant gene functional diversity and environmental adaptation. Tandem repeats and fragment repeats are the main causes of gene expansion in plants. Therefore, the chromosome distribution and collinearity of the *DfCCoAOMT* gene family were analyzed (Figure 2 and Supplemental Figure S1). In *D. farinosus*, these *DfCCoAOMT* genes were located on chromosome groups 3, 6, 7 and 11, respectively; the *DfCCoAOMT* genes were most abundant on chromosome group 7 (10/17). Furthermore, there was an obvious tandem duplication in the seventh chromosome group, *DfCCoAOMT5/6/7/8* formed a tandem repeat gene cluster on chromosome A07 and *DfCCoAOMT9/10/11/12* formed a tandem repeat gene cluster on chromosome B07. Four pairs of putative paralogous *DfCCoAOMT* genes were identified and located on different chromosomes; these results suggested that there were fragment duplication events in the *DfCCoAOMT* gene family and led to expansion of the family members.

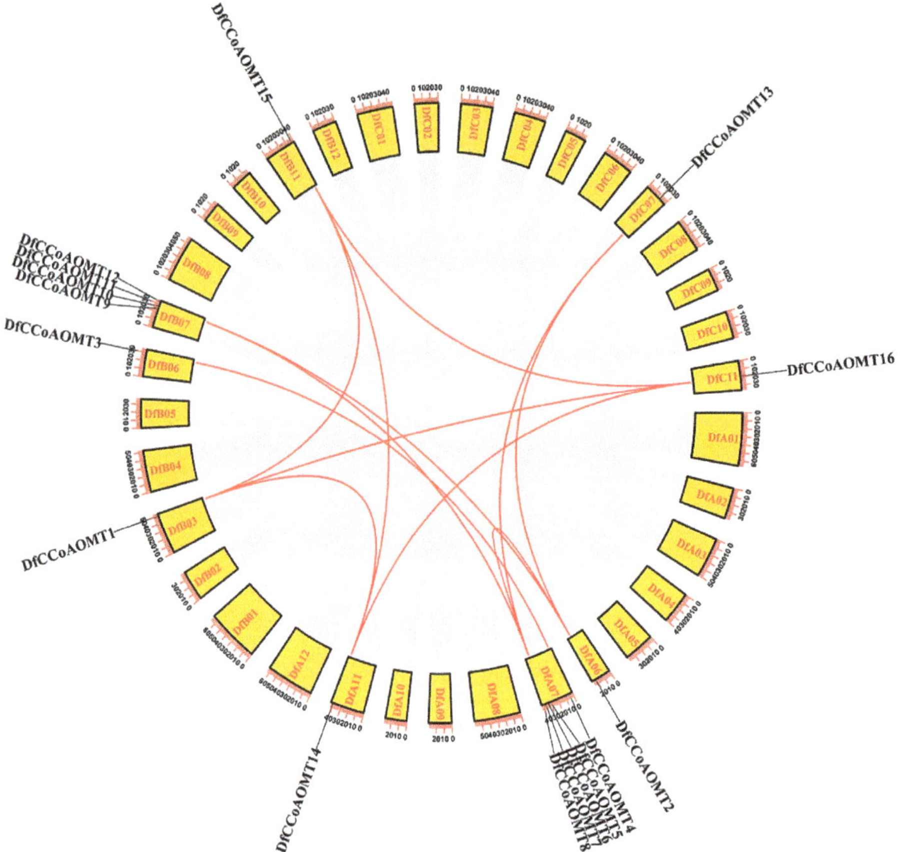

Figure 2. Chromosome distribution and collinearity analysis of *DfCCoAOMT* genes. The *DfCCoAOMT* genes are localized on different chromosomes. Chromosome numbers are indicated in the yellow boxes. The numbers on the chromosome boxes represent the sequence length in megabases. Gene pairs with sibling relationships are connected by a red line.

2.4. Tissue Expression Profile of DfCCoAOMTs and Analysis of Expression at Different Developmental Periods in D. farinosus

To better understand the function of the *DfCCoAOMT* genes in *D. farinosus*, we performed a transcriptome sequencing analysis of the expression levels of various tissues and different developmental periods in *D. farinosus* (Supplemental Table S3). As shown in Figure 3A, *DfCCoAOMT* gene expressions were different among the Young leaf (Yleaf), Mature leaf (MLeaf), Lateral bud (Labud), Sheath of bamboo shoot (ShB), Node, Rhizome bud (Rhbud), Rhizome neck (Rhne), Root, Branch, Internode (Inod) and Culm sheath (CSh). Among these 17 genes, *DfCCoAOMT10/11/12* were specifically highly expressed in the root, implying that they mainly functioned during root development. Eight genes (*DfC-CoAOMT1/2/3/4/7/8/13/17*) showed a low expression in all tissues. *DfCCoAOMT6/9/14/15/16* showed a high expression level in various tissues, especially in the Inod, Rhne, root and node. These results suggested that *DfCCoAOMT* genes function in organ development and a variety of biological processes in *D. farinosus*.

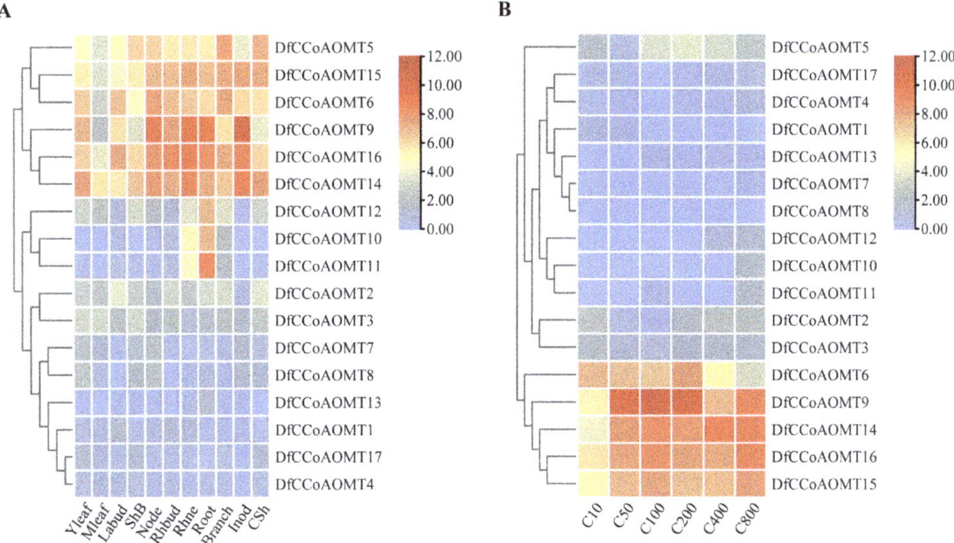

Figure 3. Expression analysis of *DfCCoAOMTs* in different tissues and developmental stages of *D. farinosus*. Hierarchical clustering analysis of expression levels of *CCoAOMTs* in different tissues. (**A**) Different developmental tissues of *D. farinosus*. (**B**) Different developmental phases of *D. farinosus*. The clustering was performed using TBtools. The red and blue colors correspond to the strong and weak expression of the genes, respectively. Dendrograms along the left sides of the heat map indicate the hierarchical clustering of genes. Young leaf (Yleaf), Mature leaf (MLeaf), Lateral bud (Labud), Sheath of bamboo shoot (ShB), Node, Rhizome bud (Rhbud), Rhizome neck (Rhne), Root, Branch, Internode (Inod) and culm sheath (CSh). C10 to C800 indicate bamboo shoots from 10 cm to 800 cm.

Since lignin deposition and secondary wall thickening usually accompany rapid growth of bamboo, we then analyzed the transcript levels of bamboo shoots during different growth periods (Figure 3B). Only *DfCCoAOMT6/9/14/15/16* showed a high expression in fast-growing culm, indicating that these five genes might be involved in lignin synthesis during the rapid growth period of bamboo culms.

2.5. Prediction of Cis-Component Analysis of DfCCoAOMTs

To explore the mechanism of *DfCCoAOMT* genes in stress response and development, the cis-regulatory elements in the 2 kb upstream sequence of *DfCCoAOMT* genes were predicted by PlantCARE and PLACE (Figure 4A,B). In the *DfCCoAOMT* genes' promoter, a variety of stress-, growth- and development-related cis-acting elements were identified, such as the low temperature response element (LTR), gibberellin (GA) response element, drought response element (MBS), light response element (G-box, Box4, MRE, I-box, ACE and LAMP element), MeJA response element (CGTCA-motif and TGACG-motif), salicylic acid (SA) response element and abscisic acid (ABA) response element (ABRE). Light- and growth-related cis-acting elements were present in almost all *DfCCoAOMT* promoter sequences. MeJA and ABA response elements were identified in the promoter of *CCoAOMT2/5/6/8/9/14/15*. Various lignin synthesis-related elements, such as MYB elements, were identified in the promoters of *CCoAOMT1/2/4/5/9/10/11/14/15/16/17*, especially *CCoAOMT14*.

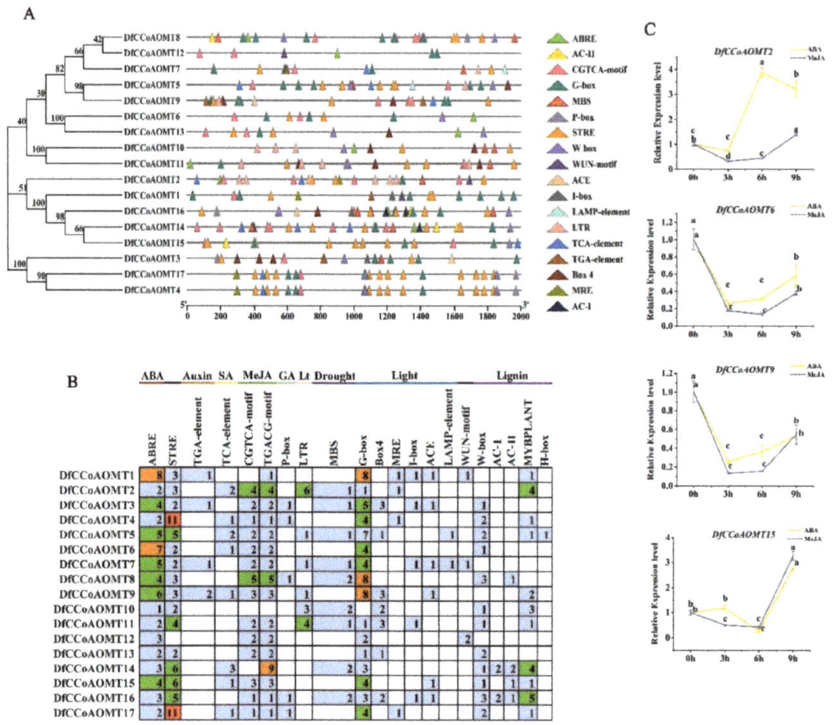

A

Legend (cis-elements): ABRE, AC-II, CGTCA-motif, G-box, MBS, P-box, STRE, W box, WUN-motif, ACE, I-box, LAMP-element, LTR, TCA-element, TGA-element, Box 4, MRE, AC-I

B

Gene	ABRE	STRE	TGA-element	TCA-element	CGTCA-motif	TGACG-motif	P-box	LTR	MBS	G-box	Box4	MRE	I-box	ACE	LAMP-element	WUN-motif	W-box	AC-1	AC-II	MYBPLANT	H-box
DfCCoAOMT1	8	3	1			1				8		1	1	1		1				1	
DfCCoAOMT2	2	3		2	4	4		6	1	1		1								4	
DfCCoAOMT3	4	2	1		2	2	1		1	5	3		1	1			1				
DfCCoAOMT4	2	11		1	1	1	1			4		1					2			1	
DfCCoAOMT5	5	5			2	2		1		7	1				1		2			1	1
DfCCoAOMT6	7	2		1	2	2				4							1				
DfCCoAOMT7	5	2	1		2	2		1	1	4			1	1	1	1					
DfCCoAOMT8	4	3			5	5	1		2	8							3		1		
DfCCoAOMT9	6	3	2	1	3	3		1		8	3		1							2	
DfCCoAOMT10	1	2					3		2		2						1			3	
DfCCoAOMT11	2	4			2	2		4	1	1	3		1				1			1	
DfCCoAOMT12	3				2	2				2					2						
DfCCoAOMT13	2	2			2	2				1	1						2				
DfCCoAOMT14	3	6		3		9			2	3							1	2	2	4	
DfCCoAOMT15	4	6		1	3	3				4				1			1			1	1
DfCCoAOMT16	3	5			1	1	1		2	3	2		1	1			3	2	1	5	
DfCCoAOMT17	2	11			1	1	1	1		4		1					2			1	

C — Relative expression levels of *DfCCoAOMT2*, *DfCCoAOMT5*, *DfCCoAOMT6*, *DfCCoAOMT8*, *DfCCoAOMT9*, *DfCCoAOMT14*, *DfCCoAOMT15* (ABA and MeJA treatments).

Figure 4. Predicted cis-element analysis in the promoter regions of *DfCCoAOMT* genes and the expression of *DfCCoAOMTs* under ABA and MeJA treatments in *D. farinosus*. (**A**) The phylogenetic tree of *DfCCoAOMT* gene family members. Putative cis-elements in *DfCCoAOMT* gene promoters. Promoter sequences (2 kb) of seventeen *DfCCoAOMT* genes were analyzed using PlantCARE and PLACE. Different color and shape boxes stand for different cis-elements. (**B**) Number and function of cis-elements. (**C**) Relative level of gene transcription of *DfCCoAOMT* in response to 100 mM ABA treatment and 100 mM MeJA treatment for 3 h, 6 h and 9 h, respectively. Error bars indicate standard deviation among three independent replicates. The values represent the mean $\pm$ SD; different letters indicate significant differences according to Duncan's multiple range test and least significance difference (LSD) ($p < 0.05$).

We then examined the expression pattern of *DfCCoAOMT2/5/6/8/9/14/15* after ABA and MeJA treatments using qRT-PCR (Figure 4C). The results showed that the expression levels of *DfCCoAOMT6/8/9/14* decreased rapidly after ABA and MeJA treatments, reaching the lowest levels at 3 h and 6 h, respectively, and then gradually bound. In contrast, the expression levels of *DfCCoAOMT2/5* were significantly increased in response to ABA after 3 h, reaching the highest levels at 6 h. These results suggested that these *DfCCoAOMT* genes might be involved not only in lignin synthesis, but also in plant biotic and abiotic stress responses.

In addition, due to the differences in lignification between different internodes of bamboo shoots, the microscopy of individual stem sections of 2 m tall bamboo shoots was performed. Phloroglucinol-Cl staining showed a gradual decrease in internode lignin deposition from base to tip in bamboo shoots and measurements of lignin content in different internodes showed consistent results (Figure 5A,B). Furthermore, the expression levels of *DfCCoAOMT* genes in different internodes were detected and showed that *DfC-CoAOMT5/6/9/14* had high expression levels in the first to third internode, consistent with the trend of lignin deposition. In contrast, *DfCCoAOMT15/16* were highly expressed in all internodes, suggesting that they are also involved in other biological functions of bamboo shoots (Figure 5C).

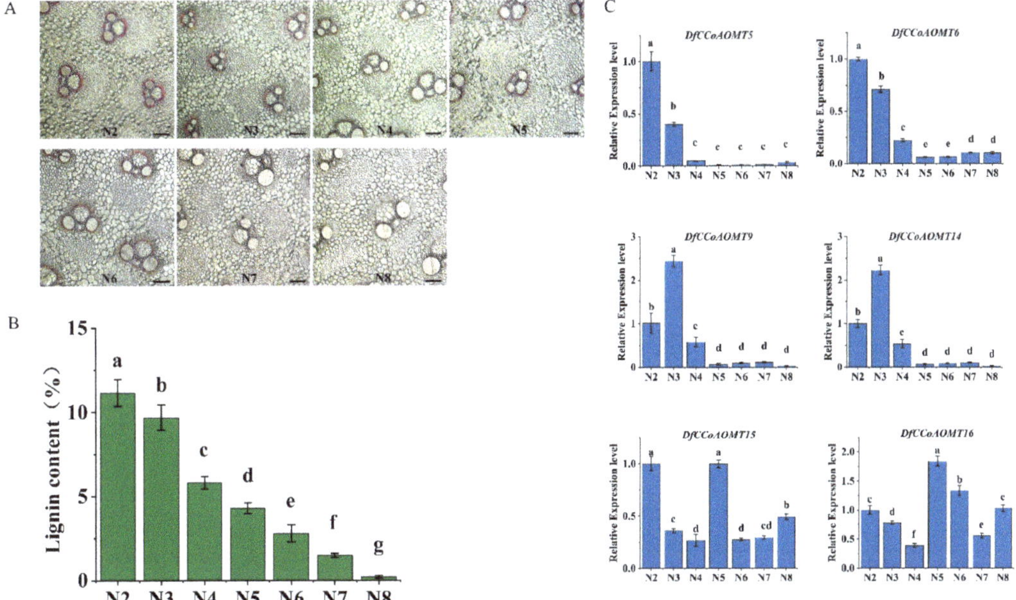

Figure 5. Shoot section diagram and lignin content determination of 2 m high bamboo shoots. (**A**) The sliced images of the 2 m high bamboo shoots from bottom to top; number 2–8 denote second to eighth internode; the slices were stained with phloroglucinol. Ten times magnification to observe sections, the scale bar = 100 μm. (**B**) The lignin content of bamboo shoots from the second to eighth internode. (**C**) The relative expression levels of *DfCCoAOMTs* in bamboo shoots from second to eighth internode. The values represent the mean ± SD, and different letters indicate significant differences according to Duncan's multiple range test and least significance difference (LSD) ($p < 0.05$).

2.6. Overexpression of DfCCoAOMT14 Improves Lignin Biosynthesis in Transgenic Tobacco

To determine the role of *DfCCoAOMTs* in lignin biosynthesis, *DfCCoAOMT14* was cloned and *CaMV 35S*-driven plant expression vectors were constructed and then transformed into tobacco. The expression of *DfCCOAOMT14* in transgenic plants and wild type was detected using qRT-PCR and we found that the expression levels of line 1, 2 and 3 were significantly higher than in wild type (Figure 6C). The phenotype of transgenic plants showed growth retardation compared with wild type (Figure 6A). In addition, the third internode of the 40-day-old transgenic plants was used to observe the anatomical cross-section; the results showed that overexpression of *DfCCOAOMT14* significantly increased xylem lignification in transgenic plants compared with wild type; the xylem widths in lines 1, 2 and 3 increased by approximately 46.5%, 9.2% and 12.3% compared with wild type, respectively (Figure 6B,D). Determination of lignin content in stems from transgenic plants also showed a significant increase (Figure 6E). Then, the expression levels of several lignin biosynthesis-related genes were detected using qRT-PCR and it was found that the expression levels of four genes (*PAL, C4H, CAD* and *COMT*) in transgenic lines were significantly higher than those of wild type (Figure 6F). These results suggested that the elevated expression levels of *DfCCoAOMT14* could enhance lignin biosynthesis in plants.

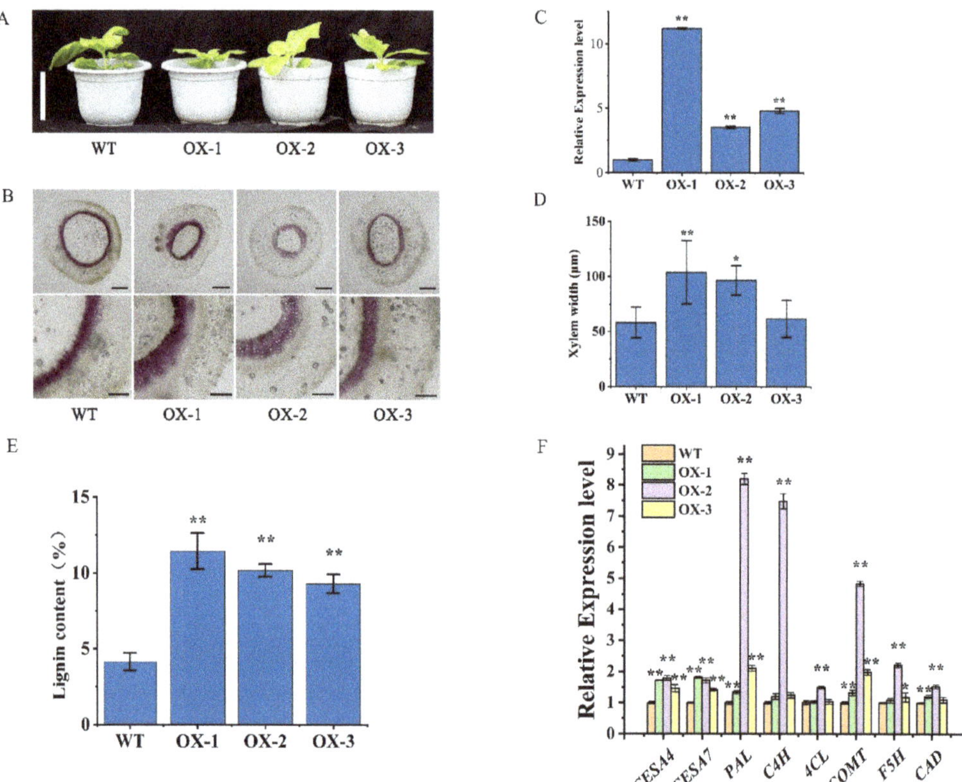

Figure 6. Overexpression of *DfCCoAOMT14* improves lignin biosynthesis in transgenic tobacco. (**A**) Phenotypic observation of WT and three transgenic plant lines. The scale bar = 10 cm. (**B**) Phloroglucinol-Cl staining of stem transverse section. The upper row images were observed using a 1.6-fold microscope, the scale bar = 250 μm, and the lower row images were magnified by 5 times, the scale bar = 100 μm. (**C**) Detection of the expression level of *DfCCoAOMT14* in overexpression tobacco and wild type. (**D**) Statistics of xylem widths of WT and three transgenic plant lines. (**E**) Determination of lignin content of the shoots. (**F**) Detection of expression level of secondary wall synthase genes including *CESA4*, *CESA7*, *PAL*, *C4H*, *4CL*, *COMT*, *F5H* and *CAD*. Data are means ± SD (n = 3). The asterisks indicate a significant difference compared to vector-transformed plants using the one-way analysis of variance (* $p < 0.05$, ** $p < 0.01$).

2.7. Overexpression of DfCCoOMT14 Improved Drought Resistance of Transgenic Plants

Previous studies have shown that lignin synthesis significantly affects the plant responses to drought [16]. We performed a drought treatment on transgenic plants for 37 days and found that the height and growth of transgenic plants performed better than wild type (Figure 7A). Moreover, we measured the drought-related physiological indicators in transgenic plants and wild type after the drought treatment; the levels of malondialdehyde (MDA), superoxide dismutase (SOD), peroxidase (POD), proline (Pro) and relative water content (RWC) in transgenic plants were significantly higher than in wild type (Figure 7C–G). qRT-PCR showed that the expression levels of the drought-related genes *DREB* and *RD29A* were also significantly higher in transgenic plants compared with wild type (Figure 7B). These results suggested that *DfCCoAOMT14* improved the drought resistance of plants.

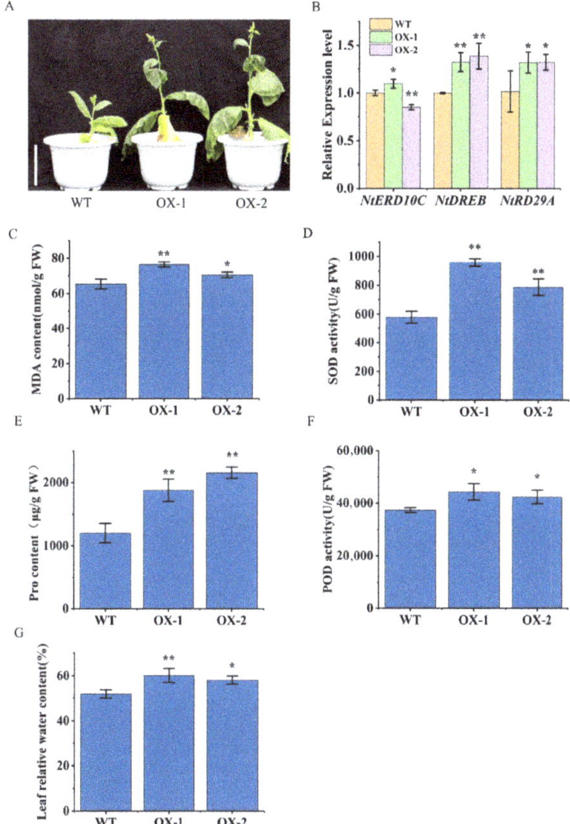

Figure 7. Overexpression of *DfCCoOMT14* enhanced drought resistance of plants. (**A**) Phenotype of wild type and transgenic plants after drought treatment. The scale bar = 10 cm. (**B**) Determination of the expression levels of drought-related genes including *NtERD10C*, *NtDREB* and *NtRD29A* in drought-treated plants. (**C–G**) The measurement of relevant physiological indices including MDA, SOD, Pro, POD and RWC. Data are means $\pm$ SD ($n = 3$). The asterisks indicate a significant difference compared to vector-transformed plants using the one-way analysis of variance (* $p < 0.05$, ** $p < 0.01$).

3. Discussion

As a cross-linked phenolic polymer, lignin is not only a major component of plant cell walls, determining the degree of lignification, but is also involved in the plant's ability to resist injury in response to abiotic and biotic stresses [21]. Lignin biosynthesis is a complex and delicate process; lignin monomers are synthesized from phenylalanine in the cytoplasm through a series of enzymatic reactions, and then transported across the membrane to participate in the synthesis of cell walls. As an important key enzyme for lignin synthesis, CCoAOMT catalyzes the methylation of caffeoyl-coenzyme A (caffeoyl-CoA) to feruloyl-coenzyme A (feruloyl-CoA), which mediates the biosynthesis of G-type lignin monomers to S-type lignin monomers, affecting the ratio of S/G lignin components [13]. We explored the number and function of *CCoAOMT* genes involved in lignin synthesis, which play important roles in improving fiber quality and lignocellulosic biomass utilization. The *CCoAOMT* gene family involved in the lignin biosynthesis pathway has been identified and analyzed in several species, including *Arabidopsis*, rice, tea tree, tobacco and poplar [22–24]. In this study, 17 *DfCCoAOMT* genes were identified based on the *D. farinosus* whole genome data, which is much more than the number of genes in *Arabidopsis* (7), poplar (6) and rice (6) (Table 1 and Figure 1A). According to the phylogenetic analysis of these CCoAOMT protein sequences from *D. farinosus* and other species, these *CCoAOMTs* could be clustered into four groups and no gene was present in Class III; *DfCCoAOMT1/14/15/16* clustered with *Arabidopsis CCoAOMT1* in Class I, 10 genes clustered in Class II and 3 genes clustered in Class IV. Proteins clustered in the same group usually exhibit similar functions; *AtCCoAOMT1* has been shown to be the true *CCoAOMT1* gene and is involved in lignin biosynthesis, suggesting that *DfCCoAOMT1/14/15/16* in Class I may also be primarily involved in this biological function [24].

In addition, the gene structure determines protein conformation and the protein structure determines function; genes with similar protein structures usually have similar functions (Figure 1B). According to our analysis, all DfCCoAOMTs have structural motif 2/3/6, which might be the key structure to maintain the functional conservation of this gene family. While motif 1 of the N-terminal region was also present in all the remaining DfCCoAOMTs except DfCCoAOMT13, since the protein sequences of the N-terminal and C-terminal regions are essential for enzyme activity and substrate binding, the deletion of motif 1 might lead to the specificity of the function of DfCCoAOMT13. DfCCoAOMT1/14/15/16 had the same protein structure, and the genes' exon–intron distribution was also consistent except for DfCCoAOMT16. DfCCoAOMT3/4/17 had a specific structural motif 17, which might lead to a functional diversification of *DfCCoAOMT* genes. Gene duplication was considered to be a major factor leading to gene family expansion and gene functional diversity. Based on the chromosomal localization analysis of the *D. farinosus* 17 *CCoAOMT* genes, 9 *DfCCoAOMT* genes were found to be densely distributed on chromosome 7, showing local tandem and proximal gene duplication; this is similar to that in *Arabidopsis* [25] and *Populus* [26], which may be due to gene evolution. The other remaining *DfCCoAOMT* genes were scattered on chromosomes 3, 6 and 11 (Figure 2). In poplar, *CCoAOMTs* were mainly evenly distributed on chromosomes 1, 8, 9 and 10 [17]. Compared with *Arabidopsis*, rice and poplar, the *DfCCoAOMT* gene family was significantly expanded, which might lead to more secondary metabolites and enhance the adaptability of *D. farinosus* to the environment. Of course, it requires more research to prove.

Cis-regulatory elements in gene promoter regions are critical for regulation of gene expression patterns and transcriptional expression [27]. Different cis-acting elements usually represent that genes are involved in different biological processes. Previous studies have found that AC elements are commonly found in the promoter regions of lignin biosynthetic enzyme genes, such as *CCoAOMT, PAL, C4H, COMT* and *CAD*, and are important for the transcriptional regulation of the lignin biosynthetic pathway [28]. H-box elements were involved in the regulation of lignin biosynthesis [29]. The Mybplant motif in the promoter of the phenylpropanoid biosynthesis genes regulates lignin biosynthesis by binding P-box elements [30]. In our study (Figure 4), all *DfCCoAOMT* gene promoters, except *DfCCoAOMT3/6/7/12/13*, contained AC elements or Mybplant elements, especially *DfCCoAOMT2/14/16*, suggesting that these genes might be involved in lignin biosynthesis. Furthermore, other cis-acting elements were identified in the *DfCoAOMT* gene promoters, which were associated with ABA, auxin, SA, MeJA, GA, LTR, drought and light. It has been well documented that lignin synthesis is closely related to environmental stress and could help plants resist mechanical damage and pathogenic bacterial attack. In switchgrass, drought and injury treatments rapidly induced an elevated expression level of *PvCCoAOMT* [31]. In *Arabidopsis*, ABA regulated lignin synthesis by phosphorylating NST1 [32]. A MeJA treatment also rapidly increased the expression of lignin synthase genes (including *CCoAOMT*) to promote lignin synthesis [18,33]; the JA core transcription factor MYC2 could regulate secondary wall synthesis by regulating the expression of *NST1* [34,35]. Therefore, we sprayed ABA and MeJA on 1-year-old plants and then detected the expression of *DfCCoAOMT2/5/6/8/9/14/15*; we found that the expression levels of *DfCCoAOMT6/8/9/14* were significantly inhibited by ABA and MeJA and reached the lowest level at 3 and 6 h, respectively, while the expression levels of *DfCCoAOMT2/5* were rapidly induced in response to the ABA treatment and reached the highest level at 6 h (Figure 4C).

These results suggested that ABA and MeJA signaling could regulate lignin synthesis by affecting the expression of *DfCCoAOMT* genes, and different *DfCCoAOMT* genes played different roles in response to ABA and MeJA, but how they coordinate environmental signals and lignin synthesis needs further investigation.

According to the analysis of transcriptome data from different growth periods/tissues of *D. farinosus* (Figure 3), the *DfCCoAOMT* genes had differential expression profiles in tissues. *DfCCoAOMT1/2/3/4/7/8/13/17* showed low expression levels in all tissues and different growth periods, suggesting that they were not involved in lignin synthesis during the normal growth and development of *D. farinosus*. *DfCCoAOMT10/11/12* were specifically highly expressed in roots, indicating that they were mainly involved in root development. *DfCCoAOMT6/9/14/15/16* showed a high expression in several tissues, especially in stems and roots, suggesting that they were involved in lignin synthesis in several tissues during plant growth and development. The elongation process of bamboo shoots is usually accompanied by secondary wall thickening and lignin accumulation. To further screen the key *DfCCoAOMT* gene in the lignin synthesis pathway [36,37], we analyzed sections of 2 m tall bamboo shoots from each internode and found that the lignin content gradually decreased from the base to the top of the bamboo shoots (Figure 5). qRT-PCR showed that the expression trends of *DfCCoAOMT5/6/9/14* were consistent with lignin accumulation, combined with the transcriptome data analysis, suggesting that these three genes might be the main enzyme genes involved in lignin synthesis during the development of *D. farinosus* shoots. According to our results, there were still a large number of *DfCCoAOMT* genes that might not be directly involved in lignin accumulation in normal development; they would be preferred to the synthesis of other secondary metabolites or participate in the stress response. For instance, *VvCCoAOMT4*, along with many *CCoAOMT* genes from other plant species [38–40], was shown to be involved in the synthesis of flavonoids, especially anthocyanins. Of course, whether the same function exists in *D. farinosus* requires further verification using transgenic technology.

Compared to *Arabidopsis* and *Populus*, the CCoAOMT gene family in *D. farinosus* has more members and more complex functions. *DfCCoAOMT14* might be a key gene that is involved in lignin synthesis in plants. To further validate our analysis, DfCCoAOMT14 was overexpressed in tobacco (Figure 6) and we found that the transgenic plants were slightly smaller than wild type, but the lignin content in stems was significantly increased and the xylem width was thickened. qRT-PCR confirmed that the expression levels of genes related to lignin synthesis were significantly up-regulated, indicating the function of *DfCCoAOMT14* involved in lignin synthesis. Drought stress experiments confirmed that overexpression of *DfCCoAOMT14* significantly improved the drought resistance of plants (Figure 7); this result is consistent with previous studies of *Pinus sylvestris CCoAOMT* [16]. Moreover, the content of lignin is crucial to the properties and quality of bamboo, which is a major limiting factor for bamboo pulp papermaking. Mutation of the *CCoAOMT* gene in *Arabidopsis* significantly down-regulated lignin content, especially G-type lignin [10]. Reverse repression of *CCoAOMT* in alfalfa significantly reduced lignin content but increased the ratio of S/G lignin components, which facilitated lignin degradation and separation [41]. So far, changing the expression levels of the key enzyme genes in the lignin biosynthesis pathway is still the main means to regulate lignin content. Therefore, this study can provide a theoretical basis for improving the wood properties of bamboo and breeding superior bamboo species with a high biomass or low lignin.

4. Materials and Methods

4.1. Identification of CCoAOMT Genes of D. farinosus

For the identification of the *CCoAOMT* family in *D. farinosus*, the methyltransferase domains (PF01596, CCoAOMT) were downloaded from Pfam (http://pfam.xfam.org, accessed on 1 May 2023) and they were used as a query in Hidden Markov Model (HMM) searches for candidate *CCoAOMT* genes in *D. farinosus*, with the cutoff at 0.01 [42].The Conserved Domain Search (https://www.ncbi.nlm.nih.gov/Structure/cdd/wrpsb.cgi, accessed on 1 May 2023) online software in NCBI was used to remove the sequences without the methylation domain. The online tool ExPASY (http://web.expasy.org/protparam/, accessed on 1 May 2023) was used in calculating the molecular weight (MW) and theoretical isoelectric point (pI) of DfCCoAOMT proteins. The subcellular localization of DfCCoAOMT proteins was predicted using Cell-PLoc 2.0 (Cell-PLoc 2.0 package (sjtu.edu.cn, accessed on 1 May 2023)). A total of 17 *CCoAOMT* family members were retained in *D. farinosus* for further analyses (Table 1). The candidate *CCoAOMT* gene sequences of *D. farinosus* were identified from the whole-genome data of *D. farinosus*, and all *DfCCoAOMT* gene sequences are attached in Supplementary File 1.

4.2. Phylogenetic Analysis and Sequence Alignment of the CCoAOMT Family of D. farinosus

The *CCoAOMT* family protein sequences of *Arabidopsis thaliana*, rice and *Populus trichocarpa* were downloaded from Phytozome (https://phytozome.jgi.doe.gov/, accessed on 1 May 2023) (Supplemental Table S1). Phylogenetic trees were constructed with IQ-TREE v1.6.12 using the Model Finder algorithm. We performed multiple sequence alignment analyses of CCoAOMT protein between *D. farinosus* and other species [43]. We constructed a phylogenetic tree between the CCoAOMT genes of *D. farinosus* by using the Clustal W program in MEGA 7.0 software and the neighbor-joining (NJ) method with 1000 bootstrap replications [44].

4.3. Chromosomal Localization and Collinearity Analysis of DfCCoAOMT Genes

The internal collinearity relationship was analyzed by using the Multiple Collinearity Scan toolkit (MCScanX) and we obtained collinearity gene pairs in the *D. farinosus* *CCoAOMT* family genes [45]. The collinearity analysis maps were then constructed using the Dual Systeny Plotter software (https://github.com/CJ-Chen/TBtools, accessed on 1 May 2023) [44]. Then, the Ka, Ks and MYA values were calculated using WGD v1.1.1 (Supplemental Tables S4 and S5).

4.4. Conserved Motifs and Gene Structure Analysis of CCoAOMT Family of D. farinosus

The conserved motifs of *DfCCoAOMTs* were determined using the MEME online tool (version 5.3.0, http://meme-suite.org/tools/meme, accessed on 1 May 2023) [46]; the number of motifs output was 10. The motif analysis and gene structure were visualized using the TBtools software. The Gene Structure Display Server 2.0 online tool (GSDS, http://gsds.gao-lab.org/index.php, accessed on 1 May 2023) [47] was used to illustrate the structure and exon/intron organization of *DfCCoAOMT* genes and the genomic length. The 2000 bp upstream promoter sequences of *DfCCoAOMT* genes were submitted to the PlantCARE database [48] (http://bioinformatics.psb.ugent.be/webtools/plantcare/html/) and PLACE (https://www.dna.affrc.go.jp/PLACE/) to predict cis-elements and to subsequently screen cis-elements manually.

4.5. Plant Materials and Hormone Treatment

All materials used in this study were from the Institute of Bamboo Research, Southwest University of Science and Technology. One-year-old young leaves of *D. farinosus* were treated with 100 mM ABA and MeJA, respectively, and sampled in liquid nitrogen before treatment and after 3 h, 6 h and 9 h of treatment.

4.6. RNA Extraction and qRT-PCR

Total RNA of *D. farinosus* leaves was extracted using the TRIzol reagent (Takara, Dalian, China). The RNA quality was monitored using 1% denaturing agarose gel and a NanoDrop 2000 spectrophotometer (Thermo Fisher Scientific, Beijing, China). Then, cDNA was synthesized from 1 to 2 μg total RNA using the FastKing cDNA First Strand Synthesis Kit (TIANGEN, Beijing, China). The specific primer used for qRT-PCR was designed by using Primer Premier 5.0 (Supplemental Table S2). *Tubulin* was used as an internal reference gene. The transcript levels of *DfCCoAOMT* were obtained using the SYBR qPCR

Master MIX kit (Vazyme, Nanjing, China) on the CFX96TM Real-Time System thermal cycler (BIO-RAD, CA, USA). The relative expression levels of *DfCCoAOMT* genes were calculated with the 2-$\Delta\Delta$Ct method [49]. Each sample had three independent biological replicates—each with three technical replicates.

4.7. RNA Sequencing

RNA sequencing was performed on 1-year-old *D. farinosus* materials in a growing chamber. The samples of *D. farinosus* included Young leaf (Yleaf), Mature leaf (MLeaf), Culm sheath (CSh), Sheath of bamboo shoot (ShB), Lateral bud (Labud), Node, Root, Branch, Internode (Inod), Rhizome bud (Rhbud), Rhizome neck (Rhne) and shoots of different height. The analysis included three biological duplications. Total RNA was extracted by using the TRIzol reagent (Takara, Dalian, China); the quantity and quality of total RNA were detected by a NanoDrop 2000 spectrophotometer (Thermo Fisher Scientific, Beijing, China). Seventeen samples were sequenced by the NovaSeq 6000 platform (Illumina, Beijing, China). The clean reads of each sample were compared with the reference genome of *Dendrocalamus farinosus*, and the alignment rate ranged from 74.47% to 91.57%. The raw reads were further processed with a bioinformatic pipeline tool, the BMKCloud (www.biocloud.net, accessed on 1 May 2023) online platform. All of these RNA-seq data were mapped to the *D. farinosus* reference genome using HISAT2 and the fragments per kilobase per million (FPKM) were calculated using StringTie. The heatmap was generated by TBtools based on the transformed data of $\log_2$ (FPKM + 1) values. The details of the sample information and the FPKM values of CCoAOMT genes are shown in Supplemental Table S3, respectively.

4.8. Overexpression of DfCCoAOMT14 in Nicotiana Tabacum

Overexpressed tobacco lines were obtained by introducing the pCAMBIA1302-*DfCCoAOMT14* with the CaMV 35S promoter recombinant vector into the *Agrobacterium tumefaciens* strain EHA105 using the leaf disc method [50]. Infected tobacco leaf discs were inoculated on a MS medium containing 0.1 mg/L NAA, 0.5 mg/L 6-BA, 400 mg/L cephalexin and 9 mg/L Hyg in the dark at 27 $\pm$ 1 °C for 2 d. The regenerated buds were transferred to a MS medium containing 0.1 mg/L NAA, 400 mg/L cephalexin and 9 mg/L Hyg for the formation of complete plants. Genomic DNA was extracted from the leaves of transgenic plants and wild type and then amplified under the following conditions: preheating at 95 °C for 5 min, followed by 30 cycles of denaturation at 95 °C for 30 s, annealing at 60 °C for 30 s, extension at 72 °C for 1 min and finally extension at 72 °C for 10 min (Supplemental Figure S2).The PCR product was checked using 1% agarose gel electrophoresis. The expression of *DfCCoAOMT14* in transgenic plants was detected by qRT-PCR and *N. tabacum NtActin* was used as the internal references for normalization [51].

4.9. Lignin Analysis

For plant cell wall preparation, the second to eighth internodes of 2 m bamboo shoots were ground to powder in liquid nitrogen. The determination method of lignin content was based on the method of Li et al. [52] and was appropriately improved as follows: weighed 0.5 g fresh sample, added liquid nitrogen to grind to fine powder, added 5 mL 0.1 M phosphate buffer (pH 7.2), incubated at 37 °C for 30 min and centrifuged for 5 min. Then, added 5 mL 80% ethanol to incubate at 80 °C for 1 h, centrifuged for 5 min and added 10 mL acetone to extract once, centrifuged for 5 min and dried at 60 °C. For the thioglycolic acid lignin analysis, approximately 20 mg of dried plant cell wall was incubated at 80 °C in 750 µL of water, 250 µL of concentrated HCl and 100 µL of thioglycolic acid for 3 h. The mixture was centrifuged and the pellet was washed with 1 mL of water and resuspended in 1 mL of 1 M NaOH on a rocking plate at room temperature overnight. The mixture was centrifuged again, and the supernatant was collected and mixed with 200 µL of concentrated HCl. After being vortexed and incubated at 4 °C for 4 h, the mixture was centrifuged and the pellet was dissolved in 1 mL of 1 M NaOH. The absorbance of a 50-fold dilution of the supernatant in 1 M NaOH was measured at 280 nm. The lignin content of *Nicotiana tabacum* overexpressing *DfCCoAOMT14* was also determined by the same method.

4.10. Histochemical Staining

The complete and thin sections were stained in 2% phloroglucinol solution for 5 min, then were stained with 30% hydrochloric acid for 1 min, washed with water, observed and photographed under a Leica microscope.

4.11. Drought Treatment in DfCCoAOMT Transgenic Tobacco

In total, 45-day-old transgenic tobacco plants and wild-type plants were treated in drought stress for 37 days. The treated plants were then subjected to the measurement of relevant physiological indices, including malondialdehyde (MDA), superoxide dismutase (SOD), proline (Pro), peroxidase (POD) and leaf relative water content (RWC). The fresh weight (FW) of tobacco leaves was weighed. The fresh tobacco leaves soaked in distilled water for 24 h in the condition of 16 h light/8 h dark at room temperature and the turgid weight (TW) was weighed. The leaves were then dried for 24 h at 80 °C to obtain the total dry weight (DW) [53]. Leaf relative water content (RWC) was estimated according to the method of Turner (1981) [54]: RWC (%) = (fresh weight − dry weight)/(turgid weight − dry weight) × 100. MDA content, Pro content, SOD activity and POD activity were all measured using the related kit (Solarbio, Beijing, China). Subsequently, qRT-PCR was used to detect the expression levels of drought-related genes including *NtERD10C*, *NtDREB* and *NtRD29A* in drought-treated plants [55]; tobacco *NtActin* served as an internal reference.

4.12. Statistical Analysis

The data in this study were statistically analyzed by using SPSS 25.0 and Origin 2018. Data are presented as means $\pm$ SD of three independent replicates. The one-way analysis of variance (ANOVA) was used for the significant difference analysis and the significant difference of $p < 0.05$ was statistically significant.

Supplementary Materials: The following supporting information can be downloaded at: https://www.mdpi.com/article/10.3390/ijms24108965/s1.

Author Contributions: S.H. and Y.C. planned and designed the research. L.W., X.G., J.P., W.S. and B.D. performed all experiments and analyzed the data. X.Z. and L.W. wrote the original manuscript. S.H. and Y.C. proofread the manuscript. All authors have read and agreed to the published version of the manuscript.

Funding: This study was supported by the National Key R & D Program of China (2021YFD2200504_1 and 2021YFD2200505_2), the Science and Technology Project of Sichuan Province, China (2022NS-FSC0093 and 2021YFYZ0006) and the Ph.D. Foundation (no. 22zx7114) from Southwest University of Science and Technology.

Institutional Review Board Statement: Not Applicable.

Informed Consent Statement: Not Applicable.

Data Availability Statement: The data which supports the findings of this study are linked in https://www.ncbi.nlm.nih.gov/, PRJNA923443.

Conflicts of Interest: The authors declare no conflict of interest.

References

1. Guo, L.; Sun, X.; Li, Z.; Wang, Y.; Fei, Z.; Jiao, C.; Feng, J.; Cui, D.; Feng, X.; Ding, Y.; et al. Morphological dissection and cellular and transcriptome characterizations of bamboo pith cavity formation reveal a pivotal role of genes related to programmed cell death. *Plant Biotechnol. J.* **2019**, *17*, 982–997. [CrossRef] [PubMed]
2. Sohel, M.; Alamgir, M.; Akhter, S.; Rahman, M. Carbon storage in a bamboo (Bambusa vulgaris) plantation in the degraded tropical forests: Implications for policy development. *Land Use Policy* **2015**, *49*, 142–151. [CrossRef]

3. Paudyal, K.; Adhikari, S.; Sharma, S.; Samsudin, Y.B.; Baral, H. *Framework for Assessing Ecosystem Services from Bamboo Forests: Lessons from Asia and Africa*; CIFOR: Bogor, Indonesia, 2019.
4. Imran, M.; Luo, X.; Hu, S.; Cao, Y.; Long, Z. Epigenetic and somaclonal divergence in Dendrocalamus farinosus for physiological augmentation and lignin degradation. *Biotechnol. Appl. Biochem.* **2022**, *69*, 1545–1556. [CrossRef] [PubMed]
5. Barros, J.; Serk, H.; Granlund, I.; Pesquet, E. The cell biology of lignification in higher plants. *Ann. Bot.* **2015**, *115*, 1053–1074. [CrossRef] [PubMed]
6. Vanholme, R.; Demedts, B.; Morreel, K.; Ralph, J.; Boerjan, W. Lignin biosynthesis and structure. *Plant Physiol.* **2010**, *153*, 895–905. [CrossRef] [PubMed]
7. Kang, X.; Kirui, A.; Dickwella Widanage, M.C.; Mentink-Vigier, F.; Cosgrove, D.J.; Wang, T. Lignin-polysaccharide interactions in plant secondary cell walls revealed by solid-state NMR. *Nat. Commun.* **2019**, *10*, 347. [CrossRef]
8. Vanholme, R.; De Meester, B.; Ralph, J.; Boerjan, W. Lignin biosynthesis and its integration into metabolism. *Curr. Opin. Biotechnol.* **2019**, *56*, 230–239. [CrossRef]
9. Zhang, G.; Zhang, Y.; Xu, J.; Niu, X.; Qi, J.; Tao, A.; Zhang, L.; Fang, P.; Lin, L.; Su, J. The CCoAOMT1 gene from jute (Corchorus capsularis L.) is involved in lignin biosynthesis in Arabidopsis thaliana. *Gene* **2014**, *546*, 398–402. [CrossRef]
10. Do, C.T.; Pollet, B.; Thévenin, J.; Sibout, R.; Denoue, D.; Barriere, Y.; Lapierre, C.; Jouanin, L. Both caffeoyl Coenzyme A 3-O-methyltransferase 1 and caffeic acid O-methyltransferase 1 are involved in redundant functions for lignin, flavonoids and sinapoyl malate biosynthesis in Arabidopsis. *Planta* **2007**, *226*, 1117–1129. [CrossRef]
11. Kwon, H.; Cho, D.J.; Lee, H.; Nam, M.H.; Kwon, C.; Yun, H.S. CCOAOMT1, a candidate cargo secreted via VAMP721/722 secretory vesicles in Arabidopsis. *Biochem. Biophys. Res. Commun.* **2020**, *524*, 977–982. [CrossRef]
12. Chun, H.J.; Lim, L.H.; Cheong, M.S.; Baek, D.; Park, M.S.; Cho, H.M.; Lee, S.H.; Jin, B.J.; No, D.H.; Cha, Y.J.; et al. Arabidopsis CCoAOMT1 Plays a Role in Drought Stress Response via ROS- and ABA-Dependent Manners. *Plants* **2021**, *10*, 831. [CrossRef]
13. Wang, G.F.; Balint-Kurti, P.J. Maize Homologs of CCoAOMT and HCT, Two Key Enzymes in Lignin Biosynthesis, Form Complexes with the NLR Rp1 Protein to Modulate the Defense Response. *Plant Physiol.* **2016**, *171*, 2166–2177. [CrossRef] [PubMed]
14. Yang, Q.; He, Y.; Kabahuma, M.; Chaya, T.; Kelly, A.; Borrego, E.; Bian, Y.; El Kasmi, F.; Yang, L.; Teixeira, P.; et al. A gene encoding maize caffeoyl-CoA O-methyltransferase confers quantitative resistance to multiple pathogens. *Nat. Genet.* **2017**, *49*, 1364–1372. [CrossRef] [PubMed]
15. Xia, Y.; Liu, J.; Wang, Y.; Zhang, X.; Shen, Z.; Hu, Z. Ectopic expression of Vicia sativa Caffeoyl-CoA O -methyltransferase (VsCCoAOMT) increases the uptake and tolerance of cadmium in Arabidopsis. *Environ. Exp. Bot.* **2018**, *145*, 47–53. [CrossRef]
16. Zhao, D.; Luan, Y.; Shi, W.; Zhang, X.; Meng, J.; Tao, J. A Paeonia ostii caffeoyl-CoA O-methyltransferase confers drought stress tolerance by promoting lignin synthesis and ROS scavenging. *Plant Sci.* **2021**, *303*, 110765. [CrossRef] [PubMed]
17. Zhao, H.; Qu, C.; Zuo, Z.; Cao, L.; Zhang, S.; Xu, X.; Xu, Z.; Liu, G. Genome Identification and Expression Profiles in Response to Nitrogen Treatment Analysis of the Class I CCoAOMT Gene Family in Populus. *Biochem. Genet.* **2022**, *60*, 656–675. [CrossRef]
18. Lin, S.J.; Yang, Y.Z.; Teng, R.M.; Liu, H.; Li, H.; Zhuang, J. Identification and expression analysis of caffeoyl-coenzyme A O-methyltransferase family genes related to lignin biosynthesis in tea plant (Camellia sinensis). *Protoplasma* **2021**, *258*, 115–127. [CrossRef]
19. Yang, G.; Pan, W.; Zhang, R.; Pan, Y.; Guo, Q.; Song, W.; Zheng, W.; Nie, X. Genome-wide identification and characterization of caffeoyl-coenzyme A O-methyltransferase genes related to the Fusarium head blight response in wheat. *BMC Genom.* **2021**, *22*, 504. [CrossRef]
20. Lu, S.; Zhuge, Y.; Hao, T.; Liu, Z.; Zhang, M.; Fang, J. Systematic analysis reveals O-methyltransferase gene family members involved in flavonoid biosynthesis in grape. *Plant Physiol. Biochem.* **2022**, *173*, 33–45. [CrossRef]
21. Xiao, Y.; Li, J.; Liu, H.; Zhang, Y.; Zhang, X.; Qin, Z.; Chen, B. The Effect of Co-Transforming Eucalyptus urophylla Catechol-O-methyltransferase and Caffeoyl-CoA O-methyltransferase on the Biosynthesis of Lignin Monomers in Transgenic Tobacco. *Russ. J. Plant Physiol.* **2020**, *67*, 879–887. [CrossRef]
22. Hamberger, B.; Ellis, M.; Friedmann, M.; de Azevedo Souza, C.; Barbazuk, B.; Douglas, C.J. Genome-wide analyses of phenylpropanoid-related genes in Populus trichocarpa, Arabidopsis thaliana, and Oryza sativa: The Populus lignin toolbox and conservation and diversification of angiosperm gene familiesThis article is one of a selection of papers published in the Special Issue on Poplar Research in Canada. *Can. J. Bot.* **2007**, *85*, 1182–1201.
23. Xu, Z.; Zhang, D.; Hu, J.; Zhou, X.; Ye, X.; Reichel, K.L.; Stewart, N.R.; Syrenne, R.D.; Yang, X.; Gao, P.; et al. Comparative genome analysis of lignin biosynthesis gene families across the plant kingdom. *BMC Bioinform.* **2009**, *10* (Suppl. S11), S3. [CrossRef] [PubMed]
24. Rakoczy, M.; Femiak, I.; Alejska, M.; Figlerowicz, M.; Podkowinski, J. Sorghum CCoAOMT and CCoAOMT-like gene evolution, structure, expression and the role of conserved amino acids in protein activity. *Mol. Genet. Genom.* **2018**, *293*, 1077–1089. [CrossRef] [PubMed]
25. Kim, B.G.; Kim, D.H.; Hur, H.G.; Lim, J.; Ahn, J.H. O-Methyltransferases from Arabidopsis thaliana. *Agric. Chem. Biotechnol.* **2005**, *48*, 113–119.
26. Barakat, A.; Choi, A.; Yassin, N.B.; Park, J.S.; Sun, Z.; Carlson, J.E. Comparative genomics and evolutionary analyses of the O-methyltransferase gene family in Populus. *Gene* **2011**, *479*, 37–46. [CrossRef]
27. Hernandez-Garcia, C.M.; Finer, J.J. Identification and validation of promoters and cis-acting regulatory elements. *Plant Sci.* **2014**, *217–218*, 109–119. [CrossRef]

28. Lacombe, E.; Van Doorsselaere, J.; Boerjan, W.; Boudet, A.M.; Grima-Pettenati, J. Characterization of cis-elements required for vascular expression of the cinnamoyl CoA reductase gene and for protein-DNA complex formation. *Plant J. Cell Mol. Biol.* **2000**, *23*, 663–676. [CrossRef]
29. Patzlaff, A. Characterisation of a pine MYB that regulates lignification. *Plant J. Cell Mol. Biol.* **2010**, *36*, 743–754. [CrossRef]
30. Tamagnone, L.; Merida, A.; Parr, A.; Mackay, S.; Culianez-Macia, F.A.; Martin, R.C. The AmMYB308 and AmMYB330 Transcription Factors from Antirrhinum Regulate Phenylpropanoid and Lignin Biosynthesis in Transgenic Tobacco. *Plant Cell* **1998**, *10*, 135–154. [CrossRef]
31. Liu, S.J.; Huang, Y.H.; Chang-Jiu, H.E.; Fang, C.; Zhang, Y.W. Cloning, bioinformatics and transcriptional analysis of caffeoyl-coenzyme A 3-O-methyltransferase in switchgrass under abiotic stress. *J. Integr. Agric.* **2016**, *15*, 636–649. [CrossRef]
32. Liu, C.; Yu, H.; Rao, X.; Li, L.; Dixon, R.A. Abscisic acid regulates secondary cell-wall formation and lignin deposition in Arabidopsis thaliana through phosphorylation of NST1. *Proc. Natl. Acad. Sci. USA* **2021**, *118*, e2010911118. [CrossRef] [PubMed]
33. Sehr, E.M.; Agusti, J.; Lehner, R.; Farmer, E.E.; Schwarz, M.; Greb, T. Analysis of secondary growth in the Arabidopsis shoot reveals a positive role of jasmonate signalling in cambium formation. *Plant J. Cell Mol. Biol.* **2010**, *63*, 811–822. [CrossRef]
34. Zhang, Q.; Xie, Z.; Zhang, R.; Xu, P.; Liu, H.; Yang, H.; Doblin, M.S.; Bacic, A.; Li, L. Blue Light Regulates Secondary Cell Wall Thickening via MYC2/MYC4 Activation of the NST1-Directed Transcriptional Network in Arabidopsis. *Plant Cell* **2018**, *30*, 2512–2528. [CrossRef] [PubMed]
35. Luo, F.; Zhang, Q.; Xin, H.; Liu, H.; Yang, H.; Doblin, M.S.; Bacic, A.; Li, L. A Phytochrome B-PIF4-MYC2/MYC4 module inhibits secondary cell wall thickening in response to shaded light. *Plant Commun.* **2022**, *3*, 100416. [CrossRef] [PubMed]
36. Chen, M.; Guo, L.; Ramakrishnan, M.; Fei, Z.; Vinod, K.K.; Ding, Y.; Jiao, C.; Gao, Z.; Zha, R.; Wang, C.; et al. Rapid growth of Moso bamboo (Phyllostachys edulis): Cellular roadmaps, transcriptome dynamics, and environmental factors. *Plant Cell* **2022**, *34*, 3577–3610. [CrossRef]
37. Yang, K.; Li, L.; Lou, Y.; Zhu, C.; Li, X.; Gao, Z. A regulatory network driving shoot lignification in rapidly growing bamboo. *Plant Physiol.* **2021**, *187*, 900–916. [CrossRef] [PubMed]
38. Liu, X.; Zhao, C.; Gong, Q.; Wang, Y.; Cao, J.; Li, X.; Grierson, D.; Sun, C. Characterization of a caffeoyl-CoA O-methyltransferase-like enzyme involved in biosynthesis of polymethoxylated flavones in Citrus reticulata. *J. Exp. Bot.* **2020**, *71*, 3066–3079. [CrossRef] [PubMed]
39. Nakamura, N.; Katsumoto, Y.; Brugliera, F.; Demelis, L.; Nakajima, D.; Suzuki, H.; Tanaka, Y. Flower color modification in Rosa hybrida by expressing the S-adenosylmethionine: Anthocyanin 3′, 5′-O-methyltransferase gene from Torenia hybrida. *Plant Biotechnol.* **2015**, *32*, 109–117. [CrossRef]
40. Widiez, T.; Hartman, T.G.; Dudai, N.; Yan, Q.; Lawton, M.; Havkin-Frenkel, D.; Belanger, F.C. Functional characterization of two new members of the caffeoyl CoA O-methyltransferase-like gene family from Vanilla planifolia reveals a new class of plastid-localized O-methyltransferases. *Plant Mol. Biol.* **2011**, *76*, 475–488. [CrossRef]
41. Guo, D.; Chen, F.; Inoue, K.; Blount, J.W.; Dixon, R.A. Downregulation of caffeic acid 3-O-methyltransferase and caffeoyl CoA 3-O-methyltransferase in transgenic alfalfa. impacts on lignin structure and implications for the biosynthesis of G and S lignin. *Plant Cell* **2001**, *13*, 73–88. [CrossRef]
42. El-Gebali, S.; Mistry, J.; Bateman, A.; Eddy, S.R.; Luciani, A.; Potter, S.C.; Qureshi, M.; Richardson, L.J.; Salazar, G.A.; Smart, A.; et al. The Pfam protein families database in 2019. *Nucleic Acids Res.* **2019**, *47*, D427–D432. [CrossRef] [PubMed]
43. Nguyen, L.-T.; Schmidt, H.A.; von Haeseler, A.; Minh, B.Q. IQ-TREE: A Fast and Effective Stochastic Algorithm for Estimating Maximum-Likelihood Phylogenies. *Mol. Biol. Evol.* **2015**, *32*, 268–274. [CrossRef] [PubMed]
44. Chen, C.; Chen, H.; Zhang, Y.; Thomas, H.R.; Frank, M.H.; He, Y.; Xia, R. TBtools: An Integrative Toolkit Developed for Interactive Analyses of Big Biological Data. *Mol. Plant* **2020**, *13*, 1194–1202. [CrossRef]
45. Wang, Y.; Tang, H.; Debarry, J.D.; Tan, X.; Li, J.; Wang, X.; Lee, T.H.; Jin, H.; Marler, B.; Guo, H.; et al. MCScanX: A toolkit for detection and evolutionary analysis of gene synteny and collinearity. *Nucleic Acids Res.* **2012**, *40*, e49. [CrossRef] [PubMed]
46. Bailey, T.L.; Williams, N.; Misleh, C.; Li, W.W. MEME: Discovering and analyzing DNA and protein sequence motifs. *Nucleic Acids Res.* **2006**, *34*, W369–W373. [CrossRef]
47. Hu, B.; Jin, J.; Guo, A.Y.; Zhang, H.; Luo, J.; Gao, G. GSDS 2.0: An upgraded gene feature visualization server. *Bioinformatics* **2015**, *31*, 1296–1297. [CrossRef]
48. Lescot, M.; Dehais, P.; Thijs, G.; Marchal, K.; Moreau, Y.; Van de Peer, Y.; Rouze, P.; Rombauts, S. PlantCARE, a database of plant cis-acting regulatory elements and a portal to tools for in silico analysis of promoter sequences. *Nucleic Acids Res.* **2002**, *30*, 325–327. [CrossRef]
49. Livak, K.J.; Schmittgen, T.D. Analysis of relative gene expression data using real-time quantitative PCR and the 2(-Delta Delta C(T)) Method. *Methods* **2001**, *25*, 402–408. [CrossRef]
50. Zhang, H.; Gao, X.; Zhi, Y.; Li, X.; Zhang, Q.; Niu, J.; Wang, J.; Zhai, H.; Zhao, N.; Li, J.; et al. A non-tandem CCCH-type zinc-finger protein, IbC3H18, functions as a nuclear transcriptional activator and enhances abiotic stress tolerance in sweet potato. *New Phytol.* **2019**, *223*, 1918–1936. [CrossRef]
51. Wang, C.; Wang, L.; Lei, J.; Chai, S.; Jin, X.; Zou, Y.; Sun, X.; Mei, Y.; Cheng, X.; Yang, X.; et al. IbMYB308, a Sweet Potato R2R3-MYB Gene, Improves Salt Stress Tolerance in Transgenic Tobacco. *Genes* **2022**, *13*, 1476. [CrossRef]
52. Li, Y.; Kim, J.I.; Pysh, L.; Chapple, C. Four Isoforms of Arabidopsis 4-Coumarate:CoA Ligase Have Overlapping yet Distinct Roles in Phenylpropanoid Metabolism. *Plant Physiol.* **2015**, *169*, 2409–2421. [PubMed]

53. Hu, W.E.I.; Huang, C.; Deng, X.; Zhou, S.; Chen, L.; Li, Y.I.N.; Wang, C.; Ma, Z.; Yuan, Q.; Wang, Y.A.N.; et al. TaASR1, a transcription factor gene in wheat, confers drought stress tolerance in transgenic tobacco. *Plant Cell Environ.* **2013**, *36*, 1449–1464. [CrossRef] [PubMed]
54. Turner, N.C. Techniques and experimental approaches for the measurement of plant water status. *Plant Soil* **1981**, *58*, 339–366. [CrossRef]
55. Liu, W.; Xiang, Y.; Zhang, X.; Han, G.; Sun, X.; Sheng, Y.; Yan, J.; Scheller, H.V.; Zhang, A. Over-Expression of a Maize N-Acetylglutamate Kinase Gene (ZmNAGK) Improves Drought Tolerance in Tobacco. *Front. Plant Sci.* **2019**, *9*, 1902. [CrossRef]

International Journal of
Molecular Sciences

Genome-Wide Analysis of the Oat (*Avena sativa*) *HSP90* Gene Family Reveals Its Identification, Evolution, and Response to Abiotic Stress

Jinghan Peng [1,2,†], Siyu Liu [1,†], Jiqiang Wu [1,2], Tianqi Liu [1], Boyang Liu [1], Yi Xiong [1], Junming Zhao [1], Minghong You [2], Xiong Lei [2,*] and Xiao Ma [1,*]

[1]   College of Grassland Science and Technology, Sichuan Agricultural University, Chengdu 611130, China
[2]   Sichuan Academy of Grassland Science, Chengdu 610097, China
*   Correspondence: lxforage@126.com (X.L.); maxiao@sicau.edu.cn (X.M.);
    Tel.: +86-028-8328-3361 (X.L.); +86-028-8629-1301 (X.M.)
[†]   These authors contributed equally to this work.

Citation: Peng, J.; Liu, S.; Wu, J.; Liu, T.; Liu, B.; Xiong, Y.; Zhao, J.; You, M.; Lei, X.; Ma, X. Genome-Wide Analysis of the Oat (*Avena sativa*) *HSP90* Gene Family Reveals Its Identification, Evolution, and Response to Abiotic Stress. *Int. J. Mol. Sci.* **2024**, *25*, 2305. https://doi.org/10.3390/ijms25042305

Academic Editor: Martin Bartas

Received: 18 December 2023
Revised: 7 February 2024
Accepted: 9 February 2024
Published: 15 February 2024

Abstract: Oats (*Avena sativa*) are an important cereal crop and cool-season forage worldwide. Heat shock protein 90 (HSP90) is a protein ubiquitously expressed in response to heat stress in almost all plants. To date, the *HSP90* gene family has not been comprehensively reported in oats. Herein, we have identified twenty *HSP90* genes in oats and elucidated their evolutionary pathways and responses to five abiotic stresses. The gene structure and motif analyses demonstrated consistency across the phylogenetic tree branches, and the groups exhibited relative structural conservation. Additionally, we identified ten pairs of segmentally duplicated genes in oats. Interspecies synteny analysis and orthologous gene identification indicated that oats share a significant number of orthologous genes with their ancestral species; this implies that the expansion of the oat *HSP90* gene family may have occurred through oat polyploidization and large fragment duplication. The analysis of cis-acting elements revealed their influential role in the expression pattern of HSP90 genes under abiotic stresses. Analysis of oat gene expression under high-temperature, salt, cadmium (Cd), polyethylene glycol (PEG), and abscisic acid (ABA) stresses demonstrated that most AsHSP90 genes were significantly up-regulated by heat stress, particularly *AsHSP90-7*, *AsHSP90-8*, and *AsHSP90-9*. This study offers new insights into the amplification and evolutionary processes of the AsHSP90 protein, as well as its potential role in response to abiotic stresses. Furthermore, it lays the groundwork for understanding oat adaptation to abiotic stress, contributing to research and applications in plant breeding.

Keywords: oat; heat shock protein; HSP90; phylogenetic analysis; expression pattern

1. Introduction

The quality of life of terrestrial organisms on Earth is increasingly impacted by climate change [1]. As sessile organisms, plants are particularly susceptible to environmental stresses such as drought, salinity, cold, and heat during growth and development [2]. Recently, heat stress has become a significant abiotic stresses affecting normal plant growth and development due to global warming, increased droughts, and extreme weather conditions [3]. In particular, prolonged growth at high temperatures severely inhibits starch synthesis and carbon assimilation, leading to a reduction in average yields and posing a major challenge to food security [4,5]. Furthermore, the exposure of plants to premature high temperatures during unsuitable phenological periods makes them more susceptible to pathogen infection, potentially affecting the quality of crop production [6].

Plants have developed regulatory mechanisms to cope with heat stress and thousands of genes are involved during the evolution of their long-term adaptation [7]. Among them, the heat shock protein (HSP) is one of the best-characterized genes and plays a significant role in regulating responses to heat stress. Generally, HSPs can be categorized into HSP20, HSP60, HSP70/DnaK, HSP90, and HSP100/ClpB families according to their molecular weight and sequence homogeneity [8], of which HSP90s are highly conserved in molecular evolution and an abundant family of ATP-dependent molecular chaperone proteins in prokaryotes and higher eukaryotes [9], which are broadly distributed in the cytoplasm, chloroplasts, mitochondria, and endoplasmic reticulum, accounting for 1–2% of total cellular proteins [10,11]. *HSP90* generally contains three structural domains: the N-terminal ATP-binding domain, the M domain, and the C-terminal substrate-binding domain [12]. *HSP90* features an unconventional Bergerat ATP-binding fold and is part of the GHKL superfamily [13]. According to previous studies, HSP90 proteins often have dual functions: HSP90s are involved in regulating and maintaining the conformation of various proteins and assisting normal cell survival under stress on the one hand [14], and act as negative feedback regulators of heat stress responses on the other [15]. *HSP90,* along with other molecular chaperones, provides a mechanism to promote protein folding [12], prevents protein aggregation, and facilitates the refolding of inactivated proteins, thereby increasing the resistance of certain cells [14]. *HSP90* expression is up-regulated when plants are stressed; it then associates with nonprotein substances to enable the repair of deformed proteins [16]. Besides its importance for protein folding, a function of *HSP90*s as negative regulators for heat stress transcription factors (Hsfs) has been proposed for Hsf1, whose activities in facilitating downstream gene expression are tightly regulated [7].

Recently, *HSP90*s have been identified and found in many plants, including seven members of the *HSP90* members in *Arabidopsis thaliana* [17], eight in *Brachypodium distachyon* [18], twenty-one in *Nicotiana tabacum* [19], and eight in *Perennial Ryegrass* [20], and directly or indirectly implicated in a host of physiological processes ranging from plant growth and development to abiotic and biotic stress responses [8]. Overexpression of *AtHSP90-2, AtHSP90-5,* and *AtHSP90-7* in *Arabidopsis thaliana* reduces tolerance to salt and drought stress but increases tolerance to high Ca^{2+} concentrations [21]. Both tobacco and *Arabidopsis* species have HSP90 members that convey resistance to pathogens by counteracting the response of signaling receptor R proteins from pathogens [22,23]. Additionally, up-regulation of the *VvHsp901a* gene was delayed when grape plants were subjected to drought and high-temperature stresses [24]. The HSP90 protein also plays a role in plant growth and developmental processes, and its mediated distribution of PIN1 regulates the distribution of auxin signaling, thereby promoting plant growth and development [25].

Oats (*Avena sativa* L.) are a global crop and one of the richest sources of protein, fat, and β-glucan among all cereals, with a low carbon footprint [26,27]. Also, oats are a widely grown annual forage worldwide and an important source of high-quality pasture for livestock [26]. Oats are a cool-season crop that is suitable for growing in humid environments [28], but due to severe global warming, temperature conditions are quickly met, which may lead to shorter growing periods, smaller plants, and lower yields in oats. Therefore, it is of great significance to understand and reveal the molecular members (e.g., HSP90s) that contribute to high temperature tolerance in oats. To date, a systematic and comprehensive study of the *HSP90* gene family in oats has not yet been reported, and the assembly of a high-quality oat genome provides the necessary information to characterize *HSP90*s at the genome-wide level [29]. In this study, we characterized the *HSP90* gene within the oat genome, including gene sequence and homology analyses, and described the evolutionary pathway of the HSP90 gene in oats. Additionally, we analyzed oat HSP90 gene expression under high-temperature, salt, cadmium (Cd), polyethylene glycol (PEG), and abscisic acid (ABA) stresses using Quantitative Real-Time PCR. Our study provides a new avenue for molecular breeding in oats, contributes to a better understanding of the heat tolerance response of oats under high-temperature conditions, and lays the foundation for future studies on the function of the AsHSP90 protein in oat stress tolerance.

2. Results

2.1. Identification of Oat HSP90 Genes and Chromosomal Distribution

A total of 20 *HSP90* gene family members has been identified in *A. sativa* (Table 1). AsHSP90 proteins were renamed *AsHSP90-1* to *AsHSP90-20* based on their molecular weight. The obtained *AsHSP90* sequences varied in length ranging from 627 to 809 amino acids, with pI values ranging from 4.6 to 5.09 and molecular weights ranging from 71,881.4 kd to 92,623.6 kd. AsHSP90 proteins are predominantly cytoplasmic and chloroplastic, except for *AsHSP90-5*, *AsHSP90-10*, *AsHSP90-1*, *AsHSP90-15*, and *AsHSP90-20*, which are localized in the nucleus and ER, respectively.

Table 1. Biophysical properties and subcellular localization of the oat HSP90 genes.

Gene	ID	Length	MW	pI	Instability Index	Aliphatic Index	GRAVY	Predicted Subcellular Location
AsHSP90-1	AVESA.00010b.r2.5DG0985750.1	627	71,881.4	4.79	41.94	74.32	−0.639	Endoplasmic reticulum
AsHSP90-2	AVESA.00010b.r2.7AG1217140.1	698	80,094.2	4.67	39.68	83.67	−0.592	Cytoplasm
AsHSP90-3	AVESA.00010b.r2.7DG1391180.1	698	80,103.2	4.69	39.7	83.81	−0.597	Chloroplast
AsHSP90-4	AVESA.00010b.r2.5CG0932690.1	699	80,214.3	4.69	39.72	83.98	−0.602	Chloroplast
AsHSP90-5	AVESA.00010b.r2.5AG0850550.1	700	80,432.6	4.67	40.22	82.17	−0.617	Nucleus
AsHSP90-6	AVESA.00010b.r2.5CG0883430.1	700	80,475.6	4.67	40.43	82.03	−0.625	Cytoplasm
AsHSP90-7	AVESA.00010b.r2.5DG0939350.1	707	80,750.8	4.67	40.95	82.48	−0.586	Chloroplast
AsHSP90-8	AVESA.00010b.r2.4CG1254200.1	713	81,379.5	4.67	41.36	82.61	−0.577	Cytoplasm mitochondrion
AsHSP90-9	AVESA.00010b.r2.6AG1070350.1	713	81,411.4	4.62	41.45	82.47	−0.577	Chloroplast
AsHSP90-10	AVESA.00010b.r2.5CG0924830.1	781	88,373.9	4.6	47.64	79.14	−0.531	Nucleus
AsHSP90-11	AVESA.00010b.r2.7DG1384250.1	781	88,429.9	4.58	46.62	80.01	−0.528	Chloroplast
AsHSP90-12	AVESA.00010b.r2.7AG1224100.1	781	88,432	4.6	46.45	79.51	−0.534	Chloroplast
AsHSP90-13	AVESA.00010b.r2.5AG0822120.1	787	88,667.1	5.05	44.63	76.24	−0.565	Cytoplasm
AsHSP90-14	AVESA.00010b.r2.5CG0884240.1	784	88,810.7	4.7	44.1	78.37	−0.544	Cytoplasm
AsHSP90-15	AVESA.00010b.r2.5DG0960220.1	785	89,044.9	4.65	44.92	78.14	−0.553	Endoplasmic reticulum
AsHSP90-16	AVESA.00010b.r2.5AG0849600.1	786	89,235.1	4.69	44.7	78.04	−0.553	Cytoplasm
AsHSP90-17	AVESA.00010b.r2.5CG0914170.1	809	91,165.9	5.09	44.57	78.74	−0.539	Chloroplast
AsHSP90-18	AVESA.00010b.r2.7DG1344440.1	806	92,426.4	4.65	37.35	79.6	−0.703	Cytoplasm
AsHSP90-19	AVESA.00010b.r2.2AG0260460.1	806	92,568.6	4.63	37.28	80.57	−0.69	Chloroplast
AsHSP90-20	AVESA.00010b.r2.6CG1147820.1	806	92,623.6	4.6	37.68	80.09	−0.699	Endoplasmic reticulum

Members of the *AsHSP90* gene family are distributed across nine chromosomes and most of them are located in positions of high gene density (Figure 1), and most *AsHSP90* genes are located on chr5 (chr5A, chr5C, chr5D). chr5 contains the most *AsHSP90* genes, although it is not the longest chromosome. Gene clusters can be observed on chr5A, chr5C. This uneven distribution might be the result of uneven replication of oat chromosome segments. In addition, it was noted that relatively less *AsHSP90* genes were located on chr2A, chr6A, and chr6C.

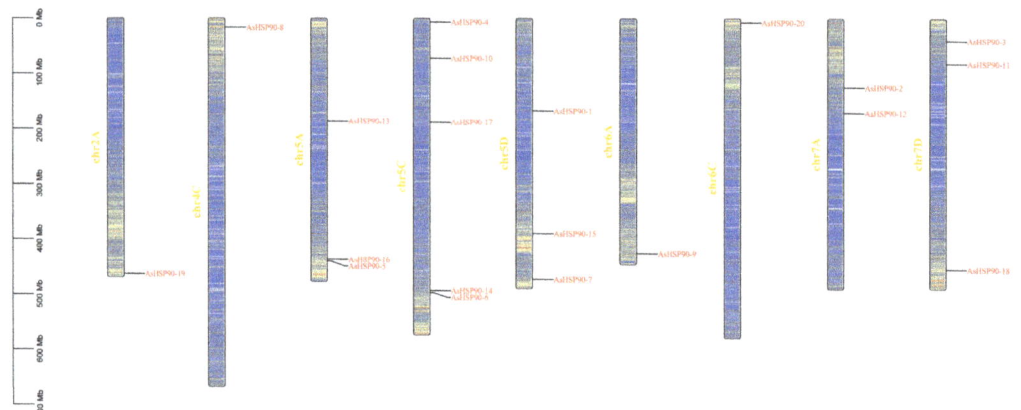

Figure 1. Chromosomal localization of members of the oat *HSP90* gene family. Blue to yellow colors within the chromosomes indicate increased gene density. Chromosome numbers are shown at the right of the vertical bar; gene locations are shown at the left of the vertical bar.

2.2. Phylogenetic Analysis of AsHSP90 Genes

In order to understand the evolutionary relationships of the *HSP90* members, a phylogenetic tree including seven *Arabidopsis* HSP90s, eight rice HSP90s, eight *Zea mays* HSP90s, nine *Brachypodium disachyon* HSP90s, seven *A. insularis* HSP90s, and thirteen *A. longiglumis* HSP90s was constructed using the maximum likelihood (ML) method with MEGA7.0 software (Figure 2). All HSP90 protein sequences were categorized into six clades (Clades 1, 2, 3, 4, 5, and 6). Clade 6 (19 members) had the most number of members, followed by Clade 3 (18 members). Seven and six species were identified in Clade 6 and Clade 3, respectively. In the phylogenetic tree, all oat HSP90 genes showed a closer evolutionary relationship with members of *A. insularis* and *A. longiglumis*. Interestingly, all *Arabidopsis* HSP90s genes were assigned to Clade 6, which may be related to the fact that it is the only dicotyledonous plant in the phylogenetic tree.

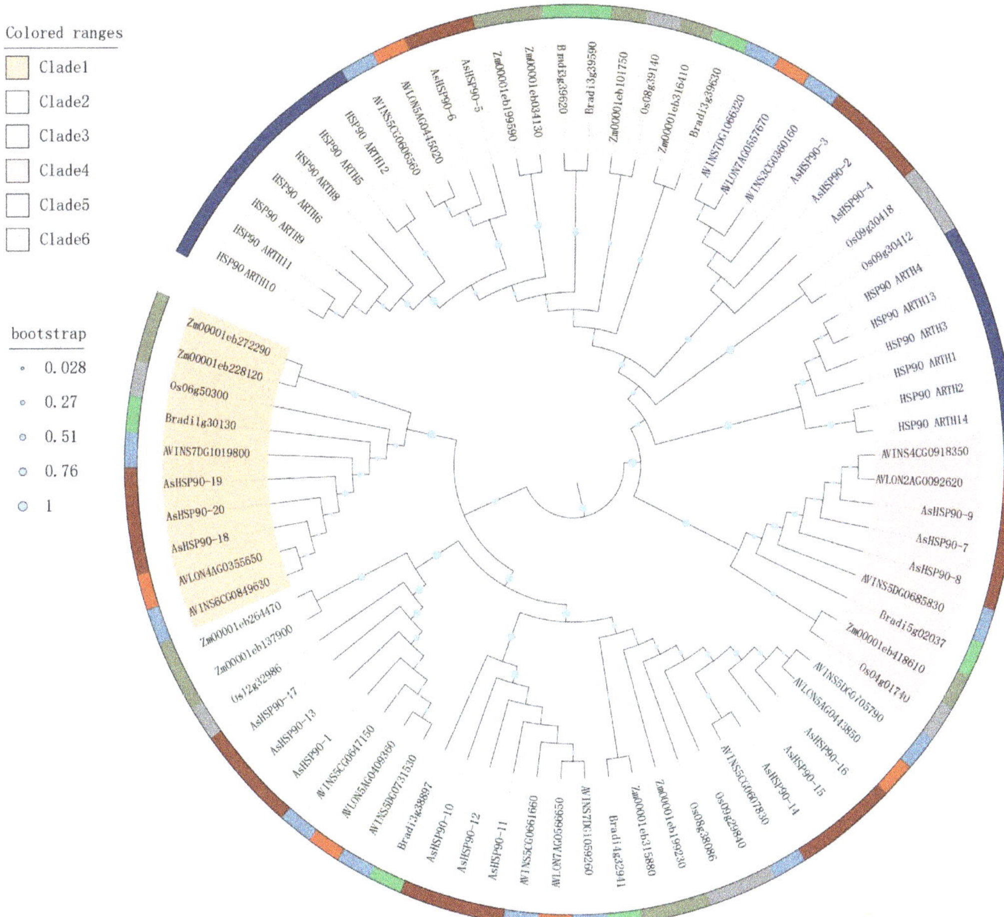

Figure 2. Phylogenetic tree analysis of the HSP90 proteins from *A. sativa*, *A. insularis*, *A. longiglumis*, *Arabidopsis thaliana*, *Brachypodium disachyon*, rice, and maize. The HSP90s were divided into six clades (Clades 1–6) based on the clustering of the protein sequence. The proteins from *A. sativa*, *A. insularis*, *A. longiglumis*, *Arabidopsis thaliana*, *Brachypodium disachyon*, rice, and maize are presented in brown, blue, dark orange, dark blue, green, gray, and dark green, respectively.

2.3. Motif Pattern and Gene Structure Analyses of AsHSP90 Members

The evolution of the oat HSP90 gene family was revealed by analyzing the gene structure and motifs of the *AsHSP90* genes. The 20 AsHSP90s proteins could be placed in six groups according to a constructed simplified phylogenetic tree. Of these, Groups 3 and 6 had the most and least members, with six and two, respectively (Figure 3A). All of these genes had between 2 and 18 introns. However, Groups 4, 5, and 6 contained 2 introns, while the remaining three groups contained 15–18 introns (Figure 3C). In addition, genes on the same branch of the evolutionary tree are similar in structure, and their CDSs have similar numbers of introns. There was little variation in the location and length of the introns within the groups, but significant variation between the groups.

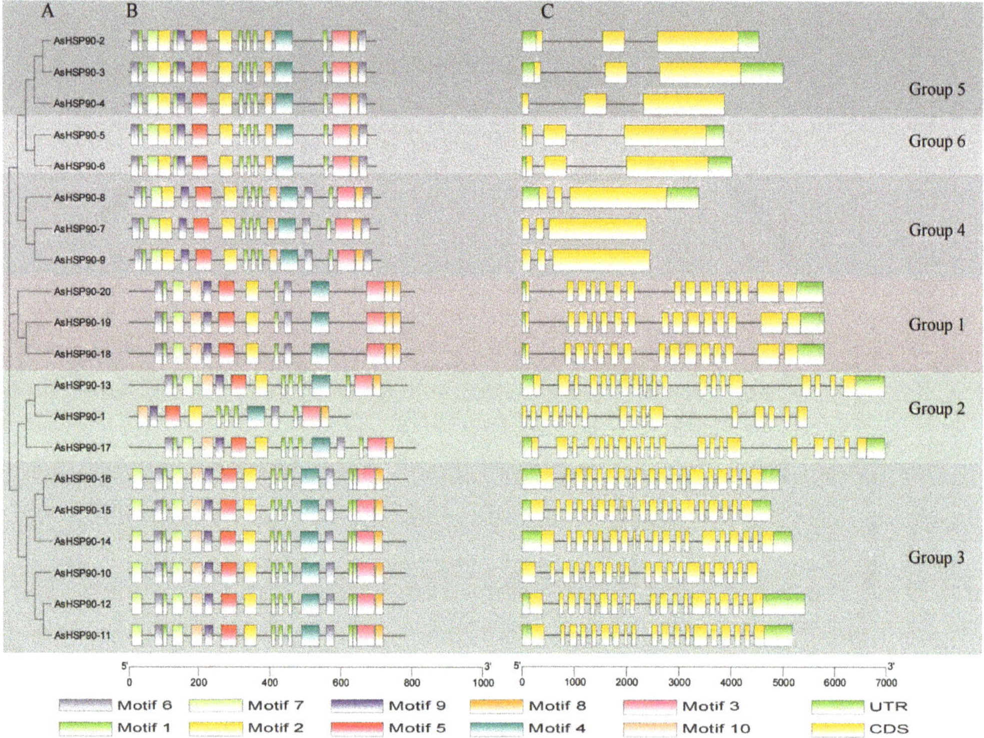

Figure 3. Phylogenetic tree, motif analysis, and gene structure of *AsHSP90*: (**A**) Phylogenetic tree analysis of the AsHSP90 protein. (**B**) Motif composition of *AsHSP90*. (**C**) Gene structure of the *AsHSP90* genes in oats.

To better understand the structural quality of the AsHSP90 protein, we identified 10 conserved motifs in the protein using the MEME [30] motif search tool and explored the distribution of these conserved motifs in the AsHSP90 protein (Figure 3B, Supplementary Figure S1). The results of this study showed that the 10 most conserved motifs detected contained 12–50 amino acids. Among them, motif 1 had the lowest amino acid content, with 12 amino acids. In addition, most of the genes consisted of 10 conserved motifs. Similar genes had similar motifs, which suggests that the *AsHSP90* gene family has similar functions. Overall, the *HSP90* gene family in oats is highly conserved, with few conserved motifs lost during evolution.

2.4. Duplication Analysis of AsHSP90 Members

Gene duplication events in the *AsHSP90* gene were analyzed using MCScanX [31]. In total, there were 10 pairs of duplication genes among *HSP90* genes (Figure 4 and Table 2). All duplicates were from Group 3 and Group 4, and the majority of them were located at the end of the chromosome. These 10 pairs of genes were defined as WGD/segmental duplicates and inter-chromosomal. Some *AsHSP90* genes had undergone more than one duplication. In addition, we calculated the non-synonymous (Ka) and synonymous (Ks) substitution rates, as well as Ka/Ks ratios (Table 2), to capture the evolutionary dynamics of the ASHSP90 protein coding sequence. Of these, 10 pairs of duplication genes had Ka/Ks values of <1, suggesting that these genes had undergone primarily purifying selection.

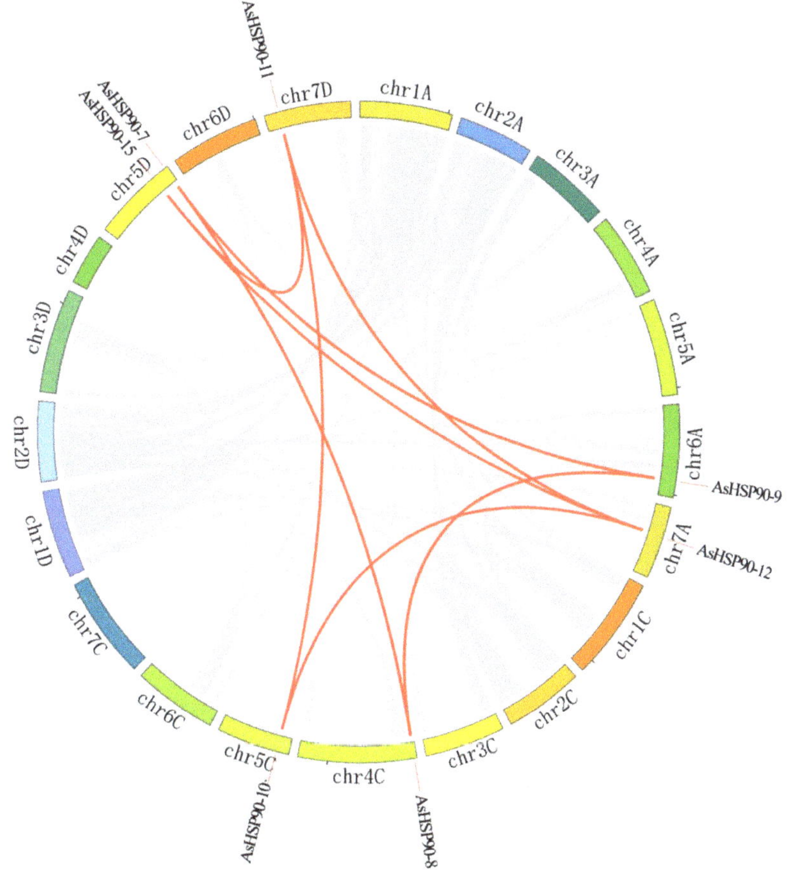

Figure 4. Gene duplications of the oat HSP90 gene family. Red lines indicate duplicated gene pairs in *AsHSP90*, and gray lines indicate co-linear gene pairs in the genome.

2.5. Collinearity Analysis of HSP90 Genes

To explore the evolutionary relationship between the oat *AsHSP90* gene and other species, synteny analyses were performed on four representative plants, including two possible ancestral species of oats (*A. insularis* and *A. longiglumis*) and two monocotyledons (*Oryza sativa* and *Brachypodium distachyon*). Thirty-seven and nineteen *AsHSP90* syntenic gene pairs were identified in *A. insularis* and *A. longiglumis*, respectively, and fifteen syntenic gene pairs each were identified in *Oryza sativa* and *Brachypodium distachyon* (Figure 5A,

Supplementary Table S1). Nine *AsHSP90*s (*AsHSP90-2*, *AsHSP90-3*, *AsHSP90-4*, *AsHSP90-12*, *AsHSP90-13*, *AsHSP90-14*, *AsHSP90-15*, *AsHSP90-16*, and *AsHSP90-17*) are present as syntenic genes in these four species (Figure 5B). In addition, syntenic genes of the oat HSP90 gene Group 1 (*AsHSP90-18*, *AsHSP90-19*, and *AsHSP-20*) were not found in either rice or *Brachypodium distachyon*, but their respective syntenic genes were found in *A. Insularis* and *A. longiglumis*. On this basis, we separately detected *AsHSP90* orthologs in rice, *A. longiglumis*, *Brachypodium distachyon*, and *A. insularis* with OrthoFinder (Supplementary Table S2). *A. longiglumis* and *A. insularis* contained 7 and 12 orthologous genes, respectively, and *Brachypodium distachyon* contained 3. Interestingly, two and one genes in *A. insularis* and *A. longiglumis*, respectively, showed to be orthologous to oat *AsHSP90* gene Group 1 (*AsHSP90-18*, *AsHSP90-19*, and *AsHSP-20*), and all of these genes were from different sub-genomes. In addition, all three genes belong to the same group and none of them were involved in duplication events, which may suggest that these three genes are more conserved during the evolutionary process.

Table 2. Segmental duplications of *AsHSP90* paralogous pairs in oats.

Paralogous HSP90 Pairs	chr. Location	Duplication Type	AsHSP90 Group	Ka	Ks	Ka_Ks
AsHSP90-8	chr4C	Segmental	Group 4	0.0036	0.0908	0.0396
AsHSP90-9	chr6A		Group 4			
AsHSP90-7	chr5D	Segmental	Group 4	0.0018	0.1265	0.0143
AsHSP90-8	chr4C		Group 4			
AsHSP90-7	chr5D	Segmental	Group 4	0.0042	0.1343	0.0316
AsHSP90-9	chr6A		Group 4			
AsHSP90-12	chr7A	Segmental	Group 3	0.0016	0.0234	0.0706
AsHSP90-11	chr7D		Group 3			
AsHSP90-12	chr7A	Segmental	Group 3	0.0080	0.0928	0.0862
AsHSP90-10	chr5C		Group 3			
AsHSP90-12	chr7A	Segmental	Group 3	0.0638	0.9539	0.0669
AsHSP90-15	chr5D		Group 3			
AsHSP90-12	chr7A	Segmental	Group 3	0.0656	0.9452	0.0694
AsHSP90-14	chr5C		Group 3			
AsHSP90-11	chr7D	Segmental	Group 3	0.0097	0.0798	0.1211
AsHSP90-10	chr5C		Group 3			
AsHSP90-11	chr7D	Segmental	Group 3	0.0647	0.9356	0.0691
AsHSP90-15	chr5D		Group 3			
AsHSP90-11	chr7D	Segmental	Group 3	0.0665	0.9406	0.0707
AsHSP90-14	chr5C		Group 3			

2.6. Cis-Element Analysis of AsHSP90 Gene Promoters

Sequences 1500 bp upstream of the promoter of each *AsHSP90* gene were isolated from the oat genome, and cis-acting elements were predicted using PlantCARE [32] to characterize the expression of each *AsHSP90* gene. A total of 32 cis-acting elements were analyzed (Figure 6, Supplementary Table S3). *AsHSP90-18* has the highest number of homeopathic acting elements and *AsHSP90-16* has the lowest number of homeopathic acting elements. The 20 *AsHSP90* promoter regions involved hormone-responsive elements, light-responsive elements, environment-responsive elements, and developmentally relevant elements (Figure 6B). These elements mainly responded to hormonal and abiotic stresses. Interestingly, there are two cis-acting regulatory elements involved in circadian control in *AsHSP90-3* and *AsHSP90-6*. Furthermore, a total of five cis-elements related to salicylic acid responsiveness were found in *AsHSP90-14*, *AsHSP90-16*, and *AsHSP90-20*.

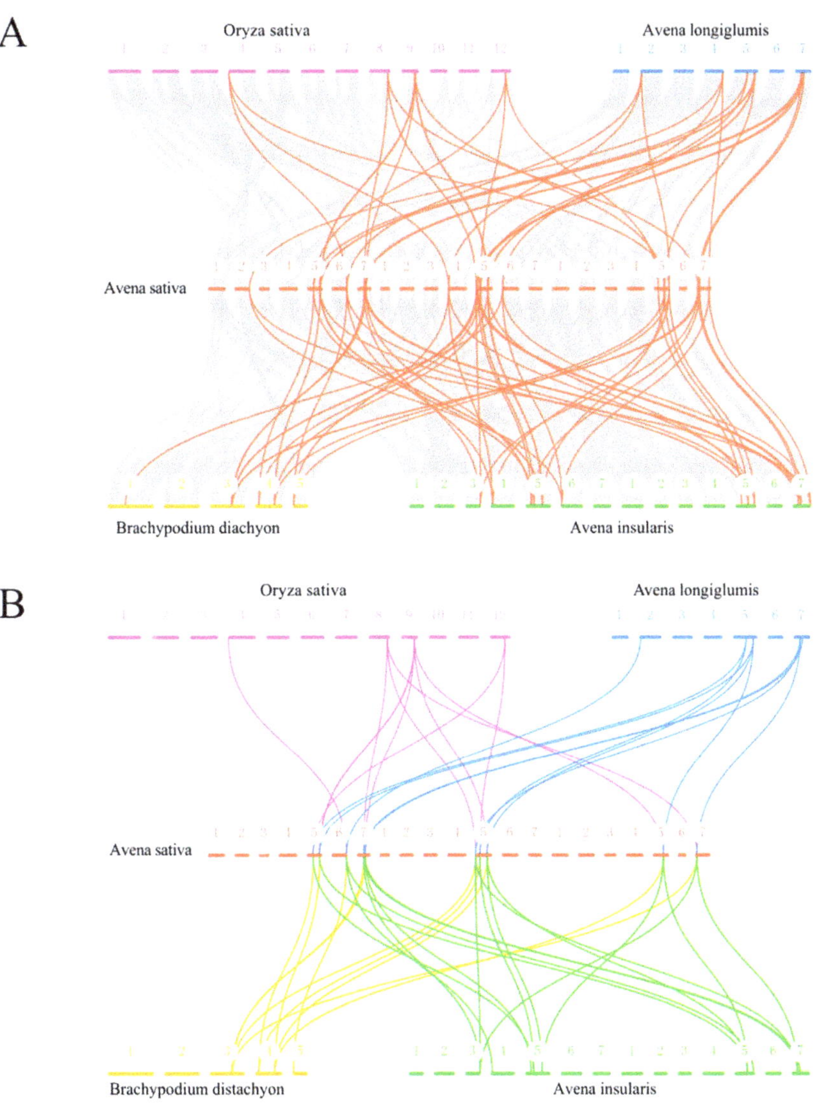

Figure 5. Synteny analysis of *AsHSP90*s in *A. sativa* and four representative plants. (**A**) All AsHSP90 synteny genes in oats and in *Oryza sativa*, *A. longiglumis*, *Brachypodium distachyon*, and *A. insularis* are indicated by red lines. The synteny blocks in the oats and the other species are shown in gray lines. (**B**) The nine *AsHSP90* genes with covariance in the four species are shown as purple, blue, yellow, and green lines.

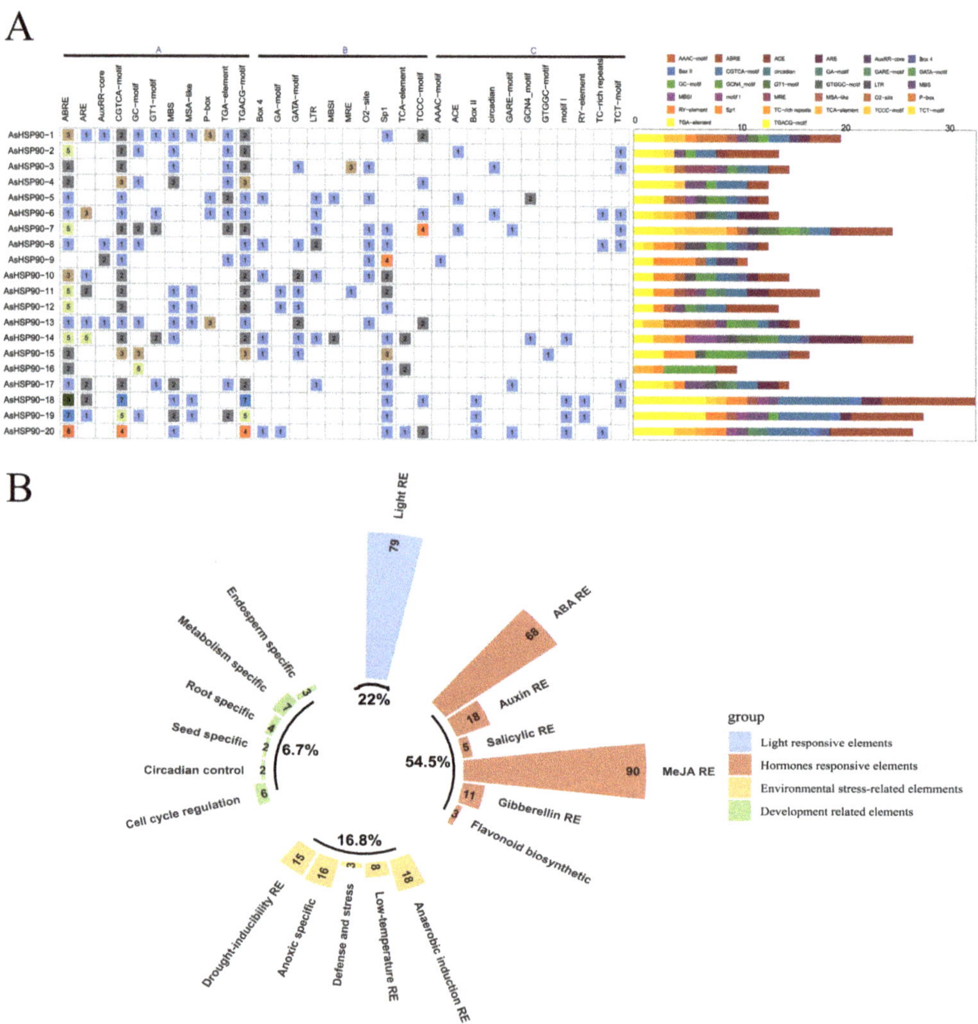

Figure 6. Distribution of cis-elements upstream of oat *AsHSP90*: (**A**) Distribution of types and numbers of cis-elements in the promoter of the oat *AsHSP90* genes. (**B**) Classification and proportion of cis-elements in the *AsHSP90* gene.

2.7. Expression Analysis of AsHSP90s in Oats under Five Abiotic Stresses

To analyze the expression pattern of *AsHSP90* under several different abiotic stresses, 20 AsHSP90 proteins were analyzed using qRT-PCR. As shown in Figure 7, different expression patterns of 20 *AsHSP90*s were observed under heat, drought, salt, Cd, and ABA stresses. Almost all AsHSP90 members were involved in expression under different abiotic stresses. Among the five treatments, heat stress elicited the most pronounced stress response, with the average expression of AsHSP90 genes being two–eight times higher than that under the other four abiotic stresses. AsHSP90-7, AsHSP90-8, and AsHSP90-9 exhibited prominent expression in response to heat stress, while other members also demonstrated varying levels of transcriptional activation.

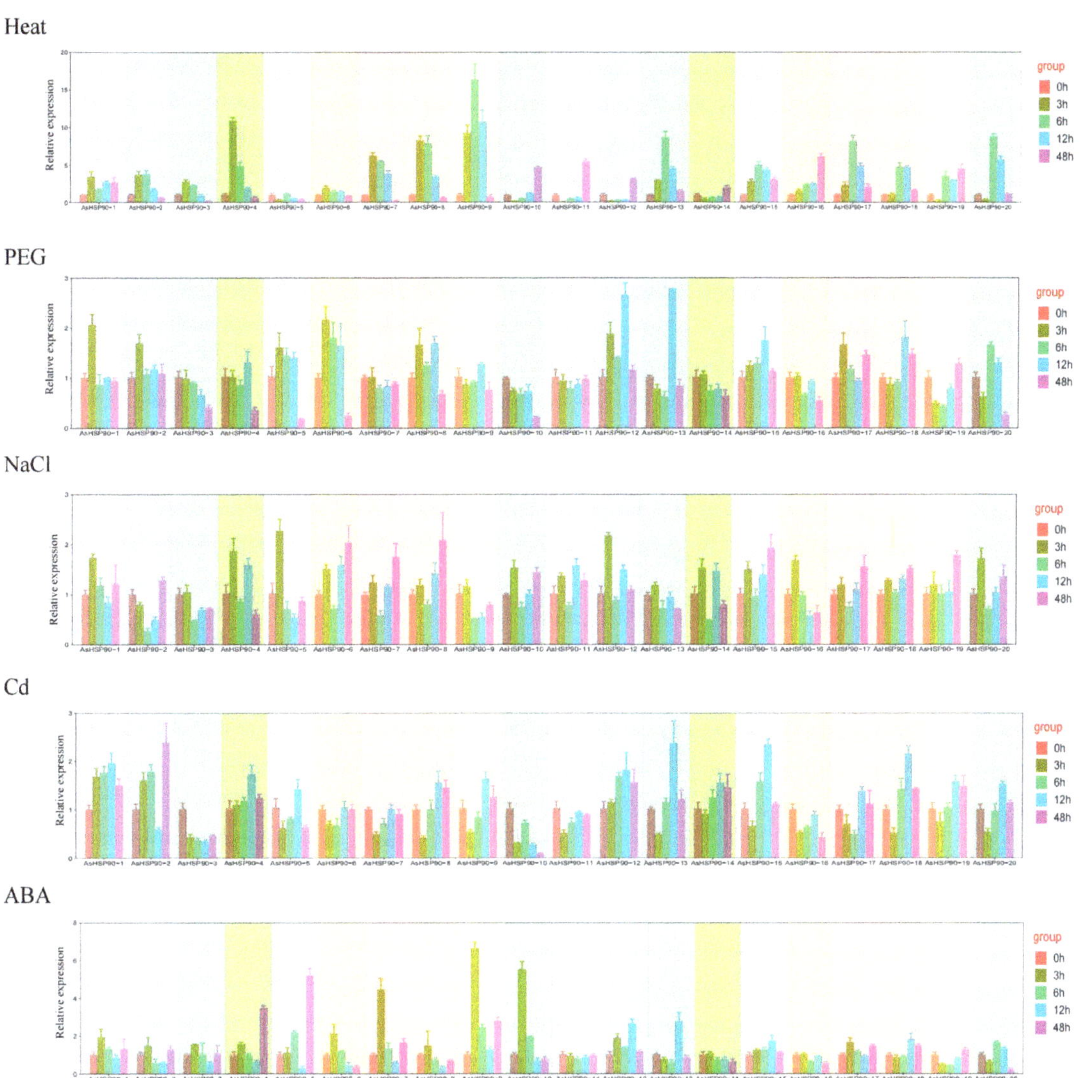

Figure 7. The relative expression of 20 *AsHSP90* genes in oat leaves was detected with qRT-PCR after treatments of 0 h (CK), 3 h, 6 h, 12 h, and 48 h under different abiotic stresses. Error bars indicate the standard error (SE) between three replicates.

Under the drought treatment (PEG treatment simulation), AsHSP90-12 and *AsHSP90-13* were significantly expressed at 12 h, but their expression decreased with increasing exposure time. *AsHSP90-1*, *AsHSP90-5*, and *AsHSP90-6* were also observed to be expressed at the 6 h point. Furthermore, the expression of *AsHSP90-3*, *AsHSP90-4*, *AsHSP90-5*, *AsHSP90-6*, *AsHSP90-10*, and *AsHSP90-20* was significantly reduced at 48 h. Under NaCl treatment, the expression of all members except *AsHSP90-3* increased at 6 h and then decreased and then increased with longer exposure time.

Under Cd treatment, *AsHSP90-13*, *AsHSP90-15*, and *AsHSP90-18* were significantly expressed at 12 h, and *AsHSP90-2* was significantly expressed at 48 h. The expression of other *AsHSP90*s showed a fluctuating pattern, suggesting that their expression may be influenced by the duration of Cd treatment. *AsHSP90-7*, *AsHSP90-9*, and *AsHSP90-10* were significantly expressed under ABA treatment at 6 h; *AsHSP90-4* and *AsHSP90-5* were significantly up-regulated at 48 h and continued to be up-regulated, while the expression of the other members was not significant.

In general, Group 5 (*AsHSP90-7*, *AsHSP90-8*, and *AsHSP90-9*) in the oat *AsHSP90* gene was induced under all five stresses. Among all of the abiotic stresses, the highest expression of *AsHSP90-9* was found after 12 h of heat stress.

2.8. Three-Dimensional Structure Prediction and Protein–Protein Interaction Network

Three-dimensional protein structures of the AsHSP90s were performed by the SWISS-MODEL [33] server and model generation was performed via PyMOL (Supplementary Figure S2). All targets had greater than 30% identity with the template, which is a threshold that is a sign of successful modeling (advances in homology protein structure modeling). The QMEAN score values of the models varied between 0.64 and 0.76, which indicated that all of the models were of better quality, while the GMQE values ranged from 0.54 to 0.76. Meanwhile, out of the 20 models, 16 models were hetero-trimer states (Supplementary Table S4).

The protein–protein interaction network was further analyzed to detect interactions between AsHSP90 and related proteins. Seven other proteins were found to interact with AsHSP90 (Supplementary Figure S3), potentially being jointly involved in certain biological processes. Protein nodes were manually rearranged based on their degree of interaction. Proteins positioned in the inner circle of the layout exhibit a higher degree of interaction. In addition to AsHSP90, proteins A0A3B5YXX4 (HSP70), A0A3B6EMP3 (calreticulin), and A0A3B6H0I5 (calreticulin) also exhibit strong associations with AsHSP90 proteins.

3. Discussion

Heat stress protein 90 (HSP90) is a highly conserved molecular chaperone within the HSP family. HSP90 proteins are rapidly synthesized in response to heat stress treatments, serving to counteract the damage caused to plants by high temperatures [10]. Here, we identified twenty HSP90 genes and assigned them to six clades. Furthermore, the number of oat HSP90 genes was higher than in some previously studied species, such as the 7 in *A. thaliana* [17] and 10 in *populus* [34], a phenomenon largely attributable to hexaploidy. These *AsHSP90* genes may play a significant role in the physiological maintenance of oats, allowing them to survive high temperatures and other abiotic stress environments. The *HSP90* family of proteins exhibits varied biophysical properties, indicative of a wide diversity among its members. This diversity lays the groundwork for further studies on the function of *HSP90* genes. In this study, the *AsHSP90* gene sequence displayed an isoelectric point (pI) ranging from 4.60 to 5.09 and was acidic, consistent with previous studies on the *HSP90* gene in *perennial ryegrass* [20]. Additionally, we observed that the *AsHSP90* gene is unevenly distributed in oats, predominantly in regions of high gene density on the chromosomes. This distributional characteristic may be linked to the uneven replication of oat chromosome segments [29].

Oats undergo complex polyploidization events and frequent translocations among subgenomes, which provide a good opportunity to study gene family formation and expansion [29,35]. Gene duplication events are thought to be an important mechanism for increasing gene family diversity [36]. Plant evolution is usually accompanied by gene fragment duplication events, considered as one of the main drivers of the expansion of plant gene families. The oat genome possesses a mosaic chromosome structure, and the chromosomal rearrangements it undergoes often result in the duplication of gene family members [29]. In this study, 10 pairs of segmental duplication genes were identified in the *AsHSP90* gene family. These genes belong to Group 3 and Group 4, comprising homologous genes with similar structural domains. Chromosomal polyploidy also significantly contributes to the expansion of the number of gene families in plants. The "A" subgenome and "CD" subgenome in oats originated from *A. insularis* and *A. longiglumis*, respectively [35].

Also, 13 and 7 HSP90 genes were identified from *A. insularis* and *A. longiglumis*, the probable ancestral species of oats, respectively, and the total number of these genes is consistent with the number of *AsHSP90* genes. Thus, chromosome doubling is likely responsible for AsHSP90 amplification in oats. This finding is consistent with the fact that *A. insularis* and *A. longiglumis* hybridized to form a heterozygous hexaploid. Furthermore, synteny analysis and orthologous gene identification reveal that not all oat *AsHSP90* genes have direct homologs in the ancestral species. For example, the *AsHSP90-9* gene located in the "A" subgenome lacks orthologous genes in *A. insularis*, whereas *AsHSP90-7* and *AsHSP90-8*, which are part of Group 3 in the "CD" subgenome, possess orthologous genes in *A. longiglumis* (Supplementary Table S2). Additionally, there is a gene duplication event between *AsHSP90-7*, *AsHSP90-8*, and *AsHSP90-9*, suggesting that *AsHSP90-9* may be a paralogous gene resulting from this duplication. This implies that the formation of paralogous genes also contributes to the expansion of the *AsHSP90* gene family. In conclusion, the formation of the *AsHSP90* gene family in oats may have arisen mainly through gene duplication after polyploidy and divergence.

Phylogenetic analyses can help to understand evolutionary relationships between species and ermine homology between and within species [37]. In this study, a phylogenetic tree was constructed using the protein sequences of the HSP90 gene from seven monocotyledonous and one dicotyledonous plant species. Based on the phylogenetic analysis, it can be categorized into six distinct clades, wherein all *Arabidopsis* HSP90 genes cluster in Clade 5, which may be attributed to *Arabidopsis* being the sole dicotyledonous plant in the phylogenetic tree [38]. Furthermore, the phylogenetic tree also reflects the closer relationship of oats and *A. insularis* to *A. longiglumis*. Intron gain or loss and intron density significantly impact the evolution of large eukaryotic genomes. According to the phylogenetic tree, the *AsHSP90* gene is categorized into six groups, each with a highly similar exon number and exon–intron structure. Similarly, the motif distribution of *AsHSP90* genes across various subgroups exhibits a consistent pattern. Consequently, these analyses further corroborate the reliability of this phylogenetic classification of *AsHSP90* genes.

Numerous studies have demonstrated that the plant HSP90 protein plays a crucial role in responding to various abiotic stresses [39,40]. According to the qRT-PCR results, it was observed that most *AsHSP90* genes were induced under all five stresses, albeit with low expression levels in some genes. HSP is a well-known protein that responds to heat stress and protects plants from high-temperature stress damage [41]. In the current study, the highest expression of all AsHSP90 proteins was observed under high-temperature stress, suggesting that HSP90 is particularly sensitive to heat stress. Group 5 (*AsHSP90-7*, *AsHSP90-8*, and *AsHSP90-9*) within the oat *AsHSP90* gene family exhibited high expression under all five stress conditions. Among them, the *AsHSP90-9* gene, a paralogous gene of *AsHSP90-7* and *AsHSP90-8* replicated, had the highest expression under heat stress. Gene duplication events usually result in an increase in the number of genes; after this, the replicated genes may undergo neofunctionalization or subfunctionalization [42]. In our study, *AsHSP90-9* had similar expression patterns to *AsHSP90-7* and *AsHSP90-8*, and we hypothesized that the *AsHSP90-9* gene may undergo subfunctionalization to carry out some of the functions of the *AsHSP90-7* and *AsHSP90-8* genes. Therefore, under high-temperature stress conditions, the expression of the *AsHSP90-9* genes involved in the response to heat stress after subfunctional differentiation was significantly up-regulated to help the oats better adapt to heat stress. Furthermore, the analysis of the cis-elements of *AsHSP90* in this study revealed a variety of cis-acting elements involved in hormone regulation and abiotic stress. For instance, *AsHSP90-18*, *AsHSP90-19*, and *AsHSP90-19* had the highest number of ABRE-binding sites within the ABA regulatory pathway. ABA-regulated genes are involved in multiple biotic and abiotic stress responses in plants [43]. However, contrary to expectations, these genes did not exhibit the highest expression levels in our RT-qPCR analysis under the five abiotic stress treatments. This discrepancy might be attributed to their status as orthologous genes from different subgenomes, suggesting that their expression is regulated through a complex mechanism of subgenomic homologous co-expression. Thus, the roles of *AsHSP90-9*, *AsHSP90-18*, *AsHSP90-19*, and *AsHSP90-20* in response to these stresses warrant further investigation. Moreover, variations were observed in gene expression at different sites, necessitating further studies to investigate the expression of the AsHSP90 protein in various tissues of oats.

4. Materials and Methods

4.1. HSP90 Gene Family Identification in Avena sativa, Avena insularis, and Avena longiglumis

The genomic resources of *Avena sativa* and its possible ancestors *Avena insularis* and *Avena longiglumis* were cv "sang_v1.1", cv "BYU209_v1.1", and cv "CN58138_v1.1" from the GrainGenes database (https://wheat.pw.usda.gov, accessed on 24 May 2023) [44], and the Hidden Markov Model (HMM) matrix file for the HSP90 structural domain was obtained from the Pfam database (PF00183). Using HMMER software version 3.0 [45], HSP90s were searched for within three genome protein sequences. A BLASTP analysis of these three genomes' genomic resources was conducted using the protein sequence of the *Brachypodium disachyon* and *Arabidopsis thaliana* HSP90 gene as a query. Based on the HMMER and BLASTP results, all candidate HSP90 proteins that possibly contained the HSP90 domain were submitted to Pfam (http://pfam.xfam.org/, accessed on 24 May 2023) and CDD (https://www.ncbi.nlm.nih.gov/Structure/bwrpsb/bwrpsb.cgi, accessed on 24 May 2023) for confirmation of the final resulted HSP90 members. *A. sativa* HSP90 genes were renamed according to their molecular weight. An in-house Perl script was used to calculate the molecular weight, instability index, and theoretical isoelectric point (pI) of *AsHSP90*. Subcellular localization was predicted using WoLF PSORT [46] (https://wolfpsort.hgc.jp/ accessed on 29 May 2023). All *HSP90* genes were mapped to *A. sativa* chromosomes using TBtools-II [47].

4.2. Phylogenetic Analysis of HSP90 Genes

The HSP90 protein sequences of *Avena sativa, Avena insularis, Avena longiglumis, Arabidopsis thaliana, Oryza sativa, Brachypodium disachyon*, and *Zea mays* were subjected to multiple-sequence alignment analysis using ClustalW [48] in order to study the evolutionary relationship between the HSP90 families of these five plants. A maximum-likelihood (ML) phylogenetic tree was construction by 1000 bootstrapping with MEGA [49].

4.3. Structure and Conserved-Motif Analysis of AsHSP90s

To extract the CDS and UTR locations corresponding to *AsHSP90*s, we used in-house Perl scripts. In addition, in the AsHSP90 protein, a phylogenetic tree was constructed using the maximum-likelihood (ML) method and 10 motifs were identified using the MEME program 5.10 [30] (https://meme-suite.org/meme/tools/fimo, accessed on 16 June 2023). The gene structures, motifs, and phylogenetic tree were mapped and modified using TBtools-II [47].

4.4. Gene Duplication and Ka and Ks Calculation

Gene duplication search for the identified HSP90 members was performed using blastall. The major criteria used for analyzing potential gene duplications included: (a) length of alignable sequence covers >75% of longer gene, and (b) similarity of aligned regions >75% [50]. The rate of synonymous substitutions (Ka) and nonsynonymous substitutions (Ks) in the *AsHSP90* gene obtained from gene duplication events was calculated using KaKs_Calculator 3.0 [51].

4.5. Synteny Analysis of AsHSP90 Genes and Selected Plants

To demonstrate the synteny relationship of the HSP90 genes obtained from *Avena sativa* and gramineous species of *Oryza sativa, Brachypodium disachyon, A. insularis*, and *A. longiglumis*, the syntenic analysis maps were constructed using the software MCScanX [31]. In addition, gene duplication events in *AsHSP90*s of oats also were visualized with MC-

ScanX. OrthoFinder 2.5.5 [52] was utilized to find directly orthologous genes between the oats and other representative species.

4.6. Identification and Analysis of cis-Elements in the Promoter Region of AsHSP90

An in-house Perl script was used to extract 1.5 kb sequences upstream of the *AsHSP90* gene as promoter regions and submit these sequences to PlantCARE [32] for the analysis of cis-regulatory elements.

4.7. Three-Dimensional Protein Structure Prediction

Protein templates in the PDB database [53] (https://www.rcsb.org/, accessed on 21 June 2023) with similar 3D structures to the AsHSP90 protein were searched using PSI-BLAST [54], and the resulting templates and AsHSP90 protein sequences were submitted to SWISS-MODEL [33] (https://swissmodel.expasy.org/, accessed on 22 June 2023) for protein 3D structure prediction. Moreover, the 3D model's quality was assessed with global model quality estimates (GMQE) and QMEAN values. The GMQE scores ranged from 0 to 1, with higher scores indicating a more reliable model, and the QMEAN scores ranged between 0 and −4, with models closer to 0 being of better quality.

4.8. Protein–Protein Interaction Network

To predict the interactions between oat HSP90 proteins and related proteins, the AsHSP90 protein sequence was submitted to the STRING database [55]. The organism was set to wheat, and advanced settings were maintained at their default values. PPI networks were visualized using Cytoscape v3.10.1 [56].

4.9. Plant Material, Growth Conditions, and Treatment

Viable seeds of oat cv "Baylor" were grown in quartz sand and grown in a greenhouse. The germinated seedlings were transferred into Hoagland's solution after 7 days. Seedlings were subjected to stress treatments after 14 days. Salt stress was simulated by dissolving a 250 mM concentration of NaCl in the culture broth. Chromium (Cr) ($K_2Cr_2O_7$) was dissolved at a concentration of 300 mg/L to simulate heavy-metal stress. For heat stress, the temperature was 40 °C during the day/30 °C at night, and the photoperiod was 12 h of light/12 h of dark. Drought stress was stimulated with 20% polyethylene glycol 6000 (PEG) after lysis. Abscisic acid (ABA) was sprayed at a concentration of 100 mM. Live biological replicates of plant leaf samples were collected at 0 h, 3 h, 6 h, 12 h, and 48 h after each application of stress. Then, the stressed seedlings were collected for RNA extraction and stored at a temperature of −8 °C.

4.10. RNA Isolation, cDNA Synthesis, and Quantitative Real-Time PCR Expression Analyses

Total RNA was isolated using the Direct-zol™ RNA MiniPrep Kit (Zymo Research, Beijing, China), according to the manufacturer's protocol. ABScript III RT Master Mix for qPCR with Gdna Remover (Abclonal, Wuhan, China) was used for the synthesis of cDNA. RT-qPCR analyses were performed using the Genious 2× SYBR Green Fast Qpcr Mix (Abclonal, China), in accordance with the manufacturer's protocol; the reactions were run using the CXF96 Connect™ Real-Time System (Bio-Rad, Singapore). The AsEIF4A gene was used as an internal reference gene to calculate the expression of 20 *AsHSP90* genes [57]. Relative expression was calculated using the $2^{-\Delta\Delta Ct}$ method [58,59]. In addition, twenty primer pairs for the *AsHSP90* gene were designed with Primer 5 software. Primers for the *AsHSP90* gene used in the qRT-PCR assay are listed in Supplementary Table S5.

5. Conclusions

We identified and localized 20 *AsHSP90* genes in the oat genome and divided them into six clades. Gene structures and motifs were highly conserved within the same groups. The formation of the *AsHSP90* gene family in oats may have arisen mainly through gene duplication after polyploidy and divergence. Under high-temperature, salt, cadmium

(Cd), polyethylene glycol (PEG), and abscisic acid (ABA) stresses, *AsHSP90* showed the strongest expression under heat stress, and members of Group 5 (*AsHSP90-7*, *AsHSP90-8*, and *AsHSP90-9*) were generally highly expressed. The function of AsHSP90 proteins remains unknown, especially the co-expression of homologous proteins among different subgenomes, and further studies are needed to determine their precise function. Our study elucidates the potential pathways for the expansion of the *AsHSP90* gene family in oats, and also lays the foundation for future functional analyses of these AsHSP90 proteins as well as studies of their synergistic expression across subgenomes.

Supplementary Materials: The following supporting information can be downloaded at: https://www.mdpi.com/article/10.3390/ijms25042305/s1.

Author Contributions: Conceptualization, J.P. and X.M.; methodology, J.P., Y.X. and X.M.; software, J.P., J.W., T.L. and B.L.; validation, X.M. and M.Y.; formal analysis, X.M. and X.L.; investigation, J.P., M.Y., J.Z. and X.M.; resources, X.M.; data curation, J.P. and S.L.; writing—original draft preparation, J.P.; writing—review and editing, X.M., J.Z. and X.L.; visualization, J.P.; supervision, X.M. and X.L.; project administration, X.M.; funding acquisition, X.M and X.L. All authors have read and agreed to the published version of the manuscript.

Funding: This research was funded by Development of SNP Fingerprints of Oat and Genetic Diversity of its Germplasm Resources: A Study of Basic Research Funding; the National Natural Science Foundation of China (32271753); Sichuan Province Beef Cattle Innovation Team (SCCXTD-20-13); Sichuan Provincial Central and Local University Regional Cooperation Project (2023YFSY0012); and supported by the earmarked fund for CARS (CARS-34) and the Sichuan Forage Innovation Team Program.

Institutional Review Board Statement: Not applicable.

Informed Consent Statement: Not applicable.

Data Availability Statement: Data are contained within the article and Supplementary Materials.

Acknowledgments: We are very grateful to the College of Grassland Science and Technology, Sichuan Agricultural University and Sichuan Academy of Grassland Science for providing us with experimental equipment and venues.

Conflicts of Interest: The authors declare no conflicts of interest.

References

1. Guihur, A.; Rebeaud, M.E.; Goloubinoff, P. How Do Plants Feel the Heat and Survive? *Trends Biochem. Sci.* **2022**, *47*, 824–838. [CrossRef]
2. Zhang, H.; Zhu, J.; Gong, Z.; Zhu, J.-K. Abiotic Stress Responses in Plants. *Nat. Rev. Genet.* **2022**, *23*, 104–119. [CrossRef]
3. Li, B.; Gao, K.; Ren, H.; Tang, W. Molecular Mechanisms Governing Plant Responses to High Temperatures. *J. Integr. Plant Biol.* **2018**, *60*, 757–779. [CrossRef]
4. Oldroyd, G.E.D.; Leyser, O. A Plant's Diet, Surviving in a Variable Nutrient Environment. *Science* **2020**, *368*, eaba0196. [CrossRef]
5. Tigchelaar, M.; Battisti, D.S.; Naylor, R.L.; Ray, D.K. Future Warming Increases Probability of Globally Synchronized Maize Production Shocks. *Proc. Natl. Acad. Sci. USA* **2018**, *115*, 6644–6649. [CrossRef] [PubMed]
6. Gray, S.B.; Brady, S.M. Plant Developmental Responses to Climate Change. *Dev. Biol.* **2016**, *419*, 64–77. [CrossRef] [PubMed]
7. Wen, J.; Qin, Z.; Sun, L.; Zhang, Y.; Wang, D.; Peng, H.; Yao, Y.; Hu, Z.; Ni, Z.; Sun, Q.; et al. Alternative Splicing of TaHSFA6e Modulates Heat Shock Protein-Mediated Translational Regulation in Response to Heat Stress in Wheat. *New Phytol.* **2023**, *239*, 2235–2247. [CrossRef] [PubMed]
8. di Donato, M.; Geisler, M. HSP90 and Co-Chaperones: A Multitaskers' View on Plant Hormone Biology. *FEBS Lett.* **2019**, *593*, 1415–1430. [CrossRef]
9. Tichá, T.; Samakovli, D.; Kuchařová, A.; Vavrdová, T.; Šamaj, J. Multifaceted Roles of HEAT SHOCK PROTEIN 90 Molecular Chaperones in Plant Development. *J. Exp. Bot.* **2020**, *71*, 3966–3985. [CrossRef] [PubMed]
10. Yadav, A.; Singh, J.; Ranjan, K.; Kumar, P.; Khanna, S.; Gupta, M.; Kumar, V.; Wani, S.; Sirohi, A. Heat Shock Proteins: Master Players for Heat-Stress Tolerance in Plants during Climate Change. In *Heat Stress Tolerance in Plants: Physiological, Molecular and Genetic Perspectives*; Wiley: New York, NY, USA, 2020; pp. 189–211. ISBN 978-1-119-43236-4.
11. Genest, O.; Wickner, S.; Doyle, S.M. Hsp90 and Hsp70 Chaperones: Collaborators in Protein Remodeling. *J. Biol. Chem.* **2019**, *294*, 2109–2120. [CrossRef] [PubMed]

12. Chiosis, G.; Digwal, C.S.; Trepel, J.B.; Neckers, L. Structural and Functional Complexity of HSP90 in Cellular Homeostasis and Disease. *Nat. Rev. Mol. Cell Biol.* **2023**, *24*, 797–815. [CrossRef] [PubMed]
13. Dutta, R.; Inouye, M. GHKL, an Emergent ATPase/Kinase Superfamily. *Trends Biochem. Sci.* **2000**, *25*, 24–28. [CrossRef] [PubMed]
14. Picard, D. Heat-Shock Protein 90, a Chaperone for Folding and Regulation. *Cell. Mol. Life Sci. CMLS* **2002**, *59*, 1640–1648. [CrossRef] [PubMed]
15. Hahn, A.; Bublak, D.; Schleiff, E.; Scharf, K.-D. Crosstalk between Hsp90 and Hsp70 Chaperones and Heat Stress Transcription Factors in Tomato. *Plant Cell* **2011**, *23*, 741–755. [CrossRef]
16. Li, W.; Chen, Y.; Ye, M.; Wang, D.; Chen, Q. Evolutionary History of the Heat Shock Protein 90 (Hsp90) Family of 43 Plants and Characterization of Hsp90s in Solanum Tuberosum. *Mol. Biol. Rep.* **2020**, *47*, 6679–6691. [CrossRef]
17. Krishna, P.; Gloor, G. The Hsp90 Family of Proteins in Arabidopsis Thaliana. *Cell Stress Chaperones* **2001**, *6*, 238. [CrossRef]
18. Zhang, M.; Shen, Z.; Meng, G.; Lu, Y.; Wang, Y. Genome-Wide Analysis of the *Brachypodium Distachyon* (L.) P. Beauv. Hsp90 Gene Family Reveals Molecular Evolution and Expression Profiling under Drought and Salt Stresses. *PLoS ONE* **2017**, *12*, e0189187. [CrossRef]
19. Song, Z.; Pan, F.; Yang, C.; Jia, H.; Jiang, H.; He, F.; Li, N.; Lu, X.; Zhang, H. Genome-Wide Identification and Expression Analysis of HSP90 Gene Family in Nicotiana Tabacum. *BMC Genet.* **2019**, *20*, 35. [CrossRef]
20. Appiah, C.; Yang, Z.-F.; He, J.; Wang, Y.; Zhou, J.; Xu, W.-Z.; Nie, G.; Zhu, Y.-Q. Genome-Wide Identification of Hsp90 Gene Family in Perennial Ryegrass and Expression Analysis under Various Abiotic Stresses. *Plants* **2021**, *10*, 2509. [CrossRef] [PubMed]
21. Song, H.; Zhao, R.; Fan, P.; Wang, X.; Chen, X.; Li, Y. Overexpression of AtHsp90.2, AtHsp90.5 and AtHsp90.7 in Arabidopsis Thaliana Enhances Plant Sensitivity to Salt and Drought Stresses. *Planta* **2009**, *229*, 955–964. [CrossRef] [PubMed]
22. Liu, Y.; Burch-Smith, T.; Schiff, M.; Feng, S.; Dinesh-Kumar, S.P. Molecular Chaperone Hsp90 Associates with Resistance Protein N and Its Signaling Proteins SGT1 and Rar1 to Modulate an Innate Immune Response in Plants. *J. Biol. Chem.* **2004**, *279*, 2101–2108. [CrossRef]
23. Hubert, D.A.; Tornero, P.; Belkhadir, Y.; Krishna, P.; Takahashi, A.; Shirasu, K.; Dangl, J.L. Cytosolic HSP90 Associates with and Modulates the Arabidopsis RPM1 Disease Resistance Protein. *EMBO J.* **2003**, *22*, 5679–5689. [CrossRef] [PubMed]
24. Banilas, G.; Korkas, E.; Englezos, V.; Nisiotou, A.A.; Hatzopoulos, P. Genome-Wide Analysis of the Heat Shock Protein 90 Gene Family in Grapevine (*Vitis Vinifera* L.). *Aust. J. Grape Wine Res.* **2012**, *18*, 29–38. [CrossRef]
25. Samakovli, D.; Roka, L.; Dimopoulou, A.; Plitsi, P.K.; Žukauskait, A.; Georgopoulou, P.; Novák, O.; Milioni, D.; Hatzopoulos, P. HSP90 Affects Root Growth in Arabidopsis by Regulating the Polar Distribution of PIN1. *New Phytol.* **2021**, *231*, 1814–1831. [CrossRef]
26. Baum, B.R. *Oats: Wild and Cultivated. A Monograph of the Genus Avena L. (Poaceae)*; Minister of Supply and Services: Ottawa, ON, Canada, 1977.
27. Fu, J.; Zhang, Y.; Hu, Y.; Zhao, G.; Tang, Y.; Zou, L. Concise Review: Coarse Cereals Exert Multiple Beneficial Effects on Human Health. *Food Chem.* **2020**, *325*, 126761. [CrossRef] [PubMed]
28. Oats | Diseases and Pests, Description, Uses, Propagation. Available online: https://plantvillage.psu.edu/topics/oats/infos (accessed on 10 December 2023).
29. Kamal, N.; Tsardakas Renhuldt, N.; Bentzer, J.; Gundlach, H.; Haberer, G.; Juhász, A.; Lux, T.; Bose, U.; Tye-Din, J.A.; Lang, D.; et al. The Mosaic Oat Genome Gives Insights into a Uniquely Healthy Cereal Crop. *Nature* **2022**, *606*, 113–119. [CrossRef]
30. Bailey, T.L.; Boden, M.; Buske, F.A.; Frith, M.; Grant, C.E.; Clementi, L.; Ren, J.; Li, W.W.; Noble, W.S. MEME SUITE: Tools for Motif Discovery and Searching. *Nucleic Acids Res.* **2009**, *37*, W202–W208. [CrossRef]
31. Wang, Y.; Tang, H.; Debarry, J.D.; Tan, X.; Li, J.; Wang, X.; Lee, T.; Jin, H.; Marler, B.; Guo, H.; et al. MCScanX: A Toolkit for Detection and Evolutionary Analysis of Gene Synteny and Collinearity. *Nucleic Acids Res.* **2012**, *40*, e49. [CrossRef]
32. Rombauts, S.; Déhais, P.; Van Montagu, M.; Rouzé, P. PlantCARE, a Plant Cis-Acting Regulatory Element Database. *Nucleic Acids Res.* **1999**, *27*, 295 296. [CrossRef]
33. Waterhouse, A.; Bertoni, M.; Bienert, S.; Studer, G.; Tauriello, G.; Gumienny, R.; Heer, F.T.; de Beer, T.A.P.; Rempfer, C.; Bordoli, L.; et al. SWISS-MODEL: Homology Modelling of Protein Structures and Complexes. *Nucleic Acids Res.* **2018**, *46*, W296–W303. [CrossRef] [PubMed]
34. Zhang, J.; Li, J.; Liu, B.; Zhang, L.; Chen, J.; Lu, M. Genome-Wide Analysis of the Populus Hsp90 Gene Family Reveals Differential Expression Patterns, Localization, and Heat Stress Responses. *BMC Genom.* **2013**, *14*, 532. [CrossRef] [PubMed]
35. Peng, Y.; Yan, H.; Guo, L.; Deng, C.; Wang, C.; Wang, Y.; Kang, L.; Zhou, P.; Yu, K.; Dong, X.; et al. Reference Genome Assemblies Reveal the Origin and Evolution of Allohexaploid Oat. *Nat. Genet.* **2022**, *54*, 1248–1258. [CrossRef] [PubMed]
36. Zhang, Y.; Zheng, L.; Yun, L.; Ji, L.; Li, G.; Ji, M.; Shi, Y.; Zheng, X. Catalase (CAT) Gene Family in Wheat (*Triticum Aestivum* L.): Evolution, Expression Pattern and Function Analysis. *Int. J. Mol. Sci.* **2022**, *23*, 542. [CrossRef] [PubMed]
37. Delsuc, F.; Brinkmann, H.; Philippe, H. Phylogenomics and the Reconstruction of the Tree of Life. *Nat. Rev. Genet.* **2005**, *6*, 361–375. [CrossRef]
38. Radoeva, T.; Vaddepalli, P.; Zhang, Z.; Weijers, D. Evolution, Initiation, and Diversity in Early Plant Embryogenesis. *Dev. Cell* **2019**, *50*, 533–543. [CrossRef]
39. Pearl, L.H.; Prodromou, C. Structure and Mechanism of the Hsp90 Molecular Chaperone Machinery. *Annu. Rev. Biochem.* **2006**, *75*, 271–294. [CrossRef]

40. Reddy, R.K.; Chaudhary, S.; Patil, P.; Krishna, P. The 90 kDa Heat Shock Protein (Hsp90) Is Expressed throughout Brassica Napus Seed Development and Germination. *Plant Sci.* **1998**, *131*, 131–137. [CrossRef]
41. Ul Haq, S.; Khan, A.; Ali, M.; Khattak, A.M.; Gai, W.-X.; Zhang, H.-X.; Wei, A.-M.; Gong, Z.-H. Heat Shock Proteins: Dynamic Biomolecules to Counter Plant Biotic and Abiotic Stresses. *Int. J. Mol. Sci.* **2019**, *20*, 5321. [CrossRef]
42. Blanc, G.; Wolfe, K.H. Widespread Paleopolyploidy in Model Plant Species Inferred from Age Distributions of Duplicate Genes. *Plant Cell* **2004**, *16*, 1667–1678. [CrossRef]
43. Nakashima, K.; Yamaguchi-Shinozaki, K. ABA Signaling in Stress-Response and Seed Development. *Plant Cell Rep.* **2013**, *32*, 959–970. [CrossRef]
44. O'Sullivan, H. GrainGenes. *Methods Mol. Biol.* **2007**, *406*, 301–314. [CrossRef]
45. Potter, S.C.; Luciani, A.; Eddy, S.R.; Park, Y.; Lopez, R.; Finn, R.D. HMMER Web Server: 2018 Update. *Nucleic Acids Res.* **2018**, *46*, W200–W204. [CrossRef]
46. Horton, P.; Park, K.-J.; Obayashi, T.; Fujita, N.; Harada, H.; Adams-Collier, C.J.; Nakai, K. WoLF PSORT: Protein Localization Predictor. *Nucleic Acids Res.* **2007**, *35*, W585–W587. [CrossRef]
47. Chen, C.; Wu, Y.; Li, J.; Wang, X.; Zeng, Z.; Xu, J.; Liu, Y.; Feng, J.; Chen, H.; He, Y.; et al. TBtools-II: A "One for All, All for One" Bioinformatics Platform for Biological Big-Data Mining. *Mol. Plant* **2023**, *16*, 1733–1742. [CrossRef] [PubMed]
48. Thompson, J.D.; Higgins, D.G.; Gibson, T.J. CLUSTAL W: Improving the Sensitivity of Progressive Multiple Sequence Alignment through Sequence Weighting, Position-Specific Gap Penalties and Weight Matrix Choice. *Nucleic Acids Res.* **1994**, *22*, 4673–4680. [CrossRef] [PubMed]
49. Kumar, S.; Stecher, G.; Li, M.; Knyaz, C.; Tamura, K. MEGA X: Molecular Evolutionary Genetics Analysis across Computing Platforms. *Mol. Biol. Evol.* **2018**, *35*, 1547–1549. [CrossRef] [PubMed]
50. Gu, Z.; Cavalcanti, A.; Chen, F.-C.; Bouman, P.; Li, W.-H. Extent of Gene Duplication in the Genomes of Drosophila, Nematode, and Yeast. *Mol. Biol. Evol.* **2002**, *19*, 256–262. [CrossRef] [PubMed]
51. Zhang, Z. KaKs_Calculator 3.0: Calculating Selective Pressure on Coding and Non-Coding Sequences. *Genom. Proteom. Bioinform.* **2022**, *20*, 536–540. [CrossRef]
52. Emms, D.M.; Kelly, S. OrthoFinder: Phylogenetic Orthology Inference for Comparative Genomics. *Genome Biol.* **2019**, *20*, 238. [CrossRef]
53. Berman, H.M.; Westbrook, J.; Feng, Z.; Gilliland, G.; Bhat, T.N.; Weissig, H.; Shindyalov, I.N.; Bourne, P.E. The Protein Data Bank. *Nucleic Acids Res.* **2000**, *28*, 235–242. [CrossRef]
54. Bhagwat, M.; Aravind, L. PSI-BLAST Tutorial. *Methods Mol. Biol.* **2007**, *395*, 177–186. [CrossRef] [PubMed]
55. Szklarczyk, D.; Gable, A.L.; Lyon, D.; Junge, A.; Wyder, S.; Huerta-Cepas, J.; Simonovic, M.; Doncheva, N.T.; Morris, J.H.; Bork, P.; et al. STRING V11: Protein-Protein Association Networks with Increased Coverage, Supporting Functional Discovery in Genome-Wide Experimental Datasets. *Nucleic Acids Res.* **2019**, *47*, D607–D613. [CrossRef] [PubMed]
56. Shannon, P.; Markiel, A.; Ozier, O.; Baliga, N.S.; Wang, J.T.; Ramage, D.; Amin, N.; Schwikowski, B.; Ideker, T. Cytoscape: A Software Environment for Integrated Models of Biomolecular Interaction Networks. *Genome Res.* **2003**, *13*, 2498–2504. [CrossRef] [PubMed]
57. Yang, Z.; Wang, K.; Aziz, U.; Zhao, C.; Zhang, M. Evaluation of Duplicated Reference Genes for Quantitative Real-Time PCR Analysis in Genome Unknown Hexaploid Oat (*Avena Sativa* L.). *Plant Methods* **2020**, *16*, 138. [CrossRef]
58. Pandey, A.; Khan, M.K.; Hamurcu, M.; Brestic, M.; Topal, A.; Gezgin, S. Insight into the Root Transcriptome of a Boron-Tolerant Triticum Zhukovskyi Genotype Grown under Boron Toxicity. *Agronomy* **2022**, *12*, 2421. [CrossRef]
59. Schmittgen, T.D.; Livak, K.J. Analyzing Real-Time PCR Data by the Comparative C(T) Method. *Nat. Protoc.* **2008**, *3*, 1101–1108. [CrossRef]

International Journal of
Molecular Sciences

Article

Transcriptome Analysis of Diurnal and Nocturnal-Warmed Plants, the Molecular Mechanism Underlying Cold Deacclimation Response in *Deschampsia antarctica*

Dariel López [1], Giovanni Larama [2], Patricia L. Sáez [1] and León A. Bravo [1,*]

[1] Departamento de Ciencias Agronómicas y Recursos Naturales, Facultad de Ciencias Agropecuarias y Medioambiente and Center of Plant, Soil Interactions and Natural Resources Biotechnology, Scientific and Technological Bioresource Nucleus, Universidad de La Frontera, Temuco 4811230, Chile; d.lopez07@ufromail.cl (D.L.); patrisaezd@gmail.com (P.L.S.)

[2] Biocontrol Research Laboratory and Scientific and Technological Bioresource Nucleus, Universidad de La Frontera, Temuco 4811230, Chile; giovanni.larama@ufrontera.cl

* Correspondence: leon.bravo@ufrontera.cl; Tel.: +56-452592821

Abstract: Warming in the Antarctic Peninsula is one of the fastest on earth, and is predicted to become more asymmetric in the near future. Warming has already favored the growth and reproduction of Antarctic plant species, leading to a decrease in their freezing tolerance (deacclimation). Evidence regarding the effects of diurnal and nocturnal warming on freezing tolerance-related gene expression in *D. antarctica* is negligible. We hypothesized that freezing tolerance-related gene (such as CBF-regulon) expression is reduced mainly by nocturnal warming rather than diurnal temperature changes in *D. antarctica*. The present work aimed to determine the effects of diurnal and nocturnal warming on cold deacclimation and its associated gene expression in *D. antarctica*, under laboratory conditions. Fully cold-acclimated plants (8 °C/0 °C), with 16h/8h thermoperiod and photoperiod duration, were assigned to four treatments for 14 days: one control (8 °C/0 °C) and three with different warming conditions (diurnal (14 °C/0 °C), nocturnal (8 °C/6 °C), and diurnal-nocturnal (14 °C/6 °C)). RNA-seq was performed and differential gene expression was analyzed. Nocturnal warming significantly down-regulated the CBF transcription factors expression and associated cold stress response genes and up-regulated photosynthetic and growth promotion genes. Consequently, nocturnal warming has a greater effect than diurnal warming on the cold deacclimation process in *D. antarctica*. The eco-physiological implications are discussed.

Keywords: Antarctic plant; asymmetric warming; CBF; climate change; cold deacclimation; gene expression; RNA-seq

Citation: López, D.; Larama, G.; Sáez, P.L.; Bravo, L.A. Transcriptome Analysis of Diurnal and Nocturnal-Warmed Plants, the Molecular Mechanism Underlying Cold Deacclimation Response in *Deschampsia antarctica. Int. J. Mol. Sci.* **2023**, 24, 11211. https://doi.org/10.3390/ijms241311211

Academic Editor: Martin Bartas

Received: 17 April 2023
Revised: 5 May 2023
Accepted: 7 May 2023
Published: 7 July 2023

1. Introduction

The phenomenon known as climate change, implying shifts in temperature, precipitation, and CO_2 levels among other factors, represents a major threat to agricultural and natural plant populations. Deforestation and fossil fuel burning have increased the concentration of greenhouse gases in the atmosphere, contributing to increases in the global average surface temperature of 1.0–3.7 °C during the past century [1,2]. It has been estimated that global warming will increase the frequency and intensity of stochastic extreme temperature events (freezing or heat waves), with changes in precipitation patterns during this century [2]. In addition, an asymmetric warming trend is expected, with a greater increase in minimum temperatures (nocturnal) with respect to maximum temperatures (diurnal) [2,3]. Furthermore, the warming process has been amplified in polar and high-mountain regions [4].

The climate of the Antarctic Peninsula and its associated islands has shown increases in the average annual temperature ca. 3.7 °C century^{-1} [5]. With longer and warmer growing seasons for the plant species that inhabit the Antarctic Peninsula, reproductive capacity and population sizes have increased [6–9]. Despite the warming trend, Antarctica is still among the coldest regions on the planet [10], and Antarctic plants must continually cope with freezing temperatures, even during the growing season [11,12]. This implies that Antarctic plants should maintain their freezing tolerance even during the growing season (summer).

Deschampsia antarctica Desv. (Poaceae), the only monocot that has naturally colonized Antarctica, is used to dealing with occasional daily warm temperatures since leaf temperatures can reach up to 20 °C on sunny days, even when the air temperature does not exceed 4 °C [13]. Although the rise of diurnal temperatures appears to have no effect on the freezing tolerance in this plant species [14], the foreseeable increase in the minimum nocturnal temperature (nocturnal warming) could lead to cold deacclimation [14].

Cold deacclimation entails the partial or total loss of previously acquired freezing tolerance traits [15], and is not just a passive reversal of cold acclimation, but rather a genetically and functionally distinct process [16,17]. Even though various studies have analyzed the cold deacclimation profile after a temperature increase in several species (*Arabidopsis thaliana*, *Brassica napus*, *Hordeum vulgare*, *Rhododendron anthopogon*, among others [17–20]), and identified up- and down-regulated genes during this process, they have not differentiated between the effects of diurnal or nocturnal temperature increments. In addition, the cold deacclimation dynamic could change among species [21], as well as the gene expression associated with this process.

The 'C-repeat binding factor' (CBF/DREB1) transcription factors are key in the regulation of a group of genes important in plants' freezing tolerance [22–24]. As plants should not be expressing their cryoprotective mechanisms during cold deacclimation, it is expected that the CBF genes would be down-regulated. However, even though some of the physiological and molecular mechanisms of freezing tolerance seem to be similar in the main angiosperm clades, the genes and genetic pathways can vary widely among species [25]. For example, *A. thaliana* has 4 CBF genes, but only 3 respond to low temperatures [26], while in pooids, the number of CBF genes can exceed 20, with variable numbers in different species [27]. Thus, given the complexity of this pathway in cereals and grasses, more studies are needed to understand its functioning and interaction with variable environmental factors.

The paradox between improved growth vs. cold acclimation faces *D. antarctica* to a likely differential response against diurnal and nocturnal warming, this prompts the question about what happens with the expression of freezing tolerance-related genes. We propose that nocturnal warming down-regulates the expression of freezing tolerance-related genes in *D. antarctica* plants, while the decrease in temperatures close to 0 °C during the night is essential for transient CBF and associated gene expression, allowing *D. antarctica* plants to maintain freezing tolerance, even with a daytime temperature increase. Differentially expressed genes during the day and night experimental warming were analyzed to assess the differential effect of nocturnal warming in cold deacclimation regulation.

2. Results

2.1. RNA-Seq Analysis of Nocturnal Expressed Genes in D. antarctica

RNA-seq transcriptome analysis was carried out using RNA from *D. antarctica* leaves collected 2 h after the beginning of the nocturnal period. Three biological replicates for each treatment with and without warming were sent to the Illumina Novaseq 6000 platform for deep sequencing. The raw sequence data can be found at BioProject ID PRJNA941125.

A total of 275.3 million paired-end reads were obtained after removing the low-quality sequence and adaptor sequence. The percentage of Q30 bases was above 95.99%, while the mapping rates of all 12 libraries ranged from 86.2% to 87.6%, regarding high-quality reads (Supplementary Table S1).

After error correction and redundancy removal, de novo transcriptome assembly resulted in 193,333 transcripts corresponding to 84,162 genes, with an average GC percentage of 47.65%. The median contig length was 1127 bp, while the average contig length was 1533 bp (Supplementary Table S2). Based on benchmarking universal single-copy ortholog (BUSCO) analysis, 1549 (95.9%) of the 1614 expected embryophyte genes were identified as complete after removing redundancy, and only 4.1% were fragmented or missing (Supplementary Table S3).

2.2. Differentially Expressed Genes (DEGs) of Warmed vs. Cold-Acclimated Plants

To analyze the differences of the *D. antarctica* transcription levels during the nocturnal period, FPKM was used to calculate the amount of differentially expressed genes. The fold change was calculated based on a comparison of the FPKM between the cold-acclimated (CA) plants and warming treatments. The genes were considered differentially expressed when the fold change of the gene expression level was at least a two-fold change and Chi-square test ($p < 0.05$) FDR < 0.01.

From the total of genes found in the transcriptome, only 7.77% (6537 genes) were differentially expressed 2 h after the drop in temperatures during the night period, in warmed plants. A total of 871, 4692 and 2403 DEGs were identified in plants with nocturnal warming (NW+), diurnal warming (DW+), and diurnal-nocturnal warming (DNW+), respectively, at 2 h after the temperature drop in the nocturnal period (Figure 1). Among these DEGs, 263 were down-regulated in plants with nocturnal warming, of which 52.1% was common for NW+ and DNW+ treatments only (Figure 2A), while 190 were up-regulated in plants with nocturnal warming, and 91.1% were also common for NW+ and DNW+ treatments (Figure 2B).

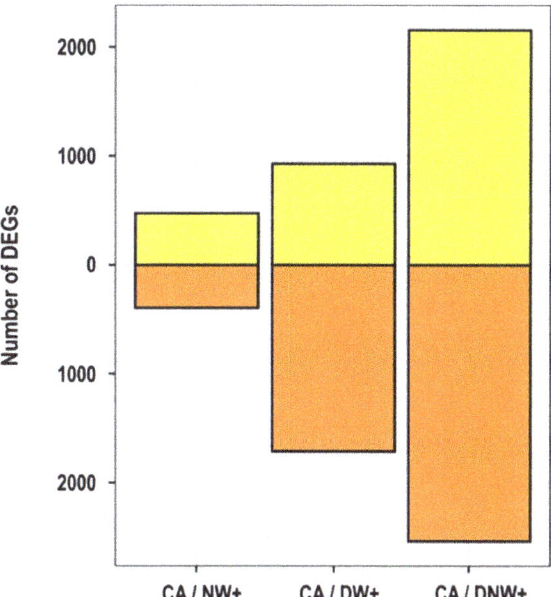

Figure 1. Number of differentially expressed genes (DEGs). The amounts of differentially expressed genes in warming treatments, up-regulated (light yellow) and down-regulated (orange) in relation to the control condition of cold-acclimated plants, 8 °C/0 °C (CA). Warming treatments received the following thermoperiods: 8 °C/6 °C, nocturnal warming (NW+); 14 °C/0 °C, diurnal warming (DW+); and 14 °C/6 °C, diurnal-nocturnal warming (DNW+). The plants were cultivated in chambers under controlled conditions at a photoperiod and thermoperiod of 18 h/6 h duration.

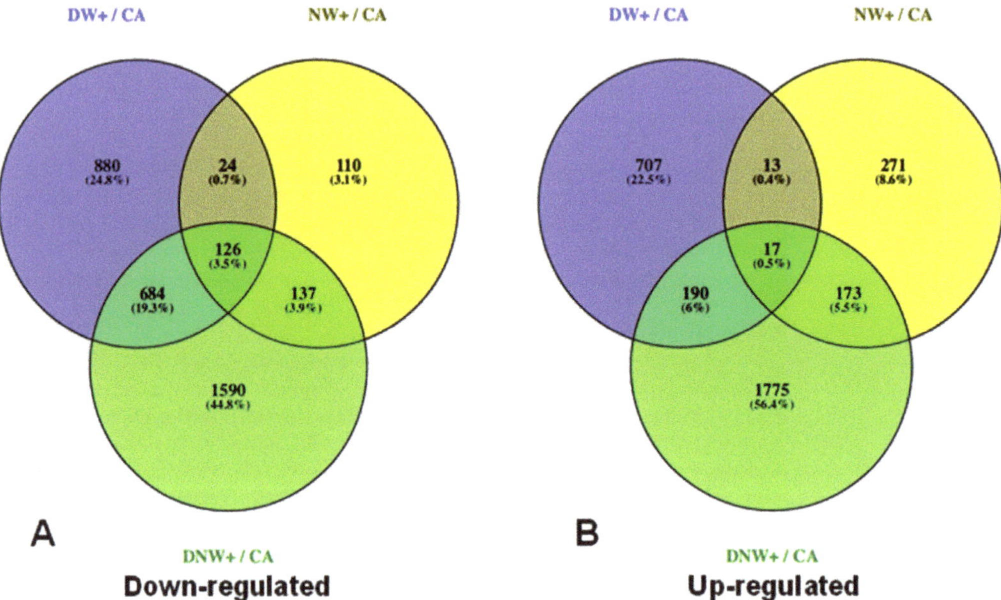

Figure 2. Venn diagram of the number of differentially expressed genes (DEGs). The number and percentage of down-regulated genes (**A**) and up-regulated genes (**B**) in warming treatments in relation to the control condition of cold-acclimated plants, 8 °C/0 °C (CA). Warming treatments received the following thermoperiods: 8 °C/6 °C, nocturnal warming (NW+); 14 °C/0 °C, diurnal warming (DW+); and 14 °C/6 °C, diurnal-nocturnal warming (DNW+). The plants were cultivated in chambers under controlled conditions at a photoperiod and thermoperiod of 18 h/6 h duration.

2.3. Analysis of Co-Expressed DEGs in Response to Nocturnal Temperatures

The analysis of principal components (PCA) and self-organized maps (SOM), together with the grouping strategy used, allowed for the creation of eight super-nodes. Among them, two super-nodes show co-expressed genes profiles related to the presence or absence of nocturnal warming (Figure S1). The gene ontology enrichment analysis in these super-nodes showed that nocturnal-warmed plants presented a significant abundance of genes related to photosynthesis, chloroplast structures, and hormonal responses 2 h after the nocturnal period began (Figure 3A). Meanwhile, the plants without nocturnal warming presented a significant enrichment in the genes related to transporter activity and membrane structure, UDP- glycosyltransferase activity, and response to the cold (Figure 3B).

Likewise, the expression patterns of 1137 genes belonging to these 2 super-nodes show abiotic stress-related genes. Those related to cold tolerance are repressed in treatments with nocturnal warming. Among them, noteworthy genes include the 'C-repeat binding factor' (CBF or CBF/DREB1), sugar metabolism, dehydrins, and other protection-related proteins (Figure 4). Moreover, genes related to photosynthesis and vegetative growth, such as photosystems and antenna complex components, photosynthetic enzymes, and auxin regulation, among others, were induced in treatments with nocturnal warming (Figure 4).

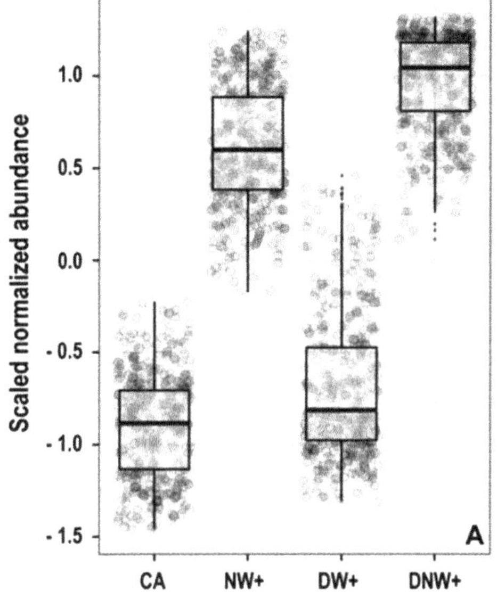

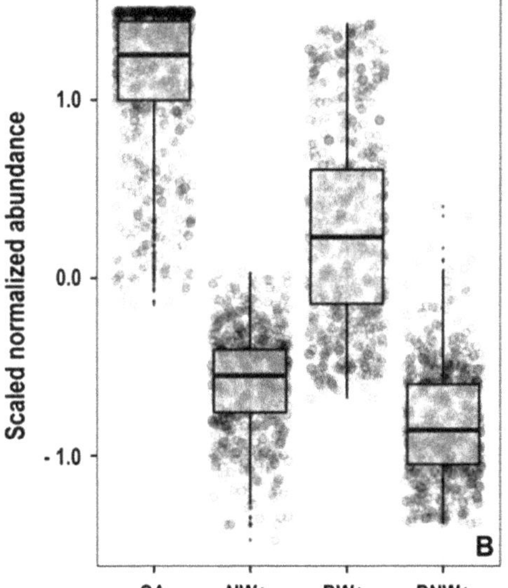

Figure 3. Clusters obtained from principal component analysis and self-organizing maps (PCA-SOM). The nodes group transcripts up-regulated (**A**) and down-regulated (**B**) by nocturnal warming, and the enriched gene ontology categories with their corresponding *p*-values are shown. The boxes represent the interquartile range (middle 50% of the data), its midline indicates the median value, and the whiskers indicate the ranges of the upper and lower 25%, excluding outliers. Corresponding to treatments with the following thermoperiods (day/night): 8 °C/0 °C, plants acclimated to cold (AF); 8 °C/6 °C, night heating (CN); 14 °C/0 °C, daytime warming (CD); and 14 °C/6 °C, daytime-night warming (CDN). The plants were cultivated in chambers under controlled conditions at a photoperiod and thermoperiod of 18 h/6 h duration.

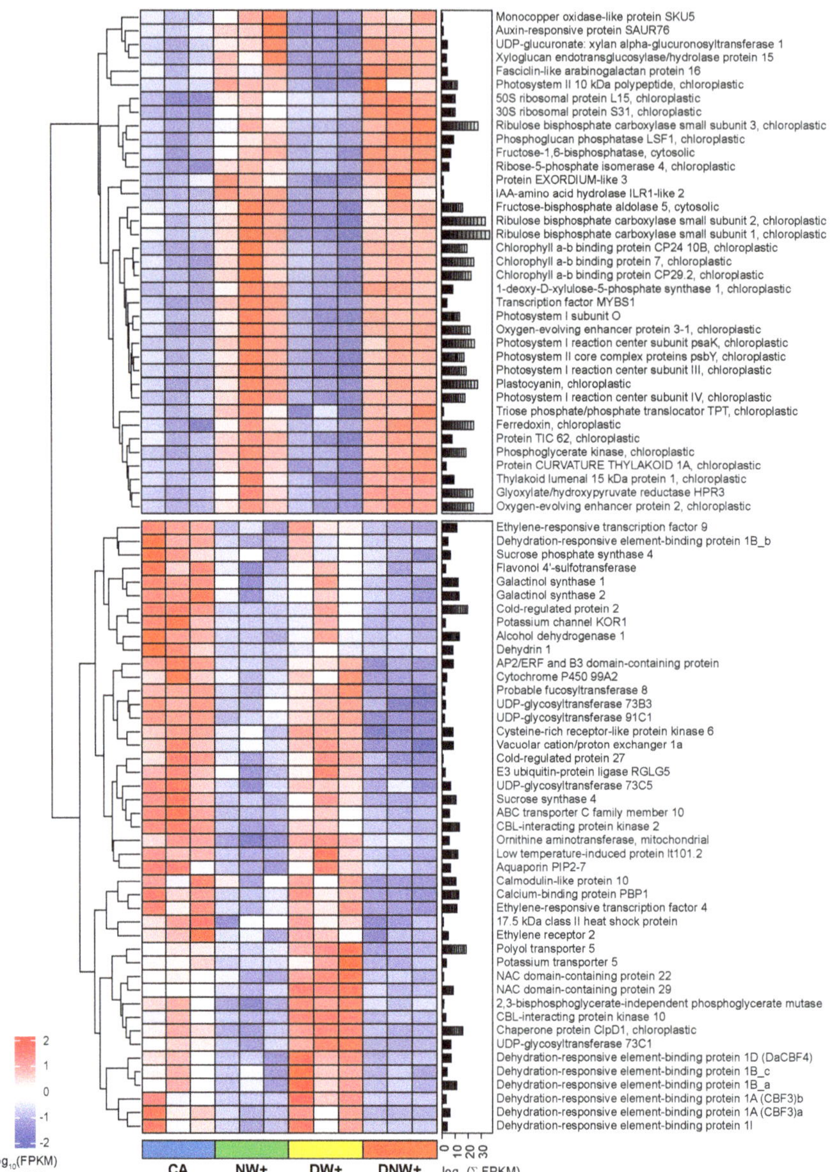

Figure 4. Heat-map of co-expressed genes' response to nocturnal warming. The genes' relative abundance is presented by the z-score of the $\log_{10}$ FPKM, after 2 h of temperature drop during the nocturnal period, of plants with a cold acclimation of 8 °C/0 °C (CA); nocturnal warming of 8 °C/6 °C (NW+); diurnal warming of 14 °C/0 °C (DW+); and diurnal-nocturnal warming of 14 °C/6 °C (DNW+). The plants were cultivated in chambers under controlled conditions at a photoperiod and thermoperiod of 18 h/6 h duration.

2.4. Validation of RNA-Seq Gene Expression by qRT-PCR

The reliability of the transcriptome data was evaluated by quantitative real-time PCR (qRT-PCR) technology. The correlation between the $\log_2$ fold change of FPKM and the relative abundance determined by qRT-PCR was highly significant ($R^2 = 0.92$; $p < 0.05$) for the 16 genes selected (Figure S2).

2.5. Nocturnal Kinetics of Interest DEGs

The relative gene expression during the nocturnal period was assessed with respect to the expression values average measured in CA plants, just before the nocturnal period started. The transcripts corresponding to 'Dehydration-responsive element-binding protein 1A' (DRE1A/CBF3), 'Dehydration-responsive element-binding protein 1D' (DRE1D/CBF4), 'Dehydrin 1' (DHN1), 'E3 ubiquitin-protein ligase RGLG5' (RGLG5), 'Ornithine amino-transferase' (OAT), 'Sucrose phosphate synthase 4' (SPS4F), 'Sucrose synthase 4' (SUS4), 'UDP-glycosyltransferase 73B3' (U73B3), and 'Vacuolar cation/proton exchanger 1' (CAX1) presented a significantly lower expression in the treatments with nocturnal warming, with respect to the CA plants, 2 h after beginning the night period (Figure 5A–I). Likewise, the aforementioned transcripts also presented significantly lower expressions in the treatments with nocturnal warming, compared to the plants with DW+, 2 h after beginning the nocturnal period, with the exception of SPS4F, (Figure 5A–I). However, in the case of SPS4F, both the CA and DW+ plants significantly increased their expression at 2 h with respect to their starting values (Figure 5I).

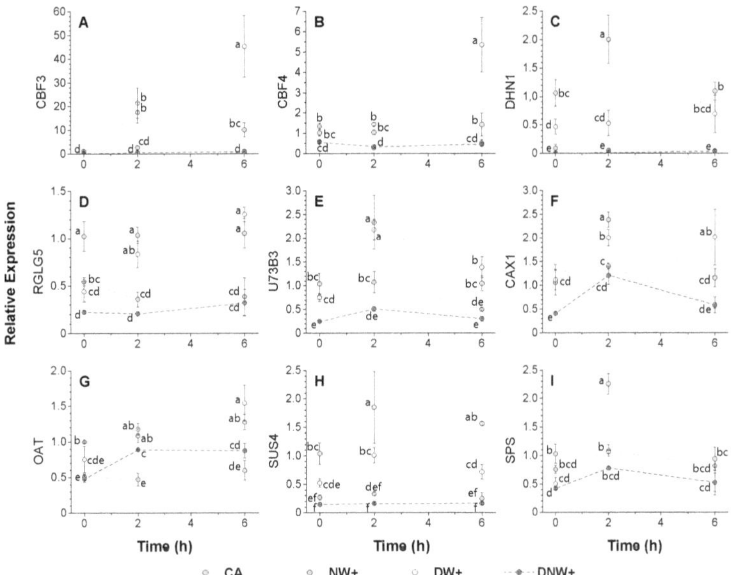

Figure 5. Relative expression of genes related to freezing tolerance. The variation on gene relative expression was evaluated by qRT-PCR. Values represent the fold change mean ± standard error (n = 3), in relation to cold-acclimated (CA) plants 8 °C/0 °C, at the beginning of the nocturnal period (0 h). Transcripts representing variation included: (**A**) 'Dehydration-responsive element-binding protein 1A' (DRE1A/CBF3), (**B**) 'Dehydration-responsive element-binding protein 1D' (DRE1D/CBF4), (**C**) 'Dehydrin 1' (DHN1), (**D**) 'E3 ubiquitin-protein ligase RGLG5' (RGLG5), (**E**) 'UDP-glycosyltransferase 73B3' (U73B3), (**F**) 'Vacuolar cation/proton exchanger 1' (CAX1), (**G**) 'Ornithine aminotransferase' (OAT), (**H**) 'Sucrose synthase 4' (SUS4), and (**I**) 'Sucrose phosphate synthase 4' (SPS4F). Warming treatments correspond to: nocturnal warming, 8 °C/6 °C (NW+); diurnal warming, 14 °C/0 °C (DW+); and diurnal-nocturnal warming, 14 °C/6 °C (DNW+), with 18 h/6 h photoperiod and thermoperiods duration. Significant differences among factors are shown as different lower cases ($p < 0.05$).

The expression of DRE1A/CBF3 was highly up-regulated during the night in CA and DW+ plants, but this was not modified significantly in the treatments with nocturnal warming (Figure 5A). On the other hand, the expression of DRE1D/CBF4 and RGLG5 remained stable overnight in CA plants, while their values increased in DW+ plants. Conversely, the above-mentioned genes in NW+ exhibited a reduced expression (only significantly for CBF4) to DNW+ similar values at 2 h, and remained low until the nocturnal period ended (Figure 5B,D). At the same time, DHN1 and SUS4 were the most down-regulated in the nocturnal warming treatments among the analyzed genes (Figure 5C,H).

In contrast, the transcripts corresponding to Fructose-bisphosphate aldolase 5 (FAB5), IAA-amino acid hydrolase ILR1-like 2 (ILL2), Ribulose bisphosphate carboxylase small subunit 1 (RBCS1), Oxygen-evolving enhancer protein 2 (PSBP2), Triose phosphate/phosphate translocator (TPT), and 1-deoxy-D-xylulose-5-phosphate synthase 1 (DXS1) significantly up-regulated their expression in treatments with nocturnal warming, at 2 h at the start of the nocturnal period, compared to CA and DW+ plants, (Figure 6A–F). Highlighting that, ILL2, DXS1, and FBA5 presented a significantly higher up-regulation in NW+ plants (Figure 6A,B,E). The transcripts corresponding to ILL2 and TPT1 maintained their significantly higher expression until the overnight period ended (Figure 6A,C). The ILL2 expression presented the higher increment during the night in plants with nocturnal warming (Figure 6A), with 9 time increments in NW+ plants at 2 h after the nocturnal period began, compared to its initial value.

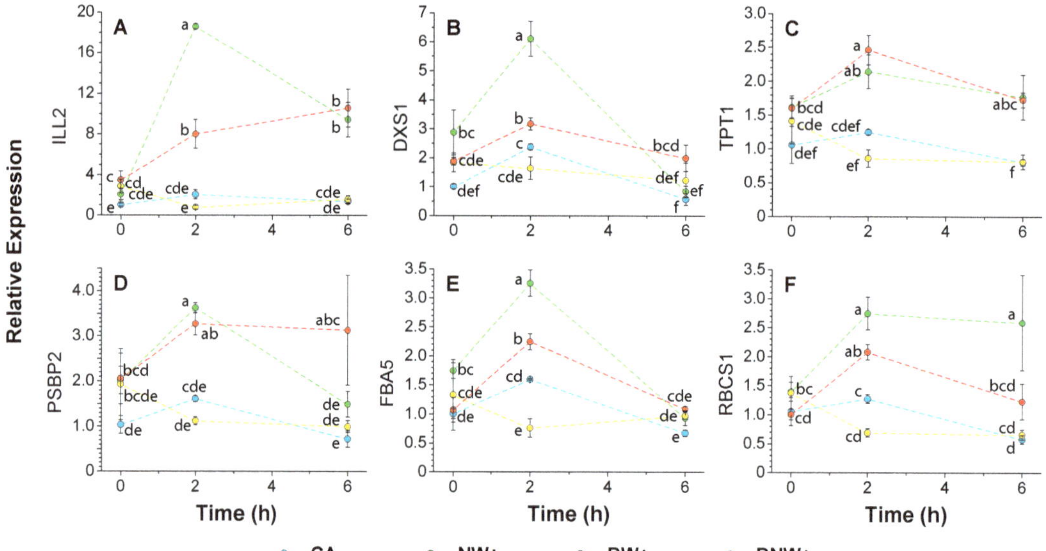

Figure 6. Relative expression of up-regulated genes by nocturnal warming. The variation in relative gene expression was evaluated by qRT-PCR. Values represent the fold change mean ± standard error (n = 3), in relation to cold-acclimated (CA) plants 8 °C/0 °C, at the beginning of the nocturnal period (0 h). Transcript variation shown includes: (**A**) 'IAA-amino acid hydrolase ILR1-like 2' (ILL2), (**B**) '1-deoxy-D-xylulose-5-phosphate synthase 1' (DXS1), (**C**) 'Triose phosphate/phosphate translocator' (TPT), (**D**) 'Oxygen-evolving enhancer protein 2' (PSBP2), (**E**) 'Fructose-bisphosphate aldolase 5' (FAB5), and (**F**) 'Ribulose bisphosphate carboxylase small subunit 1' (RBCS1). Warming treatments correspond to: nocturnal warming, 8 °C/6 °C (NW+); diurnal warming, 14 °C/0 °C (DW+); and diurnal-nocturnal warming, 14 °C/6 °C (DNW+), with 18 h/6 h photoperiod and thermoperiods duration. Significant differences among factors are shown as different lower cases ($p < 0.05$).

3. Discussion

In the current global warming context, with more frequent drastic and stochastic temperature changes, such as heat waves and spring frosts, the early cold deacclimation of plants becomes more relevant, since leaf tissues or the whole plant may become vulnerable to subsequent freezing events, threatening their survival [21,28]. Our findings suggest that only nocturnal warming is capable of down-regulating freezing tolerant-related genes and up-regulating growth promotion and carbon assimilation-related genes, inducing cold deacclimation in *D. antarctica*.

Among the freezing tolerance-related genes that were only repressed by nocturnal warming were the CBF transcription factors, which regulate the expression of a group of genes essential in plant freezing tolerance, the CBF-regulon [22–24,29]. In particular, the 'Dehydration-responsive element-binding protein 1A' (DREB1A/CBF3) transcription factor was the most induced overnight, increasing its expression 17.6 times in CA plants 2 h into the nocturnal period, and up to 2457 times in plants with CD at the end of the night. This could suggest that DREB1A/CBF3 has a very important role in maintaining freezing tolerance in plants with nocturnal temperatures below the cold deacclimation threshold, even when CBF transcription factors show a high functional redundancy [23,24]. It also has been postulated that the expression of CBFs needs to be transient, and their transcripts return to basal levels a few hours after their induction at low temperatures [30]. In accordance with the above, the treatment with CD, which presented the highest levels of transcription for most of the CBF analyzed, also presented the highest levels of coding transcripts for the 14-3-3 protein (results not shown), which has been postulated to function as an intermediate in the CBFs' ubiquitination modification for their subsequent degradation via the proteasome [30].

The CBF-regulon controls the expression of a diverse group of genes, and among them are other transcription factors, channel proteins and membrane transporters, enzymes related to sugar and proline metabolism, etc. [22,23,29]. Among these genes is the 'Vacuolar cation/proton exchanger 1' (CAX1), which has been reported to increase expression in cold-acclimated plants, such as the CA and DW+ plants in this study, and is key in the regulation of intracellular Ca^{2+} [31], one of the second messengers responsible for transmitting the cold signaling [32,33]. Some gene isoforms coding for sucrose synthase (SUS) enzymes [24,34] are also part of this regulon, and they degrade the sucrose molecule yielding UDP-glucose through a reversible reaction, allowing for its incorporation into more complex molecules synthesis, such as cellulose during the thickening of cell walls [35], or moving onto catabolic pathways as a result of an increased energy demand in the face of abiotic stress [36]. Sucrose synthase has also been found to have a predominant role in sucrose loading and unloading in phloem transport [37,38]. Both situations could be occurring in CA and DW+ plants, which would explain the induction of SUS4 in these treatments, since the cell walls' rearrangement is an expected process in *D. antarctica* cold-acclimated plants [39], and cold acclimation is also an energetically demanding process, needing to mobilize sucrose to sinks organs.

Dehydrins gene expression are also regulated by CBF transcription factors [23,24]. This agrees with dehydrins transcript's drastic reduction in plants with nocturnal warming, and with their protein levels reduction during the cold deacclimation of *D. antarctica* and other pooids (*Poa annua, Agrostis stolonifera*, and *Cynodon* spp.) [14,40–43]. Dehydrins are associated with freezing tolerance due to their ability to stabilize cell membranes, preventing the lipid bilayer transition to the hexagonal phase at freezing temperatures, as a result of the lipid interaction in a medium with less polarity, resulting leakage of electrolytes, and other essential cytoplasm components [44].

Proline accumulation has also been associated with cold acclimation and tolerance to low temperatures in *Arabidopsis thaliana, Chrysanthemum dichrum*, and *Lolium perenne*, among others [45–48]. In this sense, it has been described that the gene coding for 'δ-1-pyrroline-5-carboxylate synthase 2' (P5CS2), a key enzyme for proline synthesis, is regulated by CBFs [23]. However, the P5CS2 gene reduced its expression in all warming treatments (results not shown), while the 'Ornithine aminotransferase' (OAT) gene reduced its expression in plants that received nocturnal warming. The mitochondrial enzyme Ornithine aminotransferase is part of an alternative proline synthesis pathway usually activated by abiotic stress, dehydration, and oxidative stress mainly [49,50]. The OAT enzyme transforms ornithine into glutamyl-5-semi-aldehyde (GSA) for its subsequent conversion to pyrroline-5-carboxylate (P5C), which is exported to the cytoplasm and continues its conversion to proline [49]. Although the OAT genes' expression is not significantly affected during *Chrysanthemum dichrum* cold acclimation [47], its expression was reduced in *D. antarctica* cold-deacclimated plants. This could be related to the possible existing interspecific differences, but also reaffirms the postulate that cold deacclimation is more than a passive regression from a previous cold-acclimated state [21].

Both tissue dehydration and oxidative stress can be the result of low temperature exposure, since the freezing of apoplastic fluid results in severe cell dehydration, the formation of ice crystals that cause cell membrane rupture [51], and the release of electrolytes and free radicals [52]. In this sense, the synthesis of flavonols, flavonoids, and phenylpropanoids has been proposed in *D. antarctica* as a way to increase the antioxidant capacity of cold-acclimated plants [39]. Our results confirm that 'flavonol 4'-sulfotransferase' and various 'UDP-glycosyltransferase' genes (UGT73B3, UGT73C1, UGT73C5, UGT91C1) that encode enzymes related to these antioxidant molecules synthesis pathways [53–57] were co-expressed together in freezing tolerant plants (CA and DW+), while their expression was down-regulated during cold deacclimation as result of nocturnal warming.

Another gene whose expression is modified by the cold is 'E3 ubiquitin-protein ligase RGLG5' (RGLG5), which together with RGLG1, is responsible for phosphatase 2CA (PP2CA) ubiquitination and its degradation [58], and PP2CA is a recognized repressor of the hormonal abscisic acid (ABA) synthesis pathway [58,59]. Our results show that RGLG5 transcripts were down-regulated in cold-deacclimated plants, while their expression was higher and stable in CA plants, and it gradually increased overnight in DW+ plants. This could imply a reduction in *D. antarctica* leaf tissue ABA concentration in cold-deacclimated plants by nocturnal warming, while the ABA concentration of the leaf tissue from DW+ plants could be similar to that of cold-acclimated plants at the end of the night. However, more studies are needed to fully validate this idea. In accordance with the previous approach, although the CBF-regulon expression corresponds to an ABA-independent pathway in response to the cold [60], its expression, as well the expression of other cold-regulated genes (COR), requires ABA presence to potentiate its induction by the cold [61]. This has been related to the presence in the promoter regions of the freezing tolerance genes of the CRT/DRE elements, and the sequences called 'abscisic acid-responsive element' (ABRE) [60].

The CBFs' expression is not only related to the induction of freeze tolerance genes, but also to the repression of a group of genes related to plant growth [24]. *Arabidopsis thaliana* plants overexpressing one or more CBF genes exhibit constitutive freezing tolerance, slow growth, and a dwarf phenotype [62]. In addition, rice plants overexpressing the DaCBF4 gene presented a greater cold and drought tolerance than wild type [63], favoring the up-regulation of dehydrins and other LEA proteins, while down-regulating the expression of several proteins related to photosynthesis such as Chlorophyll A-B binding protein, 10 kDa chloroplast precursor polypeptide of photosystem II, 44 kDa reaction center protein of photosystem II, and the D2 protein of photosystem II [63]. Similar results were found in our study of plants that maintained freezing tolerance compared to those that were cold-deacclimated by nocturnal warming. This could suggest that DaCBF4 and other DaCBFs directly repress or promote the transcription of other repressors over a photosynthesis-related group of genes, given the functional redundancy found among CBFs genes [23,24].

Even though the transcription of genes related to photosynthetic activity does not necessarily imply a higher photosynthetic rate, it has been reported that night-warmed plants exhibited an increase in CO_2 assimilation during the day [64,65]. Additionally, the increased expression of carbohydrate metabolism-related genes, such as the triose phosphate/phosphate translocator and Fructose-bisphosphate aldolase 5 in plants with nocturnal warming, suggests that a remobilization process of previously accumulated photo-assimilates occurs at night to obtain energy and/or to be used as intermediates in biosynthetic processes, as proposed by Turnbull et al. [64]. This nocturnal remobilization of photo-assimilates does not seem to be directed towards sucrose synthesis, since the expression of sucrose phosphate synthase transcripts did not vary throughout the night in plants with nocturnal warming. Previous results in *D. antarctica* plants under experimental laboratory conditions with nocturnal warming show a reduction in the ratio between respiration and carbon assimilation, mainly due to a greater increase in CO_2 assimilation [65], but this increase in CO_2 assimilation is not reflected in the accumulation of soluble carbohydrates, since there is also a reduction in the amount of sucrose accumulated [14]. All the above suggests that the accumulated carbohydrates and the recently assimilated CO_2 were used in other biological processes, potentially including vegetative growth. This idea is consistent with the increased expression of genes related to the hormonal response in plants with nocturnal warming, such as 'IAA-amino acid hydrolase ILR1-like 2', which allows for the release of previously accumulated auxins in its active form, promoting plant growth [66,67].

The relationship between the consumption of accumulated soluble carbohydrates, with the activation of photosynthetic metabolism and vegetative growth, has been previously associated with the cold deacclimation process [16,21,68]. Usually, cold deacclimation occurs during the spring season, and for some species, it could be a passive process [17]. However, *D. antarctica* plants should maintain their freezing tolerance for the whole year, since they need to deal with freezing temperatures during the growing season (summer) [11,12]. Therefore, when cold deacclimation occurs in *D. antarctica* (at least 7 °C increase in LT_{50}), it was promoted by nocturnal warming [14], and this should be an active process, as was also proposed by Wójcik-Jagła et al. [17] for winter barley submitted to warm periods during the winter season. This active cold deacclimation process involves a complex gene regulation that was activated by nocturnal warming only, as has been shown in this laboratory study. Even *D. antarctica* plants cold deacclimated by nocturnal warming maintain a certain freezing resistance that prevents serious damage by the Antarctic spring and summer freezing events [14]. However, it is not possible to predict the effect that the vegetative growth promotion by nocturnal warming should have on other physiological processes, such as reproduction and seed maturity. While most of the researchers found that nocturnal warming reduces the reproductive success of cereals [69], others report opposite results [70]. Moreover, *D. antarctica* plants in the field are exposed to other variables beside temperature that can influence their physiological response, such as wind intensity, variable frequency, nutrient availability, and the intensity of freezing events, among others. Therefore, this raises questions about the relative contribution of nocturnal warming to induce the cold deacclimation process of *D. antarctica* plants under field conditions.

4. Materials and Methods

4.1. Plant Material and Warming Treatments

Deschampsia antarctica Desv. plants were collected near the Henryk Arctowski Polish Scientific Station, King George Island, Maritime Antarctic (62°09′41″ S–58°28′10″ W) during the 2018 growing season. Three individual plants (biological replicates) were vegetatively propagated, as previously described by Bravo et al. [71], in 400 mL pots with soil mixture 3:2:1 (vegetal soil:vermiculite:peat), with 18 h/6 h of light/dark photoperiod, in a controlled climate room at 13 °C constant temperature. These plants (four clones for each biological replicate) were cold-acclimated during 28 days with a gradual temperature reduction in the 18 h/6 h day/night thermoperiod, as described by López et al. [14]. Then, groups of pots were transferred to four thermoperiods treatments that were established with the following temperature regimes: (CA) maintained at 8 °C/0 °C day/night, the control treatment where plants remain in the final cold acclimation condition; (NW+) only nocturnal warming at 8 °C/6 °C day/night; (DW+) only diurnal warming at 14 °C/0 °C day/night; and (DNW+), both diurnal and nocturnal warming at 14 °C/6 °C day/night. Each treatment contained clones of the 3 biological replicates. The maximum and minimum temperatures of the above-mentioned thermoperiods were selected based on microenvironmental data recorded in the field, near the Henryk Arctowski Polish Antarctic Station, King George Island. It has been reported that an increase of 6 °C in the minimum nocturnal temperature leads to the cold deacclimation of *D. antarctica*, unlike the same increase in the maximum daytime temperature [14]. The plants remained in these conditions for 14 days.

During the fourteenth night, the plants were kept at the maximum temperature of their corresponding thermoperiod until the end of the diurnal photoperiod. Samples were collected from 3 plants for each treatment during the night. First, just before the nocturnal period began, which was considered time 0 h, the initial state before the beginning of the night and the temperature dropped. Next, the plants were moved into dark culture chambers with their corresponding nocturnal thermoperiod of each treatment (0 °C CA and DW+; 6 °C NW+ and DNW+). After 2 h, when the expression of the CBF genes should be the maximum [26], the same plants were sampled again, and also at 6 h, just before the diurnal photoperiod began and the temperature rose, in order to analyze the effect of nocturnal warming on CBFs and freezing tolerant-related genes.

4.2. Total RNA Extraction and Quantification

Total RNA extraction and purification was performed using a RNeasy ®® Plant Mini Kit (QIAGEN), according to the manufacturer's recommendations. Total RNA concentration was determined using an Infinite M200 NanoQuant spectrophotometer measured at 260 nm, and the A260/A280 absorbance ratio was used to estimate its purity.

Additionally, from extracted RNA, cDNA synthesis was performed with the commercial AffinityScript qRT-PCR kit (Stratagene, Cedar Creek, TX, USA), following the manufacturer's instructions. Reverse transcription was performed from 1 µg of total RNA, using oligos-dT and random primers as primers in a final volume of 20 µL. At the end of the reaction, the cDNA in solution was diluted in a 1:10 ratio with nuclease-free water, before being aliquoted and stored at −80 °C until use.

4.3. Sequencing and De Novo Transcriptome Assembly

A total of twelve samples were sequenced in a Novaseq 6000 platform, for 150 cycles in the paired-end mode at Novogene facilities. The resulting fastq files were processed using Trimmomatic v0.38 [72] to remove residual adapters and low-quality sequences. These high-quality reads were used to construct a de novo assembly of a *Deschampsia antarctica* transcriptome using Trinity v2.12 [73].

4.4. Transcriptome Assessment and Abundance Estimation

The raw assembly was clustered at 95% similarity using CD-HIT [74], and the completeness was evaluated using BUSCO v5 (Simao et al., 2015), which queried a collection of proteins (Embryophyta odb10) to the assembled transcripts by similarity. The relative abundance estimation was calculated using RSEM v1.2.26 [75].

4.5. Differential Gene Expression and Co-Expression Analysis

Changes in relative RNA abundance were analyzed with Bioconductor package DESeq2 [76] in the R statistical environment. The significance of the changes in gene expression were judged using a False Discovery Rate (FDR) of less than 0.05 and a minimum fold change (FC) of 2 as the threshold. The expression patterns in the differentially expressed genes were clustered using dimensionality reduction with Principal Component Analysis (PCA), followed by a 8 × 8 Self-Organizing Map (SOM) using one hundred training iterations, as implemented in the Kohonen package in R [77]. The resulting SOM nodes were grouped using a hierarchical clustering approach with a branch length cutoff of 2.2, resulting in 9 super-nodes. The abundance patterns of each super-node were visualized in a boxplot to associate their accumulation with any treatment.

4.6. Transcriptome Annotation and Enrichment Test

The resulting genes were compared by homology into the UniProt/SwissProtKB database using BLAST+ with an e-value of 1×10^{-10} as the threshold. The ontology assignment was performed using the PANTHER (version 2020_4) classification system [78], from EMBL. Each super-node was analyzed for the enrichment of gene ontologies with a false discovery rate (FDR) of 0.05 using GOSeq [79].

4.7. Quantitative Real-Time (qRT-PCR) Analysis

The expression patterns of DEGs were analyzed by qRT-PCR, performed in a Stratagene AriaMX thermocycler (Agilent, Santa Clara, CA, USA) using the previously obtained cDNA and Brilliant II SYBR Green QPCR Master mix reagent (Stratagene, Cedar Creek, TX, USA), according to the manufacturer's instructions. The gene-specific primers (Table S4) used for the qRT-PCR analysis were designed using the Primer-BLAST tool (http://www.ncbi.nlm.nih.gov/tools/primer-blast accessed on 12 October 2022). The genes Polyadenylate-binding protein2 (PAB2) and Eukaryotic translation initiation factor (IF4E4) were used as internal references to normalize the expression data. The relative expression levels were calculated according to the method proposed by Pfaffl [80], based on the gene primer efficiency of the $e^{-\Delta\Delta CT}$ (cycle threshold). To validate the repeatability of the RNA-seq data, 16 DEGs were selected for verification by qRT-PCR, using the cDNA obtained from the same sample that was sent to the sequence. The relative expression was estimated with respect to their values in the control treatment (CA), 2 h into the nocturnal period. In contrast, the kinetic variation of genes expression during the night was evaluated with respect to their values in the control treatment (CA) at 0 h just before the nocturnal temperature began.

5. Conclusions

The evidence obtained in the present study confirms that nocturnal warming is key in the cold deacclimation process through the down-regulation of freezing tolerance-related genes, and the up-regulation of photosynthetic and growth-related genes in *D. antarctica*. Furthermore, the expression of the CBF transcription factor and its regulon during the night is necessary to maintain the previously acquired freezing tolerance. However, more studies are necessary to understand the consequences that regional warming will have on *D. antarctica* field populations. In this sense, it would be interesting to evaluate how gene expression is affected by nocturnal warming under field experimental conditions.

Supplementary Materials: The following supporting information can be downloaded at: https://www.mdpi.com/article/10.3390/ijms241311211/s1.

Author Contributions: Conceptualization, D.L., P.L.S. and L.A.B.; methodology, D.L., G.L. and L.A.B.; investigation, D.L. and L.A.B.; formal analysis, D.L., G.L. and L.A.B.; resources, D.L. and L.A.B.; writing—original draft preparation and editing, D.L.; writing—review and editing, G.L., P.L.S. and L.A.B.; visualization, D.L. and G.L.; project administration, L.A.B.; funding acquisition, D.L., P.L.S. and L.A.B. All authors have read and agreed to the published version of the manuscript.

Funding: This research was funded by ANID through Fondecyt 1151173 and Scholarship Program/Beca Doctorado Nacional/2017—21170470, NEXER-UFRO NXR17-0002, and INACH RT 18-18.

Institutional Review Board Statement: Not applicable.

Informed Consent Statement: Not applicable.

Data Availability Statement: The sequencing data was deposited in the National Center for Biotechnology Information (NCBI) and can be accessed via BioProject ID PRJNA941125.

Acknowledgments: The authors thank the 42nd crew of the Henryk Arctowski Station for their tremendous help and logistical support, and the Instituto Antártico Chileno (INACH) for the support provided (including the permits to work and sample close to ASPA 128) during the 54th Antarctic Scientific Expedition (ECA). The authors acknowledge the supercomputing infrastructure of Soroban (SATREPS MACH—JPM/JSA1705) at Centro de Modelación y Computación Científica, Universidad de La Frontera. The authors thank Charles L. Guy for proofreading this manuscript, and Fernanda Meneses—Ponzini for the graphical abstract drawing.

Conflicts of Interest: The authors declare no conflict of interest.

References

1. Meehl, G.A.; Stocker, T.F.; Collins, W.D.; Friedlingstein, P.; Gaye, A.T.; Gregory, J.M.; Kitoh, A.; Knutti, R.; Murphy, J.M.; Noda, A.; et al. Global Climate Projections Coordinating. In *Climate Change 2007: The Physical Science Basis. Contribution of Working Group I to the Fourth Assessment Report of the Intergovernmental Panel on Climate Change*; Solomon, S., Qin, D., Manning, M., Chen, Z., Marquis, M., Averyt, K.B., Tignor, M., Miller, H.L., Eds.; Cambridge University Press: Cambridge, UK, 2007; pp. 747–845, ISBN 978-0521-70596-7.
2. Pachauri, R.K.; Allen, M.R.; Barros, V.R.; Broome, J.; Cramer, W.; Christ, R.; Church, J.A.; Clarke, L.; Dahe, Q.; Dasgupta, P.; et al. *Climate Change 2014: Synthesis Report. Contribution of Working Groups I, II and III to the Fifth Assessment Report of the Intergovernmental Panel on Climate Change*; IPCC: Geneva, Switzerland, 2014.
3. Vose, R.S.; Easterling, D.R.; Gleason, B. Maximum and minimum temperature trends for the globe: An update through 2004. *Geophys. Res. Lett.* **2005**, *32*, L23822. [CrossRef]
4. Xie, H.; Id, M.Z.; Yu, Y.; Zeng, X.; Tang, G.; Duan, Y.; Wang, J.; Yu, Y. Comparative transcriptome analysis of the cold resistance of the sterile rice line 33S. *PLoS ONE* **2022**, *17*, e0261822. [CrossRef]
5. Vaughan, D.G.; Marshall, G.J.; Connolley, W.M.; Parkinson, C.; Mulvaney, R.; Hodgson, D.A.; King, J.C.; Pudsey, C.J.; Turner, J. Recent Rapid Regional Climate Warming on the Antarctic Peninsula. *Clim. Change* **2003**, *60*, 243–274. [CrossRef]
6. Fowbert, J.A.; Smith, R.I.L. Rapid population increases in native vascular plants in the Argentine Islands, Antarctic Peninsula. *Arct. Alp. Res.* **1994**, *26*, 290–296. [CrossRef]
7. Smith, R.I.L. The enigma of Colobanthus quitensis and Deschampsia antarctica in Antarctica. In *Antarctic Biology in a Global Context*; Huiskes, A.H.L., Gieskes, W.W.C., Rozema, J., Schorno, R.M.L., van der Vies, S.M., Eds.; Backhuys: Leiden, The Netherlands, 2003; pp. 234–239.
8. Torres-Mellado, G.A.; Jaña, R.; Casanova-Katny, M.A. Antarctic hairgrass expansion in the South Shetland archipelago and Antarctic Peninsula revisited. *Polar Biol.* **2011**, *34*, 1679–1688. [CrossRef]
9. Cannone, N.; Guglielmin, M.; Convey, P.; Worland, M.R.; Favero Longo, S.E. Vascular plant changes in extreme environments: Effects of multiple drivers. *Clim. Change* **2016**, *134*, 651–665. [CrossRef]
10. Robinson, S.A.; Wasley, J.; Tobin, A.K. Living on the edge—Plants and global change in continental and maritime Antarctica. *Glob. Chang. Biol.* **2003**, *9*, 1681–1717. [CrossRef]
11. Convey, P. The influence of environmental characteristics on life history attributes of Antarctic terrestrial biota. *Biol. Rev.* **1996**, *71*, 191–225. [CrossRef]
12. Convey, P.; Chown, S.L.; Clarke, A.; Barnes, D.K.A.; Bokhorst, S.; Cummings, V.; Ducklow, H.W.; Frati, F.; Green, T.G.A.; Gordon, S.; et al. The spatial structure of antarctic biodiversity. *Ecol. Monogr.* **2014**, *84*, 203–244. [CrossRef]
13. Day, T.A.; Ruhland, C.T.; Grobe, C.W.; Xiong, F. Growth and reproduction of Antarctic vascular plants in response to warming and UV radiation reductions in the field. *Oecologia* **1999**, *119*, 24–35. [CrossRef]
14. López, D.; Sanhueza, C.; Salvo-Garrido, H.; Bascunan-Godoy, L.; Bravo, L.A. How Does Diurnal and Nocturnal Warming Affect the Freezing Resistance of Antarctic Vascular Plants? *Plants* **2023**, *12*, 806. [CrossRef] [PubMed]
15. Vyse, K.; Pagter, M.; Zuther, E.; Hincha, D.K. Deacclimation after cold acclimation- a crucial, but widely neglected part of plant winter survival. *J. Exp. Bot.* **2019**, *70*, 4595–4604. [CrossRef] [PubMed]
16. Zuther, E.; Juszczak, I.; Ping Lee, Y.; Baier, M.; Hincha, D.K. Time-dependent deacclimation after cold acclimation in Arabidopsis thaliana accessions. *Sci. Rep.* **2015**, *5*, 12199. [CrossRef]
17. Wójcik-Jagła, M.; Daszkowska-Golec, A.; Fiust, A.; Kopeć, P.; Rapacz, M. Identification of the genetic basis of response to de-acclimation in winter barley. *Int. J. Mol. Sci.* **2021**, *22*, 1057. [CrossRef] [PubMed]
18. Pagter, M.; Alpers, J.; Erban, A.; Kopka, J.; Zuther, E.; Hincha, D.K. Rapid transcriptional and metabolic regulation of the deacclimation process in cold acclimated Arabidopsis thaliana. *BMC Genom.* **2017**, *18*, 731. [CrossRef] [PubMed]
19. Horvath, D.P.; Zhang, J.; Chao, W.S.; Mandal, A.; Rahman, M.; Anderson, J.V. Genome-wide association studies and transcriptome changes during acclimation and deacclimation in divergent Brassica napus varieties. *Int. J. Mol. Sci.* **2020**, *21*, 9148. [CrossRef]
20. Rathore, N.; Kumar, P.; Mehta, N.; Swarnkar, M.K.; Shankar, R.; Chawla, A. Time-series RNA-Seq transcriptome profiling reveals novel insights about cold acclimation and de-acclimation processes in an evergreen shrub of high altitude. *Sci. Rep.* **2022**, *12*, 15553. [CrossRef]
21. Pagter, M.; Arora, R. Winter survival and deacclimation of perennials under warming climate: Physiological perspectives. *Physiol. Plant.* **2012**, *147*, 75–87. [CrossRef]

22. Fowler, S.; Thomashow, M.F. Arabidopsis Transcriptome Profiling Indicates That Multiple Regulatory Pathways Are Activated during Cold Acclimation in Addition to the CBF Cold Response Pathway. *Plant Cell* **2002**, *14*, 1675–1690. [CrossRef]
23. Jia, Y.; Ding, Y.; Shi, Y.; Zhang, X.; Gong, Z.; Yang, S. The cbfs triple mutants reveal the essential functions of CBFs in cold acclimation and allow the definition of CBF regulons in Arabidopsis. *New Phytol.* **2016**, *212*, 345–353. [CrossRef]
24. Zhao, C.; Zhang, Z.; Xie, S.; Si, T.; Li, Y.; Zhu, J.-K. Mutational Evidence for the Critical Role of CBF Transcription Factors in Cold Acclimation in Arabidopsis. *Plant Physiol.* **2016**, *171*, 2744–2759. [CrossRef] [PubMed]
25. Sandve, S.R.; Kosmala, A.; Rudi, H.; Fjellheim, S.; Rapacz, M.; Yamada, T.; Rognli, O.A. Molecular mechanisms underlying frost tolerance in perennial grasses adapted to cold climates. *Plant Sci.* **2011**, *180*, 69–77. [CrossRef] [PubMed]
26. Thomashow, M.F. Molecular Basis of Plant Cold Acclimation: Insights Gained from Studying the CBF Cold Response Pathway. *Plant Physiol.* **2010**, *154*, 571–577. [CrossRef]
27. Badawi, M.; Danyluk, J.; Boucho, B.; Houde, M.; Sarhan, F. The CBF gene family in hexaploid wheat and its relationship to the phylogenetic complexity of cereal CBFs. *Mol. Genet. Genom.* **2007**, *277*, 533–554. [CrossRef] [PubMed]
28. Kalberer, S.R.; Wisniewski, M.; Arora, R. Deacclimation and reacclimation of cold-hardy plants: Current understanding and emerging concepts. *Plant Sci.* **2006**, *171*, 3–16. [CrossRef]
29. Campoli, C.; Matus-Cádiz, M.A.; Pozniak, C.J.; Cattivelli, L.; Fowler, D.B. Comparative expression of Cbf genes in the Triticeae under different acclimation induction temperatures. *Mol. Genet. Genom.* **2009**, *282*, 141–152. [CrossRef]
30. Liu, Z.; Jia, Y.; Ding, Y.; Shi, Y.; Li, Z.; Guo, Y.; Gong, Z.; Yang, S. Plasma Membrane CRPK1-Mediated Phosphorylation of 14-3-3 Proteins Induces Their Nuclear Import to Fine-Tune CBF Signaling during Cold Response. *Mol. Cell* **2017**, *66*, 117–128.e5. [CrossRef]
31. Catalá, R.; Santos, E.; Alonso, J.M.; Ecker, J.R.; Martínez-Zapater, J.M.; Salinas, J. Mutations in the Ca2+/H+ Transporter CAX1 Increase CBF/DREB1 Expression and the Cold-Acclimation Response in Arabidopsis. *Plant Cell* **2003**, *15*, 2940–2951. [CrossRef]
32. Iqbal, Z.; Memon, A.G.; Ahmad, A.; Iqbal, M.S. Calcium Mediated Cold Acclimation in Plants: Underlying Signaling and Molecular Mechanisms. *Front. Plant Sci.* **2022**, *13*, 855559. [CrossRef]
33. Tong, T.; Li, Q.; Jiang, W.; Chen, G.; Xue, D.; Deng, F.; Zeng, F.; Chen, Z.-H. Molecular Evolution of Calcium Signaling and Transport in Plant Adaptation to Abiotic Stress. *Int. J. Mol. Sci.* **2021**, *22*, 12308. [CrossRef]
34. Zhang, W.; Wang, J.; Huang, Z.; Mi, L.; Xu, K.; Wu, J.; Fan, Y.; Ma, S.; Jiang, D. Effects of Low Temperature at Booting Stage on Sucrose Metabolism and Endogenous Hormone Contents in Winter Wheat Spikelet. *Front. Plant Sci.* **2019**, *10*, 498. [CrossRef] [PubMed]
35. Stein, O.; Granot, D. An Overview of Sucrose Synthases in Plants. *Front. Plant Sci.* **2019**, *10*, 95. [CrossRef] [PubMed]
36. Bilska-Kos, A.; Mytych, J.; Suski, S.; Magoń, J.; Ochodzki, P.; Zebrowski, J. Sucrose phosphate synthase (SPS), sucrose synthase (SUS) and their products in the leaves of Miscanthus × giganteus and Zea mays at low temperature. *Planta* **2020**, *252*, 23. [CrossRef] [PubMed]
37. Wang, A.; Kao, M.-H.; Yang, W.; Sayion, Y.; Liu, L.-F.; Lee, P.-D.; Jong-Ching, S. Differentially and Developmentally Regulated Expression of Three Rice Sucrose Synthase Genes. *Plant Cell Physiol.* **1999**, *40*, 800–807. [CrossRef] [PubMed]
38. Nolte, K.D.; Koch, K.E. Companion-Cell Specific Localization of Sucrose Synthase in Zones of Phloem Loading and Unloading. *Plant Physiol.* **1993**, *101*, 899–905. [CrossRef]
39. Clemente-Moreno, M.J.; Omranian, N.; Sáez, P.L.; Figueroa, C.M.; Del-Saz, N.; Elso, M.; Poblete, L.; Orf, I.; Cuadros-Inostroza, A.; Cavieres, L.A.; et al. Low-temperature tolerance of the Antarctic species Deschampsia antarctica: A complex metabolic response associated with nutrient remobilization. *Plant Cell Environ.* **2020**, *43*, 1376–1393. [CrossRef]
40. Borovik, O.A.; Pomortsev, A.V.; Korsukova, A.V.; Polyakova, E.A.; Fomina, E.A.; Zabanova, N.S.; Grabelnych, O.I. Effect of Cold Acclimation and Deacclimation on the Content of Soluble Carbohydrates and Dehydrins in the Leaves of Winter Wheat. *J. Stress Physiol. Biochem.* **2019**, *15*, 62–67.
41. Hoffman, L.; DaCosta, M.; Scott Ebdon, J. Examination of cold deacclimation sensitivity of annual bluegrass and creeping bentgrass. *Crop Sci.* **2014**, *54*, 413–420. [CrossRef]
42. Vítámvás, P.; Prášil, I.T. WCS120 protein family and frost tolerance during cold acclimation, deacclimation and reacclimation of winter wheat. *Plant Physiol. Biochem.* **2008**, *46*, 970–976. [CrossRef]
43. Zhang, X.; Wang, K.; Ervin, E.H.; Waltz, C.; Murphy, T. Metabolic changes during cold acclimation and deacclimation in five bermudagrass varieties. I. Proline, total amino acid, protein, and dehydrin expression. *Crop Sci.* **2011**, *51*, 838–846. [CrossRef]
44. Banerjee, A.; Roychoudhury, A. Group II late embryogenesis abundant (LEA) proteins: Structural and functional aspects in plant abiotic stress. *Plant Growth Regul.* **2016**, *79*, 1–17. [CrossRef]
45. Kaplan, F.; Kopka, J.; Sung, D.Y.; Zhao, W.; Popp, M.; Porat, R.; Guy, C.L. Transcript and metabolite profiling during cold acclimation of Arabidopsis reveals an intricate relationship of cold-regulated gene expression with modifications in metabolite content. *Plant J.* **2007**, *50*, 967–981. [CrossRef] [PubMed]
46. Hoermiller, I.I.; Funck, D.; Schönewolf, L.; May, H.; Heyer, A.G. Cytosolic proline is required for basal freezing tolerance in Arabidopsis. *Plant Cell Environ.* **2022**, *45*, 147–155. [CrossRef] [PubMed]
47. Chen, Y.; Jiang, J.; Chang, Q.; Gu, C.; Song, A.; Chen, S.; Dong, B.; Chen, F. Cold acclimation induces freezing tolerance via antioxidative enzymes, proline metabolism and gene expression changes in two chrysanthemum species. *Mol. Biol. Rep.* **2014**, *41*, 815–822. [CrossRef] [PubMed]
48. Hoffman, L.; DaCosta, M.; Ebdon, J.S.; Watkins, E. Physiological changes during cold acclimation of perennial ryegrass accessions differing in freeze tolerance. *Crop Sci.* **2010**, *50*, 1037–1047. [CrossRef]

49. Anwar, A.; She, M.; Wang, K.; Riaz, B.; Ye, X. Biological Roles of Ornithine Aminotransferase (OAT) in Plant Stress Tolerance: Present Progress and Future Perspectives. *Int. J. Mol. Sci.* **2018**, *19*, 3681. [CrossRef] [PubMed]
50. You, J.; Hu, H.; Xiong, L. Plant Science An ornithine δ-aminotransferase gene OsOAT confers drought and oxidative stress tolerance in rice. *Plant Sci.* **2012**, *197*, 59–69. [CrossRef]
51. Steponkus, P.L. Role of the plasma membrane in freezing injury and cold acclimation. *Annu. Rev. Plant Physiol.* **1984**, *35*, 543–584. [CrossRef]
52. Pearce, R.S. Plant freezing and damage. *Ann. Bot.* **2001**, *87*, 417–424. [CrossRef]
53. Langlois-Meurinne, M.; Gachon, C.M.M.; Saindrenan, P. Pathogen-Responsive Expression of Glycosyltransferase Genes UGT73B3 and UGT73B5 Is Necessary for Resistance to Pseudomonas syringae pv tomato. *Plant Physiol.* **2005**, *139*, 1890–1901. [CrossRef]
54. Mostek, A.; Börner, A.; Weidner, S. Plant Physiology and Biochemistry Comparative proteomic analysis of b-aminobutyric acid-mediated alleviation of salt stress in barley. *Plant Physiol. Biochem.* **2016**, *99*, 150–161. [CrossRef]
55. Pan, L.; Gao, H.; Xia, W.; Zhang, T.; Dong, L. Establishing a herbicide-metabolizing enzyme library in Beckmannia syzigachne to identify genes associated with metabolic resistance. *J. Exp. Bot.* **2016**, *67*, 1745–1757. [CrossRef] [PubMed]
56. Wang, J.; Jia, J.; Sun, J.; Pang, X.; Li, B.; Yuan, J.; Chen, E.; Li, X. Trypsin preservation: CsUGT91C1 regulates Trilobatin Biosynthesis in Cucumis sativus during Storage. *Plant Growth Regul.* **2023**. [CrossRef]
57. Poppenberger, B.; Fujioka, S.; Soeno, K.; George, G.L.; Seto, H.; Takatsuto, S.; Adam, G.; Yoshida, S.; Bowles, D. The UGT73C5 of Arabidopsis thaliana glucosylates brassinosteroids. *Proc. Natl. Acad. Sci. USA* **2005**, *102*, 15253–15258. [CrossRef] [PubMed]
58. Wu, Q.; Zhang, X.; Peirats-Llobet, M.; Belda-Palazon, B.; Wang, X.; Cui, S.; Yu, X.; Rodriguez, P.L.; Chengcai, A. Ubiquitin Ligases RGLG1 and RGLG5 Regulate Abscisic Acid Signaling by Controlling the Turnover of Phosphatase PP2CA. *Plant Cell* **2016**, *28*, 2178–2196. [CrossRef]
59. Tähtiharju, S.; Palva, T. Antisense inhibition of protein phosphatase 2C accelerates cold acclimation in Arabidopsis thaliana. *Plant J.* **2001**, *26*, 461–470. [CrossRef]
60. Thomashow, M.F. PLANT COLD ACCLIMATION: Freezing Tolerance Genes and Regulatory Mechanisms. *Annu. Rev. Plant Physiol. Plant Mol. Biol.* **1999**, *50*, 571–599. [CrossRef]
61. Xiong, L.; Ishitani, M.; Lee, H.; Zhu, J.-K. The Arabidopsis LOS5/ABA3 locus encodes a molybdenum cofactor sulfurase and modulates cold stress- and osmotic stress- responsive gene expression. *Plant Cell* **2001**, *13*, 2063–2083.
62. Gilmour, S.J.; Fowler, S.G.; Thomashow, M.F. Arabidopsis transcriptional activators CBF1, CBF2, and CBF3 have matching functional activities. *Plant Mol. Biol.* **2004**, *54*, 767–781. [CrossRef]
63. Byun, M.Y.; Cui, L.H.; Lee, J.; Park, H.; Lee, A.; Kim, W.T.; Lee, H. Identification of Rice Genes Associated With Enhanced Cold Tolerance by Comparative Transcriptome Analysis With Two Transgenic Rice Plants Isolated From Antarctic Flowering Plant Deschampsia antarctica. *Front. Plant Sci.* **2018**, *9*, 601. [CrossRef]
64. Turnbull, M.H.; Murthy, R.; Griffin, K.L. The relative impacts of daytime and night-time warming on photosynthetic capacity in Populus deltoides. *Plant Cell Environ.* **2002**, *25*, 1729–1737. [CrossRef]
65. Sanhueza, C.; Fuentes, F.; Cortes, D.; Bascuñan, L.; Saez, P.; Bravo, L.A.; Cavieres, L. Contrasting thermal acclimation of leaf dark respiration and photosynthesis of Antarctic vascular plant species exposed to nocturnal warming. *Physiol. Plant.* **2019**, *167*, 205–216. [CrossRef]
66. Korasick, D.A.; Enders, T.A.; Strader, L.C. Auxin biosynthesis and storage forms. *J. Exp. Bot.* **2013**, *64*, 2541–2555. [CrossRef] [PubMed]
67. LeClere, S.; Tellez, R.; Rampey, R.A.; Matsuda, S.P.T.; Bartel, B. Characterization of a Family of IAA-Amino Acid Conjugate Hydrolases from Arabidopsis. *J. Biol. Chem.* **2002**, *277*, 20446–20452. [CrossRef] [PubMed]
68. Wan, S.; Xia, J.; Liu, W.; Niu, S. Photosynthetic overcompensation under nocturnal warming enhances grassland carbon sequestration. *Ecology* **2009**, *90*, 2700–2710. [CrossRef] [PubMed]
69. Sadok, W.; Jagadish, S.V.K. The Hidden Costs of Nighttime Warming on Yields. *Trends Plant Sci.* **2020**, *25*, 644–651. [CrossRef] [PubMed]
70. Fan, Y.; Lv, Z.; Qin, B.; Yang, J.; Ren, K.; Liu, Q.; Jiang, F.; Zhang, W.; Ma, S.; Ma, C.; et al. Night warming at the vegetative stage improves pre-anthesis photosynthesis and plant productivity involved in grain yield of winter wheat. *Plant Physiol. Biochem.* **2022**, *186*, 19–30. [CrossRef]
71. Bravo, L.A.; Ulloa, N.; Zuñiga, G.E.; Casanova, A.; Corcuera, L.J.; Alberdi, M. Cold resistance in antarctic angiosperms. *Physiol. Plant.* **2001**, *111*, 55–65. [CrossRef]
72. Bolger, A.M.; Lohse, M.; Usadel, B. Trimmomatic: A flexible trimmer for Illumina sequence data. *Bioinformatics* **2014**, *30*, 2114–2120. [CrossRef]
73. Grabherr, M.G.; Haas, B.J.; Yassour, M.; Levin, J.Z.; Thompson, D.A.; Amit, I.; Adiconis, X.; Fan, L.; Raychowdhury, R.; Chen, Z.; et al. Trinity: Reconstructing a full-length transcriptome without a genome from RNA-Seq data. *Nat. Biotechnol.* **2013**, *29*, 644–652. [CrossRef]
74. Fu, L.; Niu, B.; Zhu, Z.; Wu, S.; Li, W. CD-HIT: Accelerated for clustering the next-generation sequencing data. *Bioinformatics* **2012**, *28*, 3150–3152. [CrossRef] [PubMed]
75. Li, B.; Dewey, C.N. RSEM: Accurate transcript quantification from RNA-Seq data with or without a reference genome. *BMC Bioinform.* **2011**, *12*, 323. [CrossRef] [PubMed]
76. Love, M.I.; Huber, W.; Anders, S. Moderated estimation of fold change and dispersion for RNA-seq data with DESeq2. *Genome Biol.* **2014**, *15*, 550. [CrossRef] [PubMed]

77. Wehrens, R.; Buydens, L.M.C. Self- and super-organizing maps in R: The kohonen package. *J. Stat. Softw.* **2007**, *21*, 1–19. [CrossRef]
78. Mi, H.; Muruganujan, A.; Huang, X.; Ebert, D.; Mills, C.; Guo, X.; Thomas, P.D. Protocol Update for Large-scale genome and gene function analysis with PANTHER Classification System (v.14.0). *Nat. Protoc.* **2019**, *14*, 703–721. [CrossRef]
79. Young, M.D.; Wakefield, M.J.; Smyth, G.K.; Oshlack, A. Gene ontology analysis for RNA-seq: Accounting for selection bias. *Genome Biol.* **2010**, *11*, R14. [CrossRef]
80. Pfaffl, M.W. A new mathematical model for relative quantification in real-time RT–PCR. *Nucleic Acids Res.* **2001**, *29*, 16–21. [CrossRef]

International Journal of
Molecular Sciences

Article

Changes in Polar Lipid Composition in Balsam Fir during Seasonal Cold Acclimation and Relationship to Needle Abscission

Mason T. MacDonald [1],*, Rajasekaran R. Lada [1], Gaye E. MacDonald [1], Claude D. Caldwell [1] and Chibuike C. Udenigwe [2]

[1] Department of Plant, Food, and Environmental Sciences, Faculty of Agriculture, Dalhousie University, Bible Hill, NS B2N 5E3, Canada; raj.lada@dal.ca (R.R.L.); gemacdon@dal.ca (G.E.M.); claude.caldwell@dal.ca (C.D.C.)
[2] School of Nutritional Sciences, Faculty of Health Sciences, University of Ottawa, Ottawa, ON K1N 6N5, Canada; cudenigw@uottawa.ca
* Correspondence: mason.macdonald@dal.ca; Tel.: +1-902-893-6626

Citation: MacDonald, M.T.; Lada, R.R.; MacDonald, G.E.; Caldwell, C.D.; Udenigwe, C.C. Changes in Polar Lipid Composition in Balsam Fir during Seasonal Cold Acclimation and Relationship to Needle Abscission. *Int. J. Mol. Sci.* 2023, 24, 15702. https://doi.org/10.3390/ijms242115702

Academic Editor: Martin Bartas

Received: 29 September 2023
Revised: 24 October 2023
Accepted: 25 October 2023
Published: 28 October 2023

Abstract: Needle abscission in balsam fir has been linked to both cold acclimation and changes in lipid composition. The overall objective of this research is to uncover lipid changes in balsam fir during cold acclimation and link those changes with postharvest abscission. Branches were collected monthly from September to December and were assessed for cold tolerance via membrane leakage and chlorophyll fluorescence changes at −5, −15, −25, −35, and −45 °C. Lipids were extracted and analyzed using mass spectrometry while postharvest needle abscission was determined gravimetrically. Cold tolerance and needle retention each significantly ($p < 0.001$) improved throughout autumn in balsam fir. There were concurrent increases in DGDG, PC, PG, PE, and PA throughout autumn as well as a decrease in MGDG. Those same lipids were strongly related to cold tolerance, though MGDG had the strongest relationship (R^2 = 55.0% and 42.7% from membrane injury and chlorophyll fluorescence, respectively). There was a similar, albeit weaker, relationship between MGDG:DGDG and needle retention (R^2 = 24.3%). Generally, a decrease in MGDG:DGDG ratio resulted in better cold tolerance and higher needle retention in balsam fir, possibly due to increased membrane stability. This study confirms the degree of cold acclimation in Nova Scotian balsam fir and presents practical significance to industry by identifying the timing of peak needle retention. It is suggested that MGDG:DGDG might be a beneficial tool for screening balsam fir genotypes with higher needle retention characteristics.

Keywords: *Abies balsamea*; cold; conifer; fluorescence; galactosyldiacylglycerol; galactolipids; phospholipids; membrane injury; needle retention

1. Introduction

Balsam fir is a mid- to late-successional conifer species native to northeastern United States and Canada [1]. In Canada, balsam fir is found as far east as Newfoundland and as far west as Alberta, though balsam fir stands are more scattered and not the dominant conifer in most western forests [2]. Balsam fir can be used for lumber or fuel, but they are also commonly used as Christmas trees. Balsam fir Christmas trees are often harvested in early October to meet export demand and are often a preferred species due to their unique fragrance, color, and high needle retention characteristics [3]. However, postharvest needle retention has decreased over time, attributed to earlier harvests and climate change limiting opportunities for cold acclimation [4].

Optimum growth of balsam fir occurs in regions with an annual temperature of 2 to 4 °C. Ideally the coldest winter temperatures would range from −18 to −12 °C, while the warmest summer temperatures would range from 16 to 18 °C [2]. Few areas in Canada provide an ideal environment for balsam fir. As an example, Nova Scotia has the warmest average temperature of any Canadian province, with an average temperature of 25.9 °C in July 2023 [5]. Yet winter temperatures the same year fell as low as -26 °C [5]. There have been fewer frost events or cold degree days in autumn in which plants could acclimate for winter conditions, possibly exacerbating the opportunity for cold stress during winter.

Cold stress and/or freezing temperatures can cause widespread metabolic dysfunction in plants, including reduced electron transport, increased oxidative stress, impaired water movement, changes in membrane fluidity, and eventual tissue death [6,7]. Like many conifers, balsam fir is adept at tolerating freezing temperatures through cold acclimation. Cold acclimation is regulated by a complex network of signaling pathways triggered by environmental cues like light and temperature [8]. Decreasing temperatures and shorter photoperiods trigger cryoprotective genes that ultimately improve protein stabilization, increase solute concentrations, and alter membrane composition [9,10].

Cold acclimation-induced membrane changes are manifested through shifts in the concentration of membrane lipids. One change is that the concentration of monogalacto-syldiacylglycerol (MGDG) tends to decrease in tandem with an increase in digalactosyl-diacylglycerol (DGDG) [11,12]. MGDG forms a single layer of lipids for greater efficiency of thylakoid membranes while DGDG forms a bilayer to create greater stability [13]. A shift towards DGDG constitutes a shift towards membrane stability in the plant and helps protect plants from cold stress. A second shift towards phospholipids (PLs) occurs after cold acclimation [14]. More specifically, there is an increase in phosphatidycholine (PC) and phoshatidylethanolamine (PE) after cold acclimation [12,15]. An increase in PLs is associated with a shift towards lipids containing unsaturated fatty acids to help maintain membrane fluidity in cold temperatures [16].

Cold acclimation and tolerance can be calculated through many different methods, but in essence all methods operate by exposing plants to cold temperatures and assessing damage [17]. The specific temperatures, exposure time, and damage assessment vary between studies and plant species [17]. One technique is to assess plants for visual damage from freezing temperatures, though this can often be a tedious task [17]. Alternatives include measuring electrolyte leakage [18,19] or chlorophyll fluorescence [20,21] to assess damage. Freezing tolerance is then often quantified using an LT50 value, or the temperature at which there is 50% electrolyte leakage or a 50% decrease in chlorophyll fluorescence. Freezing tolerance of *Abies* species ranged from -25 to -70 °C through visual observation [22]. Freezing tolerance of *Abies procera* was monitored throughout autumn and ranged from approximately -20 °C in September to -38 °C in December [23].

Cold acclimation tends to be associated with improved postharvest needle retention in balsam fir trees. Needle retention increases throughout the autumn months and reaches a peak in November or December [3]. Though improved needle retention was strongly correlated with decreasing temperatures and photoperiod, freezing tolerance of balsam fir has not been directly related to needle retention. Further, postharvest needle retention is linked to changes in lipids and fatty acids [24]. Abscising needles had lower concentrations of MGDG and DGDG, but significantly higher concentrations of PC, lysophosphatidylglyc-erol (LPG), and phosphainositol (PI) than intact needles [24]. It is reasonable to postulate that lipid changes in balsam fir are linked to both cold tolerance and postharvest needle retention. The objectives of this study are to (1) quantify changes in balsam fir cold tolerance throughout autumn, (2) determine changes in balsam fir needle polar lipids throughout autumn, and (3) to relate cold tolerance to changes in lipid concentration and needle retention in balsam fir.

2. Results

2.1. Confirming Cold Acclimation

There was a significant interactive effect ($p < 0.001$) between collection month and freezing temperatures on membrane injury (Figure 1). Branches collected in September had significantly more membrane injury once exposed to -15 °C, a trend that continued until exposure to −45 °C when there was no difference between branches collected in September and November. Exposure to −25 °C was the point where there was very clear separation in membrane injury between sample months; highest to lowest membrane injury occurred in the order of September, October, November, and December. Membrane injury was higher in all collection months when exposed to −35 °C, though the order was identical to −25 °C freezing. Ultimately, there was a clear trend of balsam fir having lower membrane injury in cold temperatures when harvested later in autumn. Blocking by genotype caused minimal improvement to the statistical model (F = 0.94, p = 0.420).

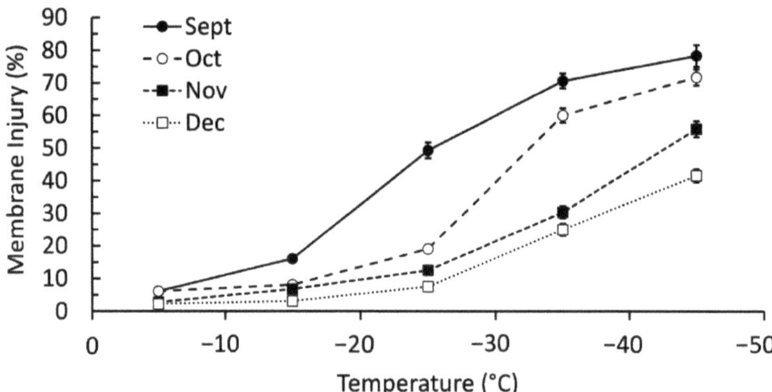

Figure 1. Membrane injury of balsam fir branches collected from September to December then subjected to freeze tests at −5, −15, −25, −35, and −45 °C. Each data point is the mean ± standard error as calculated from 20 replicates.

There was a significant interactive effect (p < 0.001) between collection month and freezing temperatures on chlorophyll fluorescence (Figure 2). September had significantly lower fluorescence than other months when exposed to −5 °C, though fluorescence was still maintained at approximately 90% of the baseline value. September was also the only sampling month with significantly lower fluorescence after −15 °C exposure. Both September and October had significantly lower fluorescence after exposure to −25 °C. December maintained significantly higher fluorescence than other months after exposure to −35 °C and −45 °C, though fluorescence had decreased compared to freeze tests at −5, −15, and −25 °C. As with membrane injury, the general trend was balsam fir maintaining their chlorophyll fluorescence in cold temperatures when harvested later in autumn. Blocking by genotype improved the statistical model with respect to chlorophyll fluorescence (F = 3.57, p = 0.014).

Sigmoidal curves fitted to membrane injury and chlorophyll fluorescence responses to freezing temperatures allowed for the determination of LT50 values. However, LT50 values were not completely consistent between LT50 calculated from membrane injury (LT50-MI) and chlorophyll fluorescence (LT50-CF) (Figure 3). A perfect relationship between LT50-MI and LT50-CF would have a slope of 1 compared to the calculated 0.7337. Instead, LT50-MI values are lower than their respective LT50-CG estimates. Sampling month had a consistent significant (p < 0.001) effect on LT50, regardless of estimation method. September always had the highest LT50, while December always had the lowest LT50 (Figure 4).

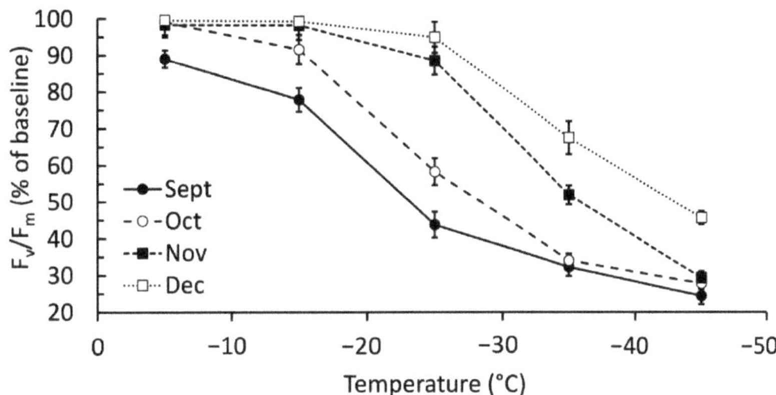

Figure 2. Chlorophyll fluorescence (F_v/F_m) of balsam fir branches collected from September to December and then subjected to freeze tests at -5, -15, -25, -35, and $-45\,°C$. Chlorophyll fluorescence is expressed as a percentage of the baseline values established by measuring branches that were not exposed to any freezing test. Each data point represents the mean $\pm$ standard error as calculated from 20 replicates.

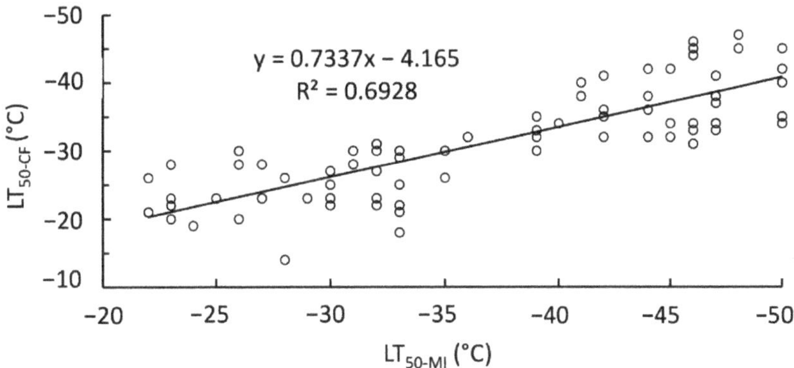

Figure 3. Linear relationship between LT50-MI and LT50-CF (N = 80).

Sampling month had a significant ($p < 0.001$) effect on needle retention duration (NRD) (Figure 4). September and October had the lowest NRD of 51 and 49 days, respectively. November and December had a 32–47% increase in NRD compared to September and October, though there was no significant difference between November and December. Although the overall trend was consistent with LT50 values per month, NRD was only weakly correlated to LT50-MI and LT50-CF (r = -0.365 and -0.290, respectively). Genotype contributed significant variation to NRD (F = 8.90, $p < 0.001$). A regression equation using genotype as a block improved the relationship between NRD and LT50-MI (R^2 = 35%) and between NRD and LT50-CF (R^2 = 36%).

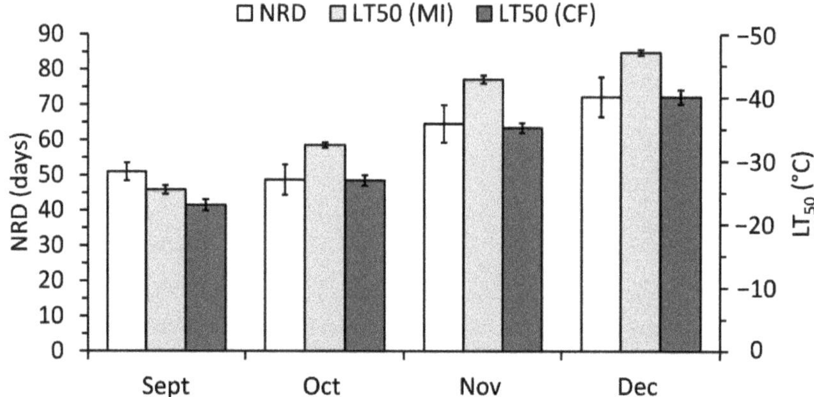

Figure 4. NRD and LT50 values estimated from membrane injury (LT50-MI) and chlorophyll fluorescence (LT50-CF) from balsam fir branches collected in September to December. Bars represent the mean ± standard error as calculated from 20 replicates.

2.2. Changes in Polar Lipids during Cold Acclimation

Sampling month had a significant effect ($p < 0.05$) on all polar lipid classes in balsam fir except for LPG and LPE (Table 1). Most polar lipids increased in relative concentration throughout autumn. There was a relative increase of 11.8% in DGDG, 30.3% in PC, 26.5% in PG, 81.7% in PE, and 23.1% in PI from September to December. The increases in the 5 lipid classes above were offset by a 26.2% decrease in MGDG over the same time span. LPG and PA did not have a consistent progression from September to December; instead, each reached their highest relative concentration in November before decreasing in December.

Table 1. Comparison of lipid classes by percentage of total lipids for balsam fir branches harvested at five different months. Values are expressed as the mean ± standard error as calculated from 20 replicates. The *p*-value denotes whether there was a significant difference in at least one of the sampling dates for each class of lipids. DGDG, digalactosyldiacylglycerol; LPG, lysophosphatidylglycerol; LPE, lysophosphatidylethanolamine; LPC, lysophosphatidylcholine; MGDG, monogalactosyldiacylglycerol; PA, phosphatidic acid; PC, phosphatidycholine; PE: phoshatidylethanolamine; PG, phosphatidylglycerol; PI, phosphainositol.

Lipid Class	September.	October	November	December	*p*-Value
DGDG	26.87 ± 0.36	27.10 ± 0.43	28.33 ± 0.38	30.04 ± 0.37	<0.001
MGDG	45.90 ± 0.87	40.98 ± 1.17	38.37 ± 0.81	33.86 ± 0.58	<0.001
PC	15.01 ± 0.39	18.12 ± 0.61	18.48 ± 0.44	19.56 ± 0.40	<0.001
PG	4.04 ± 0.20	3.89 ± 0.16	3.81 ± 0.15	5.11 ± 0.15	<0.001
PE	2.62 ± 0.17	3.74 ± 0.18	3.75 ± 0.21	4.76 ± 0.24	<0.001
PI	3.86 ± 0.15	3.99 ± 0.16	4.64 ± 0.17	4.75 ± 0.11	<0.001
PA	0.47 ± 0.06	0.93 ± 0.09	1.19 ± 0.34	0.66 ± 0.05	=0.049
LPC	0.05 ± 0.01	0.12 ± 0.02	0.14 ± 0.01	0.09 ± 0.01	=0.041
LPG	1.05 ± 0.19	0.96 ± 0.15	1.14 ± 0.15	1.00 ± 0.20	=0.344
LPE	0.11 ± 0.01	0.15 ± 0.01	0.14 ± 0.01	0.14 ± 0.01	=0.071

Sampling month had a significant effect ($p < 0.001$) on MGDG:DGDG and GL:PL ratios (Figure 5). The MGDG:DGDG was 1.69 in September and then significantly decreased each month. MGDG:DGDG decreased by 14.2% in October, 19.5% in November, and 33.1% in December all compared to September. GL:PL significantly decreased by 20.0% from September to October. There was a further significant decrease in November, but by December, GL:PL had decreased by 35.6% compared to its initial value.

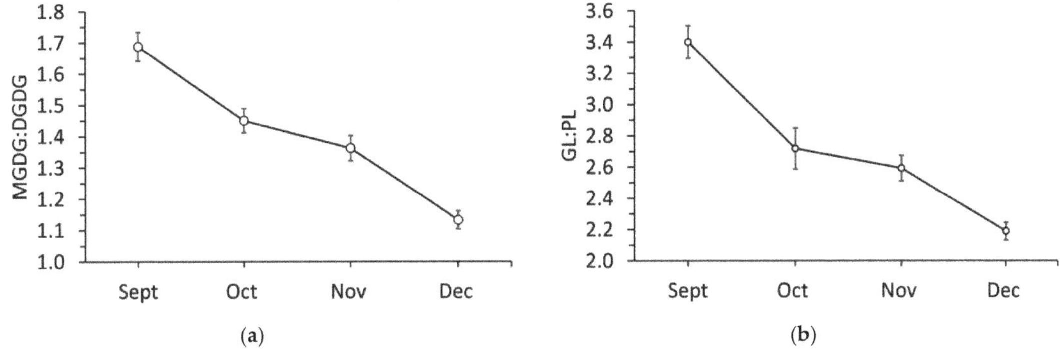

(a)   (b)

Figure 5. Ratio of (**a**) monogalactosyldiacylglycerol to digalactosyldiacylglycerol and (**b**) galactolipids to phospholipids in balsam fir collected from September to December. Each data point represents the mean ± standard error as calculated from 20 replicates.

2.3. Relationship of Polar Lipids and Lethal Temperatures

Almost all polar lipid classes had significant ($p < 0.05$), linear relationships with LT50 values, except for PA, LPC, LPG, and LPE (Table 2). MGDG had the strongest relationship with LT50, accounting for 55.0% of variation in LT50-MI and 42.7% of variation in LT50-CF. MGDG also had the only positive relationship with LT50 values. DGDG, PC, PE, and PI were weaker, accounting for 30.0–36.7% of the variation in LT50-MI and 22.9–29.5% of the variation in LT50-CF. PG accounted for less than 20% of the variation in each LT50 value.

Table 2. Strength, significance, and linear equation of relationships between relative polar lipid concentration and LT50 values. LT50 were estimated from membrane injury (LT50-MI) and chlorophyll fluorescence (LT50-CF) regression calculated from 80 data points using LT50 as the dependent variable and relative lipid concentration as the independent variable. DGDG, digalactosyldiacylglycerol; LPG, lysophosphatidylglycerol; LPE, lysophosphatidylethanolamine; LPC, lysophosphatidylcholine; MGDG, monogalactosyldiacylglycerol; PA, phosphatidic acid; PC, phosphatidycholine; PE: phoshatidylethanolamine; PG, phosphatidylglycerol; PI, phosphainositol.

Lipid Class	LT50-MI				LT50-CF			
	R^2 (%)	*p*-Value	Slope	Constant	R^2 (%)	*p*-Value	Slope	Constant
DGDG	31.5	<0.001	−2.23	25.8	29.5	<0.001	−1.87	21.3
MGDG	55.0	<0.001	1.10	−80.8	42.7	<0.001	0.86	−65.6
PC	36.7	<0.001	−1.86	−3.80	26.0	<0.001	−1.36	−7.01
PG	15.2	=0.001	−3.86	−20.6	16.6	=0.008	−3.43	−16.8
PE	35.4	<0.001	−4.40	−20.6	27.7	<0.001	−3.41	−18.6
PI	30.0	<0.001	−6.13	−10.5	22.9	<0.001	−4.56	−11.6
PA	2.5	=0.180	−1.65	−35.5	3.4	=0.411	−0.73	−30.7
LPC	5.1	=0.060	−29.0	−34.0	7.8	=0.193	24.2	−28.85
LPG	0.7	=0.498	−0.69	−36.2	3.4	=0.399	−1.10	−30.1
LPE	3.1	=0.130	−30.3	−32.8	4.6	=0.208	−28.4	−27.5

MGDG:DGDG and GL:PL ratios each had significant, linear relationship with LT50 values (Figure 6). These relationships were stronger when LT50 was determined from membrane injury as opposed to chlorophyll fluorescence. The slope of each relationship was positive so that a shift towards DGDGs or PLs was related to lower LT50.

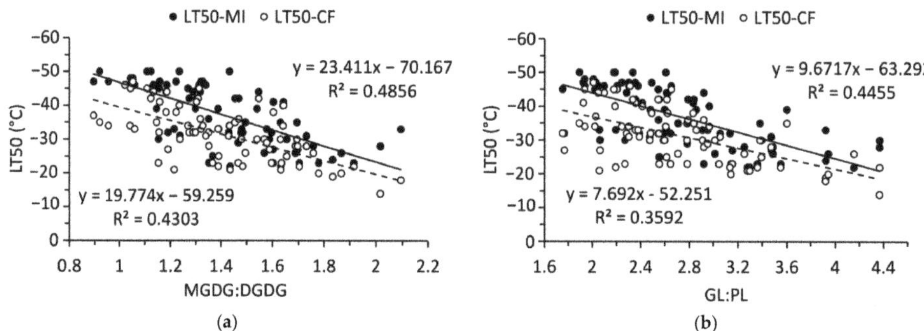

Figure 6. Relationships of (**a**) monogalactosyldiacylglycerol to digalactosyldiacylglycerol ratio and (**b**) galactolipid to phospholipid ratio to LT50 values in balsam fir collected from September to December. N = 80.

2.4. Relationship of Polar Lipids and Needle Abscission

Only DGDG and MGDG were significantly related to NRD (Figure 7). DGDG was positively related to NRD while MGDG was negatively related to NRD. Although each of the relationships was significant, these were relatively weak relationships that accounted for less than 20% of the variation in NRD. The MGDG:DGDG ratio slightly improved the model, accounting for 24% of the variation in NRD (Figure 7).

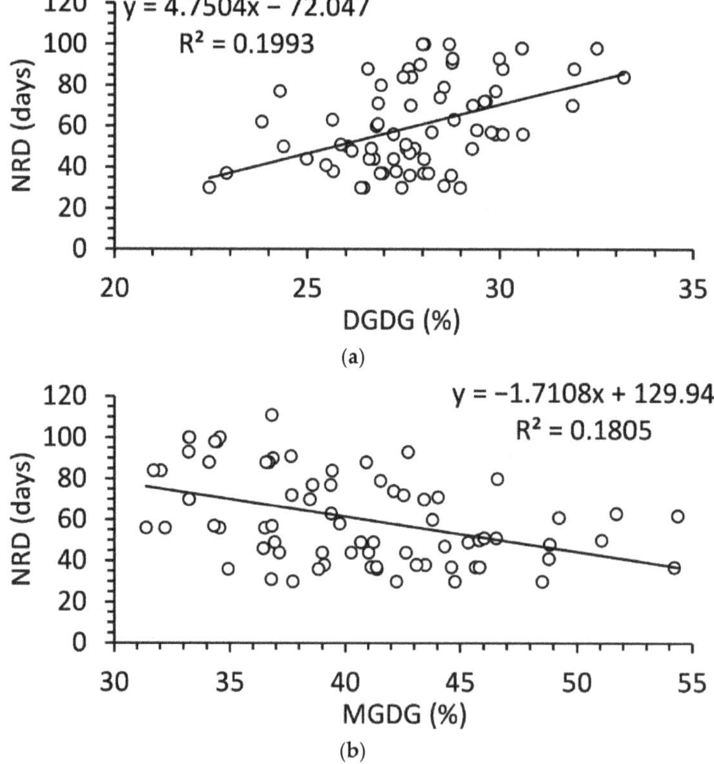

Figure 7. *Cont.*

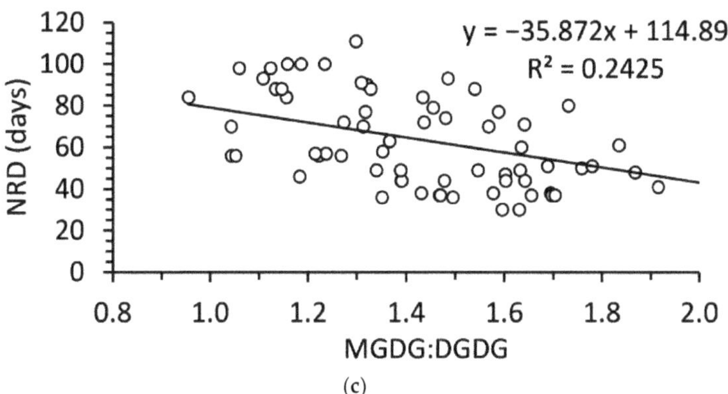

Figure 7. Relationships of (**a**) digalactosyldiacylglycerol, (**b**) monogalactosyldiacylglycerol, and (**c**) monogalactosyldiacylglycerol to digalactosyldiacylglycerol ratio to needle retention duration (NRD) in balsam fir collected from September to December. N = 80 in each.

3. Discussion

3.1. Cold Acclimation in Balsam Fir

There is sufficient evidence to support that balsam fir acclimates to cold temperatures throughout autumn in Nova Scotia. Balsam fir branches collected each month after September better resisted membrane injury and impaired chlorophyll fluorescence when subjected to freezing temperatures. The LT50 value for branches collected in December was between −40 and −50 °C, almost twice as low as branches collected in September. Freezing tolerance of balsam fir fall well within the expected range of −25 to −70 °C for *Abies* species [22].

The fact that balsam fir cold acclimates through autumn was expected. The minimum temperature observed in December during sampling was −23 °C and average temperatures continued to decrease in January and February. Although rare, it is possible to have minimum temperatures below −40 °C in Nova Scotia where these branches were collected. Balsam fir would need to adapt to such freezing temperatures to survive the winter. However, it was important to confirm and quantify their ability to acclimate to freezing temperatures. One weakness of previous cold acclimation studies in balsam fir was that it was assumed that cold acclimation occurred, but the degree of that acclimation was not measured. The LT50 values in this current study lend credence that the assumption of cold acclimation was likely correct in previous studies. It is noteworthy that accuracy of estimates in this current study could be improved with smaller increments in freezing temperatures tested [17]. Also, a few branches from December did not have 50% membrane injury or a 50% decrease in chlorophyll fluorescence even at −45 °C, although they were usually very close to that 50% mark. In those cases, LT50 values were extrapolated from a sigmoidal curve. Future evaluations of balsam fir freezing tolerance should include lower temperatures to avoid that problem.

3.2. Polar Lipids in Balsam Fir

Seasonal changes in lipid profiles have been studied extensively in other species, but there have been few studies of lipids in balsam fir. There are fewer major contributing spe-cies of polar lipids in balsam fir needles than in Arabidopsis rosettes [25]. GLs are the most abundant lipid classes in balsam fir needles, which can represent up to 80% of all polar lipids [26]. In balsam fir, galactolipids (GLs) comprise 65–70 per cent of all polar lipids. The other major contributor was PC and to a lesser extent PG. A previous study found 33% MGDG and 28% DGDG in balsam fir harvested in December [24]. Our study found 33% MGDG to 30% DGDG, resulting in a very similar ratio as previous work [24].

This ratio is higher than reported in most plants, but that could be because plants in both studies were subjected to cold acclimation [25].

The seasonal changes in balsam fir GLs are consistent with those in Pinus sylvestris [27]. MGDG decreased during the winter months and increased in the spring, while DGDG increased. When membranes of the chloroplast were isolated from rye leaves, there was a decrease in MGDG and increase in DGDG due to cold acclimation [28]. Similar re-sults were observed in Arabidopsis [25]. Studies have shown that MGDG, DGDG, sulfoqui-novosyl diacylglycerol (SQDG), and PG are all involved in maintaining the thylakoid structure and for the proper functioning of photosystem II and related proteins [11,12]. Therefore, changes in GLs were expected during cold acclimation.

MGDG and DGDG are also known to stabilize the photosystem protein complexes in chloroplasts [29]. Maintaining a constant MGDG:DGDG ratio in thylakoid membranes (at least under standard growth conditions) seems crucial for the stability and functional integrity of photosynthetic membranes [26]. An increase in the bilayer-forming galactolipid, DGDG, and a decrease in the monolayer-forming MGDG during cold acclimation should help protect the chloroplast membranes from damage during the winter keeping the membrane more fluid [26]. Due to the reduction of light and colder temperatures associated with winter, chloroplast membrane stabilization becomes ultimately important for survival. The ratio of MGDG to DGDG may vary according to the length of cold acclimation, as the biosynthesis of these compounds is tightly regulated to meet the needs of the cell under changing environmental conditions [25]. MGDG content decreased more than DGDG increased in this study. This can be explained by the fact that it takes two MGDG to form one DGDG [30]. The transition of MGDG to DGDG to maintain the appropriate ratio, which allows the reversible transition from the hexagonal II to lamellar α phase of the lipid bilayer, could be a very important factor in thylakoid biogenesis [31].

Most PLs increased during balsam fir cold acclimation, with the largest increase observed in PC. An increase in the per cent PC during cold acclimation has been reported in two other conifer species, *Pinus silvestris* and *Pinus nigra* [27,32]. As MGDG combines to form DGDG, the proportion of GL decreases while PL increases. With respect to cold acclimation, PLs are often of interest due to shifts towards longer and unsaturated fatty acid chains [28,33].

Although PC was the predominant PL, the relative increase in PLs was also a function of PG, PE, and PI. PI and PA have been identified to play a role in cell signaling [34]. PA is closely related to an increased activity of phospholipase D (PLD), which catalyzes the hydrolysis of PC to PA [35]. PA in balsam fir doubled from September to October but remained a minor contributor to overall PL concentration. Some studies with evidence of PLD activity have shown an increase from below 1 to 12 % in PA in a short time [25]. The increase in PA never exceeded 1% in balsam fir. Other PLs have been linked to a variety of stress responses. For instance, PI is the first molecule in the phosphoinositide signaling pathway [36] and increases have also been observed in drought and salinity stress [37]. The exact role of each PL in response to stress signaling needs to be further explored in plants, including balsam fir.

3.3. Needle Retention in Balsam Fir

Needle retention increased throughout autumn with peak retention occurring in November and December, consistent with several previous studies [38,39]. Previous studies had speculated that the improvement in needle retention was due to cold acclimation. This current study supports the hypothesis that cold acclimation improves needle retention through a significant relationship between LT50 values and needle retention. However, the relationship between cold acclimation and needle retention was weak unless genotype was considered. Needle retention is highly variable between genotypes [39], which at one point was attributed to differences in the rate at which certain genotypes acclimated to cold temperatures [39]. Even when genotype was included in the model, the model only explained 35–36% of the variation meaning there are likely other factors significantly affecting needle retention.

A decrease in MGDG and increase in DGDG were both associated with higher needle retention in balsam fir. As with cold acclimation in general, it is possible that increased stability of membranes due to higher proportion of DGDG also protects the plant from other stresses [26]. Membrane damage occurs prior to and during abscission in balsam fir, likely as a stress response postharvest [24]. Thus, any mechanism that protects membrane integrity might delay abscission. Although, the relationship between MGDG:DGDG ratio and needle retention was the strongest, it still only accounted for 24% of the variation. While shifting from MGDG to DGDG seems beneficial for needle retention, there must be other factors involved.

Only accounting for a portion of variation in needle retention should not diminish the importance of cold acclimation and the shift towards DGDG. Abscission of balsam fir, like other plants, is a complex physiological phenomenon. Multiple factors have been associated with abscission, with the prevailing theory being that water stress is the impetus for postharvest needle abscission [3]. Synthesis of abscisic acid and ethylene occur postharvest, transpiration and water uptake decrease, and abscission occurs [3]. Meanwhile, having a high concentration of indole-3-acetic acid can delay abscission in balsam fir. Additional factors affecting balsam fir abscission include water quality, volatiles, nutrition, and others [3,4]. With such a multitude of other factors known to influence abscission, it makes sense that a shift in polar lipids is a part of the larger puzzle.

Cold acclimation could also affect postharvest abscission through other mechanisms than polar lipid shifts. Balsam fir accumulate carbohydrates and isopentenyladenosine during autumn, which were both associated with improved needle retention [3]. Other plants have shown a variety of cold acclimation specific proteins [8,40], which have never been assessed in balsam fir but could be linked to needle retention. There is also an established link between ethylene and cold acclimation in other species. Ethylene was synthesized during exposure to cold and upregulated cold acclimation in *Arabidopsis*; plants that could not synthesize adequate amounts of ethylene did not acclimate as well as those that could [41]. Where short-term exposure to ethylene helped promote cold tolerance in *Arabidopsis*, short-term ethylene exposure delayed abscission in balsam fir [3]. Any link between cold acclimation, ethylene, and needle abscission in balsam fir remains to be explored.

Furthering our understanding of the relationships between cold acclimation, lipid composition, and needle retention could have practical significance. First, the balsam fir industry has long desired screening tools to identify trees with superior needle retention. Previous studies identified ethylene evolution rates and sensitivity as one potential screening tool [3]. However, the study was limited by only uncovering differences between genotypes categorized as having low or high needle retention. The strength of the relationship between needle retention and ethylene sensitivity was not quantified. Second, there is the possibility of artificially inducing cold tolerance to promote needle retention. Antioxidant enzymes, certain hormones, and signaling metabolites all may modify local and systemic cold tolerance [42], thus may be useful to support high needle retention. Finally, it must be noted that lipidomics is only one strategy for understanding changes in physiology. Incorporation of other omic approaches, such as proteomics, genomics, transciptomics, and others would further our understanding.

4. Materials and Methods

4.1. Sampling and Experimental Design

The plant material for this investigation was collected from the Debert clonal tree orchard (45°44′ N, −63°50′ W) located in Debert, NS, Canada from September to January. Sampling month was an explanatory variable, with specific sampling dates being 18 September, 28 October, 25 November, and 30 December. The Christmas tree germplasm collection has over 220 genotypes of balsam fir. Previous research demonstrated considerable variation in needle retention between genotypes [39], so genotype was used as a blocking factor. Four genotypes were selected for levels of the blocking factor. The general experimental design included an explanatory variable with 4 levels, block with 4 levels, and was then replicated 5 times.

Branches to be sampled were identified on 20-year-old trees at an elevation of approximately 1 m from ground level on the south facing side of the trees. Branches with any visual problems, such as pathogen or pests, were avoided. Approximately 4 g of needles were removed onsite from each tree to be used for initial lipid analysis. Needle samples that were taken on site were immediately immersed in liquid nitrogen and stored at −80 °C until analysis. Branches were then cut to include the previous 2 years growth and taken to the lab for cold tolerance, needle retention, and eventual lipid analysis. Each month required 80 branches to be collected to be submitted to freeze testing, 20 branches to be assessed for needle retention, and 80 branches to be evaluated for lipid composition.

4.2. Environmental Conditions during Sampling Months

Temperatures and photoperiod from 30 days prior to the start of the experiment were recorded from the Debert Weather Station. Photoperiod was taken on the days of sampling only. However, temperatures leading up to the sampling days were used to calculate minimum temperature (T_{min}), maximum temperature (T_{max}), days of exposure to freezing temperatures, and cold degree days (CDD), which are all shown in Table 3. CDD was determined by using the following formula:

$$\text{CDD} = \sum_{i=1}^{i=n} T_{base} - T_a, \text{if} T_{base} - T_a > 0; \text{otherwise } 0 \tag{1}$$

Table 3. Temperature parameters 30 days prior to first sampling periods and between each sampling time. Temperature-related parameters are CDD (cold degree days), T_{min} (minimum temperature in the days prior to testing), T_{max} (highest temperature experienced in the month prior and since the last sampling date), and cumulative days below 0 °C. Photoperiod represents the daylight hours on the day of sampling.

Sampling Date	CDD (Days)	T_{min} (°C)	T_{max} (°C)	Days Below 0 °C	Photoperiod (h)
18 September	0	5	21	0	12.4
28 October	9	−4	18	7	10.3
25 November	102	−8	17	28	9.2
30 December	354	−23	8.3	60	8.8

In the above equation, T_a is the daily mean air temperature calculated from the daily minimum and maximum temperatures. T_{base} is the threshold temperature of 5 °C determined as the temperature at which plant growth no longer occurs in balsam fir [43], and i is day of the period prior to sampling, with 1 being the first day and n being the last day.

4.3. Display Conditions and Needle Retention

Branches were transferred to an environmental controlled growth chamber at a constant temperature of 20 °C and light intensity of 100 µmol m-2 s-1. Light was supplied as a combination of incandescent and fluorescent. Both temperature and light conditions were selected to simulate household conditions as described by [44]. Once in the lab, branches were given a fresh aseptic cut 2.5 cm above the previous cut while submerged in water to reduce risk of cavitation. Branches were placed into amber bottles and provided 100 mL of distilled water. The neck of each flask was plugged with cotton wool to reduce direct water evaporation and provide added stability to a branch.

Needle retention was evaluated by lightly holding the branch between thumb and index finger and sliding along the length of the branch. Care was taken to slide in the same direction as the needles were attached to the stem. The goal was to dislodge needles that had abscised and not to snap off needles that were healthy. Needles were dried at 80 °C overnight and then weighed. This procedure was repeated every 2 days until complete needle shed. The day on which a branch lost 100% of its needles was referred to as its NRD.

4.4. Freeze Testing to Evaluate Cold Tolerance

Cold tolerance was assessed at 5 temperatures: -5, -15, -25, -35, and -45 °C. The temperature range was determined partially by the typical range of regional winter temperatures as well as cold tolerance estimates from other *Abies* species [22,23]. The freeze test protocol was based on previous work by [45]. Branches were placed in a programmable freezer (Thermotron SM-32-C, Holland, MI, USA), and the temperature was reduced 5 °C h^{-1} until reaching the target temperature. Branches were exposed to the target temperatures for 30 min, held at 0 °C overnight, and then returned to the growth chamber until the needles reached 20 °C for the evaluation of membrane injury and chlorophyll fluorescence.

Chlorophyll fluorescence (F_v/F_m), was determined using a MINI-PAM-II Photosynthesis Yield Analyzer (Heinz Walz, Effeltrich, Germany). Plants were dark-adapted for 20 min before fluorescence measurements with a saturating light pulse of 8000 µmol m^{-2} s^{-1}. Three measurements were taken for each experimental unit; the average of these measurements was recorded as the F_v/F_m for each branch. Chlorophyll fluorescence was measured prior to freezing test to establish a baseline for fluorescence in each branch. Chlorophyll fluorescence was measured again after the freezing test.

Membrane injury was determined from the percentage of electrolytes leaked into the solution versus the total electrolytes present. Test tubes were filled with 30 mL of distilled water and were allowed to adjust to room temperature (20 °C). The electrical conductivity of the distilled water (EC_w) alone was measured using a CDM 2e Conductivity Meter (Bach-Simpson, London, ON, Canada). Afterward, approximately 0.4 g of needles were removed from each branch and completely submerged in a centrifuge tube. The tubes were sealed and left at room temperature for 24 h. Initial conductivity (EC_0) was measured in order to determine electrolytes leaching into solution. Sealed tubes were then placed in a forced-air oven for 4 h at 90 °C to kill tissues and were then cooled to room temperature. Final conductivity measurements (EC_f) were taken after equilibrating to 20 °C to determine maximum leakage. Membrane injury was then calculated using the following equation [46]:

$$\text{Membrane injury} = \frac{EC_0 - EC_w}{EC_f - EC_w} \times 100 \tag{2}$$

LT50 values were calculated from both chlorophyll fluorescence and membrane injury by fitting a sigmoidal curve to freeze test changes for each branch. From those sigmoidal curves, LT50-MI was the temperature at which balsam fir would have 50% membrane injury, and LT50-CG was the temperature at which balsam fir would lose 50% of its initial chlorophyll fluorescence.

4.5. Lipid Extraction and Analysis

All lipids were extracted at the Kansas Lipidomic Research Center (Manhattan, KS, USA) using an extraction protocol for *Arabidopsis* leaf tissue adapted from [47]. Approximately 1 g of frozen needles were cut into smaller pieces and incubated in 1.0 mL of isopropanol with 0.01% butylated hydroxytoluene (BHT) at 75 °C for 15 min. Afterwards, 1.5 mL of chloroform and 0.6 mL of water were added to the solution. The solvent was shaken at room temperature for 1h and then transferred to a new glass tube with a Teflon-lined screw-cap using a Pasteur pipette. A total of 0.7 mL chloroform:methanol (2:1) was added, shaken for 30 min. The extraction was completed by adding 4 mL of chloroform:methanol (2:1) ten times, shaking for 1 h and collecting the solvent. The solvent extracts were washed once with 1mL KCl (1.0 M) and once with 0.66 mL water. The solvent was evaporated under nitrogen and the lipid extract was quantified and dissolved in 1.0 mL chloroform. The tissues, after lipid extraction, were dried in an oven at 105 °C and dry weights were determined (3–20 mg).

Lipid extracts were analyzed on a triple quadrupole mass spectrometer equipped for electrospray ionization (ESI–MS/MS; Applied Biosystems API 4000, Waltham, MA, USA). Acquisition and ESI-MS/MS analysis parameters are shown in Tables 4 and 5, respectively. The lipids in each class were quantified in comparison to two internal standards of the class. Lipid species within each head group were identified by total carbon number and total double bonds. The molecular species of each head class were quantified by comparrison with the signals of the internal standards [25].

Table 4. Acquisition parameters of lipid classes using Applied Biosystems API 4000 triple quadrupole mass spectrometer.

Parameter	PA	PC/LPC	PE/LPE	PG	PI	PS	DGDG	MGDG
Typical Scan Time (min)	3.51	1.28	3.34	3.21	4.00	4.01	1.67	1.67
Depolarization Potential (V)	100	100	100	100	100	100	90	90
Exit Potential (V)	14	14	14	14	14	14	10	10
Collision Energy (V)	25	40	28	20	25	26	24	21
Collision Exit Potential (V)	14	14	14	14	14	14	23	23

Table 5. ESI-MS/MS analysis parameters (using Applied Biosystems API 4000) for plant lipids. DGDG, digalactosyldiacylglycerol; ESI-MS/MS, electrospray ionization tandem mass spectrometry; MGDG, monogalactosyldiacylglycerol; PA, phosphatidic acid; PC, phosphatidylcholine; LPC, lysophosphatidylcholine; PE, phosphatidylethanolamine; LPE, lysophosphatidylethanolamine; PG, phosphatidylglycerol; LPG, lysophosphatidylglycerol; PI, phosphatidylinositol; PS, phosphatidylserine.

Class	Ion Analyzed	Positive Ion Scan Mode	m/z Range	Reference
PA	$(M + NH_4)^-$	NL of 115.00	500–850	[48]
PC/LPC	$(M + H)^-$	Pre of m/z 184.07	450–960	[49]
PE/LPE	$(M + H)^-$	NL of 141.02	420–920	[49]
PG	$(M + NH_4)^-$	NL of 189.04	650–1000	[50]
LPG	$(M-H)^-$	Prec 153	-	[25]
PI	$(M + NH_4)^-$	NL of 277.06	790–950	[50]
PS	$(M + H)^-$	NL of 185.01	600–920	[49]
DGDG	$(M + NH_4)^-$	NL of 341.13	890–1050	[51]
MGDG	$(M + NH_4)$	NL of 179.08	700–900	[51]

Average coefficient of variation (CoV) for lipid analytes is a function of corrections used in data processing. CoV is equal to the standard deviation of the measurements for each analyte divided by the average. CoV was calculated from the quality check samples, without any correction, using only the linear trend correction within each day's sample set, only the correction to the overall average across sample sets, or both corrections (as conducted on the experimental data).

4.6. Statistical Analysis

Freeze tests with membrane injury and chlorophyll fluorescence were analyzed using an analysis of variance in Minitab 19 software (Minitab, LLC., Pennsylvania State College, PA). Temperature (5 levels) and collection month (4 levels) were used an explanatory variable, and genotype (4 levels) was used as a blocking factor. Analysis of LT50 values, NRD, and lipids was also conducted using an analysis of variance with collection month as the only explanatory variable and genotype as the blocking factor. The relationships between response variables were assessed using linear regression. Genotype was included as a blocking variable in a regression model when it accounted for a significant amount of variation. Statistical assumption of normality, homogeneity, and independence was verified for each analysis.

5. Conclusions

This experiment was able to address each of the original objectives. Cold acclimation was confirmed in balsam fir, with trees able to tolerate colder temperatures throughout autumn assessed via decreased membrane injury and maintenance of chlorophyll fluorescence. LT50 values decreased from approximately $-23\,^{\circ}\mathrm{C}$ in September to $-46\,^{\circ}\mathrm{C}$ in December. Polar lipid distribution changed during cold acclimation, with a significant shift from MGDG to DGDG and from GLs to PLs. The highest increase in PLs was observed in PC, though most other PLs also increased significantly. Changes in lipids were significantly related to cold acclimation.

Needle retention improved throughout cold acclimation. Needle retention was also significantly related to cold tolerance and MGDG:DGDG. In general, higher cold acclimation or a shift from MDGD to DGDG also increased needle retention. Overall, it was concluded the cold acclimation is beneficial to needle retention, possibly through increased stability of membranes due to a relative increase in DGDG. The mechanism through which cold acclimation affects needle retention requires further work, and factors beyond polar lipid shifts should also be considered moving forward.

Author Contributions: Conceptualization, all authors; methodology, all authors; software, M.T.M. and G.E.M.; validation, G.E.M.; formal analysis, M.T.M. and G.E.M.; investigation, G.E.M.; resources, R.R.L.; data curation, M.T.M. and G.E.M.; writing—original draft preparation, M.T.M.; writing—review and editing, all authors; visualization, M.T.M.; supervision, R.R.L., C.D.C., C.C.U. and M.T.M.; project administration, R.R.L.; funding acquisition, R.R.L. All authors have read and agreed to the published version of the manuscript.

Funding: This research was funded by NSF grants MCB 0455318, MCB 092063, DBI 0521587, DBI 1228622, Kansas INBRE (NIH Grant P20 RR16475 from the INBRE program from the National Center for Research Resources, NSF EPSCoR grant EPS-0236913, Kansas Technology Enterprise Corporation, and Kansas State University).

Institutional Review Board Statement: Not applicable.

Informed Consent Statement: Not applicable.

Data Availability Statement: Data are contained within the article.

Acknowledgments: We thank the Nova Scotia Department of Natural resources for access to the clonal balsam fir orchard in Debert, NS. We also thank the Kansas Lipidomic Research Center at Kansas State University for assistance in lipid analysis.

Conflicts of Interest: The authors declare no conflict of interest.

References

1. Collier, J.; MacLean, D.A.; D'Orangeville, L.; Taylor, A.R. A Review of Climate Change Effects on the Regeneration Dynamics of Balsam Fir. *For. Chron.* **2022**, *98*, 54–65. [CrossRef]

2. Burns, R.M.; Honkala, B.H. *Silvics of North America: 1. Conifers*; US Department of Agriculture, Forest Service: Washington, DC, USA, 1990; Volume 1.

3. Lada, R.R.; MacDonald, M.T. Understanding the Physiology of Postharvest Needle Abscission in Balsam Fir. *Front. Plant Sci.* **2015**, *6*, 1069. [CrossRef] [PubMed]

4. Thiagarajan, A.; MacDonald, M.T.; Lada, R. Environmental and Hormonal Physiology of Postharvest Needle Abscission in Christmas Trees. *Crit. Rev. Plant Sci.* **2016**, *35*, 1–17. [CrossRef]

5. Environment Canada Historical Data - Climate - Environment and Climate Change Canada. Available online: https://climate.weather.gc.ca/historical_data/search_historic_data_e.html (accessed on 23 October 2023).

6. Beck, E.H.; Fettig, S.; Knake, C.; Hartig, K.; Bhattarai, T. Specific and Unspecific Responses of Plants to Cold and Drought Stress. *J. Biosci.* **2007**, *32*, 501–510. [CrossRef]

7. Thakur, P.; Kumar, S.; Malik, J.A.; Berger, J.D.; Nayyar, H. Cold Stress Effects on Reproductive Development in Grain Crops: An Overview. *Environ. Exp. Bot.* **2010**, *67*, 429–443. [CrossRef]

8. Chang, C.Y.; Bräutigam, K.; Hüner, N.P.A.; Ensminger, I. Champions of Winter Survival: Cold Acclimation and Molecular Regulation of Cold Hardiness in Evergreen Conifers. *N. Phytol.* **2021**, *229*, 675–691. [CrossRef]
9. Vogg, G.; Heim, R.; Gotschy, B.; Beck, E.; Hansen, J. Frost Hardening and Photosynthetic Performance of Scots Pine (*Pinus sylvestris*, L.). II. Seasonal Changes in the Fluidity of Thylakoid Membranes. *Planta* **1998**, *204*, 201–206. [CrossRef]
10. Crosatti, C.; Rizza, F.; Badeck, F.W.; Mazzucotelli, E.; Cattivelli, L. Harden the Chloroplast to Protect the Plant. *Physiol. Plant.* **2013**, *147*, 55–63. [CrossRef]
11. Ivanova, A.P.; Stefanov, K.L.; Yordanov, I.T. Effect of Cytokinin 4-PU-30 on the Lipid Composition of Water Stressed Bean Plants. *Biol. Plant.* **1998**, *41*, 155–159. [CrossRef]
12. Degenkolbe, T.; Giavalisco, P.; Zuther, E.; Seiwert, B.; Hincha, D.K.; Willmitzer, L. Differential Remodeling of the Lipidome during Cold Acclimation in Natural Accessions of *Arabidopsis thaliana*: Lipidomics of *Arabidopsis* Cold Acclimation. *Plant J.* **2012**, *72*, 972–982. [CrossRef]
13. Seiwert, D.; Witt, H.; Ritz, S.; Janshoff, A.; Paulsen, H. The Nonbilayer Lipid MGDG and the Major Light-Harvesting Complex (LHCII) Promote Membrane Stacking in Supported Lipid Bilayers. *Biochemistry* **2018**, *57*, 2278–2288. [CrossRef] [PubMed]
14. Yoshida, S.; Uemura, M. Protein and Lipid Compositions of Isolated Plasma Membranes from Orchard Grass (*Dactylis glomerata* L.) and Changes during Cold Acclimation. *Plant Physiol.* **1984**, *75*, 31–37. [CrossRef] [PubMed]
15. Nokhsorov, V.V.; Dudareva, L.V.; Senik, S.V.; Chirikova, N.K.; Petrov, K.A. Influence of Extremely Low Temperatures of the Pole of Cold on the Lipid and Fatty-Acid Composition of Aerial Parts of the Horsetail Family (Equisetaceae). *Plants* **2021**, *10*, 996. [CrossRef] [PubMed]
16. MacDonald, G.E.; Lada, R.R.; Caldwell, C.D.; Udenigwe, C.C.; MacDonald, M.T. Linking Changes in Fatty Acid Composition to Postharvest Needle Abscission Resistance in Balsam Fir Trees. *Forests* **2022**, *13*, 800. [CrossRef]
17. Atucha Zamkova, A.-A.; Steele, K.A.; Smith, A.R. Methods for Measuring Frost Tolerance of Conifers: A Systematic Map. *Forests* **2021**, *12*, 1094. [CrossRef]
18. Murray, M.B.; Cape, J.N.; Fowler, D. Quantification of Frost Damage in Plant Tissues by Rates of Electrolyte Leakage. *N. Phytol.* **1989**, *113*, 307–311. [CrossRef]
19. Bachofen, C.; Wohlgemuth, T.; Ghazoul, J.; Moser, B. Cold Temperature Extremes during Spring Do Not Limit the Range Shift of Mediterranean Pines into Regions with Intermittent Frost. *Funct. Ecol.* **2016**, *30*, 856–865. [CrossRef]
20. Adams, G.T.; Perkins, T.D. Assessing Cold Tolerance in Picea Using Chlorophyll Fluorescence. *Environ. Exp. Bot.* **1993**, *33*, 377–382. [CrossRef]
21. Perks, M.P.; Osborne, B.A.; Mitchell, D.T. Rapid Predictions of Cold Tolerance in Douglas-Fir Seedlings Using Chlorophyll Fluorescence after Freezing. *N. For.* **2004**, *28*, 49–62. [CrossRef]
22. Sakai, A. Comparative Study on Freezing Resistance of Conifers with Special Reference to Cold Adaptation and Its Evolutive Aspects. *Can. J. Bot.* **1983**, *61*, 2323–2332. [CrossRef]
23. Norgaard Nielsen, C.C.; Rasmussen, H.N. Frost Hardening and Dehardening in Abies Procera and Other Conifers under Differing Temperature Regimes and Warm-Spell Treatments. *Forestry* **2009**, *82*, 43–59. [CrossRef]
24. MacDonald, G.E.; Lada, R.R.; Caldwell, C.D.; Udenigwe, C.; MacDonald, M. Lipid and Fatty Acid Changes Linked to Postharvest Needle Abscission in Balsam Fir, Abies Balsamea. *Trees* **2020**, *34*, 297–305. [CrossRef]
25. Welti, R.; Li, W.; Li, M.; Sang, Y.; Biesiada, H.; Zhou, H.-E.; Rajashekar, C.B.; Williams, T.D.; Wang, X. Profiling Membrane Lipids in Plant Stress Responses. *J. Biol. Chem.* **2002**, *277*, 31994–32002. [CrossRef]
26. Dörmann, P.; Benning, C. Galactolipids Rule in Seed Plants. *Trends Plant Sci.* **2002**, *7*, 112–118. [CrossRef]
27. Oquist, G. Seasonally Induced Changes in Acyl Lipids and Fatty Acids of Chloroplast Thylakoids of *Pinus silvestris*: A Correlation between the Level of Unsaturation of Monogalactosyldiglyceride and the Rate of Electron Transport. *Plant Physiol.* **1982**, *69*, 869–875. [CrossRef] [PubMed]
28. Uemura, M.; Steponkus, P.L. Effect of Cold Acclimation on the Lipid Composition of the Lnner and Outer Membrane of the Chloroplast Envelope Lsolated from Rye Leaves. *Plant Physiol.* **1997**, *114*, 1493–1500. [CrossRef] [PubMed]
29. Sakurai, I.; Mizusawa, N.; Wada, H.; Sato, N. Digalactosyldiacylglycerol Is Required for Stabilization of the Oxygen-Evolving Complex in Photosystem II. *Plant Physiol.* **2007**, *145*, 1361–1370. [CrossRef] [PubMed]
30. Heemskerk, J.W.M.; Storz, T.; Schmidt, R.R.; Heinz, E. Biosynthesis of Digalactosyldiacylglycerol in Plastids from 16:3 and 18:3 Plants. *Plant Physiol.* **1990**, *93*, 1286–1294. [CrossRef]
31. Rocha, J.; Nitenberg, M.; Girard-Egrot, A.; Jouhet, J.; Maréchal, E.; Block, M.A.; Breton, C. Do Galactolipid Synthases Play a Key Role in the Biogenesis of Chloroplast Membranes of Higher Plants? *Front. Plant Sci.* **2018**, *9*, 126. [CrossRef]
32. Kojima, M.; Shiraki, H.; Ohnishi, N.; Ito, S. Seasonal Change Sin Glycolipids and Phospholipids in *Pinus nigra* Needles. *Res. Bull. Obihiro* **1990**, *17*, 13–19.
33. Orlova, I.V.; Serebriiskaya, T.S.; Popov, V.; Merkulova, N.; Nosov, A.M.; Trunova, T.I.; Tsydendambaev, V.D.; Los, D.A. Transformation of Tobacco with a Gene for the Thermophilic Acyl-Lipid Desaturase Enhances the Chilling Tolerance of Plants. *Plant Cell Physiol.* **2003**, *44*, 447–450. [CrossRef]
34. Van Meer, G.; Voelker, D.R.; Feigenson, G.W. Membrane Lipids: Where They Are and How They Behave. *Nat. Rev. Mol. Cell. Biol.* **2008**, *9*, 112–124. [CrossRef]
35. Selvy, P.E.; Lavieri, R.R.; Lindsley, C.W.; Brown, H.A. Phospholipase D: Enzymology, Functionality, and Chemical Modulation. *Chem. Rev.* **2011**, *111*, 6064–6119. [CrossRef] [PubMed]

36. Hou, Q.; Ufer, G.; Bartels, D. Lipid Signalling in Plant Responses to Abiotic Stress: Lipid Signalling in Plant Responses to Abiotic Stress. *Plant Cell Environ.* **2016**, *39*, 1029–1048. [CrossRef]
37. Kiełbowicz-Matuk, A.; Banachowicz, E.; Turska-Tarska, A.; Rey, P.; Rorat, T. Expression and Characterization of a Barley Phosphatidylinositol Transfer Protein Structurally Homologous to the Yeast Sec14p Protein. *Plant Sci.* **2016**, *246*, 98–111. [CrossRef] [PubMed]
38. MacDonald, M.T.; Lada, R.R. Changes in Endogenous Hormone Levels Explains Seasonal Variation in Balsam Fir Needle Abscission Patterns. *J. Plant Growth Regul.* **2017**, *36*, 723–733. [CrossRef]
39. MacDonald, M.T.; Lada, R.R.; Veitch, R.S.; Thiagarajan, A.; Adams, A.D. Postharvest Needle Abscission Resistance of Balsam Fir (*Abies Balsamea*) Is Modified by Harvest Date. *Can. J. For. Res.* **2014**, *44*, 1394–1401. [CrossRef]
40. Juurakko, C.L.; diCenzo, G.C.; Walker, V.K. Cold Acclimation and Prospects for Cold-Resilient Crops. *Plant Stress* **2021**, *2*, 100028. [CrossRef]
41. Catalá, R.; López-Cobollo, R.; Mar Castellano, M.; Angosto, T.; Alonso, J.M.; Ecker, J.R.; Salinas, J. The *Arabidopsis* 14-3-3 Protein RARE COLD INDUCIBLE 1A Links Low-Temperature Response and Ethylene Biosynthesis to Regulate Freezing Tolerance and Cold Acclimation. *Plant Cell* **2014**, *26*, 3326–3342. [CrossRef] [PubMed]
42. Baier, M.; Bittner, A.; Prescher, A.; Van Buer, J. Preparing Plants for Improved Cold Tolerance by Priming. *Plant Cell Environ.* **2019**, *42*, 782–800. [CrossRef]
43. Hassan, Q.K. Spatial Mapping of Growing Degree Days: An Application of MODIS-Based Surface Temperatures and Enhanced Vegetation Index. *J. Appl. Remote Sens.* **2007**, *1*, 013511. [CrossRef]
44. MacDonald, M.T.; Lada, R.R.; Dorais, M.; Pepin, S. Endogenous and Exogenous Ethylene Induces Needle Abscission and Cellulase Activity in Post-Harvest Balsam Fir (*Abies balsamea* L.). *Trees* **2011**, *25*, 947–952. [CrossRef]
45. Man, R.; Lu, P.; Dang, Q.-L. Cold Tolerance of Black Spruce, White Spruce, Jack Pine, and Lodgepole Pine Seedlings at Different Stages of Spring Dehardening. *N. For.* **2021**, *52*, 317–328. [CrossRef]
46. Odlum, K.D.; Blake, T.J. A Comparison of Analytical Approaches for Assessing Freezing Damage in Black Spruce Using Electrolyte Leakage Methods. *Can. J. Bot.* **1996**, *74*, 952–958. [CrossRef]
47. Bligh, E.G.; Dyer, W.J. A Rapid Method of Total Lipid Extraction and Purification. *Can. J. Biochem. Physiol.* **1959**, *37*, 911–917. [CrossRef]
48. Munnik, T.; Heilmann (Eds.) *Plant Lipid Signaling Protocols*; Methods in Molecular Biology; Humana Press: Totowa, NJ, USA, 2013; Volume 1009, ISBN 978-1-62703-400-5.
49. Brügger, B.; Erben, G.; Sandhoff, R.; Wieland, F.T.; Lehmann, W.D. Quantitative Analysis of Biological Membrane Lipids at the Low Picomole Level by Nano-Electrospray Ionization Tandem Mass Spectrometry. *Proc. Natl. Acad. Sci. USA* **1997**, *94*, 2339–2344. [CrossRef] [PubMed]
50. Taguchi, R.; Houjou, T.; Nakanishi, H.; Yamazaki, T.; Ishida, M.; Imagawa, M.; Shimizu, T. Focused Lipidomics by Tandem Mass Spectrometry. *J. Chromatogr. B* **2005**, *823*, 26–36. [CrossRef] [PubMed]
51. Isaac, G.; Jeannotte, R.; Esch, S.; Welti, R. New Mass Spectrometry-Based Strategies for Lipids. In *Genetic Engineering*; Springer: Boston, MA, USA, 2007; Volume 28.

 International Journal of
Molecular Sciences

Article

Genome-Wide Identification of *GmSPS* Gene Family in Soybean and Expression Analysis in Response to Cold Stress

Jiafang Shen [1], Yiran Xu [2], Songli Yuan [1], Fuxiao Jin [1], Yi Huang [1], Haifeng Chen [1], Zhihui Shan [1], Zhonglu Yang [1], Shuilian Chen [1], Xinan Zhou [1,*] and Chanjuan Zhang [1,*]

[1] Key Laboratory of Biology and Genetic Improvement of Oil Crops, Ministry of Agriculture and Rural Affairs, Oil Crops Research Institute of Chinese Academy of Agricultural Sciences, Wuhan 430062, China
[2] College of Life Sciences, Wuhan University, Wuhan 430072, China
* Correspondence: zhouxinan@caas.cn (X.Z.); zhangchanjuan@caas.cn (C.Z.)

Abstract: Sucrose metabolism plays a critical role in development, stress response, and yield formation of plants. Sucrose phosphate synthase (SPS) is the key rate-limiting enzyme in the sucrose synthesis pathway. To date, genome-wide survey and comprehensive analysis of the *SPS* gene family in soybean (*Glycine max*) have yet to be performed. In this study, seven genes encoding SPS were identified in soybean genome. The structural characteristics, phylogenetics, tissue expression patterns, and cold stress response of these *GmSPSs* were investigated. A comparative phylogenetic analysis of SPS proteins in soybean, *Medicago truncatula*, *Medicago sativa*, *Lotus japonicus*, Arabidopsis, and rice revealed four families. GmSPSs were clustered into three families from A to C, and have undergone five segmental duplication events under purifying selection. All *GmSPS* genes had various expression patterns in different tissues, and family A members *GmSPS13/17* were highly expressed in nodules. Remarkably, all *GmSPS* promoters contain multiple low-temperature-responsive elements such as potential binding sites of inducer of CBF expression 1 (ICE1), the central regulator in cold response. qRT-PCR proved that these *GmSPS* genes, especially *GmSPS8/18*, were induced by cold treatment in soybean leaves, and the expression pattern of *GmICE1* under cold treatment was similar to that of *GmSPS8/18*. Further transient expression analysis in *Nicotiana benthamiana* and electrophoretic mobility shift assay (EMSA) indicated that *GmSPS8* and *GmSPS18* transcriptions were directly activated by GmICE1. Taken together, our findings may aid in future efforts to clarify the potential roles of *GmSPS* genes in response to cold stress in soybean.

Keywords: soybean; sucrose phosphate synthase; genome-wide survey; gene expression; cold stress

Citation: Shen, J.; Xu, Y.; Yuan, S.; Jin, F.; Huang, Y.; Chen, H.; Shan, Z.; Yang, Z.; Chen, S.; Zhou, X.; et al. Genome-Wide Identification of *GmSPS* Gene Family in Soybean and Expression Analysis in Response to Cold Stress. *Int. J. Mol. Sci.* **2023**, *24*, 12878. https://doi.org/10.3390/ijms241612878

Academic Editor: Martin Bartas

Received: 24 May 2023
Revised: 12 August 2023
Accepted: 14 August 2023
Published: 17 August 2023

1. Introduction

Environmental factors such as cold and osmotic stresses could affect plant growth. Cold stress includes chilling (0–15 °C) and freezing (<0 °C), which have an effect on the growth and yield of plants [1,2]. Low temperature affects the absorption of water and nutrients by plants, as well as the fluidity of membrane, and affects gene expression and protein synthesis [2]. It affects cell metabolism by reducing the rate of biochemical reaction or affecting gene expression reprogramming. Cold stress will not only inhibit the metabolic reaction of plants, but also produce osmotic reactions to inhibit plant growth [2,3]. In addition to winter-habit plants, many important crops such as rice, corn, and soybean are very sensitive to cold [4]. Cold stress will seriously harm the growth of soybean, especially spring soybean cultivars, and lead to yield reduction, so it is vital to reveal the adaptation mechanism of soybean to cold stress.

In order to survive, plants have evolved complex mechanisms in response to cold stress. In previous studies, many key factors in the cold stress signal pathway have been found. *COR* (cold-responsive) genes can be upregulated by *CBFs* (C-repeat/DRE-binding factors) under cold stress [4,5], which could be induced by low temperature and had conserved functions in flowering plants in response to cold stress [6]. Overexpression of *CBF1/DREB1b* and *CBF3* can enhance the freezing resistance of Arabidopsis [7,8]. ICE transcription factor can induce the expression of *CBF* genes under normal temperature, and the *ice1* mutant blocked the expression of *CBF3* itself and its downstream genes to increase the sensitivity of plants to cold stress in Arabidopsis [9]. These results indicate that the ICE-CBF-COR signal pathway is crucial for plants in response to cold stress.

Soybean is a worldwide economical crop and a major source of oil and protein [10]. Some studies have been reported cold stress responses in soybean. Under cold treatment, transcriptome analyses identified many cold-stress-related genes, including *CBF/DREB* [11]. In addition, the plant hormone ethylene may inhibit soybean CBF/DREB1 pathway through EIN3 (ethylene-insensitive 3) [12]. The sensitivity of soybean to cold stress varies at different developmental stages. During the vegetative growth stage, *CBF* can be strongly and briefly induced under cold stress, but low-temperature stress delays the reproductive stage of soybean [13,14]. *GmTCF1a* (tolerant to chilling and freezing 1a) positively regulates cold tolerance in soybean, which is independent of the CBF pathway as *AtTCF1* [15,16].

The physiological changes of plants are closely related to freezing resistance, including the accumulation of some solutes, such as soluble sugar, proline, and other lower-molecular-weight solutes [2]. The increase of soluble sugars such as sucrose, glucose, and fructose can function as osmotic protective substances to enhance the resistance of plants to cold [17]. A previous study showed that soluble sugar accumulated significantly within 2 h after plants were exposed to low temperature [18]. Prolonged cold will lead to more sucrose accumulation, which enables *Deschampsia antarctica* to survive in the Antarctic [19]. When petunia was cold-treated, more sugar and starch were accumulated in the source leaves, while the starch content in the sink tissue decreased. Moreover, the activity of cwINV (cell wall invertase) in the sink tissue decreased, showing the reduction of sugar input and utilization [20]. Accumulating higher sugar in the source tissue can increase osmotic protection and thus enhance the ability of photosynthetic tissue to resist cold stress, which is the core element of cold stress response [21,22]. These results suggest that sugar accumulation plays an important role in plant resistance to cold stress.

Sucrose is the main transport form of photosynthate in plants and plays a crucial role in the normal growth and development of plants [23]. Sucrose phosphate synthase plays a central role in the production of sucrose in photosynthetic cells. The sucrose synthesis involves a two-step reaction, and SPS (EC 2.4.1.14) is the key rate-limiting enzyme in the sucrose synthesis pathway, which catalyzes the conversion of UDP-glucose and fructose-6-phosphate to sucrose-6-phosphate, which is then hydrolyzed to sucrose by sucrose-phosphatase (SPP) [24,25]. SPS activity has been shown to be linked with sucrose accumulation [26]. Spinach *SPS* expression in cotton promotes sucrose synthesis and improves fiber quality [27]. At the stage of fruit ripening, both *SPS* expression and activity are upregulated, which in turn promotes sucrose synthesis [28–30]. SPS in spinach has also been proposed to play significant roles in stressful environment conditions. When plants are exposed to low-temperature stress, the expression of *SPS* increases dramatically and more sucrose is synthesized in response to cold stress [31]. SPS is encoded by a multigene family. Genome-wide survey of *SPS* genes has been performed in Arabidopsis, rice, maize, wheat, tomato, cassava, Litchi, and other plants [32–36]. According to the amino acid sequences, SPS proteins can be divided into four categories (A, B, C, D), with the branch D only present in the poaceae [37]. Expression and functions of different *SPS* genes vary among different categories in different plants. In soybean leaves, drought and K-deficiency increased soluble sugar content and SPS activity [38], and accumulation of sugar content was correlated with increased SPS activity, suggesting SPS might play important roles in response to abiotic stress in soybean [39].

In this study, a genome-wide survey was performed to explore the special sequence and expression characteristics of soybean *SPS* family genes. The objective focused on the analysis of phylogenetic relationships, duplications patterns, gene structures, conserved motifs, phosphorylation sites, cis-elements, and tissue expression patterns. The expression change of *GmSPS* genes under cold stress were also examined. Furthermore, we suggest that the expression level of *GmSPS8/18* may be upregulated by GmICE1. Consequently, these findings may provide foundations for functional investigation of *GmSPS* genes in soybean.

2. Results

2.1. Genome-Wide Identification of GmSPS Genes in Soybean

HMMER (v3.3.2) searches using three conserved domains (PF00862, PF00534, PF05116) of SPS protein and BLASTP using the Arabidopsis SPS sequences were performed in the soybean protein database. Then, 82 and 18 candidate sequences were generated through HMMER (v3.3.2) search and BLASTP, respectively. Finally, a total of seven different soybean loci encoding SPS proteins were identified by removing redundant sequences. The seven putative soybean SPS proteins all contained the sucrsPsyn_pln, and were confirmed by NCBI-CDD (Table S1). Based on the annotation file of soybean genomic sequences, the seven putative soybean *SPS* genes were found to be distributed on seven different chromosomes, respectively. According to the chromosomal location, these soybean *SPS* genes were named *GmSPS4/6/8/13/14/17/18*.

Sequence alignments showed high identity and similarity in GmSPS4/6, GmSPS13/17, and GmSPS8/14/18 at both amino acid (79.29% to 97.26%) and nucleotide level (83.18% to 96.77%) (Table S2). The detailed predictions of *GmSPS* genes are described in Table 1. The full length of GmSPS-predicted proteins vary from 778 to 1064 amino acids and the molecular weight (Mw) is arranged from 87.6 to 119.3 kDa. In addition, the predicted isoelectric point (pI) ranges from 5.94 to 6.31. More than one transcript were predicted in three *GmSPS* genes. *GmSPS13* contains five transcripts, *GmSPS14* contains two transcripts, and *GmSPS17* contains four transcripts. Moreover, subcellular localization prediction suggests that GmSPS4/8/14 and GmSPS6/13/17/18 proteins are located in the cytoplasm and nucleus, respectively.

Table 1. All seven *GmSPS* genes, including their genome location and physical properties.

Gene Name	Gene ID	Genome Location	Transcript Numbers	Protein Size (aa)	MW (kDa)	pI	Subcellular Location
GmSPS4	Glyma.04g110200	Chr04: 11369193-11375676	1	778	87.58	5.99	Cytoplasm
GmSPS6	Glyma.06g323700	Gm06: 50675372-50684602	1	1038	117.09	6.31	Nucleus
GmSPS8	Glyma.08g308600	Gm08: 42128704-42135717	1	1056	118.80	6.18	Cytoplasm
GmSPS13	Glyma.13g161600	Gm13: 27135652-27142180	5	1060	118.01	6.04	Nucleus
GmSPS14	Glyma.14g029100	Gm14: 2121804-2128414	2	1064	119.33	5.94	Cytoplasm
GmSPS17	Glyma.17g109700	Gm17: 8599983-8606617	4	1060	118.08	6.09	Nucleus
GmSPS18	Glyma.18g108100	Gm18: 12235657-12242570	1	1054	118.39	6.1	Nucleus

2.2. Phylogenetic Analysis of SPS Proteins

A phylogenetic tree was constructed using the MEGA software (v11.0.11) to assess the phylogenetic relationship of SPSs from soybean with those from *M. truncatula, M. sativa, L. japonicus*, Arabidopsis, and rice, which are leguminous, dicotyledonous, and monocotyledonous model plants, respectively (Figure 1). Previous studies isolated three MsSPS members (MsSPSA, MsSPSB, MsSPSB3) in *M. sativa* [40]. However, we identified two MsSPSB3 members (named MsSPSB3-1 and MsSPSB3-2) in diploid *M. sativa*, which was sequenced in 2020 [41]. *MsSPSB3* amplified in previous studies was a partial of *MsSPSB3-1* and *MsSPSB3-2*. Thus, four SPSs (MsSPSA, MsSPSB, MsSPSB3-1, and MsSPSB3-2) from *M. sativa* were used to construct the phylogenetic tree. These SPS members from six plant species can be clustered into four distinct families (A, B, C, D). The family D is specific to rice, and the family C does not include alfalfa SPS members, which is consistent with previously studies [25,40]. Among the four families, family B had the largest number of SPS members, with twelve. The GmSPS had a closer relationship with leguminous plants than Arabidopsis and rice. In families A, B, and C, SPSs from soybean are more closely related to *L. japonicus* than two kinds of alfalfa. GmSPS13 and GmSPS17 were grouped into family A, GmSPS8, GmSPS14, and GmSPS18 belonged to family B, and GmSPS4 and GmSPS6 were grouped into family C. No SPS from soybean was grouped into family D. Both soybean and *L. japonicus* had SPS members of family C, but alfalfa did not.

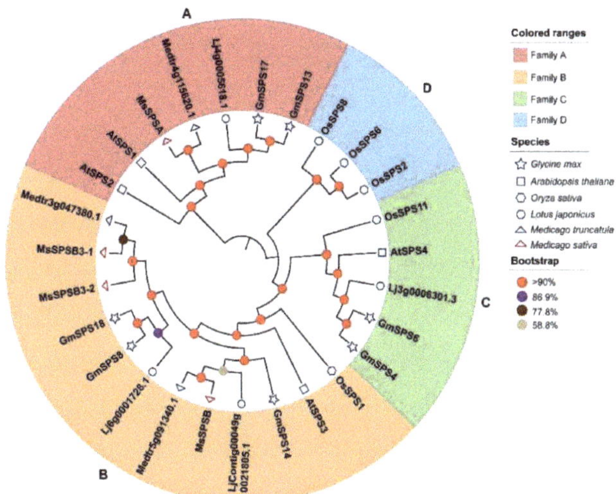

Figure 1. Phylogenetic analysis of SPS proteins. The tree was constructed and relied on the amino acid sequences by MEGA11 using neighbor joining method with JTT + G model and 1000 bootstrap replicates. Solid small circles with different colors are standard for bootstrap values. The tree can be categorized into four groups with different colors. The red, orange, green, and blue colors represent A–D groups, respectively. Different shapes indicate different species.

2.3. Collinearity Analysis of GmSPS Genes

Soybean is a paleotetraploid crop which has undergone two whole genome duplications with very high retention of homologs in the genome [42,43]. Whole genome duplication and segmental and tandem duplication are the important events during soybean genome regions [42]. Seven *GmSPS* genes were not located in the same soybean chromosome. Therefore, no tandem duplication was found in the identified *GmSPS* genes. Collinearity analyses revealed that five gene pairs underwent segmental duplication events (Figure 2).

Genome duplications occurred at approximately 59 and 13 million years ago, resulting in a highly duplicated genome with nearly 75% of the genes present in multiple copies [42]. The Ks values of segment pairs *GmSPS13/17*, *GmSPS4/6*, and *GmSPS8/18* were 0.1107, 0.0995, and 0.1182 (Table 2). The divergence times of these duplicated paralogous pairs were associated with the 13 Mya WGD events. However, the Ks of *GmSPS18/14* and *GmSPS8/14* were larger than 0.6 and the divergence time is associated with the 59 Mya early legume WGD, which indicated that both *GmSPS8* and *GmSPS18* were originated from *GmSPS14* [44]. The Ka/Ks ratio is a measure used to explore the mechanism of gene replication and evolution after ancestor differentiation [45]. A Ka/Ks value < 1 suggests purifying selection, a Ka/Ks value = 1 indicates neutral selection, and a Ka/Ks value > 1 suggests positive selection. In addition, all the *GmSPS* paralogous pairs showed Ka/Ks < 1, suggesting that *GmSPS* genes have undergone purification selection during evolution.

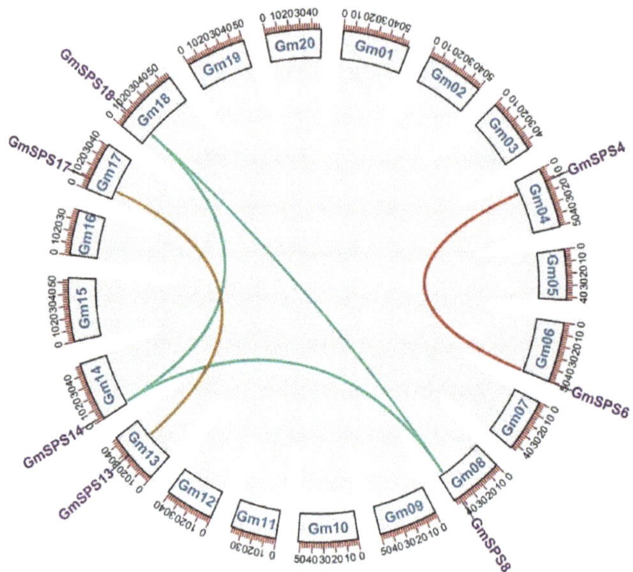

Figure 2. Collinearity analysis of *SPS* genes in soybean. The different colored lines connect two genes with collinearity. The gray lines represent other collinearity regions in the genome of soybean.

Table 2. Identification of substitution rates for homologous *GmSPS* genes.

Segmental Duplicated Genes	Method	Ka	Ks	Ka/Ks	T (Mya)
GmSPS17-GmSPS13	MA	0.012484	0.110769	0.112704	9.079426
GmSPS18-GmSPS14	MA	0.074892	0.634249	0.118079	51.98762
GmSPS6-GmSPS4	MA	0.060117	0.099571	0.60376	8.161582
GmSPS8-GmSPS14	MA	0.068167	0.640031	0.106505	52.46156
GmSPS8-GmSPS18	MA	0.014738	0.118258	0.124628	9.693279

Ka: non-syn substitution rate, Ks: syn substitution rate, T: duplication date.

2.4. Gene Structure and Conserved Motif Analyses of GmSPSs

The gene structures of seven *GmSPS* genes are shown by analyzing the sequence annotation file (Figure 3A). The number of exons and introns in *GmSPS* genes vary from twelve to seventeen and eleven to sixteen, respectively. *GmSPS* genes belonging to the same family possess similar number of exons and introns. Family A members, the putative paralogous pairs (*GmSPS13/17*), contain 13 exons and 12 introns, and family B members *GmSPS8/14/18* all contain 12 exons and 11 introns. Family C members, the putative paralogous pairs (*GmSPS4/6*), contain the most exons (17, 14) and introns (16, 13). Fifteen putative conserved motifs were identified in seven GmSPS proteins based on the amino acid sequence by the program MEME (Figure 3B). The identified multilevel consensus sequences of these motifs are shown in Table S3. All GmSPS proteins contain these fifteen motifs except GmSPS4, which not contain motif 6, motif 8, and motif 9. Most motifs are present in Glycos_transf_1, S6PP and Sucrose_synth domains (Table S3). Motif 5, motif 15, motif 9, motif 4, motif 12, and motif 3 belonged to the sucrose synthase domain. Motif 2, motif 6, motif 10, and motif 1 belonged to the Glycos_transf_1 domain. Motif 13, motif 11, motif 14, and motif 7 belonged to the S6PP domain. Glycos_transf_1 and S6PP domains existed in all GmSPS members. It is worth noting that GmSPS13 did not have the Sucrose_synth domain by using Pfam for functional domain prediction, although it had conserved motifs. This may mean that the Sucrose_synth domain function was missing in GmSPS13. The Glycos_transf_1 domain is associated with transfer of glucosyl, and the S6PP domain may mediate the interaction with SPP.

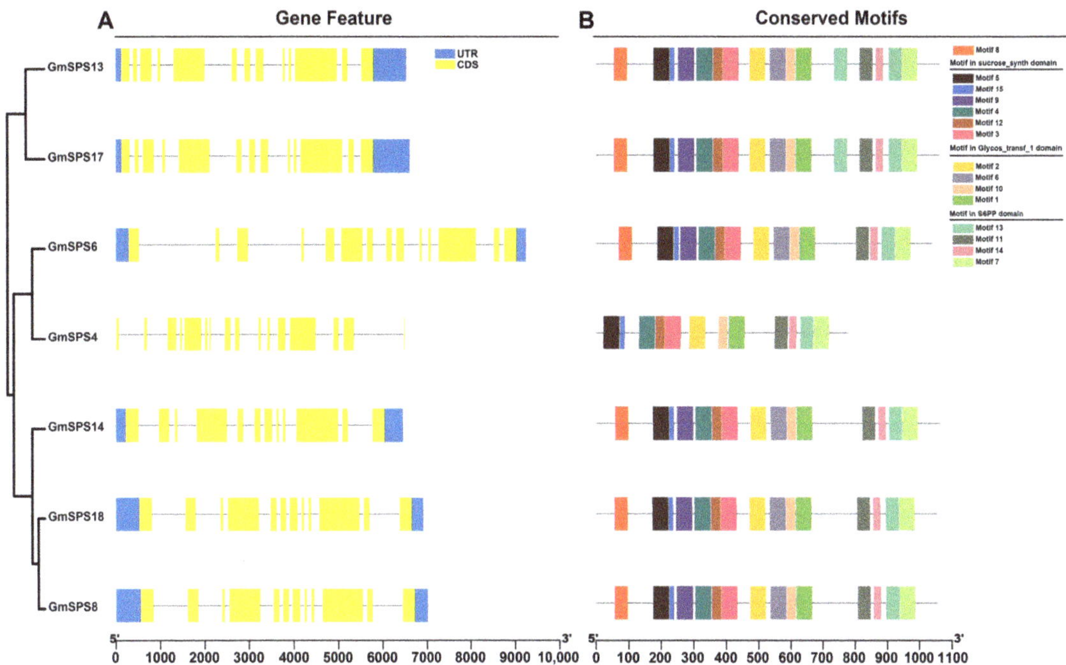

Figure 3. The gene structure and conserved motifs of *GmSPSs*. (**A**) The gene structure of *GmSPS* genes. The *GmSPSs* classified into three groups based on the phylogenetic relationships. The filled boxes and lines represent exons and introns, respectively. The blue and yellow boxes represent UTR and CDS, respectively. (**B**) Conserved motifs in GmSPS proteins. Boxes with different colors represent 15 different conserved motifs. The position and size of *GmSPS* genes and the corresponding proteins can be estimated by the scale at the bottom.

2.5. Analysis of Phosphorylation Sites in GmSPS Proteins

It has been reported that SPS proteins could be modulated by phosphorylation in response to temperature and other abiotic stresses [37]. The predicted phosphorylation sites of GmSPS proteins were analyzed using NetPhos 3.1 Server (Figures 4 and S1). The results show that the main predicted phosphorylation site of GmSPS proteins was serine. In detail, GmSPS13 had the most predicted serine phosphorylation sites with a total number of 90, and GmSPS4 had the least number of 55. GmSPS14 had the maximum number of predicted threonine phosphorylation sites and the GmSPS6 contained the maximum number of predicted tyrosine phosphorylation sites. Previous research indicated that phosphorylation sites Ser-158, Ser-229, and Ser-424 were involved in light–dark regulation, 14-3-3 protein binding, and osmotic stress activation, respectively [37]. Our results demonstrate that Ser-158 is conserved in GmSPS proteins except GmSPS4. Ser-229 is conserved in GmSPS proteins apart from family C members GmSPS4 and GmSPS6. A conversion from S to A was found in GmSPS4 and GmSPS6, which is consistent with the fact that in LcSPS4, the family C member from Litchi is observed [36]. Ser-424 is conserved in GmSPS proteins except family B members GmSPS8, GmSPS14, and GmSPS18 (Figure 4B). A conversion from S to N was found in GmSPS8, GmSPS14, and GmSPS18, which is consistent with the fact that in AtSPS3 and LcSPS3, the family B member from Arabidopsis and Litchi [36], respectively, are observed. Accordingly, our data suggest that GmSPSs contain several predicted phosphorylation sites including a part of serine, threonine, and tyrosine sites, which could possibly function in response to the abiotic stresses.

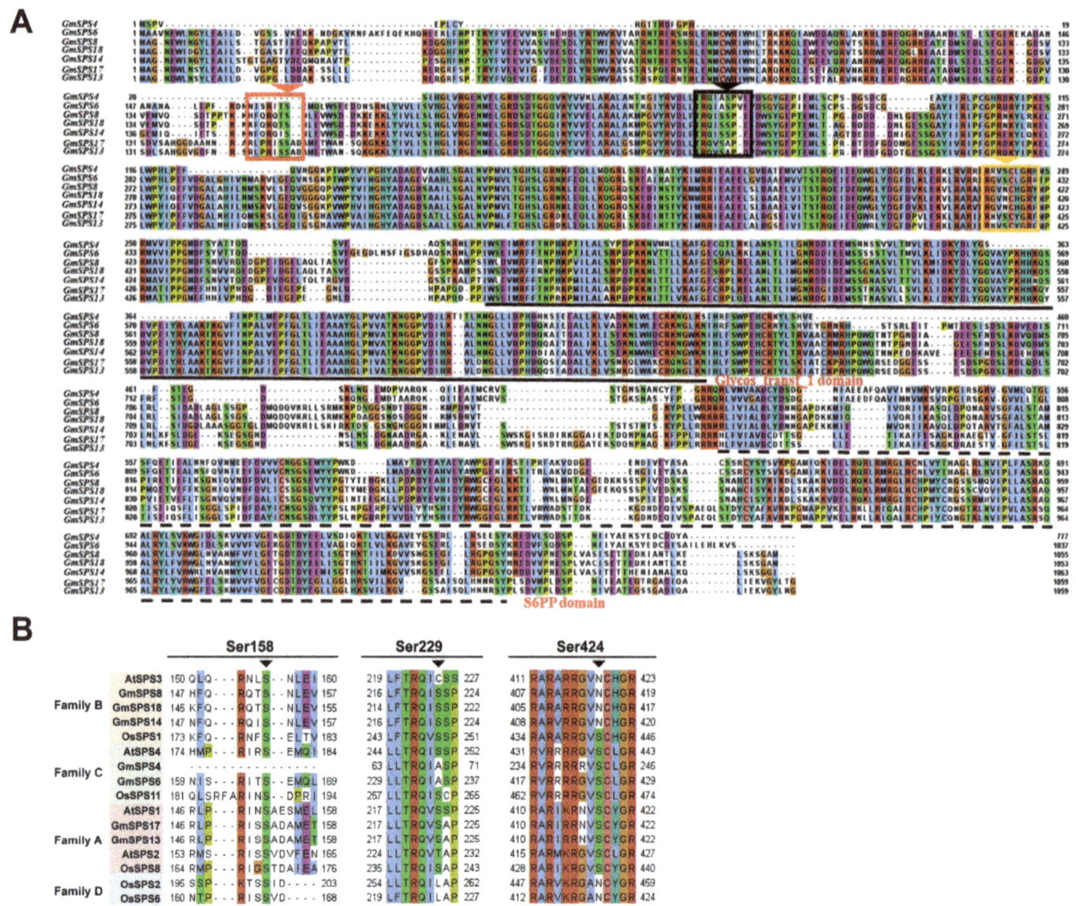

Figure 4. Phosphorylation sites analyses in SPS proteins. (**A**) The protein sequences alignment of GmSPSs. The red, black, and yellow arrows indicate Ser-158, Ser-229, and Ser-424 phosphorylation sites, respectively. Glycos transf_1 domain and S6PP domain are marked with black solid line and black dotted line, respectively. The alignment sequences are shown in jalview using Clustal colourscheme. (**B**) Ser-158, Ser-229, and Ser-424 phosphorylation sites in SPS families of soybean, Arabidopsis, and rice.

2.6. Promoter cis-Element Analysis of GmSPS

Gene promoters are essential for transcriptional regulation [46]. The cis-elements of promoters play significant roles in response to plant growth and various environmental stresses. The 3000 bp sequences upstream of all *GmSPS* genes were analyzed by PlantCARE and PlantPAN (Table S4). Some kinds of cis-elements were found in the promoter regions of *GmSPSs*, including TF binding sites, hormone response, abiotic response, and photoperiod response elements (Figure 5). Predicted cis-elements that could respond to the hormones such as auxin, gibberellin, salicylic acid, abscisic acid, and MeJA (methyl jasmonate) were identified. Some elements associated with abiotic stresses response like defense, low temperature, drought-inducibility, and anaerobic induction in upstream regions of the *GmSPS* genes were also predicted. It is worth noting that all the *GmSPS* members have many low-temperature-responsive elements, and they are closer to the predicted start codon of family B members (*GmSPS8/14/18*) than other *GmSPSs*. These results suggest that *GmSPS* genes may play important roles in response to environmental factors, especially low-temperature stress.

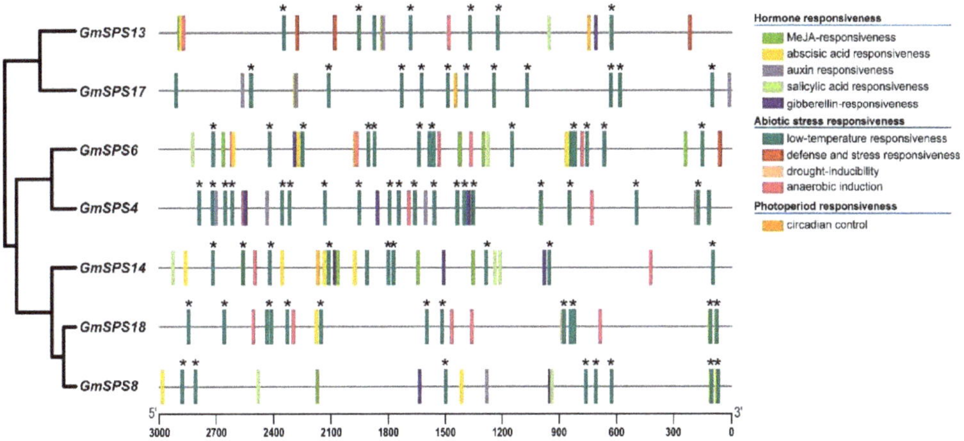

Figure 5. The cis-element analysis of *GmSPSs* promoter regions. Different colored boxes represent various elements. These asterisks represent potential binding sites of transcription factor ICE1. The positions of the cis-elements are shown at the bottom.

2.7. Tissue-Specific Expression Patterns of GmSPS Members in Soybean

To further clarify the expression pattern of *GmSPSs* in soybean, the expression levels of *GmSPSs* in eight different tissues including root, stem, leaf, petiole, flower, SAM (shoot apical meristem), pod, and mature seed were analyzed by qRT-PCR. The relative mRNA abundance of *GmSPS* genes was shown in a heatmap (Figure 6A). The results demonstrated that all *GmSPS* genes were expressed in stem, leaf, petiole, flower, and SAM, while not expressed in root and seed, except *GmSPS13* and *GmSPS17*. Our analysis revealed that *GmSPS8* had the strongest expression level in the flower. In contrast, *GmSPS6/13* and *GmSPS17* had higher expression in pod and petiole compared to other tissues, respectively. The expression patterns may lay the foundation for exploring *GmSPSs* potential functions in the future.

In our previous work, we carried out the transcriptome analysis of soybean nodules at soybean different five developmental stages (branching stage, flowering stage, fruiting stage, pod stage, and harvest stage) inoculated with *Bradyrhizobium japonicum* strain 113-2 [47]. We analyzed the expression abundance of *GmSPS* genes using the published data (Figure 6B) and found significant differences in the expression level of *GmSPS* genes in soybean nodules. Interestingly, the family A members *GmSPS13/17* were highly expressed in nodules at all five stages. This is consistent with previous research showing that *MsSPSA* revealed nodule-enhanced expression [40]. The family B members, *GmSPS8/18*, were not expressed in nodules at any stage, while *GmSPS14* had some expression level in nodules at flowering stage to harvest stage. In addition, the family C member *GmSPS4* was not expressed in nodules, while the paralogous gene *GmSPS6* had some expression level in nodules at five stages. These results suggest function divergence of some homologous genes.

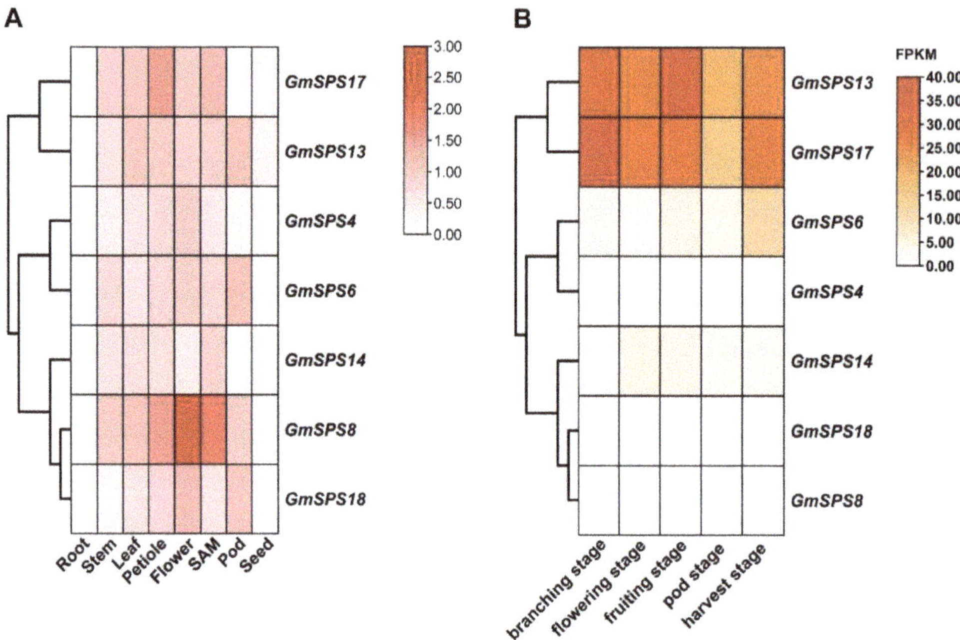

Figure 6. Expression profile of seven *GmSPS* genes in different tissues of soybean. (**A**) Expression patterns of *GmSPS* genes in soybean tissues. The expression value was obtained by qRT-PCR. At the right of the figure, different colors indicate the gene transcript abundance values. (**B**) Expression patterns of *GmSPS* genes in soybean nodules at different stages. FPKM values of *GmSPS* genes were obtained from published data [47]. Different colors of the heat map indicate the gene transcript abundance values.

2.8. Expression Analysis of GmSPS Genes in Response to Cold Stress

It was predicted that a variety of low-temperature-responsive elements would be found in the *GmSPS* promoters (Figure 5). Moreover, previous research has suggested that SPS plays a key role in abiotic stresses, including cold [31,48–50]. To verify whether *GmSPS* genes responded to cold stress, the expression levels of *GmSPSs* were examined in leaves after cold treatment at 0 h, 2 h, 4 h, 6 h, 8 h, 12 h, 24 h, and 48 h. All *GmSPSs* were found to be upregulated under cold treatment (Figure 7). The expression levels of *GmSPS4/6* increased significantly at 8 h and reached the highest value at 12 h after cold treatment. The transcript levels of *GmSPS8/18* began to accumulate after 2 h after cold treatment, and increased 8-fold at 4 h and more than 20-fold from 6 h to 12 h. The levels of *GmSPS13/17* transcripts increased at 6 h and gradually declined from 8 h after cold treatment. The transcript level of *GmSPS14* increased over 10-fold at 6 h after cold treatment. Consequently, our results reveal the potential vital biological functions of *GmSPSs* for cold response.

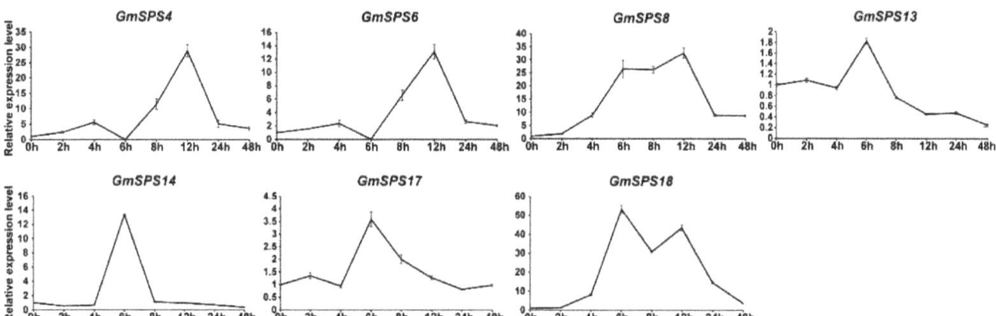

Figure 7. Relative expression levels of *GmSPS* genes in response to cold stress. RNA was extracted from treated leaves at 0 h, 2 h, 4 h, 6 h, 8 h, 12 h, 24 h, and 48 h, and qRT-PCR was performed using the specific primers of *GmSPS* genes. *GmActin11* was used as the internal reference. Treatments at each time point have their own controls, and expression at 0 h was set as "1". The results represent the mean ± SD of three independent biological repetitions.

2.9. GmSPS8/18 Were Upregulated by GmICE1 Involved in Cold Stress

The expression of *GmSPS* genes, especially *GmSPS8* and *GmSPS18*, were significantly upregulated after cold treatment. It was noted that multiple potential binding sites of transcription factor ICE1 existed in the promoter region of *GmSPS* genes. ICE1 is a central regulator in response to cold stress in plants [9], and GmICE1 is the homolog in soybean. We examined the transcript level of *GmICE1* to check whether *GmICE1* and *GmSPS* genes are in a coexpression network. We found that the transcript abundance of *GmICE1* was upregulated and reached a peak at 12 h under cold treatment (Figure 8A), which is similar to that of *GmSPS8* and *GmSPS18*. Whether GmICE1 could directly regulate the expression of *GmSPS8* and *GmSPS18* genes was investigated in transiently transformed tobacco leaves using a GUS reporter assay. The results indicated that tobacco leaves cotransformed with GmICE1 and any one of two soybean promoters produced three- to seven-fold higher value of GUS activity than the control tobacco leaves only transformed with the soybean promoters without GmICE1 (Figure 8B). To verify whether GmICE1 could bind to the promoters of *GmSPS8* and *GmSPS18* in vitro, EMSAs were conducted. The biotin labeled 50 bp DNA fragment containing CACGTG elements (the putative binding site of ICE1) in *GmSPS8* promoter (−66 bp to −115 bp) and in *GmSPS18* promoter (−68 bp to −117 bp) were used as probes. Results revealed that recombinant GmICE1-His could bind the biotin-labeled probes, which was weakened by the unlabeled probes in a dose-dependent manner. These data demonstrated that GmICE1 directly bound to *GmSPS8* and *GmSPS18* promoters (Figure 8C). Taken together, these results confirm that GmICE1 could bind to the promoter regions of *GmSPS8* and *GmSPS18* in vitro and might activate their transcriptions under cold stress.

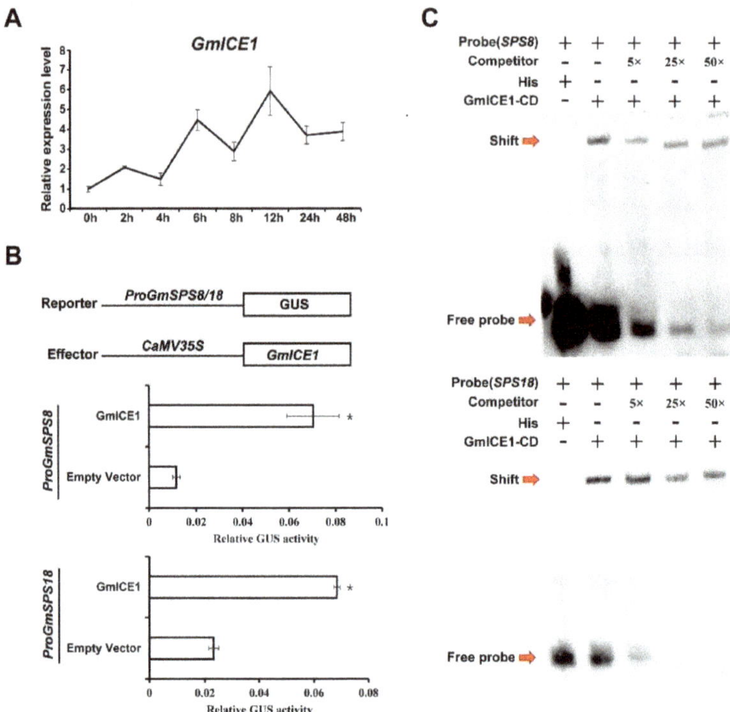

Figure 8. *GmSPS8* and *GmSPS18* transcriptions were directly activated by GmICE1. (**A**) Relative expression level of *GmICE1* in response to cold stress. The expression value was obtained by qRT-PCR. *GmActin11* was used as the internal reference. (**B**) GmICE1 promotes transcription of *GmSPS8/18*. The promoters of *GmSPS8* and *GmSPS18* are inserted in GUS reporter vector; in the meantime, the effector vector contains GmICE1. The vectors were coinfiltrated into *N. benthamiana* leaves to analyze the GUS activity. These results represent the mean ± SD of three independent biological repetitions. Asterisks represent the significant difference as determined by Student's *t*-test (* $p < 0.05$). (**C**) EMSA showed the binding of GmICE1 to the promoters of *GmSPS8* and *GmSPS18* in vitro. The conserved domain of GmICE1 was expressed in *Escherichia coli* BL21 (DE3) cells to produce His-tagged GmICE1 protein. The recombinant protein was purified by Ni-NTA Agarose. Probe indicates DNA sequence with biotin label and the competitor is the same as the DNA sequence but without the biotin label. Red arrows indicate the positions of protein-probe complexes or free probes.

3. Discussion

Extreme weather such as low temperature can have a serious impact on the growth of soybean. Previous research showed that the sensitivity of soybean to cold stress varies at different developmental stages [13,14]. Many cold-stress-related genes have been identified through transcriptome analysis in soybean, including *CBF/DREB* [13]. The ethylene signaling pathway may negatively impact soybean CBF/DREB-regulated cold response by EIN3 [12]. *GmTCF1a* regulates cold tolerance in soybean and is independent of the CBF pathway [15], but the mechanism underlying cold stress response in soybean remains unclear. In general, sucrose level is increased during cold response of plants. SPS is the key rate-limiting enzyme in the sucrose synthesis pathway in plants [27,28]. Here, we conducted the genome-wide survey of soybean SPS family genes and analyzed their response to cold stress, and our findings showed that the expression of all *GmSPS* genes increased significantly under cold stress, especially for *GmSPS8/18*, which can be directly activated by *GmICE1* in *N. benthamiana*. These results provide research ideas and clues for studying soybean cold stress response.

We assessed the phylogenetic relationships of SPS proteins between soybean and other plant species. As previously reported [33,37], all SPS proteins were divided into four families; only SPS proteins from rice belonged to family D. In families A, B, and C, the SPS proteins of legume species were clustered into a small branch, and the SPS proteins of soybean had a closer relationship with that of lotus plants than alfalfa. Different to soybean and lotus, no alfalfa SPS proteins belonged to family C, which was consistent with a previous report [40]. As an ancient tetraploid, soybean has undergone two whole genome duplication events, including an ancient duplication prior to the divergence of papilionoid (58 Mya to 60 Mya) and a Glycine-specific duplication (13 Mya), which resulted in about 75% of the genes being paralogous genes [43]. Three paralogous pairs (*GmSPS13/17, GmSPS4/6*, and *GmSPS8/18*) were identified in soybean, and their divergence time was associated with the 13 Mya WGD events. Our analysis shows that *GmSPS* genes have undergone purification selection during evolution.

The tissue-specific expression pattern was analyzed to understand the potential functions of *GmSPS* genes. All *GmSPS* genes are expressed in stem, leaf, petiole, flower, and SAM, but little in root and seed. *GmSPS8* has the highest transcript abundance in flower, indicating that itmay play an important function in flower development in soybean. Moreover, the expression level of *GmSPS* genes in nodules at different developmental stages of soybean were analyzed. The interesting finding was that members of family A (*GmSPS13/17*) were significantly expressed at branching stage, flowering stage, and fruiting stage, the critical period of symbiotic nitrogen fixation. As the ortholog gene of *GmSPS13/17* in alfalfa, *SPSA* also showed nodule-enhanced expression [40] and was involved in the synthesis of sucrose in nodules [51]. These results indicated that legume SPS members of family A may have the conserved function in carbon metabolism in nodules.

According to the cis-acting elements analysis, *GmSPS* genes may be involved in response to cold stress. The transcript abundance of all seven *GmSPS* genes was upregulated in response to cold stress. The expression level of soybean *SPS* genes in family B (*GmSPS8/14/18*) and family C (*GmSPS4/6*) was significantly increased by more than 10 times, and *GmSPS18* had the greatest increase (53 times). The transcript level of *GmSPS17* was just increased by 2.6 times, and *GmSPS13* was slightly upregulated. Our results are similar to the previous studies, which showed that low temperature increased the content and expression level of *SPS* in chilling-sensitive maize and alfalfa, respectively [52,53]. The transcription levels of many genes are affected under low-temperature stress [54]. Several studies have demonstrated that the ICE1-CBF pathway performs a key role during cold acclimation [55]. In previous studies, the ChIP-seq (chromatin immunoprecipitation sequencing) data of ICE1 indicated that it can bind to the promoter regions of many cold-responsive genes, including *CBF* and a large number of *COR* genes [56]. However, it is unknown whether ICE1 could regulate the expression level of *SPS* genes. Our data suggest that the expression levels of *GmSPS8* and *GmSPS18* could be activated by GmICE1 using transient expression GUS reporter assays in *N. benthamiana*. Additionally, the result of EMSA indicates that GmICE1 can directly bind to the promoter regions of *GmSPS8/18* in vitro. The PlantPAN prediction results showed that there was no CBF binding site in the promoters of *GmSPS8/18*. In addition, in Arabidopsis, the expression of *AtSPS3* (the orthologous gene of *GmSPS8/18*) did not significantly change in the *cbf123* triple mutant [57]. It is possible that *GmSPS8/18* participates in cold stress regulation independently of the CBF pathway in soybean. This may be a new mechanism for soybean plants to resist cold stress.

4. Materials and Methods

4.1. Identification of GmSPS Genes in Soybean

The four known Arabidopsis SPS (AtSPS1/2/3/4) amino acid sequences were extracted from the protein sequence file downloaded from the TAIR site (https://www.arabidopsis.org/) (accessed on 13 June 2022) and then were aligned to a soybean protein sequence file downloaded from the Soybase site (https://www.soybase.org/) (accessed on 13 June 2022) using BLASTP with default parameters [58,59]. In the meantime, the hidden Markov model (HMM) profile files of the SPS conserved domains Sucrose_synth (PF00862), Glycos_transf_1 (PF00534), and S6PP (PF05116) were downloaded from the Pfam site (http://pfam.xfam.org/) (accessed on 13 June 2022) [60]. HMMER (v3.3.2) was used to search the three conserved domains containing proteins in the soybean protein database with default parameter. Subsequently, the candidate soybean SPS proteins which contain the sucrsPsyn_pln were validated with NCBI-CDD (https://www.ncbi.nlm.nih.gov/cdd/) (accessed on 22 July 2022) (Table S1).

4.2. Phylogenetic Analysis of SPS Proteins

To perform the phylogenetic analysis, we integrated SPS protein sequences of soybean, *M. truncatula*, *M. sativa*, *L. japonicus*, Arabidopsis, and rice. The SPS protein sequences of *M. truncatula*, *L. japonicus*, and rice were extracted from Phytozome v12 (https://phytozome. jgi.doe.gov) (accessed on 26 August 2022) [25,61]. The *M. sativa* SPS protein sequences were obtained from https://figshare.com/articles/dataset/Medicago_sativa_genomic_ fa_zip/12859889 (accessed on 26 August 2022) [41]. The accession numbers of these *SPSs* are presented in Table S5. PRANK software (v170427) was used to generate multiple alignments of SPS protein sequences. Next, the alignment results were imported into MEGA11 [62] for phylogenetic analysis using neighbor joining method with JTT + G model and 1000 bootstrap replicates.

4.3. Analysis of Gene Structure and Conserved Motifs

The gene features of *GmSPSs* were obtained from the annotation file of soybean. In addition, the conserved motifs of seven soybean SPS protein sequences were analyzed by MEME in the MEME-Suite site (https://meme-suite.org/meme/index.html) (accessed on 16 August 2022) [63]. TBtools (v1.120) was used to show the results of gene structure and conserved motifs [64].

4.4. Phosphorylation Sites Analysis in GmSPSs

Seven soybean SPS protein sequences were uploaded to Netphos 3.1 Server (https: //services.healthtech.dtu.dk/service.php?NetPhos-3.1) (accessed on 15 August 2022) for phosphorylation sites analysis [65]. The numbers of identified serine, threonine, and tyrosine phosphorylation sites were calculated by the in-house Python script. TBtools was used to draw the heatmaps of phosphorylation sites.

4.5. Chromosomal Location and Collinearity Analysis

The chromosomal location information of all *GmSPS* genes was obtained from the annotation file of soybean. Furthermore, the chromosomal location figure of *GmSPS* genes was generated by TBtools. The longest protein-coding transcripts of *SPS* genes in soybean were screened, and the corresponding protein sequences were extracted. The genome-wide collinearity analysis was carried out by McscanX [66]. Tbtools was used to show the collinearity results. WGD segments pairs were verified using duplicate gene classifier in MCScanX. CDS alignments of each duplicate gene pair of GmSPSs were generated by ParaAT v2 with parameter "-m mafft -f axt -t -g". Next, KaKs_Calculator 3.0 was used to calculate Ka (nonsynonymous) and Ks (synonymous) using CDS alignment file [67,68]. The divergence time of duplication pairs was calculated using $T = Ks/(2 \times 6.1 \times 10^{-9}) \times 10^{-6}$ Mya [69].

4.6. cis-Regulatory Element in the Promoter Regions

The cis-regulatory element analyses of 3000 bp upstream sequences from the translation start codon of all *GmSPS* genes were generated by both PlantCARE (http://bioinformatics.psb.ugent.be/webtools/plantcare/html/) (accessed on 6 November 2022) [70] and PlantPAN (http://plantpan.itps.ncku.edu.tw/promoter_multiple.php) (accessed on 6 November 2022) sites [71]. The putative cis-regulatory elements related to resistance, hormones regulatory, transcription factor binding, etc., were displayed using TBtools.

4.7. Plant Materials and Cold Stress Treatment

At five days after germination, the seedlings of soybean cultivar Tianlong No.1 (TL-1, which was bred by our lab) were transformed to Hoagland nutrient solution under 60% relative humidity with a 12 h/12 h (light/dark) photoperiod at 25 °C. For cold treatment, seedings (V1 stage) were transferred to a low-temperature incubator at 4 °C. The first fully expanded leaves were collected at 0 h, 2 h, 4 h, 6 h, 8 h, 12 h, 24 h, and 48 h after treatment, and immediately frozen in liquid nitrogen then stored at −80 °C for RNA extraction. At each time point of treatment, six plants were harvested, with three biological replicates per sample. Untreated seedlings were used as control.

4.8. RNA Extraction and qRT-PCR Analysis

The qRT-PCR assay was used to analyze the expression pattern of *GmSPS* genes in various tissues of soybean cultivar TL-1 and soybean leaves under cold treatment. Total RNA was extracted using a TRIpure reagent (Aidlab, Beijing, China) according to the manufacturer's instruction. The quantity of the extracted RNA samples was further assessed by agarose gel and nanospectrophotometer. The Hiscript II 1st strand cDNA synthesis kit (Vazyme, Nanjing, China) was used to synthesize the first strand of cDNA. The qRT-PCR was performed using iTaq Universal SYBR Green Supermix (Bio-Rad, Hercules, CA, USA) on a Bio-Rad CFX96 Real-Time PCR system. The gene-specific primers of *GmSPSs* were designed using the primer designing tool of NCBI (https://www.ncbi.nlm. nih.gov/tools/primer-blast/index.cgi) (accessed on 15 June 2022). Meanwhile, *GmActin11* (*Glyma.18G290800*), the expression of which did not change significantly in response to cold, was used as the internal reference gene to adjust the expression level of *GmSPS* genes. The primers are listed in Table S6. All primers were synthesized by tsingke Biotech (Beijing, China). The qRT-PCR reactions were performed in a 20 μL volume and the cycling program conditions were denaturation at 95 °C for 5 min, followed by 40 cycles of 95 °C for 10 s, 60 °C for 10 s, and 72 °C for 20 s. All samples were performed with three technical and biological repeats. The $2^{-\Delta\Delta Ct}$ method [72] was utilized to analyze the relative expression level of *GmSPS* genes under cold stress.

4.9. GUS Activity Assay in Transiently Transformed N. benthamiana Leaves

The promoter sequences of *GmSPS8* (1956 bp upstream of the ATG) and *GmSPS18* (1950 bp upstream of the ATG) were cloned into the *PstI*/*BglII* and *BamHI*/*NcoI* restriction sites upstream of GUS reporter in the pCAMBIA3301 vector as the reporter constructs. The *GmICE1* CDS was constructed into PTF101 vector downstream of the *35S* promoter as the effector through homologous recombination. The constructed vectors were transformed into *Agrobacterium tumefaciens* EHA105 (Tsingke, Beijing, China). Then, the reporters and the effector were transient-expressed in *N. benthamiana* leaves as previously described [73]. These experiments were repeated at least three times independently. Primers are listed in Table S6.

4.10. Electrophoretic Mobility Shift Assay

The EMSA was performed by generally following the previous method described [74]. The conserved domain sequences (826 bp to 1398 bp) of *GmICE1* were cloned into the PET-32a vector. The constructed vector was transformed into the *E. coli* BL21 (DE3) strain. The recombined protein of GmICE1 was purified using Ni-NTA Agarose (Qiagen, Dusseldorf, Germany). The DNA probes of *GmSPS8* and *GmSPS18* with biotin label were synthesized by Tsingke. The EMSA probes were obtained through annealing and renaturation of forward and reverse primers. EMSAs were performed using an EMSA Kit (Thermo Fisher Scientific, Waltham, MA, USA). Sequences of the probes and primers are listed in Table S6.

5. Conclusions

In this study, we identified seven *SPS* genes in soybean genome via genome-wide screening. We carried out a series of analyses, including the phylogenetic relationships, duplications patterns, gene structures, conserved motifs, phosphorylation site, cis-elements, tissue expression patterns, and the expression of *GmSPS* genes in response to cold stress. We further demonstrated that *GmSPS8/18* can be directly activated by *GmICE1*. These data indicate that *GmSPS8/18* play important roles under cold stress. These investigations and analyses could increase our knowledge of the functions of SPS family genes in response to cold stress.

Supplementary Materials: The following supporting information can be downloaded at: https://www.mdpi.com/article/10.3390/ijms241612878/s1.

Author Contributions: J.S., C.Z. and X.Z. conceived and designed the experiments; J.S. and F.J. performed the experiments; J.S. and Y.X. analyzed the data; Y.H., H.C., Z.S., Z.Y. and S.C. contributed reagents/materials/analysis tools; J.S., C.Z. and S.Y. wrote the paper. All authors have read and agreed to the published version of the manuscript.

Funding: This work was supported by Knowledge Innovation Program of Wuhan-Basic Research (2022020801010294), the Agricultural Science and Technology Innovation Program of Chinese Academy of Agricultural Sciences (CAAS-ASTIP-2023-OCRI), the National Key Research and Development Program of China (2020YFD1000903-10).

Institutional Review Board Statement: Not applicable.

Informed Consent Statement: Not applicable.

Data Availability Statement: The datasets used and/or analyzed in this study are available on reasonable request from the corresponding author.

Conflicts of Interest: The authors declare no conflict of interest.

References

1. Guo, X.; Liu, D.; Chong, K. Cold signaling in plants: Insights into mechanisms and regulation. *J. Integr. Plant Biol.* **2018**, *60*, 745–756. [PubMed]
2. Ding, Y.; Shi, Y.; Yang, S. Advances and challenges in uncovering cold tolerance regulatory mechanisms in plants. *New Phytol.* **2019**, *222*, 1690–1704. [CrossRef] [PubMed]
3. Ding, Y.; Jia, Y.; Shi, Y.; Zhang, X.; Song, C.; Gong, Z.; Yang, S. OST1-mediated BTF3L phosphorylation positively regulates CBFs during plant cold responses. *EMBO J.* **2018**, *37*, e98228. [CrossRef] [PubMed]
4. Chinnusamy, V.; Zhu, J.; Zhu, J.K. Cold stress regulation of gene expression in plants. *Trends Plant Sci.* **2007**, *12*, 444–451. [CrossRef]
5. Stockinger, E.J.; Gilmour, S.J.; Thomashow, M.F. *Arabidopsis thaliana* CBF1 encodes an AP2 domain-containing transcriptional activator that binds to the C-repeat/DRE, a cis-acting DNA regulatory element that stimulates transcription in response to low temperature and water deficit. *Proc. Natl. Acad. Sci. USA* **1997**, *94*, 1035–1040. [CrossRef]
6. Jaglo, K.R.; Kleff, S.; Amundsen, K.L.; Zhang, X.; Haake, V.; Zhang, J.Z.; Deits, T.; Thomashow, M.F. Components of the *Arabidopsis* C-repeat/dehydration-responsive element binding factor cold-response pathway are conserved in *Brassica napus* and other plant species. *Plant Physiol.* **2001**, *127*, 910–917. [CrossRef]
7. Jaglo-Ottosen, K.R.; Gilmour, S.J.; Zarka, D.G.; Schabenberger, O.; Thomashow, M.F. *Arabidopsis* CBF1 overexpression induces COR genes and enhances freezing tolerance. *Science* **1998**, *280*, 104–106. [CrossRef]
8. Gilmour, S.J.; Sebolt, A.M.; Salazar, M.P.; Everard, J.D.; Thomashow, M.F. Overexpression of the *Arabidopsis* CBF3 transcriptional activator mimics multiple biochemical changes associated with cold acclimation. *Plant Physiol.* **2000**, *124*, 1854–1865. [CrossRef]
9. Chinnusamy, V.; Ohta, M.; Kanrar, S.; Lee, B.H.; Hong, X.; Agarwal, M.; Zhu, J.K. ICE1: A regulator of cold-induced transcriptome and freezing tolerance in *Arabidopsis*. *Genes Dev.* **2003**, *17*, 1043–1054. [CrossRef]
10. Wilson, R.F. *Soybean: Market Driven Research Needs*; Springer: New York, NY, USA, 2008.
11. Maruyama, K.; Takeda, M.; Kidokoro, S.; Yamada, K.; Sakuma, Y.; Urano, K.; Fujita, M.; Yoshiwara, K.; Matsukura, S.; Morishita, Y.; et al. Metabolic pathways involved in cold acclimation identified by integrated analysis of metabolites and transcripts regulated by DREB1A and DREB2A. *Plant Physiol.* **2009**, *150*, 1972–1980. [CrossRef]
12. Robison, J.D.; Yamasaki, Y.; Randall, S.K. The ethylene signaling pathway negatively impacts CBF/DREB-regulated cold response in soybean (*Glycine max*). *Front. Plant Sci.* **2019**, *10*, 121. [CrossRef] [PubMed]
13. Yamasaki, Y.; Randall, S.K. Functionality of soybean CBF/DREB1 transcription factors. *Plant Sci.* **2016**, *246*, 80–90. [CrossRef] [PubMed]

14. Wang, Z.; Reddy, V.R.; Quebedeaux, B. Growth and photosynthetic responses of soybean to short-term cold temperature. *Environ. Exp. Bot.* **1997**, *37*, 13–24. [CrossRef]
15. Dong, Z.; Wang, H.; Li, X.; Ji, H. Enhancement of plant cold tolerance by soybean RCC1 family gene *GmTCF1a*. *BMC Plant Biol.* **2021**, *21*, 369. [CrossRef]
16. Ji, H.; Wang, Y.; Cloix, C.; Li, K.; Jenkins, G.I.; Wang, S.; Shang, Z.; Shi, Y.; Yang, S.; Li, X. The *Arabidopsis* RCC1 family protein TCF1 regulates freezing tolerance and cold acclimation through modulating lignin biosynthesis. *PLoS Genet.* **2015**, *11*, e1005471. [CrossRef]
17. Rathinasabapathi, B. Metabolic engineering for stress tolerance: Installing osmoprotectant synthesis pathways. *Ann. Bot.* **2000**, *86*, 709–716. [CrossRef]
18. Wanner, L.A.; Junttila, O. Cold-induced freezing tolerance in *Arabidopsis*. *Plant Physiol.* **1999**, *120*, 391–400. [CrossRef]
19. Zúñiga-Feest, A.; Ort, D.R.; Gutiérrez, A.; Gidekel, M.; Bravo, L.A.; Corcuera, L.J. Light regulation of sucrose-phosphate synthase activity in the freezing-tolerant grass *Deschampsia antarctica*. *Photosynth. Res.* **2005**, *83*, 75–86. [CrossRef]
20. Bauerfeind, M.A.; Winkelmann, T.; Franken, P.; Druege, U. Transcriptome, carbohydrate, and phytohormone analysis of Petunia hybrida reveals a complex disturbance of plant functional integrity under mild chilling stress. *Front. Plant Sci.* **2015**, *6*, 583. [CrossRef]
21. Peng, T.; Zhu, X.; Duan, N.; Liu, J.H. PtrBAM1, a β-amylase-coding gene of *Poncirus trifoliata*, is a CBF regulon member with function in cold tolerance by modulating soluble sugar levels. *Plant Cell Environ.* **2014**, *37*, 2754–2767. [CrossRef]
22. Dong, S.; Beckles, D.M. Dynamic changes in the starch-sugar interconversion within plant source and sink tissues promote a better abiotic stress response. *J. Plant Physiol.* **2019**, *234–235*, 80–93. [CrossRef] [PubMed]
23. Rolland, F.; Moore, B.; Sheen, J. Sugar sensing and signaling in plants. *Plant Cell* **2002**, *14* (Suppl. S1), 185–205. [CrossRef] [PubMed]
24. Park, J.Y.; Canam, T.; Kang, K.Y.; Ellis, D.D.; Mansfield, S.D. Over-expression of an *Arabidopsis* family A sucrose phosphate synthase (SPS) gene alters plant growth and fibre development. *Transgenic Res.* **2008**, *17*, 181–192. [CrossRef] [PubMed]
25. Okamura, M.; Aoki, N.; Hirose, T.; Yonekura, M.; Ohto, C.; Ohsugi, R. Tissue specificity and diurnal change in gene expression of the sucrose phosphate synthase gene family in rice. *Plant Sci.* **2011**, *181*, 159–166. [CrossRef]
26. Volkert, K.; Debast, S.; Voll, L.M.; Voll, H.; Schießl, I.; Hofmann, J.; Schneider, S.; Börnke, F. Loss of the two major leaf isoforms of sucrose-phosphate synthase in *Arabidopsis thaliana* limits sucrose synthesis and nocturnal starch degradation but does not alter carbon partitioning during photosynthesis. *J. Exp. Bot.* **2014**, *65*, 5217–5229. [CrossRef]
27. Haigler, C.H.; Singh, B.; Zhang, D.; Hwang, S.; Wu, C.; Cai, W.X.; Hozain, M.; Kang, W.; Kiedaisch, B.; Strauss, R.E.; et al. Transgenic cotton over-producing spinach sucrose phosphate synthase showed enhanced leaf sucrose synthesis and improved fiber quality under controlled environmental conditions. *Plant Mol. Biol.* **2007**, *63*, 815–832. [CrossRef]
28. Miron, D.; Schaffer, A.A. Sucrose phosphate synthase, sucrose synthase, and invertase activities in developing fruit of *Lycopersicon esculentum* Mill. and the sucrose accumulating *Lycopersicon hirsutum* Humb. and Bonpl. *Plant Physiol.* **1991**, *95*, 623–627. [CrossRef]
29. Dali, N.; Michaud, D.; Yelle, S. Evidence for the involvement of sucrose phosphate synthase in the pathway of sugar accumulation in sucrose-accumulating tomato fruits. *Plant Physiol.* **1992**, *99*, 434–438. [CrossRef]
30. do Nascimento, J.R.; Cordenunsi, B.R.; Lajolo, F.M.; Alcocer, M.J. Banana sucrose-phosphate synthase gene expression during fruit ripening. *Planta* **1997**, *203*, 83–88. [CrossRef]
31. Guy, C.L.; Huber, J.L.; Huber, S.C. Sucrose phosphate synthase and sucrose accumulation at low temperature. *Plant Physiol.* **1992**, *100*, 502–508. [CrossRef]
32. Huang, T.; Luo, X.; Wei, M.; Shan, Z.; Zhu, Y.; Yang, Y.; Fan, Z. Molecular cloning and expression analysis of sucrose phosphate synthase genes in cassava (*Manihot esculenta* Crantz). *Sci. Rep.* **2020**, *10*, 20707. [CrossRef] [PubMed]
33. Lutfiyya, L.L.; Xu, N.; D'Ordine, R.L.; Morrell, J.A.; Miller, P.W.; Duff, S.M. Phylogenetic and expression analysis of sucrose phosphate synthase isozymes in plants. *J. Plant Physiol.* **2007**, *164*, 923–933. [CrossRef] [PubMed]
34. Duan, Y.; Yang, L.; Zhu, H.; Zhou, J.; Sun, H.; Gong, H. Structure and expression analysis of sucrose phosphate synthase, sucrose synthase and invertase gene families in *Solanum lycopersicum*. *Int. J. Mol. Sci.* **2021**, *22*, 4698. [CrossRef] [PubMed]
35. Sharma, S.; Sreenivasulu, N.; Harshavardhan, V.T.; Seiler, C.; Sharma, S.; Khalil, Z.N.; Akhunov, E.; Sehgal, S.K.; Röder, M.S. Delineating the structural, functional and evolutionary relationships of sucrose phosphate synthase gene family II in wheat and related grasses. *BMC Plant Biol.* **2010**, *10*, 134. [CrossRef]
36. Wang, D.; Zhao, J.; Hu, B.; Li, J.; Qin, Y.; Chen, L.; Qin, Y.; Hu, G. Identification and expression profile analysis of the sucrose phosphate synthase gene family in *Litchi chinensis* Sonn. *PeerJ* **2018**, *6*, e4379. [CrossRef]
37. Castleden, C.K.; Aoki, N.; Gillespie, V.J.; MacRae, E.A.; Quick, W.P.; Buchner, P.; Foyer, C.H.; Furbank, R.T.; Lunn, J.E. Evolution and function of the sucrose-phosphate synthase gene families in wheat and other grasses. *Plant Physiol.* **2004**, *135*, 1753–1764. [CrossRef]
38. Huber, S.C. Biochemical basis for effects of k-deficiency on assimilate export rate and accumulation of soluble sugars in soybean leaves. *Plant Physiol.* **1984**, *76*, 424–430. [CrossRef]
39. Du, Y.; Zhao, Q.; Chen, L.; Yao, X.; Zhang, H.; Wu, J.; Xie, F. Effect of drought stress during soybean R2-R6 growth stages on sucrose metabolism in leaf and seed. *Int. J. Mol. Sci.* **2020**, *21*, 618. [CrossRef]

40. Aleman, L.; Ortega, J.L.; Martinez-Grimes, M.; Seger, M.; Holguin, F.O.; Uribe, D.J.; Garcia-Ibilcieta, D.; Sengupta-Gopalan, C. Nodule-enhanced expression of a sucrose phosphate synthase gene member (MsSPSA) has a role in carbon and nitrogen metabolism in the nodules of alfalfa (*Medicago sativa* L.). *Planta* **2010**, *231*, 233–244. [CrossRef]

41. Li, A.; Liu, A.; Du, X.; Chen, J.Y.; Yin, M.; Hu, H.Y.; Shrestha, N.; Wu, S.D.; Wang, H.Q.; Dou, Q.W.; et al. A chromosome-scale genome assembly of a diploid alfalfa, the progenitor of autotetraploid alfalfa. *Hortic. Res.* **2020**, *7*, 194. [CrossRef]

42. Schmutz, J.; Cannon, S.B.; Schlueter, J.; Ma, J.; Mitros, T.; Nelson, W.; Hyten, D.L.; Song, Q.; Thelen, J.J.; Cheng, J.; et al. Genome sequence of the palaeopolyploid soybean. *Nature* **2010**, *463*, 178–183. [CrossRef] [PubMed]

43. Zhao, W.; Cheng, Y.; Zhang, C.; You, Q.; Shen, X.; Guo, W.; Jiao, Y. Genome-wide identification and characterization of circular RNAs by high throughput sequencing in soybean. *Sci. Rep.* **2017**, *7*, 5636. [CrossRef] [PubMed]

44. Fang, Y.; Cao, D.; Yang, H.; Guo, W.; Ouyang, W.; Chen, H.; Shan, Z.; Yang, Z.; Chen, S.; Li, X.; et al. Genome-wide identification and characterization of soybean GmLOR gene family and expression analysis in response to abiotic stresses. *Int. J. Mol. Sci.* **2021**, *22*, 12515. [CrossRef] [PubMed]

45. Salih, H.; Gong, W.; He, S.; Sun, G.; Sun, J.; Du, X. Genome-wide characterization and expression analysis of MYB transcription factors in *Gossypium hirsutum*. *BMC Genet.* **2016**, *17*, 129. [CrossRef] [PubMed]

46. Andersson, R.; Sandelin, A. Determinants of enhancer and promoter activities of regulatory elements. *Nat. Rev. Genet.* **2020**, *21*, 71–87. [CrossRef]

47. Yuan, S.L.; Li, R.; Chen, H.F.; Zhang, C.J.; Chen, L.M.; Hao, Q.N.; Chen, S.L.; Shan, Z.H.; Yang, Z.L.; Zhang, X.J.; et al. RNA-Seq analysis of nodule development at five different developmental stages of soybean (*Glycine max*) inoculated with *Bradyrhizobium japonicum* strain 113-2. *Sci. Rep.* **2017**, *7*, 42248. [CrossRef]

48. Liang, Y.; Zhang, M.; Wang, M.; Zhang, W.; Qiao, C.; Luo, Q.; Lu, X. Freshwater *Cyanobacterium Synechococcus* elongatus PCC 7942 adapts to an environment with salt stress via ion-induced enzymatic balance of dompatible solutes. *Appl. Environ. Microbiol.* **2020**, *86*, e02904-19. [CrossRef]

49. Yang, J.; Zhang, J.; Wang, Z.; Zhu, Q. Activities of starch hydrolytic enzymes and sucrose-phosphate synthase in the stems of rice subjected to water stress during grain filling. *J. Exp. Bot.* **2001**, *52*, 2169–2179. [CrossRef]

50. Solís-Guzmán, M.G.; Argüello-Astorga, G.; López-Bucio, J.; Ruiz-Herrera, L.F.; López-Meza, J.E.; Sánchez-Calderón, L.; Carreón-Abud, Y.; Martínez-Trujillo, M. *Arabidopsis thaliana* sucrose phosphate synthase (sps) genes are expressed differentially in organs and tissues, and their transcription is regulated by osmotic stress. *Gene Expr. Patterns* **2017**, *25–26*, 92–101. [CrossRef]

51. Padhi, S.; Grimes, M.M.; Muro-Villanueva, F.; Ortega, J.L.; Sengupta-Gopalan, C. Distinct nodule and leaf functions of two different sucrose phosphate synthases in alfalfa. *Planta* **2019**, *250*, 1743–1755. [CrossRef]

52. Bilska-Kos, A.; Mytych, J.; Suski, S.; Magoń, J.; Ochodzki, P.; Zebrowski, J. Sucrose phosphate synthase (SPS), sucrose synthase (SUS) and their products in the leaves of Miscanthus × giganteus and Zea mays at low temperature. *Planta* **2020**, *252*, 23. [CrossRef]

53. Bertrand, A.; Bipfubusa, M.; Claessens, A.; Rocher, S.; Castonguay, Y. Effect of photoperiod prior to cold acclimation on freezing tolerance and carbohydrate metabolism in alfalfa (*Medicago sativa* L.). *Plant Sci.* **2017**, *264*, 122–128. [CrossRef]

54. Zeller, G.; Henz, S.R.; Widmer, C.K.; Sachsenberg, T.; Rätsch, G.; Weigel, D.; Laubinger, S. Stress-induced changes in the Arabidopsis thaliana transcriptome analyzed using whole-genome tiling arrays. *Plant J.* **2009**, *58*, 1068–1082. [CrossRef] [PubMed]

55. Kim, Y.S.; Lee, M.; Lee, J.H.; Lee, H.J.; Park, C.M. The unified ICE-CBF pathway provides a transcriptional feedback control of freezing tolerance during cold acclimation in *Arabidopsis*. *Plant Mol. Biol.* **2015**, *89*, 187–201. [CrossRef] [PubMed]

56. Tang, K.; Zhao, L.; Ren, Y.; Yang, S.; Zhu, J.K.; Zhao, C. The transcription factor ICE1 functions in cold stress response by binding to the promoters of CBF and COR genes. *J. Integr. Plant Biol.* **2020**, *62*, 258–263. [CrossRef]

57. Shi, Y.; Huang, J.; Sun, T.; Wang, X.; Zhu, C.; Ai, Y.; Gu, H. The precise regulation of different COR genes by individual CBF transcription factors in *Arabidopsis thaliana*. *J. Integr. Plant Biol.* **2017**, *59*, 118–133. [CrossRef] [PubMed]

58. Langenkämper, G.; Fung, R.W.; Newcomb, R.D.; Atkinson, R.G.; Gardner, R.C.; MacRae, E.A. Sucrose phosphate synthase genes in plants belong to three different families. *J. Mol. Evol.* **2002**, *54*, 322–332. [CrossRef]

59. Camacho, C.; Coulouris, G.; Avagyan, V.; Ma, N.; Papadopoulos, J.; Bealer, K.; Madden, T.L. BLAST+: Architecture and applications. *BMC Bioinform.* **2009**, *10*, 421. [CrossRef]

60. Mistry, J.; Chuguransky, S.; Williams, L.; Qureshi, M.; Salazar, G.A.; Sonnhammer, E.L.L.; Tosatto, S.C.E.; Paladin, L.; Raj, S.; Richardson, L.J.; et al. Pfam: The protein families database in 2021. *Nucleic Acids Res.* **2021**, *49*, 412–419. [CrossRef]

61. Goodstein, D.M.; Shu, S.; Howson, R.; Neupane, R.; Hayes, R.D.; Fazo, J.; Mitros, T.; Dirks, W.; Hellsten, U.; Putnam, N.; et al. Phytozome: A comparative platform for green plant genomics. *Nucleic Acids Res.* **2012**, *40*, 1178–1186. [CrossRef]

62. Tamura, K.; Stecher, G.; Kumar, S. MEGA11: Molecular evolutionary genetics analysis version 11. *Mol. Biol. Evol.* **2021**, *38*, 3022–3027. [CrossRef] [PubMed]

63. Bailey, T.L.; Johnson, J.; Grant, C.E.; Noble, W.S. The MEME Suite. *Nucleic Acids Res.* **2015**, *43*, 39–49. [CrossRef] [PubMed]

64. Chen, C.; Chen, H.; Zhang, Y.; Thomas, H.R.; Frank, M.H.; He, Y.; Xia, R. TBtools: An integrative toolkit developed for interactive analyses of big biological data. *Mol. Plant* **2020**, *13*, 1194–1202. [CrossRef]

65. Blom, N.; Sicheritz-Pontén, T.; Gupta, R.; Gammeltoft, S.; Brunak, S. Prediction of post-translational glycosylation and phosphorylation of proteins from the amino acid sequence. *Proteomics* **2004**, *4*, 1633–1649. [CrossRef] [PubMed]

66. Wang, Y.; Tang, H.; Debarry, J.D.; Tan, X.; Li, J.; Wang, X.; Lee, T.H.; Jin, H.; Marler, B.; Guo, H.; et al. MCScanX: A toolkit for detection and evolutionary analysis of gene synteny and collinearity. *Nucleic Acids Res.* **2012**, *40*, e49. [CrossRef] [PubMed]

67. Zhang, Z.; Xiao, J.; Wu, J.; Zhang, H.; Liu, G.; Wang, X.; Dai, L. ParaAT: A parallel tool for constructing multiple protein-coding DNA alignments. *Biochem. Biophys. Res. Commun.* **2012**, *419*, 779–781. [CrossRef] [PubMed]
68. Zhang, Z. KaKs_calculator 3.0: Calculating selective pressure on coding and non-coding sequences. *Genom. Proteom. Bioinform.* **2022**, *20*, 536–540. [CrossRef]
69. Lynch, M.; Conery, J.S. The evolutionary fate and consequences of duplicate genes. *Science* **2000**, *290*, 1151–1155. [CrossRef]
70. Lescot, M.; Déhais, P.; Thijs, G.; Marchal, K.; Moreau, Y.; Van de Peer, Y.; Rouzé, P.; Rombauts, S. PlantCARE, a database of plant cis-acting regulatory elements and a portal to tools for in silico analysis of promoter sequences. *Nucleic Acids Res.* **2002**, *30*, 325–327. [CrossRef]
71. Chow, C.N.; Lee, T.Y.; Hung, Y.C.; Li, G.Z.; Tseng, K.C.; Liu, Y.H.; Kuo, P.L.; Zheng, H.Q.; Chang, W.C. PlantPAN3.0: A new and updated resource for reconstructing transcriptional regulatory networks from ChIP-seq experiments in plants. *Nucleic Acids Res.* **2019**, *47*, 1155–1163. [CrossRef]
72. Xia, W.; Mason, A.S.; Xiao, Y.; Liu, Z.; Yang, Y.; Lei, X.; Wu, X.; Ma, Z.; Peng, M. Analysis of multiple transcriptomes of the African oil palm (*Elaeis guineensis*) to identify reference genes for RT-qPCR. *J. Biotechnol.* **2014**, *184*, 63–73. [CrossRef] [PubMed]
73. Li, X.; Guo, W.; Li, J.; Yue, P.; Bu, H.; Jiang, J.; Liu, W.; Xu, Y.; Yuan, H.; Li, T.; et al. Histone acetylation at the promoter for the transcription factor PuWRKY31 affects sucrose accumulation in pear fruit. *Plant Physiol.* **2020**, *182*, 2035–2046. [CrossRef] [PubMed]
74. Li, M.; Zhang, H.; He, D.; Damaris, R.N.; Yang, P. A stress-associated protein OsSAP8 modulates gibberellic acid biosynthesis by reducing the promotive effect of transcription factor OsbZIP58 on *OsKO2*. *J. Exp. Bot.* **2022**, *73*, 2420–2433. [CrossRef] [PubMed]

International Journal of
Molecular Sciences

Article

Impact of Heavy Metals on Cold Acclimation of *Salix viminalis* Roots

Valentin Ambroise [1,2], Sylvain Legay [1], Marijke Jozefczak [2], Céline C. Leclercq [1], Sebastien Planchon [1], Jean-Francois Hausman [1], Jenny Renaut [1], Ann Cuypers [2] and Kjell Sergeant [1,*]

[1] Greentech Innovation Centre (GTIC), Environmental Research and Innovation (ERIN) Department, Luxembourg Institute of Science and Technology (LIST), 5 Avenue des Hauts-Fourneaux, L-4362 Esch-sur-Alzette, Luxembourg; valentin.ambroise@outlook.com (V.A.); sylvain.legay@list.lu (S.L.); celine.leclercq@list.lu (C.C.L.); sebastien.planchon@list.lu (S.P.); jean-francois.hausman@list.lu (J.-F.H.); jenny.renaut@list.lu (J.R.)

[2] Centre for Environmental Sciences, Hasselt University, Agoralaan Building D, B-3590 Diepenbeek, Belgium; marijke.jozefczak@uhasselt.be (M.J.); ann.cuypers@uhasselt.be (A.C.)

* Correspondence: kjell.sergeant@list.lu; Tel.: +352-275-888-5024

Abstract: In nature, plants are exposed to a range of climatic conditions. Those negatively impacting plant growth and survival are called abiotic stresses. Although abiotic stresses have been extensively studied separately, little is known about their interactions. Here, we investigate the impact of long-term mild metal exposure on the cold acclimation of *Salix viminalis* roots using physiological, transcriptomic, and proteomic approaches. We found that, while metal exposure significantly affected plant morphology and physiology, it did not impede cold acclimation. Cold acclimation alone increased glutathione content and glutathione reductase activity. It also resulted in the increase in transcripts and proteins belonging to the heat-shock proteins and related to the energy metabolism. Exposure to metals decreased antioxidant capacity but increased catalase and superoxide dismutase activity. It also resulted in the overexpression of transcripts and proteins related to metal homeostasis, protein folding, and the antioxidant machinery. The simultaneous exposure to both stressors resulted in effects that were not the simple addition of the effects of both stressors taken separately. At the antioxidant level, the response to both stressors was like the response to metals alone. While this should have led to a reduction of frost tolerance, this was not observed. The impact of the simultaneous exposure to metals and cold acclimation on the transcriptome was unique, while at the proteomic level the cold acclimation component seemed to be dominant. Some genes and proteins displayed positive interaction patterns. These genes and proteins were related to the mitigation and reparation of oxidative damage, sugar catabolism, and the production of lignans, trehalose, and raffinose. Interestingly, none of these genes and proteins belonged to the traditional ROS homeostasis system. These results highlight the importance of the under-studied role of lignans and the ROS damage repair and removal system in plants simultaneously exposed to multiple stressors.

Keywords: abiotic stress; heavy metals; frost; transcriptomics; proteomics; antioxidant system; integrative biology

Citation: Ambroise, V.; Legay, S.; Jozefczak, M.; Leclercq, C.C.; Planchon, S.; Hausman, J.-F.; Renaut, J.; Cuypers, A.; Sergeant, K. Impact of Heavy Metals on Cold Acclimation of *Salix viminalis* Roots. *Int. J. Mol. Sci.* **2024**, *25*, 1545. https://doi.org/10.3390/ijms25031545

Academic Editor: Martin Bartas

Received: 27 November 2023
Revised: 15 January 2024
Accepted: 19 January 2024
Published: 26 January 2024

1. Introduction

As sessile organisms, plants are exposed to changing environmental conditions. When these environmental conditions negatively impact plant growth and primary production, they are called abiotic stresses [1].

To cope with adverse conditions, plants have evolved a wide network of molecular mechanisms that allow them to respond to stress by adjusting their metabolism [2]. These metabolic changes are made possible by fine-tuning gene transcription and protein abundance and activity. During the last few decades, substantial progress has been made in the identification of stress-related genes and proteins and uncyphering the mechanisms involved in stress responses. However, a large majority of these studies have been carried out in laboratories under controlled conditions. Under natural conditions, plants are seldomly exposed to one single stress at a time [3]. An increasing amount of evidence from field and molecular studies have shown that plants respond to combinations of stresses in a non-additive manner, resulting in interacting effects that cannot be predicted from studies focusing on single stress exposure [3,4].

Another issue is the temporal scale and the degree of exposure used in stress-related studies. For example, heavy metal stress-related studies often focus on short-term responses with metal concentrations exceeding ecologically relevant concentrations [5]. However, it has been observed that plants exposed to long-term heavy metal stress present a different metabolic profile than plants exposed to acute stress [6].

Despite the specificity of stress responses at a gene and protein level, some common processes have been observed in the response to different stresses, including the production of reactive oxygen species (ROS) and the accumulation of hormones such as abscisic acid (ABA) and jasmonic acid (JA). Among those, [4] identified the antioxidant defence machinery as an underlying pathway leading to tolerance to stress combinations. Unravelling the crosstalk existing between the responses to different stresses could pave the way to the breeding of multiple-stress-resistant plants.

A few years ago, our team observed that *Salix viminalis* grown in pots in a soil polluted by a mixture of heavy metals were more winter-tolerant than plants grown in unpolluted soil (unpublished results). *S. viminalis*, known as basket willow, is a fast-growing pioneering shrub with a wide Eurasian distribution. When fully cold-acclimated, its twigs can withstand up to $-196\ °C$ [7]. Given the extreme frost tolerance of its aboveground organs, it was hypothesized that the difference in winter-tolerance was the consequence of an increased frost-tolerance at the root level.

In this present study, we investigated how exposure to a mild polymetallic stress was able to increase cold acclimation and frost hardiness in the roots of *S. viminalis*. Our hypothesis was that the exposure of the roots to a low concentration of heavy metals would prime its antioxidant system, resulting in higher antioxidative capacity, itself leading to a higher root frost hardiness, in a way analogous to how cold priming alleviates acute metallic stress [8]. In order to test this hypothesis, we characterised the long-term effects of exposure to cold and a mild polymetallic contamination, combined and separate, at the transcriptome and proteome level. We also measured the enzymatic activity and the concentration of several compounds involved in redox homeostasis in the roots of *S. viminalis*. The aim of the present study was to compare the effects of a mild polymetallic mixture and/or cold acclimation on the antioxidant defence system, with an emphasis on systems displaying interaction patterns.

2. Results

2.1. Plant Material

Plants were exposed to cold and/or a polymetallic mixture, resulting in four temperature–polymetallic contamination combinations possible: control soil temperate (CT), control soil cold acclimated (CC), metal-polluted soil temperate (MT), and metal-polluted soil cold acclimated (MC).

Plants exposed to the polymetallic mixture were smaller and showed a sickly habit (Figure 1). Their leaves were chlorotic, and some necrotic spots could be observed. In addition, some of the older leaves dried on the plants. Upon cold exposure, plant growth slowed down in plants cultivated in contaminated and uncontaminated soil. No further growth was observed after three weeks of cold acclimation.

Figure 1. Phenotype of *Salix viminalis* exposed to a polymetallic mixture and/or cold-acclimated. Plants were exposed to metals for two months before being cold-acclimated at 7 °C/5 °C for one month.

2.1.1. Chlorophyll Fluorescence

The chlorophyll fluorescence of young leaves was measured at several time points during the four weeks of cold exposure, and the F_v/F_m ratio was calculated. The F_v/F_m ratio of plants in the CT group stayed relatively stable, with an average value of 0.82 (Supplementary Table S1). The F_v/F_m ratio of MT plants was lower, with an average value of 0.71, but it fluctuated more, as shown by its standard deviation of 0.15.

Upon cold exposition, F_v/F_m values temporarily decreased in plants cultivated in control and polluted soil. The average F_v/F_m ratio of plants exposed to cold was 0.80 and 0.65 for plants cultivated in control and polluted soil, respectively.

The effects of cold and polymetallic exposure on the F_v/F_m ratio were assessed with a two-way ANOVA. The polymetallic exposure had a significant (p-value < 0.001) impact on the F_v/F_m ratio, but no effect of cold exposure (p-value = 0.340), sampling date (p-value = 0.456), or interaction between cold and polymetallic exposure (p-value = 0.590) could be detected.

2.1.2. Root Frost Tolerance

Cold acclimation significantly decreased root electrolyte leakage (REL) at −2 °C (two-way ANOVA, p-value < 0.001). Indeed, while REL was over 75% in non-cold acclimated roots exposed to −2 °C, indicating that the roots were not frost-tolerant, it remained below 40% for cold-acclimated roots (Supplementary Table S2).

On the other hand, exposure to heavy metals (HM) did not significantly impact REL at −2 °C (two-way ANOVA, p-value = 0.228). However, it must be noted that exposure to HM reduced root growth and that material from only one plant could be harvested for plants at the control temperature, while only two biological replicates could be measured for HM plants at low temperature. Although values for the latter measurements indicate that combined exposure to cold and HM further decreased REL compared to cold alone, no significant interaction could be detected (two-way ANOVA, p-value = 0.728). This needs to be reassessed.

The effects of cold and polymetallic exposure were assessed with a two-way ANOVA. The polymetallic exposure had a significant (p-value < 0.001) impact on the F_v/F_m ratio, but no effect of cold exposure (p-value = 0.340), sampling date (p-value = 0.456), or interaction between cold and polymetallic exposure (p-value = 0.590) could be detected.

2.2. Impact on the Antioxidant System

2.2.1. Antioxidant Capacity

Antioxidant capacity is divided into two components: the antioxidant capacity of the hydrophilic fraction (containing glutathione, ascorbic acid, and flavonoids compounds) and the antioxidant capacity of the lipophilic fraction (containing tocopherol, zeaxanthin, and carotene, amongst others [9]). Of the three tested conditions, none significantly affect the lipophilic antioxidant capacity (Figure 2, Table 1(1)). Nevertheless, cold-acclimated samples tend to have a higher lipophilic antioxidant capacity, while metal exposure roots have a lower lipophilic antioxidant capacity. In the hydrophilic fraction, metal exposure has a significant negative impact on antioxidant capacity (Figure 2, Table 1(1)). On the other hand, cold acclimation has no significant effect on the hydrophilic antioxidant capacity.

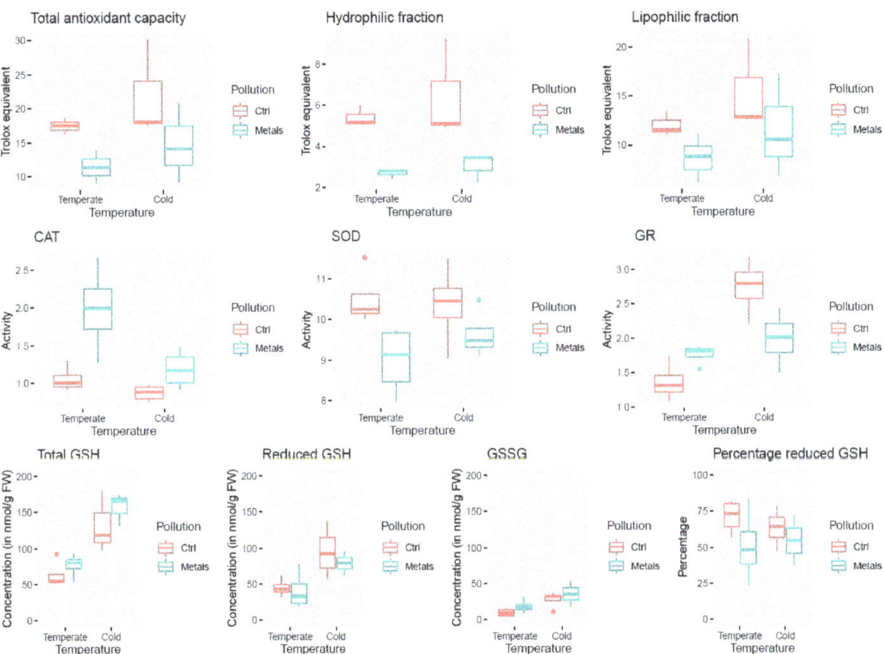

Figure 2. Impact of cold acclimation and heavy metals on root antioxidant system. First row: impact of cold and metals on the antioxidant capacity of the whole extract, hydrophilic fraction, and lipophilic fraction. Second row: impact of cold and metals on the activity of catalase (CAT), superoxide dismutase (SOD), and glutathione reductase (GR). Third row: impact of cold and metals on total glutathione (GSH), reduced glutathione, oxidised glutathione (GSSG), and the fraction of reduced glutathione.

Table 1. Impact of cold acclimation and heavy metals on root antioxidant system.

(1)

Factor	TAC	Hydro.	Lipo.
Temperature	0.175	0.366	0.157
Heavy metals	0.031 *	0.002 *	0.112
Interaction	0.84	0.668	0.914

(2)

Factor	CAT	SOD	GR
Temperature	0.017	0.562	0.003 *
Heavy metals	0.004 *	0.015 *	0.446
Interaction	0.09	0.345	0.004 *

(3)

Factor	GSH Tot.	GSH Red.	GSSG	Perc. GSH
Temperature	<0.001 *	0.007 *	0.0127 *	0.746
Heavy metals	0.258	0.529	0.1877	0.132
Interaction	0.723	0.704	0.939	0.58

Two-way ANOVA of the factors explaining antioxidative system variation. Values given are *p*-values. 1. Impact of cold acclimation (CA) and heavy metals (HM) on total antioxidant capacity (TAC), antioxidant capacity of the hydrophilic (Hydro.), and lipophilic (Lipo.) fractions. 2. Impact of CA and HM on catalase (CAT), superoxide dismutase (SOD), and glutathione reductase (GSH) activity. 3. Impact of CA and HM on total glutathione (GSH tot.), reduced glutathione (GSH red.), glutathione disulfide (GSSG) concentrations, and on the percentage of reduced glutathione over total glutathione (Perc. GSH). *: statistically significant. $p < 0.05$

When the antioxidant capacity of both fractions was summed up, only metal exposure has a significant and negative impact on the total antioxidant capacity (TAC) of the roots (Figure 2, Table 1(1)). While cold increased the average TAC of the root samples, the high variability observed makes this trend non-significant.

2.2.2. ROS Scavenging Enzymes Activity

The activity of three ROS scavenging enzymes was studied. (1) Catalase (CAT) activity is significantly and negatively affected by cold acclimation, while it increases significantly after exposure to heavy metals (Figure 2, Table 1(2)). The exposure to both conditions simultaneously does not have any significant interaction effect on the CAT activity. (2) The activity of glutathione reductase (GR) is not significantly affected by exposure to a single stress, but simultaneous exposure has a significant interaction effect (Figure 2, Table 1(2)). (3) Superoxide dismutase (SOD) activity is significantly and negatively affected by metals (Figure 2, Table 1(2)), but not by cold acclimation.

2.2.3. Glutathione

As shown in Figure 2, cold acclimation has a strong and significant effect on the total glutathione content. While metals tend to increase the total glutathione concentration, its effect is non-significant.

Similar results can be observed for glutathione (GSH) and oxidized glutathione (GSSG) (Figure 2, Table 1(3)). Cold acclimation significantly increases their concentration. However, while metals also tend to increase GSSG content, they have the opposite effect on GSH. Interestingly, no significant result could be observed for the ratio between oxidised and total glutathione content (Figure 2, Table 1(3)), but metals tend to decrease the percentage of reduced GSH.

2.3. Transcriptomics

2.3.1. De Novo Transcriptome Assembly, Functional Annotation, and Mapping

The de novo transcriptome of *S. viminalis* was obtained by merging and assembling the reads obtained by sequencing cDNA libraries of fine roots exposed to cold and/or a polymetallic mixture.

A total of 607,614,886 of 150-base pair-sequence reads were obtained, with libraries ranging from 40 to 57 million reads. After trimming and filtering, approximately 480 million reads remained. The merging and filtering of the de novo assembled transcriptomes generated 86,843 unique contigs. Of those, approximately 75% (64,173) were successfully annotated using *S. purpurea* as reference. More than 75% of the filtered reads mapped back to the de novo assembled transcriptome, indicating a good quality assembly [10]. Between 17% and 20% of the reads mapped to multiple contigs. This large proportion of reads mapped to multiple contigs likely arises from the whole-genome duplication event, known as the "salicoid" duplication, which occurred 58 Mya [11].

2.3.2. Differentially Expressed Genes

In order to determine differentially expressed genes (DEGs), the R-based package edgeR (v. 3.28.1 [12]) was used. As recommended, low-expressed contigs were filtered out because they have a low probability of playing a biological role and by necessity due to statistical approximations used by the edgeR algorithm. The filtering threshold was fixed at 0.5 counts per million (representing on average 20 counts per library) in at least three libraries. After filtering, 43,169 contigs remained.

Multidimensional scaling analysis of the data resulted in a good separation of the treatments (Figure 3a). Cold acclimation and exposure to metals appeared to have a distinct effect on gene expression. Based on this analysis, the effect of exposure to both stresses generated a distinct expression profile compared to the other conditions.

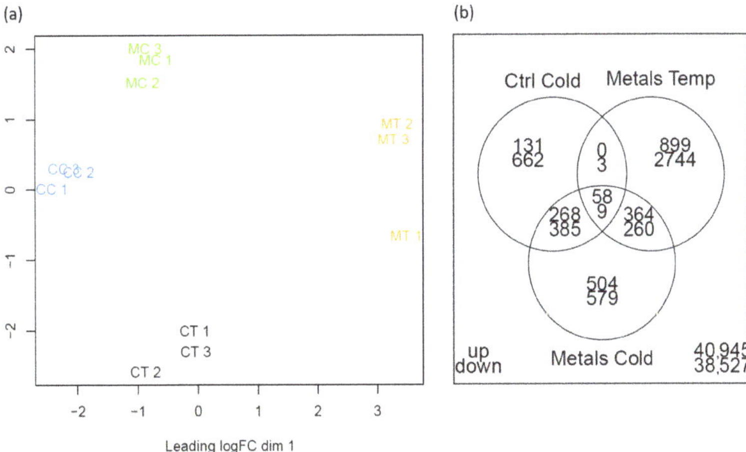

Figure 3. (**a**) Multidimensional scaling plot of the top 500 most significant DEGs. (**b**) Venn diagram of the DEGs in the three conditions investigated. Upper numbers are upregulated genes, lower numbers are downregulated genes.

Of the 43,169 contigs analysed, 6722 contigs were differentially expressed (FDR-adjusted *p*-value < 0.01) compared to the control in at least one of the conditions (Figure 3b). In CC roots, 1059 and 457 transcripts were down- and upregulated; in MT roots, 3016 and 1321 transcripts were down- and upregulated; and in MC roots, 1233 and 1194 transcripts were down- and upregulated. BLAST analysis of these 6722 unique contigs against the TAIR database resulted in 3960 annotations. For these, a TAIR name was obtained.

Only 67 contigs were significantly regulated at the intersection of the three conditions (Figure 3b). From the three conditions, exposure to heavy metals impacted the expression of the most genes (4337 genes up- or downregulated), followed by the simultaneous exposure to the polymetallic mixture and cold acclimation (2427) and cold acclimation (1516). It seems that cold acclimation reduced the impact of exposure to the polymetallic mixture on gene expression.

2.3.3. Gene Ontology Analysis

In order to identify the ontologies most involved in each condition, a gene ontology (GO) enrichment analysis was performed using clusterProfiler (v. 3.14.3 [13]). All three conditions of interest showed significant GO term enrichment ($p < 0.05$) for genes both down- and upregulated (Figure 4).

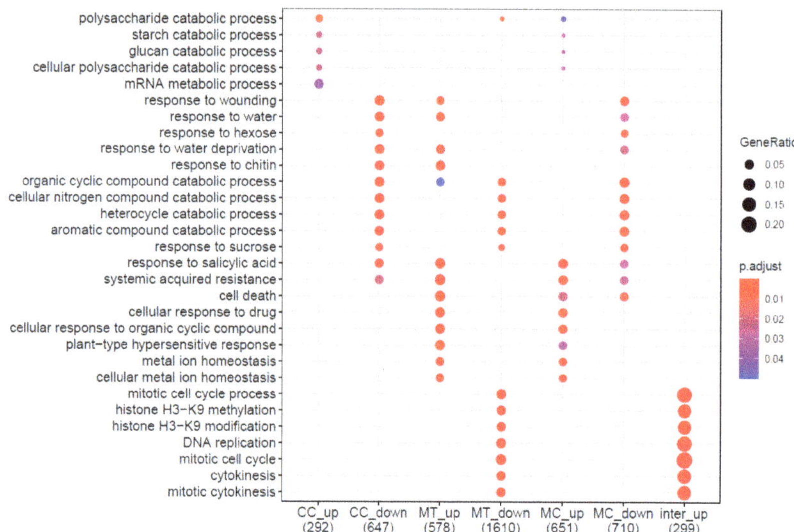

Figure 4. Dot plot of gene ontology (GO) enrichment analysis relative to the CT condition showing the five most over-represented biological processes for each category. Point size is determined by the proportion of all transcripts within a category that were annotated with the GO term (GeneRatio). Colours are based on the adjusted *p*-value of the enrichment analysis.

Using the 3960 contigs for which a GO annotation was obtained, the GO enrichment analysis revealed that processes related to the response to sucrose, catabolism of organic aromatic and cyclic compounds, and response to water deprivation were downregulated in plants exposed to cold (Figure 4). Conversely, processes related to polysaccharide catabolism were upregulated, along with mRNA metabolic processes.

In the roots of willows exposed to metals, processes related to the catabolism of organic aromatic and cyclic compounds as well as responses to sucrose and processes related to cell cycle regulation and gene silencing were downregulated. Upregulated GO terms were associated with response to water deprivation, cell death, metal ion homeostasis, and detoxification (Figure 4).

In plants exposed to both stresses, the GO terms that were significantly downregulated corresponded to those downregulated by cold acclimation alone, with the addition of the cell death GO term. Contrastingly, upregulated processes were similar to those upregulated in willows exposed to single conditions. These processes were related to metal ion homeostasis, carbohydrate catabolism, plant-type hypersensitive response, and cell death

(Figure 4). Interestingly, when filtering GO terms to levels 6 and 7, it appeared that root hair cell development was downregulated under all three conditions.

2.4. Proteomics

A total of 895 proteins (884 with TAIR annotation) were identified using our in-house database and the criteria described in the Section 4. Of those, 390 proteins (387 with TAIR annotation) were found to be significantly differentially abundant (absolute log FC $\geq$ 1.5 and FDR adjusted p-value $\leq$ 0.05).

In order to unravel the effect of cold acclimation and long-term polymetallic exposure on the root proteome, these proteins were clustered. Protein abundance was normalised by the average control abundance, then Eisen's cosine as a distance metric was used, followed by hierarchical clustering with complete linkage. Using a correlation threshold of 0.75, seven clusters were obtained (Figure 5a).

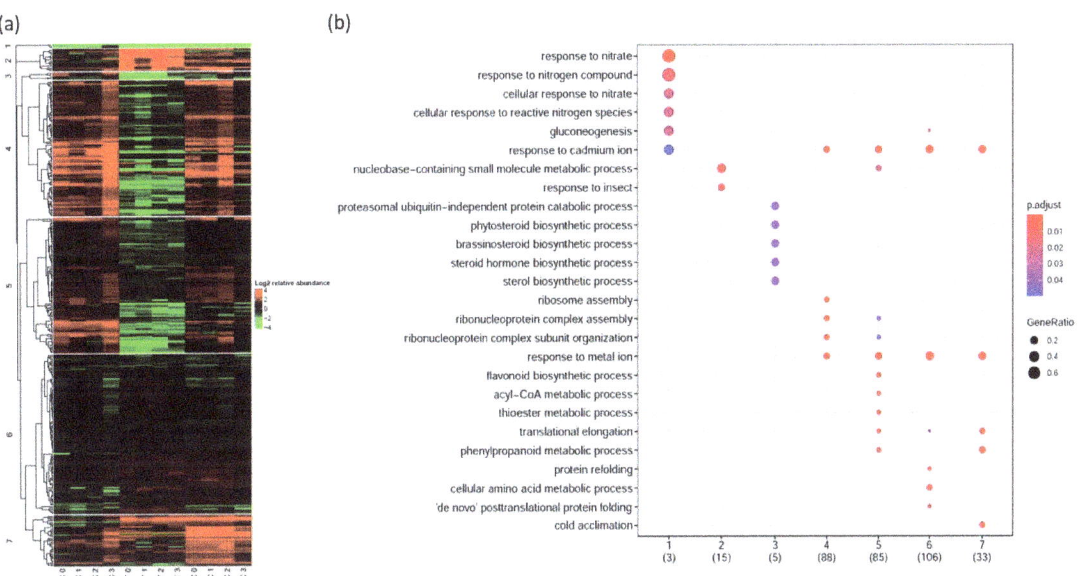

Figure 5. (**a**) Hierarchical clustering of proteins displaying interaction patterns. Clusters were made by cutting at a correlation level of 0.75. For better visualisation, protein abundance was normalised by the protein abundance of CT roots. For each condition, a technical replicate, whose name ends in 0, was made by mixing an equal protein amount of each biological replicate. (**b**) Dot plot of gene ontology (GO) enrichment analysis showing the five most over-represented biological processes for each protein cluster. Point size is determined by the proportion of all proteins within a cluster that were annotated with the GO term (GeneRatio). Colours are based on the adjusted p-value of the enrichment analysis.

Cluster 1 is composed of proteins that are downregulated in all conditions. Although it is composed of only three proteins (NRT2.1, GAPDH C2, and a calcineurin-like metallo-phosphoesterase protein), clusterProfiler determined that GO terms associated with response to nitrate, gluconeogenesis, and cadmium (Cd) ions were significantly enriched in this cluster (Figure 5b, Supplementary Figure S1).

Cluster 2 comprises proteins that are more abundant in MT roots but relatively stable in CC and MC roots. GO terms associated with response to insects, and nucleobase-containing small molecule metabolic processes, are enriched in this cluster (Supplementary Figure S2).

Opposite to cluster 2, cluster 3 is characterised by a lower protein abundance under MT and an abundance close to the control level under CC and MC. Its GO terms can be divided into three distinct groups: those related to brassinosteroid biosynthetic processes, those related to negative translation regulation, and those related to the proteasome catabolic process (Supplementary Figure S3).

The abundance of proteins in cluster 4 was significantly increased under CC and MC, while it decreased under MT. As shown in Figure 5b, GO terms related to response to Cd, ribosome assembly, and metal ion homeostasis are significantly enriched in this cluster. In addition, GO terms related to aromatic amino acid metabolism are also enriched (Supplementary Figure S4).

Cluster 5 presents a similar abundance profile as cluster 4. It is characterised by GO terms associated with response to Cd and ribosome assembly, but also to acetyl-CoA, nucleotides, flavonoid and phenylpropanoid metabolism, translation elongation, and proton transmembrane transport (Figure 5b and Supplementary Figure S5).

The abundance of proteins in cluster 6 remains relatively close to the control level. In this cluster, GO terms related to protein refolding and amino acid metabolism are significantly enriched. In addition, as shown in Supplementary Figure S6, GO terms related to response to Cd, fatty acid metabolism, response to fungus, and glucose metabolism were also enriched.

The proteins in cluster 7 have a positive interaction: while CC and MT roots have protein abundance close to the control level, protein abundance in MC roots significantly increases. GO terms enriched in this cluster can be divided into five groups: those related to cold acclimation, response to Cd, response to ROS, phenylpropanoid metabolism, and JA metabolism (Supplementary Figure S7).

3. Discussion

Salix viminalis can withstand various abiotic stresses, amongst which are exposure to high concentrations of metals and freezing. In this study, cuttings of three-month-old *S. viminalis* clones were exposed to either cold, a polymetallic mixture, or the combination of both in a controlled environment to reduce experimental variation. Aboveground organs were monitored during the cold exposure to evaluate the progression of the cold acclimation process.

3.1. Impact of Cold Acclimation

Cold exposure was sufficient to reduce and even stop plant growth, as previously observed in poplar [14], and *Juniperus chinensis* [15], amongst others. Although the F_v/F_m reduction was significant, it was smaller than in poplar exposed to 4 °C [14]. The roots were able to cold acclimate, as shown by the decrease in REL at −2 °C (Supplementary Table S2). However, the gain in frost hardiness was lower than what has been reported previously [16]. This can originate from two factors. First, the plants used in this study were young cuttings that had almost exclusively fine white roots. Younger plants tend to cold acclimate to a lesser degree than older ones, and fine roots are more frost sensitive than coarse roots [17]. Second, to focus on the effect of cold, the photoperiod was kept at 16 h/8 h, while under natural conditions the temperature decrease leading to cold acclimation coincides with a reduction of the photoperiod. Although the photoperiod does not directly impact roots, it could play a role in root cold acclimation by changing the source-sink relationship between roots and aboveground organs [18].

Cold acclimation increased the abundance of various proteins (in clusters 4, 5, 6, and 7 in Figure 5a). These include heat-shock proteins (HSP) 70 and 90, molecular chaperones during cold stress; dehydrins and proteins related to polyamines synthesis (S-adenosylmethionine synthetase), known to stabilise membranes and proteins during cold stress; and peroxidase involved in ROS detoxification. Proteins related to the energy metabolism, such as sucrose synthase, GAPDH, ATPase, and alcohol dehydrogenase, increased in abundance. These proteins are more abundant in cold-exposed chicory roots [19] and poplar [14], amongst others.

Few studies focus on long-term cold-induced transcriptomic changes. However, GO enrichment analysis is consistent with what is observed in previous studies. Starch and polysaccharide catabolism GO terms were expected as cold acclimation is associated with the conversion of starch into simple sugars [14,15,20]. Cold acclimation is also associated with a transient increase in the basal metabolism followed by a reduction of it. GO terms related to "response to water" were expected, as an essential function of cold acclimation is to mitigate upcoming frost-induced osmotic stress [16,21]. Although they play a predominant role in cold acclimation [22,23], no CBFs (genes commonly linked to cold signalling) were found in the DEGs. This might be because CBFs are highly induced upon cold perception but gradually return to control levels as exposure time increases. Kreps et al. [24] observed that *CBF1, 2,* and *3* were induced (respectively 2.5-fold, 22-fold, and 30-fold), in roots of *Arabidopsis thaliana* 3 h after cold exposure, but only 0.73-fold, 2.5-fold, and 7-fold after 27 h.

Cold acclimation did not significantly increase CAT and SOD activity. This is contrary to observations in the roots of cold acclimated *Cichorium intybus* [19], *Triticum aestivum* [25], and Cicer arietinum [26]. Nevertheless, GR activity and GSH levels significantly increased in cold acclimated roots (Figure 2), as observed in *Pinus banksiana* [27].

3.2. Impact of Long-Term Metal Exposure

Exposure to the polymetallic mixture strongly impacted plant growth and habit (Figure 1). Plants exposed to metals were shorter and chlorotic, as previously observed in multiple species [28–30]. The exposure to the polymetallic mixture also significantly lowered F_v/F_m, consistent with what has been observed previously in the same clone [31] and other *Salix* species [32].

Several protein families were more abundant in HM-exposed roots. These proteins, grouped in clusters 2, 6, and 7 (Figure 5a), were involved in protein synthesis and folding (HSPs 70 and 90 and ribosomal proteins), membrane transporters, SAM-related proteins, and Kunitz trypsin inhibitors. All these proteins were already found to be more abundant in poplar exposed to Cd for 56 days [28] and in various *S. fragilis* × *alba* clones exposed to metal-contaminated sediments [32]. Sulphite reductase, involved in sulphate assimilation, was also more abundant in MT roots. Sulphur compounds play an important role in mitigating various stresses, including metal stress, via their action on ROS detoxification and metal chelation and sequestration [33]. Interestingly, some transcripts related to sulphate assimilation were more abundant in MC roots than in MT roots and followed a positive interaction pattern. Various sulphur assimilation-related proteins have been reported to be more abundant in the roots of *A. thaliana* [34] and *Triticum aestivum* [35] under Cd stress. Peroxidase, CAT, and glutathione-S-transferases were also more abundant in MT roots, as previously reported in poplar [28]. Chromatin remodelling proteins (histones, TF jumonji family protein, pds5-related proteins) were also more abundant in MT roots. The accumulation of these proteins is consistent with the impact of metals on the expression of the genes involved in the cell cycle and histone modification, as observed in the transcriptomics analysis (Figure 4). Metals, and in particular Cd, are known to impact the cell cycle. Monteiro et al. observed that the root cells of *Lactuca sativa* exposed to 1 µM Cd tended to be blocked at the G2 checkpoint [36], and Hendrix et al. similarly observed that Cd inhibits cell division and endoreduplication in the leaves of *A. thaliana* exposed to 5 µM Cd [37]. Several proteins involved in cell wall biosynthesis (cellulose synthase 1, 4, and 6, GPAT8) were also more abundant. The cell wall can play a barrier role against Cd and reduce Cd absorption and translocation, via Cd chelation, in several plant phyla [38,39].

GO terms associated with metal ion homeostasis, detoxication, cell death, and water deprivation were enriched in MT roots. The first three terms are the direct consequence of the toxicity of the polymetallic mixture and the different mechanisms used by the plant to reduce the deleterious effects of exposure [33,40]. On the other hand, water deprivation is often observed in metal-exposed plants and could be the result of the impact of metals on root growth and aquaporins [39].

Exposure to the polymetallic mixture significantly increased CAT and GR activity and decreased SOD activity. A decrease in SOD activity was observed in Cd-exposed poplar [28], where it was hypothesised to be due to a lower Cu and Zn uptake induced by Cd exposure. However, in this study, Cu and Zn were also supplemented in the soil, indicating that the reduction in SOD activity is not linked to a decreased availability of Cu and Zn. Smeets et al. observed no significant variation in SOD activity in the roots of *A. thaliana* exposed to Cd and/or Cu [41]. In addition, they observed that CAT activity was not significantly influenced by the simultaneous exposure to Cd and Cu but that such exposure significantly reduced GR activity, contrary to what was observed here. Conversely, Tauqeer et al. [42] observed an increase in APX, CAT, POD, and SOD capacity in the roots of *Alternanthera bettzickiana* exposed to Cd for eight weeks, although this tended to be lower after exposure to 2 mM Cd than after exposure to 1 mM Cd. Similar results were observed in four cultivars of alfalfa exposed to Zn [43]. This indicates that plants have different strategies to cope with long-term metal stress. The exposure to metals also significantly reduced TAC, as observed in *Solanum tuberosum* and *Allium cepa* grown in metal-contaminated soil [44]. This decrease in TAC was not linked to reduced GSH levels as there is no difference in reduced GSH concentration between CT and MT roots.

3.3. Simultaneous Exposure to Cold and Metals

The simultaneous exposure to cold and metals did not further decrease plant growth, nor was there a further increase in chlorosis. In addition, no interaction effects could be observed in F_v/F_m. Likewise, no increase in REL at 4 °C was observed, although metals are known to increase electrolyte leakage via lipid peroxidation [42,45,46]. No significant difference in REL at −2 °C between CC and MC could be observed. This contrasts with what Talanova et al. [47] reported in Cd-exposed wheat. They observed that exposure to Cd and cold significantly increases the frost hardiness of wheat leaves after one day, but frost hardiness returned to control levels after three days.

Contrarily to what was hypothesized, the increased frost tolerance that was previously observed in *S. viminalis* roots exposed to cold and the polymetallic mixture could not be explained by a priming effect of heavy metals on the ROS detoxification capacity of the roots. Indeed, based on the measures of the antioxidant system, it seems that the polymetallic mixture impacted more strongly and more negatively the antioxidant system than cold acclimation did (Figure 2). In addition, CAT, GR, GSH1, and GSH2 were not induced at a transcriptomic or proteomic level, and while SOD transcripts were upregulated in MT and MC roots, it did not result in an increase in SOD activity. Other processes that could explain the increased frost hardiness were investigated.

Frost hardiness is a complex trait that can arise from multiple metabolic adaptations. The most commonly proposed metabolic changes leading to increased frost hardiness are (1) the increase in the concentration of small saccharides (glucose, fructose, and trehalose), (2) modification of the cell wall, (3) an increase in the unsaturation of the membrane lipids, (4) an increase in the activity of the ROS homeostasis machinery, and (5) the production of protective molecules such as polyamines and dehydrins [21]. The potential impact of these mechanisms on cold acclimation in the roots of plants exposed to a polymetallic mixture will be assessed based on our transcriptomic, proteomic, and enzymatic activity results.

With the increase in the ROS homeostasis machinery being ruled out as a possible explanation, we will now focus on the four remaining candidate processes related to frost hardiness. Given the vast amount of data generated by this study, a special emphasis will be placed on the interactions observed between cold acclimation and metal exposure. Indeed, the impact of metals on roots being stronger on the antioxidant capacity than the impact of cold acclimation, we wanted to find processes that could explain the recovery of frost hardiness despite the strong negative impact of metals on ROS homeostasis. Several nomenclatures to describe interaction patterns have been established; for the sake of clarity, the one described by Rasmusen et al. [48] will be used.

When looking at the transcriptome, different gene response patterns can be distinguished. These patterns are as follows: combinatorial (genes respond similarly for single conditions but not when exposed to both), cancelled (genes respond to at least one condition, but there is no response when exposed to both conditions), prioritised (both single conditions have an opposite effect and one response is prioritised when exposed to both conditions), independent (one single and the simultaneous exposure lead to the same response, the other condition does not influence gene response), and similar (all responses are similar).

Similarly to what Rasmusen et al. [48] observed in *Arabidopsis* exposed to multiple stresses, the cancelled, independent, and combinatorial responses were the most common, representing 63.1%, 19.2%, and 16.3% of the DEGs, respectively. The cancelled pattern corresponds to an antagonistic response at the expression level. Similarly, the prioritised (0.4%) pattern reflects an antagonist response where the response to one stress is prioritised over the other. Conversely, combinatorial and similar (1%) patterns represent positive interactions in the responses to different stresses. Finally, the independent pattern is observed in genes whose activation is entirely due to one stress and not the other. The difference between the prevalence of the cancelled response in this study and what was observed in *Arabidopsis* (30.6%) likely comes from a methodological difference. While Rasmusen et al. [48] used the top 500 DEGs for the 11 conditions studied, we used all the DEGs, even those with a low abundance. Once this difference is considered, our results are in line with the results of Rasmusen et al. [48].

While a large proportion of the DEGs is specific to simultaneous exposure to both metals and cold, no new biological process GO term arose in the GO analysis of the transcriptome. Therefore, the combinatorial pattern was not observed at the GO level, suggesting a high level of functional overlap at the gene level. Most GO terms had an independent or similar pattern (present 9 and 4 times in the 30 most significant GO terms, respectively). This is because most GO terms were enriched in only one of the single exposure conditions.

The prioritised pattern was the least represented at the transcriptome level; however, three GO terms displayed this pattern. Nine GO terms presented the cancelled expression pattern. Interestingly, out of those, seven were downregulated by metals and upregulated in the interaction group. Together, this indicates that while it is generally impossible to predict the interaction pattern of a single transcript, at the functional level, based on "biological process" GO terms, the predictability increases. Finally, four GO terms could not be classified as they were found to be simultaneously up- and downregulated.

The relationship between correlated and anti-correlated genes and proteins exhibits a similar pattern. While only five TAIR annotations were observed in both positively and negatively correlated genes and proteins, most of the GO terms found in the negatively correlated group were also found in the positively correlated group. Two exceptions to this observation are the GO terms associated with protein folding and chemical homeostasis.

When applying the same classification pattern to the clustered proteomic data, the following results were obtained: cluster 1 is similar, cluster 2 is independent, cluster 3 is cancelled, clusters 4 and 5 are prioritised, and cluster 7 is combinatorial. Proteins in cluster 6 remain close to control values in all conditions but show significant differences between the treatments; these do not show an interaction pattern. Generally, the proteome profile of MC roots was closer to that of CC roots than that of MT roots. Indeed, both conditions had similar protein abundance in clusters 1, 3, 4, 5, and 6, representing a total of 286 out of 335 proteins (Figure 5a). This might partially explain why MC roots have REL values similar to CC roots at $-2\,°C$.

Several proteins were most abundant in roots simultaneously exposed to both stresses. These proteins were found in clusters 4, 5, 6, and 7 (Figure 5a). Of those clusters, the last is of particular interest as it shows a strong synergetic interaction pattern, indicating that simultaneous exposure increased the abundance of these proteins more than expected from a mere additive effect. These proteins are of interest as their overexpression could lead to plants more resistant to frost and heavy metals simultaneously. When looking at the precise GO terms associated with this cluster (Supplementary Figure S7), five major groups can be highlighted: cold acclimation (associated with response to water and carbohydrate catabolic process), response to Cd, response to ROS, phenylpropanoid biosynthetic process, and JA biosynthetic process.

The first two GO terms, cold acclimation and response to Cd, are responses to cold and HM treatment, respectively. Since these proteins show a strong synergetic effect, they could be important in the resistance to multiple stresses. The proteins associated with the cold acclimation GO term were AT1G56070 (LOS, translation elongation factor 2-like protein), AT1G20450 (ERD10, dehydrin), AT1G10760 (SEX1, α-glucan water dikinase), and AT3G02360 (6PGD2, 6-phosphogluconate dehydrogenase). The proteins associated with the response to the Cd ion GO term were AT1G56450 (PBG1, beta subunit G1 of the 20S proteasome), AT3G48000 (ALDH2, aldehyde dehydrogenase), AT1G60710 (ATB2, aldo-keto reductase), AT1G07920 (a protein belonging to the elongation factor-tu family), AT1G77120 (ADH1, alcohol dehydrogenase), and AT4G39230 (PCBER1, a phenylcoumaran benzylic ether reductase).

The three remaining GO terms were associated with a more general abiotic stress response. The first of those was the response to oxidative stress GO term. Initial disturbance in the ROS homeostasis, followed by an increase in the ROS scavenging capacity, has been observed in response to most abiotic stresses [49,50]. The five proteins associated with this GO term were AT5G67400 (RHS19, a protein with expected peroxidase activity), AT5G58400 (a peroxidase), AT3G13160 (RPPR3B, a ribosomal pentatricopeptide repeat protein), AT1G77120 (ADH1, also related to cold acclimation), and AT2G06050 (OPR3, a 12-oxophytodienoate reductase, also involved in JA biosynthesis).

Similarly, the biosynthesis of JA has been observed as a response to multiple abiotic stresses. Furthermore, the external application of JA increases plant resistance to chilling, freezing, and heavy-metal stress, as reviewed [51]. The proteins associated with this GO term were AT2G06050 (OPR3, also found in response to oxidative stress) and AT1G55020 (LOX1, a lipoxygenase catalysing the first step of JA biosynthesis).

The least-significant GO term, phenylpropanoid biosynthesis, has also been observed in response to several stresses, including cold [52] and heavy metals [53]. The proteins linked to this GO term were AT3G13610 (F6'H1, an oxoglutarate-dependent dioxygenase), AT4G39230 (PCBER1, also found in response to Cd), AT3G53480 (ABC G3, an auxin transporter involved in the export of phenolic compounds), and AT1G64160 (DIR5, a protein involved in the synthesis of pinoresinol). Gutsch et al. observed that the cell wall proteome of *M. sativa* exposed to long-term Cd exposure is enriched in proteins related to the phenylpropanoid pathway and lignification [54]. However, they also observed that while the phenylpropanoid pathway was induced in alfalfa stems exposed to long-term Cd, this resulted in the accumulation of secondary metabolites, such as lignan and flavonoids, rather than in increased lignification [54]. In this current study, the synthesis of lignans seemed to be favoured, as shown by the overabundance of PCBER1 and DIR5. PCBER1 is known to be abundant in the xylem of poplar [55] and DIR proteins could play a role in the lignification of casparian strips [56]. In addition, based on our transcriptomics data, the genes involved in the first steps of the phenylpropanoid pathway were not upregulated in MC roots. However, the first two genes involved in the flavonoid biosynthetic pathway (*CHS* and *CHI*) were downregulated while the genes leading to the formations of lignans (including *HCT, CAD, PCBER1*, and *DIR5*) were upregulated. Altogether, this indicates a shift from the biosynthesis of flavonoids toward the production of lignans.

Of the five tolerance mechanisms mentioned above, four have been encountered in the presented data so far: the adaptation of the ROS homeostasis machinery, the modification of the cell wall (potentially via PCBER1 and DIR5), the production of protective molecules including dehydrins (ERD10), and small saccharides (through SEX1 and 6PGD2). Interestingly, transcript coding for raffinose synthesis (*DIN10* and *SIP1*), an oligosaccharide whose accumulation is related to root frost hardiness in alfalfa [57], were found to be more abundant in MC roots, while they were downregulated in the roots exposed to the other treatments, although the difference was non-significant. A similar pattern was observed in the transcripts coding for trehalose-6-phosphate synthase (*TPS*) and trehalose-6-phosphate phosphatase (*TPP*). These are involved in the production of trehalose, a disaccharide with strong frost-protecting effects [21].

In addition to these mechanisms, more general stress responses were observed in the proteome data, namely the accumulation of the 20S proteasome subunit, HSP 90, and translation elongation factors. 20S proteasome is involved in ubiquitin-independent protein degradation and plays an important role during oxidative stress [58]. HSPs are molecular chaperones that prevent protein misfolding. They are known to be important in multiple stresses, including cold and heavy metals [59]. Finally, translation elongation factors function in protein synthesis, but they also have chaperone activity and participate in the 26S proteasome [60,61]. All these proteins support the importance of the protein repair and removal mechanisms during multiple stress exposure, leading to increased ROS burden, as highlighted by Mittler [49]. Furthermore, transcripts related to sulphate assimilation were also found to be more abundant in MC roots and followed a positive interaction pattern. As mentioned above, sulphur compounds are known to play an important role in several stress-tolerance mechanisms [33].

4. Materials and Methods

4.1. Plant Materials

Cuttings of 10 cm from one individual *S. viminalis* were rooted in containers filled with a mix of 25 kg of potting soil and 17.5 kg sand under a controlled environment (23/20 °C, 60% relative humidity, 16/8 h photoperiod). After three weeks, plantlets were transferred to individual pots. Half of those pots contained soil that had been spiked with a mixture of heavy metals provided in the form of metal chlorides (final concentrations of 1.5 mg/kg of dry soil Cd, 175 mg/kg of dry soil Cu, 30 mg/kg of dry soil Ni, 500 mg/kg of dry soil Zn). After 40 days, half of the plants were randomly chosen and cold-acclimated (7/5 °C, 60% relative humidity, 16/8 h photoperiod), while the remaining plants were kept under control conditions. A total of four temperature–polymetallic contamination combinations were obtained, namely control soil temperate, control soil cold acclimated, metal-polluted soil temperate, and metal-polluted soil cold acclimated. Three biological replicates per stress combination were analysed. During the four weeks of cold treatment, chlorophyll fluorescence was measured at several time points, and the F_v/F_m ratio was calculated. After four weeks, plants were uprooted and the roots were rinsed. Approximately one gram of roots was used to determine root frost tolerance and the remaining was snap-frozen in liquid nitrogen and stored at -80 °C until subsequent use.

Root frost tolerance was assessed using root electrolyte leakage [62]. Briefly, for each biological replicate roots were washed using Elix-purified water, divided into five (100 to 350 mg), and placed into capped test tubes, and a drop of Elix-purified water was added to prevent desiccation and supercooling. For each biological replicate, one sample was kept at 4 °C and the remaining were exposed to 0 °C for 2 h. After that, the temperature was decreased at a rate of 5 °C h^{-1} to a minimum temperature of -10 °C. After reaching the selected temperature, the test tubes were removed and placed at 4 °C for thawing. After 2 h thawing, each tube was supplemented with 12 mL of Elix-purified water and left overnight at room temperature on a shaker. The conductivity of the bathing solution (E1) was measured using a conductivity probe (CyberScan CON 400, Eutech Instrument, Thermo Scientific, Villebon-Sur-Yvette, France). Root samples were then autoclaved at 121 °C for 10 min, and the conductivity of the bathing solution (E2) was re-measured the next day. The electrical conductivity of the Elix-purified water was also measured (E0). Root electrolyte leakage was calculated as:

$$REL = \frac{E1 - E0}{E2 - E0}$$

4.2. Impact on the Antioxidant System

4.2.1. Antioxidant Capacity

Analysis of the antioxidant capacity was performed using the ferric-reducing antioxidant power (FRAP) assay on hydrophilic and lipophilic fractions [63]. Briefly, 1 mL of 80% extraction buffer was added to 75 mg finely ground roots, vortexed, placed on ice, and centrifuged ($18,000\times g$, 30 min, 4 °C). Antioxidant capacity was measure at 593 nm after 10 min, using TPTZ (2,4,6 Tris (2 pyridyl) s-triazine) as a substrate and trolox (6 Hydroxy-2,5,7,8 tetramethylchrommane-2-carboxylic acid) as a standard. Total antioxidant capacity was calculated as the sum of the hydrophilic and lipophilic fractions.

4.2.2. ROS Scavenging Enzymes Activity

The enzymatic activity of catalase, glutathione reductase, and superoxide dismutase were determined spectrophotometrically. Briefly, approximately 100 mg of roots were finely ground in liquid nitrogen and added to 2 mL of extraction buffer (0.1 M Tris-HCl at pH 7.8, 1 mM DTT, 1 mM EDTA). After homogenisation, samples were filtered and centrifugated (10 min, $13,500\times g$, 4 °C). Enzyme activity was measured by colorimetry at 25 °C in the supernatant. Catalase activity was measured as the rate of decomposition of hydrogen peroxide during 10 min, measured at 240 nm, GR activity was measured at 340 nm based on the reduction of GSSG using NADPH, and the activity of SOD was measured at 550 nm using the xanthine/cytochrome c method [64].

4.2.3. Glutathione Redox Couples

Glutathione redox couples were measured by spectrophotometry, as described in [65]. Glutathione was extracted from finely ground roots under acidic conditions using 200 mM HCl. Extracts were centrifuged ($20,000\times g$, 10 min, 4 °C) and the pH of the supernatant was brought to 4.5 using 200 mM of NaH_2PO_4. Each sample was separated into two fractions; one of these was incubated with 2-vinylpyridine to determine GSSG concentrations. The other was used for total glutathione concentration, determined using the kinetics of 5,5′-dithiobis-(2-nitrobenzoic acid) reduction followed at 412 nm.

4.3. Transcriptomics

4.3.1. RNA Extraction, Library Preparation, Sequencing, and De Novo Transcriptome Assembly

The precise procedure can be found in Supplementary Materials and Methods. Briefly, root samples (250 mg) were ground into a fine powder using a mortar and pestle in liquid nitrogen. Total RNA was extracted following a modified CTAB buffer extraction protocol [66] and cleaned using the RNeasy plant mini kit (Qiagen, Leusden, The Netherlands), according to the manufacturer's instructions, using an extra DNAse I treatment to remove genomic DNA. RNA purity and quantity were assessed using a Nanodrop ND1000 spectrophotometer (Thermo Scientific) and total RNA integrity was assessed using the RNA Nano 6000 assay (Agilent Technologies, Diegem, Belgium) and a 2100 Bioanalyzer (Agilent Technologies). All RNAs had RIN above 7.

Libraries were prepared from 5 µg of total RNA using a SMARTer Stranded RNA-Seq kit (Takara Bio Inc., Mountain View, CA, USA), following the manufacturer's instructions. After ligation of the cDNA to an Illumina Indexing Primer Set (Takara Bio Inc.), an enrichment step was carried out using 13 cycles of PCR. The pooled libraries were sequenced on an Illumina NextSeq500 (Illumina NextSeq 500/550 High Output Kit v2.5 (300 cycles)) to generate 150 base-pair pair-end reads. Raw sequences have been deposited at the Gene Expression omnibus as accession GSE218490. The generated FASTQ files were imported into CLC Genomics Workbench v.11.0.1. Sequences were filtered and trimmed before transcriptome assembly.

De novo assembly was performed using the CLC genomics Workbench (v. 11.0.1) and Trinity [10,67], and the obtained assemblies were merged and filtered using cd-hit-est [68,69]) and TransDecoder (v. 5.5.0, https://github.com/TransDecoder/TransDecoder, accessed on 19 October 2021). The thus-selected transcripts were blasted against the annotated *S. purpurea* transcriptome (US Department of Energy Joint Genome Institute, Berkeley, California, USA), obtained from Phytozome (v. 12.1 [70], https://phytozome.jgi.doe.gov/, accessed on 19 October 2021). Top hit sequences were filtered at 10^{-5}. Finally, filtered reads were mapped against the de novo assembled transcriptome using the CLC genomics Workbench. A summary of the result of these steps and their optimization can be found in Supplementary Table S3.

4.3.2. Differentially Expressed Genes and Gene Ontology Analysis

To identify DEGs, count tables generated by CLC genomics were analysed using edgeR after filtering out low-expressed transcripts (count per million < 0.5). Two statistical models were used depending on the aim of the analysis. To detect DEGs, the following model was used:

$$\text{Direct effect} = \text{Condition of interest} - \text{CT}$$

where Condition of interest is either CC, MT, or MC.

Genes were considered differentially expressed when the absolute value of their adjusted fold-change was greater than 2 and their FDR-adjusted p-value was lower than 0.01. Gene ontology enrichment analysis was performed on DEGs using clusterProfiler to highlight significantly regulated biological processes (FDR-adjusted p-value < 0.05).

4.4. Proteomics

4.4.1. Protein Extraction and Gel-Free Proteomics

Soluble protein extraction was carried out as described previously [71], with modifications. The precise procedure can be found in Supplementary Materials and Methods. An amount of 400 mg of root were powdered in liquid nitrogen, suspended in ice-cold 10% TCA in acetone with 0.07% DTT, and centrifuged. The pellets were washed thrice with ice-cold acetone, dried, and resuspended in 1.4 mL SDS buffer/Tris-saturated phenol (pH 8.0, 1:1 v/v). After centrifugation, 5 volumes of ice-cold 0.1 M ammonium acetate in methanol were added to the upper phase. After 2 h at $-20\,^{\circ}\text{C}$, proteins were precipitated by centrifugation and washed twice with ice-cold 0.1 M ammonium acetate in methanol and twice with 80% ice-cold acetone before being dried. Dried samples were then re-solubilised in labelling buffer. Protein concentration was determined with the Bradford method [72].

Total proteins were loaded and separated for a short time of migration on 1D gels (Criterion, Bio-Rad, Temse, Belgium). Gels were stained with Instant Blue (Gentaur BVBA, Kampenhout, Belgium). Each sample was divided into two halves (bands of A—high molecular weight and B—low molecular weight), which were destained, reduced, alkylated, and digested using trypsin. The peptides were analysed using a NanoLC 425 Eksigent System (Sciex, Foster City, CA, USA) coupled to a TripleTOF 6600 mass spectrometer (Sciex). A MS survey scan from 300 to 1250 m/z with 250 ms of accumulation time was followed by 30 MS/MS scans (mass range 100–1500 m/z).

4.4.2. Data Analysis

Raw data were imported into Progenesis QI for Proteomics data analysis software (v. 4, Nonlinear Dynamics, Waters, Newcastle upon Tyne, UK). Spectra were processed by Mascot (v. 2.6.0, Matrix Science, London, UK) by searching against a custom-built database using our de novo-assembled transcriptome with the following search parameters: peptide tolerance of 20 ppm; fragment mass tolerance of 0.5 Da; carbamidomethylation of cysteine as fixed modification; and oxidation of methionine, N-terminal acetylation, and tryptophan to kynurenine as variable modifications. Proteins identified with a confidence of 95% were kept for further analysis. The fractions A and B were subsequently recombined. Proteins were considered significantly different between conditions when there were at least 2 significantly identified peptides per protein, of which one is unique, for the identified protein, ANOVA p-value < 0.05 and a fold change > 1.5.

The mass spectrometry proteomics data have been deposited at the ProteomeXchange Consortium via the PRIDE [73] partner repository with the dataset identifiers PXD030968 and 10.6019/PXD030968.

5. Conclusions

Our study investigated the interactions in the responses of *S. viminalis* roots to cold acclimation and/or exposure to a polymetallic mixture. The plants were exposed to ecologically relevant temperatures and heavy metal concentrations, at a temporal scale emphasising a long-term stress response.

While roots exposed to heavy metals are known to have a higher degree of electrolyte leakage, simultaneous exposure to both stresses resulted in a root electrolyte leakage level comparable to cold-acclimated roots. However, unlike the hypothesis that was the basis of this work, the low root electrolyte leakage was not linked to a metal-priming effect on the ROS scavenging capacity. Indeed, the antioxidant system of roots simultaneously exposed to both stressors was highly similar to the one of roots exposed to the polymetallic mixture alone. To understand how the roots simultaneously exposed to cold and a polymetallic mixture were still able to increase their frost tolerance despite the reduced ROS scavenging capacity, we studied their responses at the transcriptome and proteome level. A special focus was put on genes and proteins displaying interactive patterns.

At the level of the individual genes, the impact of simultaneous exposure to both stresses could not be predicted based on the analysis of the gene expression of roots exposed to each stress alone, thereby confirming that the effects of exposure to multiple stresses are not the mere addition of effects of single stresses. However, from a biological function perspective, represented by biological process GO terms, a higher degree of predictability is obtained. However, this observation requires further study and should include more diverse stress treatments.

When looking at the proteome, cold acclimation appears to be the dominant determinant in roots exposed to both conditions simultaneously. A cluster of 33 proteins was of particular interest as their abundance increased in roots exposed to both conditions simultaneously but was close to the control level in roots exposed to a single condition. Roots simultaneously exposed to the two stressors also had a higher level of PCBER and DIR5 (two proteins involved in the lignans biosynthetic pathway) and a lower level of transcripts coding for CHS and CHI, the first two proteins involved in the isoflavonoid biosynthetic pathway. This indicates a shift from the production of isoflavonoids towards the production of lignans. In addition, two transcripts of the proteins involved in the production of raffinose (*DIN10* and *SIP1*) and two transcripts involved in the biosynthesis of trehalose (*TPS* and *TPP*) had their level restored to the control level, although they were downregulated in the roots exposed to cold and metals alone. Finally, the proteins involved in the mitigation and repair of ROS damage during oxidative stress displayed a strong synergetic interaction pattern. These proteins were HSP 90, two elongation factors, and the beta subunit G1 of the 20S proteasome. This could indicate that roots simultaneously exposed to cold and metals rely more on repairing frost-induced oxidative damage when it occurs rather than going through long-term metabolic adjustment to avoid damage, as is observed in the aboveground parts of *Petunia* × *hybrida* [74]. These proteins might be specifically implicated in the tolerance to multiple stresses and could be targeted for further characterisation and targeted breeding.

Supplementary Materials: The following supporting information can be downloaded at: https://www.mdpi.com/article/10.3390/ijms25031545/s1.

Author Contributions: Conceptualization, V.A., J.-F.H., J.R., S.L., A.C. and K.S.; methodology, V.A., M.J., S.L., S.P. and K.S.; data acquisition, V.A., S.L., C.C.L., S.P. and M.J.; formal analysis, V.A., M.J., C.C.L., S.L. and S.P.; data curation, V.A., M.J., C.C.L. and S.L.; writing—original draft preparation, V.A., A.C. and K.S.; writing—review and editing, all co-authors; supervision, A.C., K.S., J.-F.H. and

J.R.; project administration, A.C., J.-F.H. and K.S.; funding acquisition, V.A. and K.S. All authors have read and agreed to the published version of the manuscript.

Funding: This work was funded by the Luxembourg National Research Fund (FNR) through the project Xpress AFR PhD/17/SR 11634190.

Institutional Review Board Statement: Not applicable.

Informed Consent Statement: Not applicable.

Data Availability Statement: Raw sequences from the transcriptome experiment have been submitted at the Gene Expression Omnibus as accession GSE218490. The mass spectrometry proteomics data have been deposited at the ProteomeXchange Consortium via the PRIDE [73] partner repository with the dataset identifiers PXD030968 and 10.6019/PXD030968. All other data is either provided in the text or as supplemental material.

Acknowledgments: The authors want to thank Aude Corvisy, Laurent Solinhac, and Ann Wijgaerds for their technical assistance.

Conflicts of Interest: The authors declare no conflicts of interest.

References

1. Lichtenthaler, H.K. The Stress Concept in Plants: An Introduction. *Ann. N. Y. Acad. Sci.* **1998**, *851*, 187–198. [CrossRef]
2. Zhu, J.K. Abiotic Stress Signaling and Responses in Plants. *Cell* **2016**, *167*, 313–324. [CrossRef]
3. Rizhsky, L.; Liang, H.; Shuman, J.; Shulaev, V.; Davletova, S.; Mittler, R. When Defense Pathways Collide. The Response of *Arabidopsis* to a Combination of Drought and Heat Stress. *Plant Physiol.* **2004**, *134*, 1683–1696. [CrossRef]
4. Suzuki, N.; Rivero, R.M.; Shulaev, V.; Blumwald, E.; Mittler, R. Abiotic and Biotic Stress Combinations. *New Phytol.* **2014**, *203*, 32–43. [CrossRef]
5. Arruda, M.A.Z.; Azevedo, R.A. Metallomics and Chemical Speciation: Towards a Better Understanding of Metal-Induced Stress in Plants. *Ann. Appl. Biol.* **2009**, *155*, 301–307. [CrossRef]
6. Hédiji, H.; Djebali, W.; Cabasson, C.; Maucourt, M.; Baldet, P.; Bertrand, A.; Boulila Zoghlami, L.; Deborde, C.; Moing, A.; Brouquisse, R.; et al. Effects of Long-Term Cadmium Exposure on Growth and Metabolomic Profile of Tomato Plants. *Ecotoxicol. Environ. Saf.* **2010**, *73*, 1965–1974. [CrossRef]
7. Burke, M.J.; Gusta, L.V.; Quamme, H.A.; Weiser, C.J.; Li, P.H. Freezing and Injury in Plants. *Plant Physiol.* **1976**, *27*, 507–528. [CrossRef]
8. Hossain, M.A.; Li, Z.G.; Hoque, T.S.; Burritt, D.J.; Fujita, M.; Munné-Bosch, S. Heat or Cold Priming-Induced Cross-Tolerance to Abiotic Stresses in Plants: Key Regulators and Possible Mechanisms. *Protoplasma* **2018**, *255*, 399–412. [CrossRef]
9. Bielen, A.; Remans, T.; Vangronsveld, J.; Cuypers, A. The Influence of Metal Stress on the Availability and Redox State of Ascorbate, and Possible Interference with Its Cellular Functions. *Int. J. Mol. Sci.* **2013**, *14*, 6382–6413. [CrossRef]
10. Haas, B.J.; Papanicolaou, A.; Yassour, M.; Grabherr, M.; Blood, P.D.; Bowden, J.; Couger, M.B.; Eccles, D.; Li, B.; Lieber, M.; et al. *De Novo* Transcript Sequence Reconstruction from RNA-Seq Using the Trinity Platform for Reference Generation and Analysis. *Nat. Protoc.* **2013**, *8*, 1494–1512. [CrossRef]
11. Dai, X.; Hu, Q.; Cai, Q.; Feng, K.; Ye, N.; Tuskan, G.A.; Milne, R.; Chen, Y.; Wan, Z.; Wang, Z.; et al. The Willow Genome and Divergent Evolution from Poplar after the Common Genome Duplication. *Cell Res.* **2014**, *24*, 1274–1277. [CrossRef]
12. Robinson, M.D.; McCarthy, D.J.; Smyth, G.K. EdgeR: A Bioconductor Package for Differential Expression Analysis of Digital Gene Expression Data. *Bioinformatics* **2009**, *26*, 139–140. [CrossRef]
13. Yu, G.; Wang, L.G.; Han, Y.; He, Q.Y. ClusterProfiler: An R Package for Comparing Biological Themes among Gene Clusters. *Omi. A J. Integr. Biol.* **2012**, *16*, 284–287. [CrossRef]
14. Renaut, J.; Lutts, S.; Hoffmann, L.; Hausman, J.F. Responses of Poplar to Chilling Temperatures: Proteomic and Physiological Aspects. *Plant Biol.* **2004**, *6*, 81–90. [CrossRef]
15. Bigras, F.J.; Paquin, R.; Rioux, J.A.; Therrien, H.P. Influence de La Photopériode et de La Température Sur l'évolution de La Tolérance Au Gel, de La Croissance et de La Teneur En Eau, Sucres, Amidon et Proline Des Rameaux et Des Racines de Genévrier (*Juniperus Chinensis* L. 'Pfitzerana'). *Can. J. For. Res.* **1989**, *69*, 305–316. [CrossRef]
16. Sakai, A.; Larcher, W. (Eds.) *Frost Survival of Plants: Responses and Adaptation to Freezing Stress*; Sakai, A.; Larcher, W. (Eds.) Springer Science & Business Media: Berlin, Germany, 2012; ISBN 9783642717475.
17. Ryyppö, A.; Repo, T.; Vapaavuori, E. Development of Freezing Tolerance in Roots and Shoots of Scots Pine Seedlings at Nonfreezing Temperatures. *Can. J. For. Res.* **1998**, *28*, 557–565. [CrossRef]
18. Smit-Spinks, B.; Swanson, B.T.; Markhart III, A.H. Effect of Photoperiod and Thermoperiod on Cold Acclimation and Growth of *Pinus Sylvestris*. *Can. J. For. Res.* **1985**, *15*, 453–460. [CrossRef]
19. Degand, H.; Faber, A.M.; Dauchot, N.; Mingeot, D.; Watillon, B.; Van Cutsem, P.; Morsomme, P.; Boutry, M. Proteomic Analysis of Chicory Root Identifies Proteins Typically Involved in Cold Acclimation. *Proteomics* **2009**, *9*, 2903–2907. [CrossRef] [PubMed]

20. Regier, N.; Streb, S.; Zeeman, S.C.; Frey, B. Seasonal Changes in Starch and Sugar Content of Poplar (*Populus Deltoides* × *Nigra* Cv. Dorskamp) and the Impact of Stem Girdling on Carbohydrate Allocation to Roots. *Tree Physiol.* **2010**, *30*, 979–987. [CrossRef] [PubMed]

21. Ambroise, V.; Legay, S.; Guerriero, G.; Hausman, J.F.; Cuypers, A.; Sergeant, K. The Roots of Plant Frost Hardiness and Tolerance. *Plant Cell Physiol.* **2020**, *61*, 3–20. [CrossRef] [PubMed]

22. Jia, Y.; Ding, Y.; Shi, Y.; Zhang, X.; Gong, Z.; Yang, S. The *Cbfs* Triple Mutants Reveal the Essential Functions of *CBFs* in Cold Acclimation and Allow the Definition of CBF Regulons in Arabidopsis. *New Phytol.* **2016**, *212*, 345–353. [CrossRef] [PubMed]

23. Park, S.; Gilmour, S.J.; Grumet, R.; Thomashow, M.F. CBF-Dependent and CBF-Independent Regulatory Pathways Contribute to the Differences in Freezing Tolerance and Cold-Regulated Gene Expression of Two *Arabidopsis* Ecotypes Locally Adapted to Sites in Sweden and Italy. *PLoS One* **2018**, *13*, e0207723. [CrossRef] [PubMed]

24. Kreps, J.A.; Wu, Y.; Chang, H.; Zhu, T.; Wang, X.; Harper, J.F.; Mesa, T.; Row, M.; Diego, S.; California, J.A.K.; et al. Transcriptome Changes for *Arabidopsis* in Response to Salt, Osmotic, and Cold Stress. *Plant Physiol.* **2002**, *130*, 2129–2141. [CrossRef] [PubMed]

25. Scebba, F.; Sebastiani, L.; Vitagliano, C. Protective Enzymes against Activated Oxygen Species in Wheat (*Triticum Aestivum* L.) Seedlings: Responses to Cold Acclimation. *J. Plant Physiol.* **1999**, *155*, 762–768. [CrossRef]

26. Nayyar, H.; Chander, S. Protective Effects of Polyamines against Oxidative Stress Induced by Water and Cold Stress in Chickpea. *J. Agron. Crop Sci.* **2004**, *190*, 335–365. [CrossRef]

27. Zhao, S.; Blumwald, E. Changes in Oxidation-Redution State and Antioxidant Enzymes in the Root of Jack Pine Seeding during Cold Acclimation. *Physiol. Plant.* **1998**, *104*, 134–142. [CrossRef]

28. Kieffer, P.; Dommes, J.; Hoffmann, L.; Hausman, J.F.; Renaut, J. Quantitative Changes in Protein Expression of Cadmium-Exposed Poplar Plants. *Proteomics* **2008**, *8*, 2514–2530. [CrossRef]

29. Luo, Z.B.; He, J.; Polle, A.; Rennenberg, H. Heavy Metal Accumulation and Signal Transduction in Herbaceous and Woody Plants: Paving the Way for Enhancing Phytoremediation Efficiency. *Biotechnol. Adv.* **2016**, *34*, 1131–1148. [CrossRef]

30. Di Baccio, D.; Galla, G.; Bracci, T.; Andreucci, A.; Barcaccia, G.; Tognetti, R.; Sebastiani, L. Transcriptome Analyses of *Populus* × *Euramericana* Clone I-214 Leaves Exposed to Excess Zinc. *Tree Physiol.* **2011**, *31*, 1293–1308. [CrossRef]

31. Ambroise, V.; Legay, S.; Guerriero, G.; Hausman, J.F.; Cuypers, A.; Sergeant, K. Selection of Appropriate Reference Genes for Gene Expression Analysis under Abiotic Stresses in *Salix Viminalis*. *Int. J. Mol. Sci.* **2019**, *20*, 4210. [CrossRef]

32. Evlard, A.; Sergeant, K.; Ferrandis, S.; Printz, B.; Renaut, J.; Guignard, C.; Paul, R.; Hausman, J.F.; Campanella, B. Physiological and Proteomic Responses of Different Willow Clones (*Salix Fragilis* × *Alba*) Exposed to Dredged Sediment Contaminated by Heavy Metals. *Int. J. Phytoremediation* **2014**, *16*, 1148–1169. [CrossRef] [PubMed]

33. Samanta, S.; Singh, A.; Roychoudhury, A. Involvement of Sulfur in the Regulation of Abiotic Stress Tolerance in Plants. In *Protective chemical agents in the amelioration of plant abiotic stress: Biochemical and Molecular Perspectives*; Wiley: Hoboken, NJ, USA, 2020; pp. 437–466.

34. Howarth, J.R.; Domínguez-Solís, J.R.; Gutiérrez-Alcalá, G.; Wray, J.L.; Romero, L.C.; Gotor, C. The Serine Acetyltransferase Gene Family in *Arabidopsis Thaliana* and the Regulation of Its Expression by Cadmium. *Plant Mol. Biol.* **2003**, *51*, 589–598. [CrossRef] [PubMed]

35. Khan, N.A.; Samiullah; Singh, S.; Nazar, R. Activities of Antioxidative Enzymes, Sulphur Assimilation, Photosynthetic Activity and Growth of Wheat (*Triticum Aestivum*) Cultivars Differing in Yield Potential under Cadmium Stress. *J. Agron. Crop Sci.* **2007**, *193*, 435–444. [CrossRef]

36. Monteiro, C.; Conceicao, S.; Pinho, S.; Oliveira, H.; Pedrosa, T.; Dias, M.C. Cadmium-Induced Cyto- and Genotoxicity Are Organ-Dependent in Lettuce. *Chem. Res. Toxicol.* **2012**, *25*, 1423–1434. [CrossRef] [PubMed]

37. Hendrix, S.; Keunen, E.; Mertens, A.I.G.; Beemster, G.T.S.; Vangronsveld, J.; Cuypers, A. Cell Cycle Regulation in Different Leaves of *Arabidopsis Thaliana* Plants Grown under Control and Cadmium-Exposed Conditions. *Environ. Exp. Bot.* **2018**, *155*, 441–452. [CrossRef]

38. Le Gall, H.; Philippe, F.; Domon, J.M.; Gillet, F.; Pelloux, J.; Rayon, C. Cell Wall Metabolism in Response to Abiotic Stress. *Plants* **2015**, *4*, 112–166. [CrossRef]

39. Poschenrieder, C.; Barceló, J. Water Relations in Heavy Metal Stressed Plants. In *Heavy Metal Stress in Plants*; Springer: Berlin, Heidelberg, 1999; pp. 207–229.

40. Cortés-Eslava, J.; Gómez-Arroyo, S.; Risueño, M.C.; Testillano, P.S. The Effects of Organophosphorus Insecticides and Heavy Metals on DNA Damage and Programmed Cell Death in Two Plant Models. *Environ. Pollut.* **2018**, *240*, 77–86. [CrossRef]

41. Smeets, K.; Opdenakker, K.; Remans, T.; Van Sanden, S.; Van Belleghem, F.; Semane, B.; Horemans, N.; Guisez, Y.; Vangronsveld, J.; Cuypers, A. Oxidative Stress-Related Responses at Transcriptional and Enzymatic Levels after Exposure to Cd or Cu in a Multipollution Context. *Plant Physiol.* **2009**, *166*, 1982–1992. [CrossRef]

42. Tauqeer, H.M.; Ali, S.; Rizwan, M.; Ali, Q.; Saeed, R.; Iftikhar, U.; Ahmad, R.; Farid, M.; Abbasi, G.H. Phytoremediation of Heavy Metals by *Alternanthera Bettzickiana*: Growth and Physiological Response. *Ecotoxicol. Environ. Saf.* **2016**, *126*, 138–146. [CrossRef] [PubMed]

43. Dai, H.P.; Shan, C.J.; Zhao, H.; Li, J.C.; Jia, G.L.; Jiang, H.; Wu, S.Q.; Wang, Q. The Difference in Antioxidant Capacity of Four Alfalfa Cultivars in Response to Zn. *Ecotoxicol. Environ. Saf.* **2015**, *114*, 312–317. [CrossRef] [PubMed]

44. Zeneli, L.; Daci-Ajvazi, M.; Daci, N.M.; Hoxha, D.; Shala, A. Environmental Pollution and Relationship between Total Antioxidant Capacity and Heavy Metals (Pb, Cd, Zn, Mn, and Fe) in *Solanum Tuberosum* L. and *Allium Cepa* L. *Hum. Ecol. Risk Assess.* **2013**, *19*, 1618–1627. [CrossRef]
45. Gajewska, E.; Bernat, P.; Dlugoński, J.; Sklodowska, M. Effect of Nickel on Membrane Integrity, Lipid Peroxidation and Fatty Acid Composition in Wheat Seedlings. *J. Agron. Crop Sci.* **2012**, *198*, 286–294. [CrossRef]
46. Taulavuori, K.; Prasad, M.N.V.; Taulavuori, E.; Laine, K. Metal Stress Consequences on Frost Hardiness of Plants at Northern High Latitudes: A Review and Hypothesis. *Environ. Pollut.* **2005**, *135*, 209–220. [CrossRef] [PubMed]
47. Talanova, V.V.; Titov, A.F.; Repkina, N.S.; Topchieva, L.V. Cold-Responsive *COR/LEA* Genes Participate in the Response of Wheat Plants to Heavy Metals Stress. *Dokl. Biol. Sci.* **2013**, *448*, 28–31. [CrossRef] [PubMed]
48. Rasmussen, S.; Barah, P.; Suarez-Rodriguez, M.C.; Bressendorff, S.; Friis, P.; Costantino, P.; Bones, A.M.; Nielsen, H.B.; Mundy, J. Transcriptome Responses to Combinations of Stresses in *Arabidopsis*. *Plant Physiol.* **2013**, *161*, 1783–1794. [CrossRef] [PubMed]
49. Mittler, R. ROS Are Good. *Trends Plant Sci.* **2017**, *22*, 11–19. [CrossRef]
50. Noctor, G.; De Paepe, R.; Foyer, C.H. Mitochondrial Redox Biology and Homeostasis in Plants. *Trends Plant Sci.* **2007**, *12*, 125–134. [CrossRef]
51. Wang, J.; Song, L.; Gong, X.; Xu, J.; Li, M. Functions of Jasmonic Acid in Plant Regulation and Response to Abiotic Stress. *Int. J. Mol. Sci.* **2020**, *21*, 1446. [CrossRef]
52. Domon, J.M.; Baldwin, L.; Acket, S.; Caudeville, E.; Arnoult, S.; Zub, H.; Gillet, F.; Lejeune-Hénaut, I.; Brancourt-Hulmel, M.; Pelloux, J.; et al. Cell Wall Compositional Modifications of *Miscanthus* Ecotypes in Response to Cold Acclimation. *Phytochemistry* **2013**, *85*, 51–61. [CrossRef]
53. Sharma, A.; Shahzad, B.; Rehman, A.; Bhardwaj, R.; Landi, M.; Zheng, B. Response of Phenylpropanoid Pathway and the Role of Polyphenols in Plants under Abiotic Stress. *Molecules* **2019**, *24*, 1–22. [CrossRef]
54. Gutsch, A.; Hendrix, S.; Guerriero, G.; Renaut, J.; Lutts, S.; Alseekh, S.; Fernie, A.R.; Hausman, J.F.; Vangronsveld, J.; Cuypers, A.; et al. Long-Term Cd Exposure Alters the Metabolite Profile in Stem Tissue of *Medicago Sativa*. *Cells* **2020**, *9*, 1–22. [CrossRef]
55. Bruegmann, T.; Wetzel, H.; Hettrich, K.; Smeds, A.; Willför, S.; Kersten, B.; Fladung, M. Knockdown of *PCBER1*, a Gene of Neolignan Biosynthesis, Resulted in Increased Poplar Growth. *Planta* **2019**, *249*, 515–525. [CrossRef]
56. Paniagua, C.; Bilkova, A.; Jackson, P.; Dabravolski, S.; Riber, W.; Didi, V.; Houser, J.; Gigli-Bisceglia, N.; Wimmerova, M.; Budínská, E.; et al. Dirigent Proteins in Plants: Modulating Cell Wall Metabolism during Abiotic and Biotic Stress Exposure. *J. Exp. Bot.* **2017**, *68*, 3287–3301. [CrossRef]
57. Cunningham, S.M.; Nadeau, P.; Castonguay, Y.; Laberge, S.; Volenec, J.J. Raffinose and Stachyose Accumulation, Galactinol Synthase Expression, and Winter Injury of Contrasting Alfalfa Germplasms. *Crop Sci.* **2003**, *43*, 562–570. [CrossRef]
58. Pena, L.B.; Azpilicueta, C.E.; Benavides, M.P.; Gallego, S.M. Protein Oxidative Modifications. In *Metal Toxicity in Plants: Perception, Signaling and Remediation*; Springer: Berlin/Heidelberg, Germany, 2012; pp. 207–225.
59. Jacob, P.; Hirt, H.; Bendahmane, A. The Heat-Shock Protein/Chaperone Network and Multiple Stress Resistance. *Plant Biotechnol. J.* **2017**, *15*, 405–414. [CrossRef]
60. Sasikumar, A.N.; Perez, W.B.; Kinzy, T.G. The Many Roles of the Eukaryotic Elongation Factor 1 Complex. *Wiley Interdiscip. Rev. RNA* **2012**, *3*, 543–555. [CrossRef] [PubMed]
61. Fu, J.; Momčilović, I.; Prasad, P.V.V. Roles of Protein Synthesis Elongation Factor EF-Tu in Heat Tolerance in Plants. *J. Bot.* **2012**, *2012*, 1–8. [CrossRef]
62. McKay, H.M. Frost Hardiness and Cold-Storage Tolerance of the Root System of *Picea Sitchensis, Pseudotsuga Menziesii, Larix Kaempferi* and *Pinus Sylvestris* Bare-Root Seedlings. *Scand. J. For. Res.* **1994**, *9*, 203–213. [CrossRef]
63. Benzie, I.F.F.; Strain, J.J. The Ferric Reducing Ability of Plasma (FRAP) as a Measure of '"Antioxidant Power"': The FRAP Assay. *Anal. Biochem.* **1996**, *239*, 70–76. [CrossRef]
64. McCord, J.M.; Fridovich, I. Superoxide Dismutase, an Enzymic Function for Erythrocuprein (Hemocuprein). *J. Biol. Chem.* **1969**, *244*, 6049–6055. [CrossRef]
65. Queval, G.; Noctor, G. A Plate Reader Method for the Measurement of NAD, NADP, Glutathione, and Ascorbate in Tissue Extracts: Application to Redox Profiling during *Arabidopsis* Rosette Development. *Anal. Biochem.* **2007**, *363*, 58–69. [CrossRef]
66. Legay, S.; Guerriero, G.; Deleruelle, A.; Lateur, M.; Evers, D.; André, C.M.; Hausman, J.F. Apple Russeting as Seen through the RNA-Seq Lens: Strong Alterations in the Exocarp Cell Wall. *Plant Mol. Biol.* **2015**, *88*, 21–40. [CrossRef]
67. Grabherr, M.G.; Haas, B.J.; Yassour, M.; Levin, J.Z.; Thompson, D.A.; Amit, I.; Adiconis, X.; Fan, L.; Raychowdhury, R.; Zeng, Q.; et al. Full-Length Transcriptome Assembly from RNA-Seq Data without a Reference Genome. *Nat. Biotechnol.* **2011**, *29*, 644–652. [CrossRef]
68. Li, W.; Godzik, A. Cd-Hit: A Fast Program for Clustering and Comparing Large Sets of Protein or Nucleotide Sequences. *Bioinformatics* **2006**, *22*, 1658–1659. [CrossRef] [PubMed]
69. Fu, L.; Niu, B.; Zhu, Z.; Wu, S.; Li, W. CD-HIT: Accelerated for Clustering the next-Generation Sequencing Data. *Bioinformatics* **2012**, *28*, 3150–3152. [CrossRef] [PubMed]
70. Goodstein, D.M.; Shu, S.; Howson, R.; Neupane, R.; Hayes, R.D.; Fazo, J.; Mitros, T.; Dirks, W.; Hellsten, U.; Putnam, N.; et al. Phytozome: A Comparative Platform for Green Plant Genomics. *Nucleic Acids Res.* **2012**, *40*, 1178–1186. [CrossRef]

71. Wang, W.; Scali, M.; Vignani, R.; Spadafora, A.; Sensi, E.; Mazzuca, S.; Cresti, M. Protein Extraction for Two-Dimensional Electrophoresis from Olive Leaf, a Plant Tissue Containing High Levels of Interfering Compounds. *Electrophoresis* **2003**, *24*, 2369–2375. [CrossRef]
72. Bradford, M.M. A Rapid and Sensitive Method for the Quantitation of Microgram Quantities of Protein Utilizing the Principle of Protein-Dye Binding. *Anal. Biochem.* **1976**, *72*, 248–254. [CrossRef]
73. Perez-Riverol, Y.; Bai, J.; Bandla, C.; García-Seisdedos, D.; Hewapathirana, S.; Kamatchinathan, S.; Kundu, D.J.; Prakash, A.; Frericks-Zipper, A.; Eisenacher, M.; et al. The PRIDE Database Resources in 2022: A Hub for Mass Spectrometry-Based Proteomics Evidences. *Nucleic Acids Res.* **2022**, *50*, D543–D552. [CrossRef]
74. Pennycooke, J.C.; Cox, S.; Stushnoff, C. Relationship of Cold Acclimation, Total Phenolic Content and Antioxidant Capacity with Chilling Tolerance in Petunia (*Petunia × Hybrida*). *Environ. Exp. Bot.* **2005**, *53*, 225–232. [CrossRef]

International Journal of
Molecular Sciences

MDPI

Article

Transcriptome and Low-Affinity Sodium Transport Analysis Reveals Salt Tolerance Variations between Two Poplar Trees

Xuan Ma [1,†,‡], Qiang Zhang [1,†], Yongbin Ou [1], Lijun Wang [1], Yongfeng Gao [1], Gutiérrez Rodríguez Lucas [1], Víctor Resco de Dios [1,2,*] and Yinan Yao [1,*]

[1] School of Life Science and Engineering, Southwest University of Science and Technology, Mianyang 621010, China
[2] Department of Crop and Forest Sciences & Agrotecnio Center, Universitat de Lleida, 25003 Leida, Spain
* Correspondence: v.rescodedios@gmail.com (V.R.d.D.); yinanyao@swust.edu.cn (Y.Y.)
† These authors contributed equally to this work.
‡ Current address: National Key Laboratory of Crop Genetic Improvement, Huazhong Agricultural University, Wuhan 430070, China.

Abstract: Salinity stress severely hampers plant growth and productivity. How to improve plants' salt tolerance is an urgent issue. However, the molecular basis of plant resistance to salinity still remains unclear. In this study, we used two poplar species with different salt sensitivities to conduct RNA-sequencing and physiological and pharmacological analyses; the aim is to study the transcriptional profiles and ionic transport characteristics in the roots of the two *Populus* subjected to salt stress under hydroponic culture conditions. Our results show that numerous genes related to energy metabolism were highly expressed in *Populus alba* relative to *Populus russkii*, which activates vigorous metabolic processes and energy reserves for initiating a set of defense responses when suffering from salinity stress. Moreover, we found the capacity of Na$^+$ transportation by the *P. alba* high-affinity K+ transporter1;2 (HKT1;2) was superior to that of *P. russkii* under salt stress, which enables *P. alba* to efficiently recycle xylem-loaded Na$^+$ and to maintain shoot K$^+$/Na$^+$ homeostasis. Furthermore, the genes involved in the synthesis of ethylene and abscisic acid were up-regulated in *P. alba* but downregulated in *P. russkii* under salt stress. In *P. alba*, the gibberellin inactivation and auxin signaling genes with steady high transcriptions, several antioxidant enzymes activities (such as peroxidase [POD], ascorbate peroxidase [APX], and glutathione reductase [GR]), and glycine-betaine content were significantly increased under salt stress. These factors altogether confer *P. alba* a higher resistance to salinity, achieving a more efficient coordination between growth modulation and defense response. Our research provides significant evidence to improve the salt tolerance of crops or woody plants.

Keywords: salt stress; *Populus*; Na$^+$ transportation; transcriptome; HKT1;2

Citation: Ma, X.; Zhang, Q.; Ou, Y.; Wang, L.; Gao, Y.; Lucas, G.R.; Resco de Dios, V.; Yao, Y. Transcriptome and Low-Affinity Sodium Transport Analysis Reveals Salt Tolerance Variations between Two Poplar Trees. *Int. J. Mol. Sci.* **2023**, 24, 5732. https://doi.org/10.3390/ijms24065732

Academic Editor: Martin Bartas

Received: 13 February 2023
Revised: 4 March 2023
Accepted: 5 March 2023
Published: 17 March 2023

1. Introduction

Soil salinity is a detrimental environmental stress that constrains plant growth and crop production. High salinity induces osmotic stress, ionic toxicity, and oxidative stress in plants [1,2]. Saline soils limit the capacity of plant roots to absorb water, promoting water-deficiency and osmotic stress, which results in the inhibition of plant growth. Furthermore, when an excess of sodium (Na$^+$) and/or chlorine (Cl$^-$) ions enter through the transpiration stream into the plant leaves, it causes ionic imbalance and cellular damage, hampering plant growth and development [3,4]. These osmotic and ionic stresses may induce further secondary oxidative effects, promoting retardation of plant growth, metabolic and developmental changes, and reduction of productivity [1,4,5].

The transport of Na$^+$ into plant roots presents a biphasic behavior [6]: when external Na$^+$ concentration is at a very low level, Na$^+$ transport is mediated by a high-affinity transport system (HATS); at higher Na$^+$ concentrations, it is mediated via a low-affinity system (LATS). Extensive research of the adverse effects of high salinity on agricultural production has, during the last decades, consolidated LATS as a well-acknowledged mechanism in the field of plant physiology [7–9]. Electrophysio- and pharmacological analyses have long demonstrated that the LATS of Na^+ uptake in glycophytes or halophytes (e.g., *Arabidopsis* and *Suaeda maritime*) mainly consist of non-selection cation channels (NSCC) [7,9] as well as several Na^+/K^+ transporters, i.e., the low-affinity cation transporter 1 (LCT1), high-affinity K^+ transporter (HKT), and K^+ channels *Arabidopsis* K^+ Transporter 1 (AKT1) [8–12]. HKT1 mediates Na^+ transport in both high- and low-affinity systems and may function as either an Na^+ uniporter or Na^+/K^+ symporter [4,13]. In *Arabidopsis*, there is only evidence of one HKT ortholog gene *AtHKT1;1* performing Na^+ exclusion from leaves and regulating Na^+ distribution between roots and shoots [14,15]. In other plant species, a great number of studies have also revealed that the HKT1 ortholog protein plays a key role in salt tolerance in both dicots and monocots via Na^+ translocation [15–17]. Although the function of HKTs in herbs or grass plants has been thoroughly studied, however, little is yet known about their function in perennial woody plants.

Poplars are widely distributed across the northern temperate and cold-temperate zones of the world. Different *Populus* species evolutionarily adapted to these diverse ecological environments have thus generated long-term variations in responsiveness and tolerance to salinity [18–20]. Differences in salt tolerance between *Populus* species possibly result from different Na^+ transporting or translocation capacities in the plant organ/tissue or cell, including the following: (1) Na^+ and/or K^+ uptake capacities are clearly different between salt-tolerant and salt-sensitive *Populus* species under salinity stress [18,19]. (2) Excess Na^+ remains in roots rather than translocating into the salt-sensitive photosynthetic tissues, which is an efficient strategy to tolerate salinity stress in plants [1,2]. For example, the AtHKT1;1 in *Arabidopsis* and HKT1;5 in rice or wheat perform a similar function in the storge of Na^+ in roots, unloading Na^+ from the xylem sap into the surrounding parenchyma cells and thereby protecting the photosynthetic organs from Na^+ toxicity [14–16,21]. (3) Differences are present in the Na^+ compartmentation and Na^+ detoxification in cells subjected to salinity stress. The sequestration of Na^+ within the vacuole through the Na^+/H^+ antiporters (NHXs) and the extrusion of Na^+ from the cytoplasm through the plasma membrane-localized Na^+/H^+ antiporter SOS1 (Salt Overly Sensitive) are key mechanisms to improve plants' salt tolerance at cellular levels [2,22,23].

Moreover, the capacity of K^+ uptake and maintaining the K^+/Na^+ ratio in plant cells is crucial to improve salt tolerance [18,19]. Since Na^+ competes with K^+ and induces K^+ efflux from cytoplasm, affecting various enzymatic processes and metabolic pathways, it can cause nutrient ion imbalance [1,24]. Our recent study shows that maintaining the K^+/Na^+ ratio plays a vital role against salinity stress under different nitrogen levels [25]. As Na^+ enters into the plant cell, K^+/Na^+ ratio maintenance is an essential route to tolerate salt-induced osmotic stress to keep water availability inside the cytoplasm via concerted regulation of metabolic processes and strengthened biosynthesis of osmoprotectants such as proline, glycine-betaine, etc. [1,2,26]. Furthermore, excess Na^+ triggers the generation of reactive oxygen species (ROS) inside the plant cell; the capacity to activate prompt redox responses and enhance the activity of antioxidant enzymes that scavenge the excess ROS is paramount to resist oxidative stress induced by high salinity [1,27].

Na^+ transportation under high salinity conditions plays an important role in plant growth and development [2,18]. However, still little is known about Na^+ transport and the underlying molecular processes of salt tolerance in poplar plants. The regulation of Na^+ transport mainly depends on the transcriptional level, which is energy-saving and highly plastic, and also depends on the capacity of Na^+/K^+ transporters and related ion channels. In this study, we performed RNA-sequencing and combined the pharmacological analysis with a hydroponic-cultivation condition to analyze the Na^+ transportation and root transcriptomic profiles in two *Populus* species (*Populus alba* and *Populus russkii*, separately belong to the Leuce section and the Aigeiros section in the *Populus* genus). Our results show distinct features of Na^+ transportation and extensive variations in physiological and molecular responses to salt stress between the two *Populus*.

2. Results

2.1. Na⁺ and K⁺ Uptake in the Two Populus and Their Distinct Physiological Responses to Salt Stress

To analyze whether there were differences in salt tolerance between *P. alba* and *P. russkii* and explore the characteristics of ion uptake/transport under salinity stress, we cultivated the cuttings of the two poplars using the hydroponic technique (see methods section) and subsequently treated them with 100 mM NaCl. We analyzed the dynamics of Na⁺ and K⁺ contents in the two *Populus* leaves before NaCl treatment (0 h) and at three treatment time points during the treatment (for 24 h, 7 days, and 12 days). Firstly, we detected that Na⁺ accumulation in *P. russkii* leaves was significantly higher than that of *P. alba* (Figure 1b,c), albeit Na⁺ concentration gradually increased in both species during the NaCl treatment. Secondly, K⁺ content was higher in *P. alba* than in *P. russkii* (Figure 1d), while the K⁺/Na⁺ ratio in *P. alba* was also significantly higher than in *P. russkii* (Figure 1e). Thirdly, we observed severe Na⁺ toxicity and osmotic stress in the leaf phenotype of *P. russkii*, whose leaves were dehydrated and shrunk after the NaCl treatment for 12 days, whereas no obvious variation was observed in *P. alba* (Figure 1a). These results altogether indicate that there was a substantial interspecific divergence in salt tolerance and Na⁺ and K⁺ transport between *P. alba* and *P. russkii*.

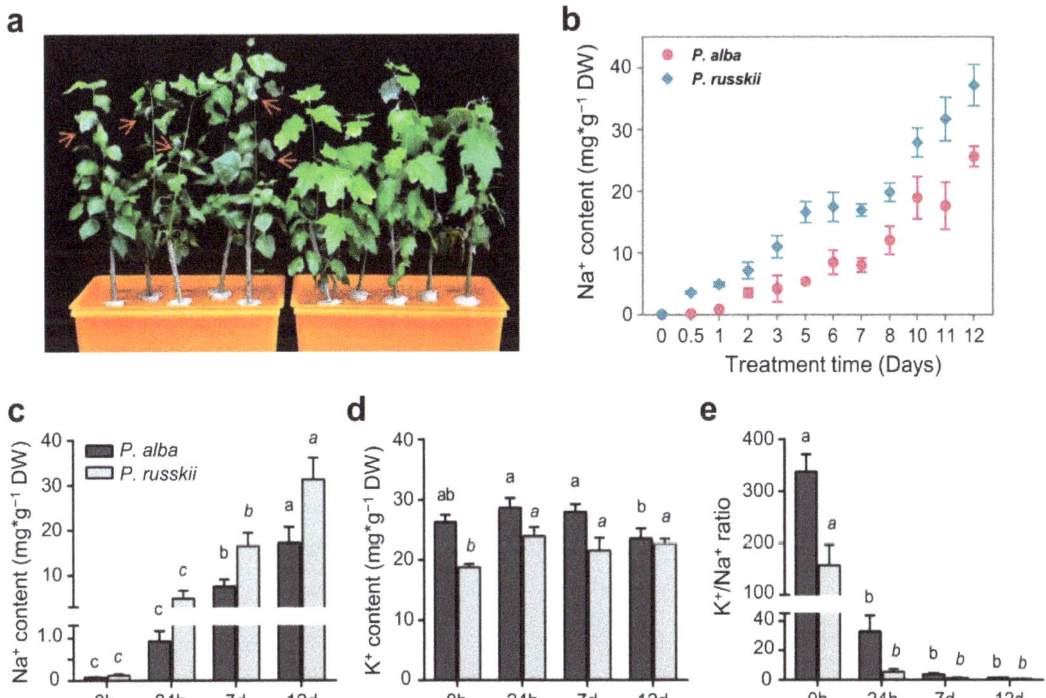

Figure 1. Phenotype and the content of Na⁺ or K⁺ in leaves of two *Populus* species under salt stress. (**a**) Morphological change of *P. russkii* (**left**) and *P. alba* (**right**) with 100 mM NaCl treatment for 12 days. Red arrows indicate the shrunken leaves of *P. russkii*. (**b**) Dynamic of Na⁺ content in the two *Populus* leaves following NaCl treatment. (**c**) Na⁺ content, (**d**) K⁺ content, and (**e**) K⁺/Na⁺ ratio in leaves of *P. alba* and *P. russkii* before treatment and after treatment for 24 h, 7 d, and 12 d. Data shown are the mean ± SD (n = 3). Different letters above the bars indicate the statistical significance at the *p* < 0.05 level among different time-points of NaCl treatment according to LSD test.

2.2. Salt Stress Effects on the Antioxidant System of the Two Populus

The activities of four anti-oxidative enzymes (POD, catalase [CAT], APX, and GR) were determined in the two *Populus* leaves under salt stress (Figure 2a–d). In contrast to the control, POD, APX, and GR activities significantly increased in *P. alba* under salt stress, but no significant changes were found for *P. russkii*; meanwhile, the CAT activity in the two *Populus* showed the opposite trend to the abovementioned three enzymes (Figure 2b). Moreover, malondialdehyde (MDA) concentration significantly increased in *P. russkii* leaves under salt stress but did not increase in *P. alba* (Figure 2e), while glycine-betaine content augmented significantly in both *Populus* leaves under salt stress and increased ca. four times in *P. alba* (Figure 2f).

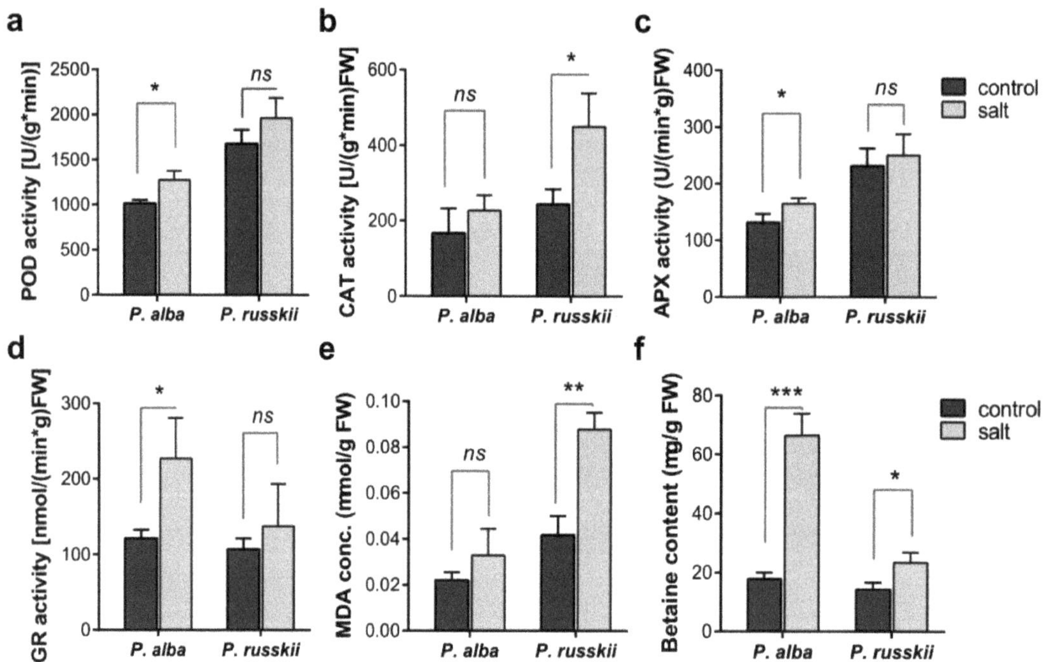

Figure 2. Antioxidant enzyme activities, MDA concentration, and glycine-betaine content in leaves of the two *Populus* species under salinity stress. (**a**) POD activity, (**b**) CAT activity, (**c**) APX activity, (**d**) glutathione reductase (GR) activity, (**e**) MDA concentration and (**f**) glycine-betaine content in *P. alba* and *P. russkii* leaves under NaCl treatment for 12 days. Bar indicates the mean ± SD (n = 3), *, $p < 0.05$, **, $p < 0.01$, ***, $p < 0.001$, ns, $p > 0.05$, two-tailed *t*-test.

2.3. Differentially Expressed Genes (DEGs) in P. alba and P. russkii Roots under Salt Stress

To investigate the molecular basis of the divergent salt tolerance in the two *Populus*, we conducted RNA-sequencing on the roots of *P. alba* and *P. russkii* just after applying the short-term (24 h) and long-term (12 d) salt stress treatments (100 mM NaCl). Two biological replicates were sequenced for each treatment group and control. We obtained an average of 23.3 million reads from twelve cDNA libraries. After removing the adapters and low-quality reads, we obtained high-quality clean reads for each library (Table S1). The clean reads were aligned on the reference genome (v4.1, PhytozomeV13) of *P. trichocarpa*, the reference model poplar species from *Populus* genus; the average mapping rate of *P. russkii* was 74.85% that higher than that of *P. alba* (60.6%), suggesting that the genetic relationship between *P. russkii* and *P. trichocarpa* was closer than that between *P. alba* and *P. trichocarpa*. The numbers of the expressed transcripts (RPKM >1) detected in each library were comparable (average of 22,391, Table S1). The correlation of the two biological replicates for each treatment group was high (Pearson's $r = 0.8\sim1.0$, Figure S1). Principal component analysis (PCA) revealed important differences between *P. alba* and *P. russkii* transcriptomes on PC1 (31.4% of variation, Figure 3a). With regard to PC2 (16.1% variation), we observed that the transcriptome of short- and long-term salt stress groups and CK was clearly separated in *P. russkii*, whereas it was not observed in the case of *P. alba* (Figure 3a). One control sample of *P. alba* was far away from other samples, which was attributed to a batch effect of RNA-seq, and therefore was removed for further analysis.

DEGs were identified in *P. alba* and *P. russkii* using a cutoff of fold change ≥ 2 and FDR < 0.01. The numbers of up-regulated genes were more than double those of downregulated genes in the *P. alba* root under either short- (CK (Control) vs. S24h, 1070/493) or long-term salt stress (CK vs. S12d, 609/179) (Figure 3b; Supplementary Data S1). In the *P. russkii* root, the number of up-regulated genes was comparable with that of downregulated genes under short-term salt stress, whereas downregulated genes were two times higher than up-regulated genes under long-term salt stress (2482/1090) (Figure 3b; Supplementary Data S1).

GO enrichment analysis of the up- or downregulated genes in four comparisons (*Pr*CK vs. *Pr*S24h, *Pr*CK vs. *Pr*S12d, *Pa*CK vs. *Pa*S24h, and *Pa*CK vs. *Pa*S12d) identified several biological processes uniquely/commonly overrepresented (Figure 3c). The GO term of the carbohydrate metabolic process (GO:0005975) was significantly enriched in nearly all comparisons (except for downregulated genes in *Pa*CK vs. *Pa*S12d). Several other metabolic processes were significantly enriched in the downregulated genes in *P. russkii* and *P. alba*, such as catabolic (GO:0009056), secondary metabolic (GO:0019748), and lipid metabolic (GO:0006629) processes (Figure 3c). The GO term of response to stress (GO:0006950) was enriched in the up-regulated genes in *P. russkii* (*Pr*CK vs. *Pr*S12d) but in the downregulated genes in *P. alba* (*Pa*CK vs. *Pa*S24h and *Pa*CK vs. *Pa*S12d) (Figure 3c).

To further dissect the characteristics of gene transcription in *P. russkii* and *P. alba* roots under salt stress, we analyzed the overlapping of DEGs that were identified in the four comparisons. We found that numerous DEGs overlapped between the short-term and long-term salt stress comparisons in *P. russkii*. For instance, 737 downregulated genes overlapped in *Pr*CK vs. *Pr*S24h and *Pr*CK vs. *Pr*S12d, and 329 up-regulated genes were also overlapped (Figure 4a). The 737 downregulated genes were significantly enriched in response to stress, while the 329 up-regulated genes were mainly enriched in metabolic processes (GO:0019784, GO:0005975, and GO:0006629) (Figure 4b). A total of 308 up-regulated genes overlapped in *Pa*CK vs. *Pa*S24h and *Pa*CK vs. *Pa*S12d that were significantly enriched in carbohydrate metabolic, biosynthetic, and nucleobase-containing compound metabolic processes. A total of 65 downregulated genes overlapped in *Pa*CK vs. *Pa*S24h and *Pa*CK vs. *Pa*S12d that were enriched for transport (Figure 4b).

Moreover, we found 41 genes were commonly up-regulated in *P. alba* and *P. russkii* under salt stress (Figure 4c), including some transcription factors/genes that were important for response to osmotic stress, abscisic acid, or abiotic stress, such as MYB43/78, WRKY48, bZIP1, Responsive to Dessication 26 (RD26a/b), Highly ABA-Induced PP2C Gene 1 (HAI1), Drought-Induced 21 (DI21), and Late Embryogenesis Abundant 4-5 (LEA4-5) [28–32]. There were 20 genes commonly downregulated in *P. alba* and *P. russkii* under salt stress, including several that were previously reported to play an important role in ion homeostasis and root development, such as Vacuolar iron Transporter-Like 2 (VTL2), Ferric Reductase Defective 3 (FRD3), Nitrate Reductase 1 (NR1), DWARF14-LIKE2 (DLK2), and GRAS [33–36].

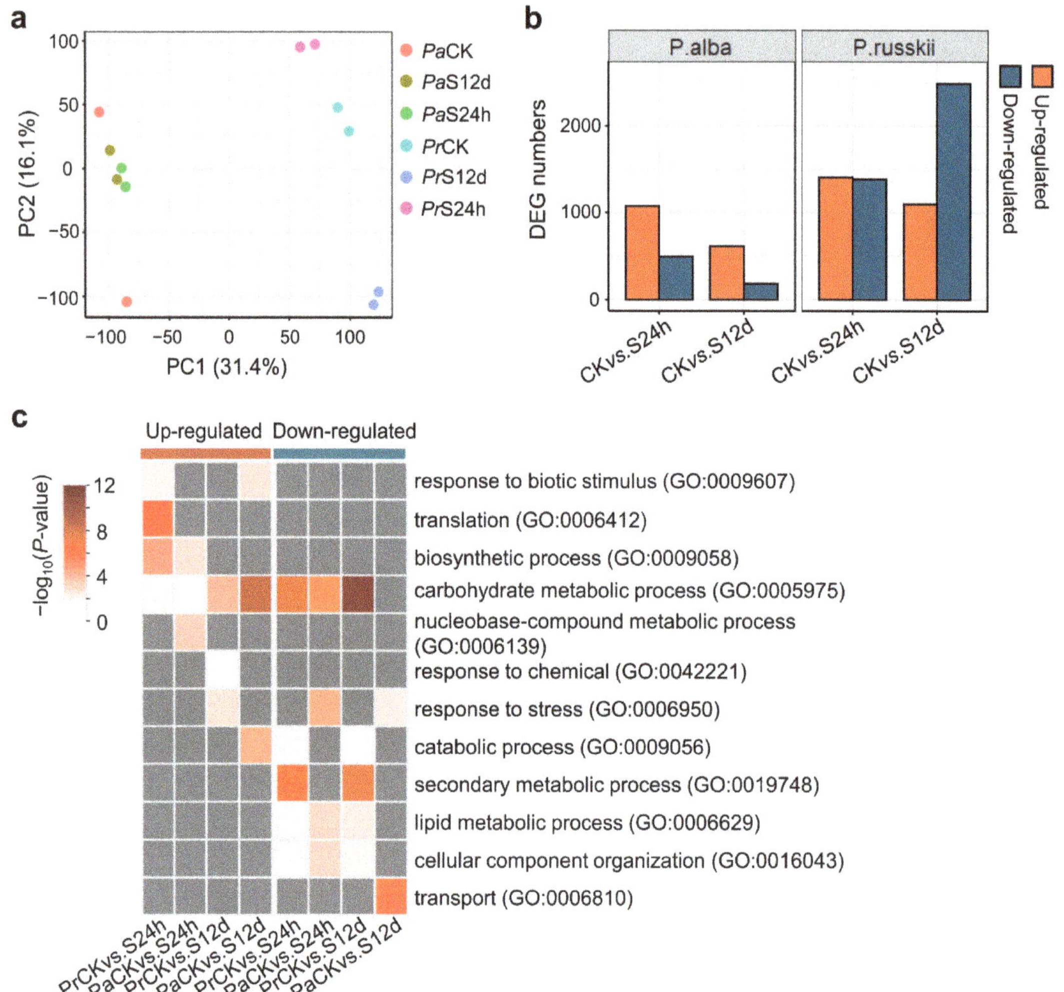

Figure 3. Differentially expressed genes (DEGs) identified in root of *P. alba* and *P. russkii* under salinity stress. (**a**) PCA analysis of the transcriptome data. (**b**) DEG numbers of the identified in *P. alba* and *P. russkii* under NaCl treatment for 24 h and 12 d versus to the control (CK), respectively. (**c**) GO enrichment of the up- and down-regulated genes identified in *P. alba* and *P. russkii* under NaCl treatment for 24 h or 12 d. Grey color indicates not available.

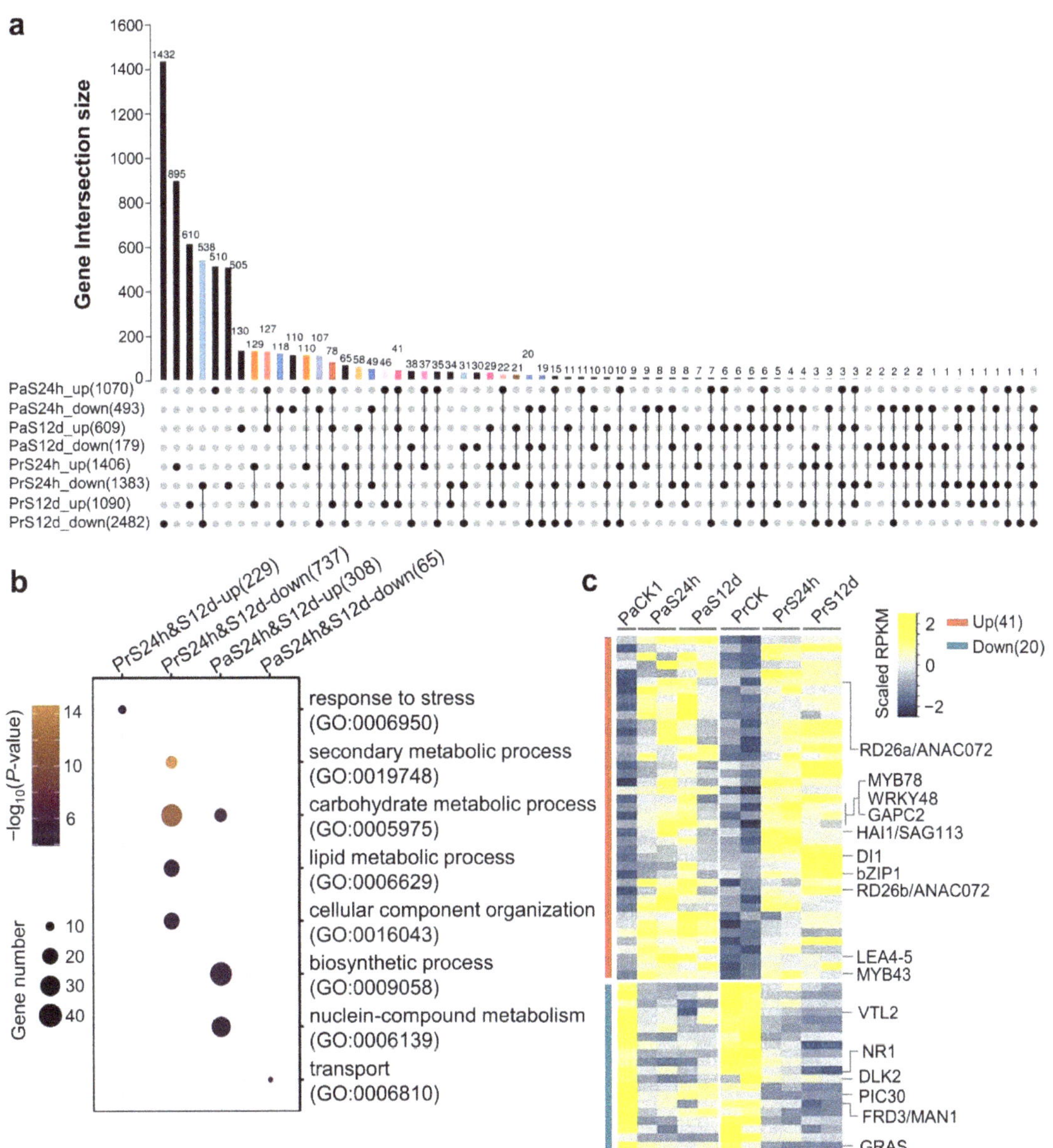

Figure 4. Characteristics of the DEGs identified in *P. russkii* and *P. alba* roots under salt stress. (**a**) Upsetplot showing the overlapping genes among the up-regulated and downregulated genes identified in *P. russkii* and *P. alba* roots. (**b**) GO enrichment of the overlapping up-regulated or downregulated genes of *P. russkii* and *P. alba* roots between short-term and long-term salt stress. (**c**) Heatmap showing the commonly up-regulated or downregulated genes in *P. russkii* and *P. alba* roots under salt stress. Two biological replicates are shown except for the control of *P. alba*.

2.4. Extensive Variation in Gene Transcription between the Two Populus under Salt Stress

We identified a large number of DEGs between *P. russkii* and *P. alba* in CK and under both short-term and long-term salt stress (Figure 5a; Supplementary Data S2). Numbers of up-regulated and downregulated genes were comparable in CK (*P. russkii* vs. *P. alba*), but under salt stress the up-regulated genes were about two times greater than downregulated genes, suggesting a substantial difference in gene expression between the two *Populus* under salt stress. In the different comparisons for these genes, GO enrichment indicated that up-regulated genes were mainly involved in metabolic and cellular processes, such as generation of precursor metabolites and energy (GO:0006091); lipid metabolic (GO:0006629), catabolic (GO:0009056), and carbohydrate metabolic (GO:0005975) processes; translation (GO:0006412); biosynthetic process (GO:0009058); and transport (GO:0006810) (Figure 5b). The downregulated genes were significantly enriched in metabolic processes (GO:0005975, GO:0019748), response to stress (GO:0006950), and response to biotic stimulus (GO:0009607) (Figure 5b). We found 1080 high-expressing genes and 598 low-expressing genes in *P. alba* versus *P. russkii* regardless of whether they were in CK or under salt stress (Figure S2a,b). These genes might be the interspecific expression genes in the two *Populus*.

Since the metabolism-related GO term was the most enriched among the DEGs between the two *Populus*, thereafter we analyzed the expression patterns of the genes corresponding to the phenylpropanoid pathway and secondary metabolism. Heatmap analysis showed that most of the key enzyme genes in flavonoids biosynthesis were differentially expressed between the two *Populus* (Figure 5c). Under short-term salt stress, most of these genes were commonly downregulated in the two *Populus*. In addition, we found that many laccases (LAC) encoding genes involved in lignin biosynthesis were differentially expressed in the two *Populus*, i.e., *AtLAC3*, *AtLAC4*, and *AtLAC5* homolog genes being up-regulated in *P. russkii* compared with *P. alba* (Figure 5c). Under short-term salt stress, the homolog genes of *AtLAC4*, *AtLAC11*, and *AtLAC17* were up-regulated in *P. russkii*, suggesting these genes were associated with the rapid response to salt stress in *Populus*.

We further analyzed the DEGs identified between the two *Populus*. Interestingly, 151 genes that were up-regulated in the *P. alba* root but were downregulated in *P. russkii* under salt stress (Figure 6a). Among these genes, several had been previously reported to function in response to oxidative stress or abiotic stress (*ERF6*, *ERF71*, *ZBDH*, *HSFA2*, *GOLS1/4*, and *HRA1*) [37–39], hormone biosynthesis (*ACO1* and *NCED3*) [40,41], and ion transport (*ALS3*, *ABCG33*, and *ABCB17*) (Figure 6b) [42,43]. Furthermore, we analyzed the transcription of the genes in the GA and auxin biosynthetic or signaling pathways, finding substantial differences between the two *Populus* and even presenting a completely reverse pattern; specifically, *GA2OXs*, a GA decomposition gene family, maintained a higher level in *P. alba* under both control and salt stress conditions relative to *P. russkii*. On the other hand, many Auxin/indole-acetic acid (Aux/IAA) repressors (*IAA7*, *IAA18*, *IAA19*, *IAA33*), small auxin up-regulated RNA (SAUR) genes (*SAUR8*, *SAUR37*, *SAUR43*, *SAUR70*, *SAUR72*, *SAUR74*), *GH3.11*, *PIN2*, etc., exhibited higher expression levels in *P. alba* than *P. russkii*, both in control and salinity conditions, whereas other IAA genes (*IAA3*, *IAA20*, *IAA29*), PINs (*PIN1*, *PIN5-8*), *ARF9*, *ARF17*, and *GH3.1* were more highly expressed in *P. russkii* than in *P. alba* (Figure 6c).

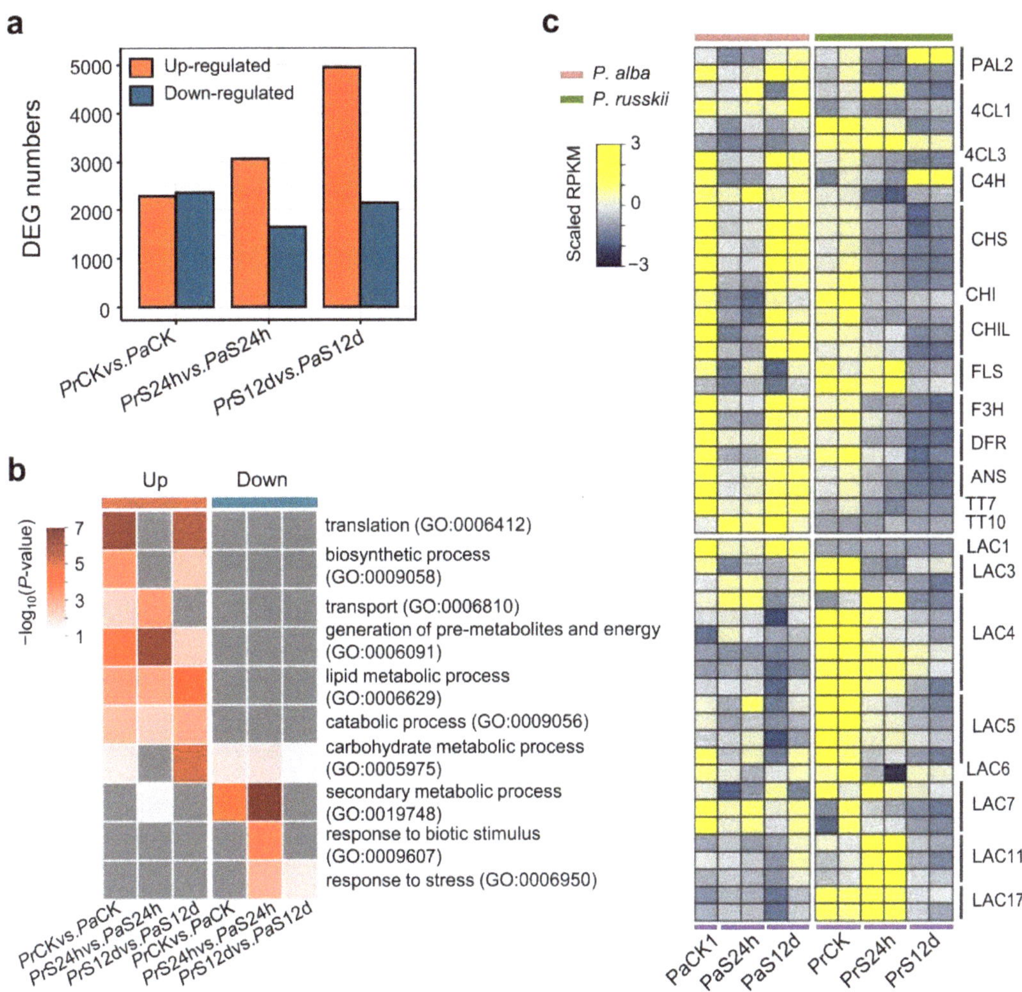

Figure 5. DEGs identified between *P. alba* and *P. russkii*. (**a**) The numbers of DEGs identified between the two *Populus* roots. (**b**) GO enrichment of the up-regulated or downregulated genes between the two *Populus*. (**c**) Heatmap showing the differential expression patterns of the genes in phenylpropanoid pathway and Laccases encoding genes between two *Populus*.

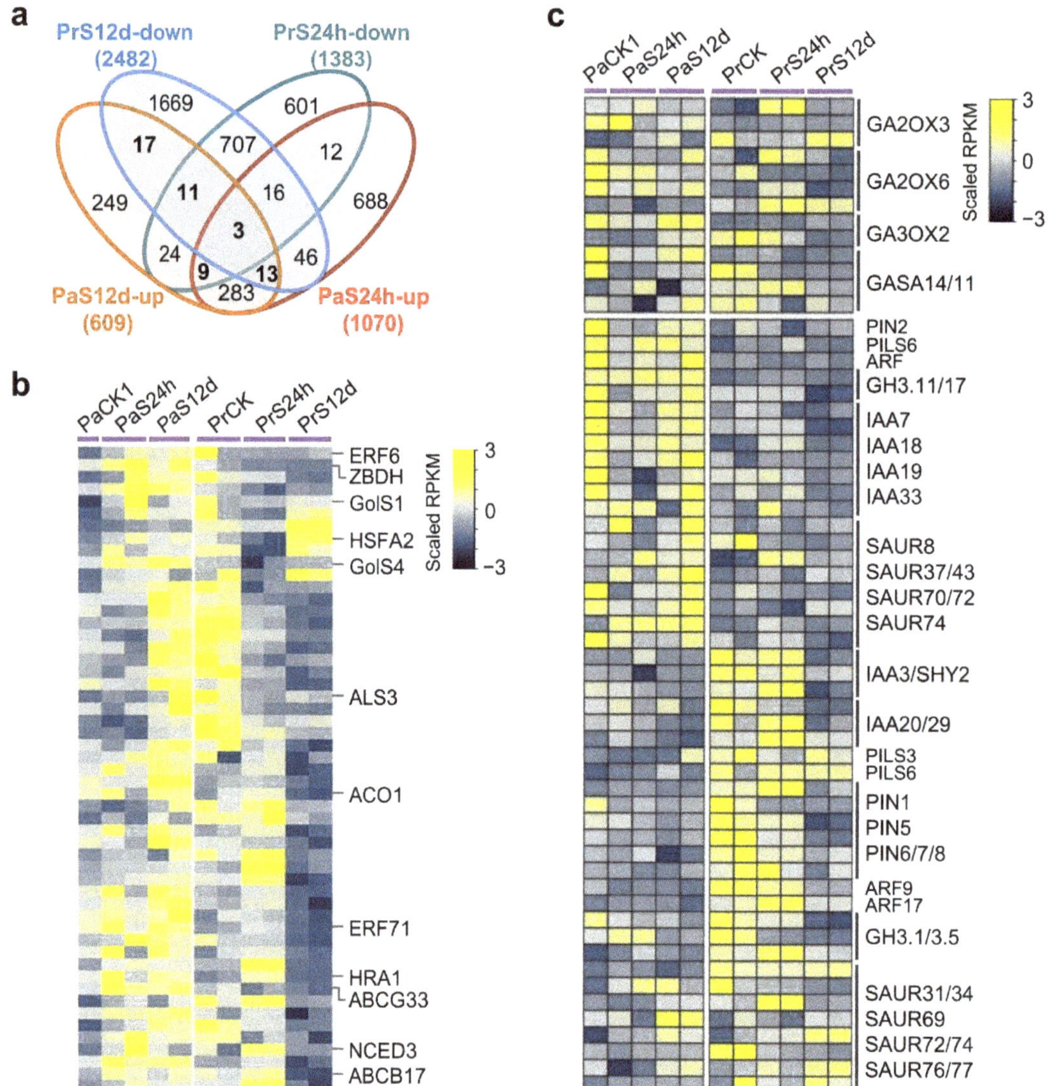

Figure 6. Identification of the expression pattern divergent genes in *P. alba* and *P. russkii* roots under salt stress. (**a**) Venn diagrams showing the overlapping of the up-regulated genes in *P. alba* and downregulated genes in *P. russkii* under salt stress. (**b**) Heatmap showing the genes that the expression patterns were reversed in the two *Populus* under salt stress. (**c**) Heatmap showing the expression of the genes involved in GA or auxin signaling pathways in the two *Populus* under salt stress.

2.5. Differential Expression Patterns of the Na⁺/K⁺ Transporters between the Two Populus under Salt Stress, and Low-Affinity Na⁺ Transport Analysis

We analyzed the transcription level of Na^+/K^+ transporters in *P. alba* and *P. russkii* root transcriptomes, the genes showing distinct expression patterns between the two *Populus* under salt stress (Figure 7). For instance, the Na^+ transporter gene *HKT1;1* (Potri.018G132200) was significantly downregulated in the *P. russkii* root under salt stress but did not change in *P. alba* under short-term salt stress. Many K^+ channels (AKT5, AKT2, KT1, KT2a/b, KCO1) and calcium-binding protein (CML15, CIPK11) encoding genes were downregulated in *P.*

russkii under salt stress, whereas they did not change in *P. alba* (Figure 7). Furthermore, the transcription levels of Na⁺/H⁺ exchanger genes (NHX1/2/6, SOS1a, SOS2/3) in the *P. alba* root had a higher expression than in *P. russkii*; meanwhile, under long-term salt stress some of them were downregulated in *P. russkii* but did not change in *P. alba* (Figure 7). We designed gene-specific primers for the above transporter genes (*HKT1;1, AKT1, NHX1/2/3, SOS1, CBL9,* and *CIPK23/25*) and conducted quantitative real-time PCR (qRT-PCR) analysis to detect their transcription levels and verify the transcriptome data (Figure S4, Table S2). The qRT-PCR results were consistent with the transcriptome data. This confirmed our transcriptome data are accurate and reliable.

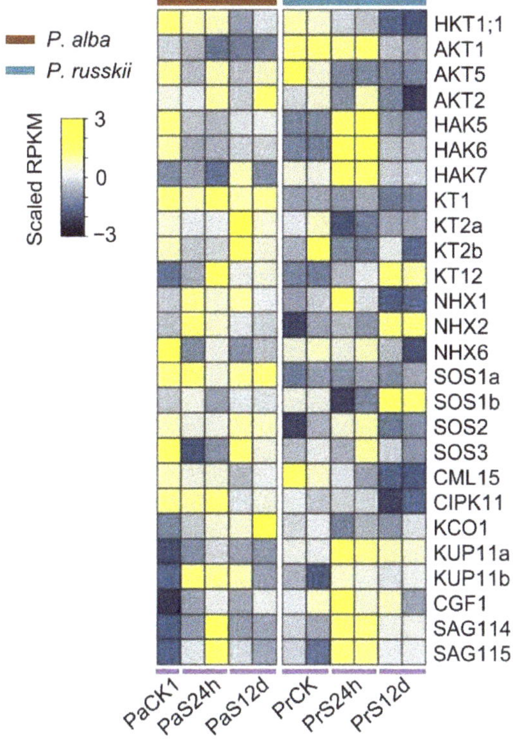

Figure 7. Heatmap showing the different expression patterns of the genes related to Na⁺ or K⁺ transportation in *P. alba* and *P. russkii* roots under salt stress. Two biological replicates were shown except for the control of *P. alba*.

To investigate the Na⁺ uptake/transport features of the two *Populus* under salt stress, we conducted a pharmacological analysis of the low-affinity transport system by the inhibitors of the K⁺ channels (tetraethylammonium chloride [TEA⁺], Cs⁺, and Ba²⁺), Na⁺/K⁺ transporter (Ba²⁺), and non-selective cation channels (Gd³⁺ and La³⁺) [8,9,44]. The results show that Ba²⁺ had strongly inhibited Na⁺ uptaking in *P. alba* but less so in *P. russkii*, that the NSCCs blockers (Gd³⁺ and La³⁺) largely constrained Na⁺ uptake in both *Populus*, and that the K⁺ channel blockers severely inhibited Na⁺ uptaking in *P. alba* more than in *P. russkii* (Table 1). TEA⁺ is a specific K⁺ inward rectifier channels blocker, and its suppression rate in the two *Populus* show that K⁺ channels play a dominant role in Na⁺ transport. Its higher inhibition in *P. alba* than in *P. russkii* suggests the K⁺ channels of *P. alba* have a stronger capacity than those of *P. russkii*. The inhibition of TEA⁺ from Ba²⁺ (i.e., Na⁺ transportation via HKT transporter), which shows a higher suppression value in *P. alba* than *P. russkii*, reveals that *P. alba* HKT has superior Na⁺ transportation than *P. russkii* (Table 1).

Table 1. Effects of pharmacological treatments on low-affinity Na$^+$ influx in *Populus* plants under 100 mM NaCl. Na$^+$ concentrations were measured before treatments and after NaCl and inhibitor treatments for 24 h (each treatment had three biological replicates). Measurements significantly different from the control are indicated with a *, $p < 0.05$ (*t*-test). Values represent mean $\pm$ SD (n = 3).

Treatment (Meaured at 100 mM NaCl)	*P. russkii*		*P. alba*	
	Na$^+$ Influx (nmol g^{-1} DW h^{-1})	Suppression	Na$^+$ Influx (nmol g^{-1} DW h^{-1})	Suppression
NaCl (100 mM)	148.44 ± 7.85	0	84.02 ± 3.94	0
TEA$^+$ (10 mM)	$38.56 * \pm 4.1$	74.02%	5.72 ± 0.16	93.19%
Cs$^+$ (10 mM)	$33.15 * \pm 2.31$	77.67%	8.57 ± 0.62	89.79%
Ba^{2+} (5 mM)	$32.87 * \pm 3.45$	77.85%	0.76 ± 0.26	99.10%
Gd^{3+} (50 μM)	$29.24 * \pm 2.03$	80.03%	9.67 ± 0.29	88.49%
La^{3+} (50 μM)	$32.63 * \pm 3.47$	78.01%	8.33 ± 1.49	90.09%

Based on the pharmacological analysis and transcriptome data, we infer that the HKT function of *P. alba* may be divergent from that of *P. russkii*. To confirm this hypothesis, we cloned the *HKT1;1* and *HKT1;2* from *P. alba* and *P. russkii* and obtained the full-length coding sequences. The alignment of the *HKT1;1* and *HKT1;2* sequences of the two *Populus* showed extensive SNP variations (Figures S5 and S6). To study whether these sequence variations affected or not the function of Na$^+$ transporting, we conducted an additional Na$^+$ transport analysis by separately transforming the *PrHKT1;1*, *PaHKT1;1*, *PrHKT1;2*, and *PaHKT1;2* to complement the yeast Na$^+$ transport loss-function mutant G19 (Figure 8a). The yeast growth curves showed that the function of PaHKT1;1 and PrHKT1;1 presented no difference, while the Na$^+$ transportation of PaHKT1;2 was significantly higher than PrHKT1;2 (Figure 8a,b), and the yeast growth was inhibited even under control conditions containing low Na$^+$ levels.

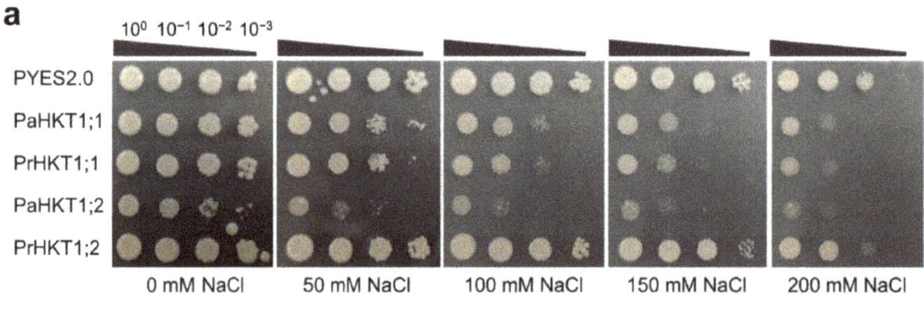

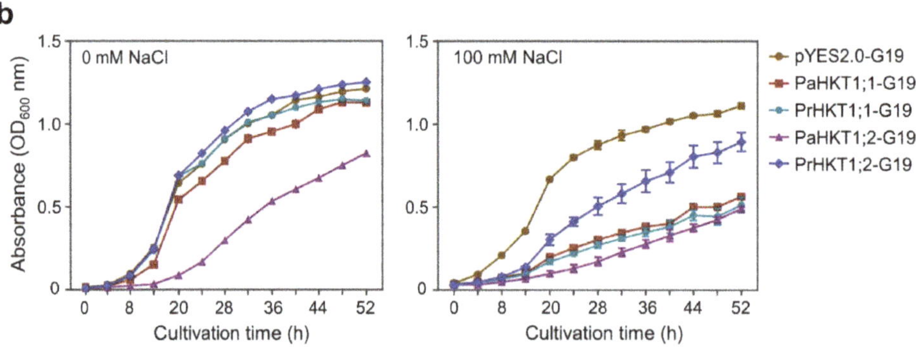

Figure 8. Difference in the Na$^+$ transport function of the HKT1;1 and HKT1;2 in *P. alba* and *P. russkii*. (**a**) Yeast mutant G19 complementary analysis of the Na$^+$ transport function of PaHKT1;1, PrHKT1;1, PaHKT1;2, and PrHKT1;2. (**b**) Yeast growth curve of the transformants of PaHKT1;1, PrHKT1;1, PaHKT1;2, and PrHKT1;2 to G19 in medium with 100 mM NaCl or without NaCl.

3. Discussion

3.1. Transcriptomic Profiles Reveal the Interspecific Difference in Salt Responsiveness between the Two Populus

Our transcriptome data comprehensively depict the salt response profiles of the two *Populus* under either short- (24 h) or long-term (12 d) salinity stress; the reliability of the data was verified by qRT-PCR (Figure S4). Our data reveal the distinct strategies of the two *Populus* responses to the short-term salt stress. The up-regulated DEG numbers substantially surpassed the downregulated in *P. alba*, whereas in *P. russkii* they were much more comparable (Figure 3b). Numerous DEGs detected between *P. alba* and *P. russkii* in the control and up-regulated genes (*P. russkii* versus *P. alba*) are mainly involved in the generation of pre-metabolites and energy, transport, and carbohydrate metabolic processes (Figure 5b). This energy reserve endows *P. alba* with a higher defense response to salt in the short term. Under long-term salt stress, phenotypic observation indicates that *P. alba* was able to resist salinity to keep growing, whereas the *P. russkii* leaves were severely dehydrated and numerous genes involved both in multiple metabolic (GO:0009056; 0019748; 0006629) and cellular processes were downregulated (Figure 3c). In contrast, in *P. alba* the up-regulated DEGs were three times as much as the downregulated, these genes being significantly enriched in response to biotic stimulus (GO:0009607) and carbohydrate metabolic (GO:0005975) and catabolic processes (GO:0009056), which suggests that *P. alba* effectively mobilizes a series of defenses and energy to combat salt stress. This response to salinity is highly similar with that of *P. euphratica*, a relative species of *P. alba* [18,19].

The transcriptomic data identified many salt-tolerant candidate genes in poplars, including among them some important transcription factors previously reported under either osmotic stress or salt stress, i.e., RD26a/ANAC072, bZIP1, MYB78, MYB43, WRKY48, and GAPC2 (Figure 4c) [29,31,32], as well as consisting of several protein-encoding genes such as DI21, ABA-induced PP2C (HAI1)/SAG113, and the LEA4-5 which related to osmotic stress [30]. These genes maybe contribute to improve salt tolerance in forest and crop plants.

3.2. Interspecific Difference in Poplar Salt Tolerance Is Related to Na$^+$ Transportation Capacity and to Its Regulation

Previous studies show that high tolerance to salinity is related to a lower Na$^+$ concentration accumulated in shoots or leaves as well as to K$^+$/Na$^+$ homeostasis in plants [10,18,45]. In our study, pharmacological analysis reveals that the capacity of Na$^+$ uptake in *P. alba* via K$^+$ inward rectifier channels and HKT transporters is higher than that in *P. russkii* (Table 1). The transcription of the K$^+$ inward rectifier channel genes, such as AKT2, AKT5, KT1, KT2a/b, and KT12, were steady expressed in *P. alba*, but their expression was reduced in *P. russkii* under salt stress (Figure 7). The downregulation of these genes contributes to a decreased Na$^+$ uptake and a decreased K$^+$ uptake, producing the negative effect of maintaining a high K$^+$/Na$^+$ cytoplasmic ratio, thereby declining salt tolerance. The HKT transporter function in Na$^+$ uptake in roots plays a key role in Na$^+$ translocation and K$^+$/Na$^+$ homeostasis maintenance in xylem parenchyma cells [14–16,21]. In *P. trichocarpa*, there is one *HKT1* ortholog gene together with a *HKT1*-like pseudogene. However, in the desert poplar *P. euhratica*, the gene has expanded to four members [20], indicating it plays a functional role in salt tolerance for poplar plants. In this study, we found there were two ortholog genes of *HKT1*, i.e., *HKT1;1* and *HKT1;2*, in both *P. alba* and *P. russkii*. We detected only *HKT1;1* expressed in the two *Populus* roots and its transcription decreased under salt stress, which serves as a defense strategy in response to salinity. *HKT1;2*, an orthology gene of *HKT1;1*, was mainly expressed in *Populus* leaves and plays a critical role in Na$^+$ transport/translocation in photosynthetic apparatus. As the yeast analysis shows, PaHKT1;2 presents a higher Na$^+$ transportation activity than PrHKT1;2, suggesting that PaHKT1;2 might play an important function in Na$^+$ transportation/translocation or salt tolerance in *P. alba*. Extensive SNP variations detected with the *HKT1;2* gene coding sequence between the two *Populus* also suggest that the HKT allele of *P. alba* has a superior performance than *P. russkii* and might have a significant role in salt tolerance; therefore, its function needs further research.

Na$^+$ uptake via the NSCCs pathway was dominant in *P. russkii* under high salinity rather than in *P. alba* (Table 1). Gd^{3+} and La^{3+} serve as the broad-spectrum inhibitors; although they can block NSCCs, they also play important roles in inhibiting Ca^{2+} channels [46]. The changes in Ca^{2+} transportation can directly impact Na$^+$ and K$^+$ channels via signal transmission [10,45,47]. The uptake of Na$^+$ mediated by NSCCs depends on the external Na$^+$ concentration and can be—to a lesser extent—in competition with K$^+$, suggesting that NSCCs may serve as a passive way to mediate Na$^+$ absorption. In this study, the NSCCs play a major role in LATS Na$^+$ uptake in *P. russkii*, which explains the weak resistance to high salinity.

The distribution of Na$^+$ in the cells and its translocation in different organs are critical to plant salt tolerance when it enters into root cells. Compartmentalizing Na$^+$ into vacuoles via the Na$^+$/H$^+$ antiporter NHX1 and squeezing out the protoplast by SOS1 is a very important strategy for Na$^+$ detoxification in plants [1,23,48,49]. Our results indicate that the transcription of *NHX1* and *SOS1a* can maintain a high level in *P. alba* while maintaining low levels in *P. russkii* under salt stress; this is another explanation for *P. russkii* being highly sensitive to salt stress.

3.3. Phytohormone-Regulated Responses Are Related to Plant Salt Tolerance

In this study, the transcriptions of NCED3 and ACO1 as rate-limiting enzymes in abscisic acid (ABA) and ethylene biosynthetic pathways were significantly downregulated in *P. russkii* but up-regulated in *P. alba* under long-term salt stress, respectively. Correspondingly, the ethylene response factors *ERF6*, *ERF71*, and *HRA1* were also significantly decreased in *P. russkii* but up-regulated in *P. alba* (Figure 6b). The expression patterns of these genes shows that *P. alba* presents a higher level of ethylene and ABA syntheses than *P. russkii*. Previous studies demonstrate that ethylene improves plants' salt tolerance, mainly via inducing antioxidant defense and ROS detoxification and by maintaining K$^+$/Na$^+$ homeostasis [50,51]. The *ERF6*, *ERF71*, and *HRA1* factors were reported to play an important role in resistance to abiotic stresses in *Arabidopsis* and crops plants [37,52,53]. ABA modulates plants responses to salt stress largely through osmotic regulation and by inducing stomatal closure. Therefore, the differences in both ethylene and ABA syntheses between *P. alba* and *P. russkii* are identified here as critical factors that cause the differential salt resistance under long-term salt stress.

GA2-oxidases (GA2oxs) regulate the deactivation of bioactive GAs and are very important to plant development and stress responses [54]. *GASA* (*Gibberellic Acid Stimulated in Arabidopsis*) genes encoding cysteine-rich peptides are involved in plant development and environmental adaption [55]. It has been reported that overexpression of *GA2ox* and *GASA14* genes in *Arabidopsis* might enhance plants' resistance to high salinity [56,57]. In our study, the *GA2ox3*, *GA2ox6*, and *GASA14s* maintained a steady high expression in *P. alba* but decreased in *P. russkii* under long-term salt stress (Figure 6c), which indicates the GAs' metabolism in *P. alba* differs from that in *P. russkii*. Furthermore, we found many genes in auxin biosynthetic or signaling pathways, such as *IAAs, ARFs, SAURs*, and *PINs*, showing a reverse expression pattern between *P. alba* and *P. russkii* (Figure 6c). Aux/IAA, ARF, and SAURs are key regulators of auxin responses that also play important roles in plant development and their responses to environmental cues [58–60]. A recent study shows that the heterologous expression of grapevine *VvIAA18* in tobacco can significantly enhance salt tolerance [61]. In this study, these auxin response genes presented a steady differential expression between the two *Populus* and were hardly affected by salinity, indicating they are mostly regulated by auxin concentration rather than directly responding to salt stress. The expression patterns of these genes reveals that the differential local auxin levels between the two *Populus* is one of the key factors shaping its salt tolerance.

Under salt stress, the activities of POD, APX, and GR increased in *P. alba* but did not significantly change in *P. russkii*, and their encoding genes had a higher expression level in

P. alba than in *P. russkii* (Figure S3), which underlines the transcriptional regulation effects on enzyme activities resulting from protein abundance. The transcription of *galactinol synthase1* (*GolS1*) and *GolS4* genes increased in *P. alba* but were diminished in *P. russkii* under salt stress. GolS is a key enzyme catalyzing the biosynthesis of raffinose family oligosaccharides (RFOs) in plants [38]. Overexpressed *GolS* can significantly enhance the tolerances to salt, cold, and drought stresses in poplar and cucumber [62–64]. In addition, most of flavonoid- or anthocyanin-related biosynthetic genes maintained a high expression under long-term salt stress in *P. alba* but were downregulated in *P. russkii*. Many studies show that overexpression of these genes can improve salt resistance [65,66]. In contrast, laccases (LACs) genes had higher expression levels in *P. russkii* than those in *P. alba* in control, whereas under long-term salt stress they were downregulated in *P. russkii* although they did not change in *P. alba*. These genes are involved in lignin synthesis and their transcription is positively related to their growth rate [67,68]. Although previous studies have found that the growth rate of *P. russkii* was higher than *P. alba* under normal conditions, its salt or drought tolerance was weaker than *P. alba* [25,27]. The expression pattern of *LACs* explains this and also reveals the different opportunistic approaches to trade-offs between growth and defense in these two poplar species: *P. russkii* adopts an opportunistic growth strategy while *P. alba* adopts an opportunistic defense strategy.

This work reveals the molecular mechanism underlying the differential salt-tolerance between two poplar species under salinity stress. Compared to the salt-sensitive poplar (*P. russkii*), the salt-tolerant poplar (*P. alba*) shows a high transcriptional level of cellular energy metabolism processes, thus effectively initiating the stress defense responses under salt stress. Moreover, PaHKT1;2 has a strong Na^+ transportation ability that effectively recycles xylem Na^+ transport in *P. abla* under salt stress. On a physiological level, it is the higher antioxidant enzyme activities, osmoregulation, and stress-related hormone levels (ABA and ethylene) that altogether confer *P. alba* high salt tolerance. This study has significant implications for improving salt resistance in crop plants.

4. Materials and Methods

4.1. Plant Material and Experimental Treatment

One-year-old saplings of *P. alba* and *P. russkii* were collected at their natural habitat (44.29° N, 87.93° E, 200 m altitude) in Xinjiang, Northwest China. The stems of *P. alba* and *P. russkii* were cut into 20 cm-long pieces (diameter with 1 cm) and grown within a hydroponic culture with a half-strength Hoagland solution containing 2.5 mM KNO_3, 2 mM $Ca(NO_3)_2 \cdot 4H_2O$, 1 mM $MgSO_4 \cdot 7H_2O$, 0.5 mM KH_2PO_3, 25 μM H_3BO_3, 1.25 μM KI, 25 μM $MnSO_4 \cdot 4H_2O$, 7.5 μM $ZnSO_4 \cdot 7H_2O$, 0.25 μM $Na_2MoO_4 \cdot 2H_2O$, 0.025 μM $CoCl_2 \cdot 6H_2O$, 0.025 μM $CuSO_4 \cdot 5H_2O$, and 25 μM Fe(III)-Ethane-1,2-diyldinitrilo tetra-acetic acid (EDTA). The hydroponic culture solution was refreshed every 3 days. An air pump supplying air to plant roots was installed for the hydroponic water culture. The cuttings were cultivated inside a controlled climate chamber at 25/18 °C with a 14/10 h light/dark photoperiod with a maximum photon flux density of 400 μmol m^{-2} s^{-1} (photosynthetically active radiation) supplied by fluorescent lights; the relative humidity was within a 65~70% range. The cuttings were rooting and sprouting after a week.

Four weeks after the start of the hydroponic culture, we selected sprouting cuttings with uniform shoots, with an average height of 32 cm for *P. alba* and 35 cm for *P. russikii*, and then initiated the experiment. The treatments included both a control (CK) and a salt stress group (with 100 mmol L^{-1} NaCl) for each of the two *Populus*. The control was cultured in the Hoagland solution without NaCl, whereas the salt stress treatment group was cultured inside a Hoagland solution containing 100 mmol L^{-1} NaCl. Each treatment group in both *Populus* species consisted of three biological replications (six plants per replicate). The fourth to sixth fully expanded leaves of the treatment group and control were collected for ion content determination and experimental analyses. Intact roots of the salt stress treatment group and control were collected and frozen immediately using liquid nitrogen and subsequently stored at $-80\,^{\circ}\text{C}$ for RNA-sequencing and physiological experimental analyses.

4.2. Measurement of Na⁺ and K⁺ Content in Poplar Leaves

Na^+ and K^+ concentration in the two *Populus* species were measured with an atomic absorption spectrophotometer (Perkin-Elmer, AA700, Waltham, MA, USA) using the method described previously [19]. Leaves of the salt-treated and control plants were oven-dried at 65 °C during more than 3 days to achieve complete dehydration. The dry samples were ground into fine powder (100 mg) and incubated into a 10 mL 0.5 M HNO_3 solution subjected to shaking extraction for 48h at room temperature. Afterwards, we used centrifugation at 12,000 rpm for 10 min to settle the solid material down into the tube bottom, and then we transferred the extracts and diluted them with deionized H_2O (Milli-Q, Merck KGaA, Darmstadt, Germany) to measure the Na^+ or K^+ content.

4.3. Determination of Antioxidant Enzymes Activities, Glycine-Betaine and Malondialdehyde (MDA) Content

Antioxidant enzymes (POD, CAT, APX, and GR) activities were assayed as described in our previous study [27]. Fresh leaves (0.5 g) were ground in liquid nitrogen and extracted with a 50 mM potassium phosphate buffer (pH 7.8) containing 0.1 mM EDTA, 1% (w/v) polyvinyl pyrrolidone (PVP), 0.1 mM phenylmethane sulfonyl fluoride (PMSF) solution, and 0.2% (v/v) Triton X-100. POD (EC1.11.1.7) activity was measured at 470 nm as described in [69]. CAT (EC1.11.1.6) activity was assayed as described in [70]. APX (EC1.11.1.11) activity was determined as described by [71]. GR (EC1.6.4.2) activity was measured using a glutathione reductase activity kit (Cominbio, Suzhou, China, Cat.# GR-2-W) following the manufacturer's instructions. Absorbance at 340 nm was monitored by spectrophotometer.

Glycine-betaine content was measured using a plant Betaine content kit (Cominbio, Suzhou, China, Cat.# TCJ-2-G) following the manufacturer's protocol. Absorbance was measured at 525 nm and expressed as $\mu g\ g^{-1}$ dry weight (DW). Cellular membrane lipid peroxidation analysis was based on MDA concentration according to our previous study [27]. Absorbance was measured at 450 nm, 532 nm, and 600 nm using a spectrophotometer. MDA concentration was calculated using the formula: MDA ($\mu mol\ L^{-1}$) = 6.45 $(A_{532} - A_{600}) - 0.56A_{450}$.

4.4. RNA Extraction, cDNA Library Preparation, and RNA Sequencing

Total RNA was extracted from the poplar roots using an RNeasy plant mini kit (Qiagen, Hilden, Germany) following the manufacturer's protocol. Each treatment group or control sample had three biological replicates. RNA concentration and purity was measured using a Qubit 2.0 fluorometer (Invitrogen, Waltham, MA, USA). High-quality RNA was processed for RNA-seq library construction. Two micrograms of total RNA were used for mRNA isolation, while mRNA fragmentation and cDNA library construction were conducted using a TruSeq Stranded mRNA Library Prep Kit (Illumina, San Diego, CA, USA) as stipulated by the manufacturer's instructions. The index codes were added to attribute sequences to each sample. The cDNA libraries were sequenced at the Beijing Genomics institution (BGI, Shenzhen, China) on the Illumina HiSeq 2500 System by 50 bp single-read sequencing.

4.5. Analysis of RNA-Seq and Identification of DEGs

RNA-seq raw reads were filtered to remove adapter sequence and low-quality reads by Trimmomatic (version 0.36) software. The clean reads were aligned to the *P. trichocarpa* reference genome (v4.1, PhytozomeV13) using HISAT2 (version 2.1.0). Gene expression quantification and differentially expressed genes (DEGs) calculation were conducted by Cufflinks (version 2.2.1) with default parameters [72]. DEGs were identified between two comparisons using the following criteria: | log2 (fold change) | > 1 and false discovery rate (FDR) < 0.01. The FDR was generated from an adjusted *P*-value using the Benjamini-

Hochberg method. GO enrichment and heatmap visualization were generated using TBtools (version 1.086) [73].

4.6. Reversal Transcription and qRT-PCR

Four micrograms of total RNA was used to synthesize first-strand cDNA with a reversal transcription reagent kit (TaKaRa). Synthesized cDNA was diluted to a final volume of 100 μL, and 1 μL was added as the template for detection in each reaction. qRT-PCR analyses were conducted using SYBR Premix ExTaq (TaKaRa) with a CFX96 real-time system (Bio-Rad Laboratories, Inc., Hercules, CA, USA). *Populus* elongation factor-1 alpha (EF-1-α) was selected as the internal control to normalize gene transcription. Three biological replicates were performed for each qRT-PCR analysis. The levels of gene expression were evaluated by calculating the threshold cycle (Ct) based on the $2^{-\Delta\Delta Ct}$ method and were measured in terms of relative quantitative variation.

4.7. Pharmacological Analysis of Low-Affinity Na⁺ Transport

After four weeks of hydroponic cultivation, sprouting cuttings of *P. alba and P. russkii*—whose shoots had heights of 32 cm and 35 cm—were used to evaluate the effect of ion channel/transporter inhibitors on Na$^+$ uptake/transport via the low-affinity transport system. The applied inhibitors in this study included K$^+$ channel blockers (TEA-Cl, CsCl, and BaCl$_2$), and non-selective cation channel (NSCCs) blockers (LaCl$_3$ and GdCl$_3$). The concentration of inhibitors was determined based on the following previous studies: [74] with 10 mM TEA-Cl; [44] with 10 mM CsCl; [75] with 5 mM BaCl$_2$; and [44] with 50 μM LaCl$_3$ and 50 μM GdCl$_3$. The inhibitor was added to the hydroponic culture solution containing 100 mM NaCl, taking the salt stress treatment group (i.e., 100 mM NaCl without any inhibitors) as the control. Each treatment had three replications and each replicate had three plants. The inhibitors treatment lasted 72 h and Na$^+$ and K$^+$ concentrations in the treated plants leaves were measured at 0 h (before treatment), 8 h, 24 h, 48 h, and 72 h.

4.8. Yeast Transformation and Growth Analysis

The yeast (*Saccharomyces cerevisiae*) strain G19 (Matα ura3 his3 trp7 ade2 ena1Δ::HIS3:: ena4Δ) [76] was used to analyze the Na$^+$ transporting function of HKT1;1 and HKT1;2 in the two *Populus*. The G19 strain was transformed with the pYES2 plasmid alone (empty vector) or by pYES2 containing the coding sequences of *PaHKT1;1, PrHKT1;1, PaHKT1;2,* and *PrHKT1;2*. Transformants were cultured on a synthetic dropout medium without uracil (SD/-Ura), and with 2% glucose as the carbon source for selection and normal growth, or with 2% galactose for inducible expression of HKT genes. Complementary analysis was performed by inoculation of 10 μL drops of cell suspension (OD600 = 1 or its dilution series) onto the surface of SD/-Ura/galactose agar medium supplied with 0, 50, 100, 150, or 200 mM NaCl. The results were recorded after 2 days of incubation at 30 °C. As for the growth curve test, yeast cells were incubated in SD/-Ura/galactose liquid medium with or without 100 mM NaCl at 30 °C with shaking. The absorbance was measured at 600 nm every 4 h.

4.9. Statistical Analysis

Both datasets were separately analyzed using SPSS software (v19.0), each bar representing the mean $\pm$ SE of at least three replicates. Different letters above the bars indicate significant differences, and values of $p < 0.05$ represented statistical significance using LSD's test.

Supplementary Materials: The supporting information can be downloaded at: https://www.mdpi.com/article/10.3390/ijms24065732/s1.

Author Contributions: Y.Y., V.R.d.D., and X.M. designed the research; Y.Y. and V.R.d.D. supervised the experiments; X.M. and Q.Z. performed most of the experiments; Y.O., L.W., and Y.G. provided technical assistance and conducted data analysis; X.M. conducted bioinformatics analysis; X.M. and Y.Y. wrote the manuscript; G.R.L., V.R.d.D., and Y.O. revised the manuscript. All authors have read and agreed to the published version of the manuscript.

Funding: This research was funded by the National Natural Science Foundation of China (grant number: 31901333 and 31270660) and the Sichuan Innovative Talent Program (grant number 2020JDRC0065) as well as by the China high-level talent project.

Institutional Review Board Statement: Not applicable.

Informed Consent Statement: Not applicable.

Data Availability Statement: The RNA-seq data generated in this study were deposited in the NCBI Sequence Read Archive (BioProject ID: PRJNA896337; https://www.ncbi.nlm.nih.gov/bioproject/PRJNA896337, accessed on 10 January 2023). The datasets supporting the conclusions of this article are included within the article and its additional files.

Conflicts of Interest: The authors declare that they have no known competing financial interest or personal relationships that could have appeared to influence the work reported in this paper.

References

1. Deinlein, U.; Stephan, A.B.; Horie, T.; Luo, W.; Xu, G.; Schroeder, J.I. Plant salt-tolerance mechanisms. *Trends Plant Sci.* **2014**, *19*, 371–379. [CrossRef] [PubMed]
2. van Zelm, E.; Zhang, Y.; Testerink, C. Salt Tolerance Mechanisms of Plants. *Annu. Rev. Plant Biol.* **2020**, *71*, 403–433. [CrossRef] [PubMed]
3. Muller, M.; Kunz, H.H.; Schroeder, J.I.; Kemp, G.; Young, H.S.; Neuhaus, H.E. Decreased capacity for sodium export out of Arabidopsis chloroplasts impairs salt tolerance, photosynthesis and plant performance. *Plant J.* **2014**, *78*, 646–658. [CrossRef] [PubMed]
4. Munns, R.; Tester, M. Mechanisms of salinity tolerance. *Annu. Rev. Plant Biol.* **2008**, *59*, 651–681. [CrossRef] [PubMed]
5. Dinneny, J.R. Traversing organizational scales in plant salt-stress responses. *Curr. Opin. Plant Biol.* **2015**, *23*, 70–75. [CrossRef] [PubMed]
6. Rains, D.W.; Epstein, E. Transport of Sodium in Plant Tissue. *Science* **1965**, *148*, 1611. [CrossRef]
7. Demidchik, V.; Tester, M. Sodium fluxes through nonselective cation channels in the plasma membrane of protoplasts from Arabidopsis roots. *Plant Physiol.* **2002**, *128*, 379–387. [CrossRef]
8. Kronzucker, H.J.; Britto, D.T. Sodium transport in plants: A critical review. *New Phytol.* **2011**, *189*, 54–81. [CrossRef]
9. Wang, S.M.; Zhang, J.L.; Flowers, T.J. Low-affinity Na+ uptake in the halophyte Suaeda maritima. *Plant Physiol.* **2007**, *145*, 559–571. [CrossRef]
10. Pantoja, O. Recent Advances in the Physiology of Ion Channels in Plants. *Annu. Rev. Plant Biol.* **2021**, *72*, 463–495. [CrossRef]
11. Tester, M.; Davenport, R. Na+ tolerance and Na+ transport in higher plants. *Ann. Bot.* **2003**, *91*, 503–527. [CrossRef] [PubMed]
12. Zhang, J.-L.; Flowers, T.J.; Wang, S.-M. Mechanisms of sodium uptake by roots of higher plants. *Plant Soil* **2009**, *326*, 45–60. [CrossRef]
13. Rodriguez-Navarro, A.; Rubio, F. High-affinity potassium and sodium transport systems in plants. *J. Exp. Bot.* **2006**, *57*, 1149–1160. [CrossRef] [PubMed]
14. Maser, P.; Brendan, E.; Schroeder, J.; Yamada, K.; Oiki, S.; Bakker, E.; Goshima, S.; Hosoo, Y.; Uozumi, N. Identification of residues essential for Na+/K+ selectivity in HKT-type K+ transporters. *Plant Cell Physiol.* **2002**, *43*, S101.
15. Sunarpi; Horie, T.; Motoda, J.; Kubo, M.; Yang, H.; Yoda, K.; Horie, R.; Chan, W.Y.; Leung, H.Y.; Hattori, K.; et al. Enhanced salt tolerance mediated by AtHKT1 transporter-induced Na unloading from xylem vessels to xylem parenchyma cells. *Plant J.* **2005**, *44*, 928–938. [CrossRef]
16. Horie, T.; Hauser, F.; Schroeder, J.I. HKT transporter-mediated salinity resistance mechanisms in Arabidopsis and monocot crop plants. *Trends Plant Sci.* **2009**, *14*, 660–668. [CrossRef]
17. Huang, S.; Spielmeyer, W.; Lagudah, E.S.; Munns, R. Comparative mapping of HKT genes in wheat, barley, and rice, key determinants of Na+ transport, and salt tolerance. *J. Exp. Bot.* **2008**, *59*, 927–937. [CrossRef]
18. Chen, S.; Polle, A. Salinity tolerance of Populus. *Plant Biol.* **2010**, *12*, 317–333. [CrossRef]
19. Ding, M.; Hou, P.; Shen, X.; Wang, M.; Deng, S.; Sun, J.; Xiao, F.; Wang, R.; Zhou, X.; Lu, C.; et al. Salt-induced expression of genes related to Na(+)/K(+) and ROS homeostasis in leaves of salt-resistant and salt-sensitive poplar species. *Plant Mol. Biol.* **2010**, *73*, 251–269. [CrossRef]
20. Ma, T.; Wang, J.; Zhou, G.; Yue, Z.; Hu, Q.; Chen, Y.; Liu, B.; Qiu, Q.; Wang, Z.; Zhang, J.; et al. Genomic insights into salt adaptation in a desert poplar. *Nat. Commun.* **2013**, *4*, 2797. [CrossRef]

21. Hamamoto, S.; Horie, T.; Hauser, F.; Deinlein, U.; Schroeder, J.I.; Uozumi, N. HKT transporters mediate salt stress resistance in plants: From structure and function to the field. *Curr. Opin. Biotechnol.* **2015**, *32*, 113–120. [CrossRef] [PubMed]
22. Ji, H.; Pardo, J.M.; Batelli, G.; Van Oosten, M.J.; Bressan, R.A.; Li, X. The Salt Overly Sensitive (SOS) pathway: Established and emerging roles. *Mol. Plant* **2013**, *6*, 275–286. [CrossRef] [PubMed]
23. Wu, Y.; Ding, N.; Zhao, X.; Zhao, M.; Chang, Z.; Liu, J.; Zhang, L. Molecular characterization of PeSOS1: The putative Na(+)/H (+) antiporter of Populus euphratica. *Plant Mol. Biol.* **2007**, *65*, 1–11. [CrossRef]
24. Maathuis, F.J. Sodium in plants: Perception, signalling, and regulation of sodium fluxes. *J. Exp. Bot.* **2014**, *65*, 849–858. [CrossRef]
25. Yao, Y.; Sun, Y.; Feng, Q.; Zhang, X.; Gao, Y.; Ou, Y.; Yang, F.; Xie, W.; Resco de Dios, V.; Ma, J.; et al. Acclimation to nitrogen × salt stress in Populus bolleana mediated by potassium/sodium balance. *Ind. Crop. Prod.* **2021**, *170*, 113789. [CrossRef]
26. Chen, T.H.; Murata, N. Glycinebetaine: An effective protectant against abiotic stress in plants. *Trends Plant Sci.* **2008**, *13*, 499–505. [CrossRef]
27. Ma, X.; Ou, Y.B.; Gao, Y.F.; Lutts, S.; Li, T.T.; Wang, Y.; Chen, Y.F.; Sun, Y.F.; Yao, Y.A. Moderate salt treatment alleviates ultraviolet-B radiation caused impairment in poplar plants. *Sci. Rep.* **2016**, *6*, 32890. [CrossRef] [PubMed]
28. Chong, G.L.; Foo, M.H.; Lin, W.D.; Wong, M.M.; Verslues, P.E. Highly ABA-Induced 1 (HAI1)-Interacting protein HIN1 and drought acclimation-enhanced splicing efficiency at intron retention sites. *Proc. Natl. Acad. Sci. USA* **2019**, *116*, 22376–22385. [CrossRef]
29. Dash, M.; Yordanov, Y.S.; Georgieva, T.; Tschaplinski, T.J.; Yordanova, E.; Busov, V. Poplar PtabZIP1-like enhances lateral root formation and biomass growth under drought stress. *Plant J.* **2017**, *89*, 692–705. [CrossRef]
30. Hundertmark, M.; Hincha, D.K. LEA (late embryogenesis abundant) proteins and their encoding genes in Arabidopsis thaliana. *BMC Genom.* **2008**, *9*, 118. [CrossRef]
31. Jiang, H.; Tang, B.; Xie, Z.; Nolan, T.; Ye, H.; Song, G.Y.; Walley, J.; Yin, Y. GSK3-like kinase BIN2 phosphorylates RD26 to potentiate drought signaling in Arabidopsis. *Plant J.* **2019**, *100*, 923–937. [CrossRef] [PubMed]
32. Zheng, P.; Cao, L.; Zhang, C.; Pan, W.; Wang, W.; Yu, X.; Li, Y.; Fan, T.; Miao, M.; Tang, X.; et al. MYB43 as a novel substrate for CRL4 PRL1 E3 ligases negatively regulates cadmium tolerance through transcriptional inhibition of HMAs in Arabidopsis. *New Phytol.* **2022**, *234*, 884–901. [CrossRef]
33. Scheepers, M.; Spielmann, J.; Boulanger, M.; Carnol, M.; Bosman, B.; De Pauw, E.; Goormaghtigh, E.; Motte, P.; Hanikenne, M. Intertwined metal homeostasis, oxidative and biotic stress responses in the Arabidopsis frd3 mutant. *Plant J.* **2020**, *102*, 34–52. [CrossRef] [PubMed]
34. Tang, X.; Peng, Y.; Li, Z.; Guo, H.; Xia, X.; Li, B.; Yin, W. The Regulation of Nitrate Reductases in Response to Abiotic Stress in Arabidopsis. *Int. J. Mol. Sci.* **2022**, *23*, 1202. [CrossRef]
35. Vegh, A.; Incze, N.; Fabian, A.; Huo, H.; Bradford, K.J.; Balazs, E.; Soos, V. Comprehensive Analysis of DWARF14-LIKE2 (DLK2) Reveals Its Functional Divergence from Strigolactone-Related Paralogs. *Front Plant Sci.* **2017**, *8*, 1641. [CrossRef] [PubMed]
36. Witt, S.N.; Gollhofer, J.; Timofeev, R.; Lan, P.; Schmidt, W.; Buckhout, T.J. Vacuolar-Iron-Transporter1-Like Proteins Mediate Iron Homeostasis in Arabidopsis. *PLoS ONE* **2014**, *9*, e110468.
37. Dubois, M.; Skirycz, A.; Claeys, H.; Maleux, K.; Dhondt, S.; De Bodt, S.; Vanden Bossche, R.; De Milde, L.; Yoshizumi, T.; Matsui, M.; et al. Ethylene Response Factor6 acts as a central regulator of leaf growth under water-limiting conditions in Arabidopsis. *Plant Physiol.* **2013**, *162*, 319–332. [CrossRef]
38. Nishizawa, A.; Yabuta, Y.; Shigeoka, S. Galactinol and raffinose constitute a novel function to protect plants from oxidative damage. *Plant Physiol.* **2008**, *147*, 1251–1263. [CrossRef]
39. Seok, H.Y.; Tran, H.T.; Lee, S.Y.; Moon, Y.H. AtERF71/HRE2, an Arabidopsis AP2/ERF Transcription Factor Gene, Contains Both Positive and Negative Cis-Regulatory Elements in Its Promoter Region Involved in Hypoxia and Salt Stress Responses. *Int. J. Mol. Sci.* **2022**, *23*, 5310. [CrossRef]
40. Datta, R.; Kumar, D.; Sultana, A.; Hazra, S.; Bhattacharyya, D.; Chattopadhyay, S. Glutathione Regulates 1-Aminocyclopropane-1-Carboxylate Synthase Transcription via WRKY33 and 1-Aminocyclopropane-1-Carboxylate Oxidase by Modulating Messenger RNA Stability to Induce Ethylene Synthesis during Stress. *Plant Physiol.* **2015**, *169*, 2963–2981.
41. Truong, H.A.; Lee, S.; Trinh, C.S.; Lee, W.J.; Chung, E.H.; Hong, S.W.; Lee, H. Overexpression of the HDA15 Gene Confers Resistance to Salt Stress by the Induction of NCED3, an ABA Biosynthesis Enzyme. *Front. Plant Sci.* **2021**, *12*, 640443. [CrossRef] [PubMed]
42. Ashraf, M.A.; Akihiro, T.; Ito, K.; Kumagai, S.; Sugita, R.; Tanoi, K.; Rahman, A. ATP binding cassette proteins ABCG37 and ABCG33 function as potassium-independent cesium uptake carriers in Arabidopsis roots. *Mol. Plant* **2021**, *14*, 664–678. [CrossRef] [PubMed]
43. Sadhukhan, A.; Agrahari, R.K.; Wu, L.; Watanabe, T.; Nakano, Y.; Panda, S.K.; Koyama, H.; Kobayashi, Y. Expression genome-wide association study identifies that phosphatidylinositol-derived signalling regulates ALUMINIUM SENSITIVE3 expression under aluminium stress in the shoots of Arabidopsis thaliana. *Plant Sci.* **2021**, *302*, 110711. [CrossRef] [PubMed]
44. Schulze, L.M.; Britto, D.T.; Li, M.; Kronzucker, H.J. A pharmacological analysis of high-affinity sodium transport in barley (Hordeum vulgare L.): A 24Na+/42K+ study. *J. Exp. Bot.* **2012**, *63*, 2479–2489. [CrossRef]
45. Demidchik, V.; Maathuis, F.J.M. Physiological roles of nonselective cation channels in plants: From salt stress to signalling and development. *New Phytol.* **2007**, *175*, 387–404. [CrossRef]

46. Biagi, B.A.; Enyeart, J.J. Gadolinium Blocks Low-Threshold and High-Threshold Calcium Currents in Pituitary-Cells. *Am. J. Physiol.* **1990**, *259*, C515–C520. [CrossRef]
47. Seifikalhor, M.; Aliniaeifard, S.; Shomali, A.; Azad, N.; Hassani, B.; Lastochkina, O.; Li, T. Calcium signaling and salt tolerance are diversely entwined in plants. *Plant Signal. Behav.* **2019**, *14*, 1665455. [CrossRef]
48. Yang, Y.; Tang, R.J.; Jiang, C.M.; Li, B.; Kang, T.; Liu, H.; Zhao, N.; Ma, X.J.; Yang, L.; Chen, S.L.; et al. Overexpression of the PtSOS2 gene improves tolerance to salt stress in transgenic poplar plants. *Plant Biotechnol. J.* **2015**, *13*, 962–973. [CrossRef]
49. Ye, C.Y.; Zhang, H.C.; Chen, J.H.; Xia, X.L.; Yin, W.L. Molecular characterization of putative vacuolar NHX-type Na(+)/H(+) exchanger genes from the salt-resistant tree Populus euphratica. *Physiol. Plant* **2009**, *137*, 166–174. [CrossRef]
50. Riyazuddin, R.; Verma, R.; Singh, K.; Nisha, N.; Keisham, M.; Bhati, K.K.; Kim, S.T.; Gupta, R. Ethylene: A Master Regulator of Salinity Stress Tolerance in Plants. *Biomolecules* **2020**, *10*, 959. [CrossRef]
51. Tao, J.J.; Chen, H.W.; Ma, B.; Zhang, W.K.; Chen, S.Y.; Zhang, J.S. The Role of Ethylene in Plants Under Salinity Stress. *Front Plant Sci.* **2015**, *6*, 1059. [CrossRef] [PubMed]
52. Li, J.; Guo, X.; Zhang, M.; Wang, X.; Zhao, Y.; Yin, Z.; Zhang, Z.; Wang, Y.; Xiong, H.; Zhang, H.; et al. OsERF71 confers drought tolerance via modulating ABA signaling and proline biosynthesis. *Plant Sci.* **2018**, *270*, 131–139. [CrossRef] [PubMed]
53. Wei, X.; Xu, H.; Rong, W.; Ye, X.; Zhang, Z. Constitutive expression of a stabilized transcription factor group VII ethylene response factor enhances waterlogging tolerance in wheat without penalizing grain yield. *Plant Cell Environ.* **2019**, *42*, 1471–1485. [CrossRef]
54. Li, C.; Zheng, L.; Wang, X.; Hu, Z.; Zheng, Y.; Chen, Q.; Hao, X.; Xiao, X.; Wang, X.; Wang, G.; et al. Comprehensive expression analysis of Arabidopsis GA2-oxidase genes and their functional insights. *Plant Sci.* **2019**, *285*, 1–13. [CrossRef]
55. Han, S.; Jiao, Z.; Niu, M.X.; Yu, X.; Huang, M.; Liu, C.; Wang, H.L.; Zhou, Y.; Mao, W.; Wang, X.; et al. Genome-Wide Comprehensive Analysis of the GASA Gene Family in Populus. *Int. J. Mol. Sci.* **2021**, *22*, 12336. [CrossRef] [PubMed]
56. Shan, C.; Mei, Z.; Duan, J.; Chen, H.; Feng, H.; Cai, W. OsGA2ox5, a gibberellin metabolism enzyme, is involved in plant growth, the root gravity response and salt stress. *PLoS ONE* **2014**, *9*, e87110. [CrossRef] [PubMed]
57. Sun, S.; Wang, H.; Yu, H.; Zhong, C.; Zhang, X.; Peng, J.; Wang, X. GASA14 regulates leaf expansion and abiotic stress resistance by modulating reactive oxygen species accumulation. *J. Exp. Bot.* **2013**, *64*, 1637–1647. [CrossRef] [PubMed]
58. Hu, W.; Yan, H.; Luo, S.; Pan, F.; Wang, Y.; Xiang, Y. Genome-wide analysis of poplar SAUR gene family and expression profiles under cold, polyethylene glycol and indole-3-acetic acid treatments. *Plant Physiol. Biochem.* **2018**, *128*, 50–65. [CrossRef]
59. Liu, W.; Li, R.J.; Han, T.T.; Cai, W.; Fu, Z.W.; Lu, Y.T. Salt stress reduces root meristem size by nitric oxide-mediated modulation of auxin accumulation and signaling in Arabidopsis. *Plant Physiol.* **2015**, *168*, 343–356. [CrossRef]
60. Strader, L.C.; Zhao, Y. Auxin perception and downstream events. *Curr. Opin. Plant Biol.* **2016**, *33*, 8–14. [CrossRef]
61. Li, W.; Dang, C.; Ye, Y.; Wang, Z.; Hu, B.; Zhang, F.; Zhang, Y.; Qian, X.; Shi, J.; Guo, Y.; et al. Overexpression of Grapevine VvIAA18 Gene Enhanced Salt Tolerance in Tobacco. *Int. J. Mol. Sci.* **2020**, *21*, 1323. [CrossRef]
62. La Mantia, J.; Unda, F.; Douglas, C.J.; Mansfield, S.D.; Hamelin, R. Overexpression of AtGolS3 and CsRFS in poplar enhances ROS tolerance and represses defense response to leaf rust disease. *Tree Physiol.* **2018**, *38*, 457–470. [CrossRef] [PubMed]
63. Liu, L.; Wu, X.; Sun, W.; Yu, X.; Demura, T.; Li, D.; Zhuge, Q. Galactinol synthase confers salt-stress tolerance by regulating the synthesis of galactinol and raffinose family oligosaccharides in poplar. *Ind. Crop. Prod.* **2021**, *165*, 113432. [CrossRef]
64. Ma, S.; Lv, J.; Li, X.; Ji, T.; Zhang, Z.; Gao, L. Galactinol synthase gene 4 (CsGolS4) increases cold and drought tolerance in Cucumis sativus L. by inducing RFO accumulation and ROS scavenging. *Environ. Exp. Bot.* **2021**, *185*, 104406. [CrossRef]
65. Fini, A.; Brunetti, C.; Di Ferdinando, M.; Ferrini, F.; Tattini, M. Stress-induced flavonoid biosynthesis and the antioxidant machinery of plants. *Plant Signal Behav.* **2011**, *6*, 709–711. [CrossRef] [PubMed]
66. Jan, R.; Kim, N.; Lee, S.H.; Khan, M.A.; Asaf, S.; Lubna; Park, J.R.; Asif, S.; Lee, I.J.; Kim, K.M. Enhanced Flavonoid Accumulation Reduces Combined Salt and Heat Stress Through Regulation of Transcriptional and Hormonal Mechanisms. *Front. Plant Sci.* **2021**, *12*, 796956. [CrossRef] [PubMed]
67. Berthet, S.; Demont-Caulet, N.; Pollet, B.; Bidzinski, P.; Cezard, L.; Le Bris, P.; Borrega, N.; Herve, J.; Blondet, E.; Balzergue, S.; et al. Disruption of LACCASE4 and 17 results in tissue-specific alterations to lignification of Arabidopsis thaliana stems. *Plant Cell* **2011**, *23*, 1124–1137. [CrossRef]
68. Qin, S.; Fan, C.; Li, X.; Li, Y.; Hu, J.; Li, C.; Luo, K. LACCASE14 is required for the deposition of guaiacyl lignin and affects cell wall digestibility in poplar. *Biotechnol. Biofuels* **2020**, *13*, 197. [CrossRef]
69. Adam, A.L.; Bestwick, C.S.; Barna, B.; Mansfield, J.W. Enzymes Regulating the Accumulation of Active Oxygen Species during the Hypersensitive Reaction of Bean to Pseudomonas-Syringae Pv Phaseolicola. *Planta* **1995**, *197*, 240–249. [CrossRef]
70. Zou, L.; Li, T.; Li, B.; He, J.; Liao, C.; Wang, L.; Xue, S.; Sun, T.; Ma, X.; Wu, Q. De novo transcriptome analysis provides insights into the salt tolerance of Podocarpus macrophyllus under salinity stress. *BMC Plant Biol.* **2021**, *21*, 489. [CrossRef]
71. Nakano, Y.; Asada, K. Hydrogen-Peroxide Is Scavenged by Ascorbate-Specific Peroxidase in Spinach-Chloroplasts. *Plant Cell Physiol.* **1981**, *22*, 867–880.
72. Trapnell, C.; Roberts, A.; Goff, L.; Pertea, G.; Kim, D.; Kelley, D.R.; Pimentel, H.; Salzberg, S.L.; Rinn, J.L.; Pachter, L. Differential gene and transcript expression analysis of RNA-seq experiments with TopHat and Cufflinks. *Nat. Protoc.* **2012**, *7*, 562–578. [CrossRef]
73. Chen, C.J.; Chen, H.; Zhang, Y.; Thomas, H.R.; Frank, M.H.; He, Y.H.; Xia, R. TBtools: An Integrative Toolkit Developed for Interactive Analyses of Big Biological Data. *Mol. Plant* **2020**, *13*, 1194–1202. [CrossRef] [PubMed]

74. Essah, P.A.; Davenport, R.; Tester, M. Sodium influx and accumulation in Arabidopsis. *Plant Physiol.* **2003**, *133*, 307–318. [CrossRef] [PubMed]
75. Garciadeblas, B.; Senn, M.E.; Banuelos, M.A.; Rodriguez-Navarro, A. Sodium transport and HKT transporters: The rice model. *Plant J.* **2003**, *34*, 788–801. [CrossRef] [PubMed]
76. Quintero, F.J.; Garciadeblas, B.; RodriguezNavarro, A. The SAL1 gene of Arabidopsis, encoding an enzyme with 3′(2′),5′-bisphosphate nucleotidase and inositol polyphosphate 1-phosphatase activities, increases salt tolerance in yeast. *Plant Cell* **1996**, *8*, 529–537.

Article

Genome-Wide Analysis of microRNAs and Their Target Genes in Dongxiang Wild Rice (*Oryza rufipogon* Griff.) Responding to Salt Stress

Yong Chen [1,†], Wanling Yang [1,†], Rifang Gao [1], Yaling Chen [1], Yi Zhou [1], Jiankun Xie [1,*] and Fantao Zhang [1,2,*]

[1] College of Life Sciences, Jiangxi Normal University, Nanchang 330022, China
[2] State Key Laboratory of Crop Gene Exploration and Utilization in Southwest China, Sichuan Agricultural University, Chengdu 611130, China
* Correspondence: 004068@jxnu.edu.cn (J.X.); 004768@jxnu.edu.cn (F.Z.)
† These authors contributed equally to this work.

Abstract: Rice (*Oryza sativa*) is a staple food for more than half of the world's population, and its production is critical for global food security. Moreover, rice yield decreases when exposed to abiotic stresses, such as salinity, which is one of the most detrimental factors for rice production. According to recent trends, as global temperatures continue to rise due to climate change, more rice fields may become saltier. Dongxiang wild rice (*Oryza rufipogon* Griff., DXWR) is a progenitor of cultivated rice and has a high tolerance to salt stress, making it useful for studying the regulatory mechanisms of salt stress tolerance. However, the regulatory mechanism of miRNA-mediated salt stress response in DXWR remains unclear. In this study, miRNA sequencing was performed to identify miRNAs and their putative target genes in response to salt stress in order to better understand the roles of miRNAs in DXWR salt stress tolerance. A total of 874 known and 476 novel miRNAs were identified, and the expression levels of 164 miRNAs were found to be significantly altered under salt stress. The stem-loop quantitative real-time PCR (qRT-PCR) expression levels of randomly selected miRNAs were largely consistent with the miRNA sequencing results, suggesting that the sequencing results were reliable. The gene ontology (GO) analysis indicated that the predicted target genes of salt-responsive miRNAs were involved in diverse biological pathways of stress tolerance. This study contributes to our understanding of DXWR salt tolerance mechanisms regulated by miRNAs and may ultimately improve salt tolerance in cultivated rice breeding using genetic methods in the future.

Keywords: wild rice; miRNA; target gene; salt tolerance; genetic resource

Citation: Chen, Y.; Yang, W.; Gao, R.; Chen, Y.; Zhou, Y.; Xie, J.; Zhang, F. Genome-Wide Analysis of microRNAs and Their Target Genes in Dongxiang Wild Rice (*Oryza rufipogon* Griff.) Responding to Salt Stress. *Int. J. Mol. Sci.* **2023**, *24*, 4069. https://doi.org/10.3390/ijms24044069

Academic Editor: Martin Bartas

Received: 27 December 2022
Revised: 15 February 2023
Accepted: 16 February 2023
Published: 17 February 2023

1. Introduction

Rice (*Oryza sativa*) is a cereal crop that feeds more than half of the world's population, especially in Asia, where approximately 80% of global rice is cultivated and consumed [1,2]. It is predicted that food production will need to increase by about 70% by 2050 to maintain sufficient food levels for the population [3]. As observed in paddy field crops, rice production decreases when subjected to abiotic stresses such as water deficiency or submergence, low or high temperatures, and high salinity [4]. After drought, salinity is the second most prevalent soil problem in rice-growing countries [5]. Approximately 30% of the world's rice lands contain too much salt to allow normal rice cultivation, and the rice yield is reduced by 68% when cultivated on such moderately salt-affected soils [6]. Even worse, as the global temperatures continue to rise, more soil in semiarid regions will be salinized by irrigation with saline water due to water scarcity and rising sea levels [7]. Therefore, salinity is considered one of the greatest environmental threats to rice production worldwide, and attaining rice cultivars that are tolerant to high salt is of the utmost necessity for the agricultural sector.

MicroRNAs (miRNAs) are short non-coding RNA molecules (20–24 nt) that regulate the expression of protein-coding genes at the post-transcriptional level by cleaving their target genes [8,9]. Almost all pathways in the eukaryotic gene regulation system are directly or indirectly regulated by miRNAs [10]. In the past few decades, a vast number of miRNAs has been identified in various plant species using high-throughput sequencing technology [11]. Meanwhile, many stress-specific miRNAs have been identified under different biotic and abiotic stress conditions such as high sanity [12], drought [13], cold [14], nutrient deficiency [15], and infection [16], suggesting that miRNAs play very important roles in various stress responses in plant species. However, previous related studies have predominantly focused on the important model plants and agricultural crops. More research needs to be done to further elucidate miRNA functions among more plant species.

Common wild rice (*Oryza rufipogon*) is thought to be the progenitor of cultivated rice, and 30–40% of its genetic variation was estimated to be lost during the domestication process [17]. Dongxiang wild rice (*Oryza rufipogon* Griff., DXWR) is the northernmost (28°14′ N) common wild rice ever found in the world, and has been found to tolerate various abiotic stresses [18,19]. Meanwhile, the previous studies revealed that DXWR has higher tolerance to salt stress than cultivated rice, making it a unique gene pool for identifying more precious salt stress response genes. Although some miRNAs have been identified in DXWR by high-throughput sequencing and a bioinformatics approach, the salt-responsive miRNAs have been little identified and characterized in DXWR, and the expression pattern of miRNAs under the salt stress of DXWR remains unclear.

In this study, a comprehensive view of known and novel miRNAs and their expression patterns under salt stress is characterized using high-throughput sequencing technology. A set of differential expression miRNAs were verified using stem-loop quantitative real-time PCR (qRT-PCR), and the target genes of salt stress-responsive miRNAs were predicted and characterized. This study helps to clarify the response of miRNAs and their target genes to salt stress, to explore the miRNA-regulated mechanism of salt stress tolerance in DXWR, and finally to help genetically improve salt stress tolerance in cultivated rice in the future.

2. Results

2.1. Overview of sRNA Library Data Sets

To investigate the possible miRNAs involved in the salt stress response in DXWR, six small RNA libraries from the control and salt treatment groups were constructed and subjected to high-throughput sequencing, named DY-CK1, DY-CK2, DY-CK3, DY-S1, DY-S2, and DY-S3. The DY-CK (1–3) libraries were three biological replicates of DXWR under normal conditions, and the DY-S (1–3) libraries were three biological replicates of DXWR after salt treatment. In total, 96.34 million raw short reads were obtained from the six libraries, with 16.06 million raw reads per library on average. After filtering out low-quality data, 3′ joint contamination data, and sequences with a length less than 18 nt or greater than 25 nt, a total of 43.88 million clean reads were obtained, with a mean of 7.31 million clean reads per library (Table 1). Meanwhile, the Q30 (sequencing error rate < 0.1%) scores of all libraries ranged from 94.05% to 95.18%, with an average of 94.54%, indicating that the sequence data were of reliable quality.

To further analyze the validity of the sequence data, a statistical analysis on the length distribution of total and unique sRNAs was performed on filtered datasets. Of all the total sRNAs, 24 nt-sRNAs were the most abundant, accounting for an average of 24.5% and 21.8% of the DY-CK and DY-S libraries, respectively (Figure 1A). Among the unique sRNAs, 24 nt-sRNAs were the most frequent, accounting for an average of 39.6% and 38.6% in the DY-CK and DY-S libraries, respectively, which was consistent with the typical size of miRNAs from Dicer-derived products (Figure 1B).

Table 1. Summary of small RNA sequencing in the DXWR at control and salt stress conditions.

Types	DY-CK1		DY-CK2		DY-CK3	
	Total	Unique	Total	Unique	Total	Unique
Raw reads	10,618,693	2,604,513	16,460,121	3,672,322	22,194,190	5,038,513
3′ adaptor & length filter	5,585,638	1,009,906	9,545,474	1,324,469	11,078,492	1,954,931
Junk reads	20,571	14,159	28,759	20,791	50,332	31,211
Clean reads	5,012,484	995,747	6,885,888	1,303,678	11,065,366	1,923,720
Rfam	637,243	14,603	859,520	15,142	1,357,192	24,867
mRNA	590,743	14,426	667,359	20,178	1,335,628	35,142
Repeats	9114	205	11,931	236	18,131	327
valid reads	3,817,501	1,552,466	5,396,359	2,292,925	8,450,225	2,994,550

Types	DY-S1		DY-S2		DY-S3	
	Total	Unique	Total	Unique	Total	Unique
Raw reads	13,215,515	2,349,112	15,702,355	2,967,602	18,147,568	3,540,661
3′ adaptor & length filter	7,629,958	1,090,528	8,986,925	1,318,486	9,451,562	1,295,788
Junk reads	19,130	11,845	24,582	15,842	38,459	24,077
Clean reads	5,566,427	1,078,683	6,690,848	1,302,644	8,657,547	1,271,711
Rfam	1,204,287	19,546	1,339,068	21,195	1,608,522	23,011
mRNA	436,033	11,146	579,789	15,631	763,307	23,762
Repeats	19,530	245	25,234	273	21,092	273
valid reads	4,012,240	1,218,038	4,863,177	1,598,664	6,395,067	2,176,500

Note: 3′ adaptor & length filter: reads removed due to 3′ adaptor not found and length with <18 nt and >25 nt were removed. Junk reads: Junk: $\geq$2 N, $\geq$7 A, $\geq$8 C, $\geq$6 G, $\geq$7 T, $\geq$10 Dimer, $\geq$6 Trimer, or $\geq$5 Tetramer. Clean reads: equal to raw reads—3′ adaptor & length filter—Junk reads. Rfam: collection of many common non-coding RNA families except microRNA, http://rfam.janelia.org, accessed on 5 September 2022. Repeats: prototypic sequences representing repetitive DNA from different eukaryotic species, http://www.girinst.org/repbase, accessed on 5 September 2022.

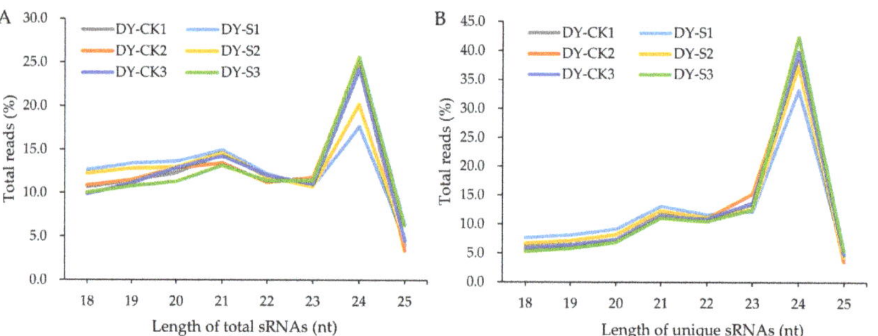

Figure 1. Length distribution and abundance of small RNAs in the libraries of DY-CK and DY-S samples. (**A**) Size distribution of total small RNAs; (**B**) Size distribution of unique small RNAs.

2.2. Identification of Known and Novel miRNAs in DXWR

To identify the known miRNAs of DXWR under normal and salt treated conditions, the unique clean reads were blasted to the miRbase database for comparison with the currently known plant precursor or mature miRNA sequences. In total, 712 pre-miRNAs corresponding to 874 known unique mature miRNAs were identified as homologues of known miRNAs from the other plants, such as *Arabidopsis thaliana*, *Cynara cardunculus*, *Glycine max*, *Medicago truncatula*, and *Triticum aestivum* (Supplementary Figure S1). Among the 874 known miRNAs, only 63 miRNAs showed high expression levels (reads greater than the average copy of the data set), and osa-miR168a-5p, osa-miR166a-3p, osa-miR1425-5p, osa-miR168a-3p_L-3, and osa-miR396e-5p were identified as the most abundantly expressed conserved miRNAs in DXWR. A majority of the known miRNAs showed middle (520 miRNAs with reads greater than 10 but below the average copy of the data set) to low

(291 miRNAs with reads less than 10) expressional levels (Supplementary Table S1). Among these known miRNAs, 523 belong to 69 families (Supplementary Table S2), whereas the families of the other 351 miRNAs were unknown. The three largest families were miR812 (54 miRNA members), miRNA166 (27 members), and miR814 (26 members), whereas most families contained less than 10 members (Supplementary Table S2 and Figure S2).

In addition, the unmapped sequences were compared with the rice genome, and mapped sequences that fulfilled the criteria for annotation of plant miRNAs were identified as novel miRNAs. Finally, a total of 476 novel miRNAs were identified from 528 pre-miRNAs (Supplementary Table S1). A majority (66.0%) of the identified novel miRNAs showed low expressional levels, 162 (34.0%) of the novel miRNAs showed middle abundance, and no novel miRNAs exhibited high abundance (Supplementary Table S1). Among the novel miRNAs, the first nucleotides of 5′ were biased toward A (adenine) (59.0%) and U (uracil) (20.6%) (Supplementary Figure S3). These pre-miRNAs range in length from 56 nt to 255 nt with an average length of 149 nt, which is consistent with the general length of pre-miRNAs. The CG percentages (CG%) of these novel pre-miRNAs range from 18.5 to 78.9%, and their minimal folding free energy index (MFEI) ranges from 0.9 to 2.3 with an average of 1.4 (Supplementary Table S1).

2.3. Differential Expression Analysis of miRNAs in DXWR under Salt Stress Condition

To identify differentially expressed miRNAs (DEMs) that responded to salt stress, the expression levels of all miRNAs in the DY-CK and DY-S libraries were normalized and analyzed. Intriguingly, 256 and 91 miRNAs were specifically expressed in the normal and salt stress conditions, respectively (Supplementary Table S1), implying that the specifically expressed miRNAs under the normal condition may play negative roles in the salt response, whereas those under the salt stress condition may play positive roles in the salt response in DXWR. Meanwhile, of the 1,350 (874 known and 476 novel) identified miRNAs, the expressions of 164 miRNAs, including 139 known and 25 novel miRNAs, were significantly altered ($p < 0.05$), and over half of the DEMs (99 out of 164) were downregulated (Supplementary Table S3). The expression levels of osa-miR399j_R-1 and osa-miR2106_R+1 significantly decreased and increased (−3.91 and 4.31 $\log_2$FC; FC means fold change), respectively (Figure 2). After similar sequences of DEMs were assigned to their family, most DEMs within their miRNA family displayed similar expression patterns, such as the family numbers of miR164 and miR166 being significantly downregulated under salt stress. Among the known DEMs, the miR166 family had the highest numbers (12), followed by the miR169_1 (9) and miR171_1 (7) (Supplementary Table S3).

To validate the high-throughput sequencing data and expression patterns of miRNAs, ten DEMs that show significant expression changes after salt treatment were randomly selected. The stem-loop qRT-PCR results showed that osa-miR5072_L-4, ath-miR8175_L-2, ptc-miR6478_1ss21GA, and osa-miR2106_R+1 were upregulated and osa-miR167d-3p, osa-miR166h-5p, osa-miR171g-p5, osa-miR1850.1, osa-miR166j-5p, and PC-5p-16439_287 were downregulated. The expression trends of these miRNAs in control libraries relative to those in salt-treated libraries detected by small RNA sequencing were basically consistent with those detected by stem-loop qRT-PCR (Figure 3, Supplementary Table S3), suggesting that the miRNA sequencing data results were credible.

2.4. Prediction and Functional Annotation of the Known and Novel DEMs Targets

Targets of known and novel DEMs were predicted using RNAplex software, and a total of 2018 transcripts from 1774 genes were identified (Supplementary Table S4). Among them, 1900 transcripts from 1666 genes were predicted to be the targets of 127 known DEMs, and 143 transcripts from 133 genes were predicted to be the targets of 21 novel DEMs (Supplementary Table S4). Of the 1,774 target genes, 115 genes are transcription factors (TFs), such as *HSFC1A*, *HSFC2A*, *OsTIFY1A*, *OsERF101*, *OsbHLH113*, and *R2R3-MYB* (Supplementary Table S4), which are DNA binding transcription factors and the regulation of stress response. The other miRNA target genes, including peptidase (such as serine carboxypeptidase-like, endoplasmic reticulum metallopeptidase, and carboxyl-terminal-processing peptidase), synthase (such as 3-ketoacyl-CoA synthase, noroxomaritidine synthase, and starch synthase), and transferase (such as acetyltransferase, O-fucosyltransferase, and nicotinate phosphoribosyltransferase) genes were involved in plant growth, development, and abiotic stress responses. These results indicated that miRNAs may play an important role in different biological processes under salt stress in DXWR.

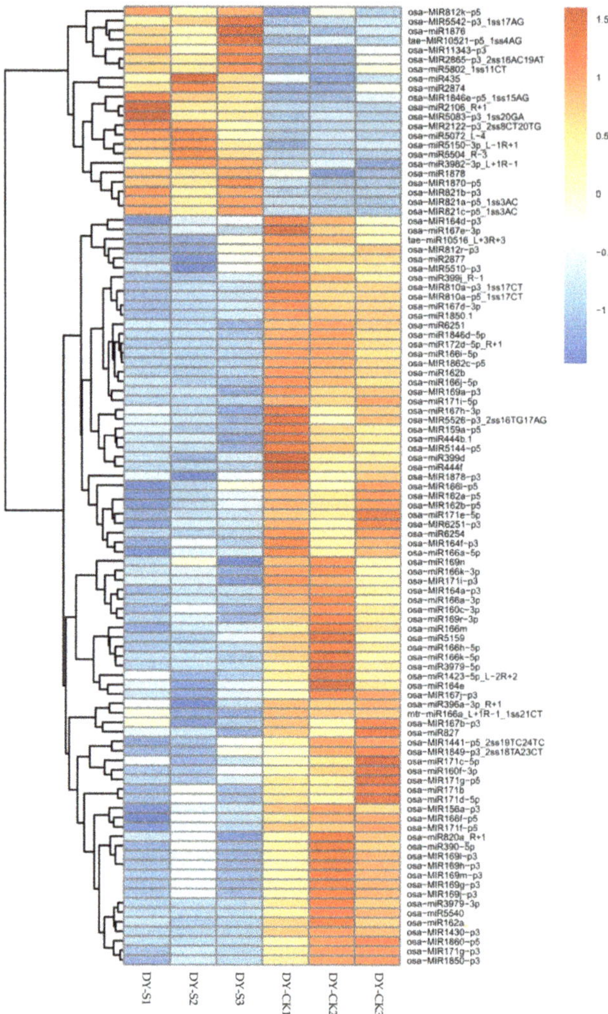

Figure 2. Heat map analysis of differentially expressed miRNAs in six libraries. Up- and downregulated genes were indicated in red and blue, respectively. Color brightness reflected the magnitude of difference.

To further analyze the specific biological functions of DXWR miRNAs under salt stress, a Gene ontology (GO) analysis was performed to explore the biological functions of the 1774 target genes for known and novel DEMs; the results indicated that the target genes were mainly annotated into 50 GO terms, which were most related to the biological process category (25 terms), followed by the cellular component (15 terms) and molecular function (10 terms) categories (Figure 4). In the classification of the cellular component category, the GO terms including nucleus (384 genes), plasma membrane (237 genes), cytoplasm (180 genes), and integral component of membrane (175 genes) have the top four number of genes; in the molecular function classification, GO terms of protein binding (176 genes) and molecular function (130 genes) have the most target genes; in the biological process category, biological process (144 genes) and regulation of transcription (114 genes) have the most gene numbers, and in this category, 56 target genes were annotated in response to stress, of which 26, 17, and 13 target genes were annotated in response to salt, oxidative, and osmotic stress, respectively (Supplementary Table S4). From the GO terms, we found that many processes, such as responses to stress (GO:0006950), abscisic acid transport (GO:0080168), auxin transport (GO:0060918), and rRNA binding (GO:0019843) were prominently down-regulated and that the processes of responses to DNA demethylation (GO:0080111), water homeostasis (GO:0030104), secondary growth (GO:0080117), and histone acetylation (GO:0016573) were up-regulated (Supplementary Table S4).

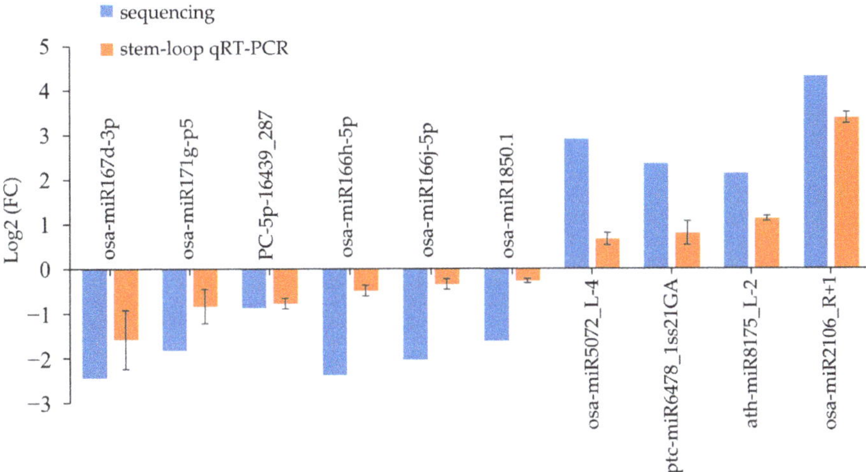

Figure 3. qRT-PCR validation of ten randomly selected salt stress-responsive miRNAs' relative expression (fold changes of sequencing reads and qRT-PCR) between control and salt-treated library. The blue bars represent the fold change (log2) in control libraries relative to that in salt-treated libraries detected by small RNA sequencing, while the orange bars represent the fold change (log2) in control libraries relative to that in salt-treated libraries detected by stem-loop qRT-PCR (normalized to U6snRNA; n = 3).

Furthermore, our previous study revealed an expression profile of genes in DXWR under salt stress: 743 genes were downregulated, while 892 were upregulated in both roots and leaves [20]. Among these genes, 81 genes coexisted in the target genes predicted in this study, and 41 miRNA–mRNA pairs showed opposite expression patterns in DXWR under salt stress (Table 2). Among them, nine genes have been named. Meanwhile, five of the nine genes have been well studied and proved to be associated with various abiotic stresses, including *OsERF101* (Os04g0398000), *OsSPX-MFS1* (Os04g0573000), *OsKOS1* (Os06g0569500), *OsPDK1* (Os07g0637300), and *OsRab16A* (Os11g0454300). The upstream miRNAs of the five genes were downregulated in this study, and when we measured the expression levels of the five genes in DXWR under salt stress, all were upregulated, consistent with the previous results and negatively correlating with the miRNA expression (Figure 5). Therefore, these miRNAs and their putative targets could be potential salt stress-associated regulators that deserve further investigation.

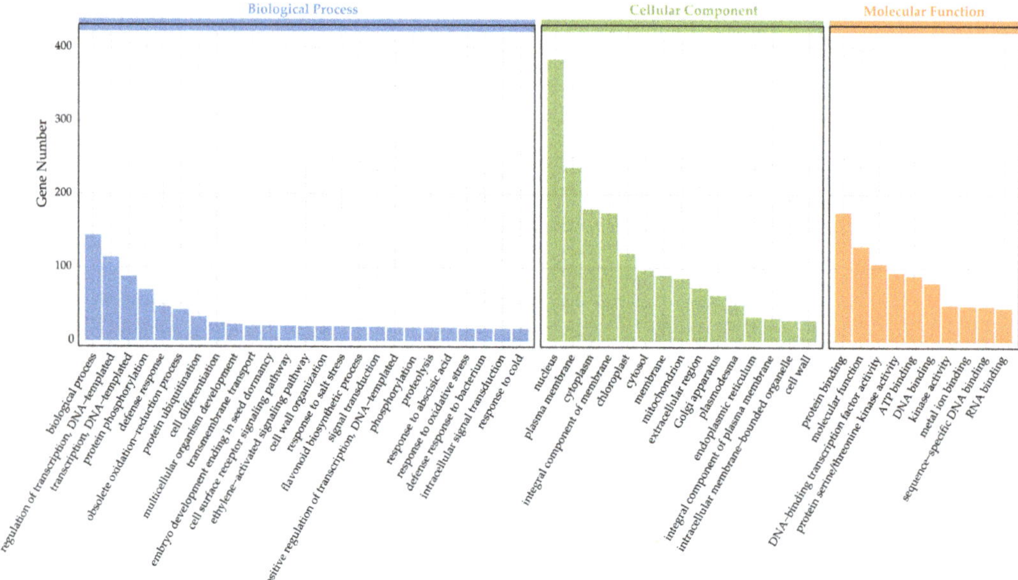

Figure 4. Gene ontology classification of the targeted genes of salt stress-responsive miRNAs.

Table 2. miRNA–mRNA pairs showed the opposing expression under salt stress condition in DXWR.

miRNAs	Expression Trends	Target Genes	Expression Trends [a]	Gene Name
bdi-miR5054_1ss10TA	up	Os01g0504100	down	*OsPUP8*
ath-miR8175_L-2	up	Os03g0130700	down	-
ath-miR8175_L-2_1ss20AT	up	Os03g0130700	down	-
gma-miR6300_1ss18GC	up	Os03g0219100	down	-
gma-MIR4995-p5_1ss18GC	up	Os03g0637900	down	-
ptc-MIR6476a-p3_2ss6AG18AC	up	Os04g0477000	down	-
bdi-miR5054_1ss10TA	up	Os05g0179300	down	-
gma-MIR6300-p5_1ss6AG	up	Os05g0219900	down	-
ath-miR8175_L-2	up	Os06g0495800	down	-
gma-miR6300_1ss18GC	up	Os07g0531500	down	-
gma-miR6300_R+1	up	Os07g0531500	down	-
bdi-miR5054_1ss10TA	up	Os08g0495500	down	-
osa-miR5072_L-4	up	Os10g0117000	down	-
ath-miR8175_L-1	up	Os10g0477900	down	-
ath-miR8175_L-2	up	Os10g0477900	down	-
ath-miR8175_L-2_1ss20AT	up	Os10g0477900	down	-
PC-5p-57749_50	up	Os10g0532200	down	-
osa-MIR1846e-p5_1ss15AG	up	Os11g0107700	down	-
gma-MIR4995-p5_1ss20GC	up	Os11g0170000	down	-
osa-MIR169g-p3	down	Os02g0596000	up	-
osa-MIR169h-p3	down	Os02g0596000	up	-
osa-MIR169j-p3	down	Os02g0596000	up	-
osa-MIR169l-p3	down	Os02g0596000	up	-
osa-MIR169m-p3	down	Os02g0596000	up	-
osa-MIR6251-p3	down	Os02g0756800	up	-
osa-MIR159a-p5	down	Os03g0130300	up	*DEFL8*
osa-miR3979-5p	down	Os03g0386500	up	-
osa-miR172d-5p_R+1	down	Os04g0398000	up	*OsERF101*

Table 2. *Cont.*

miRNAs	Expression Trends	Target Genes	Expression Trends [a]	Gene Name
osa-miR827	down	Os04g0573000	up	*OsSPX-MFS1*
osa-miR399j_R-1	down	Os04g0691900	up	-
osa-miR5540	down	Os05g0582600	up	*OsSCP30*
osa-miR172d-5p_R+1	down	Os06g0154200	up	*D3*
osa-miR169r-3p	down	Os06g0569500	up	*OsKOS1*
osa-MIR159a-p5	down	Os07g0637300	up	*OsPDK1*
vvi-MIR3638-p5_2ss17GT18CT	down	Os08g0425800	up	-
osa-MIR812r-p3	down	Os10g0181200	up	-
osa-MIR164f-p3	down	Os11g0454300	up	*OsRab16A*
osa-MIR164f-p3	down	Os11g0673000	up	-
osa-miR444b.1	down	Os12g0116100	up	-
osa-MIR1860-p5	down	Os12g0174100	up	-
PC-5p-75382_31	down	Os12g0491800	up	-

Note: [a], the expression trends of the genes were derived from the previous study [20].

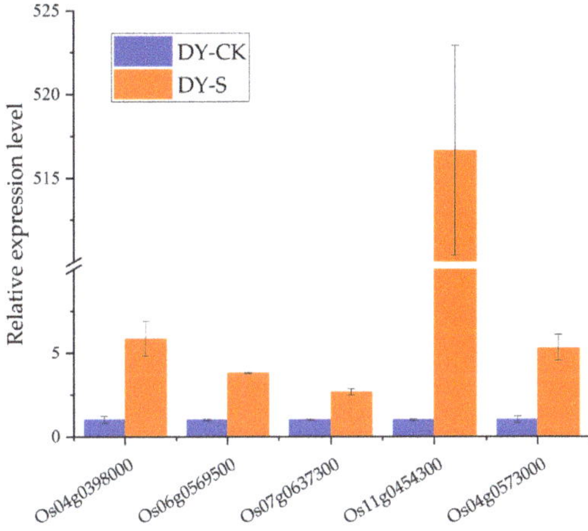

Figure 5. The expression levels of target genes under salt stress detected by qRT-PCR. The blue bars represent the samples under the normal condition, and the orange bars represent the samples under salt stress. The relative expression was calculated by the $2^{-\Delta\Delta Ct}$ method, and the standard deviation was calculated with three biological repeats.

3. Discussion

After drought, salinity is the second most common soil problem in rice-cultivating countries and has become a serious obstacle to improving global rice production [21]. More than 50% of arable lands may be lost to serious salinization by the year 2050, making it difficult to secure rice production and exacerbating food shortages [22]. Although rice salt tolerance has been improved by molecular biological techniques in recent years [23,24], surprisingly little is known about the mechanistic basis of the salt response. Wild rice is characterized by its excellent agronomic traits and tolerance to biotic and abiotic stresses [25]. During the domestication process, genetic diversity is rapidly lost, resulting in the loss of genes related to useful agronomic traits [26]. As a progenitor of cultivated rice (*Oryza sativa* L.), Dongxiang wild rice (*Oryza rufipogon* Griff., DXWR) has survived natural selection, and its gene diversity and tolerance to abiotic and biotic stress has been lost in cultivated rice [20]. Therefore, the biological mechanism underlying DXWR tolerance to stress requires further study.

MiRNAs are universally present in plants and mediate gene expression through target cleavage or translational repression [11]. Numerous stress-responsive miRNAs have been identified using sequencing technology since the first study reported miRNAs involved in the plant stress response in *Arabidopsis* [27]. Interestingly, as some miRNAs manage the metabolic pathway network in response to biotic and abiotic stress, it is important to investigate the roles of miRNAs in DXWR salt tolerance [28]. In this study, 874 known and 476 novel miRNAs were identified using high-throughput sequencing technology on a genome-wide scale in DXWR (Supplementary Table S1), of which 99 miRNAs were significantly downregulated and 65 were upregulated under salt stress (Supplementary Table S3). Among the DEMs, miR156 [29], miR159 [30], miR160 [31], miR164 [32], miR166 [33], miR167 [34], miR169 [35], miR171 [36], miR172 [37], miR399 [38], miR444 [39], and miR827 [40] have been reported to mediate stress responses in plants. Additionally, we identified known and novel miRNAs in DXWR under drought stress and found that the expression levels of miR160, miR164, miR166, miR167, miR172, miR444, miR810, miR5072, and miR1846 were significantly different [19,41]. Moreover, miR160, miR166, miR167, miR172, and miR5072 expression patterns coincided with miRNA expression in this study, suggesting that these miRNAs may retain the same properties when DXWR is subjected to drought and salinity stresses.

In addition, 2018 transcripts were predicted as candidate target genes for the salt-responsive DEMs in DXWR. The majority of these transcripts were predicted as functional genes encoding TFs, peptidases, synthases, transporters, or transferases. TFs are considered to have the greatest effects on plant salinity tolerance because these key regulators can regulate the expression levels of a range of salinity tolerance genes [42]. The bZIP family is one of the largest TF families in higher plants. Until now, many bZIP genes involved in salt stress responses have been characterized in various plant species. Gong et al. identified 48 bZIP genes in the genome of sugar beet (*Beta vulgaris* L.) and analyzed their biological functions and response patterns to salt stress [43]. Chai et al. revealed that the overexpression of a soybean bZIP gene *GmbZIP152* can enhance the tolerance to salt, drought, and heavy metal stresses in *Arabidopsis* [44]. In this study, four bZIP genes (*OsbZIP17*, *OsbZIP27*, *OsbZIP81*, and *OsbZIP85*) were predicted to be targeted by the salt-responsive DEMs. Intriguingly, Liu et al. found that *OsbZIP81* can regulate jasmonic acid (JA) levels by targeting the genes in the JA signaling and metabolism pathways in rice [45]. Many studies have revealed that JA can mediate the effect of abiotic stresses and help plants to acclimatize under stress conditions [46]. Thus, *OsbZIP81* and its corresponding miRNAs could be potential salt-associated regulators that deserve further investigation. Meanwhile, an increasing number of studies have shown that transferase genes help plants to respond and adapt to abiotic stresses. Sun et al. reported that ectopic expression of the *Arabidopsis* glycosyltransferase UGT85A5 gene can enhance salt stress tolerance in the plants of tobacco [47]. Duan et al. identified a total of 189 UDP-glycosyltransferase genes in the *Melilotus albus* genome and revealed their vital roles in abiotic stress responses [48]. In this study, 59 transferase genes were predicted to be targeted by the salt-responsive DEMs. Certainly, further investigation is required to confirm, if any, the roles of these transferase genes and their corresponding miRNAs involved in the salt tolerance of DXWR.

Meanwhile, the expression patterns of miRNAs and target genes can be used as indicators to determine the function of miRNAs, since the miRNAs and target genes were usually oppositely expressed. Previously, we analyzed the transcriptome profiles of DXWR under salt stress [20]. In this study, the expression patterns of miRNAs and target genes were evaluated by a combination analysis of miRNAs and transcriptome profiles. Finally, we screened out five stress-related genes for further investigation based on functional annotation, i.e., *OsERF101*, *OsSPX-MFS1*, *OsKOS1*, *OsPDK1*, and *OsRab16A*. *OsERF101* is an ethylene-responsive factor that is mainly expressed in rice reproductive tissues; *OsERF101*-overexpression plants were more tolerant to osmotic stress than wild-type plants, and when subjected to drought stress in the reproductive stage, transgenic plants had higher survival and seed setting rates [49]. *OsSPX-MFS1* belongs to the *SPX-MFS* family and is mainly expressed in shoots; *OsSPX-MFS1* mutant (*mfs1*) or overexpression of the upstream regulator miR827 impairs phosphate homeostasis [50]. *OsKOS1* is an *ent*-kaurene oxidase-like protein, and its expression is induced by UV irradiation and is likely to participate in phytoalexin biosynthesis [51], whereas phytoalexin is a compound that regulates the interaction between parasites and the host plant [52]. *OsPDK1* can positively regulate basal disease resistance in rice [53]. *OsRab16A* is induced when plants are subjected to stress, and overexpression of *Rab16A* increases plant salt tolerance [54,55]. The qRT-PCR analysis results showed that the expression levels of these five genes were all upregulated under salt stress in DXWR, negatively correlating with their upstream miRNAs' expression. Most notably, the expression level of *OsRab16A* (*Os11g0454300*) was the most significantly changed among the five genes, which further implies the role of this gene in conferring salt stress tolerance in plants.

Furthermore, GO enrichment analyses of target genes can help us to understand the functions of miRNAs more effectively. Our functional prediction according to GO categories showed that the target genes were significantly enriched in molecular functions, including DNA-binding TF activity, TF binding, L-alanine transmembrane transporter activity, arginine transmembrane transporter activity, and gamma-aminobutyric acid transmembrane transporter activity. As mentioned above, genes involved in DNA-binding TF activity and TF binding have shown vital roles in regulating stress tolerance in plants [42–45]. Meanwhile, according to Zhou et al. [56], miRNAs can also be involved in stress tolerance by regulating target genes that control transmembrane proteins. Thus, our findings are in agreement with the results of previous studies. In addition, plant hormones and their signal transductions play important roles in response to various abiotic stresses [57]. Based on the annotation of target genes, we identified that 14, 3, 2, 2, and 1 target genes of salt-responsive DEMs were associated with ethylene, auxin, abscisic acid, gibberellin, and cytokinin, respectively. Therefore, these results provide an abundant resource of candidate miRNAs and target genes associated with salt tolerance and their enriched regulatory networks in plants.

In the past two decades, some stress-responsive miRNAs have shown potential in stress tolerance improvement of various plant species. Moreover, miRNA modulation has been successfully developed to improve plant stress tolerance using many available biotechnological tools, such as overexpression, RNAi, and CRISPR Cas9 systems. For instance, transgenic tobacco plants expressing Zm-miR156c exhibit enhanced drought and salt stress tolerance [58]. The overexpression of sha-miR319d increases chilling and heat tolerance in tomato plants [59]. The overexpression of miR408 leads to enhanced resistance against cold, salt, and oxidative stresses in *Arabidopsis* [60]. The deletion of miR169a by CRISPR/Cas9 increases drought stress tolerance in *Arabidopsis* [61]. Therefore, we believe that the identified salt-responsive miRNAs from DXWR could provide a basis for developing salt stress-tolerant rice varieties through molecular design breeding in the future.

4. Materials and Methods

4.1. Plant Materials, Culture, and Sample Collection

The seeds of Dongxiang wild rice (*Oryza rufipogon* Griff., DXWR) were stored in our laboratory of Jiangxi Normal University. Seed germination and cultivation were carried out in accordance with the previous study [19]. Salt treatment was carried out when the rice plants were at the four-leaf stage. After two days of treatment with and without 200 mM salt, seedlings were collected and immediately frozen with liquid nitrogen.

4.2. Small RNA Library Construction and Deep Sequencing

To explore the regulatory mechanisms of miRNAs in response to salt stress in DXWR, we constructed two sample groups, namely, the salt-treated group (DY-S) and normal control group (DY-CK). Each sample group contains three biological replicates, named DY-CK1, DY-CK2, DY-CK3, DY-S1, DY-S2, and DY-S3. Library preparation and sequencing-related experiments were performed in accordance with the standard procedure provided by Illumina (San Diego, CA, USA). The TruSeq Small RNA Sample Prep Kit (Illumina, San Diego, CA, USA) was used for the preparation of the sequencing library. Then, the prepared libraries were sequenced by the Illumina Hiseq2000/2500 sequencing system (Illumina, San Diego, CA, USA) with a single-end 50 bp read length. The library construction and deep-sequencing analysis were performed by Lianchuan BioTech Co., Ltd. (Hangzhou, Zhejiang, China).

4.3. Sequencing Data Analysis and Identification of Known and Novel miRNAs

Raw sequencing reads were analyzed by the ACGT101-miR program (LC Sciences, Houston, TX, USA). After removing the $3'$ adapters and junk sequences, the remaining sequences with a length of 18–25 nt were aligned to mRNA (http://rapdb.dna.affrc.go.jp/download/irgsp1.html, accessed on 3 September 2022), Rfam (http://rfam.janelia.org, accessed on 3 September 2022), and repeat (http://www.girinst.org/repbase, accessed on 3 September 2022) databases to remove the matched sequences, respectively. The remaining valid reads were blasted against miRbase (http://www.mirbase.org/, accessed on 3 September 2022) [62] to identify the known miRNAs. The remaining unmapped sequences were compared with the rice genome (http://rapdb.dna.affrc.go.jp/download/irgsp1.html, accessed on 3 September 2022), and mapped sequences that fulfilled the criteria for the annotation of plant miRNAs were identified as novel miRNAs [63]. The p value of the Student's t-test was used to analyze the DEMs based on normalized deep-sequencing counts; the p-value ≤ 0.05 was set as the significance threshold of DEMs in this test.

4.4. Verification of Sequencing Data

To verify the reliability of the sequencing data, miRNAs were randomly selected from DEMs for qRT-PCR analysis. RNAs were extracted using a TRIzol reagent (Sangon Biotech Co., Ltd., Shanghai, China), following the manufacturer's instructions. MiRNA reverse transcription was performed using the miRNA 1st Strand cDNA Synthesis Kit (by stem-loop) (Vazyme, Nanjing, China), and mRNA reverse transcription was performed using the PrimeScript™ RT reagent Kit (Takara, Dalian, China). The qRT-PCR experiments were performed using the SYBR Premix Ex Taq II kit (Takara, Dalian, China) on an ABI 7500 Real-Time System (Applied Biosystems, Carlsbad, CA, USA). U6 small nuclear RNA (U6snRNA) and actin genes were used as endogenous controls to normalize the threshold cycle (Ct) values for the miRNAs and mRNAs detected by qRT-PCR, respectively. The relative expression levels of the miRNAs and mRNAs were calculated using the $2^{-\Delta\Delta Ct}$ method [64]. All reactions were repeated three times. The primers used in this study were listed in Table S5.

4.5. Prediction of Target Genes for Salt Stress-Responsive miRNAs

Based on miRNA sequencing, the putative target genes of differentially expressed miRNAs were predicted using RNAplex software 2.5.1 (http://www.tbi.univie.ac.at/RNA/RNAplex.1.html, accessed on 5 September 2022) [65] with a minimum free energy ratio (MFE ratio) cutoff >0.65. The putative target genes were then used for gene ontology (GO) enrichment analysis, which was conducted with a hypergeometric distribution (LC-BIO, Hangzhou, China).

5. Conclusions

In summary, we performed small RNA sequencing of DXWR during salt stress and identified 874 known and 476 novel miRNAs. Among these, 65 and 99 miRNAs were significantly upregulated and downregulated, respectively. The predicted target genes of salt-responsive miRNAs were annotated to participate in multiple biological processes. Our findings provide a comprehensive view of miRNA regulation of target genes in DXWR under salt stress, serving as a useful resource for better understanding the biological mechanisms of salt tolerance and for developing tolerant rice breeding practices.

Supplementary Materials: The supporting information can be downloaded at: https://www.mdpi.com/article/10.3390/ijms24044069/s1.

Author Contributions: Y.C. (Yong Chen), W.Y. and J.X. performed statistical analyses. R.G. performed the salt stress experiments. Y.C. (Yaling Chen) and Y.Z. performed the field experiments. Y.C. (Yong Chen), W.Y., J.X. and F.Z. drafted the manuscript. F.Z. and J.X. contributed to the experimental design and edition of the manuscript. All authors have read and agreed to the published version of the manuscript.

Funding: This research was partially supported by the National Natural Science Foundation of China (31960370, 32070374), the "Biological Breeding" project of the State Key Laboratory of Crop Gene Exploration and Utilization in Southwest China (SKL-KF202217), the Natural Science Foundation of Jiangxi Province, China (20202ACB205002), the Foundation of Jiangxi Provincial Key Lab of Protection and Utilization of Subtropical Plant Resources (YRD201913), and the Postgraduate Innovation Fund of Jiangxi Normal University (YC2019-B043).

Institutional Review Board Statement: Not applicable.

Informed Consent Statement: Not applicable.

Data Availability Statement: The data presented in this study are available in the article or Supplementary Material.

Conflicts of Interest: The authors declare no conflict of interest.

References

1. Bandumula, N. Rice Production in Asia: Key to Global Food Security. *Proc. Natl. Acad. Sci. India Sect. B Biol. Sci.* **2018**, *88*, 1323–1328. [CrossRef]
2. Kim, Y.; Chung, Y.S.; Lee, E.; Tripathi, P.; Heo, S.; Kim, K.-H. Root Response to Drought Stress in Rice (*Oryza sativa* L.). *Int. J. Mol. Sci.* **2020**, *21*, 1513. [CrossRef] [PubMed]
3. Hoang, T.M.L.; Tran, T.N.; Nguyen, T.K.T.; Williams, B.; Wurm, P.; Bellairs, S.; Mundree, S. Improvement of Salinity Stress Tolerance in Rice: Challenges and Opportunities. *Agronomy* **2016**, *6*, 54. [CrossRef]
4. Todaka, D.; Nakashima, K.; Shinozaki, K.; Yamaguchi-Shinozaki, K. Toward Understanding Transcriptional Regulatory Networks in Abiotic Stress Responses and Tolerance in Rice. *Rice* **2012**, *5*, 6. [CrossRef] [PubMed]
5. Mondal, S.; Borromeo, T.H. Screening of Salinity Tolerance of Rice at Early Seedling Stage. *J. Biosci. Agric. Res.* **2016**, *10*, 843–847. [CrossRef]
6. Zayed, B.; El-Rafaee, I.; Sedeek, S. Response of Different Rice Varieties to Phosphorous Fertilizer under Newly Reclaimed Saline Soils. *J. Plant Prod.* **2010**, *1*, 1479–1493. [CrossRef]
7. Ha-Tran, D.M.; Nguyen, T.T.M.; Hung, S.-H.; Huang, E.; Huang, C.-C. Roles of Plant Growth-Promoting Rhizobacteria (PGPR) in Stimulating Salinity Stress Defense in Plants: A Review. *Int. J. Mol. Sci.* **2021**, *22*, 3154. [CrossRef] [PubMed]
8. Llave, C.; Kasschau, K.D.; Rector, M.A.; Carrington, J.C. Endogenous and Silencing-Associated Small RNAs in Plants. *Plant Cell* **2002**, *14*, 1605–1619. [CrossRef]
9. Reinhart, B.J.; Bartel, D.P. Small RNAs Correspond to Centromere Heterochromatic Repeats. *Science* **2002**, *297*, 1831. [CrossRef]
10. Jodder, J. MiRNA-Mediated Regulation of Auxin Signaling Pathway during Plant Development and Stress Responses. *J. Biosci.* **2020**, *45*, 91. [CrossRef]
11. Phillips, J.R.; Dalmay, T.; Bartels, D. The Role of Small RNAs in Abiotic Stress. *FEBS Lett.* **2007**, *581*, 3592–3597. [CrossRef] [PubMed]
12. Gao, S.; Yang, L.; Zeng, H.Q.; Zhou, Z.S.; Yang, Z.M.; Li, H.; Sun, D.; Xie, F.; Zhang, B. A Cotton MiRNA is Involved in Regulation of Plant Response to Salt Stress. *Sci. Rep.* **2016**, *6*, 19736. [CrossRef] [PubMed]
13. Shi, G.; Fu, J.; Rong, L.; Zhang, P.; Guo, C.; Xiao, K. TaMIR1119, a MiRNA Family Member of Wheat (*Triticum aestivum*), is Essential in the Regulation of Plant Drought Tolerance. *J. Integr. Agric.* **2018**, *17*, 2369–2378. [CrossRef]
14. Thiebaut, F.; Rojas, C.A.; Almeida, K.L.; Grativol, C.; Domiciano, G.C.; Lamb, C.R.C.; De Almeida Engler, J.; Hemerly, A.S.; Ferreira, P.C.G. Regulation of MiR319 during Cold Stress in Sugarcane. *Plant Cell Environ.* **2012**, *35*, 502–512. [CrossRef] [PubMed]
15. Paul, S.; Datta, S.K.; Datta, K. MiRNA Regulation of Nutrient Homeostasis in Plants. *Front. Plant Sci.* **2015**, *6*, 232. [CrossRef]
16. Baldrich, P.; San Segundo, B. MicroRNAs in Rice Innate Immunity. *Rice* **2016**, *9*, 6. [CrossRef]
17. Liu, L.; Zhang, Y.; Yang, Z.; Yang, Q.; Zhang, Y.; Xu, P.; Li, J.; Islam, A.; Shah, L.; Zhan, X.; et al. Fine Mapping and Candidate Gene Analysis of *QHD1b*, a QTL That Promotes Flowering in Common Wild Rice (*Oryza rufipogon*) by up-Regulating *Ehd1*. *Crop J.* **2022**, *10*, 1083–1093. [CrossRef]
18. Xie, J.; Agrama, H.A.; Kong, D.; Zhuang, J.; Hu, B.; Wan, Y.; Yan, W. Genetic Diversity Associated with Conservation of Endangered Dongxiang Wild Rice (*Oryza rufipogon*). *Genet. Resour. Crop Evol.* **2010**, *57*, 597–609. [CrossRef]
19. Zhang, F.; Luo, X.; Zhou, Y.; Xie, J. Genome-Wide Identification of Conserved MicroRNA and Their Response to Drought Stress in Dongxiang Wild Rice (*Oryza rufipogon* Griff.). *Biotechnol. Lett.* **2016**, *38*, 711–721. [CrossRef]
20. Zhou, Y.; Yang, P.; Cui, F.; Zhang, F.; Luo, X.; Xie, J. Transcriptome Analysis of Salt Stress Responsiveness in the Seedlings of Dongxiang Wild Rice (*Oryza rufipogon* Griff.). *PLoS ONE* **2016**, *11*, e0146242. [CrossRef]

21. Mohammadi-Nejad, G.; Singh, R.K.; Arzani, A.; Rezaie, A.M.; Sabouri, H.; Gregorio, G.B. Evaluation of Salinity Tolerance in Rice Genotypes. *Int. J. Plant Prod.* **2010**, *4*, 199–208.

22. Yang, Y.; Ye, R.; Srisutham, M.; Nontasri, T.; Sritumboon, S.; Maki, M.; Yoshida, K.; Oki, K.; Homma, K. Rice Production in Farmer Fields in Soil Salinity Classified Areas in Khon Kaen, Northeast Thailand. *Sustainability* **2022**, *14*, 9873. [CrossRef]

23. Tang, Y.; Bao, X.; Zhi, Y.; Wu, Q.; Guo, Y.; Yin, X.; Zeng, L.; Li, J.; Zhang, J.; He, W.; et al. Overexpression of a MYB Family Gene, *OsMYB6*, Increases Drought and Salinity Stress Tolerance in Transgenic Rice. *Front. Plant Sci.* **2019**, *10*, 168. [CrossRef] [PubMed]

24. Zhang, A.; Liu, Y.; Wang, F.; Li, T.; Chen, Z.; Kong, D.; Bi, J.; Zhang, F.; Luo, X.; Wang, J.; et al. Enhanced Rice Salinity Tolerance via CRISPR/Cas9-Targeted Mutagenesis of the *OsRR22* Gene. *Mol. Breed.* **2019**, *39*, 47. [CrossRef] [PubMed]

25. Sakai, H.; Itoh, T. Massive Gene Losses in Asian Cultivated Rice Unveiled by Comparative Genome Analysis. *BMC Genom.* **2010**, *11*, 121. [CrossRef] [PubMed]

26. Caicedo, A.L.; Williamson, S.H.; Hernandez, R.D.; Boyko, A.; Fledel-Alon, A.; York, T.L.; Polato, N.R.; Olsen, K.M.; Nielsen, R.; McCouch, S.R.; et al. Genome-Wide Patterns of Nucleotide Polymorphism in Domesticated Rice. *PLoS Genet.* **2007**, *3*, e163. [CrossRef] [PubMed]

27. Jones-Rhoades, M.W.; Bartel, D.P. Computational Identification of Plant MicroRNAs and Their Targets, Including a Stress-Induced MiRNA. *Mol. Cell* **2004**, *14*, 787–799. [CrossRef] [PubMed]

28. Pagano, L.; Rossi, R.; Paesano, L.; Marmiroli, N.; Marmiroli, M. MiRNA Regulation and Stress Adaptation in Plants. *Environ. Exp. Bot.* **2021**, *184*, 104369. [CrossRef]

29. Stief, A.; Altmann, S.; Hoffmann, K.; Pant, B.D.; Scheible, W.-R.; Bäurle, I. *Arabidopsis* MiR156 Regulates Tolerance to Recurring Environmental Stress through *SPL* Transcription Factors. *Plant Cell* **2014**, *26*, 1792–1807. [CrossRef]

30. López-Galiano, M.J.; García-Robles, I.; González-Hernández, A.I.; Camañes, G.; Vicedo, B.; Real, M.D.; Rausell, C. Expression of MiR159 is Altered in Tomato Plants Undergoing Drought Stress. *Plants* **2019**, *8*, 201. [CrossRef]

31. Lin, J.-S.; Kuo, C.-C.; Yang, I.-C.; Tsai, W.-A.; Shen, Y.-H.; Lin, C.-C.; Liang, Y.-C.; Li, Y.-C.; Kuo, Y.-W.; King, Y.-C.; et al. MicroRNA160 Modulates Plant Development and Heat Shock Protein Gene Expression to Mediate Heat Tolerance in *Arabidopsis*. *Front. Plant Sci.* **2018**, *9*, 68. [CrossRef] [PubMed]

32. Shan, T.; Fu, R.; Xie, Y.; Chen, Q.; Wang, Y.; Li, Z.; Song, X.; Li, P.; Wang, B. Regulatory Mechanism of Maize (*Zea mays* L.) MiR164 in Salt Stress Response. *Russ. J. Genet.* **2020**, *56*, 835–842. [CrossRef]

33. Kitazumi, A.; Kawahara, Y.; Onda, T.S.; De Koeyer, D.; de los Reyes, B.G. Implications of MiR166 and MiR159 Induction to the Basal Response Mechanisms of an Andigena Potato (*Solanum tuberosum* Subsp. *Andigena*) to Salinity Stress, Predicted from Network Models in Arabidopsis. *Genome* **2015**, *58*, 13–24. [PubMed]

34. Kinoshita, N.; Wang, H.; Kasahara, H.; Liu, J.; MacPherson, C.; Machida, Y.; Kamiya, Y.; Hannah, M.A.; Chua, N.-H. *IAA-Ala Resistant3*, an Evolutionarily Conserved Target of MiR167, Mediates *Arabidopsis* Root Architecture Changes during High Osmotic Stress. *Plant Cell* **2012**, *24*, 3590–3602. [CrossRef] [PubMed]

35. Ni, Z.; Hu, Z.; Jiang, Q.; Zhang, H. *GmNFYA3*, a Target Gene of MiR169, is a Positive Regulator of Plant Tolerance to Drought Stress. *Plant Mol. Biol.* **2013**, *82*, 113–129. [CrossRef]

36. Hwang, E.-W.; Shin, S.-J.; Yu, B.-K.; Byun, M.-O.; Kwon, H.-B. MiR171 Family Members Are Involved in Drought Response in Solanum Tuberosum. *J. Plant Biol.* **2011**, *54*, 43–48. [CrossRef]

37. Cheng, X.; He, Q.; Tang, S.; Wang, H.; Zhang, X.; Lv, M.; Liu, H.; Gao, Q.; Zhou, Y.; Wang, Q.; et al. The MiR172/*IDS1* Signaling Module Confers Salt Tolerance through Maintaining ROS Homeostasis in Cereal Crops. *New Phytol.* **2021**, *230*, 1017–1033. [CrossRef]

38. Du, Q.; Wang, K.; Zou, C.; Xu, C.; Li, W.X. The *PILNCR1*-MiR399 Regulatory Module is Important for Low Phosphate Tolerance in Maize. *Plant Physiol.* **2018**, *177*, 1743–1753. [CrossRef]

39. Gao, S.; Guo, C.; Zhang, Y.; Zhang, F.; Du, X.; Gu, J.; Xiao, K. Wheat MicroRNA Member TaMIR444a is Nitrogen Deprivation-Responsive and Involves Plant Adaptation to the Nitrogen-Starvation Stress. *Plant Mol. Biol. Report.* **2016**, *34*, 931–946. [CrossRef]

40. Ferdous, J.; Whitford, R.; Nguyen, M.; Brien, C.; Langridge, P.; Tricker, P.J. Drought-Inducible Expression of *Hv*-MiR827 Enhances Drought Tolerance in Transgenic Barley. *Funct. Integr. Genom.* **2017**, *17*, 279–292. [CrossRef]

41. Zhang, F.; Luo, Y.; Zhang, M.; Zhou, Y.; Chen, H.; Hu, B.; Xie, J. Identification and Characterization of Drought Stress- Responsive Novel MicroRNAs in Dongxiang Wild Rice. *Rice Sci.* **2018**, *25*, 175–184.

42. Reboledo, G.; Agorio, A.; De León, I.P. Moss Transcription Factors Regulating Development and Defense Responses to Stress. *J. Exp. Bot.* **2022**, *73*, 4546–4561. [CrossRef] [PubMed]

43. Gong, Y.; Liu, X.; Chen, S.; Li, H.; Duanmu, H. Genome-Wide Identification and Salt Stress Response Analysis of the BZIP Transcription Factor Family in Sugar Beet. *Int. J. Mol. Sci.* **2022**, *23*, 11573. [CrossRef] [PubMed]

44. Chai, M.; Fan, R.; Huang, Y.; Jiang, X.; Wai, M.H.; Yang, Q.; Su, H.; Liu, K.; Ma, S.; Chen, Z.; et al. *GmbZIP152*, a Soybean BZIP Transcription Factor, Confers Multiple Biotic and Abiotic Stress Responses in Plant. *Int. J. Mol. Sci.* **2022**, *23*, 10935. [CrossRef] [PubMed]

45. Liu, D.; Shi, S.; Hao, Z.; Xiong, W.; Luo, M. OsbZIP81, A Homologue of Arabidopsis VIP1, May Positively Regulate JA Levels by Directly Targetting the Genes in JA Signaling and Metabolism Pathway in Rice. *Int. J. Mol. Sci.* **2019**, *20*, 2360. [CrossRef] [PubMed]

46. Raza, A.; Charagh, S.; Zahid, Z.; Mubarik, M.S.; Javed, R.; Siddiqui, M.H.; Hasanuzzaman, M. Jasmonic Acid: A Key Frontier in Conferring Abiotic Stress Tolerance in Plants. *Plant Cell Rep.* **2021**, *40*, 1513–1541. [CrossRef] [PubMed]

47. Sun, Y.-G.; Wang, B.; Jin, S.-H.; Qu, X.-X.; Li, Y.-J.; Hou, B.-K. Ectopic Expression of Arabidopsis Glycosyltransferase *UGT85A5* Enhances Salt Stress Tolerance in Tobacco. *PLoS ONE* **2013**, *8*, e59924. [CrossRef]
48. Duan, Z.; Yan, Q.; Wu, F.; Wang, Y.; Wang, S.; Zong, X.; Zhou, P.; Zhang, J. Genome-Wide Analysis of the UDP-Glycosyltransferase Family Reveals Its Roles in Coumarin Biosynthesis and Abiotic Stress in *Melilotus Albus*. *Int. J. Mol. Sci.* **2021**, *22*, 10826. [CrossRef]
49. Jin, Y.; Pan, W.; Zheng, X.; Cheng, X.; Liu, M.; Ma, H.; Ge, X. *OsERF101*, an ERF Family Transcription Factor, Regulates Drought Stress Response in Reproductive Tissues. *Plant Mol. Biol.* **2018**, *98*, 51–65. [CrossRef]
50. Wang, C.; Huang, W.; Ying, Y.; Li, S.; Secco, D.; Tyerman, S.; Whelan, J.; Shou, H. Functional Characterization of the Rice *SPX-MFS* Family Reveals a Key Role of *OsSPX-MFS1* in Controlling Phosphate Homeostasis in Leaves. *New Phytol.* **2012**, *196*, 139–148. [CrossRef]
51. Itoh, H.; Tatsumi, T.; Sakamoto, T.; Otomo, K.; Toyomasu, T.; Kitano, H.; Ashikari, M.; Ichihara, S.; Matsuoka, M. A Rice Semi-Dwarf Gene, *Tan-Ginbozu* (*D35*), Encodes the Gibberellin Biosynthesis Enzyme, *Ent*-Kaurene Oxidase. *Plant Mol. Biol.* **2004**, *54*, 533–547. [CrossRef] [PubMed]
52. Chowdhary, V.A.; Tank, J.G. Biomolecules Regulating Defense Mechanism in Plants. *Proc. Natl. Acad. Sci. India Sect. B Biol. Sci.* **2022**, 1–9. [CrossRef]
53. Matsui, H.; Miyao, A.; Takahashi, A.; Hirochika, H. Pdk1 Kinase Regulates Basal Disease Resistance Through the OsOxi1–OsPti1a Phosphorylation Cascade in Rice. *Plant Cell Physiol.* **2010**, *51*, 2082–2091. [CrossRef] [PubMed]
54. RoyChoudhury, A.; Roy, C.; Sengupta, D.N. Transgenic Tobacco Plants Overexpressing the Heterologous Lea Gene *Rab16A* from Rice during High Salt and Water Deficit Display Enhanced Tolerance to Salinity Stress. *Plant Cell Rep.* **2007**, *26*, 1839–1859. [CrossRef]
55. Ganguly, M.; Datta, K.; Roychoudhury, A.; Gayen, D.; Sengupta, D.N.; Datta, S.K. Overexpression of *Rab16A* Gene in Indica Rice Variety for Generating Enhanced Salt Tolerance. *Plant Signal. Behav.* **2012**, *7*, 502–509. [CrossRef]
56. Zhou, Y.; Wang, B.; Yuan, F. The Role of Transmembrane Proteins in Plant Growth, Development, and Stress Responses. *Int. J. Mol. Sci.* **2022**, *23*, 13627. [CrossRef]
57. Waadt, R.; Seller, C.A.; Hsu, P.-K.; Takahashi, Y.; Munemasa, S.; Schroeder, J.I. Plant Hormone Regulation of Abiotic Stress Responses. *Nat. Rev. Mol. Cell Biol.* **2022**, *23*, 680–694. [CrossRef]
58. Kang, T.; Yu, C.-Y.; Liu, Y.; Song, W.-M.; Bao, Y.; Guo, X.-T.; Li, B.; Zhang, H.-X. Subtly Manipulated Expression of ZmmiR156 in Tobacco Improves Drought and Salt Tolerance Without Changing the Architecture of Transgenic Plants. *Front. Plant Sci.* **2020**, *10*, 1664. [CrossRef]
59. Shi, X.; Jiang, F.; Wen, J.; Wu, Z. Overexpression of *Solanum habrochaites* MicroRNA319d (Sha-MiR319d) Confers Chilling and Heat Stress Tolerance in Tomato (*S. lycopersicum*). *BMC Plant Biol.* **2019**, *19*, 214. [CrossRef]
60. Ma, C.; Burd, S.; Lers, A. *MiR408* is Involved in Abiotic Stress Responses in Arabidopsis. *Plant J.* **2015**, *84*, 169–187. [CrossRef]
61. Zhao, Y.; Zhang, C.; Liu, W.; Gao, W.; Liu, C.; Song, G.; Li, W.-X.; Mao, L.; Chen, B.; Xu, Y.; et al. An Alternative Strategy for Targeted Gene Replacement in Plants Using a Dual-SgRNA/Cas9 Design. *Sci. Rep.* **2016**, *6*, 23890. [CrossRef] [PubMed]
62. Kozomara, A.; Birgaoanu, M.; Griffiths-Jones, S. MiRBase: From MicroRNA Sequences to Function. *Nucleic Acids Res.* **2019**, *47*, D155–D162. [CrossRef]
63. Meyers, B.C.; Axtell, M.J.; Bartel, B.; Bartel, D.P.; Baulcombe, D.; Bowman, J.L.; Cao, X.; Carrington, J.C.; Chen, X.; Green, P.J.; et al. Criteria for Annotation of Plant MicroRNAs. *Plant Cell* **2008**, *20*, 3186–3190. [CrossRef] [PubMed]
64. Wu, Y.; Xu, J.; Han, X.; Qiao, G.; Yang, K.; Wen, Z.; Wen, X. Comparative Transcriptome Analysis Combining SMRT- and Illumina-Based RNA-Seq Identifies Potential Candidate Genes Involved in Betalain Biosynthesis in Pitaya Fruit. *Int. J. Mol. Sci.* **2020**, *21*, 3288. [CrossRef] [PubMed]
65. Tafer, H.; Hofacker, I.L. RNAplex: A Fast Tool for RNA–RNA Interaction Search. *Bioinformatics* **2008**, *24*, 2657–2663. [CrossRef] [PubMed]

International Journal of
Molecular Sciences

Article

Multi-Omics Analysis of the Effects of Soil Amendment on Rapeseed (*Brassica napus* L.) Photosynthesis under Drip Irrigation with Brackish Water

Ziwei Li [1], Hua Fan [1], Le Yang [1], Shuai Wang [1], Dashuang Hong [1], Wenli Cui [1], Tong Wang [1], Chunying Wei [1], Yan Sun [1], Kaiyong Wang [1,*] and Yantao Liu [2]

[1] Agricultural College, Shihezi University, Shihezi 832000, China; liziwei09@stu.shzu.edu.cn (Z.L.); fanhua@shzu.edu.cn (H.F.); yl_agr@shzu.edu.cn (L.Y.); wangshuai@stu.shzu.edu.cn (S.W.); hongdashaugn@stu.shzu.edu.cn (D.H.); cuiwenli@xjshzu.com (W.C.); wt@stu.shzu.edu.cn (T.W.); 20201012057@stu.shzu.edu.cn (C.W.); sunvan72@xishzu.com (Y.S.)

[2] Institute of Crop Research, Xinjiang Academy of Agricultural Reclamation Sciences, Shihezi 832000, China; ziheng1979@126.com

* Correspondence: wangkaiyong@shzu.edu.cn

Abstract: Drip irrigation with brackish water increases the risk of soil salinization while alleviating water shortage in arid areas. In order to alleviate soil salinity stress on crops, polymer soil amendments are increasingly used. But the regulation mechanism of a polymer soil amendment composed of polyacrylamide polyvinyl alcohol, and manganese sulfate (PPM) on rapeseed photosynthesis under drip irrigation with different types of brackish water is still unclear. In this field study, PPM was applied to study the responses of the rapeseed (*Brassica napus* L.) phenotype, photosynthetic physiology, transcriptomics, and metabolomics at the peak flowering stage under drip irrigation with water containing 6 g·L^{-1} NaCl (S) and Na$_2$CO$_3$ (A). The results showed that the inhibitory effect of the A treatment on rapeseed photosynthesis was greater than that of the S treatment, which was reflected in the higher Na$^+$ content (73.30%) and lower photosynthetic-fluorescence parameters (6.30–61.54%) and antioxidant enzyme activity (53.13–77.10%) of the A-treated plants. The application of PPM increased the biomass (63.03–75.91%), photosynthetic parameters (10.55–34.06%), chlorophyll fluorescence parameters (33.83–62.52%), leaf pigment content (10.30–187.73%), and antioxidant enzyme activity (28.37–198.57%) under S and A treatments. However, the difference is that under the S treatment, PPM regulated the sulfur metabolism, carbon fixation and carbon metabolism pathways in rapeseed leaves. And it also regulated the photosynthesis-, oxidative phosphorylation-, and TCA cycle-related metabolic pathways in rapeseed leaves under A treatment. This study will provide new insights for the application of polymer materials to tackle the salinity stress on crops caused by drip irrigation with brackish water, and solve the difficulty in brackish water utilization.

Keywords: brackish water; metabolic pathways; energy metabolism; abiotic stresses; multi-omics

Citation: Li, Z.; Fan, H.; Yang, L.; Wang, S.; Hong, D.; Cui, W.; Wang, T.; Wei, C.; Sun, Y.; Wang, K.; et al. Multi-Omics Analysis of the Effects of Soil Amendment on Rapeseed (*Brassica napus* L.) Photosynthesis under Drip Irrigation with Brackish Water. *Int. J. Mol. Sci.* **2024**, *25*, 2521. https://doi.org/10.3390/ijms25052521

Academic Editor: Martin Bartas

Received: 5 January 2024
Revised: 8 February 2024
Accepted: 11 February 2024
Published: 21 February 2024

1. Introduction

Freshwater is scarce in arid regions, so drip irrigation with brackish water is an important measure to alleviate the contradiction between agricultural production and water scarcity [1]. At present, countries around the world mainly use brackish and fresh water mixed irrigation, brackish and fresh water rotation irrigation, and filtration technology to increase the use of brackish water [2]. However, these methods still do not meet freshwater needs due to freshwater shortages, great domestic and industrial freshwater needs, and high costs in most arid regions, and they also lead to the surface accumulation of a large amount of salt, which increases the risk of salinization. Particularly, irrigation with brackish water could significantly alter many physiological processes of plants, especially photosynthesis, resulting in retarded plant growth and great yield reduction [3,4].

The possible ways of salt stress affecting plant photosynthesis mainly include: ion poisoning, osmotic stress, and sugar accumulation-induced feedback inhibition [5]. Previous studies have shown that high soil salinity could destroy the stomatal structure and chloroplast structural integrity of plants [6], reduce net photosynthesis, the photosynthetic pigment content, the F_v/F_m (maximum photochemical efficiency of photosystem II) ratio, stomatal conductance, the transformation rate, gas exchange, the PSII (photosystem II) efficiency, and water potential, and cause osmotic stress and nutrient deficiency in plants [7,8]. Salt stress can also inhibit the carbon assimilation of plants, resulting in a reduced accumulation of photosynthates [9]. In addition, salt stress also impacts the activities of key enzymes related to photosynthesis, such as RuBPCase (ribulose-1, 5-bisphophatecarboxylase), leading to reduced photosynthetic carbon assimilation and photosynthetic rate [10].

At present, there are many methods used to improve the tolerance of crops to salt stress. Biochar, a commonly used solid soil amendment, can reduce the negative effects of drought and salt stress on plants, as well as improve the soil's physical structure, water holding capacity, and fertility and plant photosynthesis [11]. However, biochar is insoluble in water and cannot be used through a drip irrigation system that is widely applied by farmers in arid areas [12]. The artificial and mechanical application of biochar are time-consuming and costly. Other soil amendments widely studied by scholars such as zeolite, bentonite and gypsum are also insoluble in water [13]. Bioactive compounds and organic amendments such as plant growth-promoting bacteria and biofertilizers can enhance crop salt tolerance and yield by maintaining ionic homeostasis and reducing oxidative damage, but such amendments are expensive and have a slow effect and low utilization rate. Especially, a high pH environment always causes inactivation of most bacteria [14]. Therefore, bioactive compounds and organic amendments are also not suitable for large-scale application in arid areas. Therefore, it is urgent to find a soil amendment suitable for a drip irrigation system.

PPM is an independently developed soil amendment with inorganic and organic polymers as the main components. It has high-water solubility, and can be applied through drip irrigation [15]. Studies have shown that polyacrylamide has no significant negative effects on aquatic ecosystems and soil organisms when used at recommended doses $(10–20 \text{ kg·hm}^{-2})$ [16,17]. The biodegradability of PVA depends on its molecular weight and other structural properties [18]. In terms of ecological safety, polyvinyl alcohol is safer, but with increasing decomposition time, there is a need to be concerned about its long-term accumulation in soil and water bodies. The proper use of manganese sulfate avoids manganese deficiency and thus enhances plant growth and resistance, and generally the safe concentration of manganese sulphate solution is 0.1–1% [19,20]. Our previous study found that PPM had a variety of functional groups, such as carboxyl, alcohol hydroxyl, and silicone, whose main effect is to improve the soil structure and buffer saline ions in the soil [15,21]. The application of PPM could improve the resistance of cotton, wheat, and other crops to salt stress [15]. *B. napus* is a salt-tolerant plant. In contrast to salt-tolerant crops such as cotton and wheat, which resist salt damage by "salt avoidance" (avoiding uptake of salt), variety Huayouza 82 is able to take up a large amount of sodium ions and store them in the stalks and petioles [13,22]. It was found that variety Huayouza 82 was able to grow in a salt concentration of 0.8–1.2%; variety Huayouza showed the best salinity tolerance under salt stress at 0.5–1.5% NaCl concentration compared to varieties such as Zheshuang 72, Zheyou 50, Zheda 622, and Feeding Oil No.2 [23]. The cropping of *B. napus* combined with drip irrigation with brackish water can improve the utilization of brackish water, alleviate the freshwater crisis, and ensure agriculture production in arid areas. However, previous study showed that when soil salinity reached 5 g·L^{-1}, the growth of *B. napus* was significantly inhibited [24].

Therefore, in this study, the molecular regulatory mechanism of PPM on the photosynthesis of B. napus under drip irrigation with neutral (NaCl) and alkaline (Na_2CO_3) water was determined by the integrated transcriptomic and metabolomic analysis. The objectives were to reveal (1) the differences in the effects of drip irrigation with neutral and alkaline water on the physiological and photosynthetic parameters of *B. napus*, and (2) the differences in the mechanisms of PPM improving the photosynthesis of *B. napus* under drip irrigation with neutral and alkaline water. This study will have guiding significance for brackish water drip irrigation and crop yield increase in arid areas.

2. Results

2.1. Effects of Application of PPM on Biomass and Leaf Physiological Parameters of B. napus

Comparing the ACK (alkaline brackish water irrigation control treatment) group with the SCK (neutral brackish water irrigation control treatment) group, the yield, root, stem, and leaf fresh weight decreased by 13.24%, 17.88%, 9.33%, and 6.01%, respectively ($p < 0.05$). A decrease of 17.96% was observed in the leaf K^+ content, while there was a rise of 51.32% in the K^+/Na^+ ratio, and a decrease of 40.52% in the REC (relative electrical conductivity) in the ACK group ($p < 0.05$), the Na^+ content increased by 73.30% ($p < 0.05$), compared with those in the SCK group. The application of PPM obviously increased the fresh weight, leaf K^+/Na^+ ratio, K^+ content, and REC content of *B. napus* under brackish water irrigation. The yield, root, stem, and leaf fresh weight in the SPPM (neutral brackish water irrigation and PPM treatment) group increased by 30.19%, 64.38%, 61.60% and 78.98%, respectively ($p < 0.05$) compared with those in the SCK group. The leaf K^+/Na^+ ratio and K^+ content in the SPPM group increased by 29.54% and 12.22%, respectively ($p < 0.05$), while the leaf Na^+ content and REC decreased by 12.22% and 64.88%, respectively ($p < 0.05$), compared with those in the SCK group. The yield, root, stem, and leaf fresh weight in the APPM (alkaline brackish water irrigation and PPM treatment) group increased by 38.83%, 71.04%, 68.58%, and 73.86%, respectively ($p < 0.05$) compared with those in the ACK group, but it decreased by 22.68%, 13.67%, and 5.91%, respectively ($p < 0.05$) compared with those in the SPPM group. The leaf K^+/Na^+ ratio and K^+ content in the APPM group increased by 36.14% and 26.55%, respectively ($p < 0.05$), while the leaf Na^+ content and REC decreased by 7.05% and 22.23%, respectively ($p < 0.05$), compared with those in the ACK group (Figure 1).

2.2. Effects of PPM Application on Photosynthetic Parameters of B. napus Leaves

The Pn (leaf net photosynthetic rate), Gs (stomatal conductance), and Tr (transpiration rate) of *B. napus* leaves in the SPPM group increased by 10.55%, 17.01%, and 34.06%, respectively ($p < 0.05$), but the C_i (intracellular CO_2 concentration) decreased by 3.36% ($p > 0.05$), compared with those in the SCK group. The Pn, Gs, and Tr in the APPM group increased by 17.01%, 23.74%, and 42.34%, respectively ($p < 0.05$), but the C_i decreased by 18.83% ($p < 0.05$), compared with those in the ACK group. In addition, there are no differences in Pn, Tr, and Gs between the S (SCK vs. SPPM treatment) groups and A (ACK vs. APPM treatment) groups ($p > 0.05$) (Figure 2a). The leaf Chl a (chlorophyll a) in the ACK group increased by 10.00% ($p < 0.05$), but the Chl b (chlorophyll b) and Car (carotenoids) reduced by 14.49% and 16.74%, respectively, compared with those in the SCK group. The application of PPM enhanced photosynthetic pigment contents of *B. napus* leaves under drip irrigation using brackish water. The leaf Chl a, Chl b, Chl a + Chl b, and Car in the SPPM group increased by 56.34%, 40.05%, 51.68%, and 33.83%, respectively ($p < 0.05$) compared with those in the SCK group, and those in the APPM group increased by 35.00%, 43.25%, 36.97%, and 62.52%, respectively ($p < 0.05$) compared with those in the ACK group. In addition, the Chl a in the SPPM group decreased by 6.08% ($p < 0.05$) compared with that in the APPM group (Figure 2a). As compared with the SCK group, the F_v/F_0 (potential activity of photosystem II photochemistry), F_v/F_m (maximum photochemical efficiency of photosystem II), and qP (photochemical quenching coefficient) of *B. napus* leaves in the SPPM group increased by 49.91%, 10.30%, and 14.29%, respectively ($p < 0.05$). In comparison to the ACK group, the F_v/F_0, F_v/F_m, and qP of *B. napus* leaves in the APPM group increased by 183.73%, 48.07%, and 57.62%, respectively ($p < 0.05$). The PSII (PSII photochemistry) did not differ among the four groups (Figure 2a).

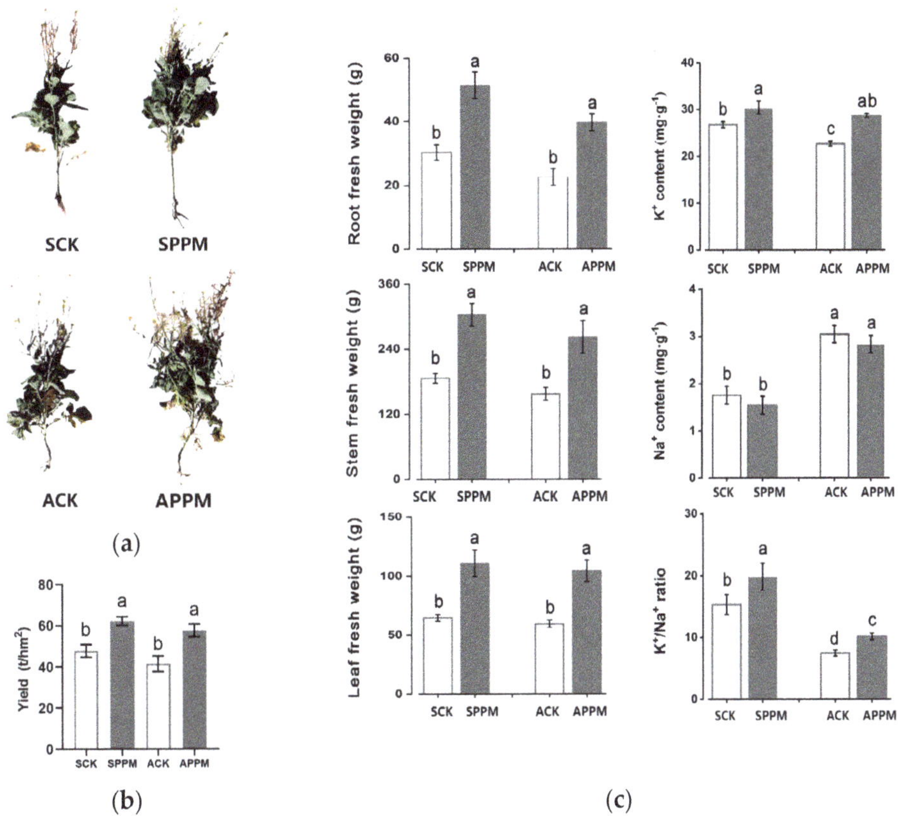

Figure 1. Effects of soil amendment (PPM) on the morphology (**a**), and yield of *Brassica napus* under different treatments (**b**). Fresh weight, and leaf K^+ and Na^+ content of *Brassica napus* under drip irrigation with water containing $6\ g \cdot L^{-1}$ NaCl and $6\ g \cdot L^{-1}$ Na_2CO_3 (**c**). Different lowercase letters indicate significant difference between groups at $p < 0.05$.

2.3. Effects of PPM Application on Antioxidant Enzyme Activity

A comparison between the ACK and SCK treatments revealed that SOD (superoxide dismutase), POD (peroxidase), and CAT (catalase) activity increased by 53.13%, 54.25%, and 77.10% ($p < 0.05$). The activity of SOD, POD, and CAT in the SPPM group increased by 29.21%, 140.17%, and 198.57%, respectively compared with those in the SCK group ($p < 0.05$), and the activity of SOD, POD, and CAT in the APPM group increased by 28.37%, 70.33%, and 77.87%, respectively compared with those in the ACK group ($p < 0.05$). The content of MDA (malondialdehyde) in the SPPM group decreased by 15.42% compared with that in the APPM group ($p < 0.05$) (Figure 2a). RDA (redundancy analysis) analysis (Figure 2b) showed that the leaf photosynthetic parameters (Tr, Ci, qP, and ΦPSII), antioxidant enzyme activities, and photosynthetic pigment content were significantly positively correlated with stem and leaf fresh weight, and they were negatively correlated with leaf REC, K^+ and Na^+ content. Meanwhile, F_v/F_m, F_v/F_0, and qP were positively correlated with root fresh weight, and negatively correlated with K^+ and Na^+ content. However, the MDA content and Gs were negatively correlated with the fresh weight of plant organs and positively correlated with the K^+ content and REC.

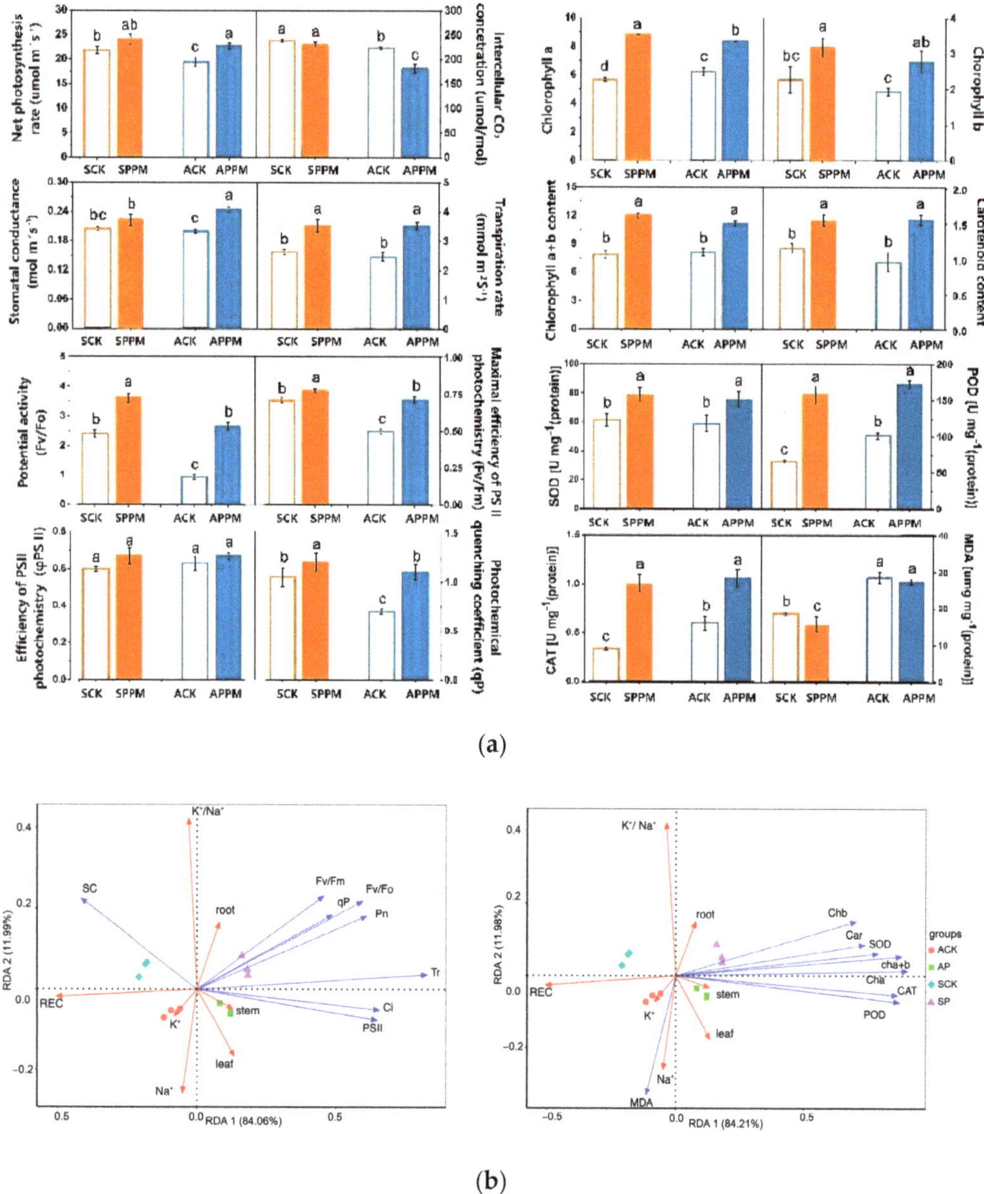

(a)

(b)

Figure 2. Effects of application of PPM on photosynthetic parameters, pigment content, fluorescence parameters, and antioxidant enzyme activities (**a**) of *Brassica napus* under drip irrigation with waters containing 6 g·L^{-1} NaCl and 6 g·L^{-1} Na$_2$CO$_3$ and RDA analysis of the parameters. Different lowercase letters indicate significant difference between groups at $p < 0.05$ (**b**).

2.4. Transcriptomic and Metabonomic Analysis

2.4.1. Determination of Photosynthetic Parameters, Chlorophyll Fluorescence Parameters and Plant Fresh Weight

Principal PCA (component analysis) was performed to assess the correlation of the transcriptomes and metabolomes. The results showed that samples of the SCK and ACK group were separated on PC2 (Principal Component 2), and those of the control (SCK and ACK) and PPM (SMMP and AMMP) groups were separated on PC1 (Principal Component 1) based on the metabolite count (Figure 3a). Samples of the S (SCK and SPPM) and A(ACK and APPM) groups were separated on PC2, and those of the control and PPM groups were separated on PC1 based on the gene counts (Figure 3a). Moreover, there were 83 shared DAMs (differentially accumulated metabolites) and 7 shared DEGs (differentially expressed genes) for the four groups (Figure 3a). The number of DEGs and DAMs in the SPPM and APPM groups were more than those in the SCK and ACK groups. In addition, the up-regulated DEGs were more than the down-regulated ones, while the down-regulated DAMs were more than the up-regulated ones (Figure 3a).

Many DEGs and DAMs for the four groups were enriched in pathways related to energy metabolism (including sulfur metabolism and oxidative phosphorylation) and carbohydrate metabolism. DEGs were also enriched in Nitrogen metabolism, carbon fixation in photosynthetic organisms, Photosynthesis-antenna proteins, and Photosynthesis pathways, and DAMs were also enriched in the amino sugar and nucleotide sugar metabolism, tricarboxylic acid (TCA) cycle, carbon metabolism, and carbon fixation pathways in photosynthetic organisms. The DEGs for the SCK vs. ACK were significantly enriched in Nitrogen metabolism and Photosynthesis -antenna proteins pathways, and those for the SCK vs. SPPM were significantly enriched in the sulfur metabolism pathway. The DEGs for the ACK vs. APPM were significantly enriched in photosynthesis-antenna proteins and photosynthesis pathways, while those for the SPPM vs. APPM were significantly enriched in the citrate cycle pathway (Figure 3b).

2.4.2. PPM's Impacts on Energy Metabolism and Carbohydrate Metabolism Pathways in *B. napus*

DEGs and DAMs with high photosynthesis-related content were enriched in pathways such as Energy metabolism, Carbohydrate metabolism, Photosynthesis-antenna proteins, and Photosynthesis (Figure 4a). Many genes (*BanA09G0675700ZS*, *BanC04G0172900ZS*, and *BanA09G0666300ZS*) were highly expressed in the four groups, but the expression of genes such as *BanA09G0675700ZS*, *BanC04G0172900ZS*, and *BanA09G0666300ZS* were at a low level. Compared with the CK (ACK and SCK), the expression of *BanC02G0540200ZS* and *BanC02G0555600ZS* in the PPM groups (APPM and SPPM) were up-regulated, while those of *BanA03G0578600ZS*, *BanC01G0424400ZS*, and *BanC07G0195800ZS* were down-regulated. Different treatments had different mechanisms in regulating the carbon metabolism pathway (Figure 4a). The expression of *ALDO* (aldolase A) and *FBP* were significantly down-regulated in the Calvin cycle, while that of *RPIA* (ribose 5-phosphate isomerase A) was significantly up-regulated. the expression of *ACLY* (ATP citrate lyase), *FH* (fumarate hydratase), and *SDHB* (Succinate dehydrogenase complex iron sulfur subunit B) were significantly down-regulated in the TCA cycle. Compared with the CK, the relative abundance of metabolites 2-oxoglutric acid and citric acid were significantly down-regulated, while that of fumarate and L-malic acid were significantly up-regulated. The abundance of metabolites related to amino acid metabolism and carbohydrate metabolism was down-regulated.

RDA analysis results (Figure 4b) showed that the expression of photosynthesis-related transcription factors *psb.O* (photosystem II subunit O), *psb.W* (photosystem II subunit W), *psa.O*, and *psa.N* (photosystem I subunit N) and the abundance of fumaric acid, L-malic acid, and 5′-deoxy-5′-(methylthio)adenosine were positively correlated with the fresh weight of plant organs and negatively correlated with REC, K^+ content, and Na^+ content. Meanwhile, the expression of *psb.D* (photosystem II subunit D), *psb.W*, *psa.H* (photosystem I subunit H), and *psa.N* and the abundance of fumaric acid, L-malic acid, and 5′-deoxy-5′-(methylthio) adenosine were negatively correlated with the fresh weight of plant organs and positively correlated with REC, K^+ content, and Na^+ content.

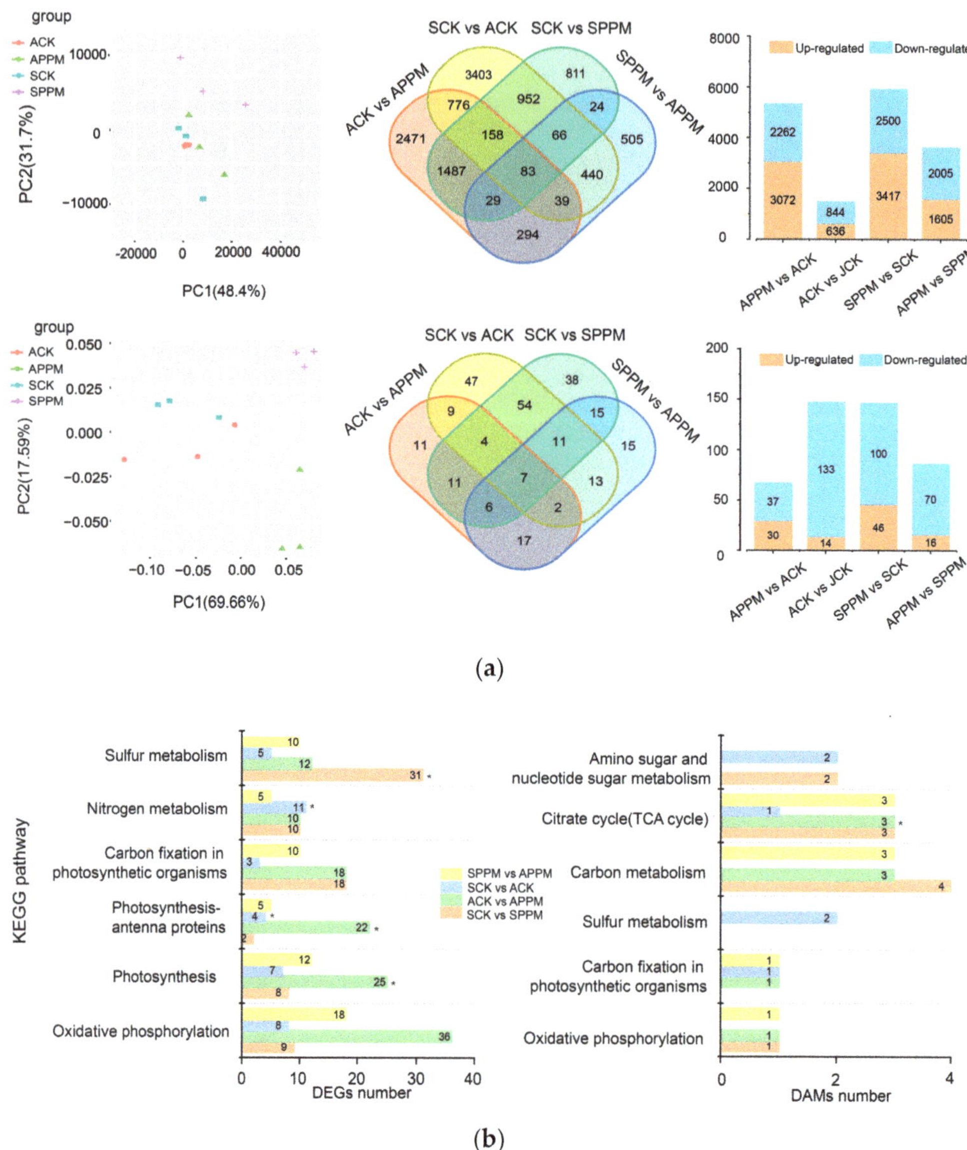

Figure 3. PCA analysis, Venn diagram, up- and down- regulation statistics (**a**), and KEGG annotation of DEGs (differentially expressed genes) and DAMs (differentially accumulated metabolites) in different groups. *, $p < 0.05$ (**b**).

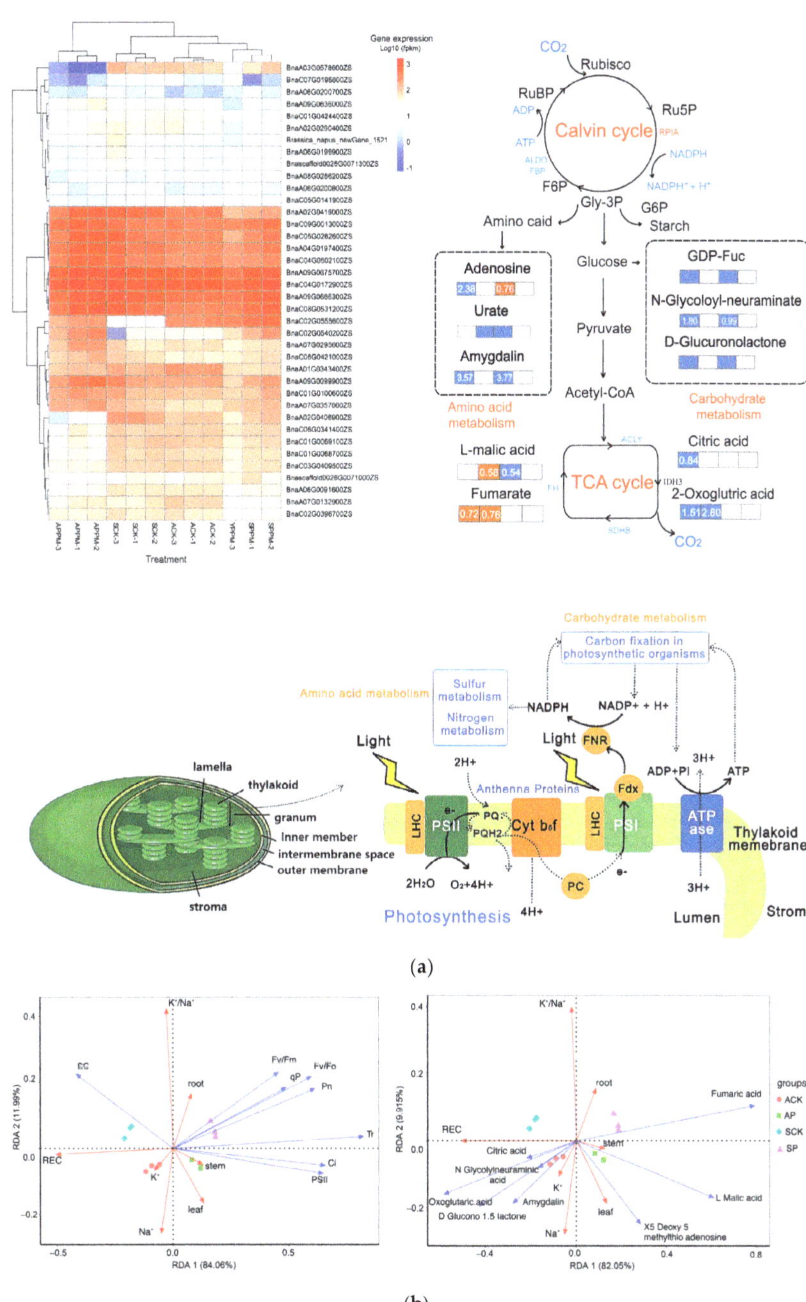

Figure 4. The expression of differentially expressed genes (DEGs) related to photosynthesis in different groups, metabolic pathway analysis, photosynthetic mechanism and energy metabolic pathway analysis (**a**), and RDA analysis of various parameters between groups (**b**). The four boxes in the metabolic pathway represent the ratios of SPPM vs. SCK, APPM vs. ACK, ACK vs. SCK and APPM vs. SPPM metabolite contents, respectively.

3. Discussion

Irrigation using brackish water with high salt concentration produces ionic toxicity on *B. napus* and inhibits its photosynthesis and cell growth [8]. This study found that the photosynthetic characteristics and antioxidant enzymes activities of *B. napus* showed differences when subjected to different types of brackish water drip irrigation. The leaf photosynthetic parameters (Pn, Tr, and Gs), fluorescence parameters (F_v/F_0, F_v/F_m and qP), and Chl b under alkaline water drip irrigation condition were lower than those under neutral water drip irrigation condition. This indicates that alkaline salts have greater inhibition on the leaf photosynthesis of *B. napus*, especially on the net photosynthetic rate and fluorescence parameters. This ultimately inhibits the dry matter accumulation in *B. napus* [25,26].

Long-term brackish water drip irrigation could also lead to massive accumulation of Na^+ and other salt ions in plants, disrupting osmotic balance, and competitively inhibiting the uptake of other ions [27]. Neutral brackish water irrigation is dominated by Na^+ and Cl^- stress, while alkaline brackish water irrigation also includes CO_3^{2-}, and high pH stress [28,29]. This study found that more Na^+ and less K^+ were accumulated in leaves under alkaline water drip irrigation compared to neutral water drip irrigation. A high Na^+ content could cause high osmotic pressure in *B. napus*, resulting in the closing of leaf stomata and inhibited photosynthesis, metabolism, and biomass accumulation. Besides, excessive Na^+ can also cause a deficiency of phosphorus and nitrogen, restricting the synthesis of chloroplasts [30,31]. The less K^+ content under alkaline water drip irrigation may hinder starch synthesis in *B. napus* and accelerate the decomposition of starch into sugar. However, Carbohydrates in leaves cannot be transported smoothly, which causes a massive accumulation of photosynthesis-synthesized sugars in cells. This ultimately causes feedback inhibition and a reduced accumulation of dry matter by photosynthesis. This study also found that alkaline water irrigation had a greater effect on the chlorophyll content of *B. napus* than neutral water irrigation. Chloroplasts are the main sites of plant photosynthesis [32,33]. So, in this study, alkaline water irrigation led to reduced leaf photosynthates and plant biomass.

In this study, the application of PPM regulated *B. napus* growth and photosynthesis during the full flowering stage under brackish water drip irrigation conditions. Photosynthesis and antioxidant defense play key roles in the response of *B. napus* to salt stress [34]. The present study showed that PPM application reduced the ionic toxicity and oxidative stress to *B. napus* under brackish water drip irrigation conditions. This may be because the acidic functional groups such as COOH-, C=O, -SH, and -CHO in PPM could combine with the salt ions brought by brackish water drip irrigation, and improve the structure of soil aggregates and soil structure [35]. This could reduce the uptake of salt ions by *B. napus*, alleviate ionic toxicity and oxidative stress, and ensure normal photosynthesis and biomass accumulation. PPM can also regulate the osmotic potential in *B. napus* leaves by regulating the accumulation of compatible solutes such as soluble sugars, organic acids, amino acids, choline, and betaine, which could protect the photosynthase system [36], thus improving the photosynthetic rate and antioxidant defense system of *B. napus* [37]. In addition, our study found an increase in the MDA content of leaves after PPM application, which may act as one of the adaptive responses of the plant to environmental changes, helping the plant survive under adversity and adapt to changing environmental conditions [38]. Therefore, PPM can maintain the intracellular stability and normal physiological activities of *B. napus* leaves under salt stress, and protect photosynthesis and other physiological processes from serious impact.

B. napus can adapt to high salinity environments by molecular and metabolic regulations. The differences in *B. napus*'s self-regulation under different types of brackish water drip irrigation were reflected in pathways related to nitrogen metabolism and photosynthesis. This study (Figure 5a,b) found that under neutral water irrigation, transcription factors *psaN* and *atpF*(ATP synthase F subunit) and metabolites malic acid, glycolyneural, and fumaric were the main responders in the photosynthesis-related pathways and played positive regulatory roles. A previous study [23] has shown that salt stress could inhibit the transcription and translation of psbA (photosystem II reaction center protein D1) encoding D1 protein and the expression of light-induced genes, so that these genes could not be transcribed, translated, and repaired to form an active PSII reaction center. However, the expression of *psbW* (photosystem II reaction center protein W) and *psbN* (photosystem II reaction center protein N) were negatively correlated with the root fresh weight, leaf fresh weight, and the abundance of oxoglutaric acid, and the leaf and root fresh weight and photosynthetic parameters were negatively regulated by Chla, F_v/F_0, and oxoglutaric abundance. This indicates that *psbW* and *psbN* could regulate plant growth and photosynthesis under neutral water irrigation conditions. In addition, in the ACK group, transcription factors *psbW* and *atpF* and the metabolites lactone, malic, and fumaric acid played a positive regulatory role on leaf and root fresh weight. In addition, the PSII and Chl a + Chl b played a positive regulatory role on the internal homeostasis of *B. napus* leaves. This indicates that these transcription factors and metabolites directly regulated the growth of *B. napus* under alkaline water irrigation condition. Therefore, under different brackish water irrigation conditions, rapeseed can change the accumulation of metabolites in photosynthesis-related metabolic pathways by regulating the expression of photosynthesis-related genes and metabolites in rapeseed leaves, thereby maintaining the normal growth of plants in adverse environments. It should be noted that compared to neutral water irrigation, alkaline water irrigation had a greater impact on plant cell homeostasis.

It was found that soil amendment could improve plant photosynthetic performance and biomass accumulation by changing the expression of genes, transcription factors and metabolites related to photosynthesis-related pathways under salt stress conditions (Figure 4). However, the application of PPM could regulate transcription factors and metabolites in pathways related to energy metabolism and carbohydrate metabolism in *B. napus* leaves to regulate photosynthates. Under neutral water irrigation, the application of PPM mainly regulated sulfur metabolism, carbon fixation pathways in photosynthetic organisms, and carbohydrate metabolism pathways. The regulatory networks (Figure 5c,d) showed that the application of PPM changed the internal homeostasis of *B. napus* leaves under brackish water drip irrigation, and *atpF*, *psbO*, *psaN*, *psaH*, malic, adenosine, and lactone were the main regulators. It has been reported that *PsbO* (photosystem II oxygen-evolving enhancer protein), *PsbP* (photosystem II oxygen-evolving enhancer protein P), and *PsbQ* (photosystem II oxygen-evolving enhancer protein Q) play a crucial role in maintaining the active site of PSII [39]. This indicates that in this study, PPM may improve the photosynthetic efficiency by maintaining the active site in PSII, thereby increasing the biomass. It was also found that under neutral water irrigation conditions, *atpF* and *psaH* mainly regulated the dry matter. Among them, *atpF* positively regulated the fresh weight and K^+/Na^+ ratio, while *psaH* negatively regulated them (Figure 5). *PsbN* and *psbO* mainly regulated the Pn, Tr, POD activity, SOD activity, and MDA content, while *psbN* had an opposite effect on these parameters. Under alkaline water irrigation, the photosynthesis, oxidative phosphorylation, and TCA cycle metabolic pathways were mainly regulated. The regulatory mechanisms of *atpF*, *psaH*, and *psbN* under alkaline water irrigation conditions were similar to those under neutral water irrigation conditions, but *psbO* did not play a good regulatory role under alkaline water irrigation conditions.

This study found that brackish water drip irrigation mainly led to changes in nitrogen metabolism, sulfur metabolism, and oxidative phosphorylation pathways in *B. napus* leaves at the full flowering stage. This is different from the self-regulation mechanism of *B. napus* under salt stress at the seedling stage. Wang reported that under salt stress, the glucose metabolism, amino acid metabolism, and glycerol metabolism pathways were mainly regulated in *B. napus* at the seedling stage [40]. In this study, the application of PPM further changed the regulatory pathways of *B. napus* in the full flowering stage, with Calvin cycle and TCA cycle metabolism as the main metabolic pathways, and enhanced the regulatory effects of transcription factors (*atpF*, *psbO*) and metabolites (fumarate, citric acid).In addition, after the application of PPM, changes in transcription factors (such as

ALDO (aldolase A), *RPIA* (ribose 5-phosphate isomerase A), and FH (fumarate hydratase)) and metabolites (such as fumarate and, citric acid,) that were positively correlated with photosynthesis were more under neutral water drip irrigation than under alkaline water drip irrigation. This indicates that PPM has a better effect on alleviating neutral salt stress. It should be noted that under neutral and alkaline water irrigation conditions, the regulatory mechanism of *B. napus* showed great differences (Figure 5), the main factors were inapparent, and the regulatory effects of transcription factors (*atpF*, *psaN*) and metabolites (adenosine, malic) related to photosynthesis were weak. However, after the application of PPM, the regulatory networks under neutral and alkaline water irrigation conditions tended to be consistent. That is, the regulatory effects of transcription factors (*atpF*, *psbO*, *psaH* and *psaN*) and metabolites (malic and adenosine) were enhanced, and the biomass and physiological indicators of *B. napus* were positively regulated.

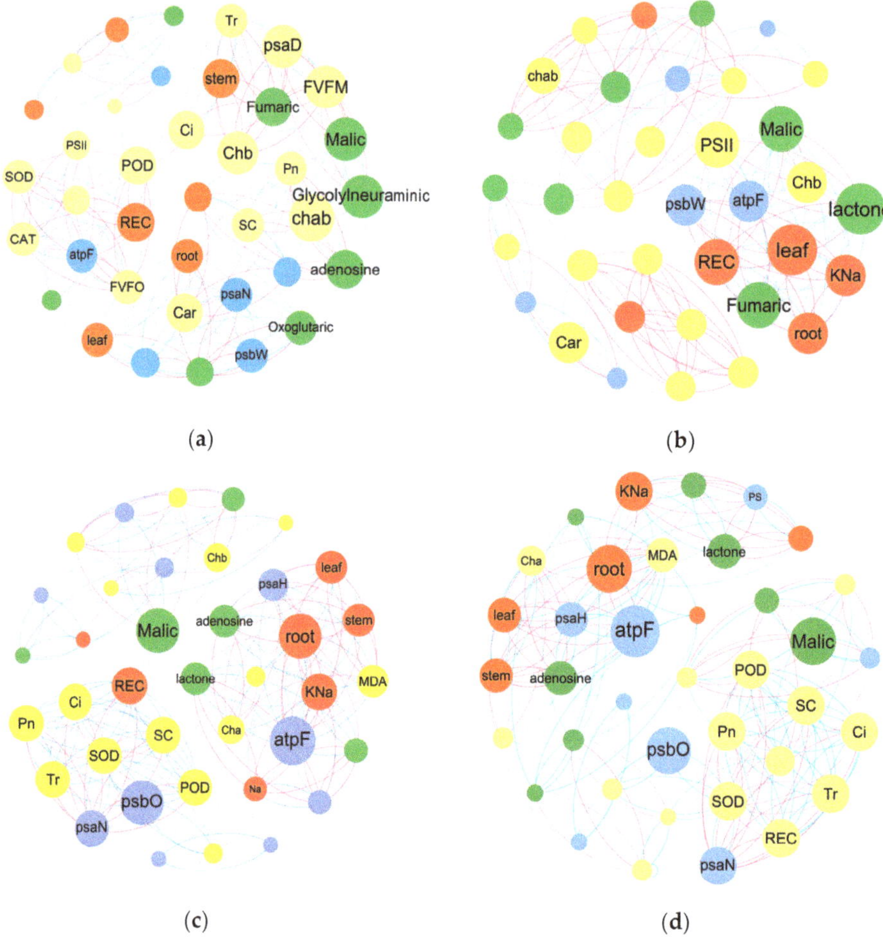

Figure 5. Relationship networks of the yield/ion content (red), photosynthetic parameters (yellow), transcription factors (blue), and DAMs (green) of *B. napus* under different treatments. (**a**) SCK treatment; (**b**) ACK treatment; (**c**) SPPM treatment; and (**d**) APPM treatment. The red lines indicate positive correlation, and the blue lines indicate negative correlation. Node size is positively correlated with the number of connections.

4. Materials and Methods

4.1. Experimental Site and Materials

This experiment was conducted in Shihezi, Xinjiang, China (44°32′44.6″ N, 85°99′877″ E). The region is a continental dry climate, with annual sunshine hours of 2300–2700 h, annual average rainfall of 220 mm and annual evaporation of 1000–1500 mm. Soil amendment (PPM), a liquid mixture of polyacrylamide, polyvinyl alcohol, and manganese sulfate prepared at 90 °C was used in this experiment. The PPM material was invented by our research team in 2020. Before application (22 July 2020), the PPM was mixed with inorganic fertilizer, using the following as the mass ratio of the PPM-type modifier:polyvinyl alcohol:polyacrylamide:manganese sulfate:inorganic fertilizer = 1:3:6:50.

4.2. Experimental Design

On 15 June 2020, 120 kg of soil (0–60 cm soil layer, pH: 8.25, cation exchange capacity: 17.32 cmol·kg^{-1}; alkaline 12ydrolysable nitrogen content: 56 mg·kg^{-1}; available phosphorus content: 10.7 mg·kg^{-1}; available potassium content: 226 mg·kg^{-1}) was collected from the experimental site. Then, the soil was transferred into a cylindrical barrel (0.3 m × 0.6 m × 0.6 m) and the original soil layers were kept. After that, the barrels were buried back to the field. This experiment adopted a randomized complete block design with four groups, namely (1) SCK group (no soil amendment and water containing 6 g·L^{-1} NaCl (neutral salt) was used for drip irrigation), (2) ACK group (no soil amendment and water containing 6 g·L^{-1} Na$_2$CO$_3$ (alkaline salt) was used for drip irrigation); (3) SPPM group (water containing 6 g·L^{-1} NaCl was used for drip irrigation and 12 g·L^{-1} of PPM was applied); (4) APPM treatment (water containing 6 g·L^{-1} Na$_2$CO$_3$ was used for drip irrigation and 12 g·L^{-1} of PPM was applied). The pH values of the irrigation solutions under different treatments of ACK, SCK, APPM and SPPM were 7.05, 11.04, 6.89 and 10.53, respectively. Each group had three replicates.

B. napus seeds (variety Huayouza 82) were sown after mixing with triple superphosphate (1:15) on 15 July 2020. After emergence, six seedlings were retained in each barrel. According to the experimental design, brackish water was irrigated at 10 d intervals throughout the growth period, and PPM was dissolved in irrigation water and applied through the drip irrigation system during the first irrigation. The irrigated soil is 0–40 cm and the irrigation rate is 750 m^3·ha^{-1}. Six leaves were collected from the plants in each group at the full flowering stage (20 October) and, stored in liquid nitrogen.

4.3. Measurement Methods

4.3.1. Determination of Photosynthetic Parameters, Chlorophyll Fluorescence Parameters and Plant Fresh Weight

At the full flowering stage (20 October), photosynthetic parameters including the net photosynthesis rate (Pn), transpiration rate (Tr), stomatal conductance (Gs), and intracellular CO$_2$ concentration (C_i) were measured with a Li-6400 portable photosynthesis instrument (LI-COR, Lincoln, Nebraska, USA) [25]. Then, the maximum fluorescence after dark adaptation (F_v), minimum fluorescence after dark adaptation (F_0), maximum fluorescence under light (F_m), and steady-state fluorescence after light adaptation ($F_m{}'$) were determined with a PAM-2100 modulated chlorophyll fluorometer (WALZ, Germany) [41]. Each measurement was repeated four times.

$$F_v/F_0 = (F_m - F_0)/F_0 \tag{1}$$

$$F_v/F_m = (F_m - F_0)/F_m \tag{2}$$

$$\Phi PSII = (F_m{}' - F_s)/F_m{}' \tag{3}$$

$$qP = (F_m{}' - F_s)/(F_m{}' - F_0) \tag{4}$$

On 23 October, three plants were collected from each group, The roots, leaves and stems were weighed after rinsing with distilled water.

4.3.2. Determination of Chlorophyll and Carotenoids in Plant Leaves

Leaves were cut into small pieces. Then, 0.2 g of leaf sample and 20 mL of extract (absolute ethanol: acetone = 1:1) were mixed evenly in a test tube, sealed, and placed in the dark until the sample turned white. After that, the chlorophyll solution was transferred into a cuvette, and the extract was used as a blank. Finally, colorimetry was conducted at 645 nm, 652 nm, 663 nm, and 440 nm [42].

4.3.3. Determination of Leaf Antioxidant Enzyme Activity and Malondialdehyde (MDA) Content

Leaf superoxide dismutase (SOD), peroxidase (POD), catalase (CAT) activities and malondialdehyde (MDA) content were determined by an NBT photochemical reduction method, guaiacol absorbance method, ultraviolet spectrophotometry and thiobarbituric acid method, respectively [43].

4.3.4. Determination of Na^+, K^+ Content and Relative Electrical Conductivity in Leaves

The Na^+ and K^+ content and REC were determined via a previous method by AP1200 flame spectrophotometer (AP1200, Shanghai, China) and DDSJ-219L conductivity meter [44,45].

4.3.5. Transcriptomic and Metabolomic Assays

First, 12 leaf samples were rapidly stored in liquid nitrogen and stored by Beijing Biomic Biotechnology Co., Ltd. (Beijing, China) for transcriptomic and metabolomic analysis. The transcriptional sequencing platform was Illumina HiSeq, and the metabolome was determined by UPLC-MS/MS [40].

4.4. Statistical Analysis

The data were processed by Excel 2016 software. SPSS 23.0 (SPSS Inc., Chicago, IL, USA) was used for one-way variance (ANOVA) analysis ($\alpha = 0.05$). Figures was drawn using Origin 8.0 (Origin Lab, Northampton, MA, USA). The co-occurrence relationships were visualized using Gephi 0.9.2 (https://Gephi.org/ (accessed on: 20 March 2022)). Fastp was used to remove low-quality sequences in adapters and reads [46]. The filtered reads were aligned to the reference genome of *B. napus* (https://www.cottongen.org/species/Gossypium_hirsutum/ZJU-AD1_v2.1 (accessed on: 15 December 2021)). The gene expression level was calculated in units of FPKM (Fragments Per Kilo bases Per Million Fragments Mapped). Differential analysis of gene expression was performed using the DESeq package in R software, and differentially expressed genes (DEGs) were selected based on $|log2FoldChange| > 1$ and $p < 0.05$. Differentially expressed genes and transcription factors were subjected to GO and KEGG enrichment analysis using the clusterProfiler package in R software [47].

5. Conclusions

Irrigation using different types of brackish water cause different ionic toxicity to *B. napus*, resulting in differences in photosynthetic characteristics. Alkaline water drip irrigation inhibited the photosynthesis of *B. napus* greater, causing lower photosynthetic (Pn, Tr, and Gs) and fluorescence (F_v/F_0, F_v/F_m and qP) parameters and chlorophyll b content in *B. napus* leaves. In addition, there were differences in the expression of photosynthesis-related genes, transcription factors, and metabolites under different types of brackish water drip irrigation conditions, leading to differences in the photosynthesis-related metabolic pathways and photosynthate content. The pathways related to nitrogen metabolism and photosynthesis were changed by brackish water drip irrigation However, the application of PPM reduced the ion toxicity and oxidative stress induced by brackish water drip irrigation, increased the photosynthetic pigment content and improved the chlorophyll fluorescence parameters, thereby enhancing the photosynthetic performance and biomass accumulation. The integrated transcriptomic and metabonomic analysis results showed that the application of PPM could regulate transcription factors and metabolites related to energy metabolism and carbohydrate metabolism-related pathways, to regulate photosynthates and increase biomass. However, these transcription factors and metabolites were different under alkaline and neutral water drip irrigation conditions. Under neutral water irrigation, the application of PPM mainly regulated sulfur metabolism, carbon fixation pathways in photosynthetic organisms, and carbohydrate metabolism pathways in *B. napus* leaves. However, under alkaline water irrigation, the photosynthesis, oxidative phosphorylation, and tricarboxylic acid (TCA) cycle metabolic pathways were mainly regulated (Figure 6). This study revealed the regulation mechanism of PPM on oilseed rape photosynthesis under different brackish water irrigation conditions, and at the same time, it provided a practical method for the use of PPM-type modifiers in agriculture using brackish water irrigation.

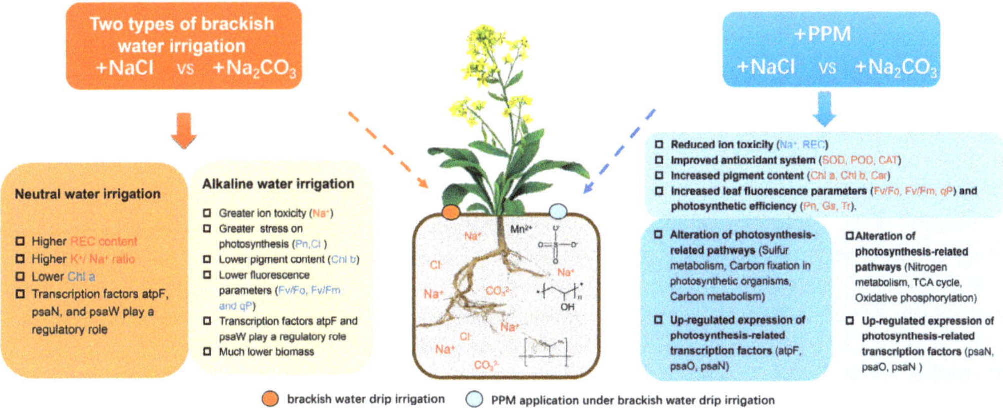

Figure 6. Regulatory mechanism of the effect of soil amendment on rapeseed (*Brassica napus* L.) photosynthesis under drip irrigation with brackish water.

Author Contributions: Conceptualization, Z.L. and K.W.; methodology, Z.L. and H.F.; software, Z.L.; validation, H.F., L.Y. and Y.L.; formal analysis, Y.S.; investigation, T.W., W.C. and C.W.; resources, Y.L.; data curation, S.W. and D.H.; writing—original draft preparation, Z.L.; writing—review and editing, K.W.; visualization, Z.L. and Y.S.; supervision, H.F.; project administration, Y.L.; funding acquisition, K.W. All authors have read and agreed to the published version of the manuscript.

Funding: The research was funded by the Special Fund for the Science and Technology Department of Xinjiang Uygur Autonomous Region, China, grant number 2022B02021-3; Shihezi University Innovative Development Special Project, China, grant number (CXFZ202207, CGZH202204); The Guiding Science and Technology Plan Project of Xinjiang production and Construction Corps (2022ZD054).

Institutional Review Board Statement: Not applicable.

Informed Consent Statement: Informed consent was obtained from all subjects involved in the study. Written informed consent has been obtained from the patient(s) to publish this paper.

Data Availability Statement: The raw reads of *Brassica Napus* Transcriptome and Gene expression (TaxID: 3708) are available under accession PRJNA772052 at NCBI Sequence Read Archive (SRA) repository. All data have been released.

Acknowledgments: Thanks are due to D.D.L. and W.L.C. for assistance with the experiments and to H.F. and C.C.J. for suggestions.

Conflicts of Interest: The authors declare no conflicts of interest.

Abbreviations

PPM	a soil amendment composed of polyacrylamide, polyvinyl alcohol, and manganese sulfate
Pn	leaf net photosynthetic rate
Gs	stomatal conductance
Tr	transpiration rate
C_i	intracellular CO_2 concentration
F_v	maximum fluorescence after dark adaptation
F_0	minimum fluorescence value after dark adaptation
$F_{0'}$	minimum fluorescence under light
F_m	variable fluorescence
$F_{m'}$	maximum fluorescence after light adaptation
F_s	steady-state fluorescence after light adaptation
F_v/F_m	maximum photochemical efficiency of photosystem II
F_v/F_0	potential activity of photosystem II photochemistry
qP	photochemical quenching coefficient
Chl a	chlorophyll a
Chl b	chlorophyll b
Car	carotenoids
SOD	superoxide dismutase
POD	peroxidase
CAT	catalase
MDA	malondialdehyde
REC	relative electrical conductivity
DEGs	differentially expressed genes
DAMs	differentially accumulated metabolites
LC/MS	liquid chromatography-mass spectrometry
PCA	principal component analysis
RDA	redundancy analysis
FPKM	fragments per kilobase of exon per million fragments mapped

References

1. Hussain, M.I.; Farooq, M.; Muscolo, A.; Rehman, A. Crop diversification and saline water irrigation as potential strategies to save freshwater resources and reclamation of marginal soils—A review. *Environ. Sci. Pollut. Res. Int.* **2020**, *27*, 28695–28729. [CrossRef] [PubMed]
2. Leogrande, R.; Vitti, C.; Lopedota, O.; Ventrella, D.; Montemurro, F. Effects of irrigation volume and saline water on maize yield and soil in Southern Italy. *Irrig. Drain.* **2016**, *65*, 243–253. [CrossRef]
3. Ahmad, H.; Hayat, S.; Ali, M.; Liu, T.; Cheng, Z.H. The combination of arbuscular mycorrhizal fungi inoculation (*Glomus versiforme*) and 28-homobrassinolide spraying intervals improves growth by enhancing photosynthesis, nutrient absorption, and antioxidant system in cucumber (*Cucumis sativus* L.) under salinity. *Evol. Ecol.* **2018**, *8*, 5724–5740.
4. Sharma, A.; Kumar, V.; Shahzad, B.; Ramakrishnan, M.; Zheng, B.S. Photosynthetic response of plants under different abiotic stresses: A review. *Plant Growth Regul.* **2020**, *39*, 509–531. [CrossRef]
5. Zhao, S.; Zhang, Q.; Liu, M.; Zhou, H.P.; Ma, C.L.; Wang, P.P. Regulation of plant responses to salt stress. *Int. J. Mol. Sci.* **2021**, *22*, 4609. [CrossRef]
6. Omoto, E.; Taniguchi, M.; Miyake, H. Effects of salinity stress on the structure of bundle sheath and mesophyll chloroplasts in NAD-malic enzyme and PCK type C4 plants. *Plant Prod. Sci.* **2010**, *13*, 169–176. [CrossRef]
7. Chaves, M.M.; Flexas, J.; Pinheiro, C. Photosynthesis under drought and salt stress: Regulation mechanisms from whole plant to cell. *Ann. Bot.* **2009**, *103*, 551–560. [CrossRef]
8. Yang, Z.; Li, J.L.; Liu, L.N.; Xie, Q.; Sui, N. Photosynthetic regulation under salt stress and salt-tolerance mechanism of sweet sorghum. *Front. Plant Sci.* **2020**, *10*, 1722. [CrossRef]
9. Chojak-Koźniewska, J.; Kuźniak, E.; Linkiewicz, A.; Sowa, S. Primary carbon metabolism-related changes in cucumber exposed to single and sequential treatments with salt stress and bacterial infection. *Plant Physiol. Biochem.* **2018**, *123*, 160–169. [CrossRef]
10. Li, N.; Zhang, Z.; Gao, S.; Lv, Y.; Chen, Z.J.; Cao, B.L.; Xu, K. Different responses of two Chinese cabbage (*Brassica rapa* L. ssp. *pekinensis*) cultivars in photosynthetic characteristics and chloroplast ultrastructure to salt and alkali stress. *Planta* **2021**, *254*, 102. [CrossRef]
11. Zhao, B.; Nan, X.; Xu, H.; Zhang, T.; Ma, F.F. Sulfate sorption on rape (*Brassica campestris* L.) straw biochar, loess soil and a biochar-soil mixture. *J. Environ. Manag.* **2017**, *201*, 309–314. [CrossRef]

12. Ali, S.; Rizwan, M.; Qayyum, M.F.; Ok, Y.S.; Ibrahim, M.; Riaz, M.; Arif, M.S.; Hafeez, F.; Al-Wabel, M.I.; Shahzad, A.N. Biochar soil amendment on alleviation of drought and salt stress in plants: A critical review. *Environ. Sci. Pollut. Res.* **2017**, *24*, 12700–12712. [CrossRef]

13. Wang, X.; Wang, J.; Wang, J. Seasonality of soil respiration under gypsum and straw amendments in an arid saline-alkali soil. *J. Environ. Manag.* **2021**, *1*, 111494. [CrossRef]

14. Lu, P.; Bainard, L.D.; Ma, B.; Liu, J. Bio-fertilizer and rotten straw amendments alter the rhizosphere bacterial community and increase oat productivity in a saline-alkaline environment. *Sci. Rep.* **2020**, *16*, 19896. [CrossRef]

15. An, M.; Wang, X.; Chang, D.; Wang, S.; Wang, K.Y. Application of compound material alleviates saline and alkaline stress in cotton leaves through regulation of the transcriptome. *BMC Plant Biol.* **2020**, *20*, 462. [CrossRef]

16. Sojka, R.E.; Bjorneberg, D.L.; Entry, J.A.; Lentz, R.D.; Orts, W.J. Polyacrylamides in agriculture and environmental land management. *Adv. Agron.* **2007**, *92*, 75–162.

17. Boya, X.; Dettam, L.R.; Derrick, S.; Taylor, P.; Richard, H.; Andrew, Z.L.; Manish, K. Polyacrylamide degradation and its implications in environmental systems. *Npj Clean Water* **2018**, *1*, 17–25.

18. Kabata-Pendias, A.; Mukherjee, A.B. *Trace Elements from Soil to Human*; Springer: Berlin/Heidelberg, Germany, 2007; Volume 5, 1. 52.

19. Kolvenbach, B.A.; Fournier, S.; Mu, Q. Polyvinyl alcohol biodegradation in wastewater treatment plants: A review. *Water Res.* **2018**, *139*, 118–128.

20. Hutton, M.; Eggett, D.; Van, H.S. Manganese for plant growth and development: A review. *Agronomy* **2020**, *2*, 279–284.

21. An, M.; Wei, C.; Wang, K.; Fan, H.; Wang, X.L.; Chang, D.D. Effects of polymer modifiers on the bacterial communities in cadmium-contaminated alkaline soil. *Appl. Soil Ecol.* **2021**, *157*, 103777. [CrossRef]

22. Wang, X.; An, M.; Wang, K.; Cheng, K. Effects of organic polymer compound material on k^+ and na^+ distribution and physiological characteristics of cotton under saline and alkaline stresses. *Front. Plant Sci.* **2021**, *12*, 636536. [CrossRef] [PubMed]

23. Allakhverdiev, S.I.; Nishiyama, Y.; Miyairi, S.; Yamamoto, H.; Inagaki, N.; Kanesaki, Y.; Murata, N. Salt stress inhibits the repair of photodamaged photosystem II by suppressing the transcription and translation of *psbA* genes in *synechocystis. Plant Physiol.* **2002**, *130*, 1443–1453. [CrossRef] [PubMed]

24. Shah, A.N.; Tanveer, M.; Abbas, A.; Fahad, S.; Baloch, M.S.; Ahmad, M.I.; Saud, S.; Song, Y.H. Targeting salt stress coping mechanisms for stress tolerance in Brassica: A research perspective. *Plant Physiol. Biochem.* **2021**, *158*, 53–64. [CrossRef] [PubMed]

25. Liu, X.; Mak, M.; Babla, M.; Wang, F.; Chen, G.; Veljanoski, F.; Wang, G.; Shabala, S.; Zhou, M.; Chen, Z.H. Linking stomatal traits and expression of slow anion channel genesHvSLAH1 and HvSLAC1 with grain yield for increasing salinity tolerance in barley. *Front. Plant Sci.* **2014**, *5*, 634. [CrossRef] [PubMed]

26. Muhammad, I.; Shalmani, A.; Ali, M.; Yang, Q.H.; Ahmad, H.; Feng, B.L. Mechanisms regulating the dynamics of photosynthesis under abiotic stresses. *Front. Plant Sci.* **2021**, *11*, 615942. [CrossRef]

27. Saddhe, A.A.; Mishra, A.K.; Kumar, K. Molecular insights into the role of plant transporters in salt stress response. *Plant Physiol.* **2021**, *173*, 1481–1494. [CrossRef] [PubMed]

28. Lu, P.; Dai, S.Y.; Yong, L.T.; Zou, B.H.; Wang, N.; Dong, Y.Y.; Liu, W.C.; Wang, F.W.; Yang, H.Y.; Li, X.W. A Soybean Sucrose Non-Fermenting Protein Kinase 1 Gene, GmSNF1, Positively Regulates Plant Response to Salt and Salt-Alkali Stress in Transgenic Plants. *Int. J. Mol. Sci.* **2023**, *24*, 12482. [CrossRef]

29. Smith, J.D.; Johnson, R.L. Saline water irrigation: Effects of soil characteristics and irrigation management on crop yield and quality. *Agric. Water Manag.* **2019**, *221*, 301–312.

30. Farooq, M.; Hussain, M.; Usman, M.; Farooq, S.; Alghamdi, S.S.; Siddique, K.H.M. Impact of abiotic stresses on grain composition and quality in food legumes. *J. Agric. Food Chem.* **2018**, *66*, 8887–8897. [CrossRef]

31. Kopriva, S.; Chu, C. Are we ready to improve phosphorus homeostasis in rice? *Exp. Bot.* **2018**, *69*, 3515–3522. [CrossRef]

32. Hameed, A.; Ahmed, M.Z.; Hussain, T.; Asis, I.; Ahmad, N.; Gul, B.; Nielsen, B.L. Effects of salinity stress on chloroplast structure and function. *Cells* **2021**, *10*, 2023. [CrossRef] [PubMed]

33. Ran, X.; Wang, X.; Gao, X.; Liang, H.Y.; Liu, B.X.; Huang, X.X. Effects of salt stress on the photosynthetic physiology and mineral ion absorption and distribution in white willow (*Salix alba* L.). *PLoS ONE* **2021**, *16*, e0260086. [CrossRef] [PubMed]

34. ElSayed, A.I.; Rafudeen, M.S.; Gomaa, A.M.; Hasanuzzaman, M. Exogenous melatonin enhances the reactive oxygen species metabolism, antioxidant defense-related gene expression, and photosynthetic capacity of *Phaseolus vulgaris* L. to confer salt stress tolerance. *Plant Physiol.* **2021**, *173*, 1369–1381. [CrossRef] [PubMed]

35. An, M.; Chang, D.; Hong, D.; Fan, H.; Wang, K.Y. Metabolic regulation in soil microbial succession and niche differentiation by the polymer amendment under cadmium stress. *J. Hazard. Mater.* **2021**, *416*, 126094. [CrossRef] [PubMed]

36. Kumar, D.; Hassan, M.A.; Naranjo, M.A.; Agrawal, V.; Boscaiu, M.; Vicente, O. Effects of salinity and drought on growth, ionic relations, compatible solutes and activation of antioxidant systems in oleander (*Nerium oleander* L.). *PLoS ONE* **2017**, *12*, e0185017. [CrossRef] [PubMed]

37. Kirsch, F.; Klähn, S.; Hagemann, M. Salt-regulated accumulation of the compatible solutes sucrose and glucosylglycerol in cyanobacteria and its biotechnological potential. *Front. Microbiol.* **2019**, *10*, 2139. [CrossRef] [PubMed]

38. Chen, Z.; Chen, W.; Dai, F. Study on the physiological response differences of Pinus yunnanensis seedlings under different drought conditions. *J. Fujian For. Sci. Technol.* **2012**, *39*, 218–222.

39. Bricker, T.M.; Frankel, L.K. Auxiliary functions of the PsbO, PsbP and PsbQ proteins of higher plant Photosystem II: A critical analysis. *Photochem. Photobiol.* **2011**, *104*, 65–178. [CrossRef]

40. Wang, W.; Pang, J.; Zhang, F.; Sun, L.P.; Yang, L.; Zhao, Y.Z.; Yang, Y.; Wang, Y.J.; Siddique, K.H.M. Integrated transcriptomics and metabolomics analysis to characterize alkali stress responses in canola (*Brassica napus* L.). *Plant Physiol. Biochem.* **2021**, *166*, 605–620. [CrossRef]

41. Zhao, R.; An, L.; Song, D.; Li, M.; Qiao, L.; Liu, N.; Sun, H. Detection of chlorophyll fluorescence parameters of potato leaves based on continuous wavelet transform and spectral analysis. *Spectrochim. Acta Part A Mol. Biomol. Spectrosc.* **2021**, *259*, 119768. [CrossRef]

42. Lichtenthaler, H.K. Chlorophylls and carotenoids: Pigments of photosynthetic biomembranes. *Meth. Enzymol.* **1987**, *148*, 350–382.

43. Martin, W.F.; Bryant, D.A.; Beatty, J.T. A physiological perspective on the origin and evolution of photosynthesis. *FEMS Microbiol. Rev.* **2018**, *42*, 205–231. [CrossRef] [PubMed]

44. Ding, X.; Jiang, Y.; Zhao, H.; Guo, D.; He, L.Z.; Liu, F.G.; Zhou, Q.; Nandwani, D.; Hui, D.F.; Yu, J.Z. Electrical conductivity of nutrient solution influenced photosynthesis, quality, and antioxidant enzyme activity of pakchoi (*Brassica campestris* L. ssp. *Chinensis*) in a hydroponic system. *PLoS ONE* **2018**, *13*, e0202090. [CrossRef]

45. Bao, S.D. *Soil Agrochemical Analysis*, 3rd ed.; China Agriculture Press: Beijing, China, 2000; pp. 99–200.

46. Chen, S.; Zhou, Y.; Chen, Y.; Gu, J. Fastp: An ultra-fast all-in-one FASTQ preprocessor. *Bioinformatics* **2018**, *34*, i884–i890. [CrossRef] [PubMed]

47. Sun, S.; Song, H.; Li, J.; Chen, D.; Tu, M.Y.; Jiang, G.L.; Yu, G.Q.; Zhou, Z.Q. Comparative transcriptome analysis reveals gene expression differences between two peach cultivars under saline-alkaline stress. *Hereditas* **2020**, *157*, 9. [CrossRef]

International Journal of
Molecular Sciences

MDPI

Article

STAY-GREEN Accelerates Chlorophyll Degradation in *Magnolia sinostellata* under the Condition of Light Deficiency

Mingjie Ren [†], Jingjing Ma [†], Danying Lu, Chao Wu, Senyu Zhu, Xiaojun Chen, Yufeng Wu and Yamei Shen *

Zhejiang Provincial Key Laboratory of Germplasm Innovation and Utilization for Garden Plants, College of Landscape and Architecture, Zhejiang Agriculture and Forestry University, Hangzhou 311300, China
* Correspondence: yameishen@zafu.edu.cn
† These authors contributed equally to this work.

Abstract: Species of the Magnoliaceae family are valued for their ornamental qualities and are widely used in landscaping worldwide. However, many of these species are endangered in their natural environments, often due to being overshadowed by overstory canopies. The molecular mechanisms of *Magnolia*'s sensitivity to shade have remained hitherto obscure. Our study sheds light on this conundrum by identifying critical genes involved in governing the plant's response to a light deficiency (LD) environment. In response to LD stress, *Magnolia sinostellata* leaves were endowed with a drastic dwindling in chlorophyll content, which was concomitant to the downregulation of the chlorophyll biosynthesis pathway and upregulation in the chlorophyll degradation pathway. The *STAY-GREEN* (*MsSGR*) gene was one of the most up-regulated genes, which was specifically localized in chloroplasts, and its overexpression in Arabidopsis and tobacco accelerated chlorophyll degradation. Sequence analysis of the *MsSGR* promoter revealed that it contains multiple phytohormone-responsive and light-responsive cis-acting elements and was activated by LD stress. A yeast two-hybrid analysis resulted in the identification of 24 proteins that putatively interact with MsSGR, among which eight were chloroplast-localized proteins that were significantly responsive to LD. Our findings demonstrate that light deficiency increases the expression of *MsSGR*, which in turn regulates chlorophyll degradation and interacts with multiple proteins to form a molecular cascade. Overall, our work has uncovered the mechanism by which *MsSGR* mediates chlorophyll degradation under LD stress conditions, providing insight into the molecular interactions network of MsSGR and contributing to a theoretical framework for understanding the endangerment of wild Magnoliaceae species.

Keywords: *Magnolia sinostellata*; *STAY-GREEN* gene (*SGR*); chlorophyll; light deficiency; transcriptome; yeast two-hybrids analysis

Citation: Ren, M.; Ma, J.; Lu, D.; Wu, C.; Zhu, S.; Chen, X.; Wu, Y.; Shen, Y. *STAY-GREEN* Accelerates Chlorophyll Degradation in *Magnolia sinostellata* under the Condition of Light Deficiency. *Int. J. Mol. Sci.* **2023**, 24, 8510. https://doi.org/10.3390/ijms24108510

Academic Editors: Martin Bartas, Koichi Kobayashi and Tibor Janda

Received: 21 February 2023
Revised: 5 May 2023
Accepted: 6 May 2023
Published: 9 May 2023

1. Introduction

Magnoliaceae species are cultivated globally due to their high ornamental and commercial values. Unfortunately, many Magnoliaceae species are on the verge of extinction due to community succession and environmental threats to their habitats [1]. Some endangered Magnoliaceae species, including *Magnolia stellata*, *M. wufengensis*, *M. officinalis*, *Sinomanglietia glauca*, and *M. sinostellata* [2–6], tend to grow in coniferous woods and have tiny populations in evergreen broad-leaved forests because of light deficiency (LD) stress. Hence, the shading from taller trees has been generally recognized as an important abiotic stress factor that significantly affects the growth and development of deciduous Magnoliaceae plants [7].

Light is a critical environmental factor that drives photosynthesis [8]. The ability of plants from various ecological niches to capture light effectively is a key factor that determines their survival adaptation in nature [9]. Multiple layers of vegetation can interact in complex ways, with the amount of light that reaches the understory vegetation depending on a variety of factors such as the height and size of the canopy, the species composition of the forest, and the density and arrangement of the trees. Trees with broad leaves generally cast more shade than those with smaller leaves, and taller trees can cast longer shadows. The light intensity perceived by the understory vegetation is influenced by the canopy density of the overstory vegetation. This can have a significant impact on the growth and development of heliophilic plants, which are plants that require a lot of light to grow. Shading caused by a dense canopy cover can be an important abiotic stress factor for these plants, and it can affect their ability to photosynthesize and grow [10,11].

Plants have evolved mechanisms to sense changes in light intensity and quality, including shading caused by the surrounding vegetation. Under a shaded environment, the hypocotyl and petiole of Arabidopsis tend to over-elongate, and the leaf blade area is often reduced [12]. The percentage of the *Vitis vinifera* flowers that fell prematurely was found to increase under shading conditions [13]. Likewise, in *Paeonia lactiflora*, shading caused delays in the initial flowering date, a reduction in flower fresh weight, and a fade in the flower's color [14]. The chlorophyll-a and -b levels in tea leaves were significantly elevated under shade conditions, allowing them to capture more light energy for photosynthesis [15]. A decline in chlorophyll concentration, which stunts plant growth, was observed as shade intensified over time [16]. Further studies are needed to better understand the connection between light intensity and chlorophyll content.

Chlorophyll is a crucial molecule for photosynthesis in plants. It is a tetrapyrrole compound that contains magnesium and has a porphyrin ring and a long aliphatic side chain (phytol), which plays an important role in the light absorption and photosynthesis of plants [17]. The regulatory networks that control chlorophyll biosynthesis and degradation are complex and involve many genes and signaling pathways [18,19]. The functions of chlorophyll synthase and the regulatory networks that govern chlorophyll biosynthesis have been extensively studied [20,21]. Mutations in genes involved in the chlorophyll biosynthesis pathway can affect the biosynthesis of chlorophyll and result in the altered pigmentation of plant tissues. For example, transgenic Arabidopsis plants that expressed antisense *HEMA1* mRNA showed a conspicuous deficiency in chlorophyll, resulting in yellow leaves [22]. Mutations in other genes involved in the biosynthesis of chlorophyll also result in reduced chlorophyll accumulation. For instance, mutations in genes such as *ChlD* and *ChlI*, which encode subunits of the Mg-chelatase enzyme that is required for chlorophyll biosynthesis, resulted in decreased chlorophyll accumulation and altered leaf pigmentation [23].

Despite the efforts made so far to elucidate the chlorophyll synthase regulatory hierarchy, the mechanisms underlying how they mediate chlorophyll degradation are still largely shrouded in mystery. Leaf yellowing is a prominent feature of leaf senescence, which is a natural process that occurs in plants as they age [24]. Chlorophyll degradation is a critical component of this process as it allows the plant to recycle valuable nutrients and prevent oxidative damage from accumulated chlorophyll in aging tissues [25,26]. The PAO (phophorbide a oxygenase) pathway is a multi-layered regulatory network that controls chlorophyll degradation in plants [27]. The *STAY-GREEN* (*SGR*) gene encodes a Mg-dechelatase, which is a key component of this pathway. The SGR protein interacts with chlorophyll catabolic enzymes (CCEs) and light-harvesting complex II (LHCII) to form the SGR-CCEs-LHCII complex, which catalyzes the conversion of chlorophyll-a to pheophytin-a [28].

The expression of the *SGR* gene is regulated by various plant hormones and abiotic stress signals. For instance, ethylene, abscisic acid, and jasmonic acid are known to play important roles in regulating plant senescence, and their signaling pathways can modulate *SGR* expression. Several transcription factors have also been identified that can bind to the promoter of the *SGR* gene and regulate its expression. For example, EIN3, ABI3, and MYC2/3/4 are transcription factors involved in ethylene, abscisic acid, and jasmonic acid signaling, respectively, and can regulate *SGR* expression in Arabidopsis [29–31]. During dark-induced senescence, phytochrome-interacting factors (PIFs) such as PIF4/5 can promote the expression of *EIN3*, *ORE1*, and *ABI5*, which in turn can promote *SGR* expression [32]. Additionally, PIF5 can bind to G-box motifs in the promoter regions of *SGR*, *NYC1*, and *ORE1* and enhance their expression in Arabidopsis [33].

M. *sinostellata* is an endangered species and is listed in the "International Union for Conservation of Nature (IUCN) red list of threatened species" [34]. Previous studies have shown that M. *sinostellata* is particularly sensitive to shading stress, which can compromise its growth and development by affecting plant biological activities such as chlorophyll accumulation, photosynthesis efficiency, and hormone signaling [7,35]. There is a clear knowledge gap about the molecular mechanism of chlorophyll degradation in *M. sinostellata* under LD stress conditions, necessitating further investigation. In this study, we found the chlorophyll degradation gene *MsSGR*, which was significantly up-regulated under LD conditions and served a chlorophyll degradation function. Analysis of the *MsSGR* promoter elucidated that it contains hormone- and light-responsive cis-acting elements and that it responded to LD stress. Yeast two-hybrid (Y2H) screening identified numerous proteins that directly interact with MsSGR, providing vital clues about the molecular pathways involved in LD adaptation in *M. sinostellata*.

2. Results

2.1. The Light Intensity Received by M. sinostellata Varys Greatly among Different Forest Communities

The overstory trees cast shadows onto the understory, reducing the amount of sunlight reaching the deciduous plant there. *M. sinostellata* plants in three distinct habitats, including coniferous forest, broadleaf forest, and mixed forest communities, were analyzed for their relative sensitivity to LD stress. There were substantial variations in light intensity among the three different habitats (Figure 1). In the coniferous forest, the light intensity was about 51,000–57,000 lx; however, in the broadleaf forest and mixed forest communities, the light intensities were in the ranges of 180–250 lx and 400–750 lx, respectively (Table 1). Meanwhile, *M. sinostellata* exhibited diverse phenotypes across different populations. In the broadleaf forest and mixed forest communities, the seedlings of *M. sinostellata* displayed larger leaves and longer stems compared to those in coniferous forest communities (Figure S1). Neither fruits were discernible in the broadleaf forest and mixed forest communities. In the mixed forest community, *M. sinostellata* branches were prostrate with adventive roots (Figure S1B). These observed variations in *M. sinostellata's* morphology suggest that LD stress caused by the shade of overstory trees exerts a far-reaching impact on the plant's growth and development.

Table 1. Geographic information and light intensity of three forest communities of *M. sinostellata*.

Type	Side	Latitude	Longitude	Altitude	Aspect	Slope	Light Intensity/(Lx)		
Coniferous forest	A	28°18′58″ N	119°49′12″ E	990	318° Northwest	16°	60,000	56,000	53,000
	B	28°18′58″ N	119°46′12″ E	1000	310° Northwest	11°	52,000	51,000	50,000
	C	28°12′59″ N	119°46′11″ E	980	20° North	22°	48,000	54,000	52,000
Coniferous-broadleaf mixed forest	A	28°12′51″ N	119°46′10″ E	960	340° North	42°	501	360	350
	B	28°12′51″ N	119°46′10″ E	960	358° North	41°	671	806	732
	C	28°12′51″ N	119°46′11″ E	970	344° North	30°	420	321	433
Broadleaf forest	A	28°10′59″ N	119°49′21″ E	990	88° East	25°	207	238	212
	B	28°10′58″ N	119°49′21″ E	980	74° East	23°	273	204	196
	C	28°10′58″ N	119°49′21″ E	990	87° East	40°	150	214	197

Figure 1. Three typical growth habitats of *M. sinostellata*: (**A**) coniferous forest; (**B**) coniferous-broadleaf mixed forest; and (**C**) broadleaf forest. The red arrow represents the seedlings of *M. sinostellata* in the forest community.

2.2. Light Deficiency Accelerated Leaves Senescence in M. sinostellata

To further clarify the specific influence caused by LD stress in *M. sinostellata*, a field environment simulation experiment was constructed. Under LD conditions, the leaf morphology of *M. sinostellata* altered significantly compared to that under normal light conditions (CK). *M. sinostellata* leaves began to fall after 15 days (d) of light deprivation on the seedlings, accompanied by the appearance of many dark spots. Following 20 d of light deprivation, the dark spots on the leaves of *M. sinostellata* became more noticeable. After 25 d, two-thirds of the leaf surfaces were brown and wilted. After 30 d, practically all of the leaves turned brown and dropped off (Figure 2A,B). Total chlorophyll contents in both LD and control conditions showed an ascending trend followed by a descending trend, but the chlorophyll content was significantly lower in the LD leaves than those in the control from 10 d on (Figure 2C).

2.3. Expression Patterns of Genes Involved in Chlorophyll Biosynthesis and Degradation Pathway in M. sinostellata under Light Deficiency Conditions

The expression patterns of genes in chlorophyll biosynthesis and degradation pathways were analyzed to elucidate the molecular processes that underpin chlorophyll depletion under LD conditions. The sequences of genes in chlorophyll biosynthesis and degradation pathways of *M. sinostellata* were similar to the homologous gene sequences in Arabidopsis and rice, and they all contained the same conserved domain, which indicated the same function (Figure S2). As shown in Figure 3A,B, the majority of genes involved in chlorophyll biosynthesis were down-regulated in response to LD treatment, but the majority of genes, with the exception of the *MsSGR* gene, did not exhibit significant expression changes. The expression pattern of genes was verified by the quantitative reverse transcriptase polymerase chain reaction (qRT-PCR) analysis (Figure 4), which revealed that the majority of genes involved in chlorophyll biosynthesis, except for *MsHEMB*, were down-regulated under LD conditions. Meanwhile, *MsSGR*, *MsPPH*, and *MsPAO* were up-regulated, and the up-regulated trend of the *MsSGR* gene was the most significant under LD conditions, indicating that *MsSGR* plays a vital role in the shade tolerance response of *M. sinostellata* under LD conditions.

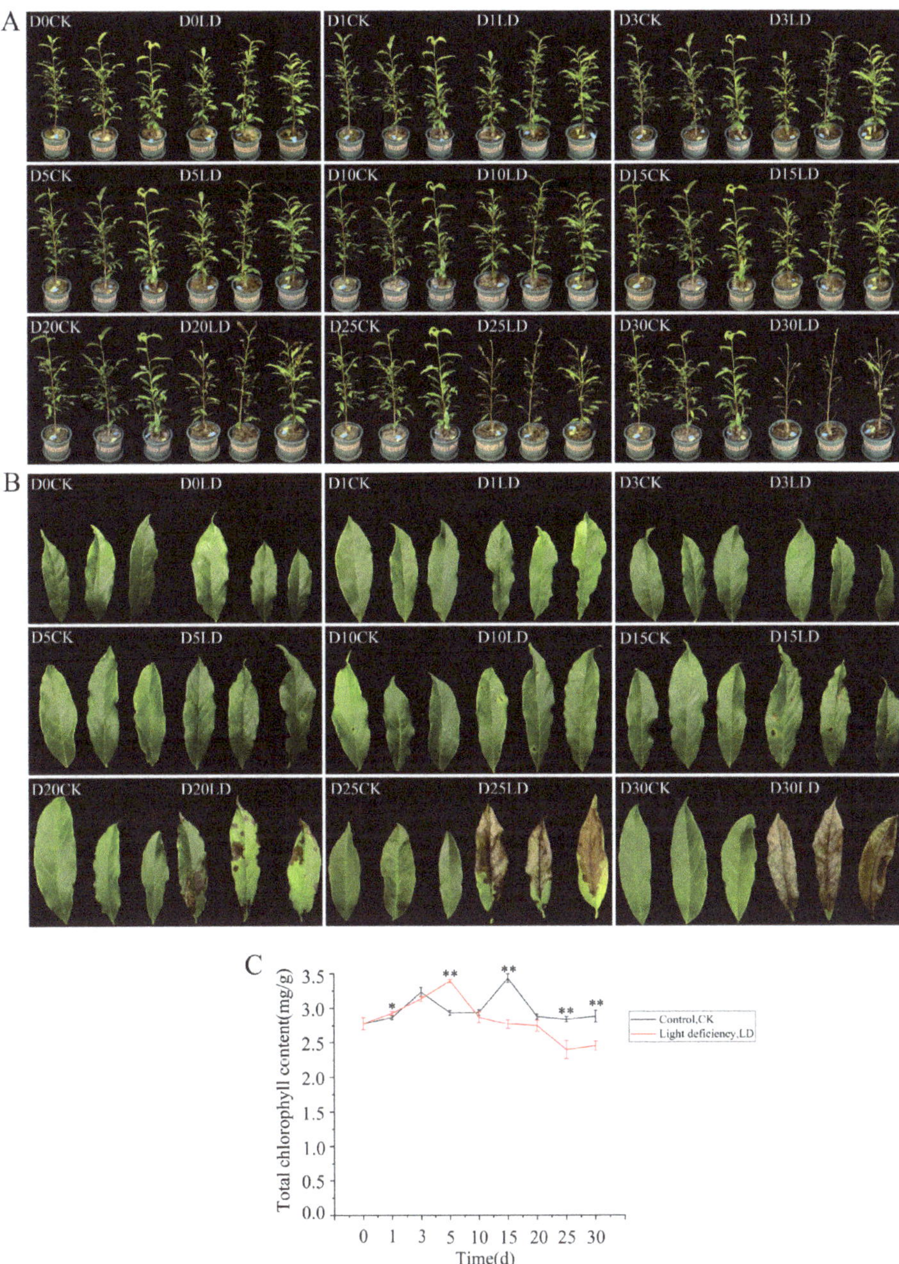

Figure 2. Phenotypic and chlorophyll content changes of *M. sinostellata* under normal and light deficiency (LD) conditions. (**A**) plant shape changes (Control, CK; Light deficiency, LD); (**B**) leaf shape changes; (**C**) chlorophyll content changes of leaves; * represents a significant difference ($p < 0.05$); ** represents a significant difference ($p < 0.01$).

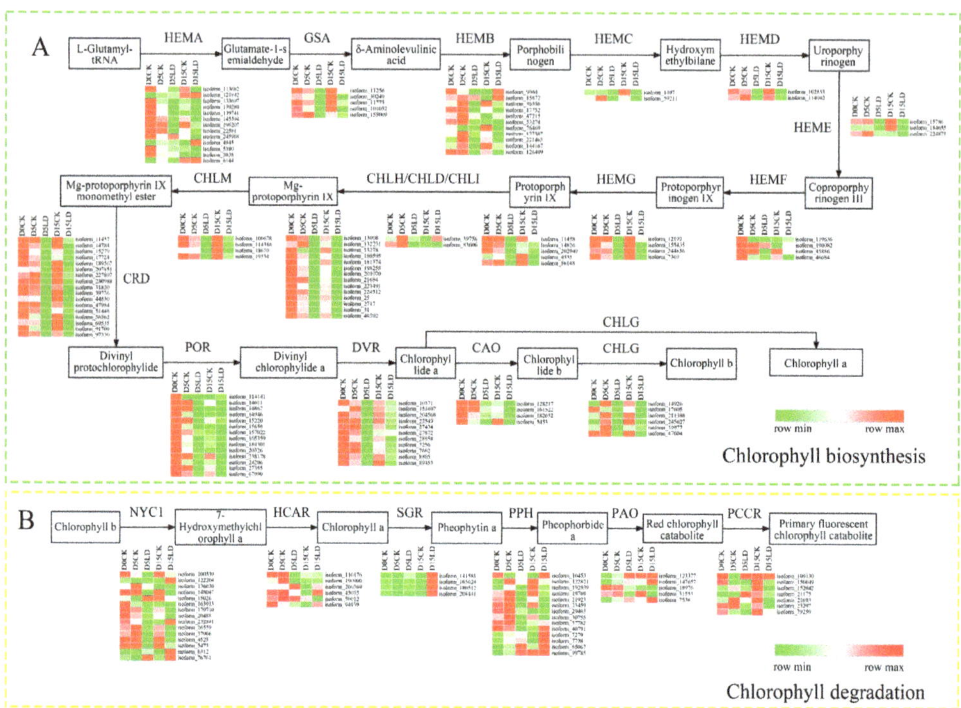

Figure 3. Expression patterns of the genes involved in chlorophyll biosynthesis and degradation pathways in *M. sinostellata* under light deficiency (LD). (**A**) Chlorophyll biosynthesis pathway; (**B**) chlorophyll degradation pathway.

2.4. Identification and Characterization of MsSGR

MsSGR gene was amplified by PCR using *M. sinostellata* leaf cDNA as a template. Multiple protein alignments of the MsSGR and its orthologs in various other plant species showed that MsSGR contains the hallmark STAY-GREEN domain and the cysteine-rich motif, indicating that MsSGR is a member of the SGR subfamily but not the SGRL subfamily (Figure S3). The phylogenetic tree revealed that SGR homologs clearly fall into two categories: monocotyledonous plants and dicotyledonous plants, indicating that SGR conforms to the laws of evolution (Figure 5A). In addition, *MsSGR* gene was expressed in the stem, leaf, leaf bud, flower, flower bud, stamen, and pistil, as determined by tissue-specific expression. Expression was the most abundant in the flower and flower bud, followed by pistil, leaf, stem, stamen, and leaf bud (Figure 5C,D). To determine if chloroplasts are the functional location of MsSGR, *Agrobacterium tumefaciens* cells harboring the 35S::GFP or 35S::MsSGR-GFP constructs were infiltrated into tobacco leaves. As illustrated in Figure 5B, in the leaf discs infiltrated with 35S::GFP, the GFP fluorescence signal was present in the nucleus and the cytomembrane but not the chloroplasts. In contrast, GFP fluorescence was only visible in the chloroplasts in the leaf discs infiltrated with 35S::MsSGR-GFP (Figure 5B). Therefore, MsSGR was conceivably located in the chloroplasts.

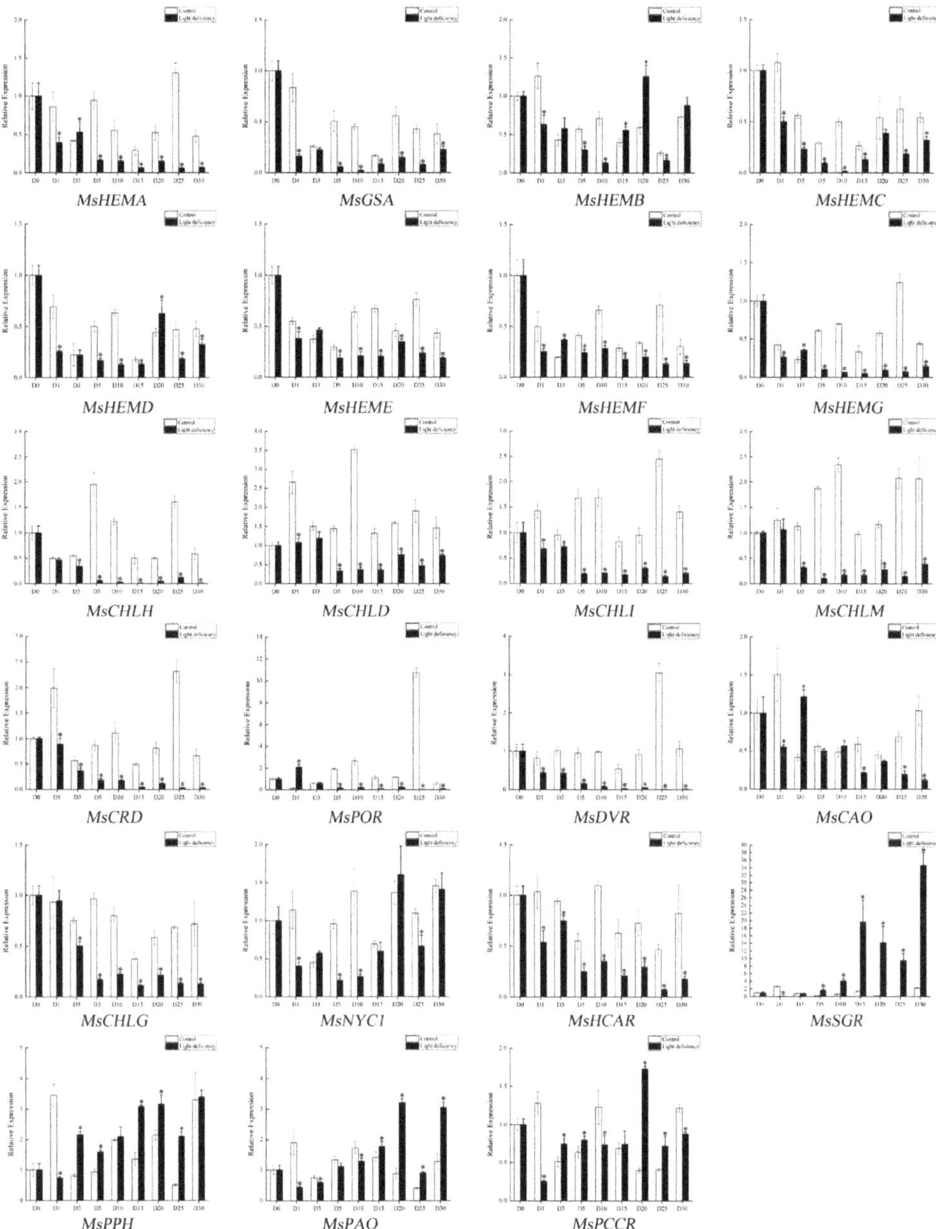

Figure 4. Quantitative reverse transcriptase PCR (qRT-PCR) analysis of chlorophyll biosynthesis and degradation gene under LD conditions; * represents a significant difference (*p* < 0.05).

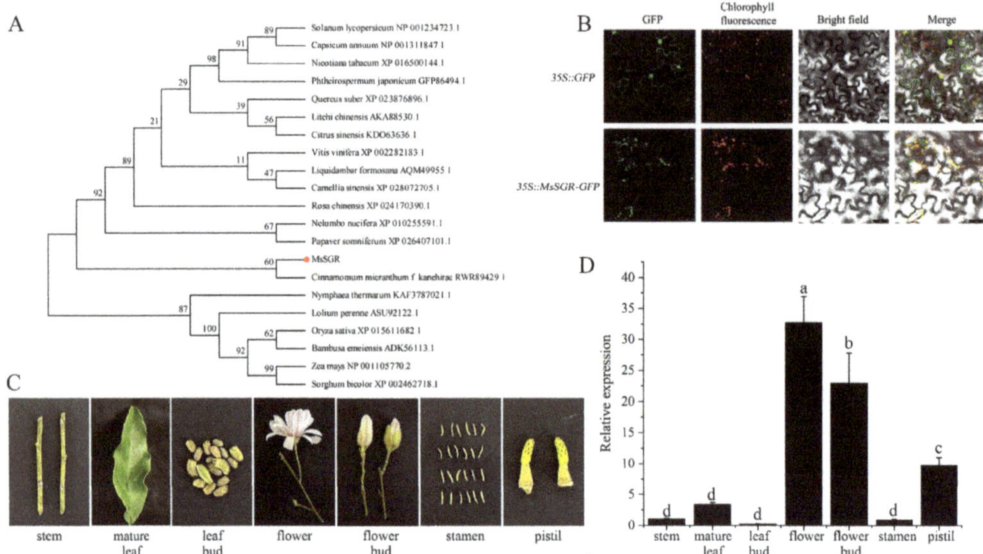

Figure 5. Identification and characterization of *MsSGR*. (**A**) Phylogenetic tree analysis of SGR protein. The amino acid sequence used in the analysis was listed as follows: *Litchi chinensis* (AKA88530.1), *Vitis vinifera* (XP_002282183.1), *Nelumbo nucifera* (XP_010255591.1), *Liquidambar formosana* (AQM49955.1), *Rosa chinensis* (XP_024170390.1), *Nicotiana tabacum* (XP_016500144.1), *Citrus sinensis* (KDO63636.1), *Cinnamomum micranthum* (RWR89429.1), *Quercus suber* (XP_023876896.1), *Phtheirospermum japonicum* (GFP86494.1), *Camellia sinensis* (XP_028072705.1), *Solanum lycopersicum* (NP_001234723.1), *Capsicum annuum* (NP_001311847.1), *Oryza sativa* (XP_015611682.1), *Lolium perenne* (ASU92122.1), *Nymphaea thermarum* (KAF3787021.1), *Papaver somniferum* (XP_026407101.1), *Zea mays* (NP_001105770.2), *Sorghum bicolor* (XP_002462718.1), and *Bambusa emeiensis* (ADK56113.1). (**B**) Subcellular localization of MsSGR in tobacco leaves; (**C**) the tissues of *M.sinostellata* seedlings; (**D**) tissue-specific expression of *MsSGR* gene, different letters represent significant differences ($p < 0.05$).

2.5. MsSGR Induced Chlorophyll Degradation in Arabidopsis and Tobacco

To investigate how *MsSGR* functions in the chlorophyll degradation process, the *MsSGR* gene was cloned and integrated into the plant expression vector pORE-R4 (Figure S4). *MsSGR* was overexpressed in transgenic Arabidopsis and transiently expressed in tobacco leaves. The transgenic Arabidopsis plants displayed conspicuous alterations in morphology compared to the wild-type (WT) lines. Notably, overexpression of the *MsSGR* gene in Arabidopsis resulted in pre-withered rosette leaves (Figure 6A,G,H). In addition, the transient expression of the *MsSGR* gene in tobacco leaves caused severely withered maculas in leaves and induced the chlorophyll degradation process (Figure S5). In addition to significant phenotypic aberrations in the leaves, transgenic Arabidopsis exhibited significantly stunted development, reaching a maximum height of 12–13 cm as opposed to 30 cm in WT (Figure 6B), and forming hypogenetic inflorescences with yellow sepals and peduncles (Figure 6C,D). The structures of hypogenetic inflorescences further developed into short and abortive siliques that bear few viable seeds (Figure 6E,F). Taken together, these findings demonstrate that *MsSGR* plays a crucial role in the chlorophyll degradation process and has a profound effect on plant growth and development.

Figure 6. *MsSGR* induces chlorophyll degradation in Arabidopsis; (**A**) represents the plant shapes of Arabidopsis at 15 days (d); (**B**) represents the plant shapes of Arabidopsis at 35 d; (**C,E,G**) represent inflorescences, pods and leaves of wild-type (WT) Arabidopsis; and (**D,F,H**) represent inflorescences, pods, and leaves of *MsSGR* overexpressed Arabidopsis, respectively, bar = 1 cm.

2.6. Light Deficiency Promotes MsSGR Promoter Activity

The putative *MsSGR* promoter, spanning 1944 bp in length, was cloned using PCR and verified using sequence analysis (Figure S6A). GUS enzyme activity analysis revealed the regulatory activity of the *MsSGR* promoter (Figure S6B,C). The analysis of the cis-acting elements by searching the PlantCARE database revealed that the *MsSGR* promoter encompasses four hormone-responsive elements, including gibberellin, abscisic acid, salicylic acid, and MeJA, as well as two light-responsive elements, including the I-box and the G-box. It is tempting to infer that *MsSGR* gene expression is regulated by light signals and various plant hormones (Figure 7A). To further investigate the regulatory mechanisms of the *MsSGR* gene in response to LD stress, the putative *MsSGR* promoter sequence was used to drive β-glucuronidase (GUS) expression by generating a MsSGR::GUS construct that was used to transform Arabidopsis. It appeared that the GUS staining was discernible in the leaf, hypocotyl, and root in the transgenic lines (Figure 7B(a,b)). In response to LD and dark treatments, the hypocotyls and petioles of the transgenic lines were overtly elongated and displayed a significantly higher level of GUS staining in leaves, hypocotyls, and petioles compared to the CK, indicating the regulatory response of the *MsSGR* promoter to LD treatment (Figure 7B(b–d)).

2.7. Potential Interacting Proteins of MsSGR by Y2H Analysis

In order to clarify the potential molecular mechanisms of MsSGR, a Y2H library with a capacity of 5.56×10^7 CFU was constructed (Figure S7). MsSGR exhibited no self-activation or cell toxicity in yeast cells (Figure S8). The screening among the Y2H library produced 74 colonies, which were growing strongly on SD-Trp-Leu-His. These *HIS*[+] positive colonies were sequenced and contained sequences encoding 41 proteins. In order to further determine the positive interaction between 41 proteins and MsSGR, these colonies were transferred to the SD-Trp-Leu-Hi-Ade and SD-Trp-Leu-His-Ade+x-α-gal to detect whether the *Ade* and *MEL1* reporter genes were activated. A total of 24 proteins directly interacted with MsSGR, eight of which were chloroplast proteins (Table 2, Figure 8A). To elucidate the potential shading response mechanisms, the expression patterns of eight interacting chloroplast proteins under light deficiency conditions were examined. The results showed that the expression levels of eight chloroplast genes significantly responded to LD stress, of which six genes were significantly downregulated and one was highly upregulated. In particular, a gene encoding polyphenol oxidase was downregulated at 5 d and accumulated highly at 15 d (Figure 8B).

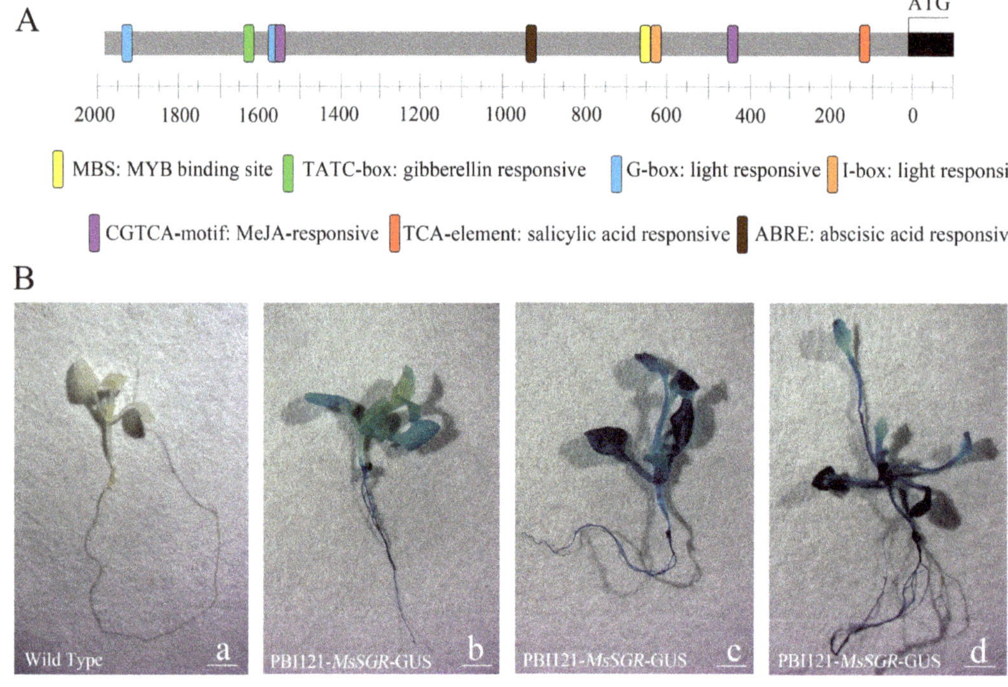

Figure 7. Analysis of *MsSGR* promoter; (**A**) cis-acting element analysis of the *MsSGR* promoter; (**B**) GUS enzyme activity analysis under light deficiency. (a) represents wild type in normal environment; (b–d) represent *MsSGR::GUS* transgenic Arabidopsis in normal environment (CK), light deficiency treatment and dark treatment, respectively, bar = 2 mm.

Table 2. Y2H Screened colonies information.

No.	GenBank No.	Homologous Protein	Clone Numbers
1	EHA8589896.1	polyphenol oxidase, chloroplastic	4
2	XP_030942398.1	peptidyl-prolyl cis-trans isomerase FKBP13, chloroplastic	4
3	XP_020695605.1	ribulose bisphosphate carboxylase small chain, chloroplastic	4
4	XP_034688430.1	PHOTOSYSTEM I ASSEMBLY 2, chloroplastic	1
5	RWR91980.1	multiple organellar RNA editing factor 8, chloroplastic/mitochondrial-like	1
6	RWR94664.1	ATP-dependent Clp protease proteolytic subunit-related protein 3, chloroplastic	1
7	RWR83419.1	protein PTST, chloroplastic	1
8	XP_029123969.1	uroporphyrinogen decarboxylase, chloroplastic	1

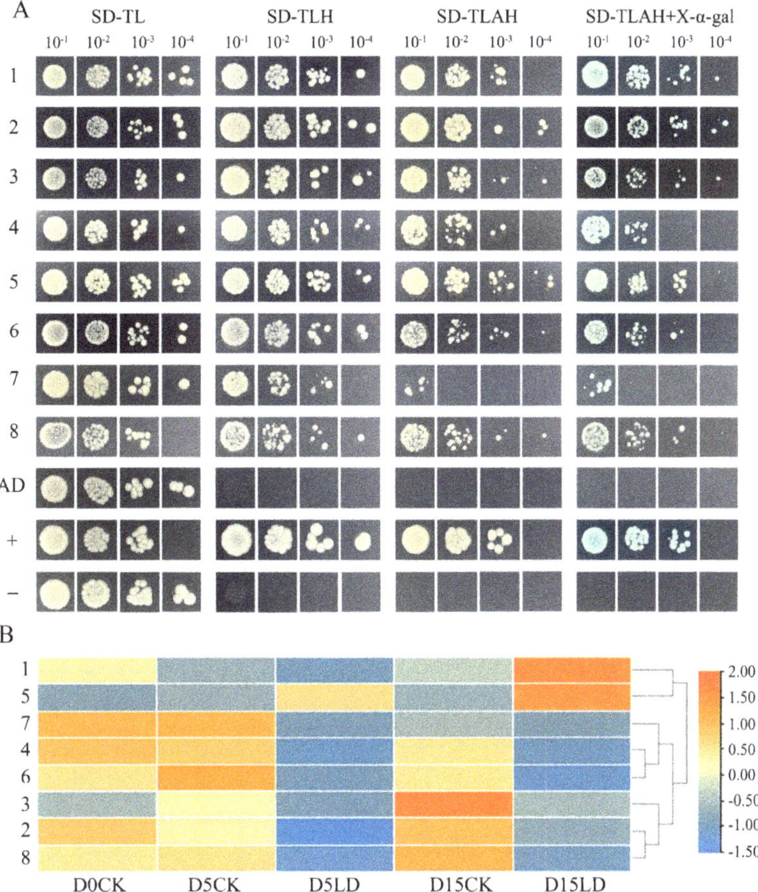

Figure 8. Yeast two-hybrid (Y2H) screening of the putative chloroplastic proteins that are interacting with MsSGR. (**A**) Screened colonies grew on medium SD-Trp-Leu, SD-Trp-Leu-His, SD-Trp-Leu-His-Ade, and SD-Trp-Leu-His-Ade+x-α-gal. "AD" represents the self-activation group (pGBKT7-MsSGR+pGADT7); "+" represents the positive control (pGBKT7-T+pGADT7-P53); and "−" represents the negative control (pGBKT7-T+pGADT7-Lam). Yeast liquid dilution concentrations: 10^{-1}, 10^{-2}, 10^{-3}, and 10^{-4}. (**B**) The heatmap of coding chloroplastic genes expression patterns under light deficiency treatment.

3. Discussion

LD is one of the major abiotic factors affecting plant growth and development as it is a crucial source of energy for photosynthesis [36,37]. Without sufficient light, plants may not be able to synthesize enough food to support their growth and development, leading to various physiological and morphological changes, as observed in *M. sinostellata* in the present study. It was emphatically shown that the light intensity of *M. sinostellata* in the mixed forest and broadleaf forest was far below that in the coniferous forest (Table 1). Compared with seedlings grown in the coniferous forest, the *M. sinostellata* seedlings grown in the mixed forest and broadleaf forest had larger leaves and longer stems (Figure S1), which is consistent with the plant shade avoidance syndrome (SAS), which is a set of physiological and morphological changes observed in plants in response to reduced light availability or shading by neighboring plants [38]. When plants are exposed to shade, they perceive changes in the light spectrum, particularly a decrease in the ratio of red to far-red light. This triggers a series of responses, including elongation of stems and petioles, increased leaf area, reduced branching, and altered flowering time, all of which are thought to enhance the plant's ability to capture more light. *M. sinostellata* did not produce any fruits in the broadleaf forest and mixed forest communities. In the mixed forest community, the branches of *M. sinostellata* were observed to be prostrate and have adventive roots (Figure S1B). Furthermore, in conditions of light deprivation, the leaves of *M. sinostellata* rapidly wilt and senesce, with almost all of the leaves falling off under an LD environment, indicating that *M. sinostellata* is particularly sensitive to LD (Figure 2A,B). Such observations are congruent with previous reports in many other plant species under LD conditions, such as hypocotyl elongation in Arabidopsis and tomato [12,39], increases in plant height, canopy perimeter, and canopy volume in lemon trees [40], reduction in the number of flower buds, and a delay in floral transition in Lisianthus [41]. Based on the evidence presented, it is conceivable that LD has a significant impact on the growth, development, and reproduction of *M. sinostellata*, which may have played a major role in the endangerment of this species in its natural habitat.

It is well-recognized that leaf yellowing is a classic symptom of leaves senescence, which is primarily caused by chlorophyll degradation [42]. Chlorophyll degradation is a natural process that occurs in plants during various stages of their life cycle, such as senescence, ripening, and stress responses. To prevent free chlorophyll from generating additional photo-oxidative damage, plants must rapidly break down these molecules as a survival strategy [43]. Numerous plants, including rice and Arabidopsis, exhibit a senescence phenotype characterized by leaf yellowing under darkness to promote chlorophyll degradation [44,45]. In this study, the total chlorophyll content showed a general increasing trend followed by a descending trend, although the descending trend of the total chlorophyll content was not obvious compared with the leaf color changes. Because the weight changes caused by water loss in senescent leaves will affect the measurement of chlorophyll content based on weight (mg g^{-1}), it is better to use leaf area-based measurement (mg cm^{-2}) to measure the chlorophyll content of senescent leaves [46]. However, the descending trend of total chlorophyll content was still significant ($p < 0.01$) based on weight measurement (Figure 2C). The decline in the content of chlorophyll content impaired *M. sinostellata*'s ability to capture light energy. In addition, LD induced the downregulation of almost all the genes involved in chlorophyll biosynthesis, whereas the *MsSGR* gene was significantly upregulated (Figures 3 and 4), indicating that *M. sinostellata* reduces leaf chlorophyll content by inhibiting chlorophyll biosynthesis and accelerating chlorophyll degradation to prevent chlorophyll from causing more photooxidative damage to cells in LD. Overall, *M. sinostellata* responded to LD by reducing the chlorophyll content in its leaves. However, there was a possibility that the decrease in chlorophyll content impaired photosynthesis in *M. sinostellata*.

As a key gene in the chlorophyll degradation pathway, the *SGR* gene encodes a Mg-dechelatase, which catalyzes the removal of the central magnesium ion from chlorophyll, leading to its breakdown and the subsequent degradation of the chloroplast [47]. In this study, *MsSGR* isolated from *M. sinostellata* was determined to be a chloroplast-specific protein through subcellular localization (Figure 5B), which is consistent with its orthologs in Arabidopsis and *Camellia sinensis* [28,48]. Furthermore, protein structure analysis revealed that MsSGR does not have any transmembrane structure (Figure S9), suggesting that it is most likely localized in the chloroplast stroma, which is the fluid-filled space inside the chloroplasts where many of the biochemical reactions associated with photosynthesis and chlorophyll degradation occur. Overexpression of *MsSGR* caused leaf yellowing in the transgenic Arabidopsis and a decrease in the chlorophyll content of tobacco that was transiently expressing *MsSGR* (Figures 6A,G,H and S5). These results were consistent with previous studies, which demonstrated the functional role of the *SGR* gene in chlorophyll degradation [48]. In addition, yellow stems, yellow inflorescences, and abortive fruit pods were also observed in transgenic Arabidopsis overexpressing *MsSGR* (Figure 6D–F), lending further credence to the assumed role of the *MsSGR* gene in accelerating chlorophyll degradation. Furthermore, the spatial differentiation of *MsSGR* in *M. sinostellata* tissues suggests its functional diversity and specificity, which may have evolved through gene duplication. Notably, *MsSGR* was observed in higher expression in flower and flower bud, which is consistent with previous reports in Arabidopsis [49]. The chlorophyll degradation gene *PAO* was also expressed highly in flowers [25]. In *Magnolia* plants, cyanidin and peonidin make the flower petals appear red-purple and purple, respectively, and the flavonols perform as auxiliary pigments [50]. It is reasonable to speculate that the high expression of *MsSGR* might promote chlorophyll degradation in flowers. Nonetheless, further research is needed.

Analysis of the *MsSGR* promoter sequence revealed the presence of four hormone-responsive elements and two light-responsive cis-acting elements (Figure 7A). The results indicated that the *MsSGR* gene is regulated by a complex regulatory network. As indicated by previous works, *SGR* is controlled by plant senescence hormone and transcription factor families [29,31,51]. Meanwhile, *SGR* also responds to changes in light signals, phytochrome interacting factors PIF4 can bind to the promoter of *SGR* to promote its expression in Arabidopsis [45]. In this study, a greater deposition of GUS stain was observed in the hypocotyl, leaf, and petiole of *MsSGR::GUS* transgenic lines under LD conditions, indicating that the promoter of *MsSGR* is responsive to LD stress (Figure 7B). Taken together, *MsSGR* was found to be an important gene in the chlorophyll degradation process of *M. sinostellata* in response to LD stress.

Previous studies have focused mostly on how SGR regulates chlorophyll degradation [52–54]. SGR is able to bind the light-harvesting complex II (LHCII) protein and recruit chlorophyll catabolic enzymes (CCEs) to form the SGR-CCEs-LHCII complex [28]. In *Camellia sinensis*, CsSGR can generate a CsSGR-CssHSP-CsLHCII protein complex to regulate albinism [48]. In addition, tomato SlSGR1 directly interacts with the carotenoid biosynthesis enzyme SlPSY1 and inhibits its activity in order to regulate tomato lycopene accumulation [55]. Therefore, not only is SGR engaged in chlorophyll degradation but it also contributes to the carotenoid biosynthesis pathway, indicating that SGR has a diverse range of functions. In the present study, a total of eight potential chloroplast proteins that interacted with MsSGR were identified (Table 2, Figure 8A). Among these proteins, polyphenol oxidase is the key enzyme in the enzymatic browning of fruits and vegetables [56]; uroporphyrinogen decarboxylase encoded by the *HEME* gene is involved in the chlorophyll biosynthesis pathway [57]. The transcriptome analysis revealed that genes encoding eight chloroplast proteins exhibited differential expression in response to LD stress (Figure 8B). It is envisaged that the eight chloroplast proteins are likely to be involved in regulating chlorophyll degradation by interacting with MsSGR in *M. sinostellata* in response to LD stress. Therefore, the findings of Y2H screening in this study can provide valuable information about protein–protein interactions and potential molecular mechanisms involved in the chlorophyll degradation process in response to LD stress. Further research could then be directed towards investigating these interactions and mechanisms to gain a better understanding of MsSGR's role in an intricate molecular network involving many other chloroplast-localized proteins under LD stress.

4. Material and Methods

4.1. Light Intensity Measurement in Habitats of M. sinostellata

Light intensity was measured in the coniferous forest, broadleaf forest, and mixed forest communities where *M. sinostellata* grew. In each community, three *M. sinostellata* seedlings were chosen as measurement stations. The coordinates for each measurement station were recorded by global positioning system (GPS) (Garmin, Olathe, KS, USA). The light intensity was measured in luminous flux using a Digital Luxmeter ZDS-10 (Shanghai Jiading Xuelian Instrument, Shanghai, China), with three replicates for each measurement.

4.2. Plants Materials and Light Deficiency Treatments

The 3-year-old grafted *M. sinostellata* seedlings were collected from the Shengzhou Magnolia base, Shaoxing, Zhejiang Province, China. In this experiment, these seedlings were evenly placed in a climate chamber at 25 °C, 50% humidity, 14/10 h light/dark photoperiod, and 648 μmol m^{-2}·s^{-1} PAR (photosynthesis active radiation). To simulate the LD stress conditions caused by shading in natural environments, one part of *M. sinostellata* seedlings was placed under an LD stress treatment condition (16.2 μmol m^{-2}·s^{-1} PAR, LD) that was set up using a black shade net and several bamboo poles, and the other part of the seedlings was placed in a normal condition (648 μmol m^{-2}·s^{-1} PAR, CK) in a climate chamber. The light intensity was measured in luminous flux with a Digital Luxmeter ZDS-10 (Shanghai Jiading Xuelian Instrument, Shanghai, China) and converted the light intensity from LUX to PAR as previously described [58]. All other experimental conditions were maintained the same for both LD stress treatment and CK, with three biological replicates. Leaf samples were collected from the seedlings in LD treatment and CK groups at 0, 1, 3, 5, 10, 15, 20, 25, and 30 d following the treatment. The samples of leaf, leaf bud, flower, flower bud, stamen, and pistil were collected from the *M. sinostellata* seedlings. Each sample was collected from three seedlings and each collection was repeated three times as biological replicates. All samples were stored at -80 °C for further experimentation.

4.3. Determination of Chlorophyll Content

The chlorophyll was extracted from 100 mg leaves using 10 mL of 95% ethanol. The extracts were filtered and analyzed with a Shimadzu UV2700 spectrophotometer (Shimadzu, Kyoto, Japan), and the absorbances were recorded at both 649 and 665 nm. Total chlorophyll content (mg g^{-1}) was estimated using the method previously described by Welburn and Lichtenthaler [59]. All the experiments were performed in triplicate.

4.4. Transcriptome Analysis of Chlorophyll Metabolism Pathway Genes

The transcriptome data are derived from our previously published study [7]. The genes involved in the chlorophyll biosynthesis and degradation pathways were mainly retrieved through the Kyoto Encyclopedia of Genes and Genomes (KEGG) (map00860, Porphyrin Metabolism) pathway enrichment analysis. Heatmaps of gene expression were produced using Morpheus (Morpheus, https://software.broadinstitute.org/morpheus, accessed on 27 May 2021).

4.5. RNA Extraction, cDNA Synthesis, and DNA Preparation

Total RNA was extracted from an RNAprep Pure Plant Plus Kit (Tiangen, Beijing, China) following the manufacturer's instructions. The integrity of RNAs was confirmed using 1% agarose gel electrophoresis, and the purity and concentration of the total RNAs were analyzed using a NanoDrop spectrophotometer (Thermo Fisher Scientific, Waltham, MA, USA). The total RNAs were converted to cDNAs using the PrimeScriptTM^{RT} Master Mix (Takara, Tokyo, Japan) following the manufacturer's instructions. Total DNA was extracted from a FastPure Plant DNA Isolation Mini Kit (Vazyme, Nanjing, China) following the manufacturer's instructions.

4.6. Expression Analysis of Chlorophyll Biosynthesis and Degradation Gene in Light Deficiency Stress and MsSGR in Different Tissues

The expression patterns of chlorophyll biosynthesis and degradation genes and *MsSGR* in different tissues were assayed using qRT-PCR and the primers were designed using Primer 5.0 (Table S1). The common isoform sequences were used to design primers for qRT-PCR (Figure S10). The qRT-PCR experiment was conducted using BCG qPCR Master Mix (Beijing Baikaiji Biotechnology, Beijing, China) on a LightCycler$^{®}$ 480 II (Roche Applied Science, Penzberg, Germany). The qRT-PCR program was as follows: 95 °C for 2 min, 40 cycles of 95 °C for 15 s, and 60 °C for 40 s. *EF1-α* was selected as the reference gene to normalize the gene expression, and the relative gene expression was determined using the

$2^{-\Delta\Delta Ct}$ method [60]. All the qRT-PCR analysis experiments were performed in triplicate. The bar charts of gene expression were generated using Origin 2018 (OriginLab Corporation, Northampton, MA, USA), and SPSS 24.0 (SPSS Inc., Chicago, IL, USA) was used to analyze statistical significance.

4.7. Cloning of the MsSGR Gene and Bioinformatics Analysis

The *MsSGR* full-length sequence was isolated from our previous transcript data, and primers were designed on both sides of the *MsSGR* open reading frame (ORF). PCR amplification was performed using the primers listed in Table S1. The PCR program was as follows: 95 °C for 3 min, 35 cycles of 95 °C for 15 s, 58 °C for 15 s, and 72 °C for 1 min; with a final extension at 72 °C for 5 min. The purified PCR product was ligated into the pMD-18T vector (Takara, Tokyo, Japan) and verified using DNA sequencing. Multiple sequence alignments of amino acid sequences were performed using DNAMAN 6.0 (Lynnon Biosoft, San Ramon, CA, USA). Phylogenetic trees were generated using MEGA 6.0 (Koichiro Tamura, Tokyo, Japan) based on the neighbor-joining method (1000 bootstrap replicates). Homologous protein sequences of SGR were all downloaded from NCBI (https://www.ncbi.nlm.nih.gov/, accessed on 18 August 2023). The protein transmembrane regions were predicted using TMHMM 2.0 (https://services.healthtech.dtu.dk/services/TMHMM-2.0/, accessed on 16 January 2023).

4.8. MsSGR Subcellular Localization and Overexpression in Arabidopsis and Tobacco

The binary vector pORE_R4 containing GFP was used in this experiment. The open reading frame (ORF) of *MsSGR* without a stop codon was amplified by using the primers listed in Table S1. The PCR products were digested with XhoI and ClaI and ligated into the corresponding sites of the pre-digested pORE_R4 to generate the expression vector 35Spro::MsSGR::GFP by using the ClonExpress® II One Step Cloning kit (Vazyme, Nanjing, China). The vector was transformed into *A. tumefaciens* strain GV3101 Chemically Competent Cell (Shanghai Weidi Biotechnology, Shanghai, China), which was used to infiltrate the tobacco leaves. After incubation for 12 h in the dark followed by 48 h under light, GFP fluorescence was observed with a Leica TCS SP8DLS confocal microscope (Leica, Nussloch, Germany). The excitation wavelength for GFP was 488 nm, and the emission wavelength was 498–538 nm. The excitation wavelength for chloroplasts was 552 nm, and the emission wavelength was 640–720 nm. Arabidopsis was transformed with the inflorescence-dip method and selected on Kanamycin-containing 1/2 MS medium.

4.9. MsSGR Promoters Clone and Activity Analysis

To isolate the promoter sequence of *MsSGR*, its orthologous sequence, *MbSGR*, was first retrieved from the *Magnolia biondii* genome database [61]. A set of PCR primers was designed based on the promoter sequence of *MbSGR* (Table S1). The putative promoter sequence of the *MsSGR* gene was amplified by PCR using the genomic DNA derived from *M. sinostellata* leaves as a DNA template and verified by DNA sequencing and sequence alignment with the *MbSGR* promoter. The cis-acting elements were predicted using PlantCARE (http://bioinformatics.psb.ugent.be/webtools/plantcare/html/, accessed on 6 August 2021). The *MsSGR* promoter fragment was directionally cloned into the *Hind III* and *XbaI* sites of the binary vector pBI121-GUS replacing the 35S CaMv promoter. The ClonExpress® II One Step Cloning kit (Vazyme) was used to generate a MsSGR::GUS vector. This vector was introduced into *A. tumefaciens* GV3101 (Shanghai Weidi Biotechnology, Shanghai, China), which was used to infiltrate tobacco leaves. *A. tumefaciens* cells containing the pBI121-GUS vector with 35S CaMv promoter served as a positive control, and the cells without any vector were used as a negative control. GUS histochemical staining was performed using the GUS Stain kit (Coolaber, Beijing, China). Arabidopsis was transformed with the inflorescence-dip method and selected on the Kanamycin-containing 1/2 MS medium. Transgenic Arabidopsis plants were subjected to LD stress of 16.2 μmol m^{-2}·s^{-1} PAR or a completely dark environment.

4.10. Construction of Yeast Two-Hybrid (Y2H) pGADT7 Library

Total RNA was extracted from the sample of leaves and leaf buds derived from *M. sinostellata* using the Trizol method (Invitrogen, Carlsbad, CA, USA), and mRNA was purified from total RNA using oligo (dT) magnetic beads (Vazyme, Nanjing, China). Subsequently, mRNA was reverse-transcribed into single-strand cDNA, and the cDNA was amplified and purified. TRIMMER-2 cDNA normalization kit (EVROGEN, Moscow, Russia) was used to construct homogenized cDNA and Clontech CHROMA SPIN™+TE-1000 Columns (Takara, Tokyo, Japan) was usedto remove small fragments. The double-stranded cDNA was recombined into pGADT7 and then transformed into TOP10 competent cells (Shanghai Weidi Biotechnology, Shanghai, China). Finally, the capacity of two-hybrid (Y2H) pGADT7 library capacity was identified, and it used HighPure Maxi Plasmid Kit (Tiangen, Beijing, China) to extract library plasmid.

4.11. Yeast Two-Hybrid (Y2H) Screen of MsSGR Protein

The ORF of *MsSGR* was inserted into the bait vector pGBKT7 to generate the BD fusion vector, pGBKT7-*MsSGR*, which was co-transformed with pGADT7 (AD) into the yeast strain AH109 (Shanghai Weidi Biotechnology) using heat shock transformation. Positive transformants containing both AD and BD were selected on a synthetic defined (SD) medium (Coolaber, Beijing, China) lacking Trp and Leu (SD-Trp-Leu). Following PCR confirmation, the positive transformants were inoculated on an SD medium lacking Trp, Leu, and His (SD-Trp-Leu-His), an SD medium lacking Trp, Leu, His, and Ade (SD-Trp-Leu-His-Ade), and an SD medium lacking Trp, Leu, His, and Ade but adding x-α-gal (SD-Trp-Leu-His-Ade+x-α-gal), to test whether MsSGR can self-activate the expression of His, Ade, and MEL1 reporter genes. Subsequently, the BD (pGBKT7-MsSGR) and AD (Y2H pGADT7 library) were co-transformed into the yeast strain AH109 (Shanghai Weidi Biotechnology), and the cells were cultured on an SD medium lacking Trp, Leu, and His (SD-Trp-Leu-His). The positive transformants were selected for PCR and sequenced. Finally, the positive transformants were inoculated on SD-Trp-Leu, SD-Trp-Leu-His, SD-Trp-Leu-His-Ade, and SD-Trp-Leu-His-Ade with added x-α-gal to further determine the potentially interacting proteins with MsSGR.

5. Conclusions

The present study revealed the effect of LD stress on the chlorophyll metabolism pathway of *M. sinostellata* and the role of *MsSGR* in regulating chlorophyll degradation in *M. sinostellata* under LD stress. We also studied the interactive network of the MsSGR protein and identified eight chloroplast proteins interacting with MsSGR. Overall, these findings highlight the fact that LD stress can have a significant impact on the growth of *M. sinostellata* by affecting the chlorophyll metabolism pathway. This study also contributes to our understanding of the endangerment mechanisms of wild Magnoliaceae species and provides a theoretical basis for developing conservation strategies. By identifying the interactive network of MsSGR, this study can provide insights into potential targets for interventions aimed at mitigating the effects of LD stress on Magnoliaceae species.

Supplementary Materials: The following supporting information can be downloaded at: https://www.mdpi.com/article/10.3390/ijms24108510/s1.

Author Contributions: M.R. and J.M.: conceptualization; M.R.: methodology; D.L.: software; M.R.: data analysis; C.W. and X.C.: validation; S.Z. and Y.W.: investigation; M.R.: original draft preparation; J.M.: revised and improved technical language; J.M. and Y.S.: funding acquisition and project administration. All authors have read and agreed to the published version of the manuscript.

Funding: This research was funded by Zhejiang Provincial Key Laboratory of Germplasm Innovation and Utilization for Garden Plants (S2020004); Key Scientific and Technological Grant of Zhejiang for Breeding New Agricultural Varieties (No. 2021C02071-3).

Institutional Review Board Statement: Not applicable.

Informed Consent Statement: Not applicable.

Data Availability Statement: The RNAseq data that support the findings of this study were deposited in the National Center for Biotechnology Information (NCBI) with the accession number PRJNA770262.

Conflicts of Interest: The authors declare no conflict of interest.

References

1. Shen, Y.; Meng, D.; McGrouther, K.; Zhang, J.; Cheng, L. Efficient isolation of *Magnolia* protoplasts and the application to subcellular localization of *MdeHSF1*. *Plant Methods* **2017**, *13*, 44. [CrossRef] [PubMed]
2. Setsuko, S.; Ishida, K.; Ueno, S.; Tsumura, Y.; Tomaru, N. Population differentiation and gene flow within a metapopulation of a threatened tree, *Magnolia stellata* (Magnoliaceae). *Am. J. Bot.* **2007**, *94*, 128–136. [CrossRef] [PubMed]
3. Yang, Y.; Jia, Z.; Chen, F.; Sang, Z.; Ma, L. Optimal Light Regime for the Rare Species *Magnolia wufengensis* in Northern China. *Plant Sci. J.* **2015**, *33*, 377–387.
4. Tan, M.; Yang, Z.; Yang, X.; Cheng, X.; Li, G.; Ma, W. Study on Seed Germination and Seedling Growth of *Houpoëa officinalis* in Different Habitats. *J. Ecol. Rural. Environ.* **2018**, *34*, 910–916.
5. Yang, Q.; Xiao, Z.; Hu, X.; Ouyang, M.; Chen, X.; Lin, G.; Xu, J.; Yang, G. Endangered mechanisms of *Sinomanglietia glauca*: Exploring and prospect. *Guihaia* **2017**, *37*, 653–660.
6. Yu, Z.; Chen, X.; Lu, L.; Liu, X.; Yin, H.; Shen, Y. Distribution and Community Structure of *Magnolia sinostellata*. *J. Zhejiang For. Sci. Technol.* **2015**, *35*, 47–52.
7. Lu, D.; Liu, B.; Ren, M.; Wu, C.; Ma, J.; Shen, Y. Light Deficiency Inhibits Growth by Affecting Photosynthesis Efficiency as well as JA and Ethylene Signaling in Endangered Plant *Magnolia sinostellata*. *Plants* **2021**, *10*, 2261. [CrossRef]
8. Gao, S.; Liu, X.; Liu, Y.; Cao, B.; Chen, Z.; Xu, K. The Spectral Irradiance, Growth, Photosynthetic Characteristics, Antioxidant System, and Nutritional Status of Green Onion (*Allium fistulosum* L.) Grown under Different Photo-Selective Nets. *Front. Plant Sci.* **2021**, *12*, 650471. [CrossRef]
9. Kim, B.; Jeong, Y.J.; Corvalan, C.; Fujioka, S.; Cho, S.; Park, T.; Choe, S. Darkness and *gulliver2/phyB* mutation decrease the abundance of phosphorylated BZR1 to activate brassinosteroid signaling in Arabidopsis. *Plant J.* **2014**, *77*, 737–747. [CrossRef]
10. Katahata, S.; Naramoto, M.; Kakubari, Y.; Mukai, Y. Photosynthetic capacity and nitrogen partitioning in foliage of the evergreen shrub *Daphniphyllum humile* along a natural light gradient. *Tree Physiol* **2007**, *27*, 199–208. [CrossRef]
11. Casal, J.J. Photoreceptor signaling networks in plant responses to shade. *Annu. Rev. Plant Biol.* **2013**, *64*, 403–427. [CrossRef] [PubMed]
12. Martínez-García, J.F.; Galstyan, A.; Salla-Martret, M.; Cifuentes-Esquivel, N.; Gallemí, M.; Bou-Torrent, J. Regulatory Components of Shade Avoidance Syndrome. *Adv. Bot. Res.* **2010**, *53*, 65–116.
13. Domingos, S.; Scafidi, P.; Cardoso, V.; Leitao, A.E.; Di Lorenzo, R.; Oliveira, C.M.; Goulao, L.F. Flower abscission in *Vitis vinifera* L. triggered by gibberellic acid and shade discloses differences in the underlying metabolic pathways. *Front. Plant Sci.* **2015**, *6*, 457. [CrossRef] [PubMed]
14. Zhao, D.; Hao, Z.; Tao, J. Effects of shade on plant growth and flower quality in the herbaceous peony (*Paeonia lactiflora* Pall.). *Plant Physiol. Biochem.* **2012**, *61*, 187–196. [CrossRef] [PubMed]
15. Liu, L.; Lin, N.; Liu, X.; Yang, S.; Wang, W.; Wan, X. From Chloroplast Biogenesis to Chlorophyll Accumulation: The Interplay of Light and Hormones on Gene Expression in *Camellia sinensis* cv. Shuchazao Leaves. *Front. Plant Sci.* **2020**, *11*, 256. [CrossRef]
16. Hashemi-Dezfouli, A.; Herbert, S.J. Intensifying Plant Density Response of Corn with Artificial Shade. *Agron. J.* **1992**, *84*, 547–551. [CrossRef]
17. Tanaka, R.; Tanaka, A. Tetrapyrrole biosynthesis in higher plants. *Annu. Rev. Plant Biol.* **2007**, *58*, 321–346. [CrossRef]
18. Bollivar, D.W. Recent advances in chlorophyll biosynthesis. *Photosynth. Res.* **2006**, *90*, 173–194. [CrossRef]
19. Meguro, M.; Ito, H.; Takabayashi, A.; Tanaka, R.; Tanaka, A. Identification of the 7-hydroxymethyl chlorophyll a reductase of the chlorophyll cycle in Arabidopsis. *Plant Cell* **2011**, *23*, 3442–3453. [CrossRef]
20. Nagata, N.; Tanaka, R.; Satoh, S.; Tanaka, A. Identification of a vinyl reductase gene for chlorophyll synthesis in *Arabidopsis thaliana* and implications for the evolution of Prochlorococcus species. *Plant Cell* **2005**, *17*, 233–240. [CrossRef]
21. Beale, S.I. Green genes gleaned. *Trends Plant Sci.* **2005**, *10*, 309–312. [CrossRef] [PubMed]
22. Kumar, A.M.; Soll, D. Antisense *HEMA1* RNA expression inhibits heme and chlorophyll biosynthesis in Arabidopsis. *Plant Physiol.* **2000**, *122*, 49–56. [CrossRef] [PubMed]
23. Zhang, H.; Li, J.; Yoo, J.H.; Yoo, S.C.; Cho, S.H.; Koh, H.J.; Seo, H.S.; Paek, N.C. Rice *Chlorina-1* and *Chlorina-9* encode ChlD and ChlI subunits of Mg-chelatase, a key enzyme for chlorophyll synthesis and chloroplast development. *Plant Mol. Biol.* **2006**, *62*, 325–337. [CrossRef] [PubMed]
24. Zhou, D.; Li, T.; Yang, Y.; Qu, Z.; Ouyang, L.; Jiang, Z.; Lin, X.; Zhu, C.; Peng, L.; Fu, J.; et al. *OsPLS4* Is Involved in Cuticular Wax Biosynthesis and Affects Leaf Senescence in Rice. *Front. Plant Sci.* **2020**, *11*, 782. [CrossRef]
25. Pruzinska, A.; Tanner, G.; Aubry, S.; Anders, I.; Moser, S.; Muller, T.; Ongania, K.H.; Krautler, B.; Youn, J.Y.; Liljegren, S.J.; et al. Chlorophyll breakdown in senescent *Arabidopsis* leaves. Characterization of chlorophyll catabolites and of chlorophyll catabolic enzymes involved in the degreening reaction. *Plant Physiol.* **2005**, *139*, 52–63. [CrossRef]

26. Kusaba, M.; Ito, H.; Morita, R.; Iida, S.; Sato, Y.; Fujimoto, M.; Kawasaki, S.; Tanaka, R.; Hirochika, H.; Nishimura, M.; et al. Rice Non-Yellow Coloring1 is involved in light-harvesting complex II and grana degradation during leaf senescence. *Plant Cell* **2007**, *19*, 1362–1375. [CrossRef]

27. Hortensteiner, S.; Krautler, B. Chlorophyll breakdown in higher plants. *Biochim. Biophys. Acta* **2011**, *1807*, 977–988. [CrossRef]

28. Sakuraba, Y.; Schelbert, S.; Park, S.Y.; Han, S.H.; Lee, B.D.; Andres, C.B.; Kessler, F.; Hortensteiner, S.; Paek, N.C. Stay-Green and chlorophyll catabolic enzymes interact at light-harvesting complex II for chlorophyll detoxification during leaf senescence in Arabidopsis. *Plant Cell* **2012**, *24*, 507–518. [CrossRef]

29. Qiu, K.; Li, Z.; Yang, Z.; Chen, J.; Wu, S.; Zhu, X.; Gao, S.; Gao, J.; Ren, G.; Kuai, B.; et al. EIN3 and ORE1 Accelerate Degreening during Ethylene-Mediated Leaf Senescence by Directly Activating Chlorophyll Catabolic Genes in Arabidopsis. *PLoS Genet.* **2015**, *11*, e1005399. [CrossRef]

30. Delmas, F.; Sankaranarayanan, S.; Deb, S.; Widdup, E.; Bournonville, C.; Bollier, N.; Northey, J.G.; McCourt, P.; Samuel, M.A. ABI3 controls embryo degreening through Mendel's I locus. *Proc. Natl. Acad. Sci. USA* **2013**, *110*, E3888–E3894. [CrossRef]

31. Zhu, X.; Chen, J.; Xie, Z.; Gao, J.; Ren, G.; Gao, S.; Zhou, X.; Kuai, B. Jasmonic acid promotes degreening via MYC2/3/4- and ANAC019/055/072-mediated regulation of major chlorophyll catabolic genes. *Plant J.* **2015**, *84*, 597–610. [CrossRef] [PubMed]

32. Sakuraba, Y.; Jeong, J.; Kang, M.Y.; Kim, J.; Paek, N.C.; Choi, G. Phytochrome-interacting transcription factors PIF4 and PIF5 induce leaf senescence in Arabidopsis. *Nat. Commun.* **2014**, *5*, 4636. [CrossRef] [PubMed]

33. Zhang, Y.; Liu, Z.; Chen, Y.; He, J.X.; Bi, Y. PHYTOCHROME-INTERACTING FACTOR 5 (PIF5) positively regulates dark-induced senescence and chlorophyll degradation in Arabidopsis. *Plant Sci.* **2015**, *237*, 57–68. [CrossRef] [PubMed]

34. *Magnolia sinostellata*. The IUCN Red List of Threatened Species 2015. Available online: https://doi.org/10.2305/IUCN.UK.2015-2.RLTS.T69306476A69306540.en (accessed on 15 February 2023).

35. Yu, Q.; Shen, Y.; Wang, Q.; Wang, X.; Fan, L.; Wang, Y.; Zhang, S.; Liu, Z.; Zhang, M. Light deficiency and waterlogging affect chlorophyll metabolism and photosynthesis in *Magnolia sinostellata*. *Trees* **2018**, *33*, 11–22. [CrossRef]

36. Chen, T.; Zhang, H.; Zeng, R.; Wang, X.; Huang, L.; Wang, L.; Wang, X.; Zhang, L. Shade Effects on Peanut Yield Associate with Physiological and Expressional Regulation on Photosynthesis and Sucrose Metabolism. *Int. J. Mol. Sci.* **2020**, *21*, 5284. [CrossRef]

37. Semchenko, M.; Lepik, M.; Götzenberger, L.; Zobel, K. Positive effect of shade on plant growth: Amelioration of stress or active regulation of growth rate? *J. Ecol.* **2012**, *100*, 459–466. [CrossRef]

38. Wang, X.; Gao, X.; Liu, Y.; Fan, S.; Ma, Q. Progress of Research on the Regulatory Pathway of the Plant Shade-Avoidance Syndrome. *Front. Plant Sci.* **2020**, *11*, 439. [CrossRef]

39. Sun, W.; Han, H.; Deng, L.; Sun, C.; Xu, Y.; Lin, L.; Ren, P.; Zhao, J.; Zhai, Q.; Li, C. Mediator Subunit MED25 Physically Interacts with phytochrome interacting factor4 to Regulate Shade-Induced Hypocotyl Elongation in Tomato. *Plant Physiol.* **2020**, *184*, 1549–1562. [CrossRef]

40. García-Sánchez, F.; Simón, I.; Lidón, V.; Manera, F.J.; Simón-Grao, S.; Pérez-Pérez, J.G.; Gimeno, V. Shade screen increases the vegetative growth but not the production in 'Fino 49' lemon trees grafted on *Citrus macrophylla* and *Citrus aurantium* L. *Sci. Hortic.* **2015**, *194*, 175–180. [CrossRef]

41. Lugassi-Ben-Hamo, M.; Kitron, M.; Bustan, A.; Zaccai, M. Effect of shade regime on flower development, yield and quality in lisianthus. *Sci. Hortic.* **2010**, *124*, 248–253. [CrossRef]

42. Lee, S.; Kim, M.H.; Lee, J.H.; Jeon, J.; Kwak, J.M.; Kim, Y.J. Glycosyltransferase-Like RSE1 Negatively Regulates Leaf Senescence through Salicylic Acid Signaling in Arabidopsis. *Front. Plant Sci.* **2020**, *11*, 551. [CrossRef] [PubMed]

43. Jin, D.; Wang, X.; Xu, Y.; Gui, H.; Zhang, H.; Dong, Q.; Sikder, R.K.; Yang, G.; Song, M. Chemical Defoliant Promotes Leaf Abscission by Altering ROS Metabolism and Photosynthetic Efficiency in *Gossypium hirsutum*. *Int. J. Mol. Sci.* **2020**, *21*, 2738. [CrossRef] [PubMed]

44. Kim, T.; Kang, K.; Kim, S.H.; An, G.; Paek, N.C. OsWRKY5 Promotes Rice Leaf Senescence via Senescence-Associated NAC and Abscisic Acid Biosynthesis Pathway. *Int. J. Mol. Sci.* **2019**, *20*, 4437. [CrossRef] [PubMed]

45. Song, Y.; Yang, C.; Gao, S.; Zhang, W.; Li, L.; Kuai, B. Age-triggered and dark-induced leaf senescence require the bHLH transcription factors PIF3, 4, and 5. *Mol. Plant.* **2014**, *7*, 1776–1787. [CrossRef] [PubMed]

46. Wang, R.; Yang, F.; Zhang, X.Q.; Wu, D.; Tan, C.; Westcott, S.; Broughton, S.; Li, C.; Zhang, W.; Xu, Y. Characterization of a Thermo-Inducible Chlorophyll-Deficient Mutant in Barley. *Front. Plant Sci.* **2017**, *8*, 1936. [CrossRef] [PubMed]

47. Shimoda, Y.; Ito, H.; Tanaka, A. Arabidopsis *Stay-Green*, Mendel's Green Cotyledon Gene, Encodes Magnesium-Dechelatase. *Plant Cell* **2016**, *28*, 2147–2160. [CrossRef]

48. Chen, X.; Li, J.; Yu, Y.; Kou, X.; Periakaruppan, R.; Chen, X.; Li, X. *Stay-Green* and *light-harvesting complex II chlorophyll a/b binding protein* are involved in albinism of a novel albino tea germplasm 'Huabai 1'. *Sci. Hortic.* **2022**, *293*. [CrossRef]

49. Ren, G.; An, K.; Liao, Y.; Zhou, X.; Cao, Y.; Zhao, H.; Ge, X.; Kuai, B. Identification of a novel chloroplast protein AtNYE1 regulating chlorophyll degradation during leaf senescence in Arabidopsis. *Plant Physiol* **2007**, *144*, 1429–1441. [CrossRef]

50. Wang, N.; Zhang, C.; Bian, S.; Chang, P.; Xuan, L.; Fan, L.; Yu, Q.; Liu, Z.; Gu, C.; Zhang, S.; et al. Flavonoid Components of Different Color *Magnolia* Flowers and Their Relationship to Cultivar Selections. *HortScience* **2019**, *54*, 404–408. [CrossRef]

51. Gao, S.; Gao, J.; Zhu, X.; Song, Y.; Li, Z.; Ren, G.; Zhou, X.; Kuai, B. ABF2, ABF3, and ABF4 Promote ABA-Mediated Chlorophyll Degradation and Leaf Senescence by Transcriptional Activation of Chlorophyll Catabolic Genes and Senescence-Associated Genes in Arabidopsis. *Mol. Plant* **2016**, *9*, 1272–1285. [CrossRef]

52. Sato, Y.; Morita, R.; Nishimura, M.; Yamaguchi, H.; Kusaba, M. Mendel's green cotyledon gene encodes a positive regulator of the chlorophyll-degrading pathway. *Proc. Natl. Acad. Sci. USA* **2007**, *104*, 14169–14174. [CrossRef] [PubMed]
53. Aubry, S.; Mani, J.; Hortensteiner, S. Stay-green protein, defective in Mendel's green cotyledon mutant, acts independent and upstream of pheophorbide a oxygenase in the chlorophyll catabolic pathway. *Plant Mol. Biol.* **2008**, *67*, 243–256. [CrossRef] [PubMed]
54. Hortensteiner, S. Stay-green regulates chlorophyll and chlorophyll-binding protein degradation during senescence. *Trends Plant Sci.* **2009**, *14*, 155–162. [CrossRef] [PubMed]
55. Luo, Z.; Zhang, J.; Li, J.; Yang, C.; Wang, T.; Ouyang, B.; Li, H.; Giovannoni, J.; Ye, Z. A Stay-Green protein SlSGR1 regulates lycopene and beta-carotene accumulation by interacting directly with SlPSY1 during ripening processes in tomato. *New Phytol.* **2013**, *198*, 442–452. [CrossRef] [PubMed]
56. Bibhuti Bhusan Mishra, B.x. Polyphonel Oxidases: Biochemical and Molecular Characterization, Distribution, Role and its Control. *Enzym. Eng.* **2016**, *05*, 141–149. [CrossRef]
57. Fu, S.; Shao, J.; Zhou, C.; Hartung, J.S. Co-infection of Sweet Orange with Severe and Mild Strains of *Citrus tristeza virus* Is Overwhelmingly Dominated by the Severe Strain on Both the Transcriptional and Biological Levels. *Front. Plant Sci.* **2017**, *8*, 1419. [CrossRef]
58. Chen, J. Conversion relationship between lx and umol·m^{-2} s^{-1} of practical light source. *Acta Agric. Univ. Henanensis* **1998**, *32*, 200–202.
59. Wellburn, A.R.; Lichtenthaler, H. Formulae and Program to Determine Total Carotenoids and Chlorophylls A and B of Leaf Extracts in Different Solvents. In *Advances in Photosynthesis Research, Proceedings of the VIth International Congress on Photosynthesis, Brussels, Belgium, 1–6 August 1983; Volume 2*; Sybesma, C., Ed.; Springer: Dordrecht, The Netherlands, 1984; pp. 9–12.
60. Livak, K.J.; Schmittgen, T.D. Analysis of relative gene expression data using real-time quantitative PCR and the 2(-Delta Delta C(T)) Method. *Methods* **2001**, *25*, 402–408. [CrossRef]
61. Dong, S.; Liu, M.; Liu, Y.; Chen, F.; Yang, T.; Chen, L.; Zhang, X.; Guo, X.; Fang, D.; Li, L.; et al. The genome of *Magnolia biondii* Pamp. provides insights into the evolution of *Magnoliales* and biosynthesis of terpenoids. *Hortic. Res.* **2021**, *8*, 38. [CrossRef]

 International Journal of *Molecular Sciences*

Article

Evaluating the Oxidation Rate of Reduced Ferredoxin in *Arabidopsis thaliana* Independent of Photosynthetic Linear Electron Flow: Plausible Activity of Ferredoxin-Dependent Cyclic Electron Flow around Photosystem I

Miho Ohnishi [1,2], Shu Maekawa [1], Shinya Wada [1,2], Kentaro Ifuku [2,3] and Chikahiro Miyake [1,2,*]

[1] Graduate School for Agricultural Science, Kobe University, 1-1 Rokkodai, Nada-Ku, Kobe 657-8501, Japan
[2] Core Research for Evolutional Science and Technology (CREST), Japan Science and Technology Agency (JST), 7 Gobancho, Tokyo 102-0076, Japan
[3] Graduate School for Agriculture, Kyoto University, Kitashirakawa Oiwake-cho, Sakyo-ku, Kyoto 606-8502, Japan
* Correspondence: cmiyake@hawk.kobe-u.ac.jp

Abstract: The activity of ferredoxin (Fd)-dependent cyclic electron flow (Fd-CEF) around photosystem I (PSI) was determined in intact leaves of *Arabidopsis thaliana*. The oxidation rate of Fd reduced by PSI (vFd) and photosynthetic linear electron flow activity are simultaneously measured under actinic light illumination. The vFd showed a curved response to the photosynthetic linear electron flow activity. In the lower range of photosynthetic linear flow activity with plastoquinone (PQ) in a highly reduced state, vFd clearly showed a linear relationship with photosynthetic linear electron flow activity. On the other hand, vFd increased sharply when photosynthetic linear electron flow activity became saturated with oxidized PQ as the net CO_2 assimilation rate increased. That is, under higher photosynthesis conditions, we observed excess vFd resulting in electron flow over photosynthetic linear electron flow. The situation in which excess vFd was observed was consistent with the previous Fd-CEF model. Thus, excess vFd could be attributed to the in vivo activity of Fd-CEF. Furthermore, the excess vFd was also observed in NAD(P)H dehydrogenase-deficient mutants localized in the thylakoid membrane. The physiological significance of the excessive vFd was discussed.

Keywords: cyclic electron flow; ferredoxin; NADH dehydrogenase; pgr5; photosynthesis; photosystem I

Citation: Ohnishi, M.; Maekawa, S.; Wada, S.; Ifuku, K.; Miyake, C. Evaluating the Oxidation Rate of Reduced Ferredoxin in *Arabidopsis thaliana* Independent of Photosynthetic Linear Electron Flow: Plausible Activity of Ferredoxin-Dependent Cyclic Electron Flow around Photosystem I. *Int. J. Mol. Sci.* **2023**, *24*, 12145. https://doi.org/10.3390/ijms241512145

Academic Editor: Martin Bartas

Received: 11 June 2023
Revised: 24 July 2023
Accepted: 24 July 2023
Published: 29 July 2023

1. Introduction

In photosynthesis, both reaction center chlorophylls (P680 in photosystem II (PSII) and P700 in photosystem I (PSI)) are excited by the photon energy absorbed by the light-harvesting systems located in PSII/I. The absorbed photon energy is converted to the electron flux starting at H_2O oxidation in PSII, and the electron flow ends in the reduction of ferredoxin (Fd) at PSI. The reduced Fd delivers electrons mainly to reactions catalyzed by Fd-NADP oxidoreductase to produce NADPH. Simultaneously, with the photosynthetic linear electron flow from H_2O to Fd, protons accumulate on the luminal side of the thylakoid membrane, forming ΔpH across the membrane. The ΔpH, as a proton motive force, drives ATP synthase to produce ATP. These energy compounds, including reduced Fd, NADPH, and ATP produced in the light reaction, drive the dark reactions, net CO_2 assimilation, and photorespiration in C_3 plants.

Many researchers have proposed that Fd could also deliver electrons to plastoquinone (PQ) in the photosynthetic electron transport system through Fd-quinone oxidoreductase (FQR) [1–3]. FQR-dependent electron flow has been called Fd-dependent cyclic electron flow around PSI (Fd-CEF). If the process works, Fd-CEF could contribute to the induction of ΔpH across the thylakoid membrane [2,4,5]. The electron flow from the reduced Fd to PQ through the cytochrome (Cyt) b_6/f complex could drive excess Q-cycle activity in the

Cyt b_6/f complex against the photosynthetic linear electron flow and form an excessive ΔpH. The ΔpH induced by Fd-CEF could contribute to the induction of nonphotochemical quenching of chlorophyll (Chl) fluorescence and the production of ATP [2,5].

In vivo, the activity of Fd-CEF has been shown as the excess quantum yield of PSI against the apparent quantum yield of PSII [Y(II)] [6,7]. The apparent quantum yield of PSI [Y(I)] becomes excessive against Y(II) with oxidized P700 [8]; these observations were artifacts [9]. The value of Y(I) was measured by the saturation-pulse illumination method [10]. The saturation-pulse illumination under actinic light illumination excites the ground state of P700 to oxidized P700 (P700$^+$) through light-excited P700 (P700*), and the ratio of induced P700$^+$ to total P700 has been estimated as Y(I). The amount of induced P700$^+$ depends on the rate-determining step of the P700 photooxidation reduction cycle in PSI [9]. In the P700 photooxidation reduction cycle, P700* is oxidized to P700$^+$ by donating electrons to the electron acceptors A_o, A_1, F_x, and F_A/F_B sequentially to Fd. P700$^+$ is reduced to the ground state of P700 by electrons from PSII through PQ, the Cyt b_6/f complex, and plastocyanin (PC). If the reduction of P700$^+$ in the cycle was the rate-determining step, in which oxidized P700 accumulated under actinic light illumination, the amount of the ground state of P700 was overestimated by the saturation-pulse illumination method [9]. On the other hand, if the oxidation of P700* in the cycle was the rate-determining step, the amount of the ground state of P700 was underestimated by the saturation-pulse illumination method [10]. It is too difficult to estimate the true Y(I) when evaluating Fd-CEF activity [9].

Furthermore, the ability of electron donation from the reduced Fd to the oxidized PQ through FQR has been evaluated using the isolated thylakoid membrane [3]. The addition of Fd/NADPH to the thylakoid membrane increased the minimal yield of Chl fluorescence, and PQ was reduced by the reduced Fd. The rate of increase in the minimal yield of Chl fluorescence was treated as the activity of FQR. However, the reduced Fd donates electrons to PSII, not PQ [11]. Furthermore, the reduced Fd donates electrons to Cyt b_{559}, which is inhibited by antimycin A [12]. The effect of antimycin A also inhibits Fd-dependent quenching of 9-aminoacridine fluorescence, which is driven by far-red light illumination [12]. The quenching of 9-aminoacridine fluorescence shows ΔpH formation across the thylakoid membrane. Far-red-driven quenching also results from the activity of Fd-CEF. However, as described above, electron donation from the reduced Fd to PSII could also induce the quenching of 9-aminoacridine fluorescence by maintaining the photosynthetic linear electron flow.

As described above, no credible methods are available to detect and evaluate Fd-CEF activity, that is, FQR activity. To elucidate the physiological function of Fd-CEF, an assay system capable of detecting Fd-CEF in vivo was required. In the present research, we monitored the redox reaction of Fd simultaneously with Chl fluorescence, P700$^+$ and PC$^+$ absorbance changes, and net CO_2 assimilation using intact leaves of *Arabidopsis thaliana*. As a result, excessive turnover of the Fd redox reaction was observed, which is not explained by the photosynthetic linear electron flow in vivo. The excessive Fd redox reaction, which followed the model of Fd-CEF activity, was characterized [1]. The redox balance between the electron donor (the reduced Fd) and the electron acceptor (the oxidized PQ) is needed to obtain the maximum Fd-CEF activity, which is a possible reason why we did not detect the excessive Fd redox reaction [13]. During limited photosynthesis, in which PQ was highly reduced and the apparent quantum yield of PSII was low, the excessive Fd redox reaction was suppressed in vivo. We proposed that the excessive Fd redox reaction reflected the activity of Fd-CEF. Next, we compared Fd-CEF activity between the wild type (WT) and an *Arabidopsis thaliana* mutant lacking NAD(P)H dehydrogenase (NDH) (*crr4*). NDH has been considered to catalyze electron donation from reduced Fd to PQ in the photosynthetic electron transport system [6,14]. *crr4* did not show an Fd-dependent increase in the minimal Chl fluorescence [15]. Surprisingly, *crr4* showed almost the same Fd-CEF activity as the wild type in vivo. The physiological function of Fd-CEF from our results was discussed.

2. Results

Both the gross CO_2 assimilation rate and the apparent quantum yield of PSII [Y(II)] were plotted against the intercellular partial pressures of CO_2 (Ci) (Figure 1). These two parameters showed the same dependencies on Ci in both WT and *crr4*.

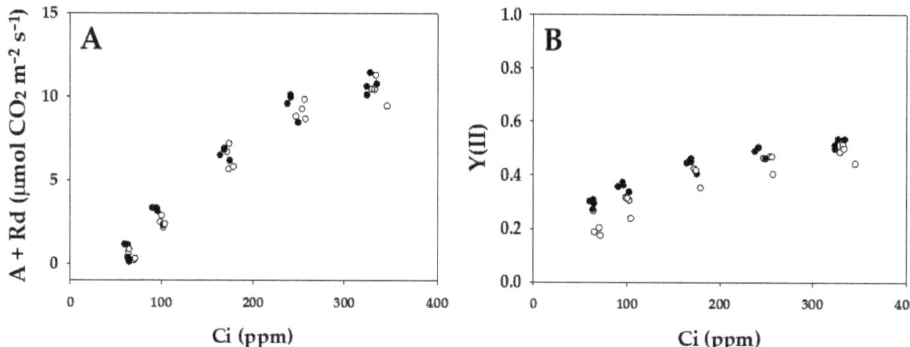

Figure 1. Effects of the intercellular partial pressure of CO_2 (Ci) on the gross CO_2 assimilation rate (A + Rd) and apparent quantum yield of photosystem II (PSII) [Y(II)] in wild-type (WT) and *crr4 Arabidopsis*. (**A**) The net CO_2 assimilation rates were measured at 400 μmol photons m^{-2} s^{-1} and 21 kPa O_2, and Y(II) was simultaneously measured. The dark respiration rates (Rd) were measured before starting actinic light illumination. The gross CO_2 assimilation rates are expressed as A + Rd and were plotted against Ci. (**B**) Y(II) was plotted against Ci. The data were obtained from four independent experiments using leaves attached to four plants of both WT and *crr4* ($N = 4$). The ambient partial pressures of CO_2 were changed from 400 ppm to 300 ppm, then 200 ppm, then 100 ppm, and finally 50 ppm at 21 kPa O_2 for the same leaves. Black symbols, WT; White symbols, *crr4*.

The parameters $P700^+$, PC^+, and Fd^- against Y(II) were plotted in Figure 2. With the decrease in Y(II) caused by lowering Ci, P700 was oxidized from approximately 10 to 40% in WT and from 10 to 50% in *crr4* (Figure 2A). Similarly, PC was oxidized from 65 to 90% in both WT and *crr4* (Figure 2B). In contrast to both $P700^+$ and PC^+, Fd^- did not change in response to the decrease in Y(II) in either WT or *crr4* (Figure 2C), due to the oxidation of P700 in PSI [16].

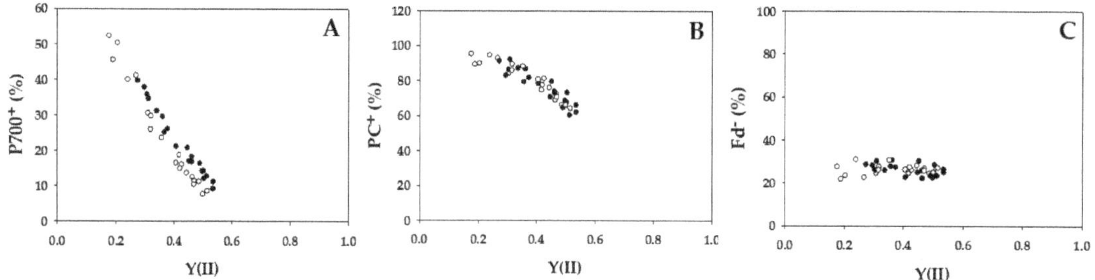

Figure 2. Relationships between $P700^+$, PC^+, Fd^-, and apparent quantum yield of photosystem II (PSII) [Y(II)]. The data for each parameter were measured in the experiments depicted in Figure 1, simultaneously with the net CO_2 assimilation rates and Y(II). (**A**) $P700^+$, (**B**) PC^+, and (**C**) Fd^- were plotted against Y(II). Those ratios of $P700^+$, PC^+, and Fd^- against the total contents are expressed. The data were obtained from four independent experiments using leaves attached to four WT and *crr4* plants ($N = 4$). Black symbols, WT; White symbols, *crr4*.

The parameters non-photochemical quenching (NPQ) and plastoquinone reduced state (1 − qP) against Y(II) were plotted in Figure 3. With the decrease in Y(II), NPQ increased from approximately 0.5 to 1.5 in WT and *crr4* (Figure 3A). Increase in NPQ showed the enhancement of heat dissipation of photon energy absorbed by PSII. WT and *crr4* also showed the same dependence of 1 − qP on the decrease in Y(II), where 1 -qP increased with the decrease in Y(II) (Figure 3B). Increase in 1 − qP showed the reduction of plastoquinone pool.

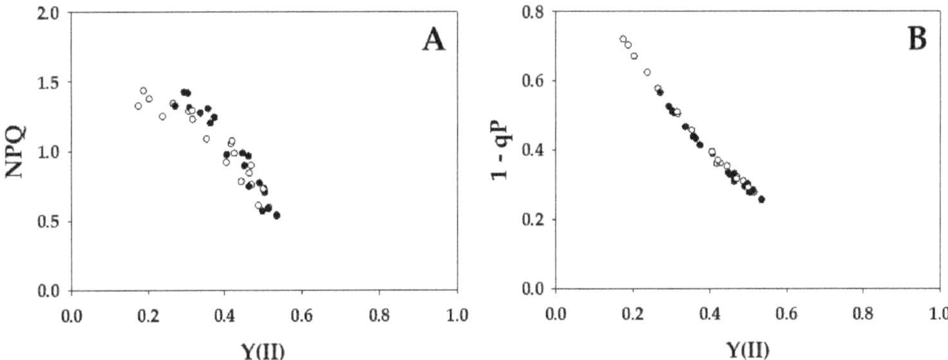

Figure 3. Relationships between non-photochemical quenching (NPQ), plastoquinone reduced state (1 − qP), and apparent quantum yield of photosystem II (PSII) [Y(II)]. The data for each parameter were measured in the experiments depicted in Figure 1, simultaneously with the net CO_2 assimilation rates and Y(II). (**A**) NPQ and (**B**) 1 − qP were plotted against Y(II). The data were obtained from four independent experiments using leaves attached to four WT and *crr4* plants (*N* = 4). Black symbols, WT; White symbols, *crr4*.

The parameters apparent quantum yield of PSI [Y(I)] and apparent quantum yield of non-photochemically energy dissipation of photoexcited P700 [Y(NA)] were plotted in Figure 4. As described in the Introduction, Y(I) and Y(NA) were estimated by illuminating the leaves with saturated-pulse light under actinic light illumination. Y(I) reflected the strength of the donor-side limitation of the P700 photooxidation reduction cycle, and Y(NA) reflected that of the acceptor-side limitation during the saturated-pulse illumination [16]. That is, if P700 was highly oxidized under actinic light, Y(I) showed a higher value, and the reverse was also true. For example, if Y(NA) was higher, Y(I) was lower. A higher Y(NA) is accompanied by a highly reduced state of Fd [16]. With the decrease in Y(II), Y(I) decreased from approximately 0.8 to 0.5 in WT and *crr4* (Figure 4A). Y(NA) in WT was slightly lower than that in *crr4* (Figure 4B). Principally, the dependencies of Y(I) and Y(NA) on Y(II) in *crr4* were almost the same as those in WT.

Finally, the oxidation rates of the reduced Fd (vFd) against Y(II) under steady-state conditions, which was estimated by DIRK analysis (see Section 4), were plotted in Figure 5. vFd did not show a linear relationship with Y(II) in either WT or *crr4* (Figure 5A). In WT, increasing Y(II) increased vFd, and a nearly linear relationship between vFd and Y(II) was found in the low range of Y(II) from 0.25 to 0.35. However, above Y(II) = 0.35, although Y(II) became saturated, vFd further increased. That is, excessive turnover of the redox reaction of Fd against Y(II) appeared in the higher range of Y(II). This behavior of vFd against Y(II) was also observed in *crr4* (Figure 5A). Furthermore, vFd showed dependence on the increase in qP (Figure 5B). The increase in qP and the enhancement of PQ oxidation stimulated the appearance of excessive vFd in WT and *crr4*. That is, the activation of photosynthetic linear electron flow oxidized PQ and induced excessive vFd, as observed in the increase in Y(II).

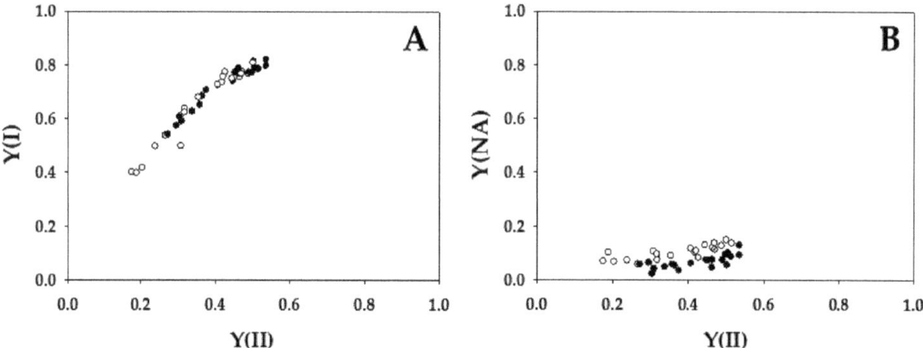

Figure 4. Relationships between apparent quantum yield of PSI [Y(I)], apparent quantum yield of non-photochemical energy dissipation of photoexcited P700 [Y(NA)], and apparent quantum yield of photosystem II (PSII) [Y(II)]. The data for each parameter were measured in the experiments depicted in Figure 1, simultaneously with the net CO_2 assimilation rates and Y(II). (**A**) Y(I) and (**B**) Y(NA) were plotted against Y(II). The data were obtained from four independent experiments using leaves attached to four WT and *crr4* plants (N = 4). Black symbols, WT; White symbols, *crr4*.

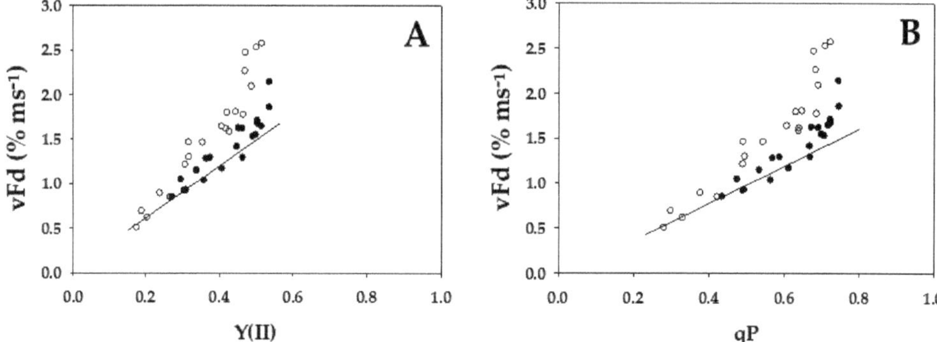

Figure 5. Relationships between apparent quantum yield of photosystem II (PSII) [Y(II)], plasto-quinone oxidized state (qP), and vFd. The data for each parameter were measured in the experiments depicted in Figure 1, simultaneously with the net CO_2 assimilation rates and Y(II). (**A**) Y(II) and (**B**) qP were plotted against vFd. In the experiments shown in Figure 1, the oxidation rate of ferredoxin (Fd) was determined by DIRK analysis (see Section 4). To determine the oxidation rate of Fd^- under illuminated conditions, actinic light was transiently turned off for 400 ms. The initial slope of the decrease in Fd^- indicates the oxidation rate of Fd^- (vFd). These data were obtained at a steady state, which was confirmed by the achievement of stable conditions for both net CO_2 assimilation and Y(II). The data were obtained from four independent experiments using leaves attached to four WT and *crr4* plants (N = 4). Black symbols, WT; White symbols, *crr4*.

3. Discussion

In the present research, we compared the activity of Fd-CEF in *Arabidopsis thaliana*, WT, and *crr4*. There was a nonlinear relationship between Y(II) and vFd (Figure 5A). The gross CO_2 assimilation rate increased with increasing Ci. On the other hand, Y(II) became saturated. Furthermore, vFd did not saturate and continued to increase (Figure 5A). That is, the increase in vFd deviated from the increase in Y(II) in both WT and *crr4*. This result indicates the presence of excessive vFd unrelated to photosynthetic linear electron flow (Figure 5A). A deviation of vFd from the linear relationship was also found between vFd and qP (Figure 5B). One parameter of Chl fluorescence, qP, reflected the reduction-oxidation state of PQ. The increase in qP showed the oxidation of the reduced PQ, which was induced by the stimulation of the photosynthetic linear electron flow. That is, the appearance of excessive vFd required the reduced PQ to be oxidized.

We propose that the excessive vFd, which is unrelated to the photosynthetic linear electron flow, reflects the electron flux in the ferredoxin-dependent cyclic electron flow around PSI (Fd-CEF), judged from the following model (Figure 6) [1]. In Fd-CEF, the reduced Fd donates electrons to PQ through Fd-quinone oxidoreductase (FQR). That is, the electron donor is the reduced Fd, and the electron acceptor is the oxidized PQ. The Fd-CEF velocity is proportional to the product of the reduced Fd and the oxidized PQ and depends on the activity of FQR [1]. If PQ was completely reduced, the activity of Fd-CEF was zero, even if Fd was reduced (Figure 6). Conversely, if PQ was completely oxidized, the activity of Fd-CEF was also zero because Fd did not possess any electrons for the reduction of PQ [1,5]. In the present research, the reduction-oxidation state of Fd was constant in the range of Y(II) (Figure 2C), due to the oxidation of P700, which reflects the limitation of the electron flow from the reduced PQ to the oxidized P700 in PSI; that is, the reduction of the oxidized P700 is the rate-determining step in the P700 photooxidation reduction cycle [9]. The constant level of reduced Fd induced Fd-CEF activity, depending on the reduced state of PQ (Figures 3B and 5B). As the reduced PQ was oxidized by the stimulation of the linear electron flow of photosynthesis, excessive vFd appeared (Figures 3B and 5B). The reverse was true. These relationships between excessive vFd and qP followed the Fd-CEF model [1]. Hereafter, excessive vFd will be called Fd-CEF activity.

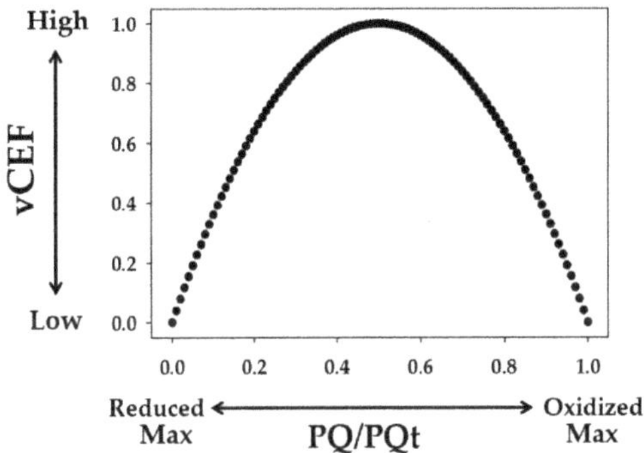

Figure 6. Model for the expression of Fd-CEF activity (vCEF): the dependence of vCEF on the ratio of the oxidized PQ to total PQ pool (PQt), (PQ/PQt). The vCEF is plotted against PQ/PQt according to the model of Allen [1]. In the extremely reduced state of PQ, in which the intensity of actinic light is higher and/or the intercellular partial pressure of CO_2 is lower, vCEF is greatly suppressed. The stimulation of net CO_2 assimilation with photosynthetic linear electron flow enhanced the oxidation of the reduced PQ, with vCEF increasing, as shown in the present research.

The expression of excessive vFd did not correspond to the enhancement in the ratio of Y(I) to Y(II) caused by the decrease in Y(II) (Supplemental Figure S1A). The increase in the ratio of Y(I) to Y(II) clearly showed a linear relationship with the increase in the oxidation of P700 (Supplemental Figure S1B). The content of the oxidized P700 (P700max') produced by saturation-pulse light illumination (see Section 4) depended on the rate-determining step of the P700 photooxidation reduction cycle in PSI, and the cycle turned over under saturation-pulse light and actinic light illumination [9]. If the reduction of $P700^+$ was the rate-determining step, as observed in the accumulation of the oxidized P700, the yield of P700max' (Y(I)) became larger than Y(II) and did not reflect the true electron flux in PSI [9].

Fd-CEF activity was not detected in our previous study [13]. The vFd observed during the induction of net CO_2 assimilation precisely followed the increase in Y(II) in wheat leaves under the higher light intensity. Furthermore, vFd followed Y(II) in the A/Ci analysis under higher light intensity. That is, vFd clearly showed a positive linear relationship with

Y(II). In these analyses, Y(II) showed a lower value in the 0.05 to 0.3 range [13]. These lower Y(II) values indicate a higher reduction level of PQ. The present research suggests that higher light intensity and lower Ci induced a higher reduction state of PQ, which suppressed Fd-CEF activity. The expression of Fd-CEF activity requires the oxidation of PQ (Figure 6) [1], which is why we could not detect Fd-CEF activity [13].

It has been proposed that Fd-CEF contributes to excess production of ΔpH across the thylakoid membrane in cooperation with the photosynthetic linear electron flow [2]. The ΔpH induced by Fd-CEF drives the induction of NPQ of Chl fluorescence and satisfies the ATP requirement for achieving a higher rate of net CO_2 assimilation [1,2]. In the present research, vFd showed a negative relationship with the NPQ of Chl fluorescence (Figures 3A and 5A). However, this finding does not indicate that Fd-CEF cannot induce ΔpH for the induction of NPQ because the excessive vFd was accompanied by the stimulation of the gross CO_2 assimilation rate as Ci increased (Figures 1 and 5). That is, the increase in ATP consumption occurred simultaneously with the increase in excessive vFd. Therefore, the ΔpH induced by Fd-CEF did not accumulate, or the total ΔpH induced by the enhanced photosynthetic linear electron flow (the light reaction in photosynthesis) driven by net CO_2 assimilation (the dark reaction in photosynthesis) and the enhanced Fd-CEF (the light reaction in photosynthesis) greatly decreased.

The reaction center Chl in PSI (P700) is oxidized in response to the suppression of net CO_2 assimilation [17]. This is a universal and robust phenomenon observed in organisms that carry out oxygenic photosynthesis [17,18]. The oxidation of P700 alleviates the accumulation of electrons in the electron acceptors of PSI: Fx, F_A/F_B, and Fd [16]. These electron acceptors, including phylloquinone, can donate electrons to O_2 to produce superoxide anion radicals, the primary product of O_2 reduction in the Mehler reaction [19,20]. The accumulation of the reduced forms of these electron carriers causes PSI photoinhibition [21–25]; therefore, P700 should be oxidized under the suppressed condition of photosynthesis and/or the suppressed utilization of photon energy for net CO_2 assimilation [16].

A molecular mechanism has been proposed for the suppression of electron accumulation in the electron carriers at the acceptor side of PSI [17,26]. To suppress the electron accumulation at the acceptor side of PSI, the electron flux to the acceptor side of PSI from PSII should be decreased, and the bottlenecked reaction is the oxidation of reduced PQ by the Cyt b_6/f complex. The downregulation of PQ oxidation activity by the Cyt b_6/f complex is induced by acidification of the luminal side of the thylakoid membrane [27] and the highly reduced state of PQ (reduction-induced suppression of electron flow, RISE) [17]. RISE occurs under extremely suppressed photosynthetic conditions, in which the induction of acidification at the luminal side of the thylakoid membrane saturates in response to the decrease in photosynthesis activity [28,29]. On the other hand, acidification of the luminal side of the thylakoid membrane, observed as the induction of ΔpH formation across the thylakoid membrane, may be triggered by Fd-CEF [2]. However, as shown in Figure 5, no enhanced activity of Fd-CEF was observed when the efficiency of the net CO_2 assimilation rate was lower, in which the photosynthetic linear electron flow was suppressed and PQ was highly reduced. That is, Fd-CEF cannot function and cannot contribute to the induction of the ΔpH under such conditions. On the one hand, it was suggested that photorespiration was a prerequisite for the induction of ΔpH formation across the thylakoid membrane [29]. Furthermore, photorespiration can function as soon as actinic light illumination starts [30,31]. The electron flux in the photosynthetic linear electron flow can be accounted for by both net CO_2 assimilation and photorespiration [17,31]. That is, the photosynthetic linear electron flow, which is observed as Y(II), reflected mainly the net CO_2 assimilation and photorespiration. Based on this knowledge, it was concluded that unless photorespiration functioned, P700 was not oxidized [32]. Under photorespiratory conditions, Y(II) increased with the enhancement in P700 oxidation as soon as the illumination started with actinic light and without CO_2 assimilation [32]. Furthermore, photorespiration induced ΔpH formation across the thylakoid membrane, which caused P700 oxidation in rice plants [28]. The rice plants show intrinsic fluctuations in the stomatal opening. This natural phenomenon occurs in all plants [28]. That is, all plants with stomata are exposed to the fluctuating condition of net CO_2 assimilation, and photorespiratory regulation of the oxidation of the electron carriers at the acceptor side of PSI is necessary at the closed state of the stomata. Then, CO_2 is decreased at the ribulose 1,5-bisphosphate carboxylase/oxygenase (Rubisco) carboxylation site, and photorespiration is activated to form ΔpH for the suppressed electron flow from PSII to the acceptor side of PSI.

To date, the precise pathway of electron flow in Fd-CEF has not been clarified. In the present research, the possibility of catalyzing Fd-CEF by NDH in *Arabidopsis thaliana* was tested. However, the Fd-CEF activity of the *crr4* mutant, which lacks NDH, was the same as that of WT *Arabidopsis thaliana* (Figure 5). Furthermore, *crr4* showed the same dependence of Fd-CEF activity on Y(II) (Figure 5). That is, NDH is not the mediator for the observed Fd-CEF. The Shikanai group proposed, as a candidate for Fd-CEF, that *pgr5/grl1* proteins mediate Fd-CEF [33–35]. However, the mutants deficient in pgr5 and/or pgrl1 showed higher H^+-conductance, which caused the ΔpH across the thylakoid membrane to decrease and suppress P700 oxidation compared to that of WT [36]. No theory has been demonstrated to explain the higher H^+-conductance to lower ΔpH across the thylakoid membrane. Recently, it was elucidated that *pgr5*, which the Shikanai group isolated, is a double mutant that showed lower CO_2 fixation activity than that of WT [37]. That is, the pure single mutant (*pgr5^{hope1}*), which is deficient in only pgr5, showed the same electron sink activity and the same CO_2 fixation rate as WT [16,37]. These results showed that pgr5 is not a factor for Fd-CEF. If pgr5 functioned in Fd-CEF to supply ATP, *pgr5^{hope1}* would not maintain the same net CO_2 assimilation rate as WT. This knowledge should not be ignored when considering the physiological function of pgr5 in photosynthesis. At present, we cannot refer to the physiological function of pgr5 in photosynthesis.

From the knowledge obtained in the present research, one possibility for the molecular mechanism of Fd-CEF in PSI was considered, which does not induce ΔpH formation across the thylakoid membrane [4]. That is, the reduced Fd donates electrons to heme c in the Cyt b_6/f complex. The reduced heme c donates electrons to the high-potential heme b through the lower-potential heme b in the Cyt b_6/f complex. This thermodynamically favored Fd-CEF route drives the fast CEF pathway [4]. If Fd-CEF functioned in the fast CEF pathway, excessive vFd was not related to the induction of NPQ of Chl fluorescence and P700 oxidation in PSI. To further elucidate the physiological function of the excessive vFd and the molecular mechanism driving Fd-CEF, further research is needed.

We developed a method to detect Fd-CEF activity around PSI in intact leaves of *Arabidopsis thaliana*. The oxidation rate of the reduced Fd is a useful indicator of Fd-CEF activity. If the oxidation rate exceeded the photosynthetic linear electron flow rate, Fd-CEF started to function. The Fd-CEF activity required the oxidation of PQ because the reduced Fd could donate electrons to PQ. Therefore, suppression of the net CO_2 assimilation rate with PQ reduction decreased Fd-CEF activity, in contrast to the expectation that Fd-CEF could induce ΔpH across the thylakoid membrane to regulate the electron flux through PQ to PSI; that is, Fd-CEF could not oxidize P700 to suppress reactive oxygen species (ROS) production in PSI. On the other hand, as the net CO_2 assimilation increased, Fd-CEF activity increased. The accelerated activity of Fd-CEF would contribute to higher CO_2 assimilation. Surprisingly, NDH did not affect the activity of Fd-CEF around PSI. Elucidation of the physiological function of NDH requires further research.

4. Materials and Methods

4.1. Plant Materials and Growth Conditions

Arabidopsis plants (*Arabidopsis thaliana* WT and *crr4*) were grown from seeds under standard air-equilibrated conditions with 10 h/14 h day–night cycles at 23 and 20 °C, respectively, and 55–60% relative humidity. The photon flux density was adjusted to 100 µmol photons m^{-2} s^{-1}, which was measured with a light meter (LI-189, LI-COR, Lincoln, NE, USA) equipped with a quantum sensor. Seeds were planted in the soil after 3 days of vernalization at 4 °C. Seedlings were kept in 0.2 (dm)^3 pots containing a 2:1.5 ratio of seeding-culture soil (TAKII Co., Ltd., Kyoto, Japan) to vermiculite and were watered daily. Plants were fertilized with 1000-fold diluted Hyponex fertilizer 8–12–6 (Hyponex Japan, Osaka, Japan) only once in the 3rd week after seeding. The plants at 10 h after the dark duration start of the light/dark cycle of growth conditions were used for all the measurements, which were conducted using rosette leaves of 5- to 6-week-old plants.

4.2. Determination of Chlorophyll and Nitrogen

Leaves after measuring photosynthetic parameters were sampled for content analysis and stored at -80 °C until use. Upon sampling, detached leaves were weighed and electronic images were acquired with a scanner for leaf area measurement by ImageJ (NIH). The Chl and nitrogen (N) contents in the leaves of WT and *crr4* were determined [38] and are shown in Supplemental Table S1. A raw leaf blade was homogenized in 50 mM sodium-phosphate buffer (pH 7.2) containing 120 mM 2-mercaptoethanol, 1 mM iodoacetic acid, and 5% (v/v) glycerol at a leaf:buffer ratio of 1:9 (g/mL) in a chilled mortar and pestle. The total Chl and leaf N contents were measured from a part of this homogenate. The absorbance at 663.6 and 646.6 nm was measured to calculate the Chl content [39]. The Chl content in the leaves is represented on a leaf-area basis [38]. The total leaf N content was determined using Nessler's reagent in a digestion solution after potassium sodium tartrate was added [38]. The homogenate was decomposed by 60% (v/v) sulfuric acid and 30% (v/v) H_2O_2 with heat. The decomposing leaf solution was mixed with distilled water, 10% (w/v) potassium sodium tartrate solution, and 2.5 N NaOH, and Nessler's reagent was immediately added to the mixture. The N content was determined by measuring the change in absorbance at 420 nm.

4.3. Simultaneous Measurements of Chlorophyll Fluorescence, P700, and Fd-Signals with Gas Exchange

Chl fluorescence, P700, Fd, and CO_2 exchange were simultaneously measured using Dual/KLAS-NIR (Heinz Walz GmbH, Effeltrich, Germany), and an infrared gas analyzer (IRGA) LI-7000 (Li-COR, Lincoln, NE, USA) measuring system equipped with a 3010-DUAL gas exchange chamber at several ambient partial pressures of CO_2 at 21 kPa O_2 (Heinz Walz GmbH) was used [16]. The gases were saturated with water vapor at 16 ± 0.1 °C. The leaf temperature was controlled at 25 ± 0.5 °C (relative humidity: 55–60%). The actinic photon flux density at the upper position on the leaf in the chamber was adjusted to the indicated intensity. The net CO_2 assimilation rate (A) and the dark respiration rate (Rd) were measured.

The Chl fluorescence parameters were calculated as follows [40]: F_o, minimum fluorescence from a dark-adapted leaf; F_o', minimum fluorescence from a light-adapted leaf; F_m, maximum fluorescence from a dark-adapted leaf; F_m', maximum fluorescence from a light-adapted leaf; Fs, fluorescence emission from a light-adapted leaf; the apparent quantum yield of PSII, $Y(II) = (F_m' - Fs)/F_m'$ [41]; non-photochemical quenching, non-photochemical quenching (NPQ) $= (F_m - F_m')/F_m'$ [42]; and PQ oxidized state $(qP) = (F_m' - Fs)/(F_m' - F_o')$ [43]. To obtain F_m and F_m', a saturating pulse light (630 nm, 8000 µmol photons m^{-2} s^{-1}, 300 ms) was applied. Red actinic light (630 nm, 400 µmol photons m^{-2} s^{-1}) was supplied using a chip-on-board LED array.

The signals for oxidized P700 (P700$^+$), oxidized plastocyanin (PC$^+$), and reduced ferredoxin (Fd$^-$) were calculated based on the deconvolution of four pulse-modulated dual-wavelength difference signals in the near-infrared region (780–820, 820–870, 840–965, and 870–965 nm) [44]. Both P700 and PC were completely reduced, and Fd was fully oxidized in the dark. To determine the total photo-oxidizable P700 (P700max) and PC (PCmax), a saturation flash was applied after 10 s of illumination with far-red light (740 nm). The following formulas were used: The apparent quantum yield of PSI, $Y(I) = (P700max' - P700^+)/P700max$; the quantum yield of oxidized P700 (P700$^+$), $Y(ND) = P700^+/P700max$; and the apparent quantum yield of nonphotochemical energy dissipation of photoexcited P700 (P700*), $Y(NA) = (P700max - P700max')/P700max$. In the present research, we showed $Y(ND)$ as

P700$^+$ (Figure 2A). The summation of these quantum yields is 1 (Y(I) + Y(ND) + Y(NA) = 1). Total photo-reducible Fd (Fdmax) was determined by illumination with red actinic light (450 μmol photons m^{-2} s^{-1}) after plant leaves were adapted to the dark for 5 min [44]. The redox states of both P700 and PC under actinic light illumination were evaluated as the ratios of P700$^+$ and PC$^+$ to total P700 and total PC, respectively. The redox state of Fd was also determined similarly. The values of P700max, PCmax, and Fdmax in the leaves of WT and crr4 are shown as relative values in Supplemental Table S1.

For the analysis of dark-interval relaxation kinetics (DIRK analysis, [45]), red actinic light (400 μmol photons m^{-2} s^{-1}) was temporarily turned off for 400 ms at steady-state photosynthesis [13]. The oxidation rate of Fd$^-$ was estimated by a Dual/KLAS-NIR spectrophotometer and expressed as the relative values by estimating the initial decay of Fd$^-$.

4.4. Statistical Analysis

Statistical analyses of the corresponding data in Supplemental Table S1 (CI, confidential interval) were performed using the commercial software JMP8 (ver. 14.2.0, SAS Institute Inc., Cary, NC, USA).

Supplementary Materials: The following supporting information can be downloaded at https://www.mdpi.com/article/10.3390/ijms241512145/s1.

Author Contributions: Conceptualization, C.M.; writing-original draft preparation, M.O. and C.M.; writing-review and editing, M.O., C.M., S.W., S.M., and K.I.; supervision, C.M.; funding acquisition, C.M.; software, M.O. All authors have read and agreed to the published version of the manuscript.

Funding: This work was supported by Core Research for Evolutional Science and Technology (CREST) of the Japan Science and Technology Agency, Japan (grant number JPMJCR15O3) to C.M.

Institutional Review Board Statement: Not applicable.

Informed Consent Statement: Not applicable.

Data Availability Statement: Data are contained within the article and Supplementary Material.

Conflicts of Interest: The authors declare no conflict of interest.

References

1. Allen, J.F. Cyclic, pseudocyclic and noncyclic photophosphorylation: New links in the chain. *Trend. Plant Sci.* **2003**, *8*, 15–19. [CrossRef]
2. Heber, U.; Walker, D. Concerning a dual function of coupled cyclic electron transport in leaves. *Plant Physiol.* **1992**, *100*, 1621–1626. [CrossRef]
3. Munekage, Y.; Hojo, M.; Meurer, J.; Endo, T.; Tasaka, M.; Shikanai, T. PGR5 is involved in cyclic electron flow around photosystem I and is essential for photoprotection in *Arabidopsis. Cell* **2002**, *110*, 361–371. [CrossRef]
4. Laisk, A.; Talts, E.; Oja, V., Eichelmann, H., Peterson, R.B. Fast cyclic electron transport around photosystem I in leaves under far-red light: A proton-uncoupled pathway? *Photosyn. Res.* **2010**, *103*, 79–95. [CrossRef]
5. Miyake, C. Alternative electron flows (water-water cycle and cyclic electron flow around PSI) in photosynthesis: Molecular mechanisms and physiological functions. *Plant Cell Physiol.* **2010**, *51*, 1951–1963. [CrossRef]
6. Yamamoto, H.; Takahashi, S.; Badger, M.R.; Shikanai, T. Artificial remodelling of alternative electron flow by flavodiiron proteins in *Arabidopsis. Nat. Plants* **2016**, *2*, 16012. [CrossRef] [PubMed]
7. Wada, S.; Yamamoto, H.; Suzuki, Y.; Yamori, W.; Shikanai, T.; Makino, A. Flavodiiron protein substitutes for cyclic electron flow without competing CO$_2$ assimilation in rice. *Plant Physiol.* **2018**, *176*, 1509–1518. [CrossRef] [PubMed]
8. Miyake, C.; Miyata, M.; Shinzaki, Y.; Tomizawa, K. CO$_2$ response of cyclic electron flow around PSI (CEF-PSI) in tobacco leaves--relative electron fluxes through PSI and PSII determine the magnitude of non-photochemical quenching (NPQ) of Chl fluorescence. *Plant Cell Physiol.* **2005**, *46*, 629–637. [CrossRef]
9. Furutani, R.; Ohnishi, M.; Mori, Y.; Wada, S.; Miyake, C. The difficulty of estimating the electron transport rate at photosystem I. *J. Plant Res.* **2022**, *135*, 565–577. [CrossRef] [PubMed]
10. Klughammer, C.; Schreiber, U. An improved method, using saturating light pulses, for the determination of photosystem I quantum yield via P700-absorbance changes at 830 nm. *Planta* **1994**, *192*, 261–268. [CrossRef]
11. Fisher, N.; Kramer, D.M. Non-photochemical reduction of thylakoid photosynthetic redox carriers in vitro: Relevance to cyclic electron flow around photosystem I? *Biochim. Biophys. Acta* **2014**, *1837*, 1944–1954. [CrossRef] [PubMed]

12. Miyake, C.; Schreiber, U.; Asada, K. Ferredoxin-dependent and antimycin A-sensitive reduction of cytochrome *b-559* by far-red light in maize thylakoids; Participation of a menadiol-reducible cytochrome *b-559* in cyclic electron flow. *Plant Cell Physiol.* **1995**, *36*, 743–748. [CrossRef]

13. Kadota, K.; Furutani, R.; Makino, A.; Suzuki, Y.; Wada, S.; Miyake, C. Oxidation of P700 induces alternative electron flow in photosystem I in wheat leaves. *Plants* **2019**, *8*, 152. [CrossRef] [PubMed]

14. Yamori, W.; Shikanai, T. Physiological functions of cyclic electron transport around photosystem I in sustaining photosynthesis and plant growth. *Ann. Rev. Plant Physiol.* **2016**, *67*, 81–106. [CrossRef]

15. Hashimoto, M.; Endo, T.; Peltier, G.; Tasaka, M.; Shikanai, T. A nucleus-encoded factor, CRR2, is essential for the expression of chloroplast ndhB in *Arabidopsis*. *Plant J.* **2003**, *36*, 541–549. [CrossRef]

16. Furutani, R.; Wada, S.; Ifuku, K.; Maekawa, S.; Miyake, C. Higher reduced state of Fe/S-signals, with the suppressed oxidation of P700, causes PSI inactivation in *Arabidopsis thaliana*. *Antioxidants* **2022**, *12*, 21. [CrossRef]

17. Miyake, C. Molecular mechanism of oxidation of P700 and suppression of ROS production in photosystem I in response to electron-sink limitations in C_3 Plants. *Antioxidants* **2020**, *9*, 230. [CrossRef] [PubMed]

18. Furutani, R.; Ifuku, K.; Suzuki, Y.; Noguchi, K.; Shimakawa, G.; Wada, S.; Makino, A.; Sohtome, T.; Miyake, C. *P700 Oxidation Suppresses the Production of Reactive Oxygen Species in Photosystem I*; Toru, H., Ed.; Acad Press: Cambridge, MA, USA, 2020; Volume 96, p. 26.

19. Asada, K.; Kiso, K.; Yoshikawa, K. Univalent reduction of molecular oxygen by spinach chloroplasts on illumination. *J. Biochem. Biol.* **1974**, *249*, 2175–2181.

20. Kozuleva, M.; Petrova, A.; Milrad, Y.; Semenov, A.; Ivanov, B.; Redding, K.E.; Yacoby, I. Phylloquinone is the principal Mehler reaction site within photosystem I in high light. *Plant Physiol.* **2021**, *186*, 1848–1858. [CrossRef]

21. Havaux, M.; Davaud, A. Photoinhibition of photosynthesis in chilled potato leaves is not correlated with a loss of Photosystem-II activity: Preferential inactivation of photosystem I. *Photosyn. Res.* **1994**, *40*, 75–92. [CrossRef]

22. Inoue, K.; Fujie, T.; Yokoyama, E.; Matsuura, K.; Hiyama, T.; Sakurai, H. The photoinhibition sites of photosystem I in isolated chloroplasts under extremely reducing conditions. *Plant Cell Physiol.* **1989**, *30*, 7. [CrossRef]

23. Satoh, K. Mechanism of photoinactivation in photosynthetic systems. III. The site and mode of photoinactivation in photosystem I. *Plant Cell Physiol.* **1970**, *11*, 187. [CrossRef]

24. Sonoike, K.; Terashima, I.; Iwaki, M.; Itoh, S. Destruction of photosystem I iron-sulfur centers in leaves of *Cucumis sativus* L. by weak illumination at chilling temperatures. *FEBS Lett.* **1995**, *362*, 235–238. [CrossRef] [PubMed]

25. Terashima, I.; Funayama, S.; Sonoike, K. The site of photoinhibition in leaves of *Cucumis sativus* L. at low temperatures in photosystem I, not photosystem II. *Planta* **1994**, *193*, 7. [CrossRef]

26. Foyer, C.; Furbank, R.; Harbinson, J.; Horton, P. The mechanisms contributing to photosynthetic control of electron transport by carbon assimilation in leaves. *Photosyn. Res.* **1990**, *25*, 83–100. [CrossRef]

27. Tikhonov, A.N. The cytochrome b_6f complex at the crossroad of photosynthetic electron transport pathways. *Plant Physiol. Biochem. PPB* **2014**, *81*, 163–183. [CrossRef]

28. Furutani, R.; Makino, A.; Suzuki, Y.; Wada, S.; Shimakawa, G.; Miyake, C. Intrinsic fluctuations in transpiration induce photorespiration to oxidize P700 in photosystem I. *Plants* **2020**, *9*, 1761. [CrossRef]

29. Wada, S.; Suzuki, Y.; Miyake, C. Photorespiration enhances acidification of the thylakoid lumen, reduces the plastoquinone pool, and contributes to the oxidation of P700 at a lower partial pressure of CO_2 in wheat leaves. *Plants* **2020**, *9*, 319. [CrossRef]

30. Hanawa, H.; Ishizaki, K.; Nohira, K.; Takagi, D.; Shimakawa, G.; Sejima, T.; Shaku, K.; Makino, A.; Miyake, C. Land plants drive photorespiration as higher electron-sink: Comparative study of post-illumination transient O_2-uptake rates from liverworts to angiosperms through ferns and gymnosperms. *Physiol. Plant.* **2017**, *161*, 138–149. [CrossRef]

31. Sejima, T.; Hanawa, H.; Shimakawa, G.; Takagi, D.; Suzuki, Y.; Fukayama, H.; Makino, A.; Miyake, C. Post-illumination transient O_2-uptake is driven by photorespiration in tobacco leaves. *Physiol. Plant.* **2016**, *156*, 227–238. [CrossRef]

32. Miyake, C.; Suzuki, Y.; Yamamoto, H.; Amako, K.; Makino, A. O_2-enhanced induction of photosynthesis in rice leaves: The Mehler-ascorbate peroxidase (MAP) pathway drives cyclic electron flow within PSII and cyclic electron flow around PSI. *Soil Sci. Plant Nutri.* **2012**, *58*, 718–727. [CrossRef]

33. Yamamoto, H.; Shikanai, T. PGR5-dependent cyclic electron flow protects photosystem I under fluctuating light at donor and acceptor sides. *Plant Physiol.* **2019**, *179*, 588–600. [CrossRef] [PubMed]

34. Yamamoto, H.; Shikanai, T. Does the Arabidopsis *proton gradient regulation 5* mutant leak protons from the thylakoid membrane? *Plant Physiol.* **2020**, *184*, 421–427. [CrossRef]

35. Suganami, M.; Konno, S.; Maruhashi, R.; Takagi, D.; Tazoe, Y.; Wada, S.; Yamamoto, H.; Shikanai, T.; Ishida, H.; Suzuki, Y.; et al. Expression of flavodiiron protein rescues defects in electron transport around PSI resulting from overproduction of Rubisco activase in rice. *J. Exp. Bot.* **2022**, *73*, 2589–2600. [CrossRef]

36. Rantala, S.; Lempiäinen, T.; Gerotto, C.; Tiwari, A.; Aro, E.M.; Tikkanen, M. PGR5 and NDH-1 systems do not function as protective electron acceptors but mitigate the consequences of PSI inhibition. *Biochim. Biophys. Acta Bioenerg.* **2020**, *1861*, 148154. [CrossRef] [PubMed]

37. Wada, S.; Amako, K.; Miyake, C. Identification of a novel mutation exacerbated the PSI photoinhibition in pgr5/pgrl1 mutants; Caution for overestimation of the phenotypes in *Arabidopsis* pgr5-1 Mutant. *Cells* **2021**, *10*, 2884. [CrossRef]

38. Ohnishi, M.; Furutani, R.; Sohtome, T.; Suzuki, T.; Wada, S.; Tanaka, S.; Ifuku, K.; Ueno, D.; Miyake, C. Photosynthetic parameters show specific responses to essential mineral deficiencies. *Antioxidants* **2021**, *10*, 996. [CrossRef]
39. Porra, R.J.; Scheer, H. Towards a more accurate future for chlorophyll *a* and *b* determinations: The inaccuracies of Daniel Arnon's assay. *Photosyn. Res.* **2019**, *140*, 215–219. [CrossRef]
40. Baker, N.R. Chlorophyll fluorescence: A probe of photosynthesis in vivo. *Annu. Rev. Plant Biol.* **2008**, *59*, 89–113. [CrossRef]
41. Genty, B.; Harbinson, J.; Briantais, J.M.; Baker, N.R. The relationship between non-photochemical quenching of chlorophyll fluorescence and the rate of photosystem 2 photochemistry in leaves. *Photosyn. Res.* **1990**, *25*, 249–257. [CrossRef]
42. Bilger, W.; Björkman, O. Relationships among violaxanthin deepoxidation, thylakoid membrane conformation, and nonphotochemical chlorophyll fluorescence quenching in leaves of cotton (*Gossypium hirsutum* L.). *Planta* **1994**, *193*, 238–246. [CrossRef]
43. Oxborough, K.; Baker, N.R. An evaluation of the potential triggers of photoinactivation of photosystem II in the context of a Stern-Volmer model for downregulation and the reversible radical pair equilibrium model. *Phil. Trans. R. Soc. Lond. B Biol. Sci.* **2000**, *355*, 1489–1498. [CrossRef] [PubMed]
44. Klughammer, C.; Schreiber, U. Deconvolution of ferredoxin, plastocyanin, and P700 transmittance changes in intact leaves with a new type of kinetic LED array spectrophotometer. *Photosyn. Res.* **2016**, *128*, 195–214. [CrossRef]
45. Sacksteder, C.A.; Kramer, D.M. Dark-interval relaxation kinetics (DIRK) of absorbance changes as a quantitative probe of steady-state electron transfer. *Photosyn. Res.* **2000**, *66*, 145–158. [CrossRef]

International Journal of
Molecular Sciences

Enhanced Reduction of Ferredoxin in PGR5-Deficient Mutant of *Arabidopsis thaliana* Stimulated Ferredoxin-Dependent Cyclic Electron Flow around Photosystem I

Shu Maekawa [1,†], Miho Ohnishi [1,2,†], Shinya Wada [1,2], Kentaro Ifuku [2,3] and Chikahiro Miyake [1,2,*]

[1] Graduate School for Agricultural Science, Kobe University, 1-1 Rokkodai, Nada-Ku, Kobe 657-8501, Japan; 2001de325@gmail.com (S.M.); miho.ohnishi02@gamil.com (M.O.); swada@penguin.kobe-u.ac.jp (S.W.)
[2] Core Research for Evolutional Science and Technology (CREST), Japan Science and Technology Agency (JST), 7 Gobancho, Tokyo 102-0076, Japan; ifuku.kentaro.2m@kyoto-u.ac.jp
[3] Graduate School for Agriculture, Kyoto University, Kitashirakawa Oiwake-cho, Sakyo-ku, Kyoto 606-8502, Japan
* Correspondence: cmiyake@hawk.kobe-u.ac.jp
† These authors contributed equally to the present work.

Abstract: The molecular entity responsible for catalyzing ferredoxin (Fd)-dependent cyclic electron flow around photosystem I (Fd-CEF) remains unidentified. To reveal the in vivo molecular mechanism of Fd-CEF, evaluating ferredoxin reduction–oxidation kinetics proves to be a reliable indicator of Fd-CEF activity. Recent research has demonstrated that the expression of Fd-CEF activity is contingent upon the oxidation of plastoquinone. Moreover, chloroplast NAD(P)H dehydrogenase does not catalyze Fd-CEF in *Arabidopsis thaliana*. In this study, we analyzed the impact of reduced Fd on Fd-CEF activity by comparing wild-type and pgr5-deficient mutants (*pgr5^{hope1}*). PGR5 has been proposed as the mediator of Fd-CEF, and *pgr5^{hope1}* exhibited a comparable CO_2 assimilation rate and the same reduction–oxidation level of PQ as the wild type. However, P700 oxidation was suppressed with highly reduced Fd in *pgr5^{hope1}*, unlike in the wild type. As anticipated, the Fd-CEF activity was enhanced in *pgr5^{hope1}* compared to the wild type, and its activity further increased with the oxidation of PQ due to the elevated CO_2 assimilation rate. This in vivo research clearly demonstrates that the expression of Fd-CEF activity requires not only reduced Fd but also oxidized PQ. Importantly, PGR5 was found to not catalyze Fd-CEF, challenging previous assumptions about its role in this process.

Keywords: cyclic electron flow; ferredoxin; NADH dehydrogenase; pgr5; photosynthesis; photosystem I

Citation: Maekawa, S.; Ohnishi, M.; Wada, S.; Ifuku, K.; Miyake, C. Enhanced Reduction of Ferredoxin in PGR5-Deficient Mutant of *Arabidopsis thaliana* Stimulated Ferredoxin-Dependent Cyclic Electron Flow around Photosystem I. *Int. J. Mol. Sci.* **2024**, 25, 2677. https://doi.org/10.3390/ijms25052677

Academic Editor: Martin Bartas

Received: 6 January 2024
Revised: 12 February 2024
Accepted: 25 February 2024
Published: 26 February 2024

1. Introduction

In oxygenic photosynthesis, the photon energy absorbed by the light-harvesting systems of both photosystem II (PSII) and photosystem I (PSI) in the photosynthetic electron transport system initiates the excitation of the reaction center chlorophylls—P680 in PSII and P700 in PSI. The excitation of these chlorophylls initiates their catalytic reactions, leading to electron flow from oxidation to reduction: H_2O oxidation in PSII results in O_2 evolution, while plastoquinone (PQ) undergoes reduction. Simultaneously, PQ is oxidized through cytochrome (Cyt) b_6/f and plastocyanin, eventually leading to the reduction of ferredoxin (Fd) via electron transport carriers, including phylloquinone, Fx, and F_A/F_B in PSI. The electrons from reduced Fd are primarily utilized in the production of NADPH catalyzed by Fd-NADP oxidoreductase. Concomitant with the photosynthetic linear electron flow from H_2O to NADPH, protons accumulate in the lumen of thylakoid membranes, creating a ΔpH across the thylakoid membranes. These protons originate from water oxidation in PSII and their transport from the stroma to the lumen facilitated by the Q-cycle during PQ oxidation in the Cyt b_6/f complex [1,2]. The ΔpH, acting as a proton motive force, drives ATP synthase to generate ATP. These energy compounds—reduced Fd, NADPH, and

ATP—produced during the light reaction, play pivotal roles in driving the dark reactions of CO_2 assimilation and photorespiration in C3 plants.

The photosynthetic linear electron flow faces a potential threat from the generation of reactive oxygen species (ROS). The rate at which NADPH is supplied to ATP, produced in the photosynthetic linear electron flow, surpasses the consumption rate of NADPH to ATP in dark reactions, even under non-photorespiratory conditions [3–10]. This excess supply of NADPH is further heightened under photorespiratory conditions, where a greater amount of ATP is consumed in dark reactions [11–13]. Additionally, as CO_2 assimilation is stimulated by the photosynthetic linear electron flow, the surplus NADPH supply becomes more pronounced. In essence, the photosynthetic linear electron flow, the sole source of electrons for NADPH production in photosynthesis, accumulates NADPH and saturates electrons in the photosynthetic electron transport system. The accumulation of electrons in PSI is evident in the reduction of electron carriers at the acceptor side of PSI, including the Fe/S-series, Fx, F_A/F_B, and Fd. This accumulation triggers the reduction of O_2 to produce O_2^-, and the O_2^- subsequently degrades Fe/S compounds, leading to the inactivation of PSI [14–23].

The issue of a supply rate of NADPH exceeding its demand, as caused by the photosynthetic linear electron flow, finds resolution in the cyclic electron flow around photosystem I (CEF) [3]. Within the CEF, Fd reduced by photosystem I donates electrons to PQ through Fd-quinone oxidoreductase (FQR). In essence, FQR catalyzes Fd-dependent CEF. This cyclic flow induces a ΔpH across thylakoid membranes, generating ATP without producing electrons for NADPH supply. Instead, Fd-CEF promotes the consumption of NADPH and actively contributes to the activation of the dark reaction. In this way, Fd-CEF possesses the potential to alleviate the challenges posed by the photosynthetic linear electron flow.

Until now, the potential threat posed by the photosynthetic linear electron flow has not been thoroughly explored, and the physiological role of the Fd-CEF has not been investigated in terms of oxidative stress. Although CO_2 assimilation could potentially proceed at an ATP supply-limited rate under photorespiratory conditions, the issue arises when NADPH starts to accumulate. The accumulation of electrons at the acceptor side of PSI, manifesting as the reduction of electron carriers such as Fe/S-clusters, is known to induce ROS production [19]. It is crucial to alleviate the accumulation of NADPH, and this is where the function of Fd-CEF becomes significant. Fd-CEF is proposed to promote the consumption of NADPH by supplying ATP in the dark reaction, thus counteracting the potential ROS production associated with the photosynthetic linear electron flow. This dual role of Fd-CEF in ATP supply and NADPH consumption could play a crucial role in maintaining redox balance and mitigating oxidative stress in the photosynthetic process.

To unravel the physiological function of Fd-CEF, it was imperative to establish an assay system capable of measuring Fd-CEF in vivo. In this study, we monitored the redox reaction of Fd concurrently with chlorophyll fluorescence, $P700^+$ and PC^+ absorbance changes, and net CO_2 assimilation using intact leaves of *Arabidopsis thaliana*. Successful measurement of the electron flux in Fd-CEF in *Arabidopsis thaliana* had been previously achieved [24]. The oxidation rate of reduced Fd, independent of the photosynthetic linear electron flow—termed the extra oxidation rate of Fd—was designated as the electron flux in Fd-CEF, denoted as vCEF. The regulation of vCEF was found to be linked to the reduction–oxidation state of PQ. Specifically, as PQ became oxidized with an increase in CO_2 assimilation, vCEF exhibited a corresponding increase [24]. These observed characteristics align with the established model of cyclic electron flow [3]. Importantly, our inability to detect the extra Fd oxidation reaction [24] might be attributed to the suppressed in vivo occurrence of this reaction under conditions of limited photosynthesis. In instances where PQ was highly reduced and the apparent quantum yield of PSII was low, the extra Fd oxidation reaction was inhibited. We posit that the extra Fd oxidation reaction serves as a reflection of Fd-CEF activity.

In this study, we further characterized the impact of the reduction–oxidation state of Fd on Fd-CEF activity in vivo. Fd-CEF necessitates Fd as the electron donor to effectively reduce oxidized PQ [3]. Typically, the redox state of Fd remains constant in response to decreases in both the net CO_2 assimilation rate and the photosynthetic linear electron flow under constant actinic light intensity [19,24–26]. Our recent findings revealed an augmented reduction of Fd in a PGR5-less mutant (*pgr5hope1*) of *Arabidopsis thaliana* due to the inhibited oxidation of P700 in PSI [19]. Subsequently, we conducted a comparative analysis of Fd-CEF activity between the wild type (WT) and *pgr5hope1* to elucidate the influence of the Fd redox state in vivo. As anticipated, *pgr5hope1* exhibited a higher Fd-CEF activity than the WT. This underscores the regulation of Fd-CEF activity by the redox states of both Fd and PQ in vivo. Our study delves into the molecular mechanisms and physiological functions of Fd-CEF in vivo, shedding light on the intricate interplay between Fd redox status and the activity of the cyclic electron flow around PSI.

2. Results

The impact of intercellular partial pressures of CO_2 (Ci) on both the gross CO_2 assimilation rate and the apparent quantum yield of photosystem II [Y(II)] was examined (Figure 1). The gross CO_2 assimilation rates exhibited a consistent dependence on Ci for both the wild type (WT) and *pgr5hope1*. While Y(II) demonstrated a similar Ci dependence in both the WT and *pgr5hope1*, the Y(II) values in *pgr5hope1* were observed to be lower than those in the WT (Figure 1). The distinctions in Y(II) between WT and *pgr5hope1* were particularly evident in the lower range of Y(II), as highlighted in Figure 2.

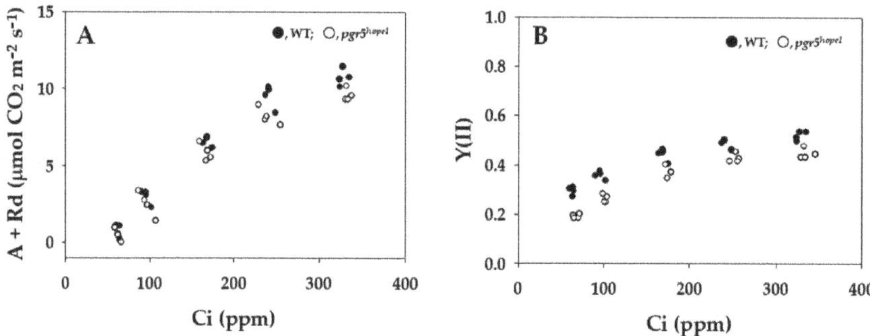

Figure 1. Effects of the intercellular partial pressure of CO_2 (Ci) on the gross CO_2 assimilation rate (A + Rd) and apparent quantum yield of photosystem II (PSII) [Y(II)] in wild-type (WT) and *pgr5hope1* *Arabidopsis*. (**A**) The gross CO_2 assimilation rates were measured at 400 µmol photons m^{-2} s^{-1} and 21 kPa O_2, and Y(II) was simultaneously measured. The dark respiration rates (Rd) were measured before starting actinic light illumination. The gross CO_2 assimilation rates are expressed as A + Rd and were plotted against Ci. (**B**) Y(II) was plotted against Ci. The data were obtained from four independent experiments using leaves attached to four plants of both WT and *pgr5hope1* (n = 4). The ambient partial pressures of CO_2 were changed from 400 ppm to 50 through 300, 200, and 100 Pa at 21 kPa O_2 for the same leaves. Black symbols, WT; white symbols, *pgr5hope1*.

In Figure 2, the relationships between the parameters $P700^+$, PC^+, and Fd^- and Y(II) were depicted. As Y(II) decreased due to lowering Ci, P700 in WT was oxidized, increasing from approximately 10% to 40% (see Figure 2A). In contrast, P700 in *pgr5hope1* was not oxidized even as Y(II) decreased (Figure 2A). Similarly, PC in WT was oxidized, ranging from 65% to 90% (Figure 2B). Conversely, the oxidized PC percentage decreased from 20% to 5% with the decrease in Y(II) in *pgr5hope1* (Figure 2B). Unlike $P700^+$ and PC^+, Fd^- in WT did not show significant changes in response to the decrease in Y(II) (Figure 2C), attributed to the oxidation of P700 in PSI [19]. In contrast, Fd^- in *pgr5hope1* surpassed that in WT and increased from 40% to 55% with the decrease in Y(II) (Figure 2C). This elevation was attributed to the suppression of P700 oxidation in *pgr5hope1*, leading to an enhanced electron flux toward the acceptor side of PS I, ultimately resulting in the reduction of Fd [19].

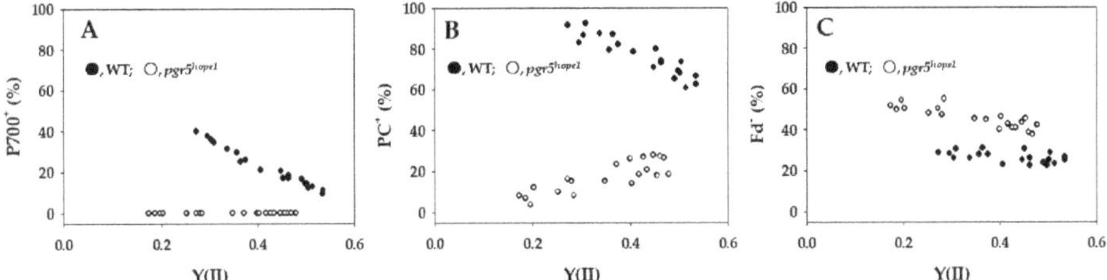

Figure 2. Relationships between P700$^+$, PC$^+$, Fd$^-$, and apparent quantum yield of photosystem II (PSII) [Y(II)]. The data for each parameter were measured in the experiments depicted in Figure 1, simultaneously with the gross CO$_2$ assimilation rates and Y(II). (**A**) P700$^+$, (**B**) PC$^+$, and (**C**) Fd$^-$ were plotted against Y(II). The ratios of P700$^+$, PC$^+$, and Fd$^-$ against the total contents are expressed. The data were obtained from four independent experiments using leaves attached to four WT and *pgr5^{hope1}* plants (*n* = 4). Black symbols, WT; white symbols, *pgr5^{hope1}*.

In Figure 3, the parameters non-photochemical quenching (NPQ) and plastoquinone reduced state (1 − qP) were plotted against Y(II). An increase in NPQ indicated the enhancement of heat dissipation of photon energy absorbed by PSII. As Y(II) decreased, NPQ in WT rose from approximately 0.5 to 1.5 (Figure 3A). Conversely, NPQ in *pgr5^{hope1}* was lower than that in WT, and it increased from 0.2 to 0.9 with the decrease in Y(II) (Figure 3A). The increase in 1 − qP reflects a reduction in the plastoquinone pool. Both WT and *pgr5^{hope1}* exhibited the same dependence of 1 − qP on the decrease in Y(II), with 1 − qP rising as Y(II) decreased (Figure 3B). In WT, 1 − qP increased from approximately 0.25 to 0.6 with the decrease in Y(II) (Figure 3B). Similarly, in *pgr5^{hope1}*, 1 − qP increased from approximately 0.35 to 0.75 with the decrease in Y(II) (Figure 3B). While the dependence of 1 − qP on Y(II) in *pgr5^{hope1}* mirrored that of WT, the values of 1 − qP in *pgr5^{hope1}* showed a further increase with lowering Y(II) from 0.25 to 0.15 compared to WT.

In Figure 4, the parameters related to photosystem I (PSI), namely Y(I) and Y(NA), were shown. As Y(II) decreased, Y(I) in WT declined from approximately 0.8 to 0.5 (Figure 4A). Conversely, in *pgr5^{hope1}*, Y(I) decreased from about 0.5 to 0.2 with the decrease in Y(II) (Figure 4A). Notably, the dependence of Y(I) on Y(II) in *pgr5^{hope1}* differed from that in WT, and the values of Y(I) in *pgr5^{hope1}* were consistently lower than those in WT. Turning to Y(NA), in WT, it maintained lower values ranging from 0.1 to 0.05 as Y(II) decreased (Figure 4B). In contrast, Y(NA) in *pgr5^{hope1}* was higher than in WT. Specifically, in *pgr5^{hope1}*, Y(NA) increased from 0.5 to 0.8 as Y(II) decreased (Figure 4B). This suggests an acceleration of the limitation of the oxidation of the excited P700 in the photo-oxidation cycle of P700 in PSI (Figure 2C). Furthermore, in *pgr5^{hope1}*, Y(NA) appeared to be related to the dependence of Fd reduction on Y(II).

To delve deeper into this relationship, a statistical comparison was conducted between the dependence of Y(NA) on the reduced state of Fd in WT and *pgr5^{hope1}* (Supplemental Figure S5). ANCOVA analysis revealed a significant interaction between plants (WT and *pgr5^{hope1}*) and the reduced Fd ($p < 0.01$). Subsequently, correlation analysis between Y(NA) and the reduced Fd was performed for each plant. In WT, ANOVA of the regression analysis showed no significant relationship between Y(NA) and the reduced Fd. However, in *pgr5^{hope1}*, ANOVA of the regression analysis demonstrated a significant relationship between Y(NA) and the reduced Fd (F value 35.29, $p < 0.01$). The regression line was Y(NA) = −0.0717 + 0.01589** × Fd$^-$ (** $p < 0.01$), indicating a significant slope. Thus, the correlation between Y(NA) and the reduced Fd was observed exclusively in *pgr5^{hope1}*.

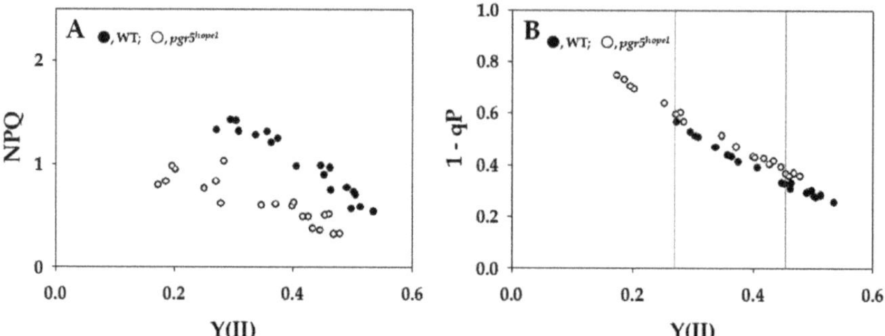

Figure 3. Relationships between non-photochemical quenching (NPQ), plastoquinone reduced state (1 − qP), and apparent quantum yield of photosystem II (PSII) [Y(II)]. The data for each parameter were measured in the experiments depicted in Figure 1 simultaneously with the gross CO_2 assimilation rates and Y(II). (**A**) NPQ and (**B**) 1 − qP were plotted against Y(II). The two vertical lines were drawn at approximately 0.27 and 0.45 of Y(II), where the values of 1 − qP were the same between WT and *pgr5^hope1^*. These characteristics were used for the comparison of the oxidation of the reduced Fd in Figure 5. The data were obtained from four independent experiments using leaves attached to four WT and *pgr5^hope1^* plants (*n* = 4). Black symbols, WT; white symbols, *pgr5^hope1^*.

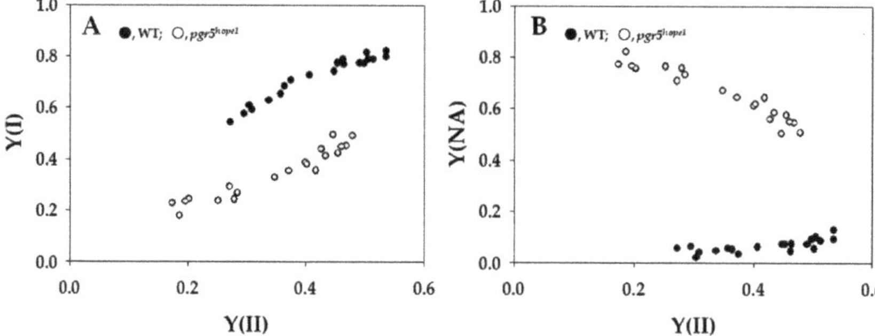

Figure 4. Relationships between the apparent quantum yield of PSI [Y(I)], apparent quantum yield of non-photochemical energy dissipation of photoexcited P700 [Y(NA)], and apparent quantum yield of photosystem II (PSII) [Y(II)]. The data for each parameter were measured in the experiments depicted in Figure 1, simultaneously with the gross CO_2 assimilation rates and Y(II). (**A**) Y(I) and (**B**) Y(NA) were plotted against Y(II). The data were obtained from four independent experiments using leaves attached to four WT and *pgr5^hope1^* plants (*n* = 4). Black symbols, WT; white symbols, *pgr5^hope1^*.

Figure 5A illustrated the relationship between the oxidation rate of reduced Fd (vFd) and Y(II) in both WT and *pgr5^hope1^*. In WT, vFd was proportional to Y(II) in the range of Y(II) below 0.4. Beyond 0.4 of Y(II), Y(II) became saturated against vFd, indicating the discovery of excess turnover of Fd, indicative of Fd-CEF activity. The behavior of vFd against Y(II) in *pgr5^hope1^* was similar to that in WT (Figure 5A), but the values of vFd in *pgr5^hope1^* were higher in the range of Y(II) compared to WT. These findings suggest that Fd-CEF was activated in *pgr5^hope1^*. Supplemental Figure S1A provided a comparison of the typical kinetics of the oxidation of Fd after turning off the actinic light in the dark-interval relaxation kinetics (DIRK) analysis between WT and *pgr5^hope1^* at approximately the same two values of Y(II). At approximately 0.45 of Y(II), the initial decay rate of the reduced Fd in *pgr5^hope1^* was larger than that in WT, indicating a higher Fd-CEF activity. The reduced level of Fd before turning off the actinic light showed a reduced level at the steady state. At approximately 0.27 of Y(II) at lower Ci, the initial decay rate of the reduced Fd in *pgr5^hope1^* was also larger than that in WT (Supplemental Figure S1B). Additionally, Figure 5B demonstrates that vFd exhibited a dependence on the increase in qP. The increase in qP, reflecting the oxidation of PQ, stimulated the oxidation rate of Fd, resulting in excessive vFd, observed with the increase in Y(II) in both WT and *pgr5^hope1^*. These observations suggest that Fd-CEF activity is induced by the oxidation of PQ, which is a consequence of the enhanced photosynthetic linear electron flow. Furthermore, the vFd in *pgr5^hope1^* was also larger than in WT, indicating that the higher reduced state of Fd stimulates Fd-CEF activity in vivo.

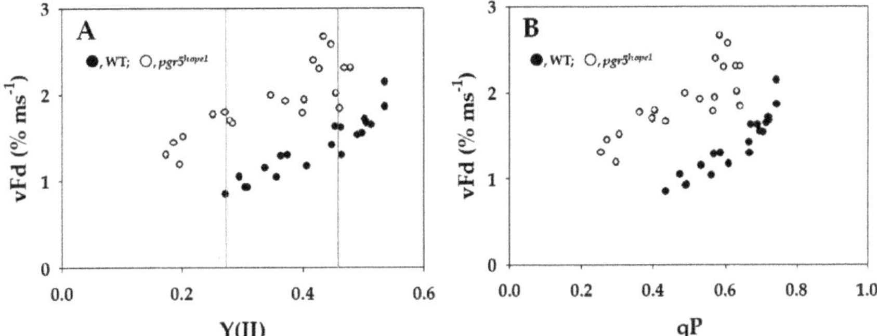

Figure 5. Relationships between the apparent quantum yield of photosystem II (PSII) [Y(II)], plasto-quinone oxidized state (qP), and vFd. The data for each parameter were measured in the experiments depicted in Figure 1 simultaneously with the gross CO_2 assimilation rates and Y(II). (**A**) Y(II) was plotted against vFd. (**B**) qP was plotted against vFd. In the experiments shown in Figure 1, the oxidation rate of Fd was determined by DIRK analysis (see "Section 4"). To determine the oxidation rate of Fd^- (vFd) under illuminated conditions, actinic light was transiently turned off for 400 ms. The initial slope of the decrease in Fd^- indicates vFd. These data were obtained at a steady state, which was confirmed by the achievement of stable conditions for both gross CO_2 assimilation and Y(II). The two vertical lines were drawn at approximately 0.27 and 0.45 of Y(II) to compare vFd between WT and *pgr5^hope1^*. The data were obtained from four independent experiments using leaves attached to four WT and *pgr5^hope1^* plants (*n* = 4). Black symbols, WT; white symbols, *pgr5^hope1^*.

3. Discussion

In this present study, we employed *pgr5^hope1^* as our experimental model due to its manifestation of a higher reduced state of Fd at the steady state. Consequently, we antici-pated that *pgr5^hope1^* would serve as a suitable material for investigating the impact of Fd on Fd-CEF activity in vivo. We conducted a comparative analysis of Fd-CEF activity in *Arabidopsis thaliana* wild type (WT) and *pgr5^hope1^*. The *pgr5^hope1^* mutant exhibited compa-rable values for both the gross CO_2 assimilation rate and the reduction–oxidation level of PQ when compared to WT (see Figure 1) [19,27]. The oxidation rate of the reduced Fd displayed a nonlinear relationship with Y(II) in both WT and *pgr5^hope1^* (Figure 5A). The increase in vFd deviated from the rise in Y(II) in both cases, indicating the presence of excessive vFd unrelated to photosynthetic linear electron flow—the electron flux in Fd-CEF (Figure 5A) [24]. Furthermore, we observed a deviation in the relationship between vFd and qP in both WT and *pgr5^hope1^* (Figure 5B). The increase in qP reflected the oxidation of the reduced PQ induced by the stimulation of photosynthetic linear electron flow. These findings align with a previous report [24], suggesting that the expression of excessive vFd corresponds to a characteristic feature of Fd-CEF. In essence, the appearance of Fd-CEF required PQ oxidation in both WT and *pgr5^hope1^*, aligning with the molecular mechanism of Fd-CEF (Supplemental Figure S2) [3].

Moreover, we observed an elevated oxidation rate of reduced Fd in *pgr5^hope1^* com-pared to WT, as illustrated in Figure 5A. This implies that the electron flux of Fd-CEF in *pgr5^hope1^* was enhanced. In the Fd-CEF process, the reduced Fd serves as the electron donor, transferring electrons to PQ through FQR. The electron acceptor in this process is the oxidized PQ. The electron flux in Fd-CEF (vCEF) is directly proportional to the product of the concentrations of both the reduced Fd ([Fd⁻]) and the oxidized PQ ([PQ]). Additionally, it depends on the activity of FQR and the rate constant (k), as expressed by the equation:

$$vCEF = k \times [PQ] \times [Fd^-] \tag{1}$$

If PQ reached complete reduction, the electron flux of Fd-CEF became zero, even if Fd was in a reduced state (Supplemental Figure S2) [3,24]. Conversely, when PQ was entirely oxidized, the activity of Fd-CEF was also zero, since Fd lacked electrons for the reduction of PQ [3]. Similarly, if Fd was fully reduced with PQ also in a completely reduced state, Fd-CEF could not function. Furthermore, if Fd was entirely oxidized with PQ completely oxidized, Fd-CEF would not be operational. In our current investigation, the reduction level of Fd in *pgr5^{hope1}* surpassed that in WT, and the further reduction of Fd was facilitated by the decline in photosynthetic linear electron transport (Figure 2C). This reduction was a consequence of the suppressed P700 oxidation in *pgr5^{hope1}* (Figure 2A). Unlike WT, where the rate-determining step in the P700 photo-oxidation reduction cycle is the oxidation of the excited P700 by the electron acceptor in PSI, *pgr5^{hope1}* demonstrated a distinct pattern. In *pgr5^{hope1}*, this rate-determining step was observed as a larger Y(NA) and a higher reduced level of Fd (Figures 2 and 4). This shift was attributed to the lower ΔpH across thylakoid membranes in *pgr5^{hope1}* compared to WT (Supplemental Figure S3). The diminished ΔpH in *pgr5^{hope1}* resulted from a higher value of H⁺-conductance (gH⁺) compared to WT [28]. However, the mechanism by which gH⁺ is decreased in *pgr5^{hope1}* has not been elucidated. Notably, the observed lower ΔpH in *pgr5^{hope1}* did not impede the oxidation of the reduced PQ by the cytochrome b_6/f-complex. Consequently, the rate-determining step in the P700 photo-oxidation reduction cycle in *pgr5^{hope1}* shifted from the reduction of the oxidized P700 to the oxidation of the excited P700. This explains the intensified reduction in Fd in *pgr5^{hope1}*, particularly in response to the suppression of photosynthetic linear electron transport. According to the Fd-CEF activity expression model (Equation (1)), the increase in [Fd⁻] results in an elevation of vCEF. Indeed, at identical qP values (e.g., 0.4 and 0.6), indicating the same [PQ], vFd in *pgr5^{hope1}* exceeded that in WT (Figure 5B). These observations align with the behaviors predicted by the Fd-CEF model (Supplemental Figure S2) [3]. In essence, Fd-CEF necessitates the presence of both oxidized PQ and reduced Fd in vivo.

The role of Fd-CEF in inducing ΔpH across thylakoid membranes has been previously explored [1–3,7,8,29–42]. The dependencies of vFd on both Y(II) and qP, as illustrated in Figure 5, indicated that the acceleration of Fd-CEF was concurrent with an increase in CO$_2$ assimilation in response to elevated Ci levels. However, despite the rise in Fd-CEF activity, ΔpH across the thylakoid membranes remains constant (Supplemental Figure S3). This implies that the ΔpH induced by Fd-CEF is dissipated by the increased CO$_2$ assimilation, where the usage of ATP is augmented. Unless Fd-CEF is stimulated by an increase in Ci, the ΔpH would not be maintained by the heightened CO$_2$ assimilation at higher Ci levels. The dependencies of ΔpH on the increase in Ci were consistent between WT and *pgr5^{hope1}* (Supplemental Figure S3). However, the ΔpH across thylakoid membranes in *pgr5^{hope1}* was lower than in WT, in line with previous reports [28,42]. The reduced ΔpH in *pgr5^{hope1}* resulted from a larger gH⁺ compared to WT (Supplemental Figure S3). The molecular mechanism underlying this increased gH⁺ in *pgr5^{hope1}* remains unclear. Despite the lower ΔpH in *pgr5^{hope1}*, the CO$_2$ assimilation rates were almost identical to WT (Figure 1). This suggests that the diminished ΔpH in *pgr5^{hope1}* was sufficient to drive CO$_2$ assimilation, a phenomenon supported by the stimulated Fd-CEF activity (Figure 5). Without the acceleration of Fd-CEF in *pgr5^{hope1}*, the ΔpH could not be sustained, jeopardizing the functionality of CO$_2$ assimilation.

In *pgr5^{hope1}*, non-photochemical quenching (NPQ) was observed to be lower compared to WT, as depicted in Figure 3. The induction of NPQ requires acidification of the luminal side of thylakoid membranes [33,43]. Consequently, the lower ΔpH in *pgr5^{hope1}* may explain the inability to induce higher NPQ. On the contrary, the behavior of NPQ in WT in response to both the increase in the gross CO$_2$ assimilation rate and the photosynthetic linear electron flow rate mirrored that of *pgr5^hope1* (Figure 3). NPQ decreased with the rise in both the gross CO_2 assimilation rate and the photosynthetic linear electron flow rate, despite ΔpH remaining unchanged in both WT and *pgr5^hope1*, as described earlier. NPQ is also influenced by the reduction–oxidation state of PQ and Y(II) [44,45]. Consequently, NPQ decreased with increases in both qP and Y(II).

In this study, we have further substantiated the expression model of Fd-CEF activity proposed in Supplemental Figure S2 [3]. The *pgr5^hope1* mutant exhibited a higher reduction in Fd compared to WT (Figure 2). As anticipated, this led to an enhancement in the electron flux of Fd-CEF in *pgr5^hope1*, as depicted in Figure 5. Both pgr5 and NDH have been recognized as potential mediators of Fd-CEF [6,7]. Mutants of these components displayed a suppression of the increase in the minimum yield of Chl fluorescence (Fo) after actinic light illumination was turned off in vivo [7,35]. Additionally, the reduced Fd-dependent increase in Fo was inhibited in the isolated thylakoid membranes from both pgr5- and NDH-less *Arabidopsis thaliana* [6]. The increase in Fo was attributed to the reduction of PQ by the reduced Fd, and thus, the Fd-dependent increase in Fo has been utilized as a measure of FQR activity [6]. However, it is noteworthy that the reduced Fd increased Fo even in the presence of the inhibitor (DCMU) of the electron transport in PSII [46]. Furthermore, Fd was observed to reduce cytochrome b_{559} in PSII, and this reduction was inhibited by antimycin A [47]. These observations imply that both pgr5 and NDH contribute to the reduction of PQ through PSII. Although the Fd-dependent reduction of PSII catalyzed by pgr5 and NDH can form the electron flow pathway in CEF around PSI, it is considered to be relatively small compared to the electron flux in photosynthetic linear electron flow. This conclusion is supported by the fact that the electron flux in PSII (Jf) estimated by Chl fluorescence, Y(II), exhibits a positive linear relationship with the electron fluxes (Jg) into both net CO_2 assimilation and photorespiration [48–51]. This suggests that no additional electron flux beyond photosynthetic linear electron flow is detected. Therefore, the observed electron flux in the Fd redox reaction, not associated with photosynthetic linear electron flow, constitutes the cyclic electron flow from PSI to PQ through the electron transport carrier localized between PSII and PSI, revealing the Fd-dependent cyclic electron flow around PSI, Fd-CEF. Our current and previous findings unequivocally demonstrate that Fd-CEF is driven by a new mediator [24]. The most compelling candidate for this mediator, FQR, is Cyt b_6/f-complex [52–54]. The Cyt b_6/f-complex possesses a potential binding site for Fd, situated close to the heme *c* location, composed of basic amino groups. The acidic region of Fd would bind to the Cyt b_6/f-complex at this site. The reduced heme *c* would then transfer electrons to the low-potential heme *b* in the Cyt *b* subunit of the Cyt b_6/f-complex [52,54]. Subsequently, the reduced heme *b* donates electrons to the oxidized PQ and/or the one-electron reduced PQ in the Q-cycle of the Cyt b_6/f-complex. This cyclic electron flow accelerates the Q-cycle and contributes to ΔpH formation. Further research is required to conclusively identify the mediator for Fd-CEF.

Subsequently, we have recognized what we consider to be the most critical insight into the physiological function of Fd-CEF in *pgr5^hope1*. The *pgr5^hope1* mutant demonstrates elevated H^+-conductance, as reported by various researchers [42,55], and this is evident in Supplemental Figures S3 and S4. Without Fd-CEF inducing acidification of the luminal space of thylakoid membranes, the proton motive force required for ATP production should not be sustained. The heightened electron flux in Fd-CEF of *pgr5^hope1* appears to play a compensatory role, counteracting the rapid loss of proton motive force by facilitating ΔpH formation across thylakoid membranes. In other words, as illustrated in Figures 1 and 5, the increased activity of Fd-CEF in *pgr5^hope1* effectively drives CO_2 assimilation at a rate comparable to that of WT.

4. Materials and Methods

4.1. Plant Materials and Growth Conditions

Arabidopsis plants (*Arabidopsis thaliana* WT and *pgr5^hope1*) were cultivated under the same condition as that reported by the previous study [24]. These plants of both WT and *pgr5^{hope1}* were analyzed with crr4, which was deficient mutant in chloroplastic NADH dehydrogenase [24]. The comparative analysis between WT and crr4 was already reported [24]. In the present research, *pgr5^{hope1}* were comparatively analyzed with WT.

4.2. Contents of Both Chlorophyll and Nitrogen

The contents of both chlorophyll and nitrogen in the leaves of *Arabidopsis* plants were determined by the method reported in the previous study [24].

4.3. Simultaneous Measurements of Chlorophyll Fluorescence, P700, Plastocyanin, and Ferredoxin with CO$_2$/H$_2$O-Exchange

One set of *Arabidopsis* plants (*Arabidopsis thaliana* WT and *pgr5^{hope1}*) grown under the above growth conditions were used for the simultaneous analysis of Chlorophyll fluorescence, P700, Plastocyanin, and Ferredoxin with CO$_2$/H$_2$O-exchange reported by the previous study [24]. The other set different from the above set were used for the simultaneous analysis of the electrochromic shift (ECS) signal with CO$_2$/H$_2$O-exchange, reported by the previous study [56].

The total photoreducible ferredoxin (Fd) signal originated from Fe/S signal [57]. The ratio of Fd to P700 in PSI was approximately 5 [58,59]. Furthermore, the leaves of tobacco plants had approximately 5 μmol Fd m^{-2} leaf area [41] and approximately 1 μmol P700 m^{-2} leaf area [60]. That is, the ratio of Fd to P700 in PSI was much closer to that of spinach leaves [58]. Then, we hypothesized that the amount of Fd in *Arabidopsis thaliana* was close to these values. The PSI complex contains Fx and F$_A$/F$_B$, in which Fe/S-clusters are the electron transfer carriers. That is, the Fe/S signal as Fd occupied less than 60% of the total Fe/S signal. Furthermore, the electron flux from Fx to NADP$^+$ through Fd is limited by the oxidation of the reduced Fd [61]. If the observed Fe/S signal was lower than 60%, we monitored the redox reaction of Fd.

4.4. Simultaneous Measurements of Electrochromic Shift Signal with CO$_2$/H$_2$O-Exchange

Electrochromic shift (ECS) signal with CO$_2$/H$_2$O-exchange in *Arabidopsis* plants (*Arabidopsis thaliana* WT and *pgr5^{hope1}*) were simultaneously analyzed by the method reported in the previous study [56]. The magnitude of the ECS signal was analyzed by DIRK analysis [62–64] and normalized as follows [65]. A single turnover flash (10 μs) was used to illuminate the leaf under far-red light. Then, the single turnover flash induced PSII-dependent production of ECS signal, which corresponds to the membrane potential induced by single-charge separation of P680 in PSII. The average value of a single turnover (ST) flash-induced ECS signal (ECS$_{ST}$) was $(4.09 \pm 0.07) \times 10^{-3}$ ΔI/Io ($n = 4$) (WT) and $(4.0 \pm 0.4) \times 10^{-3}$ ΔI/Io ($n = 4$) (*pgr5^{hope1}*). Then, the measured ECS signal was divided by ECS$_{ST}$ and was used as the normalized ECS signal (ECS$_N$) [56] (Equation (2)):

$$ECS_N = ECS/ECS_{ST} \tag{2}$$

The contribution of both ΔpH and Δψ to the total ECS signal was separately evaluated after turning off the AL illumination over longer periods of darkness [63]. The relative H$^+$ consumption rate vH$^+$ is the decay rate of ECS signal, which was evaluated by DIRK analysis [63]. The half-time of the decay reflects the H$^+$ conductance (gH$^+$) [63]. The vH$^+$ is proportional to both ECS$_N$ and gH$^+$, as described by (Equation (3)):

$$vH^+ = gH^+ \times ECS_N \tag{3}$$

4.5. Statistical Analytics

Statistical analysis of the corresponding data in both the text (ANCOVA, ANOVA, regression analysis) and Supplemental Table S1 (CI, confidential interval) was performed using the commercial software JMP8 (ver. 14.2.0, SAS Institute Inc., Cary, NC, USA).

Supplementary Materials: The following supporting information can be downloaded at https://www. mdpi.com/article/10.3390/ijms25052677/s1.

Author Contributions: Conceptualization, C.M.; writing—original draft preparation, S.M., M.O., S.W. and C.M.; writing—review and editing, M.O., C.M., S.W., S.M. and K.I.; supervision, C.M.; funding acquisition, C.M.; software, M.O. All authors have read and agreed to the published version of the manuscript.

Funding: This work was supported by Core Research for Evolutional Science and Technology (CREST) of the Japan Science and Technology Agency, Japan (grant number JPMJCR15O3), to C.M.

Institutional Review Board Statement: Not applicable.

Informed Consent Statement: Not applicable.

Data Availability Statement: Data are contained within the article and Supplemental Materials.

Conflicts of Interest: The authors declare no conflict of interest.

References

1. Tikhonov, A.N. The cytochrome b_6f complex at the crossroad of photosynthetic electron transport pathways. *Plant Physiol. Biochem. PPB* **2014**, *81*, 163–183. [CrossRef] [PubMed]
2. Tikhonov, A.N.; Vershubskii, A.V. Computer modeling of electron and proton transport in chloroplasts. *Biosystems* **2014**, *121*, 1–21. [CrossRef]
3. Allen, J.F. Cyclic, pseudocyclic and noncyclic photophosphorylation: New links in the chain. *Trends Plant Sci.* **2003**, *8*, 15–19. [CrossRef]
4. Huang, W.; Yang, Y.J.; Zhang, S.B.; Liu, T. Cyclic electron flow around photosystem I promotes ATP synthesis possibly helping the rapid repair of photodamaged photosystem II at low light. *Front. Plant Sci.* **2018**, *9*, 239. [CrossRef]
5. Sato, R.; Kawashima, R.; Trinh, M.D.L.; Nakano, M.; Nagai, T.; Masuda, S. Significance of PGR5-dependent cyclic electron flow for optimizing the rate of ATP synthesis and consumption in *Arabidopsis* chloroplasts. *Photosynth. Res.* **2019**, *139*, 359–365. [CrossRef]
6. Shikanai, T. Cyclic electron transport around photosystem I: Genetic approaches. *Annu. Rev. Plant Biol.* **2007**, *58*, 199–217. [CrossRef]
7. Yamori, W.; Shikanai, T. Physiological functions of cyclic electron transport around photosystem I in sustaining photosynthesis and plant growth. *Annu. Rev. Plant Biol.* **2016**, *67*, 81–106. [CrossRef]
8. Nawrocki, W.J.; Bailleul, B.; Picot, D.; Cardol, P.; Rappaport, F.; Wollman, F.A.; Joliot, P. The mechanism of cyclic electron flow. *Biochim. Biophys. Acta* **2019**, *1860*, 433–438. [CrossRef]
9. Walker, B.J.; Strand, D.D.; Kramer, D.M.; Cousins, A.B. The response of cyclic electron flow around photosystem I to changes in photorespiration and nitrate assimilation. *Plant Physiol.* **2014**, *165*, 453–462. [CrossRef]
10. Zhang, S.; Zou, B.; Cao, P.; Su, X.; Xie, F.; Pan, X.; Li, M. Structural insights into photosynthetic cyclic electron transport. *Mol. Plant* **2023**, *16*, 187–205. [CrossRef]
11. Foyer, C.; Furbank, R.; Harbinson, J.; Horton, P. The mechanisms contributing to photosynthetic control of electron transport by carbon assimilation in leaves. *Photosynth. Res.* **1990**, *25*, 83–100. [CrossRef]
12. von Caemmerer, S.; Farquhar, G.D. Some relationships between the biochemistry of photosynthesis and the gas exchange of leaves. *Planta* **1981**, *153*, 376–387. [CrossRef]
13. Zhang, R.; Sharkey, T.D. Photosynthetic electron transport and proton flux under moderate heat stress. *Photosynth. Res.* **2009**, *100*, 29–43. [CrossRef] [PubMed]
14. Allahverdiyeva, Y.; Suorsa, M.; Tikkanen, M.; Aro, E.M. Photoprotection of photosystems in fluctuating light intensities. *J. Exp. Bot.* **2015**, *66*, 2427–2436. [CrossRef] [PubMed]
15. Asada, K.; Kiso, K. The photo-oxidation of epinephrine by spinach chloroplasts and its inhibition by superoxide dismutase: Evidence for the formation of superoxide radicals in chloroplasts. *Agri. Biol. Chem.* **1973**, *37*, 12. [CrossRef]
16. Asada, K.; Kiso, K.; Yoshikawa, K. Univalent reduction of molecular oxygen by spinach chloroplasts on illumination. *J. Biol. Chem.* **1974**, *249*, 2175–2181. [CrossRef]
17. Flint, D.H.; Emptage, M.H.; Finnegan, M.G.; Fu, W.; Johnson, M.K. The role and properties of the iron-sulfur cluster in *Escherichia coli* dihydroxy-acid dehydratase. *J. Biol. Chem.* **1993**, *268*, 14732–14742. [CrossRef]
18. Flint, D.H.; Tuminello, J.F.; Emptage, M.H. The inactivation of Fe-S cluster containing hydro-lyases by superoxide. *J. Biol. Chem.* **1993**, *268*, 22369–22376. [CrossRef]
19. Furutani, R.; Wada, S.; Ifuku, K.; Maekawa, S.; Miyake, C. Higher reduced state of Fe/S-signals, with the suppressed oxidation of P700, causes PSI inactivation in *Arabidopsis thaliana*. *Antioxidants* **2022**, *12*, 21. [CrossRef]
20. Ivanov, A.G.; Morgan-Kiss, R.M.; Krol, M.; Allakhverdiev, S.I.; Zanev, Y.; Sane, P.V.; Huner, N.P. Photoinhibition of photosystem I in a pea mutant with altered LHCII organization. *J. Photochem. Photobiology. B Biol.* **2015**, *152 Pt B*, 335–346. [CrossRef]
21. Khorobrykh, S.; Havurinne, V.; Mattila, H.; Tyystjärvi, E. Oxygen and ROS in Photosynthesis. *Plants* **2020**, *9*, 91. [CrossRef]

22. Zivcak, M.; Brestic, M.; Kunderlikova, K.; Olsovska, K.; Allakhverdiev, S.I. Effect of photosystem I inactivation on chlorophyll *a* fluorescence induction in wheat leaves: Does activity of photosystem I play any role in OJIP rise? *J. Photochem. Photobiol. B Biol.* **2015**, *152 Pt B*, 318–324. [CrossRef]

23. Zivcak, M.; Brestic, M.; Kunderlikova, K.; Sytar, O.; Allakhverdiev, S.I. Repetitive light pulse-induced photoinhibition of photosystem I severely affects CO_2 assimilation and photoprotection in wheat leaves. *Photosynth. Res.* **2015**, *126*, 449–463. [CrossRef]

24. Ohnishi, M.; Maekawa, S.; Wada, S.; Ifuku, K.; Miyake, C. Evaluating the Oxidation Rate of Reduced Ferredoxin in *Arabidopsis thaliana* independent of photosynthetic linear electron flow: Plausible activity of ferredoxin-dependent cyclic electron flow around photosystem I. *Int. J. Mol. Sci.* **2023**, *24*, 12145. [CrossRef]

25. Kadota, K.; Furutani, R.; Makino, A.; Suzuki, Y.; Wada, S.; Miyake, C. Oxidation of P700 induces alternative electron flow in photosystem I in wheat leaves. *Plants* **2019**, *8*, 152. [CrossRef]

26. Shimakawa, G.; Miyake, C. Changing frequency of fluctuating light reveals the molecular mechanism for P700 oxidation in plant leaves. *Plant Direct* **2018**, *2*, e00073. [CrossRef]

27. Wada, S.; Amako, K.; Miyake, C. Identification of a novel mutation exacerbated the PSI photoinhibition in *pgr5/pgrl1* Mutants; Caution for overestimation of the phenotypes in *Arabidopsis pgr5-1* Mutant. *Cells* **2021**, *10*, 2884. [CrossRef]

28. Rantala, S.; Lempiäinen, T.; Gerotto, C.; Tiwari, A.; Aro, E.M.; Tikkanen, M. PGR5 and NDH-1 systems do not function as protective electron acceptors but mitigate the consequences of PSI inhibition. *Biochim. Biophys. Acta* **2020**, *1861*, 148154. [CrossRef]

29. Harbinson, J.; Foyer, C.H. Relationships between the efficiencies of photosystems I and II and stromal redox state in CO_2-free air: Evidence for cyclic electron flow in vivo. *Plant Physiol.* **1991**, *97*, 41–49. [CrossRef]

30. Harbinson, J.; Genty, B.; Baker, N.R. Relationship between the quantum efficiencies of photosystems I and II in pea leaves. *Plant Physiol.* **1989**, *90*, 1029–1034. [CrossRef]

31. Harbinson, J.; Hedley, C.L. Changes in P700 oxidation during the early stages of the induction of photosynthesis. *Plant Physiol.* **1993**, *103*, 649–660. [CrossRef]

32. Hashimoto, M.; Endo, T.; Peltier, G.; Tasaka, M.; Shikanai, T. A nucleus-encoded factor, CRR2, is essential for the expression of chloroplast *ndh*B in *Arabidopsis*. *Plant J.* **2003**, *36*, 541–549. [CrossRef]

33. Heber, U.; Walker, D. Concerning a dual function of coupled cyclic electron transport in leaves. *Plant Physiol.* **1992**, *100*, 1621–1626. [CrossRef]

34. Kono, M.; Noguchi, K.; Terashima, I. Roles of the cyclic electron flow around PSI (CEF-PSI) and O_2-dependent alternative pathways in regulation of the photosynthetic electron flow in short-term fluctuating light in *Arabidopsis thaliana*. *Plant Cell Physiol.* **2014**, *55*, 990–1004. [CrossRef]

35. Munekage, Y.; Hojo, M.; Meurer, J.; Endo, T.; Tasaka, M.; Shikanai, T. PGR5 is involved in cyclic electron flow around photosystem I and is essential for photoprotection in *Arabidopsis*. *Cell* **2002**, *110*, 361–371. [CrossRef]

36. Nandha, B.; Finazzi, G.; Joliot, P.; Hald, S.; Johnson, G.N. The role of PGR5 in the redox poising of photosynthetic electron transport. *Biochim. Biophys. Acta* **2007**, *1767*, 1252–1259. [CrossRef] [PubMed]

37. Peltier, G.; Aro, E.M.; Shikanai, T. NDH-1 and NDH-2 plastoquinone reductases in oxygenic photosynthesis. *Annu. Rev. Plant Biol.* **2016**, *67*, 55–80. [CrossRef]

38. Suorsa, M.; Grieco, M.; Järvi, S.; Gollan, P.J.; Kangasjärvi, S.; Tikkanen, M.; Aro, E.M. PGR5 ensures photosynthetic control to safeguard photosystem I under fluctuating light conditions. *Plant Signal. Behav.* **2013**, *8*, e22741. [CrossRef]

39. Suorsa, M.; Järvi, S.; Grieco, M.; Nurmi, M.; Pietrzykowska, M.; Rantala, M.; Kangasjärvi, S.; Paakkarinen, V.; Tikkanen, M.; Jansson, S.; et al. PROTON GRADIENT REGULATION5 is essential for proper acclimation of *Arabidopsis* photosystem I to naturally and artificially fluctuating light conditions. *Plant Cell* **2012**, *24*, 2934–2948. [CrossRef]

40. Teicher, H.B.; Moeller, B.L.; Scheller, H.V. Photoinhibition of photosystem I in field-grown barley (*Hordeum vulagre* L.): Introduction, recovery and acclimation. *Photosynth. Res.* **2000**, *64*, 29. [CrossRef]

41. Yamamoto, H.; Kato, H.; Shinzaki, Y.; Horiguchi, S.; Shikanai, T.; Hase, T.; Endo, T.; Nishioka, M.; Makino, A.; Tomizawa, K.; et al. Ferredoxin limits cyclic electron flow around PSI (CEF-PSI) in higher plants--stimulation of CEF-PSI enhances non-photochemical quenching of Chl fluorescence in transplastomic tobacco. *Plant Cell Physiol.* **2006**, *47*, 1355–1371. [CrossRef] [PubMed]

42. Yamamoto, H.; Shikanai, T. Does the *Arabidopsis* proton gradient regulation5 mutant leak protons from the thylakoid membrane? *Plant Physiol.* **2020**, *184*, 421–427. [CrossRef] [PubMed]

43. Heber, U. Irrungen, Wirrungen? The Mehler reaction in relation to cyclic electron transport in C3 plants. *Photosynth. Res.* **2002**, *73*, 223–231. [CrossRef] [PubMed]

44. Miyake, C.; Amako, K.; Shiraishi, N.; Sugimoto, T. Acclimation of tobacco leaves to high light intensity drives the plastoquinone oxidation system--relationship among the fraction of open PSII centers, non-photochemical quenching of Chl fluorescence and the maximum quantum yield of PSII in the dark. *Plant Cell Physiol.* **2009**, *50*, 730–743. [CrossRef]

45. Ruban, A.V.; Murchie, E.H. Assessing the photoprotective effectiveness of non-photochemical chlorophyll fluorescence quenching: A new approach. *Biochim. Biophys. Acta* **2012**, *1817*, 977–982. [CrossRef]

46. Fisher, N.; Kramer, D.M. Non-photochemical reduction of thylakoid photosynthetic redox carriers in vitro: Relevance to cyclic electron flow around photosystem I? *Biochim. Biophys. Acta* **2014**, *1837*, 1944–1954. [CrossRef]

47. Miyake, C.; Schreiber, U.; Asada, K. Ferredoxin-dependent and antimycin A-sensitive reduction of cytochrome *b*-559 by far-red light in Maize Thylakoids; Participation of a menadiol-reducible cytochrome *b*-559 in cyclic electron flow. *Plant Cell Physiol.* **1995**, *36*, 743–748.

48. Driever, S.M.; Baker, N.R. The water-water cycle in leaves is not a major alternative electron sink for dissipation of excess excitation energy when CO_2 assimilation is restricted. *Plant Cell Environ.* **2011**, *34*, 837–846. [CrossRef]

49. Genty, B.; Harbinson, J.; Briantais, J.M.; Baker, N.R. The relationship between non-photochemical quenching of chlorophyll fluorescence and the rate of photosystem 2 photochemistry in leaves. *Photosynth. Res.* **1990**, *25*, 249–257. [CrossRef]

50. Ruuska, S.A.; Badger, M.R.; Andrews, T.J.; von Caemmerer, S. Photosynthetic electron sinks in transgenic tobacco with reduced amounts of Rubisco: Little evidence for significant Mehler reaction. *J. Exp. Bot.* **2000**, *51*, 357–368. [CrossRef] [PubMed]

51. Sejima, T.; Hanawa, H.; Shimakawa, G.; Takagi, D.; Suzuki, Y.; Fukayama, H.; Makino, A.; Miyake, C. Post-illumination transient O_2-uptake is driven by photorespiration in tobacco leaves. *Physiol. Plant.* **2016**, *156*, 227–238. [CrossRef]

52. Cramer, W.A.; Zhang, H. Consequences of the structure of the cytochrome b_6f complex for its charge transfer pathways. *Biochim. Biophys. Acta* **2006**, *1757*, 339–345. [CrossRef]

53. Cramer, W.A.; Zhang, H.; Yan, J.; Kurisu, G.; Smith, J.L. Transmembrane traffic in the cytochrome b_6f complex. *Annu. Rev. Biochem.* **2006**, *75*, 769–790. [CrossRef]

54. Sarewicz, M.; Pintscher, S.; Pietras, R.; Borek, A.; Bujnowicz, Ł.; Hanke, G.; Cramer, W.A.; Finazzi, G.; Osyczka, A. Catalytic reactions and energy conservation in the cytochrome bc_1 and b_6f complexes of energy-transducing membranes. *Chem. Rev.* **2021**, *121*, 2020–2108. [CrossRef]

55. Yamamoto, H.; Shikanai, T. PGR5-dependent cyclic electron flow protects photosystem I under fluctuating light at donor and acceptor sides. *Plant Physiol.* **2019**, *179*, 588–600. [CrossRef]

56. Shimakawa, G.; Miyake, C. Photosynthetic linear electron flow drives CO_2 assimilation in maize leaves. *Int. J. Mol. Sci.* **2021**, *22*, 4894. [CrossRef]

57. Klughammer, C.; Schreiber, U. Deconvolution of ferredoxin, plastocyanin, and P700 transmittance changes in intact leaves with a new type of kinetic LED array spectrophotometer. *Photosynth. Res.* **2016**, *128*, 195–214. [CrossRef]

58. Böhme, H. Quantitative determination of ferredoxin, ferredoxin-NADP$^+$ reductase and plastocyanin in spinach chloroplasts. *Eur. J. Biochem.* **1978**, *83*, 137–141. [CrossRef]

59. Kozuleva, M.; Goss, T.; Twachtmann, M.; Rudi, K.; Trapka, J.; Selinski, J.; Ivanov, B.; Garapati, P.; Steinhoff, H.J.; Hase, T.; et al. Ferredoxin:NADP(H) Oxidoreductase abundance and location influences redox poise and stress tolerance. *Plant Physiol.* **2016**, *172*, 1480–1493. [CrossRef] [PubMed]

60. Oja, V.; Eichelmann, H.; Peterson, R.B.; Rasulov, B.; Laisk, A. Deciphering the 820 nm signal: Redox state of donor side and quantum yield of Photosystem I in leaves. *Photosynth. Res.* **2003**, *78*, 1–15. [CrossRef] [PubMed]

61. Cherepanov, D.A.; Milanovsky, G.E.; Petrova, A.A.; Tikhonov, A.N.; Semenov, A.Y. Electron transfer through the acceptor side of photosystem I: Interaction with exogenous acceptors and molecular oxygen. *Biochem. Biokhimiia* **2017**, *82*, 1249–1268. [CrossRef]

62. Sacksteder, C.A.; Kramer, D.M. Dark-interval relaxation kinetics (DIRK) of absorbance changes as a quantitative probe of steady-state electron transfer. *Photosynth. Res.* **2000**, *66*, 145–158. [CrossRef]

63. Avenson, T.J.; Cruz, J.A.; Kramer, D.M. Modulation of energy-dependent quenching of excitons in antennae of higher plants. *Proc. Natl. Acad. Sci. USA* **2004**, *101*, 5530–5535. [CrossRef]

64. Cruz, J.A.; Sacksteder, C.A.; Kanazawa, A.; Kramer, D.M. Contribution of electric field ($\Delta\Psi$) to steady-state transthylakoid proton motive force (*pmf*) in vitro and in vivo. control of *pmf* parsing into $\Delta\Psi$ and ΔpH by ionic strength. *Biochemistry* **2001**, *40*, 1226–1237. [CrossRef]

65. Klughammer, C.; Siebke, K.; Schreiber, U. Continuous ECS-indicated recording of the proton-motive charge flux in leaves. *Photosynth. Res.* **2013**, *117*, 471–487. [CrossRef]

International Journal of
Molecular Sciences

Article

Zinc Oxide Nanoparticles Affect Early Seedlings' Growth and Polar Metabolite Profiles of Pea (*Pisum sativum* L.) and Wheat (*Triticum aestivum* L.)

Karolina Stałanowska [1], Joanna Szablińska-Piernik [1], Adam Okorski [2] and Lesław B. Lahuta [1,*]

[1] Department of Plant Physiology, Genetics and Biotechnology, University of Warmia and Mazury in Olsztyn, Oczapowskiego 1A, 10-719 Olsztyn, Poland; karolina.stalanowska@uwm.edu.pl (K.S.); joanna.szablinska@uwm.edu.pl (J.S.-P.)

[2] Department of Entomology, Phytopathology and Molecular Diagnostics, University of Warmia and Mazury in Olsztyn, Pl. Łódzki 5, 10-727 Olsztyn, Poland

* Correspondence: lahuta@uwm.edu.pl

Abstract: The growing interest in the use of zinc oxide nanoparticles (ZnO NPs) in agriculture creates a risk of soil contamination with ZnO NPs, which can lead to phytotoxic effects on germinating seeds and seedlings. In the present study, the susceptibility of germinating seeds/seedlings of pea and wheat to ZnO NPs of various sizes ($\leq$50 and $\leq$100 nm) applied at concentrations in the range of 100–1000 mg/L was compared. Changes in metabolic profiles in seedlings were analyzed by GC and GC-MS methods. The size-dependent harmful effect of ZnO NPs on the seedling's growth was revealed. The more toxic ZnO NPs (50 nm) at the lowest concentration (100 mg/L) caused a 2-fold decrease in the length of the wheat roots. In peas, the root elongation was slowed down by 20–30% only at 1000 mg/L ZnO NPs. The metabolic response to ZnO NPs, common for all tested cultivars of pea and wheat, was a significant increase in sucrose (in roots and shoots) and GABA (in roots). In pea seedlings, an increased content of metabolites involved in the aspartate–glutamate pathway and the TCA cycle (citrate, malate) was found, while in wheat, the content of total amino acids (in all tissues) and malate (in roots) decreased. Moreover, a decrease in products of starch hydrolysis (maltose and glucose) in wheat endosperm indicates the disturbances in starch mobilization.

Keywords: pea; wheat; seedling; zinc oxide nanoparticles; polar metabolite profiles

Citation: Stałanowska, K.; Szablińska-Piernik, J.; Okorski, A.; Lahuta, L.B. Zinc Oxide Nanoparticles Affect Early Seedlings' Growth and Polar Metabolite Profiles of Pea (*Pisum sativum* L.) and Wheat (*Triticum aestivum* L.). *Int. J. Mol. Sci.* **2023**, *24*, 14992. https://doi.org/10.3390/ijms241914992

Academic Editor: Martin Bartas

Received: 14 September 2023
Revised: 5 October 2023
Accepted: 6 October 2023
Published: 8 October 2023

1. Introduction

Zinc oxide nanoparticles (ZnO NPs) have a lot of applications in many industry sectors, e.g., in electronics and electrical engineering, in medicine (due to their anti-bacterial, anti-tumor, and anti-inflammation properties), and can also serve as UV blockers in clothes or sunscreens [1–3]. Moreover, due to their biocidal properties, ZnO NPs have great potential for agricultural applications, especially as a novel compound of pesticides. ZnO NPs decrease *Fusarium proliferatum* growth in maize (*Zea mays* L.) grains [4], restrain *Fusarium oxysporum* growth, and suppress Fusarium wilt disease in eggplant (*Solanum melongena* L.) [5], and they can also be used as a foliar spray to suppress infections of *Erysiphe betae*, a causal agent of powdery mildew of sugar beet (*Beta vulgaris* L.) [6]. It has also been reported that ZnO NPs significantly inhibit the growth of some plant pathogenic bacteria, such as *Xanthomonas oryzae* pv. *oryzae*, *Xanthomonas axonopodis* pv. *citri*, and *Ralstonia solanacearum* [7]. ZnO NPs can also protect plants against various abiotic stresses, e.g., heavy metal, drought, or salinity stress, alleviating their negative effects [8]. Exposure of cotton (*Gossypium hirsutum* L.) to ZnO NPs alleviated the phytotoxic effect of cadmium (Cd) and lead (Pb), whereas foliar application of ZnO NPs alleviated Cd stress in tobacco (*Nicotiana tabacum* L.) [9]. In maize, ZnO NPs promoted drought stress tolerance and increased the activity of enzymes involved in primary carbohydrate metabolism (UDP-glucose pyrophosphorylase, phosphoglucose isomerase, and cytoplasmic invertase) [10].

Fazian et al. [11] showed that the foliar application of ZnO NPs to tomato (*Lycopersicon esculentum* Mill.) can promote salt-stress tolerance. An increase in antioxidative enzyme activities, such as peroxidase (POX), superoxide dismutase (SOD), and catalase (CAT), was also observed. It has been suggested that ZnO NPs improve plant tolerance to various stresses by increasing the activity of antioxidant enzymes and regulating primary carbohydrate metabolism, which has a positive impact on nutrient uptake and plant growth [8]. ZnO NPs also have a positive effect on seed germination, seedling growth, and an increase in plant biomass and yield [12–14]. Thus, it is considered a potential nano-fertilizer, which can also counteract zinc deficiency in the soil [15]. Nano-fertilizers are a step forward to precision and less expensive agriculture. They are more effective than standard fertilizers (less amount is needed to achieve the same results), and the release of nutrients is slower and more controlled [1,8,16].

However, adverse effects of ZnO NPs are also observed, depending on their concentration and physicochemical properties like size, shape, surface charge, surface coating, dissolution, and agglomeration [17–19]. Moreover, their phytotoxicity is plant species-specific and connected with nanoparticle uptake and transfer capacity in plants [7,19]. ZnO NPs can inhibit seedling growth, reduce plant biomass, decrease chlorophyll content, and cause oxidative damage, DNA damage, chromosome aberration, disruption of the cell membrane, and disturbance in the cell cycle [20–22]. There is a fine line between the toxic and beneficial effects of ZnO NPs due to the easy metabolization of Zn ions by plants [22].

ZnO NPs can be simply absorbed by plants, which leads to their overaccumulation and phytotoxic effect, presumably due to the release of excess Zn^{2+} ions. However, the toxicity of nanoparticles themselves was also reported [20,23]. ZnO NPs and zinc ions released from nanoparticles can be transported axially through the phloem and xylem but also radially—via symplastic or apoplastic pathway [24,25]. Da Cruz et al. [24] revealed that in common bean (*Phaseolus vulgaris* L.), elevated expression of genes encoding metal tonoplast-localized carriers was observed after ZnO NPs and $ZnSO_4$ exposure. This suggests that an excess of ZnO NPs and/or Zn^{2+} released from nanoparticles can be stored in vacuoles to maintain the Zn balance.

Optimal concentrations of Zn are crucial for the proper functioning of plant cells. It is a prosthetic group of many enzymes, such as dehydrogenases, aldolases, isomerases, transphosphorylases, RNA, and DNA polymerases. In addition, zinc ions stabilize the zinc finger domain, which is observed in many proteins (such as transcription factors) and is responsible for proper protein–nucleic acid and protein–protein interactions [26,27]. Thus, zinc is essential for gene expression and regulation. It has also been reported that zinc is required for auxin production, as it is needed for the synthesis of tryptophan, which is a precursor of auxin (indolyl-acetic acid, IAA). Zinc is also involved in signal transduction, maintaining the integrity of the cellular membranes, carbohydrate and lipid metabolism, photosynthetic metabolism, and plant defense responses [26–28]. Therefore, zinc deficiency causes inhibition of RNA and protein synthesis, as well as chlorosis, growth retardation, wilting, and rolling of leaves and stems [15].

Due to the increasingly common use of ZnO NPs, it is particularly important to test the sensitivity of important crop plants to nanoparticles, including wheat (*Triticum aestivum* L.) and pea (*Pisum sativum* L.), one of the most important cereals and legume species in the human diet and livestock feed [29,30]. The application of ZnO NPs (13 nm) at concentrations of 10–100 mg/L had a positive effect on wheat seed germination and seedling development [31]. A more toxic effect was observed for larger ZnO NPs (20 nm)—seedlings root and coleoptile length were inhibited after nanoparticle treatment at doses over 50 and 100 mg/L, respectively. However, $ZnSO_4$ used at the corresponding concentration range was more toxic than nanoparticles [32]. Pea cultivation in soil enriched with ZnO NPs at concentrations 125, 250, and 500 mg/kg resulted in pea root increased elongation. Some alterations in the antioxidant system were also observed [29]. Huang et al. [33] showed that pea seeds exposure to ZnO NPs at concentrations 500–1000 mg/L did not influence seed germination but negatively affected root elongation. Nanoparticles also act on pea-rhizobia symbiosis by decreasing nodulation (ZnO NPs at concentrations 250 and 750 mg/L), which affects nitrogen fixation [33].

There are still limited published data regarding the changes in pea and wheat metabolome/metabolite profiles under ZnO NPs treatment [34], especially at the earliest stages of development (seed germination/seedling growth), which are crucial for further plant growth and yield. Considering the potential use of ZnO NPs in agriculture as nano-fertilizers or nano-pesticides, used as foliar sprays, or in seed nano-priming, such investigations are needed to better understand the phytotoxic properties of nanoparticles and their possible mechanisms of influence on seed germination, seedlings growth, and their primary metabolism. Therefore, in the present study, the effects of commercially available, chemically synthesized ZnO NPs on (a) germination and early seedlings development of garden pea (*Pisum sativum* L.) and wheat (*Triticum aestivum* L.), two important crops of different clades (dicots and monocots, respectively), and (b) changes in the polar metabolite profiles of seedlings were compared. To the best of our knowledge, this is the first study to analyze the changes in the polar metabolite profiles of pea and wheat seedlings induced by continuous ZnO NPs treatment. These changes may underlie the different sensitivities of wheat and pea to ZnO NPs.

2. Results

2.1. The Preliminary Study

In the first part of the preliminary study, the phytotoxic concentrations of ZnO NPs to seed germination and early seedling development of peas and wheat were investigated. The 3 and 4 days of incubation (for wheat and peas, respectively) in the suspension of ZnO NPs ($\leq$50 nm) at concentrations 20, 50, 250, 500, and 1000 mg/L did not affect the germination of pea and wheat seeds (of both cultivars). No statistical differences ($p < 0.05$) were observed in the length, FW, and DW of seedlings of cv. Nemo (Supplementary Materials, Table S1), whereas they were found in cv. Tarchalska, at elevated concentrations of ZnO NPs ($\geq$250 mg/L, Table S1). In wheat (both cultivars), the root length and FW decreased with increasing concentrations of ZnO NPs, whereas no differences were found in coleoptiles (Table S2).

Subsequently, the effects of ZnO NPs of different sizes (ø < 50 nm and <100 nm) were compared in pea cv. Tarchalska and wheat cv. Ostka Strzelecka. Nanoparticles (of both sizes) at concentrations of 100, 250, and 1000 mg/L did not affect the germination of pea and wheat seeds (Tables S3 and S4). Moreover, regardless of their size, ZnO NPs inhibited the growth of the pea seedlings. However, smaller nanoparticles caused stronger inhibition of epicotyl elongation, compared with root. The opposite effect was observed in seedlings grown in the suspension of larger ZnO NPs (Table S3). In wheat, smaller ZnO NPs caused stronger inhibition of root growth—the length of the primary root was decreased by 52, 66, and 71% (at concentrations of 100, 250, and 1000 mg/L, respectively), whereas under treatment with larger ZnO NPs (<100 nm) the length of the root was lowered by 50, 59, and 68% (Table S4).

The above results led to the selection of smaller ZnO NPs (<50 nm) at concentrations of 100, 250, and 1000 mg/L for the main experiment focusing on the comparisons of changes in metabolome in pea and wheat seedlings in response to ZnO NPs.

2.2. The Effect of ZnO NPs on Seed Germinability and Seedling Growth

ZnO nanoparticles at concentrations in the range of 100–1000 mg/L did not negatively affect the germination of pea and wheat seeds. Wheat grains (both cultivars) finished germination at 95–100% after the first day, both in water and suspensions of ZnO NPs. The rate of pea seed germination was slightly slowed—after 24 h of hydration in water, 60% of the seeds completed germination (the radicle protruded the seed coat and reached a length of 2–3 mm), while after another 24 h, the germinability increased to 100%. Zinc oxide nanoparticles at a concentration of 100 mg/L accelerated the completion of seed germination in both pea cultivars (from 60 to 80%), while higher concentrations had no significant effect on this process. However, the growth and development of seedlings of both species were significantly ($p < 0.05$) affected by ZnO NPs (along with increasing concentrations). In peas, ZnO NPs slightly decreased the elongation of root (cv. Nemo) or epicotyl (cv. Tarchalska, Figure 1A,B), whereas in wheat, nanoparticles dramatically inhibited the elongation and fresh weight gain in both roots and coleoptiles (Figures 1C,D,G,H and S1).

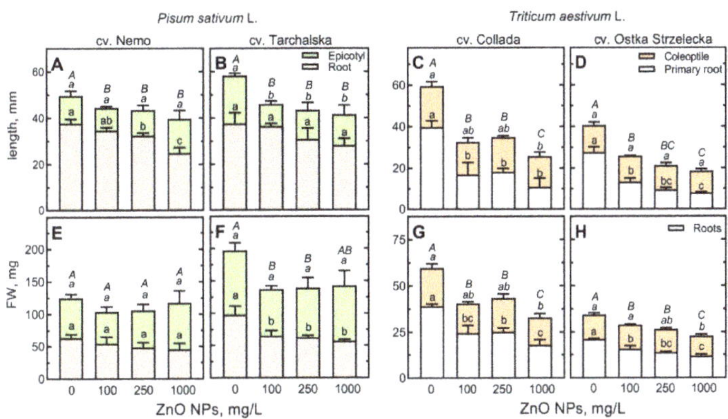

Figure 1. The effects of ZnO NPs (at concentrations of 0, 100, 250, and 1000 mg/L) on the length and fresh weight (FW) of roots and shoots (epicotyl of pea and coleoptile of wheat) of 4-day-old seedlings of pea (*Pisum sativum* L.) cv. Nemo (**A,E**) and cv. Tarchalska (**B,F**) and 3-day-old seedlings of wheat (*Triticum aestivum* L.) cv. Collada (**C,G**) and cv. Ostka Strzelecka (**D,H**). Values are means of 3 replicates + SD. The same lowercase letters (for shoots and roots, separately) and uppercase letters (for total seedlings length) above the bars indicate no significant ($p < 0.05$) differences according to the ANOVA test and post hoc Tukey's corrections.

The morphological deformations of seedlings were not observed. Noticeably, in Petri dishes with wheat seeds germinating in ZnO NP suspensions, white precipitates appeared along with increasing concentrations of ZnO NPs (Figure S1). Such precipitates were not noted in dishes with germinating pea seeds (data not presented).

2.3. The Composition and Content of Polar Metabolites in Control Seedlings of Pea and Wheat

In seedlings and cotyledon of peas, 39 and 34 polar metabolites were identified, respectively (Tables S5–S7), whereas in seedlings and endosperms of wheat, 34 and 35 were identified, respectively (Tables S8–S10). In both analyzed species, the same set of polar metabolites was identified in the tissues of growing roots and shoots, i.e., 5 soluble carbohydrates (fructose, glucose, galactose, *myo*-inositol, and sucrose), 17 amino acids (alanine, asparagine, aspartic acid, γ-aminobutyric acid, glutamic acid, glutamine, hydroxyproline, isoleucine, leucine, lysine, phenylalanine, proline, serine, threonine, tyrosine, and valine), 6 organic acids (citric, fumaric, lactic, malic, and propionic acids), phosphoric acid, and urea. In seedlings of peas, non-proteinogenic amino acids (homoserine and b-alanine) and an additional six less abundant organic acids (acetic, butyric, glutaric, malonic, oxalic, and succinic acid) were detected. Pea cotyledons also contained raffinose and stachyose (oligosaccharides absent in wheat seedlings). The polar metabolites present in seedlings of wheat, but not detected in peas, were 1-kestose (tri-saccharide), methionine, maltose, and maltotriose (in endosperm only).

The concentration of total identified polar metabolites (TIPMs) was similar in the corresponding pea and wheat organs, remaining 2-fold lower in storage tissues (cotyledons and endosperm) than in the growing parts of seedlings, regardless of seedling treatment (Figure 2, Tables S5–S10). The most prevalent fraction in the polar metabolites of both species was soluble carbohydrates (Figure 2).

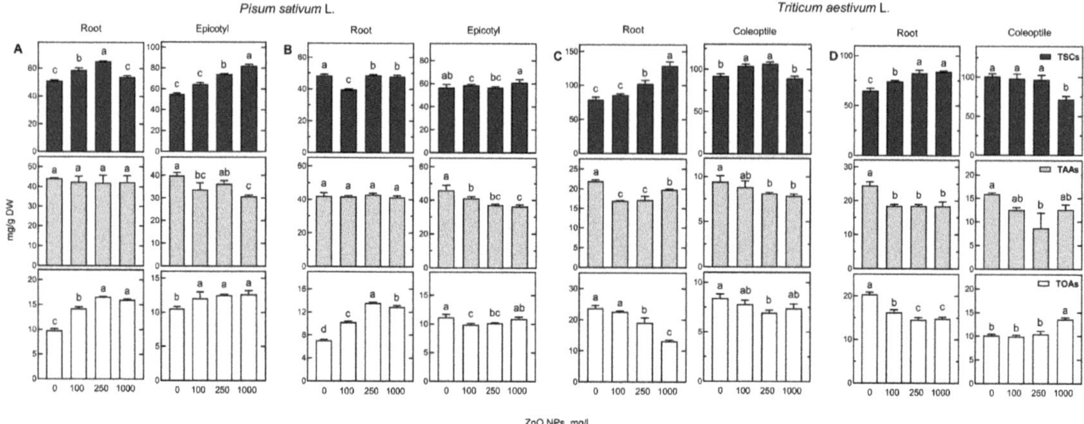

Figure 2. The concentration of total soluble carbohydrates (TSCs), total amino acids (TAAs), and total organic acids (TOAs) in the roots and epicotyls of 4-day-old seedlings of pea (*Pisum sativum* L.) cv. Nemo (**A**) and cv. Tarchalska (**B**) and roots and coleoptiles of 3-day-old seedlings of wheat (*Triticum aestivum* L.) cv. Collada (**C**) and cv. Ostka Strzelecka (**D**). Values (in mg/g DW) are means of 3 replicates + SD. The same letters above the bars indicate no significant ($p < 0.05$) differences according to the ANOVA test and post hoc Tukey's corrections.

The concentration of total soluble carbohydrates (TSCs) in epicotyls and roots of control seedlings of pea was slightly higher than that in cotyledons (48–56 and 41–46 mg/g DW, respectively) and accounted for 44 to 48% of TIPMs, while the participation of sugars in TIPMs in cotyledons was much higher (71–77%). In epicotyl and root, the second major fraction among polar metabolites was amino acids (total 42–46 mg/g DW), accounting for 35–40% of TIPMs (Figure 2A,B; Tables S5 and S6). In cotyledons, the concentrations of total amino acids (TAAs) and total organic acids (TOAs) were much lower (8–11 and 3–4.5 mg/g DW, respectively) than in growing tissues (Table S7). Moreover, in cotyledons, the concentration of phosphoric acid was much lower than in seedlings (ca. 2 and 10 mg/g DW, respectively, Tables S5–S7).

In control wheat seedlings, soluble carbohydrates (65–109 mg/g DW) constituted over half of the polar metabolites in roots, sharing 55–60% of TIPMs, and 76–80% of TIPMs in coleoptiles (92–109 mg/g DW, Figure 2C,D; Tables S8 and S9). In the endosperm, the participation of TSCs in TIPMs was much higher, up to 91–93% (Table S10). The concentration of TAAs in roots was similar in both wheat cultivars (22–25 mg/g DW) and higher than that in coleoptiles (9.5 and 16 mg/g DW). The concentration of TOAs in roots was as high (22–24 mg/g DW) as TAAs and more than twice as high as in the coleoptile (Figure 2). In endosperms, the levels of TAAs and TOAs were much lower (2.8–2.9 and 0.91–0.95 mg/g DW, respectively) than in growing tissues (Table S10).

Among the polar metabolites detected in the growing tissues of control seedlings of pea and wheat (developed in the water), sucrose, glucose, galactose, fructose, asparagine, citrate, and malate were the predominant metabolites. A high content of homoserines was detected in peas, whereas the roots of wheat seedlings contained much more glutamine and hydroxyproline than pea seedlings. Additionally, 1-kestose (a short-chain fructan) was found in wheat tissues (Table 1). Tissues also contained a considerable amount of phosphoric acid (5–10 mg/g DW). Moreover, in wheat coleoptile, elevated levels of glucose and fructose coincided with a lowered level of sucrose. Such a negative relationship was not found in peas.

Table 1. The concentration of predominant polar metabolites in roots and epicotyl/coleoptile of 4-day-old control seedlings of pea (*Pisum sativum* L.) of cultivars Nemo (N) and Tarchalska (T) and 3-day-old seedlings of wheat (*Triticum aestivum* L.) of cultivars Collada (C) and Ostka Strzelecka (OS). Values (in mg/g DW) are means of 3 replicates.

| Metabolite | *Pisum sativum* L. | | | | *Triticum aestivum* L. | | | |
| | Root | | Epicotyl | | Roots | | Coleoptile | |
	N	T	N	T	C	OS	C	OS
Sucrose	34.61	21.68	28.22	27.81	21.36	23.31	2.10	12.97
Glucose	8.70	16.34	13.72	14.73	24.38	21.57	46.59	41.83
Galactose	3.23	5.79	5.38	6.69	13.46	4.25	7.94	9.18
Fructose	0.84	1.27	3.73	3.43	10.61	10.68	32.21	28.12
Asparagine	3.00	2.22	3.95	4.88	5.80	5.83	2.45	2.70
Citrate	3.32	2.29	4.47	3.98	5.61	k4.90	5.33	6.22
Malate	4.83	3.22	4.34	5.43	16.05	12.48	2.38	2.92
Homoserine	26.59	30.33	24.36	30.36	- *	-	-	-
1-Kestose	- *	-	-	-	4.02	4.63	1.84	7.49
Glutamine	0.11	0.10	0.07	0.08	4.26	3.87	0.48	0.70
Hydroxyproline	0.48	0.27	0.29	0.18	2.30	2.88	0.72	2.55
Phosphoric acid	10.28	6.79	9.36	9.01	6.87	7.74	5.49	6.37

- * not detected.

In cotyledons of a pea, the major metabolite was sucrose (37–42 mg/g DW, Table S7), whereas in wheat endosperm, maltose (34 and 26 mg/g DW in cv. Collada and Ostka Strzelecka, respectively) and glucose (9.3 and 6.5 mg/g DW) dominated (Table S10). Small amounts of raffinose family oligosaccharides (raffinose and stachyose) were present (<2 mg/g DW, Table S3) only in pea cotyledons.

In seedlings developing in ZnO NP suspensions, the concentrations of many metabolites changed (Tables S5–S10), leading to differences in the metabolic profiles and the concentrations of polar metabolite fractions (Figure 2). However, the shift in changes was not uniform, except for a similar trend in the increase of TIPMs in the roots of both wheat cultivars (Figure 2). This was a result of the elevated levels of TSCs, along with the increasing concentration of ZnO NPs.

2.4. The PCA of Metabolic Profiles of Seedlings under ZnO NPs Treatment

The principal component analysis (PCA) of polar metabolites in root/epicotyl of pea and roots/coleoptile of wheat seedlings showed a clear separation of control samples from those treated with ZnO NPs (Figure 3). The shift in the distribution of the samples changed with the increasing concentration of ZnO NPs. This was especially noted for the roots of pea cv. Tarchalska and roots of both wheat cultivars, where PC1 shared 91.74, 86.70, and 95.76% of variability, respectively (Figure 3B–D). In wheat, the control samples of roots and those treated with 100 mg/L ZnO NPs were not only separated from those treated with ZnO NPs at higher concentrations but also from each other according to PC2 (10.20 and 2.67% of variability for cv. Collada and Ostka Strzelecka, respectively) (Figure 3C,D). Pea epicotyl cv. Nemo samples and wheat coleoptile samples of both cultivars, after treatment with ZnO NPs at concentrations 100 and 250 mg/L, were grouped together, apart from the control and samples treated with higher nanoparticle concentrations (Figure 3E,G,H). In contrast, pea epicotyl cv. Tarchalska samples treated with 250 and 1000 mg/L ZnO NPs were grouped together (Figure 3F).

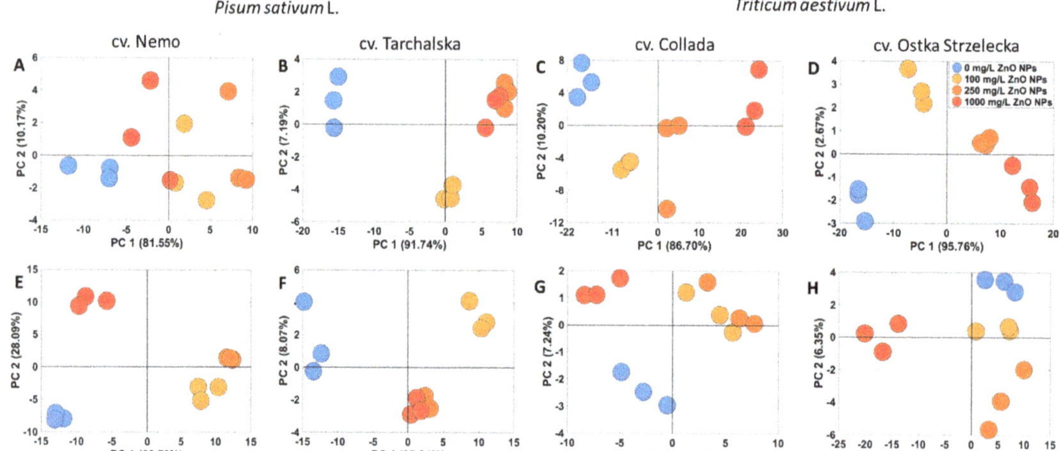

Figure 3. PCA of metabolic profiles of roots (**A,B**) and epicotyls (**E,F**) of 4-day-old seedlings of pea (*Pisum sativum* L.) and roots (**C,D**) and coleoptile (**G,H**) of 3-day-old seedlings of wheat (*Triticum aestivum* L.) developed in suspension of ZnO NPs at 0, 100, 250, and 1000 mg/L.

The distribution of seedling samples of both species was influenced mainly by changes in the concentrations of sucrose and glucose, as revealed by the PCA loading plots (Figure S2). Additionally, samples of pea seedlings were differentiated by galactose, homoserine, citric acid, and malic acid, while those of wheat by fructose and 1-kestose (Figure S2). In cotyledons of peas, sucrose and β-alanine were the major metabolites affecting the separation of samples (Figure S3). However, the pattern of sample distribution was not so clear (Figure S3), compared with those for seedlings (Figure 3). In wheat, endosperms of control seedlings were focused on the right from PC1 (sharing 96–98% of variability), whereas those treated with ZnO NPs were mostly on the left (Figure S4). This was a result of changes in the concentrations of maltose and sucrose (Figure S4).

2.5. Changes of Polar Metabolites in Seedlings' Response to ZnO NPs

2.5.1. Metabolites Mostly Differentiate Samples According to the Results of PCA

The concentration of sucrose, the carbohydrate most differentiating pea samples (Figure S2A,B,E,F), and the quantitatively major soluble carbohydrate in seedlings of both pea cultivars, was significantly ($p < 0.05$) higher in the roots and epicotyls of seedlings developing in ZnO NPs compared with the control (Tables S1 and S2). The highest concentration of sucrose was found in roots treated with ZnO NPs at 250 mg/L (49.70 and 40.63 mg/g DW in cv. Nemo and Tarchalska, respectively), while in epicotyls at 100 mg/L (46.94 mg/g DW in cv. Tarchalska) or 250 mg/L (52.74 mg/g DW in cv. Nemo, Figure 4).

Although the elevated level of sucrose was accompanied by a decrease in monosaccharides, i.e., fructose, galactose, and glucose (Tables S5 and S6), the negative correlations between changes in the concentrations of sucrose and monosaccharides (separately glucose, galactose and fructose) were found only in cv. Tarchalska in both roots ($r = -0.88$, $r = -0.88$ and $r = -0.86$, Table S11) and epicotyls ($r = -0.90$, $r = -0.87$ and $r = -0.80$, Table S11).

Among the amino acids, homoserine was dominant, regardless of pea seedlings treatment (Figure 4A, Tables S5 and S6). Its concentration was only slightly decreased in the roots of seedlings developing in ZnO NPs. However, it was significantly lowered by ZnO NPs in epicotyls, by more than 35 and 22% in cv. Nemo and cv. Tarchalska, respectively (Figure 4B, Tables S5 and S6).

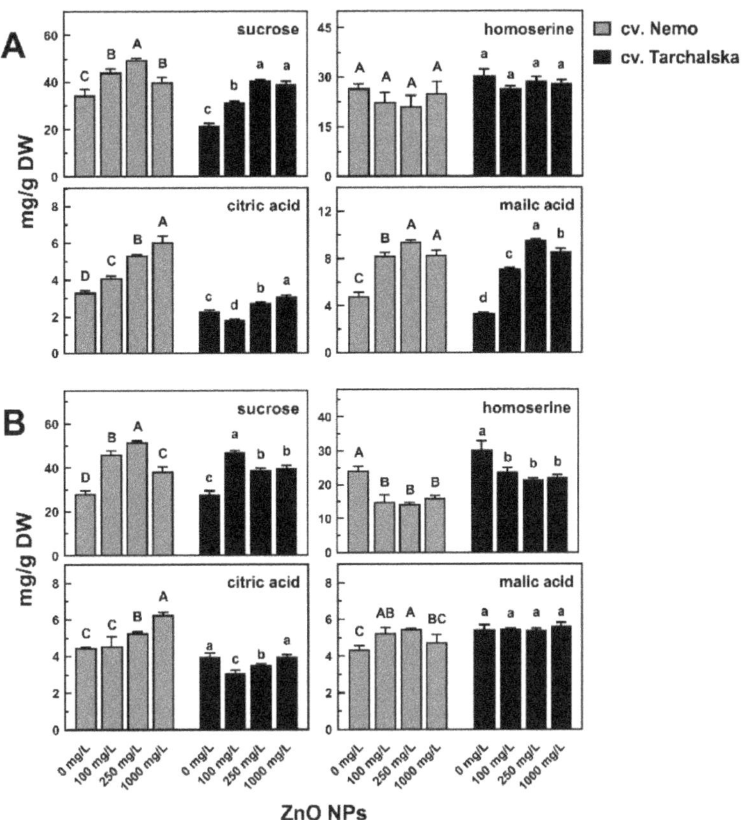

Figure 4. The effect of ZnO NPs on the concentration of metabolites mostly differentiating root (**A**) and epicotyl (**B**) samples (according to PCA analysis—Figures 3 and S2) of 4-day-old seedlings of pea (*Pisum sativum* L.) cv. Nemo and cv. Tarchalska. Means of 3 replicates + SD. The same letters (A–D and a–d separately for cv. Nemo and Tarchalska, respectively) by the values indicate no significant ($p < 0.05$) differences based on ANOVA analysis and Tukey's post hoc corrections.

The most abundant organic acids were citric and malic acids. Their concentrations significantly increased in the roots of both pea cultivars (Figure 4A) and in epicotyls of cv. Nemo (Figure 4B). The concentration of citric acid in roots and epicotyls of cv. Nemo increased along with the increasing concentrations of ZnO NPs (up to ca. 6 mg/g DW). The maximum levels of malic acid in roots and epicotyls (9.39 and 5.45 mg/g DW, respectively) were found in seedlings developed in ZnO NPs at 250 mg/L. In cv. Tarchalska, roots of seedlings treated with ZnO NPs (at 250 and 1000 mg/L) contained only slightly more citric acid than the control, but the concentration of malic acid was dramatically increased (2–3-fold, Figure 4A).

In wheat roots, the most dominant soluble sugars were sucrose and glucose, whereas in coleoptile—glucose and fructose (Figure 5; Tables S8 and S9). Changes in their concentrations were the major reasons for shifts in the metabolic profiles of seedling samples under ZnO NPs treatment (Figure S2C,D,G,H). In the roots of both wheat cultivars, the amount of sucrose increased along with the increasing concentrations of ZnO NPs (from 21.36 to 56.61 mg/g DW in cv. Collada and from 23.31 to 51.61 mg/g DW in cv. Ostka Strzelecka) (Figure 5A).

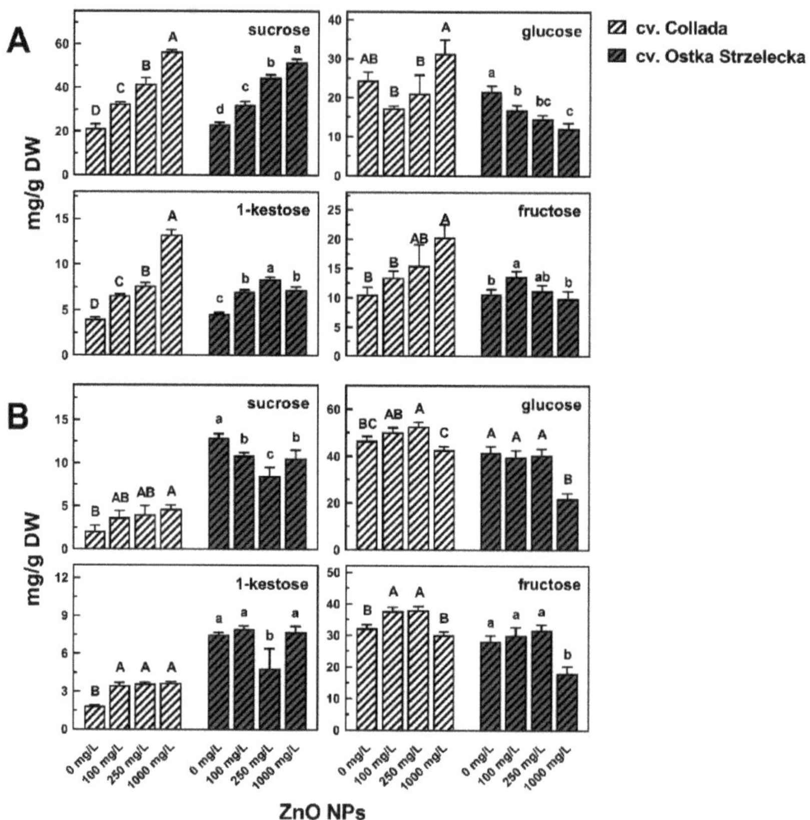

Figure 5. The effect of ZnO NPs on the concentration of metabolites mostly differentiating roots (**A**) and coleoptile (**B**) samples (according to PCA analysis—Figures 4 and S1) of 3-day-old wheat seedlings cv. Collada and cv. Ostka Strzelecka. Means of 3 replicates + SD. The same letters (A–D and a–d separately for cv. Collada and Ostka Strzelecka, respectively) by the values indicate no significant differences ($p < 0.05$) based on ANOVA analysis and Tukey's post hoc corrections.

The effect of ZnO NPs on the concentration of sucrose in coleoptiles was different, causing a 2-fold increase in cv. Collada, while a decrease in cv. Ostka Strzelecka (Figure 5B). Besides sucrose, root tissues (both cultivars) under ZnO NPs treatment accumulated 1-kestose. The concentration of this fructan was also duplicated in coleoptiles of cv. Collada (Figure 5). However, the correlations between changes in the concentrations of sucrose plus 1-kestose and monosaccharides (fructose plus glucose) in roots were opposite, i.e., positive in cv. Collada and negative in cv. Ostka Strzelecka ($r = 0.72$ and $r = -0.90$, respectively, Table S12). No correlations were found in coleoptiles (Table S12).

2.5.2. Other Metabolites

The concentrations of some amino acids were significantly affected by ZnO NPs in seedlings of both studied species. In peas, the roots of seedlings developed in ZnO NPs contained more aspartic acid, β-alanine, γ-aminobutyric acid (GABA), and glutamic acid than the roots of control seedlings (Tables S5 and S6). Similar differences were found in epicotyls, but only in seedlings that developed in ZnO NPs at concentrations of 100 and 250 mg/L. Moreover, a decrease in asparagine was found (Tables S5 and S6). It should be noted that ZnO NPs (at applied concentrations) had no effect on the concentration of TSCs (among them, sucrose, raffinose, stachyose, and *myo*-inositol) in pea cotyledons (Table S3).

In wheat, the concentrations of asparagine and glutamine (the most dominant amino acids) in roots decreased with increasing concentrations of ZnO NPs. Asparagine decreased by ca. 45 and 50% and glutamine by 40 and 75% in cv. Collada and cv. Ostka Strzelecka, respectively. In coleoptiles, the concentrations of asparagine and valine also decreased, as in roots (asparagine by ca. 30% in cv. Collada and by 50–75% in cv. Ostka Strzelecka), whereas the concentration of valine decreased by 20% in coleoptiles of both cultivars (Tables S8 and S9). Additionally, the decrease in isoleucine and valine (by ca. 20–45% and by 30–40%, respectively) and accumulation of GABA were also observed in the roots of both wheat cultivars (Tables S8 and S9).

In wheat endosperms, ZnO NPs caused a decrease in most fractions of the identified metabolites (Table S10). The concentration of maltose (predominant soluble carbohydrate, sharing more than 50% of TSCs in endosperm) decreased with increasing ZnO NPs concentration. Concomitantly, a decrease in maltotriose and a slight increase in sucrose were observed (Table S10).

3. Discussion

3.1. Physiological Effects of ZnO NPs

The dual effects of zinc oxide nanoparticles on seed germination and plant growth depend on the nanoparticles' physicochemical properties, concentration, and plant species [19]. Our findings suggest that ZnO NPs at concentrations up to 1000 mg/L have no negative impact on the germination of wheat and pea seeds. Consistent results with ZnO NPs have been documented for various plant species, i.e., pea [33], wheat [30], radish, rape, ryegrass, lettuce, corn, and cucumber [35].

The inhibitory effect of ZnO NPs on the growth of pea and wheat seedlings presented in this study (Figures 1 and S1) was probably a result of the over-uptake of nanoparticles by the roots. In our study, the uptake of ZnO NPs was not analyzed, but the reactions of seedlings to increasing concentrations of ZnO NPs (at both physiological and metabolomic levels, Figures 1 and S1, Tables S5–S10) seem to be an indirect confirmation of this process. Similar findings were presented by Huang et al. [33] in peas and by Srivastav et al. [30] in wheat. In their investigations, they used the same ZnO NPs manufactured by Sigma-Aldrich as in the present study. Huang et al. [33] showed that 24 h of pea seed imbibition in ZnO NPs at 250 mg/L resulted in inhibition of pea seedling root elongation during the next few days of seedling development in the absence of ZnO NPs. In our study, seeds germinated, and seedlings grew in the constant presence of ZnO NPs; thus, this long exposure was a reason for seedlings' growth retardation at the lower concentration of ZnO NPs (100 mg/L). Similarly, the ZnO NPs at 100 mg/L dramatically decreased the growth of wheat seedlings (Figure 1). However, Srivastav et al. [30] found a remarkable decrease in growth and biomass accumulation in seedlings of maize and wheat even at a 2-fold higher concentration of ZnO NPs (200 mg/L). Moreover, in the above-mentioned studies [30,33], an accumulation of higher amounts of zinc in roots than in shoots was documented.

In our study, ZnO NPs showed harmful effects at all tested concentrations on both species, but it was the most noticeable at the concentration of 1000 mg/L (Figure 1). Moreover, the inhibitory impact on root elongation was evident in wheat seedlings—the radicle length of both wheat cultivars was decreased by over 50% at 100 mg/L (Figure 1C,D). In peas, the root length was decreased by 25 and 34% (in cv. Tarchalska and Nemo, respectively) even at a 10-fold higher concentration of ZnO NPs (Figure 1A,B). The high susceptibility of wheat roots (and to a lesser extent also coleoptile) to other metal nanoparticles was documented earlier [36–39]. In our previous study, dead cells and accumulation of reactive oxygen species (ROS) were observed in the wheat root meristem and in the elongation zone under treatment with silver nanoparticles, applied at several-fold lower concentrations. Moreover, the visible symptom of cell deterioration was the browning of the root tip [38,40]. In the present study, such an effect was not observed, presumably due to the lower toxicity of ZnO NPs than AgNPs.

The inhibition of root growth (Figure 1) was not obviously accompanied by a decrease in the elongation of coleoptile (found only in cv. Collada) and epicotyl (found in cv. Tarchalska). The absorbed ZnO NPs (and zinc ions Zn^{2+} released from nanoparticles) can be transported from roots to shoot not only through the xylem but also via symplastic or apoplastic pathways [24,25]. Thus, the different effects of ZnO NPs on the elongation of roots/shoots of peas and wheat could be a result of differences in the transport of ZnO NPs/Zn^{2+} ions in seedlings. Moreover, zinc or specific metal transporters might be involved in ZnO NPs transport in both species [24].

On the other hand, zinc is crucial for auxin synthesis, and its excess can reduce IAA accumulation and transport in a dose-dependent manner. In *Arabidopsis*, zinc at high concentrations (over 100 µM $ZnSO_4$), besides the reduction of primary root length (caused by inhibition of meristematic cell division), also negatively affected the total number and the density of lateral roots [41]. Therefore, the effects of ZnO NPs on seedlings' growth can be also a result of an imbalance in hormonal homeostasis [42].

3.2. Changes in Polar Metabolite Profile after ZnO NPs Treatment

In the first stages of seedling development, reserve materials stored in seeds (carbohydrates, proteins, and lipids) are mobilized and utilized by growing seedlings [43]. In pea cotyledons, the content of all fractions of identified polar metabolites did not change under ZnO NPs treatment, which suggests that mobilization and utilization of storage materials in cotyledons were not affected by nanoparticles (Table S7). However, some changes in wheat endosperm (especially, soluble sugar content) occurred under ZnO NPs treatment in a dose-dependent manner (Table S10). The decreased content of maltose and maltotriose in endosperm might be related to the inhibition of α-amylase activity [30] and, therefore, inhibition of starch degradation [44]. Such inhibitory properties of ZnO NPs were previously reported in germinating seeds/seedlings of wheat [30] and also in vitro study of fibroblast cell lines [45].

Shifts in the TIPMs profiles of wheat and pea seedlings caused by ZnO NPs were related to changes in the concentration of soluble carbohydrates, especially sucrose (Figures 2, 4 and 5, Tables S5–S10), accumulated under ZnO NPs treatment in roots and epicotyl/coleoptile (Figures 4 and 5). Sucrose accumulation was also observed in the roots of wheat seedlings germinated and developed in Ag NPs solutions [38] and in tomato roots after foliar spray with ZnO NPs [46]. It should be noted that the sources of sucrose in germinating seeds of wheat and peas are not the same. In wheat, sucrose is synthesized in the scutellum (from glucose and maltose, a final product of starch hydrolysis in endosperm) and then transported to growing roots and coleoptile [44,47]. In germinating seeds of peas, sucrose is released from quickly degraded raffinose family oligosaccharides (RFOs), i.e., raffinose, stachyose, and verbascose, present at elevated concentrations in both the embryonic axis and cotyledons of mature seeds. The hydrolysis of RFOs starts just during seed imbibition and finishes in the embryonic axis at the end of seed germination (ca. 28–32 h after the start of seed imbibition), whereas in cotyledons, this process undergoes up to 7–10 days of germination [48,49]. Importantly, both products of RFOs hydrolysis, galactose and sucrose, are necessary for proper seedling growth, as documented in pea [50] and winter vetch [51]. Sucrose in sink tissues is hydrolyzed by sucrose synthase (Sus) and/or invertase (INV) to glucose and fructose—both serve as substrates for respiration and a source of carbon skeletons for other metabolic pathways [47].

Sucrose accumulation in roots and epicotyl/coleoptile under ZnO NPs treatment, found in our study (Figure 5), might be related to the inhibition of Sus and/or INV activity [41]. The increased activity of sucrose-phosphate synthase (SPS), an enzyme resynthesizing sucrose from monosaccharides, is also possible [47]. Although small-sized zinc oxide nanoparticles (ZnO NPs)—pyramids, plates, and spheres—possess the ability to inhibit the activity of a typical enzyme β-galactosidase in a biomimetic way [52], our results exclude a profound effect of ZnO NPs on α-galactosidase activity—the concentration of raffinose and stachyose in cotyledons (in 4-day-old seedlings they were absent) under ZnO

NPs treatment was as low as in the control (Table S7). Thus, an excess of ZnO NPs seems to not disturb RFOs degradation during pea seed germination and seedling growth.

It could be expected that changes in sucrose and monosaccharides (fructose and glucose) would be related. Only in pea cv. Tarchalska (in both roots and epicotyls) and wheat cv. Ostka Strzelecka (in roots), a negative correlation between sucrose and monosaccharides level was observed (cv. Tarchalska $r \geq -0.86$, cv. Ostka Strzelecka $r = -0.97$; Tables S11 and S12). However, such an effect caused by ZnO NPs was not uniform—the mentioned changes were not found in pea seedlings of cv. Nemo (Table S11) and in wheat coleoptiles (both cultivars; Tables S8 and S9). Moreover, in the roots of cv. Collada, the accumulation of sucrose was accompanied by the accumulation of monosaccharides (Table S8).

The seed germination and fast rate of early seedlings' growth are accompanied by intensive respiration [43,48]. Thus, the changes in tricarboxylic acids observed in growing seedlings under ZnO NPs treatment, as found in our study in roots of peas (Figure 4) and wheat (Tables S8 and S9), can be a result of the impact on the TCA cycle [30]. A decrease in malic acid content in wheat roots was revealed earlier, under Ag NPs treatment [38] and in cucumber roots after foliar spray with ZnO NPs [53]. On the other hand, the accumulation of organic acids, such as malate and citrate (as in peas, Figure 4), is linked with heavy-metal tolerance [54–56]. They act as metal chelators, bind them in the cell cytosol, and immobilize in vacuoles to maintain their non-toxic level but also play an important role in metals transport through the xylem [54,56,57]. Moreover, organic acids can increase the expression of heavy-metal transporters and antioxidative enzymes [56]. Citric and malic acids are the main ligands of zinc in the xylem, participate in transport, and are involved in zinc sequestration in vacuoles, especially malate [55,56]. Therefore, malate and citrate accumulation in pea seedlings might contribute to their less noticeable sensitivity to ZnO NPs than in wheat seedlings.

Pea and wheat seedlings differ also in the case of changes in the concentration of amino acids caused by ZnO NPs exposure. In pea seedlings, homoserine (a predominant amino acid) decreased after ZnO NPs treatment, but it was more pronounced in epicotyls (Figure 4B, Tables S5 and S6). Homoserine takes part in the mobilization and transport of storage reserves from seeds and is synthesized in pea seedlings after germination [58]. Homoserine is synthesized from aspartate in a series of reactions catalyzed by aspartate kinase, aspartate semialdehyde dehydrogenase, and homoserine dehydrogenase [59]. Homoserine content decreased after ZnO NPs treatment, whereas aspartate increased (Tables S5 and S6). This suggests the inhibition of homoserine synthesis, which might be related to the inhibitory activity of ZnO NPs against dehydrogenase [30]. Regarding an accumulation of aspartic acid, glutamic acid, β-alanine, and GABA in pea roots and epicotyls (Tables S5 and S6), it could be suggested that ZnO NPs affect the aspartate–glutamate pathway. The above-mentioned proteinogenic (aspartate, glutamate) and non-proteinogenic (β-alanine, GABA) amino acids are important for the proper functioning of plants, and all of them were reported to accumulate in response to various abiotic stresses [60–64]. Glutamate and aspartate accumulation was also observed in tomato roots and leaves after foliar spray with ZnO NPs [46]. The accumulation of glutamate, aspartate, and β-alanine in peas, which are involved in the TCA cycle, might be also related to the accumulation of malate and citrate.

In wheat, ZnO NPs caused a decrease in the concentration of total amino acids (Figure 5), including asparagine and valine (in roots and coleoptile), glutamine (in roots only), and isoleucine (in coleoptile). ZnO NPs treatment increased the mobilization of amino acids from the endosperm, but their amount also decreased in seedlings, which suggests an increased nitrogen requirement and consumption. Asparagine can be synthesized from glutamine, and both are considered transport and storage molecules of reduced nitrogen in plants [65]. Isoleucine and valine, branched-chain amino acids, provide energy during germination and the later phase of seedling development [66].

GABA accumulation in the roots of both wheat and peas was a common response of both species to ZnO NPs. Accumulation of this amino acid has also been observed in pea roots after treatment with Ag NPs [67] and in maize leaves after growth in soil with SiO_2, TiO_2, and Fe_3O_4 engineered nanomaterials [68]. GABA is a signaling molecule, osmolyte, and also enhances the activity of antioxidant enzymes, important in plants' response to various stresses [61]. Plants' reaction to ZnO NPs might be related to heavy-metal stress, including zinc, which is associated with proline accumulation [69]. However, in our study, changes in proline levels were not related to ZnO NPs treatment.

4. Materials and Methods

4.1. Plant Material

Pea (*Pisum sativum* L.) seeds cultivars Nemo and Tarchalska, purchased from Danko Hodowla Roślin (Choryń, Poland) and spring wheat (*Triticum aestivum* L.) grains cultivars KWS Collada and Ostka Strzelecka, purchased from KWS and Hodowla Roślin Strzelce (Strzelce, Poland), respectively, were used in the experiment.

4.2. Preparation of ZnO NPs Suspension

ZnO NPs with a diameter size < 50 nm (cat. no. 677450-5G, Sigma-Aldrich, St. Louis, MO, USA) and <100 nm (cat. no. 721077-100G, Sigma-Aldrich, St. Louis, MO, USA) were used in the experiments. Nanoparticles were suspended in the double-distilled water by sonication 2 times for 30 min (Sonic-3, 310 W, 40 KHz, POLSONIC Pałczyński, Warszawa, Poland) to obtain ZnO NPs suspensions at specific concentrations for further experiments.

4.3. Preliminary Study

In the preliminary study, the phytotoxic concentrations of ZnO NPs for seed germination and early seedling development of pea and wheat were selected. In the first experiment, pea seeds (cv. Nemo and cv. Tarchalska) and wheat grains (cv. Collada and cv. Ostka Strzelecka) were incubated in petri dishes (ø 12 cm) in water and ZnO NPs (diameter <50 nm; cat. no. 677450-5G, Sigma-Aldrich) suspensions at concentrations 20, 50, 250, 500, and 1000 mg/L for 4 and 3 days (pea and wheat, respectively) at 22 °C in the dark. After incubation, germination rate, seedlings' length, and fresh and dry weight (FW and DW) were measured.

In the second experiment, pea seeds (cv. Tarchalska) and wheat grains (cv. Ostka Strzelecka) were incubated in petri dishes (ø 12 cm) in water and ZnO NPs of diameter <50 nm (cat. no. 677450-5G, Sigma-Aldrich) and <100 nm (cat. no. 721077-100G, Sigma-Aldrich) suspensions at concentrations 100, 250, and 1000 mg/L for 4 and 3 days (pea and wheat, respectively) at 22 °C in the dark. After incubation, germination rate, seedlings' length, FW, and DW were measured.

4.4. The Effect of ZnO NPs on Seed Germination and Seedling Development

Pea seeds (cv. Nemo and cv. Tarchalska) and wheat grains (cv. Collada and cv. Ostka Strzelecka) were incubated in petri dishes (ø 12 cm) containing water and ZnO NPs (diameter <50 nm) suspensions at concentrations 100, 250, and 1000 mg/L for 4 and 3 days (pea and wheat, respectively) at 22 °C in the dark. The germinability of seeds was monitored daily. After incubation, the seedlings' length and fresh and dry weight (FW and DW) were measured. Next, the tissues of pea seedlings (root, epicotyl, and cotyledons) and wheat seedlings (roots with scutellum, coleoptile, and endosperm) were frozen in liquid nitrogen and stored at −80 °C for the polar metabolite extraction.

4.5. Polar Metabolites Analyzes

Tissue samples of pea (root, epicotyl, and cotyledons) and wheat (roots, coleoptile, and endosperm) seedlings were freeze-dried and pulverized. The extraction was performed according to Szablińska-Piernik and Lahuta [70]. The polar metabolites were extracted with a mixture of methanol:water (1:1, *v*/*v*) at 70 °C for 30 min with continuous shaking. After cooling and centrifugation, cold chloroform was used to remove the non-polar compounds. The samples were dried and derivatized in two steps—with O-methoxamine hydrochloride and with a mixture of N-methyl-N-trimethylsilyl-trifluoroacetamide (MSTFA) and pyridine (1:1, v/v). The mixtures of trimethylsilyl (TMS)-derivatives were separated on a ZEBRON ZB-5MSi Guardian capillary column (Phenomenex, Torrance, CA, USA) in the gas chromatograph GC2010 Nexia (Shimadzu, Japan) with the flame ionization detector (FID). To confirm accurate metabolite identification, the gas chromatograph coupled with a quadrupole mass spectrometer (QP-GC-2010, Shimadzu, Japan) was used. Metabolites were identified and characterized by comparison of their retention time (RT), retention indices (RI, determined according to the saturated hydrocarbons), and mass spectra of original standards derived from Sigma-Aldrich (Sigma-Aldrich, Merck, Burlington, MA, USA) and from the NIST library (National Institute of Standards and Technology), as previously described [40,70].

4.6. Statistics

The results are the mean of 3 independent replicates and were subjected to one-way ANOVA with a post hoc test (Tukey, if overall $p < 0.05$) using the Statistica software (version 12.0; StatSoft, Tulsa, OK, USA). Graphs were prepared using GraphPad Prism (version 8; GraphPad Software, San Diego, CA, USA). Principal component analysis (PCA) for multivariate statistics was performed using COVAIN [71], a MATLAB toolbox including a graphical user interface (MATLAB version 2013a; Math Works, Natick, MA, USA). Additionally, Pearson's correlation was performed for selected metabolites using Statistica software (version 12.0; StatSoft, Tulsa, OK, USA). Calculated Pearson correlation coefficients are presented in Tables S11 and S12 in Supplementary Materials.

5. Conclusions

The obtained results revealed species-, cultivar-, and organ-specific changes in the physiology and polar metabolite profiles of pea and wheat seedlings treated with ZnO NPs. Nanoparticles did not affect pea and wheat seeds' germination. However, harmful effects on seedling growth, especially roots, were observed at all tested concentrations. Inhibition of root elongation was observed, particularly noticeable in wheat seedlings—reduction of radicle length over 50% at 100 mg/L ZnO NPs. Shoots were less affected, which might be related to the differences in the transport of ZnO NPs/Zn^{2+} ions in seedlings. This effect on seedlings' growth could be a result of the over-uptake of ZnO NPs by the roots, ROS accumulation, and disturbance in phytohormone homeostasis. Our findings suggest that zinc oxide nanoparticles can affect sucrose metabolism in seedlings' roots and cause oxidative stress, as indicated by the common reaction in pea and wheat roots—accumulation of sucrose and GABA.

Pea and wheat differ in their sensitivity to ZnO NPs, which may be related to differences in changes in the metabolomic profiles caused by nanoparticles. In wheat, ZnO NPs disrupted sugar mobilization in endosperms, probably by inhibiting α-amylase activity, as indicated by the decreasing content of maltose and maltotriose in a dose-dependent manner. Moreover, in wheat roots, malate decreased, and in seedlings and endosperms, the total amino acids content decreased, which indicates a disturbance in the TCA cycle and increased nitrogen requirement and consumption. In peas, the mobilization of storage materials in cotyledons was uninterrupted, but in seedlings, the accumulation of citric acid, malic acid, glutamate, aspartate, and β-alanine in roots, and citric and malic acid in epicotyls cv. Nemo, was observed. Therefore, the accumulation of such metabolites, which indicates changes in the aspartate–glutamate pathway and the TCA cycle, might contribute to the less noticeable sensitivity of pea seedlings to ZnO NPs than wheat seedlings. Further studies are required to verify this hypothesis. Our findings confirm plant species-specific reactions to ZnO NPs, which should be considered in potential agricultural applications. Using nanoparticles for seed nano-priming seems to be an optimistic solution—shorter exposure of seeds to ZnO NPs (for a few hours instead of a few days) can eliminate the harmful effects of ZnO NPs (presumably even at the highest concentrations tested here) while maintaining their beneficial properties (nutritional and biocidal properties against phytopathogens).

Supplementary Materials: The supporting information can be downloaded at https://www.mdpi.com/article/10.3390/ijms241914992/s1.

Author Contributions: Conceptualization, K.S. and L.B.L.; methodology, K.S.; software, K.S. and J.S.-P.; validation, K.S., J.S.-P. and L.B.L.; formal analysis, K.S., J.S.-P., L.B.L. and A.O.; investigation, K.S.; resources, L.B.L.; data curation, K.S. and L.B.L.; writing—original draft preparation, K.S.; writing—review and editing, K.S. and L.B.L.; visualization, K.S.; supervision, L.B.L. and A.O.; project administration, L.B.L. All authors have read and agreed to the published version of the manuscript.

Funding: This work was financially supported by the project "Advanced Biocomposites for Tomorrow's Economy BIOG-NET", FNP POIR.04.04.00-00-1792/18-00; the project was carried out within the TEAM NET program funded by the Foundation for Polish Science and co-financed by the European Union under the European Regional Development Fund.

Institutional Review Board Statement: Not applicable.

Informed Consent Statement: Not applicable.

Data Availability Statement: Not applicable.

Conflicts of Interest: The authors declare no conflict of interest.

References

1. Dikshit, P.K.; Kumar, J.; Das, A.K.; Sadhu, S.; Sharma, S.; Singh, S.; Gupta, P.K.; Kim, B.S. Green synthesis of metallic nanoparticles: Applications and limitations. *Catalysts* **2021**, *11*, 902. [CrossRef]
2. Czyżowska, A.; Barbasz, A. A review: Zinc oxide nanoparticles–friends or enemies? *Int. J. Environ. Health Res.* **2022**, *32*, 885–901. [CrossRef]
3. Xie, J.; Li, H.; Zhang, T.; Song, B.; Wang, X.; Gu, Z. Recent advances in ZnO nanomaterial-mediated biological applications and action mechanisms. *Nanomaterials* **2023**, *13*, 1500.
4. Pena, G.A.; Cardenas, M.A.; Monge, M.P.; Yerkovich, N.; Planes, G.A.; Chulze, S.N. Reduction of *Fusarium proliferatum* growth and fumonisin accumulation by ZnO nanoparticles both on a maize based medium and irradiated maize grains. *Int. J. Food Microbiol.* **2022**, *363*, 109510. [CrossRef] [PubMed]
5. Abdelaziz, A.M.; Salem, S.S.; Khalil, A.M.A.; El-Wakil, D.A.; Fouda, H.M.; Hashem, A.H. Potential of biosynthesized zinc oxide nanoparticles to control Fusarium wilt disease in eggplant (*Solanum melongena*) and promote plant growth. *BioMetals* **2022**, *35*, 601–616. [CrossRef] [PubMed]
6. Amin, H.H.; Elsayed, A.B.; Maswada, H.F.; Elsheery, N.I. Enhancing sugar beet plant health with zinc nanoparticles: A sustainable solution for disease management. *Plant Soil Environ.* **2023**, *2*, 1–20. [CrossRef]
7. Jaithon, T.; Ruangtong, J.; T-Thienprasert, J.; T-Thienprasert, N.P. Effects of waste-derived ZnO nanoparticles against growth of plant pathogenic bacteria and epidermoid carcinoma cells. *Crystals* **2022**, *12*, 779. [CrossRef]
8. Liu, L.; Nian, H.; Lian, T. Plants and rhizospheric environment: Affected by zinc oxide nanoparticles (ZnO NPs). A review. *Plant Physiol. Biochem.* **2022**, *185*, 91–100.
9. Zou, C.; Lu, T.; Wang, R.; Xu, P.; Jing, Y.; Wang, R.; Xu, J.; Wan, J. Comparative physiological and metabolomic analyses reveal that Fe_3O_4 and ZnO nanoparticles alleviate Cd toxicity in tobacco. *J. Nanobiotechnol.* **2022**, *20*, 302. [CrossRef]
10. Sun, L.; Song, F.; Zhu, X.; Liu, S.; Liu, F.; Wang, Y.; Li, X. Nano-ZnO alleviates drought stress via modulating the plant water use and carbohydrate metabolism in maize. *Arch. Agron. Soil Sci.* **2021**, *67*, 245–259. [CrossRef]
11. Faizan, M.; Bhat, J.A.; Chen, C.; Alyemeni, M.N.; Wijaya, L.; Ahmad, P.; Yu, F. Zinc oxide nanoparticles (ZnO-NPs) induce salt tolerance by improving the antioxidant system and photosynthetic machinery in tomato. *Plant Physiol. Biochem.* **2021**, *161*, 122–130. [CrossRef]
12. Landa, P. Positive effects of metallic nanoparticles on plants: Overview of involved mechanisms. *Plant Physiol. Biochem.* **2021**, *161*, 12–24. [CrossRef] [PubMed]
13. Rai-Kalal, P.; Jajoo, A. Priming with zinc oxide nanoparticles improve germination and photosynthetic performance in wheat. *Plant Physiol. Biochem.* **2021**, *160*, 341–351. [CrossRef]
14. Sarkhosh, S.; Kahrizi, D.; Darvishi, E.; Tourang, M.; Haghighi-Mood, S.; Vahedi, P.; Ercisli, S. Effect of zinc oxide nanoparticles (ZnO-NPs) on seed germination characteristics in two Brassicaceae family species: *Camelina sativa* and *Brassica napus* L. *J. Nanomater.* **2022**, *2022*, 1892759. [CrossRef]
15. Hamzah Saleem, M.; Usman, K.; Rizwan, M.; Al Jabri, H.; Alsafran, M. Functions and strategies for enhancing zinc availability in plants for sustainable agriculture. *Front. Plant Sci.* **2022**, *13*, 1033092. [PubMed]

16. Akhtar, N.; Ilyas, N.; Meraj, T.A.; Pour-Aboughadareh, A.; Sayyed, R.Z.; Mashwani, Z.U.R.; Poczai, P. Improvement of plant responses by nanobiofertilizer: A step towards sustainable agriculture. *Nanomaterials* **2022**, *12*, 965. [CrossRef]
17. Chen, H. Metal based nanoparticles in agricultural system: Behavior, transport, and interaction with plants. *Chem. Speciat. Bioavailab.* **2018**, *30*, 123–134. [CrossRef]
18. Yusefi-Tanha, E.; Fallah, S.; Rostamnejadi, A.; Pokhrel, L.R. Zinc oxide nanoparticles (ZnONPs) as a novel nanofertilizer: Influence on seed yield and antioxidant defense system in soil grown soybean (*Glycine max* cv. Kowsar). *Sci. Total Environ.* **2020**, *738*, 140240. [PubMed]
19. Bhattacharjee, R.; Kumar, L.; Mukerjee, N.; Anand, U.; Dhasmana, A.; Preetam, S.; Bhaumik, S.; Sihi, S.; Pal, S.; Khare, T.; et al. The emergence of metal oxide nanoparticles (NPs) as a phytomedicine: A two-facet role in plant growth, nano-toxicity and anti-phyto-microbial activity. *Biomed. Pharmacother.* **2022**, *155*, 113658.
20. Chen, J.; Dou, R.; Yang, Z.; You, T.; Gao, X.; Wang, L. Phytotoxicity and bioaccumulation of zinc oxide nanoparticles in rice (*Oryza sativa* L.). *Plant Physiol. Biochem.* **2018**, *130*, 604–612. [CrossRef]
21. Sun, Z.; Xiong, T.; Zhang, T.; Wang, N.; Chen, D.; Li, S. Influences of zinc oxide nanoparticles on *Allium cepa* root cells and the primary cause of phytotoxicity. *Ecotoxicology* **2019**, *28*, 175–188. [CrossRef]
22. Leopold, L.F.; Coman, C.; Clapa, D.; Oprea, I.; Toma, A.; Iancu, Ş.D.; Barbu-Tudoran, L.; Suciu, M.; Ciorîţă, A.; Cadiş, A.I.; et al. The effect of 100–200 nm ZnO and TiO$_2$ nanoparticles on the in vitro-grown soybean plants. *Colloids Surf. B Biointerfaces* **2022**, *216*, 112536. [CrossRef] [PubMed]
23. Zhang, R.; Zhang, H.; Tu, C.; Hu, X.; Li, L.; Luo, Y.; Christie, P. Phytotoxicity of ZnO nanoparticles and the released Zn(II) ion to corn (*Zea mays* L.) and cucumber (*Cucumis sativus* L.) during germination. *Environ. Sci. Pollut. Res.* **2015**, *22*, 11109–11117. [CrossRef] [PubMed]
24. da Cruz, T.N.M.; Savassa, S.M.; Montanha, G.S.; Ishida, J.K.; de Almeida, E.; Tsai, S.M.; Lavres Junior, J.; Pereira de Carvalho, H.W. A new glance on root-to-shoot in vivo zinc transport and time-dependent physiological effects of ZnSO$_4$ and ZnO nanoparticles on plants. *Sci. Rep.* **2019**, *9*, 10416. [CrossRef] [PubMed]
25. Lv, Z.; Sun, H.; Du, W.; Li, R.; Mao, H.; Kopittke, P.M. Interaction of different-sized ZnO nanoparticles with maize (*Zea mays*): Accumulation, biotransformation and phytotoxicity. *Sci. Total Environ.* **2021**, *796*, 148927.
26. Tsonev, T.; Lidon, F.J.C. Zinc in plants-An overview. *Emir. J. Food Agric.* **2012**, *24*, 322–333.
27. Cabot, C.; Martos, S.; Llugany, M.; Gallego, B.; Tolrà, R.; Poschenrieder, C. A role for zinc in plant defense against pathogens and herbivores. *Front. Plant Sci.* **2019**, *10*, 1171. [PubMed]
28. Hacisalihoglu, G. Zinc (Zn): The last nutrient in the alphabet and shedding light on Zn efficiency for the future of crop production under suboptimal Zn. *Plants* **2020**, *9*, 1471. [CrossRef]
29. Mukherjee, A.; Peralta-Videa, J.R.; Bandyopadhyay, S.; Rico, C.M.; Zhao, L.; Gardea-Torresdey, J.L. Physiological effects of nanoparticulate ZnO in green peas (*Pisum sativum* L.) cultivated in soil. *Metallomics* **2014**, *6*, 132–138. [CrossRef]
30. Srivastav, A.; Ganjewala, D.; Singhal, R.K.; Rajput, V.D.; Minkina, T.; Voloshina, M.; Srivastava, S.; Shrivastava, M. Effect of ZnO nanoparticles on growth and biochemical responses of wheat and maize. *Plants* **2021**, *10*, 2556. [CrossRef]
31. Awasthi, A.; Bansal, S.; Jangir, L.K.; Awasthi, G.; Awasthi, K.K.; Awasthi, K. Effect of ZnO nanoparticles on germination of *Triticum aestivum* Seeds. *Macromol. Symp.* **2017**, *376*, 1700043. [CrossRef]
32. Du, W.; Yang, J.; Peng, Q.; Liang, X.; Mao, H. Comparison study of zinc nanoparticles and zinc sulphate on wheat growth: From toxicity and zinc biofortification. *Chemosphere* **2019**, *227*, 109–116. [CrossRef] [PubMed]
33. Huang, Y.C.; Fan, R.; Grusak, M.A.; Sherrier, J.D.; Huang, C.P. Effects of nano-ZnO on the agronomically relevant *Rhizobium*-legume symbiosis. *Sci. Total Environ.* **2014**, *497–498*, 78–90. [CrossRef]
34. Silva, S.; Dias, M.C.; Pinto, D.C.G.A.; Silva, A.M.S. Metabolomics as a tool to understand nano-plant interactions: The case study of metal-based nanoparticles. *Plants* **2023**, *12*, 491. [CrossRef]
35. Lin, D.; Xing, B. Phytotoxicity of nanoparticles: Inhibition of seed germination and root growth. *Environ. Pollut.* **2007**, *150*, 243–250. [CrossRef]
36. Yasmeen, F.; Razzaq, A.; Iqbal, M.N.; Jhanzab, H.M. Effect of silver, copper and iron nanoparticles on wheat germination. *Int. J. Biosci.* **2015**, *6*, 112–117.
37. Karimi, J.; Mohsenzadeh, S. Effects of silicon oxide nanoparticles on growth and physiology of wheat seedlings. *Russ. J. Plant Physiol.* **2016**, *63*, 119–123. [CrossRef]
38. Lahuta, L.B.; Szablińska-Piernik, J.; Głowacka, K.; Stałanowska, K.; Railean-Plugaru, V.; Hrobowicz, M.; Pomastowski, P.; Buszewski, B. The effect of bio-synthesized silver nanoparticles on germination, early seedling development, and metabolome of wheat (*Triticum aestivum* L.). *Molecules* **2022**, *27*, 2303. [CrossRef] [PubMed]
39. Lahuta, L.B.; Szablińska-Piernik, J.; Stałanowska, K.; Horbowicz, M.; Górecki, R.J.; Railean, V.; Pomastowski, P.; Buszewski, B. Exogenously applied cyclitols and biosynthesized silver nanoparticles affect the soluble carbohydrate profiles of wheat (*Triticum aestivum* L.) Seedling. *Plants* **2023**, *12*, 1627. [CrossRef]
40. Lahuta, L.B.; Szablińska-Piernik, J.; Stałanowska, K.; Głowacka, K.; Horbowicz, M. The Size-Dependent Effects of silver nanoparticles on germination, early seedling development and polar metabolite profile of wheat (*Triticum aestivum* L.). *Int. J. Mol. Sci.* **2022**, *23*, 13255. [CrossRef]
41. Wang, J.; Moeen-ud-din, M.; Yang, S. Dose-dependent responses of *Arabidopsis thaliana* to zinc are mediated by auxin homeostasis and transport. *Environ. Exp. Bot.* **2021**, *189*, 104554. [CrossRef]

42. EL Sabagh, A.; Islam, M.S.; Hossain, A.; Iqbal, M.A.; Mubeen, M.; Waleed, M.; Reginato, M.; Battaglia, M.; Ahmed, S.; Rehman, A.; et al. Phytohormones as growth regulators during abiotic stress tolerance in plants. *Front. Agron.* **2022**, *4*, 765068.

43. Rosental, L.; Nonogaki, H.; Fait, A. Activation and regulation of primary metabolism during seed germination. *Seed Sci. Res.* **2014**, *24*, 1–15. [CrossRef]

44. Aoki, N.; Scofield, G.N.; Wang, X.D.; Offler, C.E.; Patrick, J.W.; Furbank, R.T. Pathway of sugar transport in germinating wheat seeds. *Plant Physiol.* **2006**, *141*, 1255–1263. [CrossRef] [PubMed]

45. Dhobale, S.; Thite, T.; Laware, S.L.; Rode, C.V.; Koppikar, S.J.; Ghanekar, R.K.; Kale, S.N. Zinc oxide nanoparticles as novel alpha-amylase inhibitors. *J. Appl. Phys.* **2008**, *104*, 094907. [CrossRef]

46. Sun, L.; Wang, Y.; Wang, R.; Wang, R.; Zhang, P.; Ju, Q.; Xu, J. Physiological, transcriptomic, and metabolomic analyses reveal zinc oxide nanoparticles modulate plant growth in tomato. *Environ. Sci. Nano* **2020**, *7*, 3587–3604. [CrossRef]

47. Ruan, Y.L. Sucrose metabolism: Gateway to diverse carbon use and sugar signaling. *Annu. Rev. Plant Biol.* **2014**, *65*, 33–67. [CrossRef] [PubMed]

48. Górecki, R.J.; Obendorf, R.L. Galactosyl cyclitols and raffinose family oligosaccharides in relation to desiccation tolerance of pea and soyabean seedlings. In *Basic and Applied Aspects of Seed Biology. Current Plant Science and Biotechnology in Agriculture*; Ellis, R.H., Black, M., Murdoch, A.J., Hong, T.D., Eds.; Springer: Dordrecht, The Netherlands; Reading, UK, 1997; Volume 30, pp. 119–128.

49. Górecki, R.J.; Fordoński, G.; Halmajan, H.; Horbowicz, M.; Jones, R.G.; Lahuta, L.B.; Horbowicz, M. Seed physiology and biochemistry. In *Carbohydrates in Grain Legume Seeds: Improving Nutritional Quality and Agronomic Characteristics*; Hedley, C.L., Ed.; CABI Publishing: Wallingford, UK, 2000; pp. 117–143.

50. Blöchl, A.; Peterbauer, T.; Richter, A. Inhibition of raffinose oligosaccharide breakdown delays germination of pea seeds. *J. Plant Physiol.* **2007**, *164*, 1093–1096. [CrossRef]

51. Lahuta, L.B.; Goszczyńska, J. Inhibition of raffinose family oligosaccharides and galactosyl pinitols breakdown delays germination of winter vetch (*Vicia villosa* roth.) seeds. *Acta Soc. Bot. Pol.* **2009**, *78*, 203–208. [CrossRef]

52. Cha, S.H.; Hong, J.; McGuffie, M.; Yeom, B.; Vanepps, J.S.; Kotov, N.A. Shape-dependent biomimetic inhibition of enzyme by nanoparticles and their antibacterial activity. *ACS Nano* **2015**, *9*, 9097–9105. [CrossRef]

53. Li, S.; Liu, J.; Wang, Y.; Gao, Y.; Zhang, Z.; Xu, J.; Xing, G. Comparative physiological and metabolomic analyses revealed that foliar spraying with zinc oxide and silica nanoparticles modulates metabolite profiles in cucumber (*Cucumis sativus* L.). *Food Energy Secur.* **2021**, *10*, e269. [CrossRef]

54. Osmolovskaya, N.; Dung, V.V.; Kuchaeva, L. The role of organic acids in heavy metal tolerance in plants. *Biol. Commun.* **2018**, *63*, 9–16. [CrossRef]

55. Balafrej, H.; Bogusz, D.; Triqui, Z.E.A.; Guedira, A.; Bendaou, N.; Smouni, A.; Fahr, M. Zinc hyperaccumulation in plants: A review. *Plants* **2020**, *9*, 562. [CrossRef]

56. Vega, A.; Delgado, N.; Handford, M. Increasing heavy metal tolerance by the exogenous application of organic acids. *Int. J. Mol. Sci.* **2022**, *23*, 5438. [CrossRef] [PubMed]

57. Stanton, C.; Sanders, D.; Krämer, U.; Podar, D. Zinc in plants: Integrating homeostasis and biofortification. *Mol. Plant* **2022**, *15*, 65–85.

58. Melcher, I.M. Homoserine synthesis in dark-grown and light-grown seedlings of *Pisum sativum*. *New Phytol.* **1985**, *100*, 157–162. [CrossRef]

59. Lea, P.J. Primary nitrogen metabolism. In *Plant Biochemistry*; Dey, P.M., Harborne, J.B., Eds.; Academic Press: Cambridge, MA, USA, 1997; pp. 273–313.

60. Azevedo, R.A.; Lancien, M.; Lea, P.J. The aspartic acid metabolic pathway, an exciting and essential pathway in plants. *Amino Acids* **2006**, *30*, 143–162. [CrossRef]

61. Parthasarathy, A.; Savka, M.A.; Hudson, A.O. The synthesis and role of β-alanine in plants. *Front. Plant Sci.* **2019**, *10*, 921. [CrossRef]

62. Sita, K.; Kumar, V. Role of Gamma Amino Butyric Acid (GABA) against abiotic stress tolerance in legumes: A review. *Plant Physiol. Rep.* **2020**, *25*, 654–663. [CrossRef]

63. Qiu, X.M.; Sun, Y.Y.; Ye, X.Y.; Li, Z.G. Signaling role of glutamate in Plants. *Front. Plant Sci.* **2020**, *10*, 1743. [CrossRef] [PubMed]

64. Lei, S.; Rossi, S.; Huang, B. Metabolic and physiological regulation of aspartic acid-mediated enhancement of heat stress tolerance in perennial ryegrass. *Plants* **2022**, *11*, 199. [CrossRef] [PubMed]

65. Lea, P.J.; Sodek, L.; Parry, M.A.J.; Shewry, P.R.; Halford, N.G. Asparagine in plants. *Ann. Appl. Biol.* **2007**, *150*, 1–26. [CrossRef]

66. Gipson, A.B.; Morton, K.J.; Rhee, R.J.; Simo, S.; Clayton, J.A.; Perrett, M.E.; Binkley, C.G.; Jensen, E.L.; Oakes, D.L.; Rouhier, M.F.; et al. Disruptions in valine degradation affect seed development and germination in *Arabidopsis*. *Plant J.* **2017**, *90*, 1029–1039. [CrossRef]

67. Szablińska-Piernik, J.; Lahuta, L.B.; Stałanowska, K.; Horbowicz, M. The imbibition of pea (*Pisum sativum* L.) seeds in silver nitrate reduces seed germination, seedlings development and their metabolic profile. *Plants* **2022**, *11*, 1877. [CrossRef]

68. Zhao, L.; Zhang, H.; White, J.C.; Chen, X.; Li, H.; Qu, X.; Ji, R. Metabolomics reveals that engineered nanomaterial exposure in soil alters both soil rhizosphere metabolite profiles and maize metabolic pathways. *Environ. Sci. Nano* **2019**, *6*, 1716–1727. [CrossRef]

69. Sharma, S.S.; Dietz, K.J. The significance of amino acids and amino acid-derived molecules in plant responses and adaptation to heavy metal stress. *J. Exp. Bot.* **2006**, *57*, 711–726. [CrossRef]

70. Szablińska-Piernik, J.; Lahuta, L.B. Metabolite profiling of semi-leafless pea (*Pisum sativum* L.) under progressive soil drought and subsequent re-watering. *J. Plant Physiol.* **2021**, *256*, 153314. [CrossRef]
71. Sun, X.; Weckwerth, W. COVAIN: A toolbox for uni- and multivariate statistics, time-series and correlation network analysis and inverse estimation of the differential Jacobian from metabolomics covariance data. *Metabolomics* **2012**, *8*, 81–93. [CrossRef]

International Journal of
Molecular Sciences

Article

The Root-Colonizing Endophyte *Piriformospora indica* Supports Nitrogen-Starved *Arabidopsis thaliana* Seedlings with Nitrogen Metabolites

Sandra S. Scholz [1], Emanuel Barth [2], Gilles Clément [3], Anne Marmagne [3], Jutta Ludwig-Müller [4], Hitoshi Sakakibara [5], Takatoshi Kiba [5], Jesús Vicente-Carbajosa [6,7], Stephan Pollmann [6,7], Anne Krapp [3] and Ralf Oelmüller [1,*]

[1] Department of Plant Physiology, Matthias-Schleiden-Institute, Friedrich-Schiller-University Jena, 07743 Jena, Germany; s.scholz@uni-jena.de
[2] Bioinformatics Core Facility, Friedrich-Schiller-University Jena, 07743 Jena, Germany; bicj@uni-jena.de
[3] Université Paris-Saclay, INRAE, AgroParisTech, Institut Jean-Pierre Bourgin (IJPB), 78000 Versailles, France; anne.marmagne@inrae.fr (A.M.); anne.krapp@inrae.fr (A.K.)
[4] Institute of Botany, Technische Universität Dresden, 01217 Dresden, Germany; jutta.ludwig-mueller@tu-dresden.de
[5] Graduate School of Bioagricultural Sciences, Nagoya University, Nagoya 464-8601, Japan; sakaki@agr.nagoya-u.ac.jp (H.S.); takatoshi.kiba@riken.jp (T.K.)
[6] Centro de Biotechnología y Genómica de Plantas, Instituto Nacional de Investigación y Tecnología Agraria y Alimentación (INIA), Universidad Politécnica de Madrid (UPM), Campus de Montegancedo, 28223 Madrid, Spain; jesus.vicente@upm.es (J.V.-C.); stephan.pollmann@upm.es (S.P.)
[7] Departamento de Biotecnología-Biología Vegetal, Escuela Técnica Superior de Ingeniería Agronómica, Alimentaria y de Biosistemas, Universidad Politécnica de Madrid (UPM), 28040 Madrid, Spain
* Correspondence: ralf.oelmueller@uni-jena.de

Citation: Scholz, S.S.; Barth, E.; Clément, G.; Marmagne, A.; Ludwig-Müller, J.; Sakakibara, H.; Kiba, T.; Vicente-Carbajosa, J.; Pollmann, S.; Krapp, A.; et al. The Root-Colonizing Endophyte *Piriformospora indica* Supports Nitrogen-Starved *Arabidopsis thaliana* Seedlings with Nitrogen Metabolites. *Int. J. Mol. Sci.* **2023**, 24, 15372. https://doi.org/10.3390/ijms242015372

Academic Editor: Martin Bartas

Received: 9 September 2023
Revised: 12 October 2023
Accepted: 13 October 2023
Published: 19 October 2023

Abstract: The root-colonizing endophytic fungus *Piriformospora indica* promotes the root and shoot growth of its host plants. We show that the growth promotion of *Arabidopsis thaliana* leaves is abolished when the seedlings are grown on media with nitrogen (N) limitation. The fungus neither stimulated the total N content nor did it promote $^{15}NO_3^-$ uptake from agar plates to the leaves of the host under N-sufficient or N-limiting conditions. However, when the roots were co-cultivated with ^{15}N-labelled *P. indica*, more labels were detected in the leaves of N-starved host plants but not in plants supplied with sufficient N. Amino acid and primary metabolite profiles, as well as the expression analyses of N metabolite transporter genes suggest that the fungus alleviates the adaptation of its host from the N limitation condition. *P. indica* alters the expression of transporter genes, which participate in the relocation of NO_3^-, NH_4^+ and N metabolites from the roots to the leaves under N limitation. We propose that *P. indica* participates in the plant's metabolomic adaptation against N limitation by delivering reduced N metabolites to the host, thus alleviating metabolic N starvation responses and reprogramming the expression of N metabolism-related genes.

Keywords: *Piriformospora indica*; nitrogen starvation; nitrogen metabolism; nitrate transporter; ammonium transporter; amino acid transporter; endophyte

1. Introduction

Nitrogen is a key mineral nutrient playing a crucial role in plant growth and development [1–5]. The soil microbiome contributes to nitrogen acquisition, and among the best studied endosymbiotic interactions are those with N-fixing rhizobia and arbuscular mycorrhizal (AM) fungi. Legumes gain access to N through symbiotic association with rhizobia, which convert N_2 gas into ammonia in nodules. Although several efforts have been made to incorporate biological N fixation capacity into non-legume plants [6], agricultural crop production without N fertilization is currently not conceivable. AM fungi help plants in nutrient acquisition and much progress has been made in understanding the molecular basis of P and N transfer from the fungal partner to the host plant (cf. [7]). Less is known about endophytes, although they show relatively little host specificity and have therefore a great potential for agricultural applications [8,9].

A well-studied endophytic fungus is *Piriformospora (Serendipita) indica*, which interacts with numerous host plants and promotes their growth and resistance against biotic and abiotic stresses [10,11]. The stimulation of the growth of its hosts suggests that the fungus promotes nutrient acquisition, including nitrogen. An effect of *P. indica* on nitrate uptake and the nitrogen metabolism in the hosts has been reported repeatedly. On a full medium, the fungus promotes nitrogen accumulation and the expression of nitrate reductase in *Arabidopsis thaliana* [12]. In sunflower, *P. indica* increases the absorption of nitrogen by the root [13]. Strehmel et al. [14] showed that the concentration of nitrogen-rich amino acids decreased in inoculated *A. thaliana* plants. Ghaffari et al. [15] proposed that the nitrogen metabolism plays an important role in systemic salt-tolerance in leaves of *P. indica*-colonized barley. Furthermore, Lahrmann et al. [16] showed that the *P. indica* ammonium transporter Amt1 functions as a nitrogen sensor mediating the signal that triggers the in planta activation of the saprotrophic program. In Chinese cabbage, the amino acid γ-amino butyrate in particular is de novo synthesized in colonized roots [17]. Bandyopadhyay et al. [18] demonstrated that *P. indica*, together with *Azotobacter chroococcum*, facilitates the higher acquisition of N and P in rice. *P. indica* also improves chickpea productivity and N metabolism in a tripartite combination with Mesorhizobium [19]. Finally, *Serendipita williamsii* does not affect P status but C and N dynamics in AM tomato plants [20]. These examples highlight the importance of the N metabolism on numerous beneficial effects of *P. indica* for different plant species; however, how the fungus influences the host N metabolism is not clear. In this study, we use the model plant *A. thaliana* to investigate how *P. indica* interferes with N uptake and metabolism under N-limiting conditions.

2. Results

2.1. Shoot Growth Promotion by P. indica Requires External N Supply

P. indica colonizes *A. thaliana* roots and induces the visible growth promotion of *A. thaliana* seedlings after 4–7 days in full medium [21]. After 5 days, the fresh weight of the shoots was significantly increased (+41.9%) by the fungus, while barely any growth promotion was detectable on N-limited medium (Figure 1A). Root growth was neither affected by the fungus nor by N availability (Figure 1B). We conclude that under these experimental conditions, the shoot growth but not the root growth of *A. thaliana* seedlings is promoted by *P. indica*, and this requires N in the medium.

2.2. P. indica Colonisation Did Not Change the Total N Content in the Shoots and Transfer of ^{15}N from the Medium to the Shoots

To test whether *P. indica* interferes with N accumulation or uptake into the plant under N-limiting conditions, the total N content in the shoots and the amount of ^{15}N from $^{15}NO_3{}^-$-labelled growth medium in the shoots were compared for uncolonized and colonized seedlings, grown on either full or N-limiting media. As expected, the total N content in the shoots of seedlings that were exposed to N limitation was lower than in the shoots of seedlings grown on full medium (Figure 2A). Furthermore, the accumulation of ^{15}N in the shoots was much higher on full medium than on medium with low N (Figure 2B). However, we did not observe significant differences for uncolonized and colonized seedlings. This suggests that the fungus does not stimulate nitrate uptake from the medium under N-sufficient and N-limiting growth conditions.

N limitation might influence the colonisation of the roots. We observed that roots on N-limiting conditions were around two times more colonized than roots on full medium (Supplementary Figure S1), although the difference was not significant. This indicates that in spite of a higher colonisation rate, the transport of ^{15}N label from the $^{15}NO_3{}^-$-containing medium to the leaves was not stimulated by the fungus under N-limiting conditions.

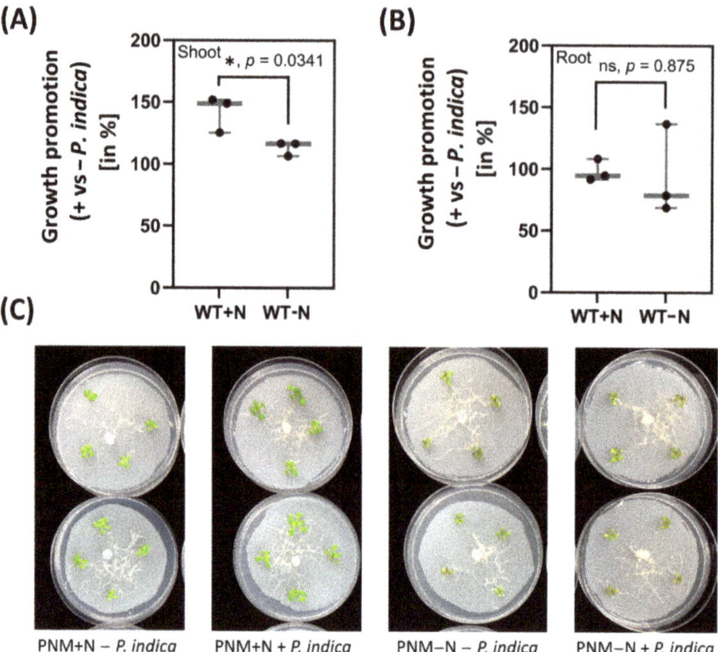

Figure 1. Shoot (**A**) and root (**B**) growth promotion of *A. thaliana* seedlings, which were either grown on full medium (+N) or N-limited medium (−N), in the presence of *P. indica* for 5 days. The % of growth promotion by the fungus was determined for 20 shoots and roots, the plant material grown on the respective media without the fungus was set as 100%. The percentage was determined for each replicate separately since the starting weight of plants was slightly different for the three independent replicates. All three replicates show the same trend of growth promotion. (**C**) shows representative pictures of the co-cultures at harvesting time. Statistic significant differences were analysed using one-way ANOVA (Holm–Sidak test). * significant, ns = not significant.

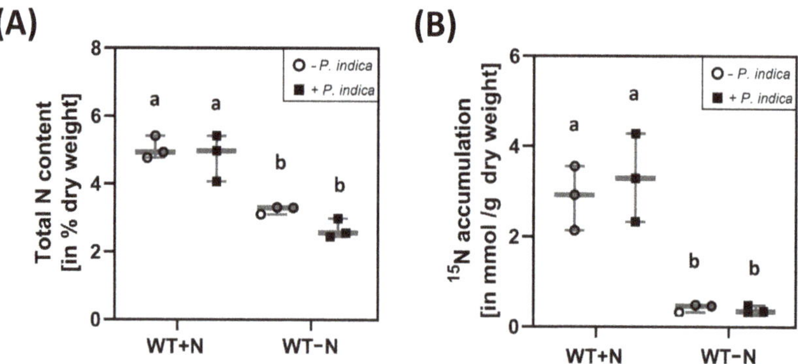

Figure 2. Total N in the shoots of uncolonized (circles) and colonized (squares) *A. thaliana* seedlings. (**A**) Total N in the shoots of seedlings grown with or without *P. indica* on either full or N-limiting conditions for 5 days. The % of N was determined in the dried material of 20 shoots. (**B**) ^{15}N accumulation in the shoots of seedlings grown with or without *P. indica* on either full or N-limiting conditions. ^{15}N was determined in dried material of 20 shoots. Based on three independent experiments. Statistic significant differences were analysed via one-way ANOVA (Holm–Sidak test). Different small letters indicate statistic significant differences.

2.3. ^{15}N Label Is Transferred from P. idica to the Host under N-Limiting Conditions

Since *P. indica* did not promote NO_3^- uptake, we tested whether labelled ^{15}N metabolites are translocated from the fungus to the plant. As shown in Figure 3A, *P. indica* was cultured on ^{15}N-containing medium for 14 days before co-culturing with *A. thaliana* seedlings on full or N-limited media.

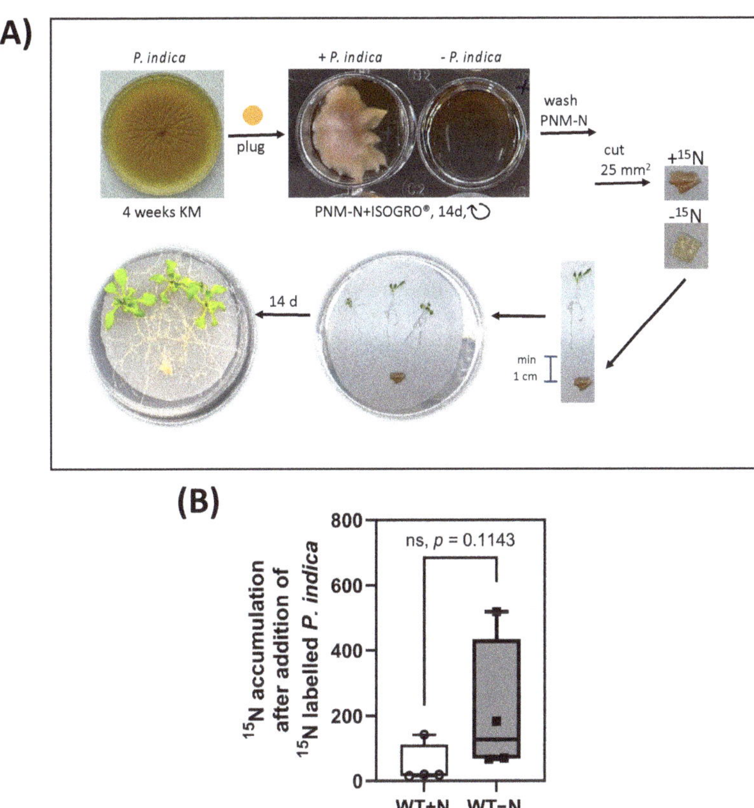

Figure 3. (**A**) Experimental set-up of sterile *A. thaliana* seedlings co-cultivated with ^{15}N-labelled *P. indica*. To obtain labelled fungal material, fungal plugs grown on KM plate were transferred to a modified liquid N-free KM medium supplemented with 10 g/L ISOGRO®-^{15}N and incubated for 14 days in a well plate. The fungal material was separated from the medium, washed carefully with PNM−N and cut into 25 mm² pieces. The fungus was placed onto the nylon membranes with 1 cm distance to the roots. The co-cultures were incubated for 14 days. For further details, see the Section 4. (**B**) ^{15}N label in the shoots of seedlings that were exposed to the ^{15}N-labelled hyphae on either full (white) or N-limited (grey) media. The accumulation of ^{15}N was determined in the dried leaf material of 20 colonized shoots, 14 days after the beginning of the co-culture. For experimental details, see the Section 4. Based on three independent experiments. ns, not significant; analysed via ranked *t*-test (Mann–Whitney test).

The ^{15}N-labelled fungal mycelium was positioned about 1 cm away from the roots. Establishing contact between the two partners and the initiation of root colonisation started approximately 24 h later, after the growing hyphae have reached the roots (Figure 3A). Since the label could be detected in the aerial parts of all analysed seedlings, which were in contact with *P. indica*, the fungus transfers N-containing metabolites to the roots of its host, and the label is further translocated to the aerial parts of the seedlings (Figure 3B).

Interestingly, more ^{15}N accumulated in the aerial parts of the plants under N-limiting conditions although the data were not significantly different (Figure 3B). This suggests that the fungus helps the host with reduced N metabolites to compensate N limitation during growth on NO_3^--limiting medium.

2.4. Reprogramming of the Metabolite Profiles to N Limitation Conditions Is Alleviated by P. indica

Next, we tested whether the fungus affects a host's N metabolism under sufficient N and N-limitation conditions. We measured the levels of primary metabolites using GC-MS in the rosettes after 2 days of transfer to N-limitation conditions compared to N-sufficient conditions in the absence and presence of *P. indica* (Supplementary Table S1). We then calculated the metabolite ratios for plants grown under limiting versus sufficient N and compared these ratios for plants grown in the absence and the presence of the fungus. Although the amino acid profiles were comparable in colonized and uncolonized shoots, we observed slight differences for several amino acids (Table 1A). The content of aspartate and alanine decreased under N-limiting conditions in both the absence and the presence of the fungus, and these decreases were less pronounced in colonized shoots (Supplementary Figure S2). Similar tendencies were observed for amino acid content that increased under N-limitation conditions; these increases were less distinct in the presence of the fungus in the case of isoleucine, lysine, tryptophan, phenylalanine, leucine, and arginine (Supplementary Figure S2). The alterations in serin contents that were triggered by N limitation varied strongly in colonized plants in comparison to uncolonized plants. We then analysed the effect of N-limitation on soluble sugars (Table 1B). In the presence of *P. indica*, N-limitation triggered stronger increases in monosaccharides—in particular in glucose and fructose. The stress-related sugars trehalose and raffinose showed strong variations between the three independent replicates. However, raffinose tended to accumulate at higher levels under N limitation when the roots were colonized (see the Section 3). Overall, the slight alteration of the metabolite profiles in response to N-limitation by the colonisation with *P. indica* suggest a lessening of the effects of N limitation on several steps of the central metabolism.

Table 1. Differentially accumulated metabolites (DAMs) in *A. thaliana* shoots. DAMs regulated by N limitation in *A. thaliana* rosettes without or with *P. indica* colonization. (**A**) amino acids; (**B**) soluble carbohydrates.

(A)	Metabolite Ratio Limiting vs. Sufficient N Supply			
	without *P. indica*		with *P. indica*	
Compound	Mean	SE	Mean	SE
Aspartate	0.36	0.07	0.44	0.17
Alanine	0.50	0.06	0.66	0.14
Homoserine	0.69	0.17	0.55	0.10
Glutamine	0.89	0.24	0.64	0.19
Glutamate	0.98	0.21	0.76	0.07
Glycine	1.10	0.46	0.72	0.12
Asparagine	1.22	0.39	0.67	0.11
Proline	1.45	0.19	1.42	0.05
Cystein	1.45	0.30	1.20	0.04
Methionine	1.51	0.71	0.72	0.04
Agmatine(-NH3)	1.64	0.27	1.40	0.52
beta-Alanine	1.67	0.46	1.25	0.12
Threonine	1.86	0.57	2.64	0.73
Valine	2.05	0.59	1.55	0.22
Arginine	2.08	0.86	0.98	0.33
Leucine	2.31	0.50	1.82	0.41
Histidine	2.46	0.88	2.13	0.97
Tyrosine	2.64	0.38	2.46	0.65
Phenylalanine	2.65	0.99	1.46	0.22

Table 1. *Cont.*

(A)	Metabolite Ratio Limiting vs. Sufficient N Supply			
	without *P. indica*		with *P. indica*	
Compound	Mean	SE	Mean	SE
Tryptophan	2.70	0.99	2.02	0.48
Lysine	2.78	0.76	2.03	0.65
Isoleucine	3.24	1.10	2.04	0.45
Serine	4.40	0.42	6.00	2.75

(B)	Metabolite Ratio Limiting vs. Sufficient N Supply			
	without *P.indica*		with *P. indica*	
Compound	Mean	SE	Mean	SE
Rhamnose	1.22	0.08	1.20	0.09
Arabinose	1.24	0.19	1.56	0.32
Gentiobiose	1.29	0.17	1.41	0.13
Ribose	1.32	0.07	1.61	0.24
Xylose	1.42	0.32	2.21	0.58
Mannose	1.43	0.28	1.89	0.15
Galactose	1.44	0.34	2.02	0.22
Sucrose	1.55	0.37	2.30	0.77
Glucose	1.59	0.67	3.34	0.73
Fructose	1.72	0.63	3.32	0.75
Maltose	2.22	1.20	1.45	0.29
Trehalose	2.88	2.21	2.02	0.54
Raffinose	3.46	1.55	10.21	3.77

0.25	0.50	0.75	1.00	1.50	2.00	4.00	10.00

Values are given as the ratio of the (relative) content in N-limiting to N-sufficient growth conditions, as measured by GC-MS profiling. Data are the means $+/-$ SE of three replicates from independent cultures on pools of 20 plantlets. There is no statistical difference with and without *P. indica* with $p < 0.1$. The gradual colour scale is indicated below the table (red, metabolite contents that decreased under limiting N supply; blue, metabolite contents that increased under limiting N supply).

2.5. P. indica Stimulates Expression of Specific Host's Transporter Genes under N Limitation

The incorporation of ^{15}N from the agar plate into the aerial parts of colonized seedlings is lower under N starvation when compared to seedlings grown on full medium (Figure 2), while a stimulation is observed for the translocation of labelled ^{15}N from *P. indica* to the leaves under N limitation (Figure 3). To test whether genes for N metabolite transporters are regulated by *P. indica*, we performed expression profiles with RNA from the roots and shoots of seedlings, which were either grown on full or N-limited medium in the presence or absence of *P. indica* (Table 2).

Of the 56 investigated genes, which code for NO_3^-, NH_4^+, amino acid or peptide transporters, 33 genes were differentially expressed in either the roots and shoots or both of colonized and uncolonized seedlings grown on full or N-limited media (Table 2). In the shoots, this included genes for two NH_4 transporters (*AMT1-3* and *AMT1-5*), three NO_3^- transporters (*NRT2.2*, *NRT2.4* and *NRT2.5*), five members of the NITRATE TRANS-PORTER 1/PEPTIDE TRANSPORTERs gene family (*NPF2.6*, *NPF2.13*, *NPF5.3*, *NPF5.12* and *NPF5.14*) as well as the urea transporter *DUR3*. Furthermore, 21 amino acid transporters, including members of the LHT and AAP families, as well as 12 UmamiT putative amino acid transporters responded to the fungus. In contrast, seven transporter genes were downregulated by *P. indica* in the shoots of N-starved seedlings (Table 2). In the roots, six of these genes showed the same regulation (Table 2). This clearly demonstrates that the expression of genes for NO_3^-, NH_4^+, amino acid and peptide transporters are major targets of the fungus under N-limited conditions (see the Section 3).

Table 2. Differentially regulated transporters in *A. thaliana*. DEGs regulated by nitrogen limitation (−N) in *A. thaliana* tissues without (w/o) or with (w) *P. indica* colonization.

DEGs			Shoot		Root	
Category	Gene	Atg Number	−N vs. +N w/o *P. indica*	−N vs. +N w *P. indica*	−N vs. +N w/o *P. indica*	−N vs. +N w *P. indica*
Nitrate (NRT2 family)	*AtNRT2.2*	At1g08100	x	3.68	x	x
Nitrate (NRT2 family)	*AtNRT2.3*	At5G60780	−1.55	−5.8	x	−5.22
Nitrate (NRT2 family)	*AtNRT2.4*	At5g60770	x	3.61	2.9	2.58
Nitrate (NRT2 family)	*AtNRT2.5*	At1g12940	2.17	3.85	5.63	3.61
Nitrate (NRT2 family)	*AtNRT2.6*	At3g45060	x	x	−3.85	−5.14
Nitrate (NPF family)	*NPF2.6*	At3g45660	x	3.34	x	x
Nitrate (NPF family)	*NPF2.8/NRT1.9*	At5g28470	x	−4.49	x	x
Nitrate (NPF family)	*NPF2.13/NRT1.7*	At1g69870	x	2.48	x	x
Nitrate (NPF family)	*NPF4.1/AIT3*	At3g25260	x	x	x	2.3
Nitrate (NPF family)	*NPF5.3/NRT1.8*	At5g46040	x	4.1	x	x
Nitrate (NPF family)	*NPF5.6*	At2g37900	x	x	−3.57	x
Nitrate (NPF family)	*NPF5.12*	At1g72140	−2.03	x	x	x
Nitrate (NPF family)	*NPF5.14/NRT1.15*	At1g72120	x	1.83	x	x
Nitrate (NPF family)	*NPF6.2/NRT1.4*	At2g26690	x	x	−2.11	−1.61
Ammonium (AMT family)	*AMT1−3*	At3g24300	x	2.42	x	X
Ammonium (AMT family)	*AMT1−4*	At4g28700	x	x	x	2.51
Ammonium (AMT family)	*AMT1−5*	At3g24290	x	4.18	4.12	2.51
Urea	*DUR3*	At5g45380	x	2.46	2.87	2.04
Amino acid (GDU family)	*GDU1*	At4g31730	x	x	−2.1	x
Amino acid (GDU family)	*GDU4*	At2g24762	−1.77	−1.99	−2.54	−1.96
Amino acid (GDU family)	*GDU5*	At5g24920	x	x	−1.8	x
Amino acid (GDU family)	*GDU6*	At3g30725	x	x	−2.89	−2.03
Amino acid (GDU family)	*GDU7*	At5g38770	x	x	−1.81	x
Amino acid (LHT family)	*LHT1*	At5g40780	x	2.16	x	x
Amino acid (LHT family)	*LHT2/AATL2*	At1g24400	x	x	x	−2.06
Amino acid (LHT family)	*LHT3*	At1g61270	x	x	x	1.53
Amino acid (LHT family)	*LHT7*	At4g35180	2.09	2.52	x	X
Amino acid (AAP family)	*AAP3*	At1g77380	x	1.84	x	X
Amino acid (AAP family)	*AAP4*	At5g63850	x	2.15	x	X
Amino acid (AAP family)	*AAP6*	At5g49630	x	1.9	x	X
Amino acid (AAP family)	*AAP7*	At5g23810	x	x	x	1.53
Amino acid (AVT family)	*AVT1E*	At5g02170	x	−4.51	−1.77	X
Amino acid (AVT family)	*AVT1H*	At5g16740	6.41	7.5	2.25	2.72
Amino acid (AVT family)	*AVT3B*	At2g42005	−2.89	−1.65	x	X
Amino acid (CAT family)	*GAT1/BAT1*	At1g08230	x	2.02	x	X
Amino acid (CAT family)	*CAT1/AAT1*	At4g21120	1.65	3.39	x	X
Amino acid (CAT family)	*CAT5*	At2g34960	x	2.18	x	X
Amino acid (UmamiT family)	*UmamiT 4*	At3G18200	x	4.31	x	X
Amino acid (UmamiT family)	*UmamiT 8*	At4G16620	x	1.99	−1.69	X
Amino acid (UmamiT family)	*UmamiT 10*	At3G56620	x	1.89	x	X
Amino acid (UmamiT family)	*UmamiT 13*	At2G37450	x	−2.07	x	−1.74
Amino acid (UmamiT family)	*UmamiT 14*	At2G39510	x	x	−1.54	X
Amino acid (UmamiT family)	*UmamiT 17*	At4G08300	x	1.76	x	X
Amino acid (UmamiT family)	*UmamiT 19*	At1G21890	x	2.27	x	3.02
Amino acid (UmamiT family)	*UmamiT 20*	At4G08290	x	2	−2.01	−1.57
Amino acid (UmamiT family)	*UmamiT 25*	At1G09380	x	2.4	x	X
Amino acid (UmamiT family)	*UmamiT 26*	At1G11460	x	−1.9	x	X
Amino acid (UmamiT family)	*UmamiT 29*	At4G01430	x	1.57	x	X
Amino acid (UmamiT family)	*UmamiT 35*	At1G60050	x	1.75	−2.5	X
Amino acid (UmamiT family)	*UmamiT 36*	At1G70260	x	x	x	1.95
Amino acid (UmamiT family)	*UmamiT 40*	At5G40240	x	2.14	x	X
Amino acid (UmamiT family)	*UmamiT 42*	At5G40210	x	1.84	x	X
Amino acid (UmamiT family)	*UmamiT 43*	At3G28060	x	−2.34	x	X
Amino acid (UmamiT family)	*UmamiT 45*	At3G28100	x	1.85	x	X
Amino acid (UmamiT family)	*UmamiT 46*	At3G28070	x	x	−5.76	X
Amino acid (UmamiT family)	*UmamiT 47*	At3G28080	x	x	−2.94	X

-6	-4	-2	0	2	4	6	8

Values are given as log2-fold differential expression identified by RNAseq analysis, $p < 0.05$. x = not differentially expressed compared to full N (+N). The gradual colour scale from red to blue is indicated (red, transcript level decreased under limited N; blue, transcript level increased under limited N supply).

3. Discussion

N limitation has severe consequences for plant performance [22], and endophytes may help plants to better adapt to the shortage. We used the well-investigated symbiotic interaction between the model plant *A. thaliana* and *P. indica* to address this question. We demonstrate that under severe N limitation, the fungus does not stimulate the uptake of nitrate into the host plant but rather the N-label from fungal metabolites appears in the leaves of the host. Since our N-limiting medium contains barely any nitrate, the absence of a detectable stimulatory effect of the fungus on nitrate uptake into the host is not surprising. The N metabolites that are translocated from the hyphae to the plants under N limitation conditions did not results in fungus-induced growth promotion (Figure 1), suggesting that the N supply to the host by the fungus might only compensate deficits. Furthermore, N-translocation from the fungus to the host occurs only under N limitation conditions, suggesting the involvement of an N-sensing system (cf. [16]). The successful transfer of ^{15}N by arbuscular mycorrhizal fungi to host plants has been shown previously [23–25]. More recently, Hoysted et al. [26] investigated clover (*Trifolium repens*) colonized by Mucoromycotina fungi and showed that the host gained both ^{15}N and ^{33}P tracers directly from the fungus in exchange for plant-fixed C. Whether the N supply to the host in our study system has comparable symbiotic features with profit for both partners or whether it is just the stress-related withdrawal of N from the fungus by the plant without any profit for the microbe remains to be investigated. However, since the fungus can grow and propagate on the host under our –N conditions, the N translocation to the host does not restrict hyphal growth. It appears that the conditions are not strong enough to induce changes in the symbiotic interaction [16]. It is also not clear which metabolites are transported from the microbe to the plant or how this occurs. In *Medicago truncatula*, three AMT2 family ammonium transporters (AMT2;3, AMT2;4, and AMT2;5) are involved in the uptake of N in form of ammonium from the periarbuscular space between the fungal plasma membrane and the plant-derived periarbuscular membrane [27]. In exchange, host plants transfer reduced carbon to the fungi [28–30]. Additionally, Cope et al. [31] showed that the colonization of *M. truncatula* with *R. irregularis* led to the elevated expression of the mycorrhiza-induced *AMT2;3* and the nitrate transporter *NPF4.12* as well as the putative ammonium transporter *NIP1;5* in the roots. A dipeptide transporter from the arbuscular mycorrhizal fungus *Rhizophagus irregularis* is upregulated in the intraradical phase [32]. To investigate how the fungus manipulates the host N metabolism, we performed a comprehensive metabolome and transcriptome analysis for N-related metabolites and genes (Tables 1 and 2).

No major impact of the colonization by *P. indica* on the changes of shoot metabolite levels in response to N limitation has been observed in this study. Liu et al. [33] demonstrated that raffinose positively regulates maize drought tolerance by reducing leaf transpiration. The raffinose family oligosaccharides are associated with various abiotic and biotic stress responses in different plant species (e.g., [34–38]). It is conceivable that the stimulatory effect of *P. indica* on the raffinose level in N-limited leaves reduces stress.

Nitrate transporter genes are often upregulated under N starvation; however, the role of endophytic microorganisms in nitrate acquisition is not fully understood. In rice, the arbuscular mycorrhizal fungus *R. irregularis* remarkably promoted growth and N acquisition, and about 42% of the overall N could be delivered via the symbiotic route under nitrate-limiting conditions [39]. Nitrate uptake occurs via NITRATE TRANSPORTER1/PEPTIDE TRANSPORTER FAMILY (NPF)4.5, a member of the low affinity nitrate transporter family, which is exclusively expressed in the arbuscles of the Gramineous species [39]. A comparable mechanism does not exist in our endophyte/*A. thaliana* model, and the putative *A. thaliana* NPF4.5 homolog is not upregulated in colonized roots under N-limited conditions. However, we observed a highly specific response of several NPF/NRT1 and NRT2 family members to *P. indica* colonisation, which suggest conclusions regarding how the fungus interferes with the plant N metabolism. The nitrate transporters NRT2.2 and NRT2.4 [40] are only upregulated in the rosettes when the roots are colonized by *P. indica*, while their expression in the roots does not respond to the fungus. This suggests that the fungus promotes nitrate scavenging, which is released from the vacuole in response to N starvation. In fact, NRT2.4 has been shown to be expressed close to the phloem in rosettes and to contribute to nitrate homeostasis in the phloem under limiting nitrate supply, since in nitrate-starved *nrt2.4* mutants, nitrate content in shoot phloem exudates was decreased [40]. Likewise, *NRT1.7* (*NPF2.13*) and *NRT1.8* (*NPF5.3*) are upregulated by *P. indica* in leaves, but not in roots. NRT1.7 loads excess nitrate stored in source leaves into phloem and facilitates nitrate allocation to sink leaves. Under N starvation, the *nrt1.7* mutant exhibits growth retardation, indicating that the NRT1.7-mediated source-to-sink remobilization of stored nitrate is important for sustaining growth in plants [41]. *NRT1.8* is expressed predominantly in xylem parenchyma cells within the vasculature and functional disruption of *NRT1.8* significantly increased the nitrate concentration in xylem sap [42]. In contrast, *NRT2.3 and -2.6* are downregulated under N-limiting conditions and this is further promoted by the fungus. *NRT2.6* has been linked to biotic and abiotic stress responses [43], and it appears that the downregulation of *NRT2.6* expression by *P. indica* alleviates the stress responses in the roots. Finally, *NRT1.9 (NPF2.9)* is strongly downregulated by *P. indica* in the leaves. *NRT1.9* is expressed in the companion cells of phloem. In *NRT1.9* mutants, downward nitrate transport was reduced, suggesting that *NRT1.9* facilitates the loading of nitrate into the phloem and enhances downward nitrate transport to the roots [44], apparently a process that is restricted by the fungus. Taken together, the analysis of the regulation of the *A. thaliana* nitrate transporter genes by *P. indica* suggests that the root-colonizing fungus supports nitrate transport to and availability in the aerial parts of the host under our nitrate-limiting conditions. This is further supported by the upregulation of *NRT1.15* (*NPF5.14*) by *P. indica* in the leaves. *NRT1.15* is a tonoplast-localized low-affinity nitrate transport [45] and the overexpression of the gene significantly decreased vacuolar nitrate contents and nitrate accumulation in *A. thaliana* shoots. *NRT1.15* regulates vacuolar nitrate efflux, and the reallocation might also contribute to osmotic stress responses other than mineral nutrition [45].

Since the medium does not contain NH_4^+, the plant can only receive NH_4^+ from the fungus via ammonium transporters (AMTs) [41,46–48]. *AMT1-4* expression is upregulated by *P. indica* in roots under N-limited conditions. Since AMT1-4 is root-specific [47], this suggests that the plant tries to compensate its N limitation by stimulating NH_4^+ uptake. NH_4^+ might also originate from the fungus, and it is conceivable that withdrawal of this ion from the fungus might ultimately result in a change of the symbiotic interaction towards saprophytism (cf. [16]). Furthermore, the expression of *AMT1-3* and *AMT1-5* as well as *DUR3*-coding for a urea transporter [49,50] is stimulated by *P. indica* in the shoots. Root colonization might create a metabolite environment in the host that requires these transporters for the proper distribution of the N metabolites in the aerial parts.

Seven amino acid transporters are regulated >log2-fold by *P. indica* colonisation in nitrate-deprived *A. thaliana* seedlings. In roots, the fungus prevents the downregulation of the gene for glutamine secreting GLUTAMINE DUMPER (GDU)1 [51], suggesting that the microbe wants to access to the plant glutamine. Furthermore, the broad-specificity high affinity amino acid transporter LYSINE HISTIDINE TRANSPORTER (LHT)1 [52], AMINO ACID PERMEASE (AAP)4, γ-AMINOBUTYRIC ACID TRANSPORTER (GAT)1, and CATIONIC AMINO ACID TRANSPORTER (CAT)5 are upregulated in the leaves of *P. indica*-colonized seedlings. These transporters have been proposed to be involved in nitrogen recycling in plants [53]. Apparently, a better or different N metabolism management is required for the plant when the roots are associated with the endophyte. LHT1 and -2 are also involved in the transport of 1-aminocyclopropane carboxylic acid, a biosynthetic precursor of ethylene [54], which might indicate an increased stress caused by the interaction with the fungus under N-limiting conditions. An involvement in nitrogen recycling has also been proposed for 5 of the 10 USUALLY MULTIPLE ACIDS MOVE IN AND OUT TRANSPORTERS (UMANIT13, −20, −40, −45 and −47) [53], which are regulated >log2-fold in either the roots or shoots of *P. indica*-colonized seedlings under N-limitations.

4. Materials and Methods

4.1. Plant and Fungus Material and Corresponding Growth Conditions

A. thaliana seeds (*Col-0*) were surface-sterilized and sown on N-free MGRL medium supplemented with 2.5 mM NH_4NO_3 and 3 g/L gelrite [55]. The KNO_3 and $Ca(NO_3)_2$ in the MGRL medium were replaced by KCl and $CaCl_2$ to ensure ion equilibrium. After 48 h of stratification at 4 °C in the dark, the seeds were transferred to long-day conditions with 22 °C, 16 h light/8 h dark, 80 µmol m^{-2} s^{-1} for 10 days.

Piriformospora indica was cultured on Kaefer's medium as described previously [56,57]. As described previously, plugs of a 4-week-old fungal culture were used for co-cultures with the seedlings. The fungus was pre-grown for 7 days on PNM medium (PNM+N) with a nylon membrane in the dark at 22 °C. For N-limiting conditions (0 mM total N, PNM−N), KNO_3 and $Ca(NO_3)_2$ were replaced by KCl and $CaCl_2$. For control plates without fungus, only empty KM plugs were placed on top of the membrane.

4.2. Plant-Fungus Co-Cultures and the Determination of Growth Promotion

For plant-fungus co-cultures for 5 days, 4 plants (per Petri dish) were placed on top of the pre-grown fungal lawn, as described previously [57], with some adaptations. Plates were sealed with 3M^{TM} Micropore tape to reduce the condensation and 10-day-old plants were used for co-cultivation to reduce the amount of N, which accumulated in the plants on MGRL medium before the co-culture. In pilot experiments, we showed that the reduced age did not affect the establishment of the symbiosis with the fungus. The co-cultures were incubated at 22 °C, 16 h light/8 h dark and 80 µmol m^{-2} s^{-1} with light from the top.

After 5 days, the roots and shoots of the plants were harvested separately. For that, 5 plates (=20 plants) were harvested as 1 sample. Both roots and shoots were washed in sterile distilled water and carefully dried before weighing and direct freezing in liquid nitrogen. Samples were stored in −80 °C until further use. These experiments were repeated 3–4 times independently.

To determine growth promotion by the fungus, the weight of the sample with fungus was normalized (divided) to the weight without fungus. This was performed for the total weights sampled from the full medium (PNM+N) as well as from the N-limited medium (PNM−N). Final growth promotion values presented in the figures are averages of 3 replicates from independent cultures.

4.3. ^{15}N Labelling Experiments in the Medium

To analyse the uptake of nitrogen by the plant, 2.5% of the total KNO_3 (which equals 0.125 mM KNO_3) of the PNM medium was replaced by $K^{15}NO_3$ (Eurisotop, Saint-Aubin, France) dissolved in distilled water. For proper comparison, the 2.5% of $K^{15}NO_3$ was also added to the N-free medium (PNM−N) resulting in a final concentration of 0.125 mM nitrate. Finally, PNM−N control plates without ^{15}N were used and contained 0.125 mM unlabelled KNO_3. Plants grown on these plates were compared to those grown on PNM+N plates to analyse the natural abundance of ^{15}N in the plant tissue. As described before, the fungus or control plug was pre-incubated on the PNM with nylon membrane for 1 week before plants were placed on the plates. The co-cultures were incubated for 5 days to ensure that enough ^{15}N was taken up by the plant.

4.4. ^{15}N Fungus-Labelling Experiments

To analyse whether the fungus can directly transfer N or N-containing metabolites to the plant, it was labelled with ^{15}N before the co-culture. A modified KM medium without the N-containing components (20 g/L dextrose, 50 mL/L macronutrients, 10 mL/L micronutrients and 1 mL/L Fe-EDTA, 1 mL/L vitamin mix, pH 6.5) was prepared and supplemented with 10 g/L ISOGRO®-15N (CortecNet, Les Ulis, France) according to the manufacturers protocol. *P. indica* plugs of 2 mm diameter were incubated (23 °C, 50 rpm, dark) in 2 mL of KM^{ISOGRO} in Greiner CELLSTAR® 12-well plates (Greiner Bio-One, Frickenhausen, Germany) sealed with 3M^{TM} Micropore tape. After 14 days of growth, the fungal tissue was separated from the remaining medium and carefully washed 3 times with N-free liquid PNM to remove the ^{15}N bound to the hyphal surface. A 76.66% enrichment in ^{15}N was achieved using this protocol. The fungus was carefully cut in 5 × 5 mm pieces and placed on PNM−N and PNM+N plates to start the co-cultures. To minimize ^{15}N uptake by the plant from dead fungal material during the washing and handling procedure, the fungal plugs were placed in minimum 1 cm distance from the roots. Under these conditions, contact between the two symbionts requires the active growth of the hyphae towards the roots. Co-cultivation was performed with 3 plants per plate for 14 days to ensure that enough ^{15}N was taken up by the plant.

4.5. Isolation and Clean-Up of RNA

RNA was isolated from the roots or shoots of 10-day-old seedlings, which were either co-cultured with the fungus for additional 5 days (root colonization results), or for an additional 4 days for expression profiling. Samples of root or shoot material were stored in −80 °C. For homogenization, the samples were ground with mortar and pistil in liquid nitrogen. A maximum of 100 mg material was used for RNA extraction. RNA was extracted with Trizol^{TM} (ThermoFisher Scientific, Waltham, MA, USA) and chloroform according to the manufacturers protocol. Briefly, the plant material was mixed with 1 mL of Trizol^{TM} and incubated on a shaker at room temperature for 15 min. After the addition of 250 µL chloroform and a second incubation phase, the sample was centrifuged (30 min, 4 °C). The supernatant was mixed with isopropanol and incubated on ice, followed by centrifugation. The pellet was washed twice with 80% ethanol, dried and resuspended in RNAse-free water. The RNA isolation was followed by an additional cleaning step to remove access salts originating from the fungus tissue. For this, the sample was mixed with 3 M sodium acetate (1/10 (v/v) in RNAse-free water, pH = 5.2) and 600 µL of ice-cold 100% ethanol and incubated at −20 °C for at least 1 hr. After centrifugation and 2 cleaning steps with 80% ethanol, the sample was resuspended in RNAse-free water. The quality and concentration of the extracted RNA was tested via absorbance analysis using a NanoVue (GE Healthcare, Uppsala, Sweden).

4.6. RNAseq and Data Analysis

After the transfer of samples to Novogene Genomics Service (Cambridge, UK), the RNA sample integrity was checked with a Bioanalyzer 2100 (Agilent). After samples passed the quality check, the service laboratory proceeded with the library construction and RNA sequencing (PE150) on Illumina NovaSeq^{TM} 6000 platforms, as described in a previous study [58].

The RNAseq libraries were filtered and quality-trimmed with fastp (v0.23.2) [59], i.e., read ends were truncated to achieve a Phred quality score of 30 or more. Reads below 15 nt length or those comprising at least 2 ambiguous N bases were removed from the dataset. Read qualities were monitored by FastQC (v0.11.3; https://www.bioinformatics. babraham.ac.uk/projects/fastqc/, accessed on 15 December 2022). Hisat2 (v2.2.1) [60] was used with default parameters to map the quality-trimmed RNAseq libraries to the *A. thaliana* reference genome (TAIR10, Ensembl release 51). The mapping allowed spliced reads and single-read mapping to multiple best-fitting locations. FeatureCounts (v1.5.3) [61] was applied to perform read-counting based on the *A. thaliana* reference annotation (TAIR10, Ensembl release 51). The Bioconductor DESeq2 (v1.10.0) package [62] was utilized to identify DEGs in the different pairwise mutant and wild type comparisons. Benjamini and Hochberg's false discovery rate (FDR) approach [63] was employed to adjust the calculated p-values for multiple testing.

To identify DEGs of transporters predicted to transport major N compounds, the obtained results were initially filtered according to their p-value ($p < 0.05$). Next, DEGs were sorted according to their log2-fold change—here only changes with numbers $\geq +1.5$ and ≤ -1.5 were further analysed. This list was cross-checked with targets identified from a search in the UniProt database (https://www.uniprot.org, accessed on 15 December 2022) using keywords like "NH$_4$ transport".

4.7. Analysis of Fungal Colonization via qPCR

A total of 1 mg of RNA was used for the synthesis of cDNA. The Omniscript RT Kit (Qiagen, Hilden, Germany) was used according to the manufacturers protocol with the oligo(dT)$_{18}$ primer (ThermoFisher Scientific, Waltham, USA). qPCR was performed with fifty nanograms of the synthesized cDNA as template in a Bio-Rad CFX96 Real-Time PCR Detection System (Feldkirchen, Germany) via DreamTaq Polymerase (ThermoFisher Scientific, Waltham, USA) and Evagreen (Biotium, Fremont, CA, USA). The data were normalized with respect to the *A. thaliana RPS18B* (*At*1g34030) gene using the $2^{-\Delta\Delta CT}$ method [64]. To quantify the *P. indica* colonization level of *A. thaliana* roots, the expression of *PiTEF1* [65] was analysed in comparison to the plant´s housekeeping gene *RPS18B* (At1g 34030). The following primers were used: *PiTEF1*: CGCAGAATACAAGGAGGCC and CGTATCGTAGCTCGCCTGC; *RPS18B*: GTCTCCAATGCCCTTGACAT and TCTTTCCTCT-GCGACCAGTT [66]. The colonization was compared between plants grown on PNM−N and PNM+N media (set as 1.0) using the $2^{-\Delta\Delta CT}$ method.

4.8. Determination of Total Nitrogen and ^{15}N Enrichment

Total N and ^{15}N contents were quantified on 1–2 mg aliquots of dry tissue, after drying a ground tissue aliquot at 65–70 °C for at least 48 h. N elements were detected using gas chromatography on a FLASH 2000 Organic Elemental Analyzer (Thermo Fisher Scientific, Villebon, France). The ^{15}N/^{14}N isotopic ratio was subsequently quantified using a coupled mass spectroscope (Delta V advantage IRMS; Thermo Fisher Scientific, Villebon, France). The total N content was only determined in plant shoots because the discrimination of N from plant or fungus was not possible in colonized root material.

4.9. Metabolomic Analysis

For GC-MS-based quantifications, 25 mg of finely ground plant material was resus-pended in 1 mL of frozen (−20 °C) water:acetonitrile:isopropanol (2:3:3, $v/v/v$) containing Ribitol at 4 ug/mL and analysed as described in [67].

5. Conclusions

We observed an unexpected complexity in the plant N metabolism when N-deprived *A. thaliana* seedlings were colonized by *P. indica*. Our data suggest that the fungus neither stimulated the total N content nor promoted ^{15}NO$_3{}^-$ uptake from agar plates to the host. Rather, reduced N metabolites were transported from the fungus to the plant. Further-more, gene expression and metabolite profiles suggest that N-containing metabolites were redistributed by *P. indica* in *A. thaliana* seedlings exposed to N-limitation.

Our initial observations highlight a few aspects that need to be investigated in greater detail. (1) Which N metabolites are transported from the fungus to a plant suffering under N limitation? (2) The plant appears to adapt its N metabolism under N limitation by transporting N metabolites shootwards, a process that is supported by the fungus. Is the fungal support for the plant specific for the symbiotic phase of the interaction? (3) Are our observations *P. indica*-specific or do they occur also in other endophyte/plant interactions?

Supplementary Materials: The following supporting information can be downloaded at: https://www.mdpi.com/article/10.3390/ijms242015372/s1.

Author Contributions: S.S.S., E.B., G.C., A.M. and A.K. performed the experiments and analysed the data, J.L.-M., H.S., T.K., J.V.-C., S.P., A.K. and R.O. developed the concept and wrote the manuscript. All authors have read and agreed to the published version of the manuscript.

Funding: This project was financially supported by the collaborative ICPS research project executed in the framework of the EIG CONCERT-Japan joint call on Food Crops and Biomass Production Technologies and the related national funding agencies: grants 01DR17007A and 01DR17007B from the Federal Ministry of Education and Research (BMBF), Germany, to R.O.; grant EIG_JC1JAPAN-045 from the Centre National de la Recherche Scientifique (CNRS), France, to A.K.; grant PCIN-2016-037 from the Ministry of Economy and Competitiveness (MINECO), Spain, to J.V.-C. and S.P.; and grant JPMJSC16C3 from the Japan Science and Technology Agency (JST) to H.S. This work was further supported by the Deutsche Forschungsgemeinschaft (DFG, German Research Foundation) CRC1127 ChemBioSys (project ID: 239748522) for R.O. This work benefited from the support of IJPB's Plant Observatory technological platforms. The IJPB benefits from additional support of Saclay Plant Sciences-SPS (ANR-17-EUR-0007).

Institutional Review Board Statement: Not applicable.

Informed Consent Statement: Not applicable.

Data Availability Statement: The dataset is available in the NCBI GEO repository, under accession number GSE239281.

Acknowledgments: We thank Sarah Mußbach, Claudia Röppischer and Christin Weilandt for the excellent technical support in the lab.

Conflicts of Interest: The authors declare no conflict of interest.

References

1. Engels, C.; Marschner, H. Plant uptake and utilization of nitrogen. In *Nitrogen Fertilization in the Environment*; Bacan, P.E., Ed.; Marcel Dekker, Inc.: New York, NY, USA, 1995; pp. 41–81.
2. Vidal, E.A.; Alvarez, J.M.; Araus, V.; Riveras, E.; Brooks, M.D.; Krouk, G.; Ruffel, S.; Lejay, L.; Crawford, N.M.; Coruzzi, G.M.; et al. Nitrate in 2020: Thirty Years from Transport to Signaling Networks. *Plant Cell* **2020**, *32*, 2094–2119. [CrossRef] [PubMed]
3. Lyzenga, W.J.; Liu, Z.; Olukayode, T.; Zhao, Y.; Kochian, L.V.; Ham, B.-K. Getting to the roots of N, P, and K uptake. *J. Exp. Bot.* **2023**, *74*, 1784–1805. [CrossRef] [PubMed]
4. Wang, Y.-Y.; Cheng, Y.-H.; Chen, K.-E.; Tsay, Y.-F. Nitrate Transport, Signaling, and Use Efficiency. *Annu. Rev. Plant Biol.* **2018**, *69*, 85–122. [CrossRef] [PubMed]
5. Aluko, O.O.; Kant, S.; Adedire, O.M.; Li, C.; Yuan, G.; Liu, H.; Wang, Q. Unlocking the potentials of nitrate transporters at improving plant nitrogen use efficiency. *Front. Plant Sci.* **2023**, *14*, 1074839. [CrossRef] [PubMed]
6. Guo, K.; Yang, J.; Yu, N.; Luo, L.; Wang, E. Biological nitrogen fixation in cereal crops: Progress, strategies, and perspectives. *Plant Commun.* **2023**, *4*, 100499. [CrossRef]
7. Rui, W.; Mao, Z.; Li, Z. The Roles of Phosphorus and Nitrogen Nutrient Transporters in the Arbuscular Mycorrhizal Symbiosis. *Int. J. Mol. Sci.* **2022**, *23*, 11027. [CrossRef]
8. Chaudhary, P.; Agri, U.; Chaudhary, A.; Kumar, A.; Kumar, G. Endophytes and their potential in biotic stress management and crop production. *Front. Microbiol.* **2022**, *13*, 933017. [CrossRef]
9. Zhang, Y.; Feng, H.; Druzhinina, I.S.; Xie, X.; Wang, E.; Martin, F.; Yuan, Z. Phosphorus/nitrogen sensing and signaling in diverse root-fungus symbioses. *Trends Microbiol.* **2023**. [CrossRef]
10. Saleem, S.; Sekara, A.; Pokluda, R. *Serendipita indica*—A Review from Agricultural Point of View. *Plants* **2022**, *11*, 3417. [CrossRef]
11. Del Barrio-Duque, A.; Ley, J.; Samad, A.; Antonielli, L.; Sessitsch, A.; Compant, S. Beneficial Endophytic Bacteria-Serendipita indica Interaction for Crop Enhancement and Resistance to Phytopathogens. *Front. Microbiol.* **2019**, *10*, 2888. [CrossRef]
12. Sherameti, I.; Shahollari, B.; Venus, Y.; Altschmied, L.; Varma, A.; Oelmüller, R. The Endophytic Fungus Piriformospora indica Stimulates the Expression of Nitrate Reductase and the Starch-degrading Enzyme Glucan-water Dikinase in Tobacco and Arabidopsis Roots through a Homeodomain Transcription Factor That Binds to a Conserved Motif in Their Promoters. *J. Biol. Chem.* **2005**, *280*, 26241–26247. [CrossRef]
13. Eliaspour, S.; Sharifi, R.S.; Shirkhani, A. Evaluation of interaction between *Piriformospora indica*, animal manure and NPK fertilizer on quantitative and qualitative yield and absorption of elements in sunflower. *Food Sci. Nutr.* **2020**, *8*, 2789–2797. [CrossRef] [PubMed]
14. Strehmel, N.; Mönchgesang, S.; Herklotz, S.; Krüger, S.; Ziegler, J.; Scheel, D. Piriformospora indica Stimulates Root Metabolism of Arabidopsis thaliana. *Int. J. Mol. Sci.* **2016**, *17*, 1091. [CrossRef]
15. Ghaffari, M.R.; Ghabooli, M.; Khatabi, B.; Hajirezaei, M.R.; Schweizer, P.; Salekdeh, G.H. Metabolic and transcriptional response of central metabolism affected by root endophytic fungus Piriformospora indica under salinity in barley. *Plant Mol. Biol.* **2016**, *90*, 699–717. [CrossRef] [PubMed]
16. Lahrmann, U.; Ding, Y.; Banhara, A.; Rath, M.; Hajirezaei, M.R.; Döhlemann, S.; von Wirén, N.; Parniske, M.; Zuccaro, A. Host-related metabolic cues affect colonization strategies of a root endophyte. *Proc. Natl. Acad. Sci. USA* **2013**, *110*, 13965–13970. [CrossRef] [PubMed]
17. Hua, M.D.-S.; Kumar, R.S.; Shyur, L.-F.; Cheng, Y.-B.; Tian, Z.; Oelmüller, R.; Yeh, K.-W. Metabolomic compounds identified in Piriformospora indica-colonized Chinese cabbage roots delineate symbiotic functions of the interaction. *Sci. Rep.* **2017**, *7*, 9291. [CrossRef] [PubMed]

18. Bandyopadhyay, P.; Yadav, B.G.; Kumar, S.G.; Kumar, R.; Kogel, K.-H.; Kumar, S. *Piriformospora indica* and *Azotobacter chroococcum* Consortium Facilitates Higher Acquisition of N, P with Improved Carbon Allocation and Enhanced Plant Growth in *Oryza sativa*. *J. Fungi* **2022**, *8*, 453. [CrossRef]

19. Mansotra, P.; Sharma, P.; Sharma, S. Bioaugmentation of Mesorhizobium cicer, Pseudomonas spp. and Piriformospora indica for Sustainable Chickpea Production. *Physiol. Mol. Biol. Plants* **2015**, *21*, 385–393. [CrossRef]

20. Hallasgo, A.M.; Spangl, B.; Steinkellner, S.; Hage-Ahmed, K. The Fungal Endophyte *Serendipita williamsii* Does Not Affect Phosphorus Status But Carbon and Nitrogen Dynamics in Arbuscular Mycorrhizal Tomato Plants. *J. Fungi* **2020**, *6*, 233. [CrossRef]

21. Abdelaziz, M.E.; Kim, D.; Ali, S.; Fedoroff, N.V.; Al-Babili, S. The endophytic fungus *Piriformospora indica* enhances *Arabidopsis thaliana* growth and modulates Na^+/K^+ homeostasis under salt stress conditions. *Plant Sci.* **2017**, *263*, 107–115. [CrossRef]

22. Krapp, A.; Berthomé, R.; Orsel, M.; Mercey-Boutet, S.; Yu, A.; Castaings, L.; Elftieh, S.; Major, H.; Renou, J.-P.; Daniel-Vedele, F. Arabidopsis Roots and Shoots Show Distinct Temporal Adaptation Patterns toward Nitrogen Starvation. *Plant Physiol.* **2011**, *157*, 1255–1282. [CrossRef]

23. Hodge, A.; Stewart, J.; Robinson, D.; Griffiths, B.S.; Fitter, A.H. Competition between roots and soil microorganisms for nutrientsfrom nitrogen-rich patches of varying complexity. *J. Ecol.* **2000**, *88*, 150–164. [CrossRef]

24. Hodge, A.; Campbell, C.D.; Fitter, A.H. An arbuscular mycorrhizal fungus accelerates decomposition and acquires nitrogen directly from organic material. *Nature* **2001**, *413*, 297–299. [CrossRef]

25. Thirkell, T.J.; Cameron, D.D.; Hodge, A. Resolving the 'nitrogen paradox' of arbuscular mycorrhizas: Fertilisation with organic matter brings considerable benefits for plant nutrition and growth. *Plant Cell Environ.* **2016**, *39*, 1683–1690. [CrossRef] [PubMed]

26. Hoysted, G.A.; Field, K.J.; Sinanaj, B.; Bell, C.A.; Bidartondo, M.I.; Pressel, S. Direct nitrogen, phosphorus and carbon exchanges between Mucoromycotina 'fine root endophyte' fungi and a flowering plant in novel monoxenic cultures. *New Phytol.* **2023**, *238*, 70–79. [CrossRef] [PubMed]

27. Breuillin-Sessoms, F.; Floss, D.S.; Gomez, S.K.; Pumplin, N.; Ding, Y.; Levesque-Tremblay, V.; Noar, R.D.; Daniels, D.A.; Bravo, A.; Eaglesham, J.B.; et al. Suppression of Arbuscule Degeneration in *Medicago truncatula phosphate transporter4* Mutants Is Dependent on the Ammonium Transporter 2 Family Protein AMT2;3. *Plant Cell* **2015**, *27*, 1352–1366. [CrossRef]

28. Paul, E.A.; Kucey, R.M.N. Carbon flow in plant microbial associations incriminated in perinatal morbidity and mortality. *Science* **1981**, *213*, 473–474. [CrossRef]

29. Wright, D.P.; Read, D.J.; Scholes, J.D. Mycorrhizal sink strength influences whole plant carbon balance of *Trifolium repens* L. *Plant Cell Environ.* **1998**, *21*, 881–891. [CrossRef]

30. Lendenmann, M.; Thonar, C.; Barnard, R.L.; Salmon, Y.; Werner, R.A.; Frossard, E.; Jansa, J. Symbiont identity matters: Carbon and phosphorus fluxes between Medicago truncatula and different arbuscular mycorrhizal fungi. *Mycorrhiza* **2011**, *21*, 689–702. [CrossRef]

31. Cope, K.R.; Kafle, A.; Yakha, J.K.; Pfeffer, P.E.; Strahan, G.D.; Garcia, K.; Subramanian, S.; Bücking, H. Physiological and transcriptomic response of Medicago truncatula to colonization by high- or low-benefit arbuscular mycorrhizal fungi. *Mycorrhiza* **2022**, *32*, 281–303. [CrossRef]

32. Belmondo, S.; Fiorilli, V.; Pérez-Tienda, J.; Ferrol, N.; Marmeisse, R.; Lanfranco, L. A dipeptide transporter from the arbuscular mycorrhizal fungus Rhizophagus irregularis is upregulated in the intraradical phase. *Front. Plant Sci.* **2014**, *5*, 436. [CrossRef] [PubMed]

33. Liu, Y.; Li, T.; Zhang, C.; Zhang, W.; Deng, N.; Dirk, L.M.A.; Downie, A.B.; Zhao, T. Raffinose positively regulates maize drought tolerance by reducing leaf transpiration. *Plant J.* **2023**, *114*, 55–67. [CrossRef] [PubMed]

34. ElSayed, A.I.; Rafudeen, M.S.; Golldack, D. Physiological aspects of raffinose family oligosaccharides in plants: Protection against abiotic stress. *Plant Biol.* **2014**, *16*, 1–8. [CrossRef] [PubMed]

35. Gu, L.; Jiang, T.; Zhang, C.; Li, X.; Wang, C.; Zhang, Y.; Li, T.; Dirk, L.M.A.; Downie, A.B.; Zhao, T. Maize HSFA2 and HSBP2 antagonistically modulate raffinose biosynthesis and heat tolerance in Arabidopsis. *Plant J.* **2019**, *100*, 128–142. [CrossRef]

36. Han, Q.; Qi, J.; Hao, G.; Zhang, C.; Wang, C.; Dirk, L.M.A.; Downie, A.B.; Zhao, T. ZmDREB1A Regulates RAFFINOSE SYNTHASE Controlling Raffinose Accumulation and Plant Chilling Stress Tolerance in Maize. *Plant Cell Physiol.* **2020**, *61*, 331–341. [CrossRef]

37. Li, C.; Tien, H.; Wen, M.; Yen, H.E. *Myo*-inositol transport and metabolism participate in salt tolerance of halophyte ice plant seedlings. *Physiol. Plant.* **2021**, *172*, 1619–1629. [CrossRef]

38. Yang, J.; Ling, C.; Liu, Y.; Zhang, H.; Hussain, Q.; Lyu, S.; Wang, S.; Liu, Y. Genome-Wide Expression Profiling Analysis of Kiwifruit *GolS* and *RFS* Genes and Identification of *AcRFS4* Function in Raffinose Accumulation. *Int. J. Mol. Sci.* **2022**, *23*, 8836. [CrossRef]

39. Wang, S.; Chen, A.; Xie, K.; Yang, X.; Luo, Z.; Chen, J.; Zeng, D.; Ren, Y.; Yang, C.; Wang, L.; et al. Functional analysis of the OsNPF4.5 nitrate transporter reveals a conserved mycorrhizal pathway of nitrogen acquisition in plants. *Proc. Natl. Acad. Sci. USA* **2020**, *117*, 16649–16659. [CrossRef]

40. Kiba, T.; Feria-Bourrellier, A.-B.; Lafouge, F.; Lezhneva, L.; Boutet-Mercey, S.; Orsel, M.; Bréhaut, V.; Miller, A.; Daniel-Vedele, F.; Sakakibara, H.; et al. The *Arabidopsis* Nitrate Transporter NRT2.4 Plays a Double Role in Roots and Shoots of Nitrogen-Starved Plants. *Plant Cell* **2012**, *24*, 245–258. [CrossRef]

41. Chen, H.-Y.; Chen, Y.-N.; Wang, H.-Y.; Liu, Z.-T.; Frommer, W.B.; Ho, C.-H. Feedback inhibition of AMT1 NH4+-transporters mediated by CIPK15 kinase. *BMC Biol.* **2020**, *18*, 196. [CrossRef]

42. Li, J.-Y.; Fu, Y.-L.; Pike, S.M.; Bao, J.; Tian, W.; Zhang, Y.; Chen, C.-Z.; Zhang, Y.; Li, H.-M.; Huang, J.; et al. The *Arabidopsis* Nitrate Transporter NRT1.8 Functions in Nitrate Removal from the Xylem Sap and Mediates Cadmium Tolerance. *Plant Cell* **2010**, *22*, 1633–1646. [CrossRef] [PubMed]
43. Dechorgnat, J.; Patrit, O.; Krapp, A.; Fagard, M.; Daniel-Vedele, F. Characterization of the Nrt2.6 Gene in Arabidopsis thaliana: A Link with Plant Response to Biotic and Abiotic Stress. *PLoS ONE* **2012**, *7*, e42491. [CrossRef] [PubMed]
44. Wang, Y.-Y.; Tsay, Y.-F. *Arabidopsis* Nitrate Transporter NRT1.9 Is Important in Phloem Nitrate Transport. *Plant Cell* **2011**, *23*, 1945–1957. [CrossRef] [PubMed]
45. Lu, Y.-T.; Liu, D.-F.; Wen, T.-T.; Fang, Z.-J.; Chen, S.-Y.; Li, H.; Gong, J.-M. Vacuolar nitrate efflux requires multiple functional redundant nitrate transporter in Arabidopsis thaliana. *Front. Plant Sci.* **2022**, *13*, 926809. [CrossRef]
46. Wu, X.; Liu, T.; Zhang, Y.; Duan, F.; Neuhäuser, B.; Ludewig, U.; Schulze, W.X.; Yuan, L. Ammonium and nitrate regulate NH4+ uptake activity of Arabidopsis ammonium transporter AtAMT1;3 via phosphorylation at multiple C-terminal sites. *J. Exp. Bot.* **2019**, *70*, 4919–4930. [CrossRef]
47. Yuan, L.; Loqué, D.; Kojima, S.; Rauch, S.; Ishiyama, K.; Inoue, E.; Takahashi, H.; von Wirén, N. The Organization of High-Affinity Ammonium Uptake in *Arabidopsis* Roots Depends on the Spatial Arrangement and Biochemical Properties of AMT1-Type Transporters. *Plant Cell* **2007**, *19*, 2636–2652. [CrossRef]
48. Yuan, L.; Graff, L.; Loqué, D.; Kojima, S.; Tsuchiya, Y.N.; Takahashi, H.; von Wirén, N. AtAMT1;4, a Pollen-Specific High-Affinity Ammonium Transporter of the Plasma Membrane in Arabidopsis. *Plant Cell Physiol.* **2009**, *50*, 13–25. [CrossRef]
49. Zanin, L.; Tomasi, N.; Wirdnam, C.; Meier, S.; Komarova, N.Y.; Mimmo, T.; Cesco, S.; Rentsch, D.; Pinton, R. Isolation and functional characterization of a high affinity urea transporter from roots of *Zea mays*. *BMC Plant Biol.* **2014**, *14*, 222. [CrossRef]
50. Wang, W.-H.; Köhler, B.; Cao, F.-Q.; Liu, G.-W.; Gong, Y.-Y.; Sheng, S.; Song, Q.-C.; Cheng, X.-Y.; Garnett, T.; Okamoto, M.; et al. Rice DUR3 mediates high-affinity urea transport and plays an effective role in improvement of urea acquisition and utilization when expressed in Arabidopsis. *New Phytol.* **2012**, *193*, 432–444. [CrossRef]
51. Yu, S.; Pratelli, R.; Denbow, C.; Pilot, G. Suppressor mutations in the Glutamine Dumper1 protein dissociate disturbance in amino acid transport from other characteristics of the Gdu1D phenotype. *Front. Plant Sci.* **2015**, *6*, 593. [CrossRef]
52. Hirner, A.; Ladwig, F.; Stransky, H.; Okumoto, S.; Keinath, M.; Harms, A.; Frommer, W.B.; Koch, W. *Arabidopsis* LHT1 Is a High-Affinity Transporter for Cellular Amino Acid Uptake in Both Root Epidermis and Leaf Mesophyll. *Plant Cell* **2006**, *18*, 1931–1946. [CrossRef] [PubMed]
53. Havé, M.; Marmagne, A.; Chardon, F.; Masclaux-Daubresse, C. Nitrogen remobilisation during leaf senescence: Lessons from Arabidopsis to crops. *J. Exp. Bot.* **2017**, *68*, 2513–2529. [CrossRef] [PubMed]
54. Choi, J.; Eom, S.; Shin, K.; Lee, R.-A.; Choi, S.; Lee, J.-H.; Lee, S.; Soh, M.-S. Identification of Lysine Histidine Transporter 2 as an 1-Aminocyclopropane Carboxylic Acid Transporter in Arabidopsis thaliana by Transgenic Complementation Approach. *Front. Plant Sci.* **2019**, *10*, 1092. [CrossRef] [PubMed]
55. Fujiwara, T.; Hirai, M.Y.; Chino, M.; Komeda, Y.; Naito, S. Effects of Sulfur Nutrition on Expression of the Soybean Seed Storage Protein Genes in Transgenic Petunia. *Plant Physiol.* **1992**, *99*, 263–268. [CrossRef]
56. Johnson, J.M.; Sherameti, I.; Ludwig, A.; Nongbri, P.L.; Sun, C.; Varma, A.; Oelmüller, R. Protocols for Arabidopsis thaliana and Piriformospora indica co-cultivation: A model system to study plant beneficial traits. *Endocyt. Cell Res.* **2011**, *21*, 101–113.
57. Johnson, J.M.; Sherameti, I.; Nongbri, P.L.; Oelmüller, R. Standardized conditions to study beneficial and nonbeneficial traits in the *Piriformospora indica/Arabidopsis thaliana* interaction. In *Piriformospora indica: Sebacinales and Their Biotechnological Applications*; Varma, A., Kost, G., Oelmüller, R., Eds.; Soil Biology; Springer: Berlin/Heidelberg, Germany, 2013; Volume 33, pp. 325–343.
58. Pérez-Alonso, M.; Guerrero-Galán, C.; Ortega-Villaizán, A.G.; Ortiz-García, P.; Scholz, S.S.; Ramos, P.; Sakakibara, H.; Kiba, T.; Ludwig-Müller, J.; Krapp, A.; et al. The calcium sensor CBL7 is required for *Serendipita indica* -induced growth stimulation in *Arabidopsis thaliana*, controlling defense against the endophyte and K$^+$ homoeostasis in the symbiosis. *Plant Cell Environ.* **2022**, *45*, 3367–3382. [CrossRef]
59. Chen, A.; Gu, M.; Wang, S.; Chen, J.; Xu, G. Transport properties and regulatory roles of nitrogen in arbuscular mycorrhizal symbiosis. *Semin. Cell Dev. Biol.* **2018**, *74*, 80–88. [CrossRef]
60. Kim, D.; Paggi, J.M.; Park, C.; Bennett, C.; Salzberg, S.L. Graph-based genome alignment and genotyping with HISAT2 and HISAT-genotype. *Nat. Biotechnol.* **2019**, *37*, 907–915. [CrossRef]
61. Liao, Y.; Smyth, G.K.; Shi, W. feature Counts: An efficient general purpose program for assigning sequence reads to genomic features. *Bioinformatics* **2014**, *30*, 923–930. [CrossRef]
62. Love, M.I.; Huber, W.; Anders, S. Moderated estimation of fold change and dispersion for RNA-seq data with DESeq2. *Genome Biol.* **2014**, *15*, 550. [CrossRef]
63. Benjamini, Y.; Hochberg, Y. Controlling the False Discovery Rate: A Practical and Powerful Approach to Multiple Testing. *J. R. Stat. Soc. Ser. B Methodol.* **1995**, *57*, 289–300. [CrossRef]
64. Pfaffl, M.W. A new mathematical model for relative quantification in real-time RT-PCR. *Nucleic Acids Res.* **2001**, *29*, e45. [CrossRef] [PubMed]
65. Bütehorn, B.; Rhody, D.; Franken, P. Isolation and Characterisation of Pitef1 Encoding the Translation Elongation Factor EF-1α of the Root Endophyte Piriformospora indica. *Plant Biol.* **2000**, *2*, 687–692. [CrossRef]

66. Scholz, S.S.; Vadassery, J.; Heyer, M.; Reichelt, M.; Bender, K.W.; Snedden, W.A.; Boland, W.; Mithöfer, A. Mutation of the Arabidopsis Calmodulin-Like Protein CML37 Deregulates the Jasmonate Pathway and Enhances Susceptibility to Herbivory. *Mol. Plant* **2014**, *7*, 1712–1726. [CrossRef]
67. Forzani, C.; Duarte, G.T.; Van Leene, J.; Clément, G.; Huguet, S.; Paysant-Le-Roux, C.; Mercier, R.; De Jaeger, G.; Leprince, A.-S.; Meyer, C. Mutations of the AtYAK1 Kinase Suppress TOR Deficiency in Arabidopsis. *Cell Rep.* **2019**, *27*, 3696–3708.e5. [CrossRef]

International Journal of
Molecular Sciences

Article

MsSPL9 Modulates Nodulation under Nitrate Sufficiency Condition in *Medicago sativa*

Vida Nasrollahi [1,2], Gamalat Allam [1,2], Susanne E. Kohalmi [2] and Abdelali Hannoufa [1,2,*]

[1] Agriculture and Agri-Food Canada, 1391 Sandford Street, London, ON N5V 4T3, Canada; vnasroll@uwo.ca (V.N.); gamalat.allam@agr.gc.ca (G.A.)
[2] Department of Biology, University of Western Ontario, 1151 Richmond Street, London, ON N6A 3K7, Canada; skohalmi@uwo.ca
* Correspondence: abdelali.hannoufa@agr.gc.ca

Abstract: Nodulation in *Leguminous* spp. is induced by common environmental cues, such as low nitrogen availability conditions, in the presence of the specific *Rhizobium* spp. in the rhizosphere. *Medicago sativa* (alfalfa) is an important nitrogen-fixing forage crop that is widely cultivated around the world and relied upon as a staple source of forage in livestock feed. Although alfalfa's relationship with these bacteria is one of the most efficient between rhizobia and legume plants, breeding for nitrogen-related traits in this crop has received little attention. In this report, we investigate the role of Squamosa-Promoter Binding Protein-Like 9 (SPL9), a target of miR156, in nodulation in alfalfa. Transgenic alfalfa plants with *SPL9*-silenced (*SPL9*-RNAi) and overexpressed (*35S::SPL9*) were compared to wild-type (WT) alfalfa for phenotypic changes in nodulation in the presence and absence of nitrogen. Phenotypic analyses showed that silencing of *MsSPL9* in alfalfa caused an increase in the number of nodules. Moreover, the characterization of phenotypic and molecular parameters revealed that *MsSPL9* regulates nodulation under a high concentration of nitrate (10 mM KNO_3) by regulating the transcription levels of the nitrate-responsive genes *Nitrate Reductase1* (*NR1*), *NR2*, *Nitrate transporter 2.5* (*NRT2.5*), and a shoot-controlled autoregulation of nodulation (AON) gene, *Super numeric nodules* (*SUNN*). While *MsSPL9*–overexpressing transgenic plants have dramatically increased transcript levels of *SUNN*, *NR1*, *NR2*, and *NRT2.5*, reducing *MsSPL9* caused downregulation of these genes and displayed a nitrogen-starved phenotype, as downregulation of the *MsSPL9* transcript levels caused a nitrate-tolerant nodulation phenotype. Taken together, our results suggest that *MsSPL9* regulates nodulation in alfalfa in response to nitrate.

Keywords: *Medicago sativa*; *SPL*; *SUNN*; nodulation; nitrate; miRNA

Citation: Nasrollahi, V.; Allam, G.; Kohalmi, S.E.; Hannoufa, A. *MsSPL9* Modulates Nodulation under Nitrate Sufficiency Condition in *Medicago sativa*. *Int. J. Mol. Sci.* **2023**, *24*, 9615. https://doi.org/10.3390/ijms24119615

Academic Editor: Martin Bartas

Received: 16 May 2023
Revised: 29 May 2023
Accepted: 30 May 2023
Published: 1 June 2023

1. Introduction

Unlike animals, the vast majority of plants have to acquire nitrogen, usually in the form of nitrates and ammonium, from the soil. Although nitrogen gas (N_2) is plentiful in the atmosphere, the biologically active forms of nitrogen are often so limited that they can constrain plant growth. For nodule-forming plants, however, the limitation of nitrogen fixation can be overcome to some extent by acquiring nitrogen from the rhizosphere [1]. While some species-specific factors may be involved, in general, the development of nitrogen-fixing root nodules is controlled by two parallel processes that are initiated by the host plant: first, nodule organogenesis, which is formed from the reinitiation of cell division in the root cortex [1,2], and second, rhizobia infecting the inside of the root hair cells that curl around rhizobia to entrap bacteria, which eventually grow and form infection threads (ITs) [1]. ITs are plant-derived conduits that are capable of crossing cell boundaries to direct rhizobia into the root cortex targets, the site of developing primordia [2,3]. Finally, the rhizobia are released from the ITs into the inner cells in the nodule while remaining encapsulated within a plant membrane. In these organelle-like structures, called symbiosomes, rhizobia are responsible for the reduction of atmospheric di-nitrogen to ammonia by expressing the nitrogenase enzyme [4].

The symbiotic nitrogen fixation of legumes takes place in nodules [2], and signal exchanges for bacterial entry to the nodules take place between host plants and rhizobia [5]. Plant roots release phenolic compounds which attract bacteria to the rhizosphere and subsequently stimulate the secretion of lipo-chito-oligosaccharides, known as nod factors (NF) [1,5]. Recognition of NFs by receptor-like kinases such as nod factor perception (NFP) leads to the induction of a signaling pathway that activates a leucine-rich repeats receptor-like kinase, known as Does not Make Infections2 (DMI2) in *Medicago truncatula* [6]. Secondary signals initiate calcium oscillation in the nuclear region, a process known as calcium spiking [7]. Activation of this signaling pathway requires three components of the nuclear pore—NUP85, NUP133, and NENA [8–10]—and the cation channels located on the nuclear envelope, encoded by a single inner-membrane-localized channel, DMI1 [11,12]. Perception of the calcium spiking signature is decoded by a nuclear calcium/calmodulin-dependent protein kinase, DMI3. DMI3 interacts with and subsequently phosphorylates the Interacting Protein of DMI3 (IPD3) [13,14]. DMI3 interacts with the nuclear protein IPD3 and other downstream components, such as two GRAS family proteins—Nodulation Signaling Pathway1 (NSP1) and NSP2—to activate expression of Nodule Inception (NIN), its downstream genes that encode Nuclear Factor YA1 (NF-YA1)/YA2, and ERF Required for Nodulation (ERN2), which are essential for rhizobium infection and nodule organogenesis [15–20].

Forming and maintaining nodules is an energy-demanding process, and consequently, excessive nodulation (super-nodulation) can negatively affect plant growth and development [21]. The host plant, therefore, tightly regulates the total root nodule number depending on the metabolic status of the shoot (carbon source) and root (nitrogen source) [22]. To that end, legumes have evolved a negative regulatory pathway called autoregulation of nodulation (AON) that functions systemically through the shoot to maintain an optimal number of nodules [23–25]. The nitrogen regulation pathway is activated in root cortical cells during rhizobial infection and nodule development to inhibit nodulation under nitrogen-rich conditions, helping the plant to conserve energy resources [25,26]. Following the initial rhizobial infection events, root-derived nodulation-specific Clavata3/Embryo surrounding region (CLE) peptides, including CLE12 and CLE13 in *M. truncatula* [27], CLE Root Signal1 (CLE-RS1) and CLE-RS2 in *Lotus japonicus*, or Rizobia-induced CLE1 (RIC1) and RIC2 in soybean (*Glycine max*) [28,29], are triggered to activate AON. Following processing, these small functional CLE peptides translocate from the root to the shoot through the xylem [30], where they bind to a specific homodimeric or heterodimeric receptor complex that includes Hypernodulation Aberrant Rant Root Formation1 (HAR1) in *L. japonicus* [30–32], *SUNN* in *M. truncatula* [33], or Nodule Autoregulation Receptor Kinase (NARK) in soybean [34]. In *L. japonicus*, *Lj*CLE-RS2 binds to *Lj*HAR1, and the application of *Lj*CLE-RS2 peptide through the xylem was found to inhibit nodulation in wild-type but not in *har1* mutants, showing that the *Lj*HAR1 receptor kinase is required for regulating the AON pathway through the *Lj*CLE peptide [30].

Nodules are induced by common environmental cues such as low nitrogen availability conditions in the presence of the specific *Rhizobium* spp. in the rhizosphere [25]. In legumes, nitrogen is utilized through assimilation regardless of whether it enters the plant as nitrate and ammonium from soil or by fixation of atmospheric nitrogen [35]. Nitrate is absorbed by the root from the external environment using two nitrate transporters, Nitrate Transporter1 (NRT1) and NRT2, which function as low-affinity and high-affinity nitrate transporters, respectively [36]. The nitrate imported into the cells is sequentially reduced into nitrite by Nitrate Reductase (NR) and into ammonium by Nitrite Reductase (NiR) [37]. Ammonium is assimilated into amino acids through the glutamine synthase (GS) and glutamine oxoglutarate aminotransferase (GOGAT) cycle [38]. High levels of nitrogen in soil reduce nodulation and inhibit nitrogen fixation in mature nodules after the addition of nitrogen fertilizers [39]. This regulation of nodulation by nitrate is a part of the AON signaling pathway [40,41]. In *M. truncatula,* Lagunas et al. [42] showed that the AON signaling pathway regulates nitrogen uptake and metabolism even when plants are not nodulating, suggesting that SUNN is involved in controlling nitrogen mobilization even in the absence of rhizobia.

The role of miRNAs in nitrate-regulated root architecture has been reported in the literature. For example, miR167 and its target Auxin Response Factor 8 (ARF8) control lateral root growth in response to nitrate in *Arabidopsis* [43,44]. In addition, miR172 positively regulates nodulation in legumes, as shown in soybean, where overexpression of miR172 resulted in plants with increased nodule number and nitrogen fixation [45]. Nova-Franco et al. [46] also showed similar results in the common bean (*Phaseolus vulgaris*). The miR2111/TML module is also involved in regulating nodulation in legumes, as overexpression of miR2111 or mutations in *TML* caused hypernodulation in *L. japonicus* [47]. In soybean, Yan et al. [48] showed that miR2606b and miR4416 affect nodulation, whereby miR2606b overexpression in roots increased nodule numbers, but overexpression of miR4416 decreased them [48]. Furthermore, De Luis et al. [49] reported that miR171 is involved in the early stages of nodulation (i.e., bacterial infection) in *L. japonicus*. When the expression of miR171 was examined in wild type (WT) and mutant *spontaneous nodule formation1* (*snf1*) and *snf2 L. japonicus* plants, it showed an increase in *Mesorhizobium loti*-inoculated *snf1* and *snf2* mutants compared to WT [49]. It was also found that miR171 targets the *NSP2*, which is an important transcription factor in the nodulation signaling pathway in *M. truncatula* [50]. In addition, Yan et al. [51] showed that miR393 negatively regulates nodule formation in soybean by targeting *Early nodulin 93* (*ENOD93*). Moreover, overexpression of miR319 in *M. truncatula* and common bean resulted in a reduction in the number of nodules in transgenic roots [52,53].

Our previous study found that overexpression of miR156 increased the number of root nodules in alfalfa [54]. Most recently, we showed that miR156-targeted *MsSPL12* has a negative effect on nodulation, as down-regulation of *MsSPL12* increased the number of nodules and nitrogen fixation in alfalfa [55]. Furthermore, it has been shown that there is a regulatory relationship between *MsSPL12* and *CLE13* and that *MsSPL12* is involved in the AON symbiotic process in alfalfa [55]. Additionally, Yun et al. [56] reported that the miR156-SPL9 regulatory system in soybean acts as an upstream master regulator of nodulation by targeting and regulating the transcript levels of nodulation genes in this plant. *Gm*SPL9 is a positive regulator of soybean nodulation which directly targets the nodulation master regulator gene, *GmNINa*, and the nodulation marker gene, *GmENOD40*, during nodule formation and development [56]. *Ms*SPL9 in alfalfa has been investigated for its role in drought response in alfalfa, where its downregulation led to improved tolerance to this stress [57]. These findings highlight the role that the miR156/SPL regulatory system plays in nodulation in legume plants.

In the current study, we conducted an investigation into the role of *MsSPL9* in nodulation in alfalfa using overexpression (OE) (*35S::SPL9*) and RNAi-silenced *SPL9* (RNAi-*SPL9*) alfalfa plants. We also investigated the role of *Ms*SPL9 in nodulation in response to high concentrations of nitrate.

2. Results

2.1. Silencing of MsSPL9 Enhances Nodulation

Overexpression of miR156 was reported earlier to increase root length and enhance nodulation in alfalfa [54]. As *MsSPL9* was one of the genes that are targeted for silencing by miR156 in this plant, we decided to investigate whether root-related traits are regulated by miR156 through *MsSPL9* silencing. For that purpose, we studied the root phenotypes in WT, *SPL9*-RNAi (R1, R2, and R3), and *35S::SPL9* (OE-1, OE-2, and OE-3) genotypes (Figure 1). To determine the ability of *SPL9*-RNAi and *35S::SPL9* transgenic plants to form symbiotic nodules, three-week-old (three weeks post-cutting) rooted plants were inoculated with *Sinorhizobium meliloti* for a period of 14 or 21 days after inoculation (dai). At 14 dai, *SPL9*-RNAi plants showed an increase in nodulation of 2.5-, 2.9-, and 2.4-fold in R3, R2, and R1, respectively, compared to WT (Figure 1a,b). The total nodule number also was increased in *SPL9*-RNAi genotypes compared to WT at 21 dai, showing 1.4-, 1.6-, and 1.5-fold in R3, R2, and R1, respectively, compared to WT (Figure 1b). Although no significant differences in nodule numbers were observed between *SPL9*-RNAi genotypes at 14 dai and 21 dai, nodulation was increased in WT at 21 dai compared to 14 dai (Figure 1b).

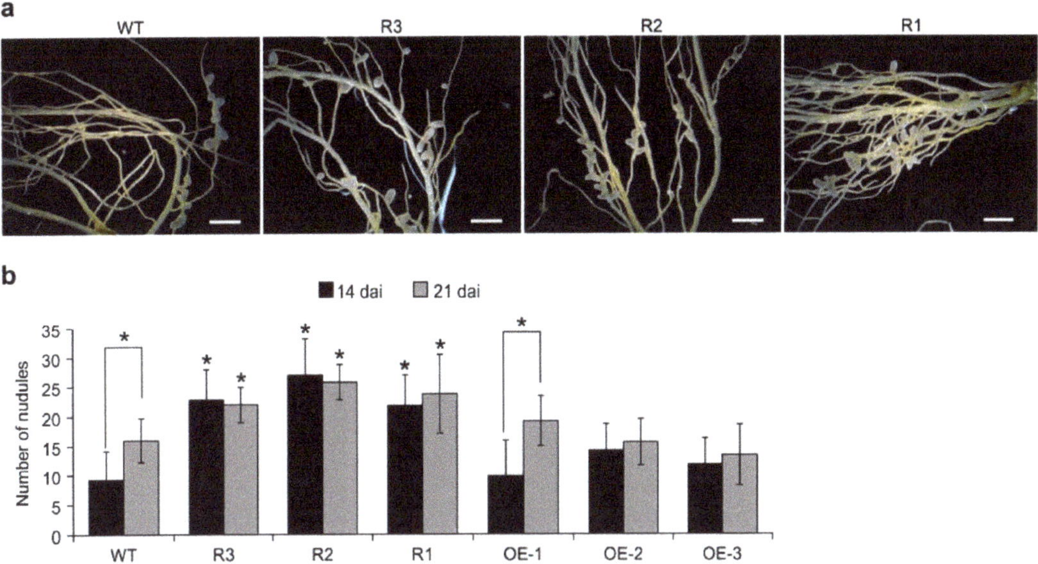

Figure 1. The effect of *MsSPL9* silencing on nodulation. (**a**) Nodule phenotypes of WT and the *SPL9*-RNAi genotypes at 14 dai. Scale bar, 1 mm. (**b**) The number of nodules in WT, *SPL9*-RNAi, and *35S::SPL9* at 14 dai and 21 dai (*n* = 10–14). * Indicates significant differences within time points (14 dai and 21 dai) between WT and *SPL9*-RNAi or *35S::SPL9* plants, and bars indicate significant differences between the two time points using Student's *t*-test *p* < 0.05. Error bars indicate standard deviation.

To determine the ability of *35S::SPL9* transgenic plants to form symbiotic nodules, three weeks after cutting, the rooted transgenic and WT plants were inoculated with *S. meliloti* for 14 and 21 days. No significant differences in nodule numbers were observed between *35S::SPL9* genotypes and WT at 14 dai nor 21 dai (Figure 1b). However, among the *35S::SPL9* genotypes, the total nodule number was significantly increased in OE-1 at 21 dai compared with this genotype at 14 dai (Figure 1b).

These results suggest that the transcript levels of *MsSPL9* are negatively correlated to nodulation in alfalfa.

2.2. MsSPL9 Silencing Affects Nodulation-Related Genes

Given the above finding that *SPL9*-RNAi alfalfa plants have enhanced nodulation in the symbiotic relationship with *S. meliloti*, we examined transcript levels of several nodulation-related genes at 14 dai in inoculated roots of alfalfa (Figure 2). These genes include *NIN* [17], *CRE1* [58], *IPD3* [13], *DELLA* [59], *DMI1* [11], *DMI2* [6], and *DMI3* [13].

Of the tested genes, *NIN*, *CRE1*, *IPD3* and *DELLA* were significantly upregulated in all the *SPL9*-RNAi genotypes (R1, R2 and R3) at 14 dai (Figure 2a–d). While *DMI1* and *DMI3* were upregulated in only one of the *SPL9*-RNAi genotypes (R3) (Figure 2e,g), significant effects of *MsSPL9* silencing on *DMI2* transcript levels were observed in roots of two *SPL9*-RNAi genotypes, R3 and R1, at 14 dai (Figure 2f). These findings suggest the involvement of *MsSPL9* in nodulation in alfalfa-*S. meliloti* symbiosis.

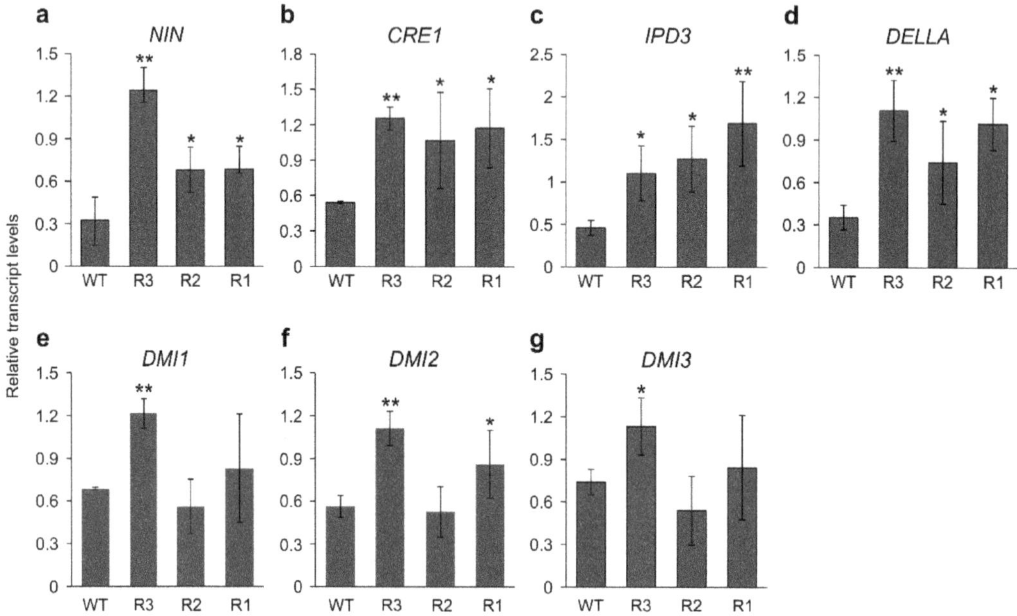

Figure 2. Transcript analysis of nodulation-related genes in *SPL9*-RNAi genotypes. Relative transcript levels of (**a**) *NIN*, (**b**) *CRE1*, (**c**) *IPD3*, (**d**) *DELLA*, (**e**) *DMI1*, (**f**) *DMI2*, and (**g**) *DMI3* in WT and *SPL9*-RNAi roots at 14 dai. Transcript levels are shown relative to WT after being normalized to ADF1 and *elF4A* reference genes. *, ** Indicate significant differences relative to WT using Student's *t*-test ($n = 3$) $p < 0.05$, $p < 0.01$, respectively. Error bar indicates standard deviation.

2.3. MsSPL9 Silencing Attenuates the Effect of Nitrate on Nodulation

Nitrogen abundance in the soil inhibits nodulation, and this regulatory process is a part of the AON pathway [41,60]. Given the effects of *Ms*SPL9 on nodulation, we assessed whether the nodulation capacity of *S. meliloti*-inoculated *SPL9*-RNAi transgenic plants, R1, R2, and R3, was affected by nitrate treatment. The number of nodules was compared between WT and *SPL9*-RNAi plants treated with 10 mM KNO_3 or KCl at 21 dai. Under watering with 10 mM KCl, *SPL9*-RNAi transgenic plants produced significantly more nodules compared to WT. When watered with 10 mM KNO_3, WT plants showed a reduction in the nodule number compared to when treated with KCl (Figure 3a), but transgenic *SPL9*-RNAi plants maintained nodulation under both treatments. In fact, nodule numbers were not noticeably affected by KNO_3 treatment in these transgenic plants (Figure 3a). These results indicate that silencing of *MsSPL9* prevents nitrate inhibition of nodulation in alfalfa.

To further investigate the role of *MsSPL9* in nodulation under nitrate treatment, we analyzed the nodulation phenotype of transgenic alfalfa plants overexpressing *MsSPL9* in the presence of either 10 mM KCl or KNO_3 (Figure 3b). The number of nodules was only significantly decreased in one of the *35S::SPL9* transgenic plants, OE-1, which expresses the highest level of *MsSPL9* compared to WT, in both conditions (watering with KCl or KNO_3) (Figure 3b). In fact, there was no significant difference in the nodule numbers of OE-2 and OE-3 compared to WT. However, when plants were watered with 10 mM KNO_3, WT and all *35S::SPL9* transgenic plants produced significantly fewer nodules than plants watered with 10 mM KCl (Figure 3b). This result suggests that overexpression of *MsSPL9* causes hypersensitivity to nitrate inhibition of nodulation in alfalfa.

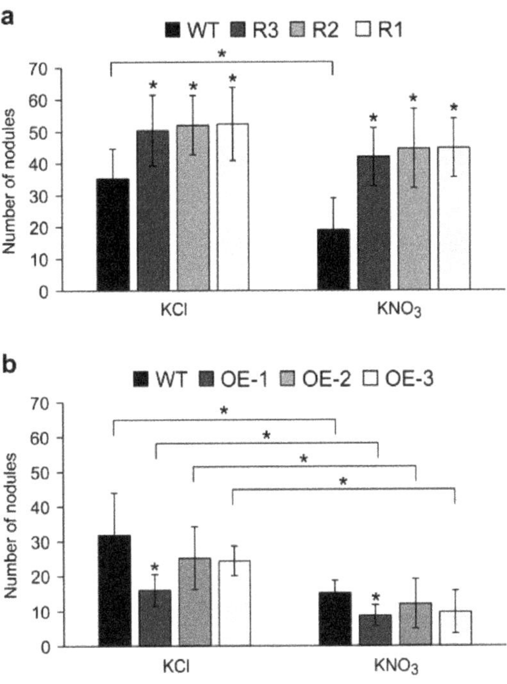

Figure 3. Effect of 10 mM KNO_3 on nodulation phenotype in *SPL9*-RNAi and *35S::SPL9* genotypes. The average numbers of nodules at 21 dai in (**a**) WT and *SPL9*-RNAi, and (**b**) WT and *35S::SPL9* genotypes (n = 10–14). * Indicates significant differences within conditions between WT and *SPL9*-RNAi or *35S::SPL9* plants, and bars indicate significant differences between conditions using Student's *t*-test $p < 0.05$. Error bars indicate standard deviation.

2.4. Differential Gene Expression in SPL9-RNAi and 35S::SPL9 Genotypes

To shed light on the molecular events associated with *Ms*SPL9 function in nodulation in response to nitrate, we investigated the transcript levels of three nitrate-responsive genes: *NR1*, *NR2*, and a high-affinity *NRT2.5* in inoculated roots of alfalfa plants with altered expression of *MsSPL9* (*SPL9*-RNAi and *35S::SPL9*) and WT. Overexpression of *MsSPL9* resulted in higher transcript levels of *NR1*, *NR2*, and *NRT2.5*, except for OE-2, which showed no significant changes for *NR2* compared to WT (Figure 4a–c).

By contrast, reduced *MsSPL9* in the *SPL9*-RNAi transgenic plants caused lower transcript levels of *NR1* and *NR2* (Figure 4a,b), except for one *SPL9*-RNAi plant, R2, that did not show any significant change for these two genes compared to WT. *NRT2.5* transcript levels were lower in all *SPL9*-RNAi plants (Figure 4c). We also analyzed the transcript levels of a shoot-controlled AON gene, *SUNN*, in *SPL9*-RNAi transgenic plants. In *M. truncatula*, *SUNN* is involved in control of nitrogen mobilization, as control plants appear to be able to transport more nitrogen to the shoot, compared with *sunn-1* loss of function mutant plants [42]. Our results showed that silencing *MsSPL9* was accompanied by lower transcript levels of *SUNN* (Figure 4d). We also detected the expression of this AON gene in *35S::SPL9*. Expression levels of *SUNN* were markedly higher in the *MsSPL9* overexpressing transgenic plants than in WT (Figure 4d). These results are consistent with the increased number of nodules in *SPL9*-RNAi compared to those of *35S::SPL9* and WT. Since SUNN is an important component in the AON signaling pathway, these results suggest that *MsSPL9* is involved in regulating AON in alfalfa.

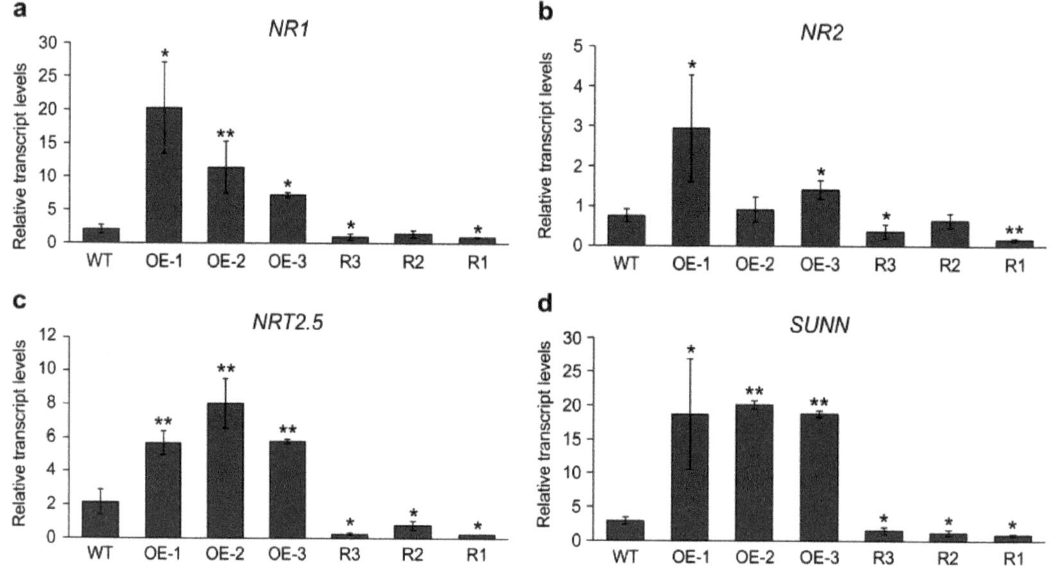

Figure 4. Transcript analysis of nitrate signaling pathway genes in *SPL9*-RNAi and *35S::SPL9* genotypes. Relative transcript levels of (**a**) *NR1*, (**b**) *NR2*, (**c**) *NRT2.5*, and (**d**) *SUNN* in *SPL9*-RNAi and *35S::SPL9* at 14 dai. Transcript levels are shown relative to WT after being normalized to ADF1 and *elF4A* reference genes. *, ** Indicate significant differences relative to WT using Student's *t*-test ($n = 3$) $p < 0.05$, $p < 0.01$, respectively. Error bar indicates standard deviation.

2.5. Root Development Is Balanced Differently in WT and SPL9-RNAi

To gain an insight into the function of *Ms*SPL9 in root architecture and nodulation in response to nitrate, we analyzed root system architecture (RSA) in *SPL9*-RNAi and WT alfalfa plants. The transgenic and WT plants were grown in the absence of nitrate and treated with either *S. meliloti* ("rhizobia") or mock for 14 days. The plants were then treated with either 10 mM KNO_3 or sterile water for 14 more days to study the individual and combinatorial effects of these treatments on RSA. We found that between the four conditions (Mock, Rhizobia, Mock-KNO_3, and Rhizobia-KNO_3), there was no significant difference in primary root (PR) and lateral root (LR) length in WT plants (Figures 5a and S1a,b), although these traits were significantly different in *SPL9*-RNAi plants in response to nitrate and inoculation with rhizobia (Figures 5a and S1a,b).

In the presence of KNO_3, *SPL9*-RNAi plants had significantly longer PR than WT, either with or without rhizobia inoculation (Figure 5b), while all plants had similar PR lengths in the absence of nitrate. Inoculated *SPL9*-RNAi did not show any significant change in LR length compared to WT in the absence of nitrate, but they showed longer LR compared to WT in the other three conditions (Figure 5c). In addition, LRs were shorter in inoculated R2 and R3 in the absence of KNO_3 than when it was present (Figure S1b).

When *SPL9*-RNAi plants were treated with KNO_3, they showed significantly more lateral roots than WT regardless of inoculation status (Figure 5d). In the absence of KNO_3, all plants had similar LR numbers, except for R2 in mock and R1 in rhizobia-inoculated conditions that showed an increase in the number of LRs compared to WT (Figure 5d). Furthermore, the number of PRs was significantly higher in *SPL9*-RNAi plants in all the conditions, except for mock plants growing in the absence of KNO_3; in fact, only R3 from this group showed an increase in the number of PRs compared to those for WT (Figure 5e).

These results suggest the involvement of *Ms*SPL9 in the regulation of root architecture in response to nitrate and inoculation with rhizobia.

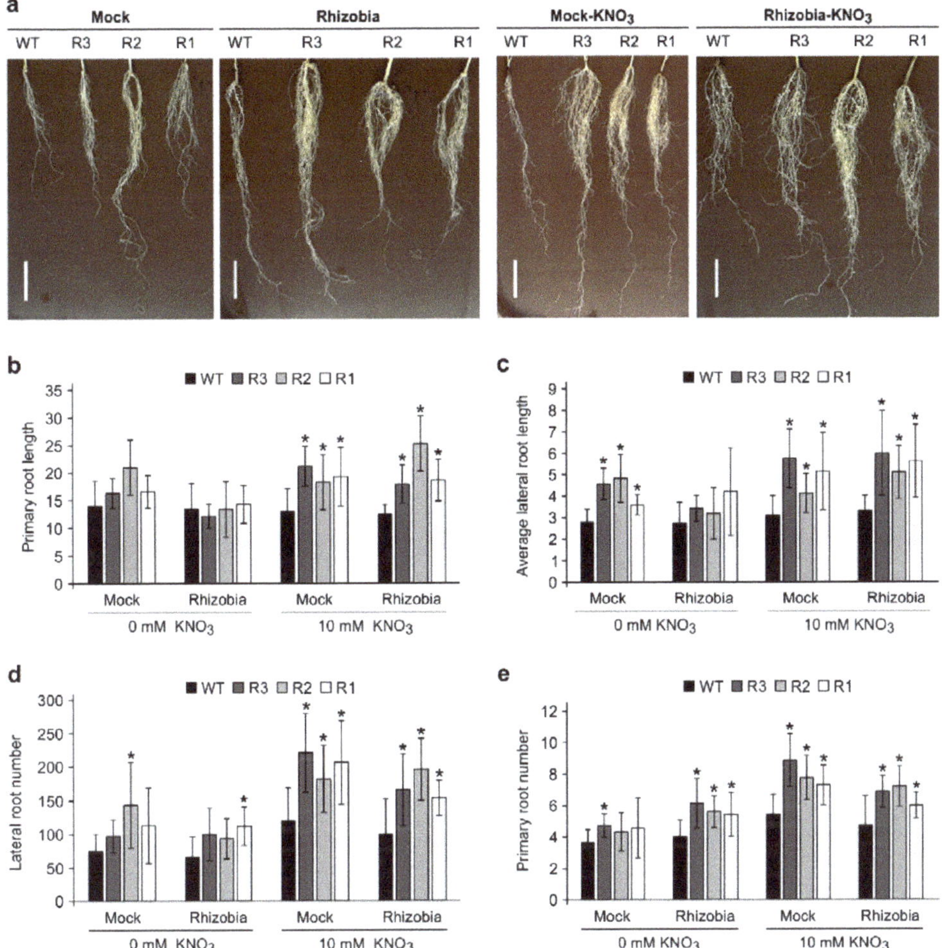

Figure 5. WT and *SPL9*-RNAi root system architecture changes after KNO_3 treatment, with effects dependent on rhizobia-inoculation status. (**a**) Root architecture phenotype of inoculated WT and the *SPL9*-RNAi genotypes with *S. meliloti* at 21 dai growing in nitrate-starved substrate or watered with 10 mM KNO_3. Scale bar, 4 cm. (**b**) Primary root length, (**c**) average lateral root length, (**d**) numbers of the lateral root, and (**e**) numbers of the primary root. * Indicates significant differences relative to WT using Student's *t*-test ($n = 10$) $p < 0.05$. Error bar indicates standard deviation.

2.6. Plant Biomass in SPL9-RNAi under KNO₃ Treatment Are Affected by Inoculation Status

Based on the differential expression of some nitrate signaling genes in the *SPL9*-RNAi plants (Figure 4), we decided to determine whether this factor also had an effect on whole plant biomass. We measured the root and shoot fresh and dry weight of WT and *SPL9*-RNAi plants grown in vermiculite. We inoculated plants with *S. meliloti*, and 14 days later, the plants were treated with either 10 mM KNO_3 or sterile water for 14 more days to study the individual and combinatorial effects of treatments on root and shoot biomass. Increased root fresh weight was observed in KNO_3-treated *SPL9*-RNAi plants in addition to the untreated R2 plants compared to WT in each treatment (Figure 6a).

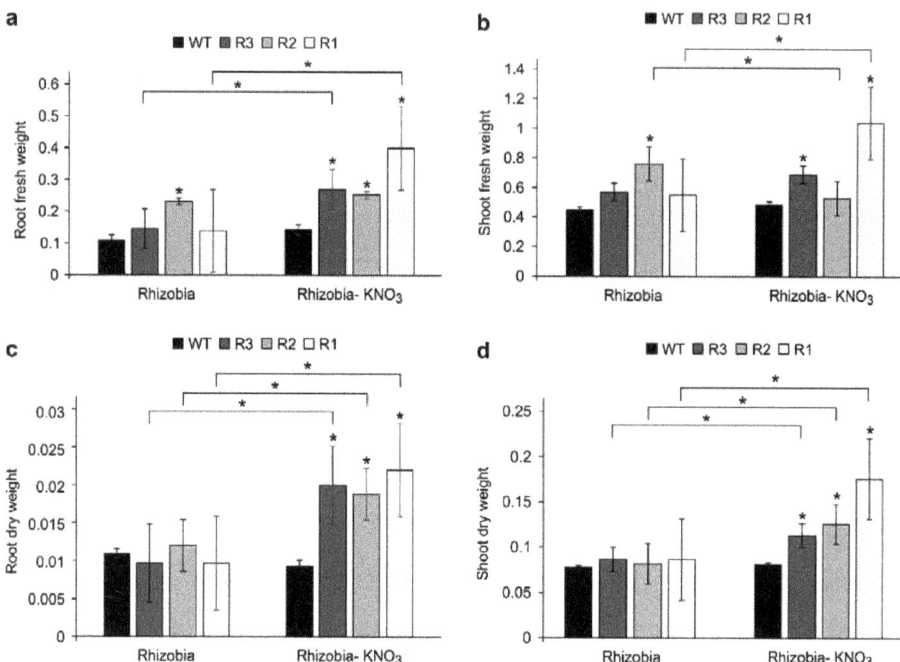

Figure 6. WT and *SPL9*-RNAi root and shoot weight changes after KNO_3 treatment. (**a**) Root fresh weight (RFW), (**b**) shoot fresh weight (SFW), (**c**) root dry weight (RDW), and (**d**) shoot dry weight (SDW). * Indicates significant differences within conditions between WT and *SPL9*-RNAi plants, and bars indicate significant differences between conditions using Student's *t*-test ($n = 10$) $p < 0.05$. Error bar indicates standard deviation.

Differences in root fresh weight were not observed between treatments in WT and R2, while R1 and R3 root fresh biomass were increased in response to KNO_3 compared to untreated plants (Figure 6a). WT and transgenic plants were also distinguishable between and within conditions when examining shoot fresh biomass. In the absence of KNO_3, there were no significant changes between *SPL9*-RNAi plants and WT in shoot fresh weight, except for R2, which showed an increase (Figure 6b). In the presence of KNO_3, two of the *SPL9*-RNAi plants, R1 and R3, showed an increase in shoot fresh weight compared to WT. When comparing the shoot fresh weight between the treatments, only R2 and R3 showed an increase in response to KNO_3 (Figure 6b).

We also measured root and shoot dry biomass in the KNO_3-treated and untreated plants. In the absence of KNO_3, WT and *SPL9*-RNAi plants were indistinguishable when examining root and shoot dry biomass (Figure 6c,d). However, in the presence of KNO_3, root and shoot dry weight were significantly increased in all the *SPL9*-RNAi plants compared to WT (Figure 6c,d). When comparing the root and shoot dry biomass between the two conditions (KNO_3-treated and untreated plants), only WT did not show any changes, while the dry biomass was significantly increased in *SPL9*-RNAi plants in response to KNO_3 (Figure 6c,d).

3. Discussion

The role of miR156 in plant growth and development has been well documented [61–63]. Previous research has shown that overexpression of *miR156* in alfalfa (miR156-OE) resulted in increased nodulation, improved nitrogen fixation, and enhanced root regenerative capacity during vegetative propagation [64]. Whereas the role of the miR156-regulated

MsSPL9 in drought stress response has been well characterized in alfalfa [57], no studies have been conducted on the possible role of miR156/SPL9 module in nodule development in alfalfa. In the present study, we analyzed transgenic alfalfa plants with altered transcript levels of *MsSPL9*, including *SPL9*-RNAi and *35S::SPL9*, to investigate the role of *MsSPL9* in root architecture.

As *MsSPL9* is one of the genes regulated by miR156 in alfalfa, we decided to investigate its involvement in nodulation. Formation of root nodules in association with rhizobia in leguminous plants is a complex process that governs the mutually beneficial relationship between the plants and their compatible rhizobia and includes the downstream components of signaling pathways that trigger changes in gene expression in both partners [1,65,66]. The signals that provide bacterial access to the plant, leading eventually to nodule organogenesis, have been well studied in legume species [65,66]. miR156/SPL was shown to play a role in nodulation in legume plants, including alfalfa, where overexpression of miR156 increased the number of root nodules [54]. Most recently, we showed that miR156-targeted *MsSPL12* and its downstream target *AGL6* are involved in the regulation of nodulation in this plant [55]. In fact, *MsSPL12* plays a negative role in nodulation in alfalfa, as its down-regulation was concomitant with changes in gene expression in both partners, alfalfa roots and *S. meliloti* [55], resulting in enhanced nodulation.

In the current study, we showed that *MsSPL9* has a negative effect on nodulation in alfalfa. While overexpression of *MsSPL9* resulted in no evident change in nodulation, silencing of this gene (*SPL9*-RNAi) increased nodulation in these plants. Based on these results, miR156/SPL was shown to play a role in nodulation in legume plants. However, the role of miR156/SPL9 in nodulation may be species-specific, as an increase in nodulation was also reported in the other study for *GmSPL9* overexpression in soybean plants. Yun et al. [56] reported that *Gm*SPL9 in soybean, which is phylogenetically close to *MsSPL9* in alfalfa [67], acts as an upstream positive regulator of nodulation by targeting and regulating the expression of nodulation genes. As such, overexpression of *GmSPL9* in soybean reduced nodule numbers [56]. *GmSPL9* regulates nodulation in soybean by directly binding to the promoter of miR172c and activating its expression [56]. *Gm*SPL9 also directly targets *GmNINa* and *GmENOD40*, which are the nodulation master regulators and nodulation marker genes, respectively, during nodule formation and development [56]. In addition to the potential species specificity of SPL9 function in nodulation, the discrepancy could be explained by the dose-dependent effects of SPLs. Hanly et al. [57] showed that only two of the *SPL9*-RNAi alfalfa plants (R2 and R3), the genotypes with the lowest *MsSPL9* transcript abundance, improved tolerance to drought. Similar findings were reported by Feyissa et al. [68] for *MsSPL13*, where only *SPL13*-RNAi alfalfa plants with decreased *MsSPL13* transcript levels but over a certain threshold showed significant drought tolerance.

Intriguingly, we also found that *MsSPL9* regulates the transcription levels of a shoot-controlled AON gene, *SUNN*, in alfalfa. We found that overexpression and silencing of *MsSPL9* resulted in up- and down–regulation of *SUNN*, respectively (Figure 4d). These results suggest a link between the increased and decreased number of nodules in *SPL9*-RNAi and *35S::SPL9*, respectively, compared to WT, and the expression level of *SUNN* in these plants. Moreover, because SUNN is an important component in the AON signaling pathway, these results also suggest that *MsSPL9* is involved in regulating the AON pathway in alfalfa and that *MsSPL9* may function upstream of SUNN to regulate alfalfa nodulation. It is noteworthy that *NIN*, *CRE1*, *IPD3*, and *DELLA* transcript levels were significantly higher in all the *SPL9*-RNAi genotypes, which supports the hypothesis of their possible regulation by *MsSPL9* acting as an essential regulator of nodule organogenesis in alfalfa. However, further experiments are required to identify the downstream targets of *MsSPL9* in roots.

Legume plants can fix atmospheric N_2 into ammonia and assimilate inorganic nitrogen sources due to the symbiotic relationship between rhizobia and root nodules. However, to conserve energy, plants inhibit nodulation under conditions of nitrate abundance in the rhizosphere, resulting in decreased nodule number, nodule mass, nitrogenase activity, and accelerated nodule senescence [60]. Repressing nodulation, sufficient nitrogen status is part of the AON signaling pathway [40,41]. As the *SPL9*-RNAi plants showed an increase in nodulation, we tested the relationship between nitrate and the miR156/SPL9 regulatory system. We found that while silencing *MsSPL9* has enhanced tolerance to nitrate during nodulation, overexpression of *MsSPL9* causes hypersensitivity to nitrate inhibition of nodulation. In fact, under nitrate-sufficient conditions, nodule numbers were not noticeably affected by nitrate treatment in rhizobia-inoculated roots of *SPL9*-RNAi plants. On the other hand, WT and all *35S::SPL9* transgenic plants produced significantly fewer nodules under nitrate-sufficient conditions than plants grown under control conditions. In our previous work, we showed that *SPL12*-RNAi plants developed more active nodules relative to WT under nitrate-sufficient conditions, demonstrating the role of miR156/SPL12 modules in controlling symbiotic nodulation in alfalfa [69]. Given that *MsSPL9* and *MsSPL12* affect nodulation under nitrate treatment during nodule development [69], it is conceivable that miR156 dynamically controls the process of nodulation by regulating *MsSPL* transcription factors in alfalfa. The observations that nodulation was regulated by miR156 overexpression in alfalfa [54], *L. japonicus* [62], and soybean [56] support the notion that the central regulatory role of *miR156*-SPL modules on symbiotic nodulation may be conserved in leguminous plants, such as alfalfa, *L. japonicus*, and soybean. In *Arabidopsis*, *At*SPL9 acts as a potential nitrate regulatory hub and *AtSPL9* expression is affected by nitrate. In addition, the transcript levels of *AtNRT1.1*, *AtNIA2*, and *AtNiR* significantly increased in response to nitrate in *AtSPL9* overexpressed *Arabidopsis* plants [70]. Furthermore, miR172c was shown to act as a signaling component of the nitrate-dependent AON in common bean, where it decreases the sensitivity of nodulation inhibition by nitrate [46]. Common bean plants overexpressing miR172 showed an increase in active nodules in the presence of nitrate [46]. In tomato (*Solanum lycopersicum*), it was reported that an *SlSPL* transcription factor, LeSPL-CNR, directly binds to the promoter of *SlNIA*, resulting in repressing its expression [71]. LeSPL-CNR was further shown to negatively regulate *SlNIA* transcription levels in response to cadmium (cd) stress [71]. In *M. truncatula*, *NIN-like protein 1* (*nlp1*) mutants had dramatically reduced induction of *NiR1* and *NRT2.1* and displayed nitrogen-starved phenotypes, including reduced nodule formation and nitrogen fixation in response to nitrate [40]. In the current study, we showed that *MsSPL9* overexpressing transgenic plants have dramatically reduced transcript levels of the nitrate-responsive genes *NR1*, *NR2*, and *NRT2.5*. By contrast, reduced *MsSPL9* in the *SPL9*-RNAi transgenic plants caused downregulation of *NR1* and *NR2*, and *NRT2.5* and displayed a nitrogen-starved phenotype, as downregulation of the *MsSPL9* expression caused a nitrate-tolerant nodulation phenotype. In fact, nodulation did not decrease in *SPL9*-RNAi plants under nitrate-sufficient conditions.

Moreover, we found *MsSPL9* to be involved in the regulation of root architecture in response to nitrate and inoculation with rhizobia. *SPL9*-RNAi plants had increased number and length of PR and LR in response to high concentrations (10 mM) of nitrate. This increase could be explained by the regulation of the *SUNN* gene (an important component in the AON signaling pathway) by *MsSPL9*, as described earlier (Figure 4d). This gene could be key in the regulation of root architecture in response to the internal and external nitrogen signals to mount an appropriate developmental response. In *M. truncatula*, *sunn-1* mutants had significantly more LRs and greater lateral root density than WT plants when treated with 5 mM NH_4NO_3 [42]. In addition, *sunn-1* mutants showed shorter LRs compared to WT plants when inoculated with rhizobia [42]. The *har1* mutant in *L. japonicus*, also had higher LR density and a shorter root system in the absence of rhizobia [72]. Lagunas et al. [42] showed that in *M. truncatula* the AON signaling pathway regulates nitrogen uptake and metabolism, as many nitrogen transporter genes were found to respond to nitrogen in *sunn-1* mutants but were not repressed by nitrogen in WT plants. It has also been shown that the dry weight in *sunn-1* rhizobia-inoculated plants increased more quickly than in WT [42]. In the current study, we showed that root and shoot dry weight were significantly increased in all *SPL9*-RNAi plants compared to WT in the presence of nitrate (Figure 6c,d).

Based on our results and the results from previous studies, it could be argued that SUNN is involved in the control of nitrogen mobilization and that *Ms*SPL9 may function upstream of SUNN.

4. Materials and Methods

4.1. Plant Material and Growth Conditions

Medicago sativa L. (alfalfa) clone N4.4.2 [73] was obtained from Daniel Brown (Agriculture and Agri-Food Canada, London, ON, Canada) and was used as a wild-type (WT) genotype. Alfalfa genotypes with reduced expression levels of *MsSPL9*, *SPL9-RNAi* (R1, R2, and R3) and plants overexpressing *SPL9*, *35S::SPL9* (OE-1, OE-2, and OE-3) were obtained from our previous study [57]. WT and transgenic alfalfa plants were grown under greenhouse conditions at 21–23 °C, 16-h light/8-h dark per day, a light intensity of 380–450 W/m^2 (approximately 500 W/m^2 at high noon time), and a relative humidity of 56% for the duration of all experiments. Because of the obligate outcrossing nature of alfalfa, WT and transgenic alfalfa were propagated by rooted stem cuttings to maintain the genotype throughout the study. Stem-cutting propagation and morphological characterization of alfalfa plants were carried out as described in [54].

4.2. Determination of Nodule Numbers

To determine the number of nodules, plants were examined at 14 and 21 days after inoculation (dai) with *Sinorhizobium meliloti* Sm1021. To eliminate potential microbial contamination, equipment was surface-sterilized using 1% sodium hypochlorite, while vermiculite and water were sterilized by autoclaving for 1 h. *S. meliloti* Sm1021 strain was cultured on a Yeast Extract Broth agar [74] for two days at 28 °C. A single colony was then inoculated in liquid TY medium and incubated at 28 °C to an optical density OD$_{600}$ nm of 1.5. The 3-week-old rooted stems were inoculated by applying 5 mL of the bacterial suspension or sterilized water (non-inoculated control) into each pot containing rooted alfalfa stem. The plants were then kept on a bench in the greenhouse and watered with distilled water once a week. Two and three weeks after inoculation with *S. meliloti*, the roots were removed from the soil and the total number of nodules from each stem was counted. At least 10 biological replicates per genotype were used, and the experiment was repeated three times.

4.3. Nitrate Treatment

To explore if *MsSPL9*-related regulation of nodulation is affected by nitrate, the nodulation test was performed upon treatment with this nutrient. WT and *SPL9-RNAi* alfalfa stem cuttings were grown on vermiculite for 21 days and were then inoculated with *S. meliloti* Sm1021 for 14 days. The 14-day-old inoculated transgenic and WT plants were watered with 10 mM potassium nitrate (KNO$_3$) or potassium chloride (KCl) twice a week for three weeks. Effects on nodulation were studied by counting the number of nodules. Other phenotypic characterizations, including fresh and dry weight of root and shoot, number of main and lateral roots, length of primary roots, and length of lateral roots on the longest root were also performed and from this the average LR length was calculated. These phenotypic characterizations were performed on nitrate-treated and non-treated WT and *SPL9*-RNAi alfalfa with approximately 10 biological replicates of each genotype. Aboveground tissue was determined by decapitating the plant approximately above the media line, and any tissue below this point was considered roots. Tissue fresh weight (FW) was measured at the time of harvest, and dry weight (DW) was determined after the tissue was baked at 65 °C for 5 days. Root length was considered the length from the top of the root crown to the tip of the longest root. The roots directly emerging from the stem were considered main roots, while those that emerged from the main roots were counted as lateral roots. The entire experiment was repeated twice under the same growth and nitrate treatment conditions to test the reproducibility of the results.

4.4. RNA Extraction, Reverse Transcription, and RT-qPCR

Alfalfa roots were collected, flash-frozen in liquid nitrogen, and stored at $-80\ °C$ until further use. Approximately 100 mg fresh weight was used for total RNA extraction using Total RNA Purification Kit (Norgen Biotek, Thorold, ON, Canada, Cat #25800) for roots. Tissue was homogenized using a PowerLyzer®24 bench top bead-based homogenizer (Cat #13155) according to the manufacturer's manual. Approximately 500 ng of Turbo DNase (Invitrogen, Waltham, MA, USA, Cat #AM1907) treated RNA was used to generate cDNA using the iScript cDNA synthesis kit (Bio-Rad, Hercules, CA, USA, Cat # 1708891). Transcript levels were analyzed by RT-qPCR using a CFX96 TouchTM Real-Time PCR Detection System (Bio-Rad) and SsoFast™ EvaGreen® Supermixes (Bio-Rad Cat # 1725204) using gene-specific primers. Each reaction consisted of 2 µL of cDNA template, 0.5 µL forward and reverse gene-specific primers (10 µM each) (Table S1), and 5 µL SsoFast Eva green Supermix, then topped up to 10 µL with ddH$_2$O. For each sample three or four biological replicates were analyzed, and each biological replicate was tested using three technical replicates. Transcript levels were analyzed relative to two reference genes: *actin–depolymerizing protein 1* (*ADF1*) and *elongation initiation factor 4A* (*elF4A*) (primers are listed in Table S1).

4.5. Statistical Analysis

Statistical analyses were performed using Microsoft Excel spreadsheet software. Pairwise comparisons were made using Student's *t*-test with either equal or unequal variance. The significant differences between sample means for three or more data sets were calculated using the one-way analysis of variance (ANOVA) where appropriate.

Supplementary Materials: The following supporting information can be downloaded at: https://www.mdpi.com/article/10.3390/ijms24119615/s1.

Author Contributions: Conceptualization of the study, experimental design, and securing the funding, A.H.; Performing the experiments and analyzing the data, V.N.; Supervision, S.E.K. and A.H.; Designing the research, V.N. and A.H.; Writing, revision, and approving the manuscript, V.N., G.A., S.E.K. and A.H. All authors have read and agreed to the published version of the manuscript.

Funding: This research was funded by Natural Sciences and Engineering Research Council of Canada, grant number RGPIN-2018-04241, and the APC was funded by Agriculture and Agri-Food Canada.

Institutional Review Board Statement: Not applicable.

Informed Consent Statement: Not applicable.

Data Availability Statement: All relevant data can be found within the article and its supporting materials.

Acknowledgments: This work was supported by grants from the Natural Sciences and Engineering Research Council of Canada and Agriculture and Agri-Food Canada to AH. We thank Alex Molnar for help with graphics.

Conflicts of Interest: The authors declare that they have no conflict of interest associated with this work.

References

1. Oldroyd, G.E.D.; Murray, J.D.; Poole, P.S.; Downie, J.A. The rules of engagement in the legume-rhizobial symbiosis. *Annu. Rev. Genet.* **2011**, *45*, 119–144. [CrossRef]
2. Madsen, L.H.; Tirichine, L.; Jurkiewicz, A.; Sullivan, J.T.; Heckmann, A.B.; Bek, A.S.; Ronson, C.W.; James, E.K.; Stougaard, J. The molecular network governing nodule organogenesis and infection in the model legume *Lotus japonicus*. *Nat. Commun.* **2010**, *1*, 1. [CrossRef]
3. Held, M.; Hossain, M.S.; Yokota, K.; Bonfante, P.; Stougaard, J.; Szczyglowski, K. Common and not so common symbiotic entry. *Trends Plant Sci.* **2010**, *15*, 540–545. [CrossRef]
4. Oldroyd, G.E.D.; Downie, J.A. Coordinating nodule morphogenesis with rhizobial infection in legumes. *Annu. Rev. Genet.* **2008**, *59*, 519–546. [CrossRef]
5. Ferguson, B.J.; Indrasumunar, A.; Hayashi, S.; Lin, M.H.; Lin, Y.H.; Reid, D.E.; Gresshoff, P.M. Molecular analysis of legume nodule development and autoregulation. *J. Integr. Plant Biol.* **2010**, *52*, 61–76. [CrossRef]

6. Bersoult, A.; Camut, S.; Perhald, A.; Kereszt, A.; Kiss, G.B.; Cullimore, J.V. Expression of the *medicago truncatula DMI2* gene suggests roles of the symbiotic nodulation receptor kinase in nodules and during early nodule development. *Mol. Plant-Microbe Interact.* **2005**, *18*, 869–875. [CrossRef]

7. Charpentier, M.; Sun, J.; Martins, T.V.; Radhakrishnan, G.V.; Findlay, K.; Soumpourou, E.; Thouin, J.; Véry, A.A.; Sanders, D.; Morris, R.J.; et al. Nuclear-localized cyclic nucleotide-gated channels mediate symbiotic calcium oscillations. *Science* **2016**, *352*, 1102–1105. [CrossRef]

8. Groth, M.; Takeda, N.; Perry, J.; Uchid, H.; Dräxl, S.; Brachmann, A.; Sato, S.; Tabata, S.; Kawaguchi, M.; Wang, T.L.; et al. NENA, a *Lotus japonicus* homolog of sec13, is required for rhizodermal infection by arbuscular mycorrhiza fungi and rhizobia but dispensable for cortical endosymbiotic development. *Plant Cell* **2010**, *22*, 2509–2526. [CrossRef]

9. Kanamori, N.; Madsen, L.H.; Radutoiu, S.; Frantescu, M.; Quistgaard, E.M.H.; Miwa, H.; Downie, J.A.; James, E.K.; Felle, H.H.; Haaning, L.L.; et al. A nucleoporin is required for induction of Ca^{2+} spiking in legume nodule development and essential for rhizobial and fungal symbiosis. *Proc. Natl. Acad. Sci. USA* **2006**, *103*, 359–364. [CrossRef]

10. Saito, K.; Yoshikawa, M.; Yano, K.; Miwa, H.; Uchida, H.; Asamizu, E.; Sato, S.; Tabata, S.; Imaizumi-Anraku, H.; Umehara, Y.; et al. Nucleoporin85 is required for calcium spiking, fungal and bacterial symbioses, and seed production in *Lotus japonicus*. *Plant Cell* **2007**, *19*, 610–624. [CrossRef]

11. Ané, J.M.; Kiss, G.B.; Riely, B.K.; Penmetsa, R.V.; Oldroyd, G.E.D.; Ayax, C.; Lévy, J.; Debellé, F.; Baek, J.M.; Kalo, P.; et al. *Medicago truncatula DMI1* required for bacterial and fungal symbioses in legumes. *Science* **2004**, *303*, 1364–1367. [CrossRef]

12. Capoen, W.; Sun, J.; Wysham, D.; Otegui, M.; Venkateshwaran, M.; Hirsch, S.; Miwa, H.; Downie, J.A.; Morris, R.J.; Ané, J.M.; et al. Nuclear membranes control symbiotic calcium signaling of legumes. *Proc. Natl. Acad. Sci. USA* **2011**, *108*, 14348–14353. [CrossRef]

13. Messinese, E.; Mun, J.H.; Li, H.Y.; Jayaraman, D.; Rougé, P.; Barre, A.; Lougnon, G.; Schornack, S.; Bono, J.J.; Cook, D.R.; et al. A novel nuclear protein interacts with the symbiotic DMI3 calcium- and calmodulin-dependent protein kinase of *medicago truncatula*. *Mol. Plant-Microbe Interact.* **2007**, *20*, 912–921. [CrossRef]

14. Yano, K.; Yoshida, S.; Müller, J.; Singh, S.; Banba, M.; Vickers, K.; Markmann, K.; White, C.; Schuller, B.; Sato, S.; et al. CYCLOPS, a mediator of symbiotic intracellular accommodation. *Proc. Natl. Acad. Sci. USA* **2008**, *105*, 20540–20545. [CrossRef]

15. Andriankaja, A.; Boisson-Dernier, A.; Frances, L.; Sauviac, L.; Jauneau, A.; Barker, D.G.; De Carvalho-Niebel, F. AP2-ERF transcription factors mediate nod factor-dependent *Mt ENOD11* activation in root hairs via a novel cis-regulatory motif. *Plant Cell* **2007**, *19*, 2866–2885. [CrossRef]

16. Hirsch, S.; Kim, J.; Muñoz, A.; Heckmann, A.B.; Downie, J.A.; Oldroyd, G.E.D. GRAS proteins form a DNA binding complex to induce gene expression during nodulation signaling in *Medicago truncatula*. *Plant Cell* **2009**, *21*, 545–557. [CrossRef]

17. Marsh, J.F.; Rakocevic, A.; Mitra, R.M.; Brocard, L.; Sun, J.; Eschstruth, A.; Long, S.R.; Schultze, M.; Ratet, P.; Oldroyd, G.E.D. *Medicago truncatula NIN* is essential for rhizobial-independent nodule organogenesis induced by autoactive calcium/calmodulin-dependent protein kinase. *Plant Physiol.* **2007**, *144*, 324–335. [CrossRef]

18. Middleton, P.H.; Jakab, J.; Penmetsa, R.V.; Starker, C.G.; Doll, J.; Kaló, P.; Prabhu, R.; Marsh, J.F.; Mitra, R.M.; Kereszt, A.; et al. An ERF transcription factor in *Medicago truncatula* that is essential for nod factor signal transduction. *Plant Cell* **2007**, *19*, 1221–1234. [CrossRef]

19. Schauser, L.; Roussis, A.; Stiller, J.; Stougaard, J. A plant regulator controlling development of symbiotic root nodules. *Nature* **1999**, *402*, 191–195. [CrossRef]

20. Smit, P.; Raedts, J.; Portyanko, V.; Debellé, F.; Gough, C.; Bisseling, T.; Geurts, R. NSP1 of the GRAS protein family is essential for rhizobial nod factor-induced transcription. *Science* **2005**, *308*, 1789–1791. [CrossRef]

21. Matsunami, T.; Kaihatsu, A.; Maekawa, T.; Takahashi, M.; Kokubun, M. Characterization of vegetative growth of a supernodulating soybean genotype, Sakukei 4. *Plant Prod. Sci.* **2004**, *7*, 165–171. [CrossRef]

22. Suzaki, T.; Yoro, E.; Kawaguchi, M. Leguminous Plants: Inventors of root nodules to accommodate symbiotic bacteria. *Int. Rev. Cell Mol. Biol.* **2015**, *316*, 111–158.

23. Caetano-Anollés, G.; Gresshoff, P.M. Plant genetic control of nodulation. *Annu. Rev. Microbiol.* **1991**, *45*, 345–382. [CrossRef] [PubMed]

24. Kosslak, R.M.; Bohlool, B.B. Suppression of nodule development of one side of a split-root system of soybeans caused by prior inoculation of the other side. *Plant Physiol.* **1984**, *75*, 125–130. [CrossRef]

25. Reid, D.E.; Ferguson, B.J.; Hayashi, S.; Lin, Y.H.; Gresshoff, P.M. Molecular mechanisms controlling legume autoregulation of nodulation. *Ann. Bot.* **2011**, *108*, 789–795. [CrossRef]

26. Lim, C.W.; Lee, Y.W.; Lee, S.C.; Hwang, C.H. Nitrate inhibits soybean nodulation by regulating expression of *CLE* genes. *Plant Sci.* **2014**, *229*, 1–9. [CrossRef] [PubMed]

27. Mortier, V.; den Herder, G.; Whitford, R.; van de Velde, W.; Rombauts, S.; D'Haeseleer, K.; Holsters, M.; Goormachtig, S. CLE peptides control *Medicago truncatula* nodulation locally and systemically. *Plant Physiol.* **2010**, *153*, 222–237. [CrossRef] [PubMed]

28. Magori, S.; Kawaguchi, M. Analysis of two potential long-distance signaling molecules, *LjCLE-RS1/2* and jasmonic acid, in a hypernodulating mutant *too much love*. *Plant Signal. Behav.* **2010**, *5*, 403–405. [CrossRef]

29. Reid, D.E.; Ferguson, B.J.; Gresshoff, P.M. Inoculation- and nitrate-induced CLE peptides of soybean control NARK-dependent nodule formation. *Mol. Plant-Microbe Interact.* **2011**, *24*, 606–618. [CrossRef]

30. Okamoto, S.; Shinohara, H.; Mori, T.; Matsubayashi, Y.; Kawaguchi, M. Root-derived CLE glycopeptides control nodulation by direct binding to HAR1 receptor kinase. *Nat. Commun.* **2013**, *4*, 2191. [CrossRef]
31. Krusell, L.; Madsen, L.H.; Sato, S.; Aubert, G.; Genua, A.; Szczyglowski, K.; Duc, G.; Kaneko, T.; Tabata, S.; De Bruijn, F.; et al. Shoot control of root development and nodulation is mediated by a receptor-like kinase. *Nature* **2002**, *420*, 422–426. [CrossRef]
32. Nishimura, R.; Hayashit, M.; Wu, G.J.; Kouchi, H.; Imaizumi-Anrakull, H.; Murakami, Y.; Kawasaki, S.; Akao, S.; Ohmori, M.; Nagasawa, M.; et al. HAR1 mediates systemic regulation of symbiotic organ development. *Nature* **2002**, *420*, 426–429. [CrossRef]
33. Schnabel, E.; Journet, E.P.; De Carvalho-Niebel, F.; Duc, G.; Frugoli, J. The *Medicago truncatula SUNN* gene encodes a CLV1-like leucine-rich repeat receptor kinase that regulates nodule number and root length. *Plant Mol. Biol.* **2005**, *58*, 809–822. [CrossRef]
34. Searle, I.R.; Men, A.E.; Laniya, T.S.; Buzas, D.M.; Iturbe-Ormaetxe, I.; Carroll, B.J.; Gresshoff, P.M. Long-distance signaling in nodulation directed by a CLAVATA1-like receptor kinase. *Science* **2003**, *299*, 109–112. [CrossRef]
35. Murray, J.D.; Liu, C.W.; Chen, Y.; Miller, A.J. Nitrogen sensing in legumes. *J. Exp. Bot.* **2017**, *68*, 1919–1926. [CrossRef]
36. Tsay, Y.F.; Chiu, C.C.; Tsai, C.B.; Ho, C.H.; Hsu, P.K. Nitrate transporters and peptide transporters. *FEBS Lett.* **2007**, *581*, 2290–2300. [CrossRef]
37. Glass, A.D.M.; Britto, D.T.; Kaiser, B.N.; Kinghorn, J.R.; Kronzucker, H.J.; Kumar, A.; Okamoto, M.; Rawat, S.; Siddiqi, M.Y.; Unkles, S.E.; et al. The regulation of nitrate and ammonium transport systems in plants. *J. Exp. Bot.* **2002**, *53*, 855–864. [CrossRef]
38. Potel, F.; Valadier, M.H.; Ferrario-Méry, S.; Grandjean, O.; Morin, H.; Gaufichon, L.; Boutet-Mercey, S.; Lothier, J.; Rothstein, S.J.; Hirose, N.; et al. Assimilation of excess ammonium into amino acids and nitrogen translocation in *Arabidopsis thaliana*- roles of glutamate synthases and carbamoylphosphate synthetase in leaves. *FEBS J.* **2009**, *276*, 4061–4076. [CrossRef] [PubMed]
39. Chaulagain, D.; Frugoli, J. The regulation of nodule number in legumes is a balance of three signal transduction pathways. *Int. J. Mol. Sci.* **2021**, *22*, 1117. [CrossRef] [PubMed]
40. Lin, J.S.; Li, X.; Luo, Z.L.; Mysore, K.S.; Wen, J.; Xie, F. NIN interacts with NLPs to mediate nitrate inhibition of nodulation in *Medicago truncatula*. *Nat. Plants* **2018**, *4*, 942–952. [CrossRef] [PubMed]
41. Moreau, C.; Gautrat, P.; Frugier, F. Nitrate-induced CLE35 signaling peptides inhibit nodulation through the SUNN receptor and miR2111 repression. *Plant Physiol.* **2021**, *185*, 1216–1228. [CrossRef] [PubMed]
42. Lagunas, B.; Achom, M.; Bonyadi-Pour, R.; Pardal, A.J.; Richmond, B.L.; Sergaki, C.; Vázquez, S.; Schäfer, P.; Ott, S.; Hammond, J.; et al. Regulation of resource partitioning coordinates nitrogen and rhizobia responses and autoregulation of nodulation in *Medicago truncatula*. *Mol. Plant* **2019**, *12*, 833–846. [CrossRef] [PubMed]
43. Gifford, M.L.; Dean, A.; Gutierrez, R.A.; Coruzzi, G.M.; Birnbaum, K.D. Cell-specific nitrogen responses mediate developmental plasticity. *Proc. Natl. Acad. Sci. USA* **2008**, *105*, 803–808. [CrossRef] [PubMed]
44. Wu, M.-F.; Tian, Q.; Reed, J.W. Arabidopsis microRNA167 controls patterns of *ARF6* and *ARF8* expression, and regulates both female and male reproduction. *Development* **2006**, *133*, 4211–4218. [CrossRef] [PubMed]
45. Yan, Z.; Hossain, M.S.; Wang, J.; Valdés-López, O.; Liang, Y.; Libault, M.; Qiu, L.; Stacey, G. miR172 regulates soybean nodulation. *Mol. Plant-Microbe Interact.* **2013**, *26*, 1371–1377. [CrossRef] [PubMed]
46. Nova-Franco, B.; Íñiguez, L.P.; Valdés-López, O.; Alvarado-Affantranger, X.; Leija, A.; Fuentes, S.I.; Ramírez, M.; Paul, S.; Reyes, J.L.; Girard, L.; et al. The micro-RNA172c-APETALA2-1 node as a key regulator of the common bean-*Rhizobium etli* nitrogen fixation symbiosis. *Plant Physiol.* **2015**, *168*, 273–291. [CrossRef]
47. Tsikou, D.; Yan, Z.; Holt, D.B.; Abel, N.B.; Reid, D.E.; Madsen, L.H.; Bhasin, H.; Sexauer, M.; Stougaard, J.; Markmann, K. Systemic control of legume susceptibility to rhizobial infection by a mobile microRNA. *Science* **2018**, *362*, 233–236. [CrossRef]
48. Yan, Z.; Hossain, M.S.; Valdés-López, O.; Hoang, N.T.; Zhai, J.; Wang, J.; Libault, M.; Brechenmacher, L.; Findley, S.; Joshi, T.; et al. Identification and functional characterization of soybean root hair microRNAs expressed in response to *Bradyrhizobium japonicum* infection. *Plant Biotechnol. J.* **2016**, *14*, 332–341. [CrossRef]
49. De Luis, A.; Markmann, K.; Cognat, V.; Holt, D.B.; Charpentier, M.; Parniske, M.; Stougaard, J.; Voinnet, O. Two microRNAs linked to nodule infection and nitrogen-fixing ability in the legume *Lotus japonicus*. *Plant Physiol.* **2012**, *160*, 2137–2154. [CrossRef]
50. Hofferek, V.; Mendrinna, A.; Gaude, N.; Krajinski, F.; Devers, E.A. MiR171h restricts root symbioses and shows like its target *NSP2* a complex transcriptional regulation in *Medicago truncatula*. *BMC Plant Biol.* **2014**, *14*, 199. [CrossRef]
51. Yan, Z.; Hossain, M.S.; Arikit, S.; Valdés-López, O.; Zhai, J.; Wang, J.; Libault, M.; Ji, T.; Qiu, L.; Meyers, B.C.; et al. Identification of microRNAs and their mRNA targets during soybean nodule development: Functional analysis of the role of miR393j-3p in soybean nodulation. *New Phytol.* **2015**, *207*, 748–759. [CrossRef]
52. Martín-Rodríguez, J.Á.; Leija, A.; Formey, D.; Hernández, G. The microRNA319d/TCP10 node regulates the common bean–rhizobia nitrogen-fixing symbiosis. *Front. Plant Sci.* **2018**, *9*, 1175. [CrossRef] [PubMed]
53. Wang, H.; Wang, H.; Liu, R.; Xu, Y.; Lu, Z.; Zhou, C. Genome-wide identification of TCP family transcription factors in *Medicago truncatula* reveals significant roles of miR319-targeted TCPs in nodule development. *Front. Plant Sci.* **2018**, *9*, 774. [CrossRef] [PubMed]
54. Aung, B.; Gruber, M.Y.; Amyot, L.; Omari, K.; Bertrand, A.; Hannoufa, A. MicroRNA156 as a promising tool for alfalfa improvement. *Plant Biotechnol. J.* **2015**, *13*, 779–790. [CrossRef] [PubMed]
55. Nasrollahi, V.; Yuan, Z.C.; Lu, Q.S.M.; McDowell, T.; Kohalmi, S.E.; Hannoufa, A. Deciphering the role of *SPL12* and *AGL6* from a genetic module that functions in nodulation and root regeneration in *Medicago sativa*. *Plant. Mol. Biol.* **2022**, *110*, 511–529. [CrossRef]

56. Yun, J.; Sun, Z.; Jiang, Q.; Wang, Y.; Wang, C.; Luo, Y.; Zhang, F.; Li, X. The miR156b-GmSPL9d module modulates nodulation by targeting multiple core nodulation genes in soybean. *New Phytol.* **2022**, *233*, 1881–1899. [CrossRef]
57. Hanly, A.; Karagiannis, J.; Lu, Q.S.M.; Tian, L.; Hannoufa, A. Characterization of the role of *SPL9* in drought stress tolerance in *Medicago sativa*. *Int. J. Mol. Sci.* **2020**, *21*, 6003. [CrossRef]
58. Gonzalez-Rizzo, S.; Crespi, M.; Frugier, F. The *Medicago truncatula* CRE1 cytokinin receptor regulates lateral root development and early symbiotic interaction with *Sinorhizobium meliloti*. *Plant Cell* **2006**, *18*, 2680–2693. [CrossRef]
59. Jin, Y.; Liu, H.; Luo, D.; Yu, N.; Dong, W.; Wang, C.; Zhang, X.; Dai, H.; Yang, J.; Wang, E. DELLA proteins are common components of symbiotic rhizobial and mycorrhizal signalling pathways. *Nat. Commun.* **2016**, *7*, 12433. [CrossRef]
60. Streeter, J. Inhibition of legume nodule formation and N$_2$ fixation by nitrate. *Crit. Rev. Plant Sci.* **1988**, *7*, 1–23. [CrossRef]
61. Cardon, G.; Höhmann, S.; Klein, J.; Nettesheim, K.; Saedler, H.; Huijser, P. Molecular characterisation of the *Arabidopsis SBP-box* genes. *Gene* **1999**, *237*, 91–104. [CrossRef] [PubMed]
62. Wang, H.; Wang, H. The miR156/SPL module, a regulatory hub and versatile toolbox, gears up crops for enhanced agronomic traits. *Mol. Plant* **2015**, *8*, 677–688. [CrossRef] [PubMed]
63. Xu, M.; Hu, T.; Zhao, J.; Park, M.Y.; Earley, K.W.; Wu, G.; Yang, L.; Poethig, R.S. Developmental functions of miR156-regulated *SQUAMOSA PROMOTER BINDING PROTEIN-LIKE (SPL)* genes in *Arabidopsis thaliana*. *PLoS Genet.* **2016**, *12*, 8. [CrossRef] [PubMed]
64. Aung, B.; Gao, R.; Gruber, M.Y.; Yuan, Z.C.; Sumarah, M.; Hannoufa, A. MsmiR156 affects global gene expression and promotes root regenerative capacity and nitrogen fixation activity in alfalfa. *Transgenic Res.* **2017**, *26*, 541–557. [CrossRef] [PubMed]
65. Mergaert, P.; Kereszt, A.; Kondorosi, E. Gene expression in nitrogen-fixing symbiotic nodule cells in *Medicago truncatula* and other nodulating plants. *Plant Cell* **2020**, *32*, 42–68. [CrossRef]
66. Roy, S.; Liu, W.; Nandety, R.S.; Crook, A.; Mysore, K.S.; Pislariu, C.I.; Frugoli, J.; Dickstein, R.; Udvardi, M.K. Celebrating 20 years of genetic discoveries in legume nodulation and symbiotic nitrogen fixation. *Plant Cell* **2020**, *32*, 15–41. [CrossRef]
67. Feyissa, B.A.; Amyot, L.; Nasrollahi, V.; Papadopoulos, Y.; Kohalmi, S.E.; Hannoufa, A. Involvement of the miR156/SPL module in flooding response in *Medicago sativa*. *Sci. Rep.* **2021**, *11*, 1. [CrossRef]
68. Feyissa, B.A.; Arshad, M.; Gruber, M.Y.; Kohalmi, S.E.; Hannoufa, A. The interplay between *miR156/SPL13* and *DFR/WD40-1* regulate drought tolerance in alfalfa. *BMC Plant Biol.* **2019**, *19*, 434. [CrossRef]
69. Nasrollahi, V.; Yuan, Z.C.; Kohalmi, S.E.; Hannoufa, A. SPL12 regulates *AGL6* and *AGL21* to modulate nodulation and root regeneration under osmotic stress and nitrate sufficiency conditions in *Medicago sativa*. *Plants* **2022**, *11*, 3071. [CrossRef]
70. Krouk, G.; Mirowski, P.; LeCun, Y.; Shasha, D.E.; Coruzzi, G.M. Predictive network modeling of the high-resolution dynamic plant transcriptome in response to nitrate. *Genome Biol.* **2010**, *11*, 12. [CrossRef]
71. Chen, W.W.; Jin, J.F.; Lou, H.Q.; Liu, L.; Kochian, L.V.; Yang, J.L. LeSPL-CNR negatively regulates Cd acquisition through repressing nitrate reductase-mediated nitric oxide production in tomato. *Planta* **2018**, *248*, 893–907. [CrossRef] [PubMed]
72. Wopereis, J.; Pajuelo, E.; Dazzo, F.B.; Jiang, Q.; Gresshoff, P.M.; De Bruijn, F.J.; Stougaard, J.; Szczyglowski, K. Short root mutant of *lotus japonicus* with a dramatically altered symbiotic phenotype. *Plant J.* **2000**, *23*, 97–114. [CrossRef] [PubMed]
73. Badhan, A.; Jin, L.; Wang, Y.; Han, S.; Kowalczys, K.; Brown, D.C.W.; Ayala, C.J.; Latoszek-Green, M.; Miki, B.; Tsang, A.; et al. Expression of a fungal ferulic acid esterase in alfalfa modifies cell wall digestibility. *Biotechnol. Biofuels* **2014**, *7*, 39. [CrossRef] [PubMed]
74. Beringer, J.E. R factor transfer in *Rhizobium leguminosarum*. *J. Gen. Microbiol.* **1974**, *84*, 188–198. [CrossRef] [PubMed]

International Journal of
Molecular Sciences

Article

Inferring the Regulatory Network of miRNAs on Terpene Trilactone Biosynthesis Affected by Environmental Conditions

Ying Guo [1,*,†], Yongli Qi [1,†], Yangfan Feng [1], Yuting Yang [1], Liangjiao Xue [1], Yousry A. El-Kassaby [2], Guibin Wang [1] and Fangfang Fu [1]

[1]   State Key Laboratory of Tree Genetics and Breeding, Key Laboratory of Forest Genetics & Biotechnology of Ministry of Education, Co-Innovation Center for Sustainable Forestry in Southern China, Nanjing Forestry University, Nanjing 210037, China; qiyongli22@njfu.edu.cn (Y.Q.); yangfanfeng@njfu.edu.cn (Y.F.)

[2]   Department of Forest and Conservation Sciences, Faculty of Forestry, The University of British Columbia, Vancouver, BC V6T 1Z4, Canada; y.el-kassaby@ubc.ca

*   Correspondence: yingguo@njfu.edu.cn

†   These authors contributed equally to this work.

Citation: Guo, Y.; Qi, Y.; Feng, Y.; Yang, Y.; Xue, L.; El-Kassaby, Y.A.; Wang, G.; Fu, F. Inferring the Regulatory Network of miRNAs on Terpene Trilactone Biosynthesis Affected by Environmental Conditions. *Int. J. Mol. Sci.* **2023**, *24*, 17002. https://doi.org/10.3390/ijms242317002

Academic Editor: Tomotsugu Koyama

Received: 23 September 2023
Revised: 28 November 2023
Accepted: 29 November 2023
Published: 30 November 2023

Abstract: As a medicinal tree species, ginkgo (*Ginkgo biloba* L.) and terpene trilactones (TTLs) extracted from its leaves are the main pharmacologic activity constituents and important economic indicators of its value. The accumulation of TTLs is known to be affected by environmental stress, while the regulatory mechanism of environmental response mediated by microRNAs (miRNAs) at the post-transcriptional levels remains unclear. Here, we focused on grafted ginkgo grown in northwestern, southwestern, and eastern-central China and integrally analyzed RNA-seq and small RNA-seq high-throughput sequencing data as well as metabolomics data from leaf samples of ginkgo clones grown in natural environments. The content of bilobalide was highest among detected TTLs, and there was more than a twofold variation in the accumulation of bilobalide between growth conditions. Meanwhile, transcriptome analysis found significant differences in the expression of 19 TTL-related genes among ginkgo leaves from different environments. Small RNA sequencing and analysis showed that 62 of the 521 miRNAs identified were differentially expressed among different samples, especially the expression of miRN50, miR169h/i, and miR169e was susceptible to environmental changes. Further, we found that transcription factors (ERF, MYB, C3H, HD-ZIP, HSF, and NAC) and miRNAs (miR319e/f, miRN2, miRN54, miR157, miR185, and miRN188) could activate or inhibit the expression of TTL-related genes to participate in the regulation of terpene trilactones biosynthesis in ginkgo leaves by weighted gene co-regulatory network analysis. Our findings provide new insights into the understanding of the regulatory mechanism of TTL biosynthesis but also lay the foundation for ginkgo leaves' medicinal value improvement under global change.

Keywords: *Ginkgo biloba* L.; ginkgolides; bilobalide; miRNA; regulatory network; environment response

1. Introduction

Ginkgo biloba L., with a long history of more than 200 million years, is one of the oldest living plant species in nature [1,2]. Ginkgo trees have significant medical value as they contain many active ingredients, such as terpene trilactones (TTLs), flavonoids, and ascorbic acid [3]. TTLs, a unique terpenoid in Ginkgo, play an essential role in treating cardiovascular and neurological diseases [4]. TTLs are diterpenes with a cage skeleton consisting of six five-membered rings that differ only in the number and position of hydroxyl groups [5]. To date, more than ten ginkgolides and bilobalide have been isolated from ginkgo leaves and root barks [6], while the content of TTL components in ginkgo leaves is too low to sustain the market demands. In the past 20 years, several attempts have been made to improve TTLs contents. For example, Crimmins et al. succeeded in synthesizing ginkgolide B through chemical routes [7]; however, this method was not amenable for commercial-scale production [8]. In vitro cultures were also attempted to promote ginkgolide and bilobalide production, but the method has shortcomings, such as a low yield and long cultivation period. Notably, the synthesis of TTLs is affected by many factors, including internal developmental genetic circuits and external environmental factors [9], which can be integrated as tools for increasing their accumulation.

It was shown that the methylerythritol 4-phosphate (MEP) pathway is mainly responsible for the biosynthesis of ginkgolides [10], while bilobalide synthesis was thought to take place through the cytosolic mevalonate (MVA) pathway [11], with cross-talks between these two pathways. Environmental stress could increase the content of TTLs in ginkgo by inducing the expression of TTL-related genes. For instance, under UV exposure for 48 h, there was a 10% surge in TTL content, attributed to the significant upregulation of *GbDXS* (1-deoxy-D-xylulose 5-phosphate synthase), *GbGGPS* (geranylgeranyl diphosphate synthase), and *GbLPS* (levopimaradiene synthase) genes. Similarly, cold stress for a period of 8 days led to a 14.5% boost in TTL content, again due to the significant increase in the expression levels of *GbDXS*, *GbGGPS*, *GbLPS*, and *GbMVD* genes [12]. While studies related to TTL biosynthetic pathways have made significant strides, the detailed molecular mechanisms underpinning TTL metabolism in responses to varying environmental conditions remain to be fully elucidated.

MicroRNAs (miRNAs) play a crucial role in mediating fundamental processes for plant survival in response to environment stress. MiRNAs are post-transcriptional regulators that target genes to negatively regulate them by cleavage and/or translational inhibition [13]. A few miRNAs have been identified to participate in TTL metabolism regulation. For instance, 25 miRNAs potentially played a vital role in TTL accumulation in ginkgo leaves by targeting key transcription factor genes [14]; Singh et al. found that miR156 could regulate the TTL metabolism pathway in *Mentha* spp. by targeting the *DXS* gene [15], and miR5021 could regulate this pathway by targeting *DXS*, *GPS*, and *GGPS* genes in *Curcuma longa* [16]. As endogenous gene modulators, moreover, several core miRNAs have been confirmed to participate in the biological processes of plant environmental adaptation. For example, miR396c/d could inhibit the expression of *GRF1* in maize to enhance drought stress tolerance [17]; the overexpression of miR319b regulated *OsPCF6* and *OsTCP21* to increase proline accumulation, thereby improving the cold tolerance ability of rice [18]; and the study in *Panicum virgatum* L. found that the down-regulation of miR166 expression could mitigate salt stress [19]. Thus, deciphering the complex molecular networks regulated by miRNAs is essential to improve TTLs synthesis and accumulation in response to changing environments.

In this study, we integrated RNA-seq and small RNA-seq high-throughput sequencing data as well as metabolomics data from leaf samples of ginkgo clones grown in different environments. Our specific objectives were to (1) investigate the response of TTL accumulation to different environments; (2) annotate miRNAs at the ginkgo genome-wide level and identify miRNAs involved in the regulation of TTL metabolism therein; and (3) reveal the regulatory mechanism mediated by miRNAs in response to TTL metabolism in different environments. This study is expected to provide new insights into the regulatory mechanism of TTL biosynthesis and to lay a theoretical foundation for ginkgo leaves' medicinal value improvement.

2. Results

2.1. Changes in TTLs Content under Different Environments

Four TTLs contents (ginkgolide A, B, C, and bilobalide) in ginkgo leaves from three different habitats were determined by HPLC-ELSD, and the molecular structural formula of TTLs is shown in Figure 1. The content of bilobalide was highest among the four TTLs, and there were significant differences ($p < 0.05$) in the accumulation of bilobalide under the three environments. The content of ginkgolide C decreased gradually from S1 to S3, and significant differences were found between S1 and S3 samples, while there were no significant differences in the content of ginkgolide A and B among leaf samples from the three different environments. Collectively, the total contents of the four TTLs in S1 (ranging from 6.15 to 7.95 mg/g) were significantly higher than the other two sites (S2 and S3), suggesting that TTLs accumulation was affected by environmental conditions (Figure 1).

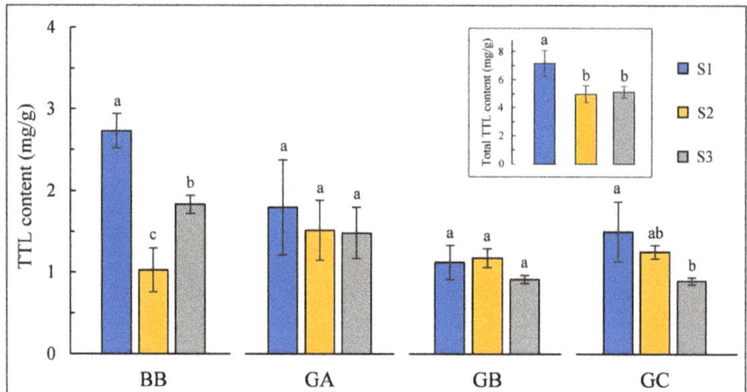

Figure 1. Variation in TTL content in ginkgo leaves from three test sites (S1–S3). The total TTL content is the sum of bilobalide (BB), ginkgolide A (GA), ginkgolide B (GB), and ginkgolide C (GC) contents. Error bars represent standard deviations, and small letters indicate the significant difference ($p < 0.05$).

The metabolome database was collected from our previous study [20]. There were 13,742 substances detected from the mass spectrums of nine selected ginkgo samples, including 8 substances in the terpenoid backbone biosynthesis pathway, such as Acetyl-CoA, DXP, CDP-ME, CDP-MEP, ME-cPP, HMBPP, Isopentenyl-PP, and FPP (Figure 2). Compared with S2 and S3 samples, the abundance of CDP-ME, HMBPP, and Isopentenyl-PP was higher in S1 samples, which were essential precursors for the biosynthesis of TTL (Figure 2).

2.2. Analysis of Environment-Responsive mRNAs

The 25,328 genes that could be expressed (TPM > 0) were found in 27,832 genes annotated in the reference genome via transcriptome analysis (NCBI BioProject number: PRJNA649066). We found a total of 8496 differentially expressed genes (DEGs) between the S1 and S3 samples, which was approximately twice the number of DEGs between S1 and S2 samples (Figure S1A and Table S1). Next, we carried out GO and KEGG enrichment analysis, and 7049 and 863 DEGs were enriched in GO terms (Figure S1B) and KEGG pathways, respectively (Figure S1C). For example, larger numbers of DEGs were observed in GO terms of the cellular polysaccharide metabolic process, fatty acid biosynthetic process, and external encapsulating structure organization, and some DEGs were enriched in the biosynthesis of amino acids, carbon metabolism, and glycolysis or gluconeogenesis pathways. Moreover, we identified 56 structural-related genes on the terpenoid backbone biosynthesis pathway, 19 of which were differentially expressed in the three environments, such as *GbGGPS*, *GbHMGR*, and *GbHDR* (Figure 2).

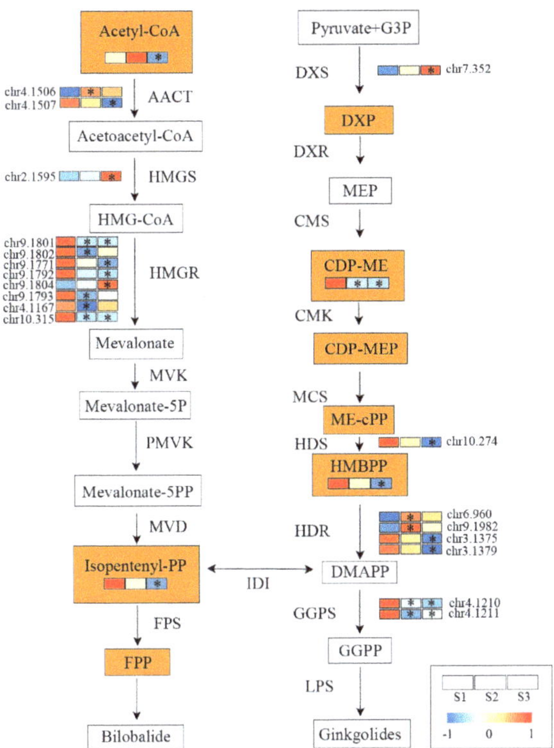

Figure 2. The pathway related to TTL metabolism in ginkgo and the heatmap showing the expression patterns of differentially expressed genes (DEGs) involved in TTL metabolism. The color change from red to blue indicates a gradual decrease in gene expression. The stars (*) in rectangular showed that there were significant differences between S2/S3 and S1 samples ($p < 0.05$). Enzyme names are abbreviated as follows: Acetyl-CoA C-acetyltransferase (AACT); Hydroxymethylglutaryl-CoA synthase (HMGS); Hydroxymethylglutaryl-CoA reductase (HMGR); Mevalonate kinase (MVK); Phosphomevalonate kinase (PMVK); Diphosphomevalonate decarboxylase (MVD); 1-deoxy-D-xylulose-5-phosphate synthase (DXS); 1-deoxy-D-xylulose-5-phosphate reductoisomerase (DXR); 2-C-methyl-D-erythritol 4-phosphate cytidylyltransferase (CMS); 4-diphosphocytidyl-2-C-methyl-D-erythritol kinase (CMK); 2-C-methy-D-erythritol 2,4-cyclodiphosphate synthase (MCS); (E)-4-hydroxy-3-methylbut-2-enyl-diphosphate synthase (HDS); 4-hydroxy-3-methylbut-2-en-1-yl diphosphate reductase (HDR); Farnesyl diphosphate synthase (FPS); Geranylgeranyl diphosphate synthase (GGPS); Isopentenyl-diphosphate Delta-isomerase (IDI); Levopimaradiene synthase (LPS).

2.3. Annotation of miRNAs and Their Target Genes

To explore the response of miRNAs to environmental changes, we first identified ginkgo miRNA loci at the genome-wide level using nine sRNA-seq datasets (NCBI Bio-Project number: PRJNA903548) and 28 public ginkgo sRNA libraries. Here, 521 miRNAs (from 337 miRNA families) were annotated, including 206 known and 315 novel miRNAs (Figure 3A and Table S2). As shown in Figure 3B, the size of miRNAs ranged between 20 nt and 24 nt, with 21 nt having the highest proportion of identified miRNAs (77.5%). The first nucleotide of the 5' end of mature miRNAs had a significant bias towards U (Figure 3C). Additionally, we found that 206 known miRNAs came from 54 conserved miRNA families. In summary, the family analysis showed that the detected miRNA family miR11534 was the most abundant with 18 members, followed by miR396 with 14 members and miR169 with 13 members (Figure 3D). Of the remaining miRNA families, 278 were represented by only one family member. Further, we predicted the targeted regulation of 20,446 genes by 521 miRNAs using psRNATarget software (http://plantgrn.noble.org/psRNATarget/ accessed on 22 September 2023), covering about 73% of the total genes (Table S3). The genome-wide annotation of miRNAs in ginkgo could provide support for further exploring the diversity and regulatory functions of miRNAs in gymnosperms.

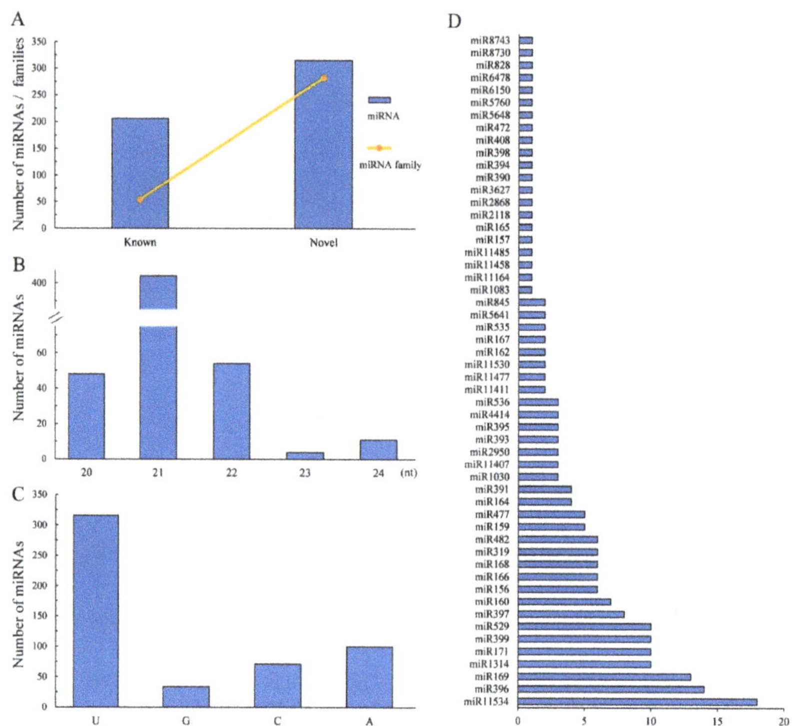

Figure 3. Annotation of miRNAs in ginkgo genome. (**A**) Number of miRNAs and miRNA families, (**B**) length distribution of identified miRNAs, (**C**) distribution of their 5′ end bases, and (**D**) number of conserved miRNA family members.

Among the 521 miRNAs annotated, 62 differentially expressed miRNAs (DEMs) were found among three environments (Figure 4A and Table S4). Consistently, more DEMs were identified between S1 and S3 samples, with the largest differences in environmental conditions. As shown in the Venn diagram (Figure 4B), three DEMs (gbi-miRN50, gbi-miR169h/i, and gbi-miR169e) were shared in the three comparison groups. There are 6346 genes predicted to be the target genes of 62 DEMs. Furthermore, we performed gene ontology (Figure S2A) and KEGG pathway (Figure S2B) enrichment analysis to explore the biological functions of target genes. Our gene ontology analysis showed that DEM target genes were involved in the biological processes of copper ion transport, transition metal ion transmembrane transporter activity, and ADP binding. Remarkably, we found that some target genes of DEMs were enriched in a few of the secondary metabolite biosynthesis pathways, such as flavonoid and terpenoid backbone biosynthetic pathways (Figure S2B). Five target genes of nine DEMs were related to TTL biosynthesis (Figure 4C), such as *GbAACT* (chr4.1506), *GbPMVK* (chr12.788), *GbHDS* (chr12.1656), *GbHMGR* (chr9.1771), and *GbGGPS* (chr8.380).

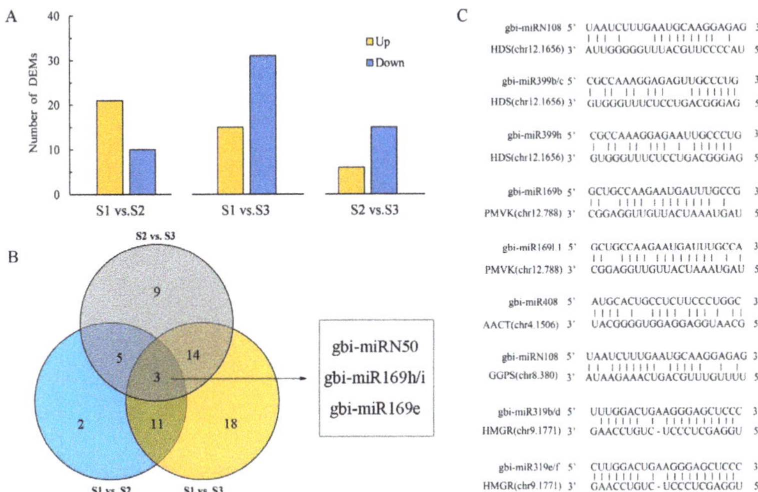

Figure 4. Analysis of the differentially expressed miRNAs (DEMs) among the samples from three environments. (**A**) The number of up- and down-regulated DEMs. Yellow rectangles represent up-regulated DEMs; blue rectangles represent down-regulated DEMs, (**B**) Venn plot shows the overlap of all identified DEMs in pairwise comparison. The blue circle represents the DEMs between S1 and S2 samples; yellow circle represents the DEMs between S1 and S3 samples; and gray circle represents the DEMs between S2 and S3 samples. The number represents the number of identified DEMs, (**C**) Prediction of DEM target sites of genes involved in TTL metabolism.

2.4. Construction of TTL-Related Gene Regulatory Network

Integrating omics datasets, the regulation mechanism of the TTL biosynthesis was explored through weighted gene co-expression network analysis (WGCNA). As shown in Figure 5A, thirty-two modules with similar expression patterns were identified (Table S5), and the number of genes in modules ranged from 4 to 8781 (Figure S3A). There were four modules with more than one thousand genes, such as Yellow, Brown, Blue, and Turquoise modules. We identified 634 transcription factor (TF) coding genes in all modules by using the PlantTFDB online website (Figure S3B). To find out the modules related to TTL biosynthesis, we quantified the correlation between each module and TTL traits. Notably, the Blue module eigengene was positively related to the content of various TTL and their precursors (r > 0.8, p < 0.05), such as CDP-ME, HMBPP, bilobalide, and total TTL (Figure 5B). In the Blue module, we found 12 genes encoding enzymes in the TTL biosynthesis pathway and extracted 30 genes closely related to their expression from the module based on weight values. In order to find the key regulators, we identified TF coding genes from the extracted gene set and retrieved miRNAs targeting these structural genes and TF coding genes. Finally, we constructed a sub-network based on the correlation coefficients of miRNA, TF gene, and structural gene expression. As shown in Figure 5C, the expression of 10 genes encoding TTL-related enzymes (*GbMVD*, *GbHMGS*, *GbHMGRs*, and *GbGGPSs*) and 18 TF encoding genes presented a certain degree of positive correlation, 13 of which were ERF encoding genes. Meanwhile, we observed the negative regulatory relationships between six miRNAs and their targets (one TF encoding gene and five structural genes), such as miR319e/f, miRN2, miRN54, miR157, miR185, and miRN188. Hence, we suggested that these TFs and miRNAs could play an important role in regulating TTL biosynthesis in ginkgo leaves, activating or inhibiting the expression of key enzyme genes. The qRT-PCR analysis of ten TTL-related genes showed the same expression patterns as the RNA sequencing results, albeit with some subtle differences, confirming the reliability of these results (| r | > 0.65 | , p < 0.005; Figure S4A,B). Furthermore, 37 Cytochrome P450 (CYP450) genes were found in the Blue module, of which the expression of 36 CYP450 family members was significantly correlated with that of TTL-related structural genes (r > 0.6, *p* < 0.05), indicating that *GbCYP450* may be involved in regulating the biosynthesis of terpene trilactone.

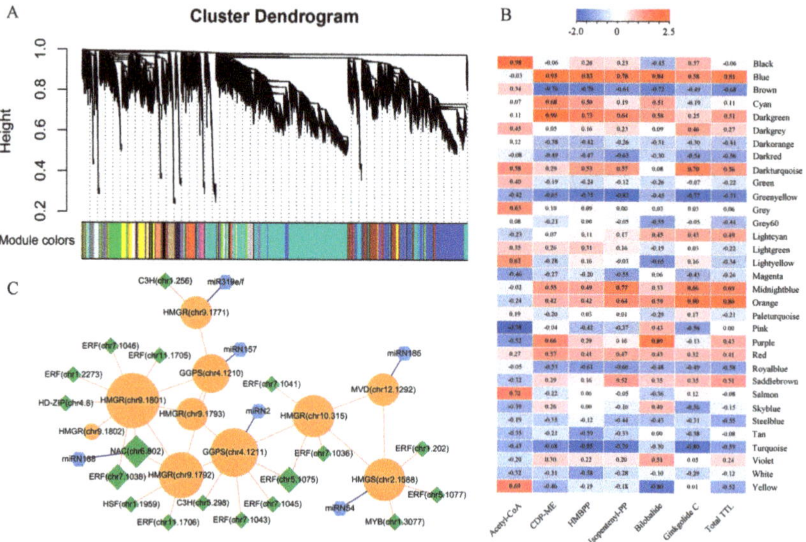

Figure 5. Construction of TTL-related gene regulatory network. (**A**) Clustering dendrogram showing modules of expressed genes identified by the WGCNA, (**B**) heatmap showing correlation of module eigengene and TTL metabolite content, (**C**) the sub−network of miRNAs and genes related to TTL metabolism. Orange circular: structural genes involved in TTL metabolism pathway; green quadrangle: TF coding genes; blue hexagon: miRNAs. The size of the shapes represents the level of connectivity. The larger the size, the higher the connectivity. The orange and blue lines represent positive and negative correlations, respectively.

3. Discussion

3.1. Effects of Environmental Condition on TTLs Accumulation in Ginkgo

Ginkgolides (GA, GB, GC, GM, GJ, GP, GQ, GK, GL, GN) and bilobalide (BB) are TTLs isolated from ginkgo leaf and its root bark. TTLs have a high medicinal value and play an essential role in curing cardiovascular diseases [4]. As an important secondary metabolite, TTLs accumulation could be influenced by ecological factors such as water, temperature, and light [21]. This study determined GA, GB, and GC and BB contents in leaf samples from three different environments. We found that the total contents of four TTLs in S1 samples were obviously higher than the other two samples (Figure 1). According to the analysis of climate data, S1 has the highest precipitation (S1, 316.3 mm; S2, 101.2 mm; S3, 4.0 mm) and the lowest evapotranspiration (107.7, 118.6, 199.5 mm) in sampling month (Table S6). A previous study found that moderate drought conditions could promote TTLs biosynthesis in ginkgo leaves [9], whose accumulation could eliminate reactive oxygen species (ROS) during abiotic stress, blocking oxidation and protecting plant growth [22]. It is speculated that moderate drought could induce additional TTLs accumulation in S1 samples by regulating the gene expression of the abscisic acid biosynthesis enzyme, transcription factors, and TTL biosynthesis-related enzyme [23]. A previous study showed that TTLs in ginkgo leaves treated with a high temperature increased compared with the control group [24], while the average monthly maximum temperature of S2 is 4 °C, which is 6 °C lower than that of S1 and S3, respectively (Table S6). Wang et al. indicated that there was an optimum light intensity in a specific light range, and when light intensity was higher or lower than this optimum, a pronounced decrease in TTLs content was observed in ginkgo leaves [25]. The average sunlight hour in August of S3 is more than 9 h (9.16 h), while that of S1 (5.76 h) and S3 (5.91 h) is less than 6 h (Table S6). Excessive radiation may be a substantial reason for the decreased TTLs accumulation in S3 samples. According to our previous research, the test sites (S1, S2, and S3) were located in high-, medium-, and low-suitability habitats, respectively [20]. Here, we suggest that priority should be given to establishing the Ginkgo plantations in high-suitability habitats to maximize TTLs production in harvested leaves.

3.2. Key Genes Regulating TTL Biosynthesis and Accumulation

Combined with previous plants' TTL metabolism studies [5] and mRNA functional annotation, we found 56 structural genes involved in the TTL metabolism pathway, including 19 DEGs among samples from different environments (Figure 2). Focused on the connection between gene expression and metabolite abundance, we found that *GbHMGRs* were highly expressed in the S1 samples, which had a consistent pattern of downstream Isopentenyl-PP content. In addition, *GbHDS* was highly expressed in S1, thus resulting in a high abundance of HMBPP. GGPS is a vital enzyme associated with the biosynthesis of ginkgolides in ginkgo, and the expression of *GbGGPSs* was enhanced in S1, which may increase the abundance of precursors for the biosynthesis of ginkgolides (Figures 1 and 2).

In ginkgo leaves, the expression of TTL-related genes was reported to be regulated by transcription factors from the bHLH, WRKY, and AP2 families [14]. In this study, through WGCNA, we identified 18 TF coding genes closely related to the TTL metabolism pathway (Figure 5C). Among these TF genes, 12 genes from the ERF family played a vital role in regulating the largest number of structural genes in this pathway. The ERF family is one of the AP2/ERF family factors found to regulate terpene biosynthesis in many plants, such as *Artemisia annua*, *Citrus sinensis*, and *Zea mays* [26–28]. A previous study of *Litsea cubeba* showed that *LcERF19* could enhance *LcTPS42* expression to improve geranial and neral biosynthesis [29]. The study of *Catharanthus roseus* found that AP2/ERF transcription factor coding gene, *ORCA3*, and its regulatory factor *CrMYC2* play a key role in terpenoid indole alkaloids biosynthesis [30]. In addition, it was found that *MYB* and *NAC* genes play important roles in regulating TTL biosynthesis, which is consistent with previous studies [31].

3.3. MicroRNA Functions in Environmental Stress Responses

In this study, we identified miRNA using a recently published high-quality reference genome to obtain an accurate annotation of miRNAs in ginkgo [32]. In addition, our miRNA method identification was robust and comprehensive. First, we selected 37 sRNA sequencing samples from different organs of ginkgo, ensuring that the miRNA library we built was not affected by sampling bias. Second, we used the ShortStack program to identify miRNAs with a rigorous set of structure- and expression-based parameter criteria. We identified 521 miRNAs belonging to 337 miRNA families (Figure 3A).

The adaptive response of plants to sudden environmental changes is a complex phenomenon, and an increasing number of studies have revealed that miRNAs are able to regulate a new gene expression program to help restore homeostasis [33]. Here, we found 62 DEMs among the three environments; in particular, the expression of miRN50, miR169h/i, and miR169e was vulnerable to environmental changes (Figure 4B). The miR169 family is a large and conserved family in plants, whose members are thought to be environmental-responsive miRNAs. The response of miR169 family members to drought stress was found in a series of studies, such as maize, *Echinacea purpurea*, and *Phaseolus vulgaris* [34–36]. An *Arabidopsis* study found that miR169, as an ambient temperature-responsive microRNA, played a vital role in stress responses and the floral transition [37].

As shown in Figure 4C, we found that the differentially expressed miR169l.1 directly binds to the promoter of *PMVK* (chr12.788), which was probably involved in the regulation of TTL biosynthesis. In the co-expression regulation sub-network, we identified four novel miRNAs and one known miR319 associated with the TTL metabolism pathway by targeting structural genes (Figure 5C). Previous studies found that miR319 played an important role in plant development and stress responses. Sun et al. found that miR319 controlled secondary cell wall formation during plant development by regulating TCP4 TF [38]. Furthermore, miR319 is highly correlated with plant drought resistance [39]. Interestingly, we found that miR319 differentially expressed in three environments can negatively regulate the expression of its target gene *HMGR* (DEG), which may further affect the synthesis and accumulation of downstream TTLs (Figures 4C and 5C).

4. Materials and Methods

4.1. Plant Materials

Ginkgo leaves were collected from 2-year-old clonally propagated (grafted) ginkgo trees grown in three different test sites. We implemented randomized block experiments, where each test site was set up with three blocks, with 20 ginkgo seedlings planted in each block and a row spacing of 40×60 cm. The first site (S1) is located in Jiangsu Province, central China, with a longitude and latitude of 34.21 °N and 117.58 °E, respectively, with a mean annual temperature (MAT) of 14.5 °C and mean annual precipitation (MAP) of 845 mm. The second site (S2) is located in Yunnan Province, southern China (latitude: 25.52 °N, longitude: 103.58 °E), characterized by a warm and humid environment (MAT: 14.1 °C, MAP: 1067 mm). And the third site (S3) is located in Xinjiang Uygur Autonomous region, northwest of China with typical continental semi-arid climate characteristics (latitude: 43.41 °N, longitude: 81.11 °E; MAT: 5.2 °C, MAP: 331 mm). Further, we obtained daily climate data of the three test sites in the sampling month from the National Meteorological Science Data Center [40], including ten climatic indicators such as ground temperature, precipitation, and sunlight hours (Table S6). At the three test sites, three vigorous ginkgo trees with a similar level of growth were randomly selected from each block (three biological replications) to collect 3–7 leaves at the upper end of the main branch for the sequencing of transcriptome, small RNA, and metabolome, and then 10 leaves were randomly collected for TTL content determination.

4.2. Extraction and Determination of TTL-Related Components

The extraction and determination of TTLs in ginkgo leaves were carried out according to the method in the People's Republic of China protocol [41]. The fresh leaves were first dried at 105 °C for 15 min and then dried further (70 °C, 48 h), crushed using a micro-high-speed universal pulverizer (JC-FW200), and screened using a 100-mesh sieve. We wrapped 1.0 g of dry leaf powder in filter paper and immersed it in a Soxhlet extractor containing 90 mL of petroleum ether, and this was refluxed for one hour to remove impurities at 70 °C; the leaf powder was then immersed in 70 mL of methanol and refluxed for 6 h at the same temperature. Then, the extract was evaporated on the Yarong rotary evaporator (Shanghai, China) under the conditions of the relative vacuum of 95 kPa, heating water bath temperature of 65 °C, cooling medium temperature of 20 °C, and rotating speed of 60 r/min, and the residue was dissolved with 10 mL methanol. After that, the eluate was sonicated for 30 min at 50 °C with 300 W power and 40 KHz frequency using Skymen ultrasonic cleaner (Shenzhen, China), and then 5 mL of the supernatant was withdrawn and filtered through an aluminum oxide column. Next, the aluminum oxide column was washed using 25 mL methanol and the eluate was evaporated on a rotary evaporator, after which the residue was dissolved with 5 mL methanol. Then, the tube containing the eluent was sonicated for 30 min after adding 4.5 mL of deionized water to the tube. After cooling to room temperature, we diluted the eluent with a 50 mL volumetric flask of methanol to a constant volume and pipetted 1 mL of the solution for TTL content determination using high-performance liquid chromatography (HPLC). TTLs determination was performed using an evaporative light scattering detector (ELSD) at ambient temperature and carrier gas pressure 40 psi. The mobile phase was tetrahydrofuran solution (10%), methanol (25%), and deionized water (65%) at 1.0 mL·min^{-1}. Standard ginkgolide (GA, GB, and GC) and bilobalide (BB) were provided by Shanghai yuanye Bio-Technology Co., Ltd., Shanghai, China. The HPLC chromatograms of the four standards are shown in Figure S5.

According to the previously reported method [20], metabolites were extracted from 9 freeze-dried samples of ginkgo leaves (three biological replicates) from different experimental sites, respectively. All of the samples' extracts were mixed as the quality control (QC) samples, which could monitor deviations of the analytical results and evaluate potential errors from analytical instruments. An UHPLC system (1290, Agilent Technologies, Santa Clara, CA, USA) was used for the LC–MS/MS analyses. The peak intensities were batch normalized to the total spectral intensity, and the mass spectrum of the QC samples is shown in Figure S6. The identification of metabolites is based on the exact molecular formula (molecular formula error < 20 ppm). For compound identification, the XCMS online program (The Scripps Research Institute, San Diego, CA, USA) with OSISMMS (version 1.0, Dalian Chem Data Solution Information Technology Co., Ltd., Dalian, China) was used for peak annotation. We mapped the identified metabolites to the Kyoto Encyclopedia of Genes and Genomes (KEGG) databases [42] to determine the TTL precursors in the terpenoid backbone biosynthesis pathway (ko00900).

4.3. RNA Sequencing (RNA-seq) and Analysis

Total RNA was extracted from nine freeze-dried leaf samples using the Trizol reagent kit (Invitrogen, Carlsbad, CA, USA) following the manufacturer's protocol. Follow the previous method for library preparation and transcriptional sequencing [20]. To obtain high-quality reads, adaptors of raw reads were removed by Trimmomatic (version 0.39) [40]. The processed reads were mapped to the ginkgo reference genome using Bowtie2 (version 2.3.0) [32,43]. To quantify mRNA expression, the TPM (Transcripts per million) value of each transcription region was calculated using RSEM (version 1.3.3) [44]. Differentially expressed genes (DEGs) were identified using DESeq2 [45] with a threshold of FDR < 0.05 and FC (fold change) > 1.5. All mRNA sequences were compared with the Swiss-Prot [46], Gene Ontology (GO) [47], and KEGG databases [42] using BLASTx software (https://blast. ncbi.nlm.nih.gov/ accessed on 22 September 2023) to perform gene functional analysis. Genes encoding key enzymes in TTL metabolism pathway were extracted according to KEGG annotation. Then, genes encoding transcription factors (TFs) were identified using PlantTFDB [48].

4.4. Small RNA Sequencing (sRNA-seq) and Analysis

Consistent with RNA-seq, sRNA libraries of nine freeze-dried leaf samples were constructed through TruSeq Small RNA Sample Prep Kits and sequenced using the Illumina Hiseq2000/2500 sequencer (50-bp single-end reads). An additional 28 small RNA data of ginkgo were downloaded from NCBI Sequence Read Archive (SRA) for analysis (Table S7). Referring to the previous method for identifying miRNAs in poplar [49], we used the ShortStack program (version 3.3.3) [50] to comprehensively annotate the miRNA loci in the ginkgo reference genome [32]. In short, we sequentially used Cutadapt software (version 2.10) [51], Bowtie software (version 1.3.0) [52], ShortStack program, and PatMaN software (version 1.2) [53] for sequencing data filtering, alignment, and miRNA recognition. Additionally, the quantification of miRNA expression levels was based on TPM. Differentially expressed miRNAs (DEMs) were identified using R package DESeq2 [45] with the threshold of FDR < 0.05 and FC > 1.5. Mature miRNA and transcript sequences were submitted to the online tool psRNATarget to conduct target prediction with default parameters (maximum cutoff of score = 5) [54].

4.5. Constructing Co-Expression Network

After the filtration of unexpressed genes (TPM = 0) from the nine RNA samples, 25,328 expressed genes were preserved for weighted gene co-regulatory network analysis (WGCNA) [55]. The dynamic decision-making tree and adjacency matrix method were used to identify similar modules using R package WGCNA with the following parameters:

min module size = 30; merge cut height = 0.25. In addition, a correlation analysis for the content of TTL-related components and epigengene was performed. In the module highly related to TTL accumulation, TTL-related structural genes were used as the hub genes to extract 30 genes close to them based on the weight value. We identified the transcription factor coding genes in the extracted gene set to select TF genes with a greater than 0.8 correlation with structural genes. Next, we retrieved miRNAs targeting these structural and TF coding genes, and miRNAs with a negative correlation between their expression ($r < -0.6$) were retained. Finally, a co-expression regulation sub-network was constructed based on the correlation among miRNAs, TF genes, and TTL-related structural genes, which were visualized using CytoScape (version 3.7.1) [56].

4.6. Quantitative Real-Time PCR (qRT-PCR) Analysis

Ten genes with a high expression involved in the regulation of TTL synthesis in ginkgo leaves were chosen for validation by qRT-PCR. Following the manufacturer's guidelines, cDNA was synthesized using MonScript RTIII All-in-One Mix with dsDNase kits (Monad, Suzhou, China). Then, the qRT-PCR analysis was conducted using cDNA templated on an Applied Biosystems™ 7500 Real-Time PCR System. Primer Premier 5 was used to design the primers, and the sequences are listed in Table S8. The glyceraldehyde 3-phosphate dehydrogenase (*GAPDH*) gene served as a reference gene. Each sample consisted of three biological replicates and three independent technical replicates. The relative gene expression was calculated using the $2^{-\Delta\Delta C^t}$ approach [57].

5. Conclusions

Ginkgo trees have the ability to biosynthesize diverse metabolites as defense substances to protect their long-term survival under complex environmental conditions. Here, we observed notable differences in the TTL content of ginkgo leaves from different environments, suggesting that some environmental factors are conducive to (moderate drought and high temperature) or not conducive to (excessive radiation) TTL biosynthesis and accumulation. Using the constructed regulatory network of miRNA-mRNA, we found several transcription factors and microRNAs to be essential regulators participating in the TTL biosynthesis pathway, which could provide new targets to enhance TTL accumulation at the molecular level. In a future study, we will investigate the relationship between miRNAs and target genes using a dual luciferase reporter assay to offer a new approach to improve TTLs accumulation in ginkgo leaves.

Supplementary Materials: The following supporting information can be downloaded at: https://www.mdpi.com/article/10.3390/ijms242317002/s1.

Author Contributions: Conceptualization, Y.G., L.X. and Y.Q.; methodology, Y.Q., Y.Y. and Y.A.E.-K.; software, Y.Q.; validation, F.F., Y.G. and Y.Q.; formal analysis, Y.Q., Y.F. and Y.G.; investigation, G.W.; resources, G.W.; data curation, Y.Q.; writing original draft preparation, Y.Q.; writing review and editing, Y.G. and Y.F.; visualization, Y.F. and Y.G.; supervision, Y.A.E.-K. and Y.G.; project administration, Y.G.; funding acquisition, L.X. All authors have read and agreed to the published version of the manuscript.

Funding: This research was funded by the Natural Science Foundation of Jiangsu Province, grant number BK20220411 (Y.G.), National Natural Science Foundation of China, grant number 32201595 (Y.G.), and National Natural Science Foundation of China, grant number 32101559 (F.F.).

Data Availability Statement: The sRNA sequencing data generated as part of this study are deposited in the NCBI Sequence Read Archive (SRA), accession BioProject number PRJNA903548.

Conflicts of Interest: The authors declare no conflict of interest.

Abbreviations

TTL	terpene trefoil lactone
miRNA	microRNA
WGCNA	weighted gene co-regulatory network
TF	transcription factor
IPP	isopentenyl diphosphate
DMAPP	dimethyl allyl pyrophosphate
MVA	mevalonate
MEP	methylerythritol 4-phosphate
DXS	1-deoxy-D-xylulose-5-phosphate synthase
DXR	1-deoxy-D-xylulose-5-phosphate reductoisomerase
HMGR	hydroxymethylglutaryl-CoA reductase
GA	ginkgolide A
GB	ginkgolide B
GC	ginkgolide C
BB	bilobalide
AACT	acetyl-CoA C-acetyltransferase
HMGS	hydroxymethylglutaryl-CoA synthase
MVK	mevalonate kinase
PMVK	phosphomevalonate kinase
MVD	diphosphomevalonate decarboxylase
CMS	2-C-methyl-D-erythritol 4-phosphate cytidylyltransferase
CMK	4-diphosphocytidyl-2-C-methyl-D-erythritol kinase
MCS	2-C-methyl-D-erythritol 2,4-cyclodiphosphate synthase
HDS	(E)-4-hydroxy-3-methylbut-2-enyl-diphosphate synthase
HDR	4-hydroxy-3-methylbut-2-en-1-yl diphosphate reductase
FPS	farnesyl diphosphate synthase
GGPS	geranylgeranyl diphosphate synthase
IDI	isopentenyl-diphosphate Delta-isomerase
LPS	levopimaradiene synthase
CYP450	Cytochrome P450
DEG	differentially expressed gene
DEM	differentially expressed miRNA
FC	fold change
GO	gene ontology
KEGG	Kyoto encyclopedia of genes and genomes
TPM	transcripts per million
sRNA	small RNA
ELSD	evaporative light scattering detector

References

1. Jacobs, B.P.; Browner, W.S. *Ginkgo biloba*: A living fossil. *Am. J. Med.* **2000**, *108*, 341–342. [CrossRef] [PubMed]
2. Li, F.; Boateng, I.D.; Yang, X.M.; Li, Y.; Liu, W. Effects of processing methods on quality, antioxidant capacity, and cytotoxicity of *Ginkgo biloba* leaf tea product. *J. Sci. Food Agric.* **2023**, *103*, 4993–5003. [CrossRef] [PubMed]
3. Chen, X.; Zhong, W.; Shu, C.; Yang, H.; Li, E. Comparative analysis of chemical constituents and bioactivities of the extracts from leaves, seed coats and embryoids of *Ginkgo biloba* L. *Nat. Prod. Res.* **2021**, *35*, 5498–5501. [CrossRef] [PubMed]
4. Dubey, A.; Marabotti, A.; Ramteke, P.W.; Facchiano, A. Interaction of human chymase with ginkgolides, terpene trilactones of *Ginkgo biloba* investigated by molecular docking simulations. *Biochem. Biophys. Res. Commun.* **2016**, *473*, 449–454. [CrossRef] [PubMed]
5. Liu, X.G.; Lu, X.; Gao, W.; Li, P.; Yang, H. Structure, synthesis, biosynthesis, and activity of the characteristic compounds from *Ginkgo biloba* L. *Nat. Prod. Rep.* **2022**, *39*, 474–511. [CrossRef] [PubMed]
6. Jaracz, S.; Malik, S.; Nakanishi, K. Isolation of ginkgolides A, B, C, J and bilobalide from *G. biloba* extracts. *Phytochemistry* **2004**, *65*, 2897–2902. [CrossRef]
7. Crimmins, M.T.; Pace, J.M.; Nantermet, P.G.; Kim-Meade, A.S.; Thomas, J.B.; Watterson, S.H.; Wagman, A.S. The total synthesis of (±)-ginkgolide B. *J. Am. Chem. Soc.* **2000**, *122*, 8453–8463. [CrossRef]

8. Sabater-Jara, A.B.; Souliman-Youssef, S.; Novo-Uzal, E.; Almagro, L.; Belchí-Navarro, S.; Pedreño, M.A. Biotechnological approaches to enhance the biosynthesis of ginkgolides and bilobalide in *Ginkgo biloba*. *Phytochem. Rev.* **2013**, *12*, 191–205. [CrossRef]

9. Zhu, C.; Cao, F.; Wang, G.; Geng, G. Effects of drought stress on annual dynamic changing pattern of the terpene lactones content in *Ginkgo biloba* leaves. *China For. Sci. Technol.* **2011**, *25*, 15–20.

10. Lange, B.M.; Ghassemian, M. Genome organization in *Arabidopsis thaliana*: A survey for genes involved in isoprenoid and chlorophyll metabolism. *Plant Mol. Biol.* **2003**, *51*, 925–948. [CrossRef]

11. Kang, S.M.; Min, J.Y.; Kim, Y.D.; Park, D.J.; Jung, H.N.; Karigar, C.S.; Ha, Y.L.; Kim, S.W.; Choi, M.S. Effect of supplementing terpenoid biosynthetic precursors on the accumulation of bilobalide and ginkgolides in *Ginkgo biloba* cell cultures. *J. Biotechnol.* **2006**, *123*, 85–92. [CrossRef] [PubMed]

12. Zheng, J.; Zhang, X.; Fu, M.; Zeng, H.; Ye, J.; Zhang, W.; Liao, Y.; Xu, F. Effects of different stress treatments on the total terpene trilactone content and expression levels of key genes in *Ginkgo biloba* Leaves. *Plant Mol. Biol. Report.* **2020**, *38*, 521–530. [CrossRef]

13. Carbonell, A.; Fahlgren, N.; Garcia-Ruiz, H.; Gilbert, K.B.; Montgomery, T.A.; Nguyen, T.; Cuperus, J.T.; Carrington, J.C. Functional analysis of three *Arabidopsis* ARGONAUTES using slicer-defective mutants. *Plant Cell* **2012**, *24*, 3613–3629. [CrossRef] [PubMed]

14. Ye, J.; Zhang, X.; Tan, J.; Xu, F.; Cheng, S.; Chen, Z.; Zhang, W.; Liao, Y. Global identification of *Ginkgo biloba* microRNAs and insight into their role in metabolism regulatory network of terpene trilactones by high-throughput sequencing and degradome analysis. *Ind. Crops Prod.* **2020**, *148*, 112289. [CrossRef]

15. Singh, N.; Srivastava, S.; Shasany, A.K.; Sharma, A. Identification of miRNAs and their targets involved in the secondary metabolic pathways of *Mentha* spp. *Comput. Biol. Chem.* **2016**, *64*, 154–162. [CrossRef] [PubMed]

16. Singh, N.; Sharma, A. Turmeric (*Curcuma longa*): miRNAs and their regulating targets are involved in development and secondary metabolite pathways. *Comptes Rendus Biol.* **2017**, *340*, 481–491. [CrossRef]

17. Aravind, J.; Rinku, S.; Pooja, B.; Shikha, M.; Kaliyugam, S.; Mallikarjuna, M.G.; Kumar, A.; Rao, A.R.; Nepolean, T. Identification, characterization, and functional validation of drought-responsive microRNAs in subtropical maize inbreds. *Front. Plant Sci.* **2017**, *8*, 941. [CrossRef]

18. Yang, C.; Li, D.; Mao, D.; Liu, X.U.E.; Ji, C.; Li, X.; Zhao, X.; Cheng, Z.; Chen, C.; Zhu, L. Overexpression of microRNA319 impacts leaf morphogenesis and leads to enhanced cold tolerance in rice (*Oryza sativa* L.). *Plant Cell Environ.* **2013**, *36*, 2207–2218. [CrossRef]

19. Wen, M.; Xie, M.; He, L.; Wang, Y.; Shi, S.; Tang, T. Expression Variations of miRNAs and mRNAs in Rice (*Oryza sativa*). *Genome Biol. Evol.* **2016**, *8*, 3529–3544. [CrossRef]

20. Guo, Y.; Gao, C.; Wang, M.; Fu, F.-f.; El-Kassaby, Y.A.; Wang, T.; Wang, G. Metabolome and transcriptome analyses reveal flavonoids biosynthesis differences in *Ginkgo biloba* associated with environmental conditions. *Ind. Crops Prod.* **2020**, *158*, 112963. [CrossRef]

21. Li, F.; Boateng, I.D.; Chen, S.; Yang, X.M.; Soetanto, D.A.; Liu, W. Pulsed light irradiation improves degradation of ginkgolic acids and retainment of ginkgo flavonoids and terpene trilactones in *Ginkgo biloba* leaves. *Ind. Crops Prod.* **2023**, *204*, 117297. [CrossRef]

22. Li, L.; Stanton, J.D.; Tolson, A.H.; Luo, Y.; Wang, H. Bioactive terpenoids and flavonoids from *Ginkgo biloba* extract induce the expression of hepatic drug-metabolizing enzymes through pregnane X receptor, constitutive androstane receptor, and aryl hydrocarbon receptor-mediated pathways. *Pharm. Res.* **2009**, *26*, 872–882. [CrossRef] [PubMed]

23. Yu, W.; Cai, J.; Liu, H.; Lu, Z.; Hu, J.; Lu, Y. Transcriptomic analysis reveals regulatory networks for osmotic water stress and rewatering response in the leaves of *Ginkgo biloba*. *Forests* **2021**, *12*, 1705. [CrossRef]

24. Zhang, C.; Guo, J.; Chen, G.; Xie, H. Effects of high temperature and/or drought on growth and secondary metabolites in *Ginkgo biloba* leaves. *J. Ecol. Rural Environ.* **2005**, *21*, 11–15.

25. Wang, H.; Xie, B.; Jiang, Y.; Wang, M. Effect of solar irradiation intensity on leaf development and flavonoid and Terpene content in *Ginkgo biloba* leaves. *Acta Agric. Univ. Jiangxiensis* **2002**, *24*, 617–622.

26. Cardenas, P.D.; Sonawane, P.D.; Pollier, J.; Vanden Bossche, R.; Dewangan, V.; Weithorn, E.; Tal, L.; Meir, S.; Rogachev, I.; Malitsky, S.; et al. GAME9 regulates the biosynthesis of steroidal alkaloids and upstream isoprenoids in the plant mevalonate pathway. *Nat. Commun.* **2016**, *7*, 10654. [CrossRef] [PubMed]

27. Li, S.; Wang, H.; Li, F.; Chen, Z.; Li, X.; Zhu, L.; Wang, G.; Yu, J.; Huang, D.; Lang, Z. The maize transcription factor *EREB58* mediates the jasmonate-induced production of sesquiterpene volatiles. *Plant J.* **2015**, *84*, 296–308. [CrossRef] [PubMed]

28. Yu, Z.; Li, J.; Yang, C.; Hu, W.; Wang, L.; Chen, X. The jasmonate-responsive AP2/ERF transcription factors AaERF1 and AaERF2 positively regulate artemisinin biosynthesis in *Artemisia annua* L. *Mol. Plant* **2012**, *5*, 353–365. [CrossRef]

29. Wang, M.; Gao, M.; Zhao, Y.; Chen, Y.; Wu, L.; Yin, H.; Xiong, S.; Wang, S.; Wang, J.; Yang, Y.; et al. LcERF19, an AP2/ERF transcription factor from Litsea cubeba, positively regulates geranial and neral biosynthesis. *Hortic. Res.* **2022**, *9*, uhac093. [CrossRef]

30. Paul, P.; Singh, S.K.; Patra, B.; Sui, X.; Pattanaik, S.; Yuan, L. A differentially regulated AP2/ERF transcription factor gene cluster acts downstream of a MAP kinase cascade to modulate terpenoid indole alkaloid biosynthesis in *Catharanthus roseus*. *New Phytol.* **2017**, *213*, 1107–1123. [CrossRef]

31. Nieuwenhuizen, N.J.; Chen, X.; Wang, M.Y.; Matich, A.J.; Perez, R.L.; Allan, A.C.; Green, S.A.; Atkinson, R.G. Natural variation in monoterpene synthesis in kiwifruit: Transcriptional regulation of terpene synthases by NAC and ETHYLENE-INSENSITIVE3-like transcription factors. *Plant Physiol.* **2015**, *167*, 1243–1258. [CrossRef] [PubMed]

32. Liu, H.; Wang, X.; Wang, G.; Cui, P.; Wu, S.; Ai, C.; Hu, N.; Li, A.; He, B.; Shao, X.; et al. The nearly complete genome of *Ginkgo biloba* illuminates gymnosperm evolution. *Nat. Plants* **2021**, *7*, 748–756. [CrossRef] [PubMed]
33. Jatan, R.; Lata, C. Role of microRNAs in abiotic and biotic stress resistance in plants. *Proc. Indian Natl. Sci. Acad.* **2019**, *85*, 553–567.
34. Luan, M.; Xu, M.; Lu, Y.; Zhang, L.; Fan, Y.; Wang, L. Expression of zma-miR169 miRNAs and their target *ZmNF-YA* genes in response to abiotic stress in maize leaves. *Gene* **2015**, *555*, 178–185. [CrossRef] [PubMed]
35. Moradi, K.; Khalili, F. Assessment of pattern expression of miR172 and miR169 in response to drought stress in *Echinacea purpurea* L. *Biocatal. Agric. Biotechnol.* **2018**, *16*, 507–512. [CrossRef]
36. Wu, J.; Wang, L.; Wang, S. MicroRNAs associated with drought response in the pulse crop common bean (*Phaseolus vulgaris* L.). *Gene* **2017**, *628*, 78–86. [CrossRef] [PubMed]
37. Serivichyaswat, P.T.; Susila, H.; Ahn, J.H. Elongated Hypocotyl 5-Homolog (HYH) negatively regulates expression of the ambient temperature-responsive microRNA gene MIR169. *Front. Plant Sci.* **2017**, *8*, 2087. [CrossRef]
38. Sun, X.; Wang, C.; Xiang, N.; Li, X.; Yang, S.; Du, J.; Yang, Y.; Yang, Y. Activation of secondary cell wall biosynthesis by miR319-targeted *TCP4* transcription factor. *Plant Biotechnol. J.* **2017**, *15*, 1284–1294. [CrossRef]
39. Liu, Y.; Li, D.; Yan, J.; Wang, K.; Luo, H.; Zhang, W. MiR319 mediated salt tolerance by ethylene. *Plant Biotechnol. J.* **2019**, *17*, 2370–2383. [CrossRef]
40. National Meteorological Science Data Center. Available online: https://data.cma.cn/ (accessed on 2 January 2022).
41. Chinese Pharmacopoeia Commission. *Pharmacopoeia of the People's Republic of China*; Chinese Medical Science and Technology Press: Beijing, China, 2010.
42. Kanehisa, M.; Goto, S. KEGG: Kyoto encyclopedia of genes and genomes. *Nucleic Acids Res.* **2000**, *28*, 27–30. [CrossRef]
43. Langdon, W.B. Performance of genetic programming optimised Bowtie2 on genome comparison and analytic testing (GCAT) benchmarks. *Biodata Min.* **2015**, *8*, 1. [CrossRef] [PubMed]
44. Li, B.; Dewey, C.N. RSEM: Accurate transcript quantification from RNA-Seq data with or without a reference genome. *BMC Bioinform.* **2011**, *12*, 323. [CrossRef] [PubMed]
45. Love, M.I.; Huber, W.; Anders, S. Moderated estimation of fold change and dispersion for RNA-seq data with DESeq2. *Genome Biol.* **2014**, *15*, 550. [CrossRef] [PubMed]
46. Bairoch, A.; Apweiler, R. The SWISS-PROT protein sequence database and its supplement TrEMBL in 2000. *Nucleic Acids Res.* **2000**, *28*, 45–48. [CrossRef] [PubMed]
47. Aleksander, S.A.; Balhoff, J.; Carbon, S.; Cherry, J.M.; Drabkin, H.J.; Ebert, D.; Feuermann, M.; Gaudet, P.; Harris, N.L. The Gene Ontology knowledgebase in 2023. *Genetics* **2023**, *224*, iyad031. [PubMed]
48. Jin, J.; Tian, F.; Yang, D.C.; Meng, Y.Q.; Kong, L.; Luo, J.; Gao, G. PlantTFDB 4.0: Toward a central hub for transcription factors and regulatory interactions in plants. *Nucleic Acids Res.* **2017**, *45*, 1040–1045. [CrossRef]
49. Qi, Y.; Xue, L.; El-Kassaby, Y.; Guo, Y. Identification and comparative analysis of conserved and species-specific microRNAs in four *Populus* sections. *Forests* **2022**, *13*, 873. [CrossRef]
50. Axtell, M.J. ShortStack: Comprehensive annotation and quantification of small RNA genes. *RNA* **2013**, *19*, 740–751. [CrossRef]
51. Martin, M. Cutadapt removes adapter sequences from high-throughput sequencing reads. *Embnet* **2011**, *17*, 10–12. [CrossRef]
52. Langmead, B. Aligning short sequencing reads with Bowtie. *Curr. Protoc. Bioinform.* **2010**, *32*, 11.7.1–11.7.14. [CrossRef]
53. Prufer, K.; Stenzel, U.; Dannemann, M.; Green, R.E.; Lachmann, M.; Kelso, J. PatMaN: Rapid alignment of short sequences to large databases. *Bioinformatics* **2008**, *24*, 1530–1531. [CrossRef]
54. Dai, X.; Zhuang, Z.; Zhao, P.X. psRNATarget: A plant small RNA target analysis server (2017 release). *Nucleic Acids Res.* **2018**, *46*, 49–54. [CrossRef] [PubMed]
55. Langfelder, P.; Horvath, S. WGCNA: An R package for weighted correlation network analysis. *BMC Bioinform.* **2008**, *9*, 559. [CrossRef] [PubMed]
56. Smoot, M.E.; Ono, K.; Ruscheinski, J.; Wang, P.L.; Ideker, T. Cytoscape 2.8: New features for data integration and network visualization. *Bioinformatics* **2010**, *27*, 431–432. [CrossRef] [PubMed]
57. Schmittgen, T.D.; Livak, K.J. Analyzing real-time PCR data by the comparative C(T) method. *Nat. Protoc.* **2008**, *3*, 1101–1108. [CrossRef]

International Journal of
Molecular Sciences

Article

Transcriptomic, Physiological, and Metabolomic Response of an Alpine Plant, *Rhododendron delavayi*, to Waterlogging Stress and Post-Waterlogging Recovery

Xi-Min Zhang [1,2,3,*], Sheng-Guang Duan [3], Ying Xia [1,3], Jie-Ting Li [3,4], Lun-Xian Liu [1,3], Ming Tang [3,4], Jing Tang [1,3], Wei Sun [1,3] and Yin Yi [1,4]

[1] Key Laboratory of Plant Physiology and Development Regulation, Guizhou Normal University, Guiyang 550025, China; llunxian@163.com (L.-X.L.); tangjing2016@gznu.edu.cn (J.T.); sunwei889@163.com (W.S.); gzklppdr@gznu.edu.cn (Y.Y.)
[2] Key Laboratory of Environment Friendly Management on Alpine Rhododendron Diseases and Pests of Institutions of Higher Learning in Guizhou Province, Guizhou Normal University, Guiyang 550025, China
[3] School of Life Sciences, Guizhou Normal University, Guiyang 550025, China; mingtang@gznu.edu.cn (M.T.)
[4] Key Laboratory of State Forestry Administration on Biodiversity Conservation in Karst Area of Southwest, Guizhou Normal University, Guiyang 550025, China
* Correspondence: author: zhxm409@163.com

Abstract: Climate change has resulted in frequent heavy and prolonged rainfall events that exacerbate waterlogging stress, leading to the death of certain alpine *Rhododendron* trees. To shed light on the physiological and molecular mechanisms behind waterlogging stress in woody *Rhododendron* trees, we conducted a study of *Rhododendron delavayi*, a well-known alpine flower species. Specifically, we investigated the physiological and molecular changes that occurred in leaves of *R. delavayi* subjected to 30 days of waterlogging stress (WS30d), as well as subsequent post-waterlogging recovery period of 10 days (WS30d-R10d). Our findings reveal that waterlogging stress causes a significant reduction in CO_2 assimilation rate, stomatal conductance, transpiration rate, and maximum photochemical efficiency of PSII (Fv/Fm) in the WS30d leaves, by 91.2%, 95.3%, 93.3%, and 8.4%, respectively, when compared to the control leaves. Furthermore, the chlorophyll a and total chlorophyll content in the WS30d leaves decreased by 13.5% and 16.6%, respectively. Both WS30d and WS30d-R10d leaves exhibited excessive H_2O_2 accumulation, with a corresponding decrease in lignin content in the WS30d-R10d leaves. At the molecular level, purine metabolism, glutathione metabolism, photosynthesis, and photosynthesis-antenna protein pathways were found to be primarily involved in WS30d leaves, whereas phenylpropanoid biosynthesis, fatty acid metabolism, fatty acid biosynthesis, fatty acid elongation, and cutin, suberin, and wax biosynthesis pathways were significantly enriched in WS30d-R10d leaves. Additionally, both WS30d and WS30d-R10d leaves displayed a build-up of sugars. Overall, our integrated transcriptomic, physiological, and metabolomic analysis demonstrated that *R. delavayi* is susceptible to waterlogging stress, which causes irreversible detrimental effects on both its physiological and molecular aspects, hence compromising the tree's ability to fully recover, even under normal growth conditions.

Keywords: *Rhododendron delavayi*; waterlogging; photosynthesis; lignin; hydrogen peroxide; soluble sugar

Citation: Zhang, X.-M.; Duan, S.-G.; Xia, Y.; Li, J.-T.; Liu, L.-X.; Tang, M.; Tang, J.; Sun, W.; Yi, Y. Transcriptomic, Physiological, and Metabolomic Response of an Alpine Plant, *Rhododendron delavayi*, to Waterlogging Stress and Post-Waterlogging Recovery. *Int. J. Mol. Sci.* **2023**, 24, 10509. https://doi.org/10.3390/ijms241310509

Academic Editor: Martin Bartas

Received: 2 May 2023
Revised: 17 June 2023
Accepted: 20 June 2023
Published: 22 June 2023

1. Introduction

Climate change has had a profound impact on the global hydrological cycle, resulting in a marked increase in heavy or recurrent precipitation events in certain regions. Such rainfall frequently exceeds the soil's water-holding capacity, resulting in oversaturation of terrestrial ecosystems, especially in zones with poor drainage, clay-rich soils, and low-lying areas. This oversaturation can trigger severe waterlogging stress as a result of soil ponding [1–3].

Waterlogging stress impedes the gas exchange between plant tissues and soil, causing anoxic or hypoxic conditions in the waterlogged tissues. These conditions can cause an "energy crisis" due to the inhibition of ATP generation [1,4]. Plant responses to waterlogging stress involve complex signal transduction pathways that alter the synthesis and balance of plant hormones, resulting in stomatal closure and morphological changes [5,6]. Additionally, waterlogging stress induces the accumulation of plant metabolites such as flavonoids, vitamin B6, and sugars that play a crucial role in the response to waterlogging stress [7]. Previous studies have shown that hormone mediated transcription factors, including MYBs, ERFs, WRKYs, and NACs, regulate plant responses to waterlogging stress [8,9]. Moreover, waterlogging triggers anaerobic metabolism in roots, leading to the production of ethanol, acetaldehyde, and other harmful substances that can poison root cells and damage protein structure [10,11]. Ultimately, this leads to accelerated root decay, damage, and even death [12].

Currently, extensive research has been conducted on the molecular and physiological responses to waterlogging stress in model plants such as Arabidopsis and crops such as *Solanum lycopersicum, Cucumis sativus*, and *Glycine max* [2,13]. However, trees play a crucial ecological role in global climate change and are essential for urban environments, such as greening streets and parks. Therefore, it is crucial to understand the mechanisms of underlying tree response to waterlogging stress. Although some studies have reported on the physiological response mechanisms of waterlogging-sensitive species or varieties [3,14–16], only a few have investigated changes in the transcriptome and metabolome of woody economic species, such as poplar, cotton, oak, kiwifruit, and *Guazuma ulmifolia*, exposed to waterlogging stress [17–20]. Thus, there is a need to investigate the response mechanisms of specific woody trees to waterlogging stress, as given that their tolerance mainly depends on the species and degree of waterlogging [21].

Rhododendron, the largest genus in the Ericaceae family and one of the most notable alpine flowers, comprises approximately 1143 species of evergreen and deciduous woody plants. In the northwest Guizhou province of China, wild *Rhododendron* trees in the Baili Azalea Nature Reserve (BANR) represent significant potential tourism resources for local communities. Although these trees typically grow in high mountain areas, some low-lying areas or clay-rich soils have experienced long-term ponding in recent years, hindering the growth of *Rhododendron* species and leading to the death of some trees. Despite the importance of Rhododenron trees in this region, there is currently a scarcity of research on the physiological and molecular response to waterlogging stress and post-waterlogging recovery in this genus.

Rhododendron delavayi, a crucial member of the *Rhododendron* genus in the BANR, has unfortunately experienced mortality in low-lying areas or clay-rich soils due to continuous rainfall. This has prompted us to urgently investigate the response mechanisms of *R. delavayi* to waterlogging stress. In this study, *R. delavayi*, an alpine plant, has been utilized as a model system to explore the physiological and molecular responses to waterlogging stress and post-waterlogging recovery. This approach aims to provide key insights into how alpine plants respond to waterlogging stress both at physiological and molecular levels. Ultimately, our findings can contribute a fundamental understanding of the death of *R. delavayi* caused by waterlogging stress.

2. Results

2.1. Waterlogging Stress Caused the Aged Leaf Wilting in R. delavayi Seedlings

To explore the physiological and molecular responses of *R. delavayi* to waterlogging stress, *R. delavayi* seedlings were subjected to various durations of waterlogging stress (WS10d, WS20d, and WS30d) followed by a post-waterlogging recovery period (WS10d-R10d, WS20d-R10d, and WS30d-R10d). During WS10d, WS20d, WS10d-R10d, and WS20d-R10d, no stress symptoms were observed when compared to their pre-waterlogging state (Figure 1). However, prolonged waterlogging stress (WS30d) led to the wilting of aged leaves in some seedlings (as indicated by white arrow in Figure 1) and even 10 days of post-waterlogging recovery was insufficient to fully restore the wilted leaves to their pre-waterlogging state (as indicated by red arrow in Figure 1). These results suggested that waterlogging stress lasting for 30 days can severely damage *R. delavayi* seedlings.

Figure 1. The leaf phenotype of *R. delavayi* after waterlogging stress and post-waterlogging recovery. Control (CK), waterlogging stress for 10 days (WS10d), 20 days (WS20d), and 30 days (WS30d) were indicated. Post-waterlogging recovery for 10 days was indicated by WS10d-R10d, WS20d-R10d, WS30d-R10d. Wilting of leaves after WS30d and WS30d-R10d was represented by white and red arrow, respectively. Bar = 5.8 cm.

2.2. Transcriptome Assembly, Analysis of Differentially Expressed Genes (DEGs), and qRT-PCR Validation

To investigate the molecular mechanisms underlying the response of *R. delavayi* to waterlogging stress at the transcriptome level, RNA sequencing was performed on leaves from CK, WS30d, and WS30d-R10d plants, based on the stress symptoms of *R. delavayi* leaves caused by waterlogging stress. Over 20,163,917 clean reads were obtained from each sample, with the lowest mapped ratio being 76.95% in WS30d-R10d-2. The guanine and cytosine/total base (GC content) and the quality of base calling accuracy at 99.9% (Q30) were more than 47.42% and 92.33%, respectively (Table S1). After assembly of the high-quality reads, a total of 97,152 unigenes and 402,244 trancripts with mean lengths of 870 bp and 1705 bp, respectively, were obtained (Table S2).

To determine the variation and similarity of the gene expression profiles among the samples, we conducted principal component analysis (PCA) of all detected genes using normalized Fragments Per Kilobase of transcript per Million mapped reads (FPKM) values. The difference in gene expression among the three samples was statistically significant ($p = 0.032$), with PC1 and PC2 accounting for 58.5% and 26.4% of the total variation, respectively (Figure 2A). The PCA plot showed that the data from the three biological replicates were closely clustered together and separated between CK and WS30d and WS30d-R10d (Figure 2A), demonstrating that the sequencing data from stressed leaves differed significantly from those of the CK leaves.

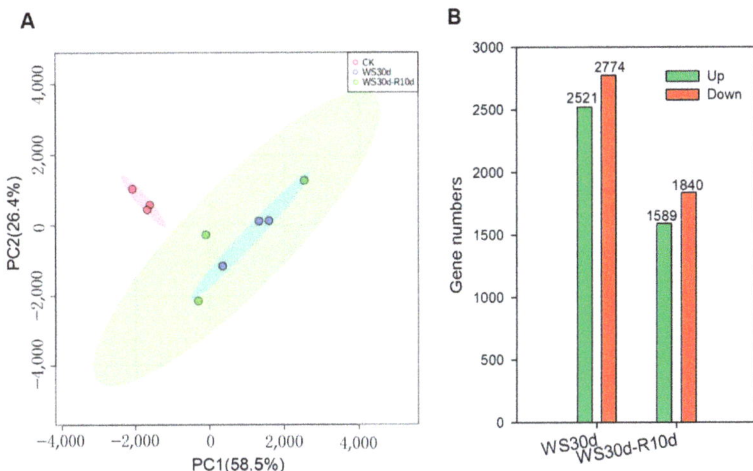

Figure 2. The principal component analysis (PCA) and differentially expressed genes (DEGs). (**A**) PCA of all detected genes; (**B**) DEGs in WS30d and WS30d-R10d leaves in *R. delavayi* compared to CK.

We subsequently identified differentially expressed genes (DEGs) among the WS30d, WS30d-R10d, and CK samples. Comparion of WS30d leaves to the CK leaves revealed 2521 up-regulated and 2774 down-regulated DEGs, while comparison of WS30d-R10d to CK leaves identified 1589 up-regulated and 1840 down-regulated DEGs (Figure 2B, Tables S3 and S4). DEGs between WS30d and WS30d-R10d were also obtained (Table S5). These results suggested that waterlogging stress and post-waterlogging recovery significantly induced gene transcription in *R. delavayi* leaves.

To verify the reliability of DEGs identified by RNA sequencing, we randomly selected six genes and quantified their expression levels using quantitative Real-Time PCR (qRT-PCR) analysis. The results demonstrated the consistency between the qRT-PCR data and the RNA sequencing data, thereby confirming the reliability of the DEG screened by RNA sequencing (Figure S1).

2.3. KEGG Pathways by DEGs of Waterlogging Stress Versus Control

To further comprehend the biological functions of the 5295 DEGs induced by 30 days of waterlogging stress, we performed Kyoto Encyclopedia of Genes and Genomes (KEGG) pathway analysis. The up-regulated DEGs were significantly enriched in the "purine metabolism" and "glutathione metabolism" pathways ($p < 0.05$) (Figure 3A), while the down-regulated DEGs were significantly enriched in "photosynthesis", "photosynthesis-antenna proteins", "fat acid biosynthesis", and "glycosaminoglycan degradation" pathways ($p < 0.05$) (Figure 3B).

In the "purine metabolism" pathway, the enzymes encoded by the up-regulated DEGs directly or indirectly catalyzed ATP production (Table 1), indicating their crucial role in providing energy. In the "Glutathione metabolism" pathway, 21 up-regulated DEGs were annotated as glutamate-cysteine ligase, glutathione peroxidase, glutathione reductase, and isocitrate dehydrogenase, which participate in the glutathione cycle (Table 1) and facilitate the accumulation of glutathione, as evidenced by metabolome detection (Table 2). Furthermore, four DEGs were up-regulated and annotated as L-ascorbate peroxidase and ribonucleoside-diphosphate reductase (Table 1), suggesting their involvement in the ascorbate cycle.

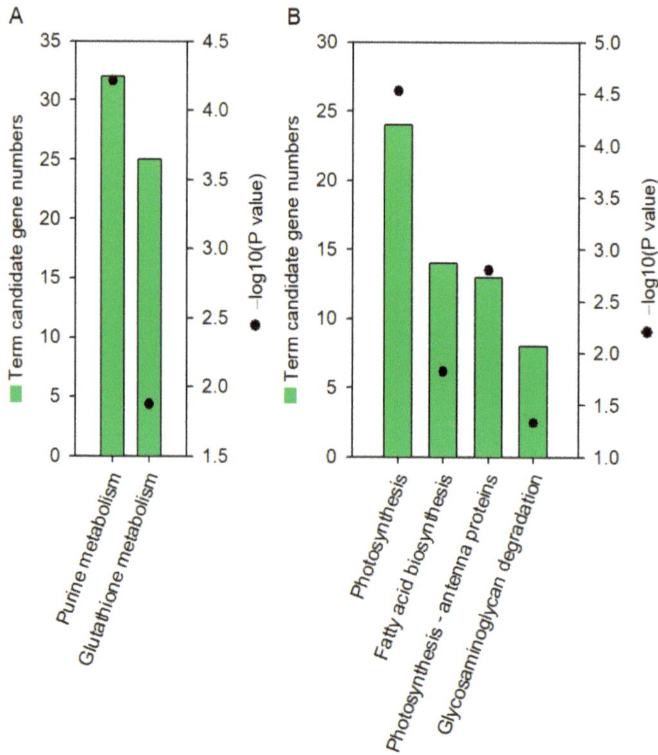

Figure 3. KEGG pathways of DEGs in WS30d leaves response to waterlogging stress. (**A**) up-regulation DEGs enriched in KEGG; (**B**) down-regulation DEGs enriched in KEGG.

Table 1. DEGs of WS30d versus CK enriched in KEGG pathways ($p < 0.05$).

#ID	FPKM		FDR	$\log_2$FC	KEGG_Annotation
	CK	WS30d			
Purine Metabolism					
c61670.graph_c0	24	121.11	5.95×10^{-8}	2.32	3′-phosphoadenosine 5′-phosphosulfate synthase
c68596.graph_c0	21.65	222.9	1.81×10^{-10}	2.88	3′-phosphoadenosine 5′-phosphosulfate synthase
c48065.graph_c0	24.05	66.58	0.009568452	1.19	adenylate kinase
c55902.graph_c0	5.9	17.61	0.002497335	1.31	adenylate kinase
c61967.graph_c0	31.45	45.22	0.001423826	1.30	amidophosphoribosyltransferase
c54389.graph_c0	0.69	2.94	5.56×10^{-8}	1.77	AMP deaminase
c56269.graph_c0	5.46	9.79	0.002969771	0.68	DNA polymerase delta subunit 3
c61104.graph_c0	3.61	7.69	3.14×10^{-5}	0.94	DNA polymerase I
c67065.graph_c1	11.34	23.74	2.70×10^{-5}	0.72	DNA polymerase I
c61560.graph_c1	7.9	16.7	0.000143442	1.37	DNA primase small subunit
c66963.graph_c1	7.71	15.06	0.002704535	0.80	DNA-directed RNA polymerase I subunit RPA1
c66139.graph_c0	6.19	15.02	4.30×10^{-5}	1.08	DNA-directed RNA polymerase I subunit RPA2
c65110.graph_c0	0.12	1.13	0.00607105	2.12	DNA-directed RNA polymerase II subunit RPB1
c54215.graph_c1	4.71	10.21	0.001854865	0.87	DNA-directed RNA polymerase II subunit RPB4
c55954.graph_c1	3.04	5.27	0.003121169	0.73	DNA-directed RNA polymerase III subunit RPC1
c63522.graph_c0	14.72	29.28	1.94×10^{-6}	0.84	DNA-directed RNA polymerase III subunit RPC2
c61042.graph_c0	0.61	2.01	0.006880565	1.56	DNA-directed RNA polymerase subunit
c61639.graph_c0	32.03	62.06	7.63×10^{-5}	0.80	DNA-directed RNA polymerase subunit
c44354.graph_c0	3	10.15	1.84×10^{-5}	1.53	DNA-directed RNA polymerases I and III subunit RPAC1
c66477.graph_c0	40.5	123.45	2.42×10^{-9}	1.43	hydroxyisourate hydrolase
c61783.graph_c1	53.95	124.98	0.004522204	1.06	nucleoside-diphosphate kinase

Table 1. *Cont.*

#ID	FPKM		FDR	log$_2$FC	KEGG_Annotation
	CK	WS30d			
c56191.graph_c0	15.05	37.97	2.23×10^{-5}	1.14	phosphoglucomutase
c66317.graph_c0	0.56	2.54	5.01×10^{-8}	1.88	polyribonucleotide nucleotidyltransferase
c46598.graph_c0	12.84	30.73	0.001705575	1.06	pyruvate kinase
c56447.graph_c0	11.28	30.88	0.000148805	1.24	pyruvate kinase
c46598.graph_c1	13.17	35.15	0.005231922	1.16	pyruvate kinase
c67771.graph_c1	20.81	59.57	0.001272097	1.27	pyruvate kinase
c65971.graph_c0	3.1	8.33	1.43×10^{-7}	1.27	ribonucleoside-diphosphate reductase subunit M1
c65971.graph_c2	3.17	7.96	1.74×10^{-5}	1.16	ribonucleoside-diphosphate reductase subunit M1
c51808.graph_c0	2.91	8.08	1.11×10^{-6}	1.32	ribonucleoside-diphosphate reductase subunit M2
c59718.graph_c0	2.19	7.62	1.01×10^{-6}	1.60	ribose-phosphate pyrophosphokinase
c65143.graph_c1	35.14	88.48	0.001053369	1.18	urate oxidase
Glutathione metabolism					
c58797.graph_c0	28.29	59.89	0.000787329	0.82	glutamate-cysteine ligase
c49129.graph_c0	17.02	64.83	0.000110947	1.66	glutathione peroxidase
c56863.graph_c0	45.09	111.71	3.73×10^{-13}	1.11	glutathione reductase
c59612.graph_c0	94.14	241.87	0.002044808	1.18	glutathione reductase
c64812.graph_c0	18.45	130.44	2.77×10^{-5}	1.92	glutathione S-transferase
c59164.graph_c0	1.5	11.4	3.70×10^{-6}	2.53	glutathione S-transferase
c63277.graph_c1	25.21	729.15	3.79×10^{-9}	3.91	glutathione S-transferase
c64156.graph_c0	11.59	23.7	0.000578537	0.95	glutathione S-transferase
c60034.graph_c0	1.33	89.58	0.000390662	3.12	glutathione S-transferase
c47361.graph_c0	2.25	13.34	0.000218054	2.10	glutathione S-transferase
c61754.graph_c0	1.38	3.79	0.006001931	1.04	glutathione S-transferase
c51875.graph_c0	102.67	346.18	0.00021695	1.49	glutathione S-transferase
c50282.graph_c0	18.24	139.8	9.15×10^{-7}	2.53	glutathione S-transferase
c67680.graph_c0	149.26	673.04	0.000191146	1.90	glutathione S-transferase
c65078.graph_c0	19.02	45.97	2.39×10^{-5}	1.14	glutathione S-transferase
c60993.graph_c0	4.01	18.46	3.33×10^{-6}	2.04	glutathione S-transferase
c65367.graph_c0	37.95	463.28	1.92×10^{-11}	3.32	glutathione S-transferase
c50305.graph_c0	1.07	8.46	0.002469129	2.14	glutathione S-transferase
c64290.graph_c0	9.52	84.97	2.20×10^{-6}	2.72	glutathione S-transferase
c52719.graph_c0	5.53	21.94	6.37×10^{-5}	1.82	glutathione synthase
c62411.graph_c0	5.21	15.07	7.17×10^{-8}	1.36	isocitrate dehydrogenase
c55430.graph_c0	287.94	520.51	0.00599229	0.64	L-ascorbate peroxidase
c65971.graph_c0	3.1	8.33	1.43×10^{-7}	1.27	ribonucleoside-diphosphate reductase subunit M1
c65971.graph_c2	3.17	7.96	1.74×10^{-5}	1.16	ribonucleoside-diphosphate reductase subunit M1
c51808.graph_c0	2.91	8.08	1.11×10^{-6}	1.32	ribonucleoside-diphosphate reductase subunit M2
Photosynthesis					
c68574.graph_c2	46.26	13.35	5.24×10^{-8}	−2.1001	cytochrome b6/f
c50907.graph_c1	158.23	86.54	1.12×10^{-6}	−1.078	ferredoxin
c44849.graph_c0	40.66	26.51	0.000278672	−0.7568	ferredoxin-NADP+ reductase
c63271.graph_c3	62.05	18.73	5.44×10^{-11}	−1.7936	F-type H+-transporting ATPase subunit a
c57998.graph_c0	67.2	20.66	1.48×10^{-12}	−2.0903	F-type H+-transporting ATPase subunit alpha
c61006.graph_c0	469.97	258.81	0.00101548	−0.9944	F-type H+-transporting ATPase subunit delta
c55205.graph_c0	1.32	0.29	0.003579127	−1.9289	F-type H+-transporting ATPase subunit epsilon
c67974.graph_c4	112.42	21.02	3.21×10^{-8}	−2.3965	photosystem I P700 chlorophyll a apoprotein A1
c66205.graph_c2	1099.4	676.68	0.005253358	−0.8404	photosystem I subunit II
c44286.graph_c0	1447.6	631.02	0.000768555	−1.3079	photosystem I subunit IV
c43761.graph_c0	1209.2	551.7	0.000426669	−1.2402	photosystem I subunit PsaN
c50596.graph_c0	603.5	317.38	5.96×10^{-5}	−1.0638	photosystem I subunit V
c47764.graph_c0	2073.3	1148.2	0.001696486	−0.9827	photosystem I subunit VI
c50541.graph_c0	1059.1	428.24	0.00013698	−1.3976	photosystem I subunit X
c46857.graph_c0	1711	892.21	0.000848878	−1.0656	photosystem I subunit XI
c63847.graph_c0	28.52	4.57	3.13×10^{-15}	−2.6784	photosystem II CP43 chlorophyll apoprotein
c63896.graph_c0	32.78	14.67	7.23×10^{-7}	−1.3887	photosystem II CP43 chlorophyll apoprotein
c55414.graph_c0	37.83	8.26	1.30×10^{-10}	−2.3996	photosystem II CP47 chlorophyll apoprotein
c48666.graph_c0	42.45	22.47	9.08×10^{-5}	−1.0589	photosystem II oxygen-evolving enhancer protein 2
c56596.graph_c0	1509.8	767.15	0.000796919	−1.1323	photosystem II oxygen-evolving enhancer protein 3
c62948.graph_c3	391.53	79.36	9.84×10^{-12}	−2.3642	photosystem II P680 reaction center D1 protein
c59404.graph_c0	46.69	10.13	5.63×10^{-11}	−2.2111	photosystem II P680 reaction center D2 protein
c53605.graph_c0	1317.5	669.41	2.04×10^{-6}	−1.1168	photosystem II PsbW protein chloroplastic-like
c65097.graph_c0	49.88	25.08	8.91×10^{-5}	−1.1178	photosystem II PsbY protein
Fatty acid biosynthesis					
c67210.graph_c2	5.34	3.78	0.005595171	−0.8695	3-oxoacyl-[acyl-carrier-protein] synthase II (FabF)

Table 1. *Cont.*

#ID	FPKM		FDR	$\log_2$FC	KEGG_Annotation
	CK	WS30d			
c57955.graph_c0	44.1	21.27	1.60×10^{-8}	−1.0614	3-oxoacyl-[acyl-carrier-protein] synthase II (FabF)
c58354.graph_c0	12.57	0.62	2.47×10^{-12}	−3.9695	3-oxoacyl-[acyl-carrier-protein] synthase II (FabF)
c62774.graph_c0	32.83	20.68	5.66×10^{-5}	−0.7974	3-oxoacyl-[acyl-carrier-protein] synthase III (FabH)
c53077.graph_c0	29.14	9.74	2.39×10^{-11}	−1.6865	acetyl-CoA carboxylase biotin carboxyl carrier protein
c67708.graph_c1	60.53	43.04	0.001805159	−0.638	acetyl-CoA carboxylase carboxyl transferase subunit alpha
c46335.graph_c0	50.52	31.1	4.92×10^{-5}	−0.8391	acetyl-CoA carboxylase, biotin carboxylase subunit
c46455.graph_c0	31.28	20.51	0.002644038	−0.7592	enoyl-[acyl-carrier protein] reductase I (FabI)
c50059.graph_c0	36.92	11.05	2.67×10^{-9}	−1.8325	enoyl-[acyl-carrier protein] reductase I (FabI)
c60574.graph_c0	55.26	7.04	5.88×10^{-24}	−2.7932	fatty acyl-ACP thioesterase A
c64735.graph_c0	242.08	108.24	3.08×10^{-12}	−1.3285	fatty acyl-ACP thioesterase B
c66291.graph_c1	252.71	32.02	6.43×10^{-28}	−3.0528	long-chain acyl-CoA synthetase
c59094.graph_c0	71.72	20.79	6.86×10^{-23}	−2.2218	long-chain acyl-CoA synthetase
c56069.graph_c0	29.9	17.27	1.13×10^{-5}	−1.0826	S-malonyltransferase
Photosynthesis—antenna proteins					
c51141.graph_c0	1625.3	463.16	1.29×10^{-5}	−1.838	light-harvesting complex I chlorophyll a/b binding protein 1
c62753.graph_c1	949.2	442.35	0.000170897	−1.2306	light-harvesting complex I chlorophyll a/b binding protein 2
c65754.graph_c0	3672.8	2120.8	0.008139475	−0.9203	light-harvesting complex I chlorophyll a/b binding protein 3
c41195.graph_c0	2546.7	785.55	2.25×10^{-5}	−1.7575	light-harvesting complex I chlorophyll a/b binding protein 4
c65915.graph_c0	66.33	39.48	4.72×10^{-5}	−0.8903	light-harvesting complex I chlorophyll a/b binding protein 5
c32187.graph_c0	23.74	0.61	2.53×10^{-8}	−4.0809	light-harvesting complex II chlorophyll a/b binding protein 1
c63116.graph_c0	18325	7434.7	3.70×10^{-6}	−1.5994	light-harvesting complex II chlorophyll a/b binding protein 1
c68500.graph_c1	294.76	87.78	2.92×10^{-6}	−1.8158	light-harvesting complex II chlorophyll a/b binding protein 2
c49775.graph_c0	30.49	8.1	6.84×10^{-5}	−1.9016	light-harvesting complex II chlorophyll a/b binding protein 2
c52982.graph_c0	1074.5	316.12	6.09×10^{-5}	−1.7878	light-harvesting complex II chlorophyll a/b binding protein 3
c56361.graph_c0	1720.6	769.22	0.000795542	−1.2611	light-harvesting complex II chlorophyll a/b binding protein 4
c65425.graph_c0	1478	735.17	0.000311123	−1.1253	light-harvesting complex II chlorophyll a/b binding protein 5
c52323.graph_c0	1566.1	601.7	0.000807719	−1.4493	light-harvesting complex II chlorophyll a/b binding protein 6
Glycosaminoglycan degradation					
c38948.graph_c0	10.24	3.43	9.09×10^{-6}	−1.6616	heparanase
c38920.graph_c0	36.54	25.82	8.92×10^{-6}	−0.6415	heparanase
c65686.graph_c0	15.85	6.55	1.99×10^{-9}	−1.4827	heparanase
c49429.graph_c0	2.2	0.05	4.01×10^{-11}	−4.5862	heparanase
c68546.graph_c0	28.62	14.23	1.60×10^{-12}	−1.2804	heparanase
c59153.graph_c0	66.93	39.86	4.05×10^{-6}	−0.9714	hexosaminidase
c64000.graph_c0	40.48	14.83	2.73×10^{-5}	−1.5638	hexosaminidase

Table 2. Differential metabolites in *R. delavayi* leaves under waterlogging stress for 30 days (WS30d).

Class	Putative Metabolites	CK	WS30d	VIP	*p* Value	FC	Log$_2$FC
	Glucose	1.020305382	1.885810307	1.414	0.026	1.848	0.886
	Sedoheptulose	0.039939754	0.384491416	1.674	0.026	9.627	3.267
	Galactose	0.242448413	0.678832861	1.637	0.004	2.800	1.485
Sugars	Sucrose	0.513948256	1.015253352	1.634	0.001	1.975	0.982
	Lyxose	0.01002245	0.014270754	1.459	0.034	1.424	0.510
	Galactonic acid	0.027194517	0.054436394	1.452	0.039	2.002	1.001
	N-Acetyl-beta-D-mannosamine	0.008378467	0.016720616	1.480	0.023	1.996	0.997
	L-Malic acid	0.506445001	1.243612412	1.505	0.044	2.456	1.296
Organic acids	Citric acid	0.145507649	0.462412497	1.373	0.019	3.178	1.668
	Glucoheptonic acid	0.180491598	0.395956908	1.627	0.001	2.194	1.133
	Gluconic lactone	0.22506794	0.492767481	1.616	0.002	2.189	1.131
Esters	beta-Mannosylglycerate	0.028496106	0.056197656	1.354	0.049	1.972	0.980
	L-Gulonolactone	0.005848757	0.050757515	1.277	0.006	8.678	3.117
Fatty acid	Linolenic acid	0.087711244	0.205478671	1.505	0.017	2.343	1.228
Alcohols	Diglycerol	0.055957005	0.12707336	1.501	0.025	2.271	1.183
	Sorbitol	0.008703194	0.024159015	1.537	0.020	2.776	1.473
Flavonoids	Arbutin	12.28242997	24.66825634	1.521	0.008	2.008	1.006
	Phenyl beta-D-glucopyranoside	0.007043534	0.018592753	1.623	0.011	2.640	1.400
	p-Coumaric acid	0.104651639	0.363258482	1.620	0.001	3.471	1.795
Others	Glycocyamine	0.005322693	0.028323088	1.261	0.016	5.321	2.412
	Glutathione	0.006044401	0.011154309	1.370	0.048	1.845	0.884

In the "photosynthesis" pathway, the down-regulated DEGs were annotated to encode proteins such as cytochrome b6/f complex, electron transport chain components, ATPase, and subunits of photosystem I (PSI) or II (PSII) (as shown in Table 1). Further analysis revealed a significant reduction in CO_2 assimilation rate, stomatal conductance, transpiration rate, and maximum photochemical efficiency by 91.2%, 95.3%, 93.3%, and 8.4%, respectively, in the WS30d leave in comparison with CK (Figure 4). Additionally, the chlorophyll a and total chlorophyll content in the WS30d leaves decreased by 13.5% and 16.6%, respectively (Figure 4). These observations suggested that excessive light energy might lead to the accumulation of reactive oxygen species (ROS) in the chloroplast. This was further supported by the significant increase in hydrogen peroxide content in WS30d leaves, which was found to be 3.39 times higher than the control (Figure 4).

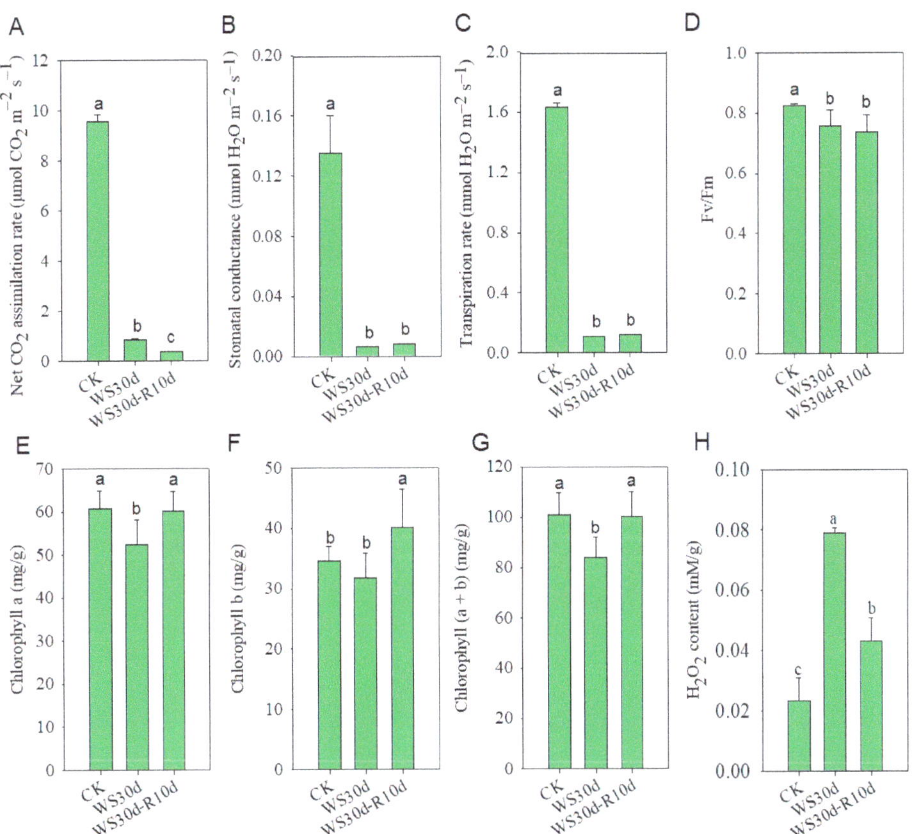

Figure 4. Photosynthesis, chlorophyll content and H_2O_2 content in CK, WS30d and WS30d-R10d leaves. (**A**) Net CO_2 assimilation rate; (**B**) stomatal conductance; (**C**) transpiration rate; (**D**) maximum photochemical efficiency; (**E**) Chlorophyll a content; (**F**) Chlorophyll b content; (**G**) Chlorophyll (a + b); and (**H**) H_2O_2 content. Different lowercase letters represented significant differences between treated groups and CK ($p < 0.05$).

In the "fatty acid biosynthesis" pathway, down-regulated DEGs were annotated as FabF, FabH, FabI, fatty acyl-ACP thioesterase, and long-chain acyl-CoA synthetase (Table 1). These genes are involved in synthesizing of long-chain acyl-CoA. Similarly, in the "glycosaminoglycan degradation" pathway, down-regulated DEGs were annotated as heparanase and hexosaminidase (Table 1).

2.4. KEGG Pathways by Common DEGs between Waterlogging Stress and Post-Waterlogging Recovery

Our objective was to identify genes that are difficult to recover after waterlogging stress. To accomplish this, we analyzed the common DEGs between WS30d and WS30d-R10d using a Venn diagram, which revealed 2417 common DEGs (Figure 5A). Heat map analysis of these DEGs indicated that 1037 DEGs were up-regulated, and 1380 DEGs were down-regulated (Figure 5B). Subsequently, we examined the biological relevance of these DEGs by using KEGG pathway analysis. The up-regulated DEGs were significantly enriched in "selenocompound metabolism" ($p < 0.05$) (Figure 5C), whereas the down-regulated DEGs were significantly enriched in "phenylpropanoid biosynthesis", "fatty acid metabolism", "fatty acid biosynthesis", "fatty acid elongation", and "cutin, suberin, and wax biosynthesis" ($p < 0.05$) (Figure 5D).

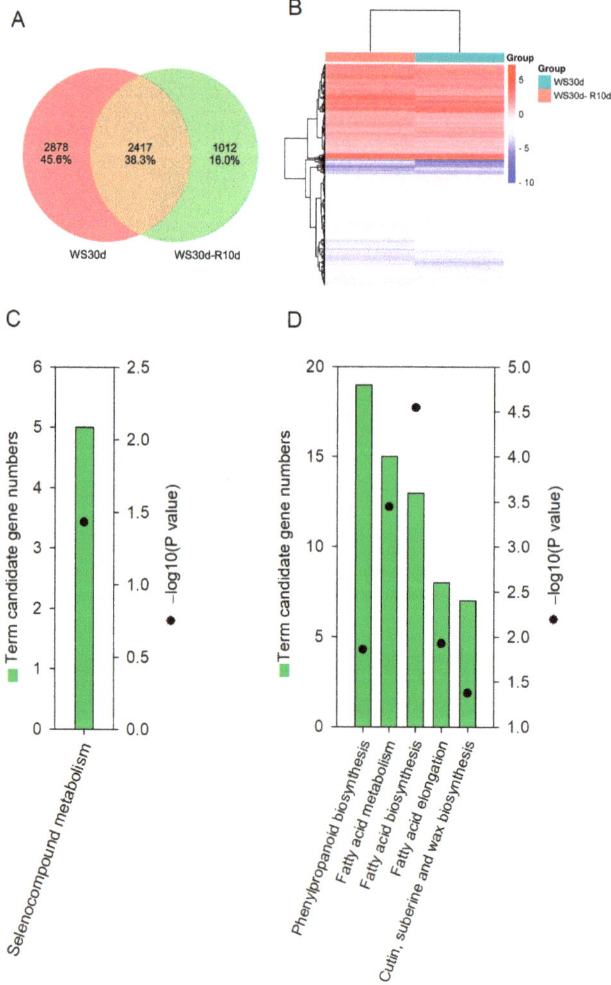

Figure 5. KEGG pathways of common DEGs in *R. delavayi* leaves between WS30d and WS30d-R10d. (**A**) Venn diagram of DEGs with annotations identified by RNA-Seq between WS30d and WS30d-R10d; (**B**) heat maps of 2417 common DEGs in (**A**); (**C**) up-regulation DEGs enriched in KEGG; (**D**) down-regulation DEGs enriched in KEGG.

In the "selenocompound metabolism" pathway, we observed up-regulated DEGs annotated as 3′-phosphoadenosine 5′-phosphosulfate synthase and methionyl-tRNA synthase (Table S6). Conversely, in the "phenylpropanoid biosynthesis" pathway, we identified down-regulated DEGs annotated as 4-coumarate-CoA ligase, beta-glucosidase, caffeic acid 3-O-methyltransferase, cinnamyl-alcohol dehydrogenase, coumaroylquinate 3′-monooxygenase, peroxidase, and shikimate O-hydroxycinnamoyltransferase (Table S6). These genes are mainly involved in the biosynthesis of lignin. As expected, the lignin content of WS30d-R10d leaves decreased significantly when compared to CK (Figure 6).

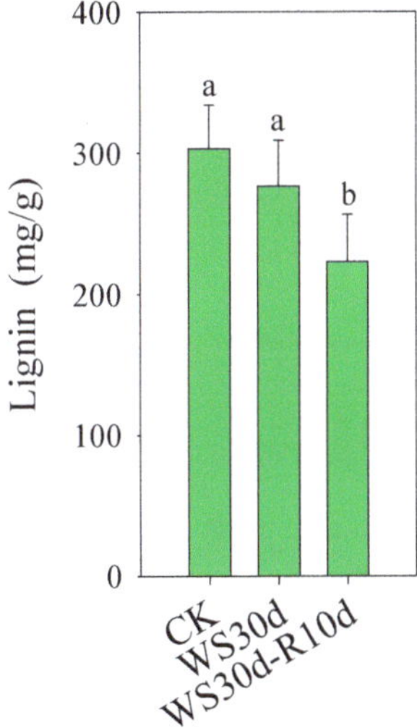

Figure 6. Lignin content in CK, WS30d and WS30d-R10d leaves. Different lowercase letters represented significant differences between treated groups and CK ($p < 0.05$).

In the "fatty acid metabolism" and "fatty acid biosynthesis" pathways, the proteins encoded by DEGs were consistent with those in the "fatty acid biosynthesis" pathway mentioned above (Table S6). The down-regulated DEGs may impede the synthesis of long-chain acyl-CaA. In the "fatty acid elongation" pathway, DEGs were annotated as 3-ketoacyl-CoA synthase, very-long-chain (3R)-3-hydroxyacyl-CoA dehydratase, and very-long-chain 3-oxoacyl-CoA reductase (Table S6). The decreased expression of these DEGs suggested that the synthesis of long-chain fatty acids with long-chain acyl-CaA as a precursor may be inhibited. Long-chain fatty acids are utilized for the biosynthesis of cutin, suberin, and wax. The down-regulated DEGs involved in these pathways may impede the biosynthesis of long-chain esters (Table S6).

2.5. Major Transcription Factors Families Active during Waterlogging Stress and Post-Waterlogging Recovery

Transcription factors (TFs) play a crucial role in the transcriptional reprogramming that occurs in response to abiotic stress. To identify the enriched TF families in response to waterlogging stress, we screened the DEGs in WS30d leaves. Our analysis revealed that 20 TF families were encoded by 118 up-regulated DEGs and 117 down-regulated DEGs (Figure 7). Among the up-regulated DEGs, the WRKY family was the most dominant, while the bHLH family was the most dominant among the down-regulated DEGs. Interestingly, the DEGs encoding most TF families, such as MYB and AP2/ERF-ERF, were both up-regulated and down-regulated under waterlogging stress (Figure 7). In the common DEGs between WS30d and WS30d-R10d, the MYB-related and bHLH families were the most dominant TF families encoded by the up-regulated and down-regulated DEGs, respectively (Figure S2).

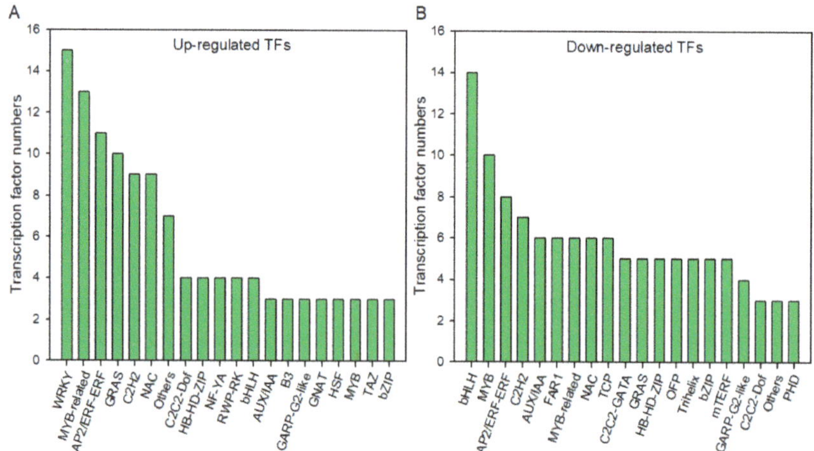

Figure 7. The predicting transcription factors (TFs) encoded by DEGs in WS30d leaves. (**A**) the TFs encoded by up-regulated DEGs; (**B**) the TFs encoded by down-regulated DEGs.

2.6. Metabolites Accumulations in Response to Waterlogging Stress and Post-Waterlogging Recovery

To investigate the accumulation of metabolites in response to waterlogging stress and post-waterlogging recovery, we utilized the GC-MS platform to determine the metabolite levels in CK, WS30d, and WS30d-R10d leaves. To summarize the similarities and differences between these groups, we employed principal component analysis (PCA) and orthogonal projections to latent structures-discriminant analysis (OPLS-DA). Our PCA analysis revealed that the data from the three duplicate samples were closely clustered, with PC1 and PC2 accounting for 20.6% and 16.7% of the total variation, respectively (Figure S3A). In the OPLS-DA score plot, component 1 and component 2 accounted for 19.8% and 12.2% of the total variation, respectively (Figure S3B). We identified a total of 149 putative metabolites and screened for differential metabolites between treatment groups (VIP > 1, and p < 0.05). Compared to CK, WS30d leaves showed significant upregulation in 22 metabolites, including seven sugars (glucose, sedoheptulose, galactose, sucrose, lyxose, galactonic acid, N-Acetyl-beta-D-mannosamine), three organic acids, three lipids, two alcohols, two flavonoids, and three others (Table 2). WS30d-R10d leaves showed upregulation in six sugars, three lipids, two alcohols and flavonoids, and five others (Table S7). Interestingly, the common metabolites in WS30d and WS30d-R10d leaves revealed that soluble sugars remained highly accumulated during post-waterlogging recovery (Table S8).

3. Discussion

3.1. Waterlogging Stress Inhibited Photosynthesis in R. delavayi Leaves

Gas exchange is usually used to assess the tolerance of plant species to waterlogging [21]. Tolerant species show a slight reduction in their photosynthetic rate [3], whereas sensitive species experience a significant reduction [22,23]. In the case of *R. delavayi*, the CO_2 assimilation rate in WS30d leaves decreased by 91.2%, suggesting that it is a waterlogging-sensitive species (Figure 4A), similar to *Quercus petraea* and *Persea Americana* [23,24]. This decrease in CO_2 assimilation rate in WS30d leaves was accompanied by a simultaneous reduction in stomatal conductance and transpiration rate (Figure 4B,C), indicating that photosynthesis may be affected by stomatal limitation [25]. Although the precise mechanism behind the decrease in stomatal conductance caused by waterlogging stress remains unclear, it has been widely reported. In addition, the decrease in maximum photochemical efficiency of PSII (Fv/Fm) in WS30d leaves (Figure 4D) suggested that photoinhibition had occurred in the stressed leaves [26,27].

The suppression of photosynthesis could also be affected by both chlorophyll content and photosynthetic enzyme activity. In the case of WS30d leaves, there was a decrease of 13.5% and 16.6% in chlorophyll a and total chlorophyll content, respectively (Figure 4), which is consistent with previous studies [3,27]. Additionally, the decrease in chlorophyll a and chlorophyll (a + b) content in stressed leaves indicates that the inhibition of photosynthesis, which results in more damage to PSII, is also more limiting [3]. This, in turn, may lead to the production of excessive ROS. RNA-sequencing analysis of stressed leaves revealed that down-regulated DEGs were enriched in "photosynthesis" and "photosynthesis-antenna proteins" pathways (Figure 3B). The DEGs annotated in these pathways were found to be involved in PSI, PSII, cytochrome *b6-f* complex, and ATP synthase (Table 1), suggesting that the activity of proteins in the photosynthesis pathway was inhibited by waterlogging stress. Although the chlorophyll a and chlorophyll (a + b) content in WS30d-R10d leaves could be restored to the control level, the CO_2 assimilation rate was still reduced (Figure 4). These findings are similar to previous studies on cotton, a woody plant that is sensitive to waterlogging stress [28,29].

Transcription factors play a crucial role in regulating gene expression, particularly in the synthesis of plant secondary metabolites such as anthocyanins, flavonoids, and lignin. MYB and bHLH transcription factors, in particular, have been identified as key regulators of this process [30,31]. In waterlogged leaves, it is posited that the down-regulation of MYB and bHLH genes may exert negative regulation on the expression of genes involved in flavonoid biosynthesis, leading to flavonoid accumulation, or alternatively, positively regulate genes involved in lignin biosynthesis, thereby inhibiting lignin synthesis (Figure 6). Further, WRKY and MYB-related transcription factors are instrumental in regulating ABA biosynthesis [30,32], a process that has been shown to play a significant role in the waterlogged plants [33]. It is hypothesized that the up-regulation of WRKY- and MYB-related transcription factors in waterlogged leaves may regulate ABA biosynthesis, ultimately causing leaf stomatal closure and reduction in stomatal conductance (Figure 4). Additionally, plant hormones induce the expression of ethylene response factors, such as *RAP2.3*, improving expression levels of downstream genes and enhancing waterlogging or post-waterlogging resistance [32,34]. Based on our findings, it is postulated that the up-regulation of ERF transcription factors may stimulate H_2O_2 accumulation in the stressed leaves by ethylene biosynthesis (Figure 4). Notably, the observed alterations in the expression levels of both up-regulated and down-regulated ERF family members in conjunction with the varying expression patterns observed in other transcription factors and plant hormone signaling pathways indicates a highly intricate regulatory network governing the plant response to waterlogging stress [17].

3.2. Waterlogging Stress Induced Oxidative Stress in R. delavayi Seedlings

The existing body of evidence indicates that short-term waterlogging stress has the potential to generate ROS in plant roots. In leaf tissues, several abiotic stressors can result in a reduction in CO_2 assimilation rate, resulting in the accumulation of excess H_2O_2 and consequent oxidative stress [35]. In addition, waterlogging stress appears to alter the accumulation of ethylene and provoke the production of H_2O_2, as evidenced by the increased H_2O_2 accumulation in the WS30d leaves (Figure 4H). Such H_2O_2 accumulation may disrupt normal metabolic processes by promoting lipid peroxidation, compromising membrane integrity, as well as protein and DNA oxidation [36–38]. Analysis of gene expression patterns revealed that the up-regulated DEGs in the stressed leaves were significantly enriched in the "purine metabolism" pathways (Figure 3A), suggesting that the plants may initiate gene expression mechanisms to repair DNA damage caused by H_2O_2 accumulation (Table 1).

In the context of ROS accumulation under waterlogging stress, the ascorbate-glutathione cycle plays a critical role in enabling plants to cope with oxidative stress to a certain extent [39]. Notably, the observed increases in ascorbate or glutathione concentrations in the WS30d leaves or WS30d-R10d (Tabels 2 and S9) were found to be consistent with the up-regulated DEGs in the glutathione metabolism pathway (Figure 3A). This strongly suggests that the accumulation of glutathione or ascorbate could serve as non-enzymatic antioxidants to preserve the balance of H_2O_2. However, despite the fact that waterlogged seedlings were able to recover, their H_2O_2 content remained at a higher level than the post-waterlogging recovery (Figure 4H). As such, oxidative stress was sustained and even accelerated the damage inflicted on the plants during the recovery, as shown in Figure 1.

3.3. Lignin and Cuticle Biosynthesis was Continuously Inhibited during Post-Waterlogging Recovery

Lignin, a phenolic biopolymer, is primarily synthesized through the phenylpropanoid biosynthesis pathway and is typically concentrated in the secondary cell wall of vascular plants. Lignin plays a vital role in providing mechanical support to plant tissues, as well as facilitating the transportation of nutrients or carbohydrates [40]. Notably, the accumulation of lignin is known to enhance cell wall reinforcement, thereby improving resistance to waterlogging stress [41]. Previous studies have shown that waterlogging stress can reduce lignin content by inhibiting the expression of genes involved in lignin biosynthesis [42]. Our transcriptome data analysis revealed that the DEGs in WS30d-R10d leaves were significantly enriched in the phenylpropanoid biosynthesis pathway (Figure 5D), with their expression being noticeably inhibited (Table S6). These results were strongly aligned with the observed decrease in lignin content in WS30d-R10d leaves (Figure 6). As such, we hypothesize that waterlogging may reduce the biosynthesis of lignin in the stressed leaves, ultimately leading to leaf wilting (Figure 1) and decreased nutrient or carbohydrate transportation (Figure 8). A similar result was reported in poplar trees [17].

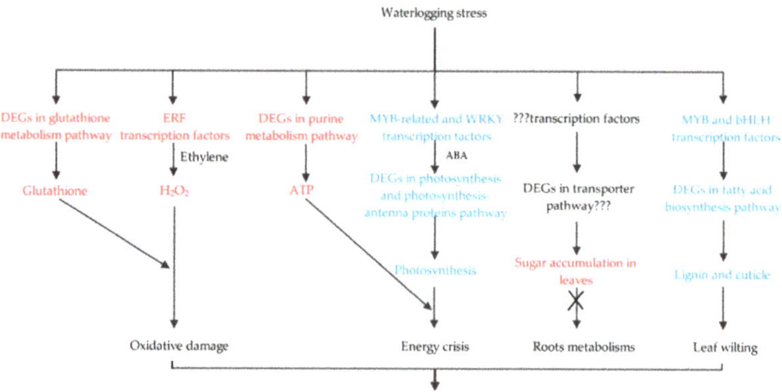

Figure 8. Simplified scheme of the response in *R. delavayi* leaves to waterlogging stress. Genes and metabolites indicated in blue are usually down-regulated and decreased in abundance, respectively. Genes and metabolites indicated in red are usually up-regulated and increased in abundance, respectively. ??? represents unknown transcription factors and pathways.

The cuticle is a layer of fatty substances that envelops above-ground plant organs, composed mainly of cutin and waxes [43]. As reported by Tellechea-Robles et al. (2019), wetland plants have adapted to waterlogging stress by augmenting the thickness of their cuticle on leaves [44]. In this study, transcriptome analysis revealed that the DEGs were significantly enriched in several pathways during post-waterlogging recovery, namely "fatty acid metabolism", "fatty acid biosynthesis", "fatty acid elongation", and "cutin, suberin, and wax biosynthesis" (Figure 5D). These pathways may reduce the production of cutin and wax, leading to thinner leaves and a diminished support function.

3.4. Waterlogging Stress Prevented Transportation of Soluble Sugar from Leaves

Maintaining a stable supply of glycolysis, particularly in roots, is a crucial factor in waterlogging tolerance. In contrast, waterlogging-sensitive plants often fail to uphold sufficient carbohydrate levels, leading to tissue death under stress conditions [21]. Previous studies have established that waterlogging-sensitive species frequently experience substantial disruptions in sugar transportation from phloem to root cells during periods of waterlogging, resulting in increased sugar accumulation in stressed leaves [22,28,45,46]. Our current study confirmed that soluble sugars, such as glucose, sedoheptulose, galactose, and sucrose, were significantly accumulated in waterlogged leaves as compared to CK leaves (Table 2). In addition, we noted excessive accumulation of soluble sugar in the stressed leaves during post-waterlogging recovery (Tables S9 and S7). These findings suggest that sugar transport in *R. delavayi* seedlings may have been severely impaired by waterlogging stress, which ultimately resulted in a deficiency of available carbohydrates in the roots [28].

4. Materials and Methods

4.1. Plant Material and Waterlogging Treatment

Three-year-old *Rhododendron delavayi* seedlings were cultivated in plastic flowerpots (8 cm in diameter and 8 cm in height), filled with a mixture of nutrient soil and humus soil (1:1 in volume). To induce waterlogging stress, the seedlings were placed within a greenhouse (with a photoperiod 16 h/8 h, temperature 22 °C, light intensity 400 μmol m^{-2} s^{-1}, relative humidity 60–70%) at Guizhou Normal University.

The flowerpots with the seedlings were placed in trays with length, width, and height of 55 cm, 30 cm, and 4.5 cm, respectively. The trays were filled with water up to a height of 4 cm, submerging the root area of the *R. delavayi* seedlings. The water level was sustained at 4 cm by daily additions, creating a waterlogging stress that was imposed for 10 days (WS10d), 20 days (WS20d), and 30 days (WS30d). After the waterlogging stress period, the water within the trays was removed, and the seedlings were allowed to recover for 10 days, marked as WS10d-R10d, WS20d-R10d, and WS30d-R10d, respectively. *R. delavayi* seedlings before the waterlogging stress were used as the control (CK). Three flowerpots were placed within each tray as a repeat, with three biological repeats set for each treatment.

4.2. Measurement of Photosynthesis Parameters and Chlorophyll Fluorescence

To measure photosynthesis parameters, we utilized the portable photosynthesis system (LI-6400, LI-COR Corporate, Lincoln, NE, USA), which was coupled with the 6400-02B chamber (6400-02B, LI-COR Corporate, Lincoln, NE, USA) providing stable and activated intensities. The third young leaf from three individuals was selected and measured between 8.00 and 12.00 a.m. Prior to measurement, the leaves were activated using a light intensity of 1000 μmol quanta m^{-2} s^{-1} for 20 s. Subsequently, the light intensity (400 μmol quanta m^{-2} s^{-1}) and CO_2 concentration (ambient CO_2 concentration) were set, and the net CO_2 assimilation rate, stomatal conductance, and transpiration rate were recorded. Additionally, we measured chlorophyll fluorescence using the portable photosynthesis system (LI-6400, LI-COR Corporate, Lincoln, NE, USA) with the 6400-40 chamber. After dark adaptation at night, we conducted measurements before dawn, recording the maximum quantum efficiency of photosystem II (Fv/Fm).

4.3. Measurement of Hydrogen Peroxide and Lignin

In this study, we extracted 0.5 g of leaves and subsequently homogenized them with 3 mL of 50 mmol/L phosphate buffer (pH = 6.5). Next, we centrifuged the mixture at 1000 rpm for 5 min at 4 °C, following which we added 1 mL of 0.1% titanium sulfate solution to 3 mL of supernatant and mixed it thoroughly. The solution was then centrifuged at 1000 rpm for 10 min at 4 °C. The resulting precipitate was washed with cooled acetone and subsequently dissolved in 2 mmol/L H_2SO_4 solution. Subsequently, we measured the absorbance at 410 nm and the hydrogen peroxide content was calculated using the standard curve. Meanwhile, lignin was extracted using an assay kit (BC4200, Solarbio Technology Co., Ltd., Beijing, China) and determined according to the instructions.

4.4. RNA Extraction, Library Preparation, and Sequencing

After a 30-day period of exposure to waterlogging stress, the aged leaves of the plant under investigation exhibited signs of wilting (Figure 1). To gain insight into the effects of this stress, we selected CK, WS30d, and WS30d-R10d leaves and roots for RNA sequencing (RNA-Seq). Total RNA was extracted from the samples using Trizol reagent (Thermo Fisher Scientific, Pleasanton, CA, USA) and evaluated for RNA degradation and contamination using agarose gelelectrophoresis (1%). RNA purity was checked using the Nano Photometer spectrophotometer (Thermo Fisher Scientific, Wilmington, NC, USA), and RNA concentration was measured using the Qubit RNA Assay Kit in Qubit 2.0Fluorometer (Thermo Fisher Scientific, Wilmington, NC, USA). RNA integrity was assessed using the RNA Nano6000 Assay Kit of the Agilent Bioanalyzer 2100 system (Agilent Technologies, Santa Clara, CA, USA). The NEBNext UltraRNA Library Prep Kit for Illumina (New England Biolabs, Inc., Beijing, China) was used to generate sequencing libraries according to the manufacturer's recommendations. However, due to the short roots and lignification of *R. delavayi* seedlings, RNA extraction from these roots deemed unsuitable for database construction (Table S9), and therefore, RNA-Seq from waterlogged root was abandoned. The index-coded samples were clustered using TruSeq PE Cluster Kit v3-cBot-HS (Illumia) on a Cluster Generation System, following the manufacturer's instructions. Once clusters had been generated, library preparations were sequenced on an Illumina Hiseq 2000 platform, with paired-end reads produced. Each sample yielded more than 6.03 Gb of clean data. Sequencing was completed by the Beijing Biomarker Biotechnology Company (Biomarker biotech, Beijing, China). The RNA-Seq data were obtained in three biological replicates.

4.5. Transcriptome Assembly and Differentially Expressed Genes (DEGs) Analysis

Transcriptome assembly was accomplished using the Trinity method [47] with min_kmer_cov set to 2 by default, with all other parameters set to default values. The expression abundance of gene was calculated using the Fragments Per Kilobase of transcript per Million mapped reads (FPKM) method [48]. To analyze DEGs, this was performed using the DESeq R package (1.10.1) between CK, WS30d, and WS30d-R10d. The resulting p values were adjusted using Benjamini and Hochberg's approach to control the false discovery rate (FDR). Genes were considered differentially expressed when FDR was less than 0.01 and fold change (FC) was more than |1.5| between the treatments.

4.6. Annotation and Classification of DEGs

To annotate the functions of DEGs, we employed BLAST with an E-value $\leq 1 \times 10^{-10}$- as the cutoff to align these sequences against the NCBI non-redundant database, COG, GO, and KEGG databases. This approach allowed us to assign putative functional annotations to each DEG, based on their sequence similarity to sequences in known databases.

4.7. Quantitative Real-Time PCR (qRT-PCR) Validation

To validate the RNA-seq results, we selected six genes from differentially expression genes for quantitative real-time PCR (qRT-PCR) analysis. Total RNA was extracted from the samples using the OmniPlant RNA Kit (CW2598S, Cwbiotech, Beijing, China) and reverse-transcribed using the TransScript All-in-One First-Strand cDNA Synthesis SuperMix for qPCR Kit (AT341-01, Transgen Biotech, Beijing, China), following the manufacturer's protocol. The primer sequences used for qRT-PCR, including β-actin as an internal control, were listed in Table S10. The qRT-PCR was performed using a Rotor-Gene Q real-time PCR system (Rotor-Gene, Qiagen, Germany). A total of 10 μL 2 X TransStart Top/TipGreen qPCR SuperMix (AQ141-02, Transgen Biotech, Beijing, China) was added to the reaction mixture, following the manufacturer's instructions. The relative expression levels of the selected genes were normalized to the expression level of the β-actin gene. Cycle threshold values were then used to calculate expression levels using the $2^{-\Delta\Delta Ct}$ method [49].

4.8. Gas Chromatograph Coupled with a Time-of-Flight Mass Spectrometer (GC-TOF-MS) Analysis

The leaves from three individual seedlings were sampled and mixed as a duplicate, and the three duplicate samples were used for GC-MS detection. Metabolites were extracted from leave samples following the method described in [50], with some modifications. Approximately 20 mg of the sample was transferred into a 2 mL EP tubes and extracted with 500 µL of pre-cold extraction liquid ($V_{Methanol}$:V_{dH_2O} = 3:1). An aliquot of 10 µL of internal standard (adonitol) was added and mixed by vortexing for 30 s. The sample was then homogenized in a ball mill for 4 min at 35 Hz and subsequently treated with ultrasound for 5 min (incubated in ice water bath). The mixture was then centrifuged at 12,000 rpm at 4 °C for 15 min. The supernatant (100 µL) was transferred to a fresh 1.5 mL EP tube before being dried completely in a vacuum concentrator without heating. Next, the extracts were dissolved in 40 µL of methoxyamine hydrochloride (20 mg mL^{-1} in pyridine) and then incubated at 80 °C for 30 min. Finally, the samples were derivatized with 60 µL of N,O-bis(trimethylsilyl) trifluoroacetamide (BSTFA) reagent (with 1% TMCS, v/v) at 70 °C for 1.5 h. Quality control (QC) samples (a mixture of all samples to be analyzed) were also processed for detection.

Metabolite detection was conducted using an Agilent 7890 gas chromatography system coupled with a Pegasus HT time-of-flight mass spectrometer (Agilent Technologies, Santa Clara, CA, USA). The analysis was conducted in a randomized order after the addition of fatty acid methylesters (FAMEs). ADB-5MS capillary column coated with 5% diphenyl cross-linked with 95% dimethylpolysiloxane (30 m × 250 µm inner diameter, 0.25 µm film thickness; J&W Scientific, Folsom, CA, USA) was employed. An aliquot of 1 µL of the analyte was injected in splitless mode, while helium was used as a carrier gas. The front inlet purge flow was 3 mL min^{-1}, and the gas flow rate through the column was 1 mL min^{-1}. The initial temperature was held at 50 °C for 1 min, then raised to 310 °C at a rate of 10 °C min^{-1} and held at 310 °C for 8 min. The injection, transfer line, and ion source temperatures were maintained at 280 °C, 280 °C, and 250 °C, respectively. The energy was -70 eV in electron impact mode, and the mass spectrometry data were acquired in full-scan mode with an m/z range of 50–500 at a rate of 12.5 spectra per second after a solvent delay of 6.27 min.

4.9. Data Preprocessing and Compound Identification

Chroma TOF 4.3X software and the LECO-Fiehn Rtx5 database (LECO Corporation, Benton Harbor, MI, USA) were used for raw peak extraction, database line filtering and calibration, peak alignment, deconvolution analysis, peak identification, and integration of peak area. Both mass spectrum match and retention index match were considered during metabolite identification. Additionally, peaks detected in less than 50% of the QC samples or for which relative standard deviation (RSD) greater than 30% in the QC samples were removed [50,51]. Metabolites separated by GC-TOF-MS were identified using LECO ChromaTOF 4.3X software and the LECO/Fiehn Rtx5 metabolite mass spectral library by matching the mass spectrum and retention index.

4.10. Data Analysis

A three-dimensional data matrix was constructed, consisting of metabolite names (tentatively identified by GC-TOF-MS), sample information (three biological replicates of each treatment), and raw abundance (peak area for each tentatively identified metabolite) This matrix was uploaded to MetaboAnalyst 5.0 and analyzed in accordance with the provided instructions. Raw data were normalized using three categories of normalization: none, log transformation, and auto scaling. Multivariate analyses between CK, WS30d, and WS30d-R10d groups were performed using MetaboAnalyst 5.0, including orthogonal projections to latent structures-discriminant analysis (OPLS-DA) and principal component analysis (PCA). Moreover, the value of the variable importance in the projection (VIP) of the first principal component in OPLS-DA analysis was determined. Metabolites with VIP > 1 and $p < 0.05$ (student t-test) were considered significantly changed between groups.

4.11. Statistical Analysis

The data were presented as mean $\pm$ standard deviation (SD) and analyzed using either a one-way analysis of variance (ANOVA) or SNK t-test. Analysis was carried out in at least three replicates, with a p value < 0.05 considered statistically significant. The statistical analysis was performed using SPSS 18.0 for Windows. The difference test of gene expression among the three samples was performed using the R version 4.2.1 vegan package.

5. Conclusions

We utilized *R. delavayi* seedlings as a system to elucidate the response mechanism of alpine *Rhododendron* to waterlogging stress. Waterlogging stress was observed to reduce CO_2 assimilation rate, stomatal conductance, transpiration rate, and maximum photo-chemical efficiency of PSII in the leave. The stress generated excessive H_2O_2, leading to oxidative stress. Additionally, waterlogging stress negatively affected lignin and cuticle biosynthesis, leading to leaf wilting, decreased carbohydrate transportation, and impeded sugar transport from leaves to roots through phloem, which ultimately results in sugar accumulation in the stressed leaves (Figure 8). Taken together, waterlogging stress has an irreversible effect on the *R. delavayi* seedlings in both physiological and molecular aspects. Regrettably, our study only shed light on the response mechanism of leaves, prompting further studies to develop into the stress response mechanism of young *R. delavayi* roots.

Supplementary Materials: The following supporting information can be downloaded at: https://www.mdpi.com/article/10.3390/ijms241310509/s1.

Author Contributions: Conceptualization: X.-M.Z. and S.-G.D.; Methodology: S.-G.D. and J.-T.L.; Software: Y.X.; Formal analysis: S.-G.D. and J.-T.L.; Investigation: S.-G.D. and L.-X.L.; Data curation: X.-M.Z. and S.-G.D.; Writing-original draft: X.-M.Z. and S.-G.D.; Writing-review and editing: X.-M.Z., M.T., J.T., W.S. and Y.Y. All authors contributed to the article and approved the submitted version. All authors have read and agreed to the published version of the manuscript.

Funding: This study was financially supported by the Joint Fund of the Natural Science Foundation of China and the Karst Science Research Center of Guizhou Province (Grant No. U1812401); the Natural Science Foundation of China (NSFC) (32260393); Key Laboratory of Environment Friendly Management on Alpine Rhododendron Diseases and Pests of Institutions of Higher Learning in Guizhou Province ([2022]044); Guizhou Science and Technology Support Plan Project [2021]224; the Science Foundation of Forestry Bureau of Guizhou Province ([2019]10); Guizhou Science and Technology Foundation ([2020]1Y130), Higher Education Science and Research Youth Project of Guizhou Education Department ([2022]130).

Institutional Review Board Statement: Not applicable.

Informed Consent Statement: Not applicable.

Data Availability Statement: The data presented in this study are available on request from the corresponding author.

Acknowledgments: We thank the staff of Biomarker Technology Co., Ltd. (Beijing, China) for their support during transcriptome data analysis.

Conflicts of Interest: The authors declare no conflict of interest.

References

1. Herzog, M.; Striker, G.G.; Colmer, T.D.; Pedersen, O. Mechanisms of waterlogging tolerance in wheat—A review of root and shoot physiology. *Plant Cell Environ.* **2016**, *39*, 1068–1086. [CrossRef] [PubMed]
2. Pan, J.; Sharif, R.; Xu, X.; Chen, X. Mechanisms of waterlogging tolerance in plants: Research progress and prospects. *Front. Plant Sci.* **2021**, *11*, 627331. [CrossRef] [PubMed]
3. Bhusal, N.; Kim, H.S.; Han, S.G.; Yoon, T.M. Photosynthetic traits and plant–water relations of two apple cultivars grown as bi-leader trees under long-term waterlogging conditions. *Environ. Exp. Bot.* **2020**, *176*, 104111. [CrossRef]

4. Gibbs, J.; Greenway, H. Review: Mechanisms of anoxia tolerance in plants. I. Growth, survival and anaerobic catabolism. *Funct. Plant Biol.* **2003**, *30*, 353. [CrossRef]

5. Phukan, U.J.; Mishra, S.; Shukla, R.K. Waterlogging and submergence stress: Affects and acclimation. *Crit. Rev. Biotechnol.* **2016**, *36*, 956–966. [CrossRef]

6. Feng, Z.; Wang, D.Y.; Zhou, Q.G.; Zhu, P.; Luo, G.M.; Luo, Y.J. Physiological and transcriptomic analyses of leaves from *Gardenia jasminoides* Ellis under waterlogging stress. *Braz. J. Bio.* **2022**, *84*, e263092. [CrossRef]

7. Hong, B.; Zhou, B.; Peng, Z.; Yao, M.; Wu, J.; Wu, X.; Guan, C.; Guan, M. Tissue-specific transcriptome and metabolome analysis reveals the response mechanism of *Brassica napus* to waterlogging stress. *Int. J. Mol. Sci.* **2023**, *24*, 6015. [CrossRef]

8. Cao, M.; Zheng, L.; Li, J.; Mao, Y.; Zhang, R.; Niu, X.; Geng, M.; Zhang, X.; Huang, W.; Luo, K.; et al. Transcriptomic profiling suggests candidate molecular responses to waterlogging in cassava. *PLoS ONE* **2022**, *17*, e0261086. [CrossRef] [PubMed]

9. Hess, N.; Klode, M.; Anders, M.; Sauter, M. The hypoxia responsive transcription factor genes ERF71/HRE2 and ERF73/HRE1 of Arabidopsis are differentially regulated by ethylene. *Physiol. Plant.* **2011**, *143*, 41–49. [CrossRef]

10. Ruperti, B.; Botton, A.; Populin, F.; Eccher, G.; Brilli, M.; Quaggiotti, S.; Trevisan, S.; Cainelli, N.; Guarracino, P.; Schievano, E.; et al. Flooding responses on grapevine: A physiological, transcriptional, and metabolic perspective. *Front. Plant Sci.* **2019**, *10*, 339. [CrossRef]

11. Xu, X.W.; Wang, H.H.; Qi, X.H.; Xu, Q.; Chen, X.H. Waterlogging-induced increase in fermentation and related gene expression in the root of cucumber (*Cucumis sativus* L.). *Sci. Hortic.* **2014**, *179*, 388–395.

12. Le, P.G.; Lesur, I.; Lalanne, C.; Da, S.C.; Labadie, K.; Aury, J.M.; Leple, J.C.; Plomion, C. Implication of the suberin pathway in adaptation to waterlogging and hypertrophied lenticels formation in pedunculate oak (*Quercus robur* L.). *Tree Physiol.* **2016**, *36*, 1330–1342.

13. León, J.; Castillo, M.C.; Gayubas, B. The hypoxia-reoxygenation stress in plants. *J. Exp. Bot.* **2021**, *72*, 5841–5856. [CrossRef]

14. Cisse, E.M.; Zhang, J.; Li, D.D.; Miao, L.F.; Yin, L.Y.; Yang, F. Exogenous ABA and IAA modulate physiological and hormonal adaptation strategies in *Cleistocalyx operculatus* and *Syzygium jambos* under long-term waterlogging conditions. *BMC Plant Biol.* **2022**, *22*, 523. [CrossRef] [PubMed]

15. Jia, L.; Qin, X.; Lyu, D.; Qin, S.; Zhang, P. ROS production and scavenging in three cherry rootstocks under short-term waterlogging conditions. *Sci. Hortic.* **2019**, *257*, 108647. [CrossRef]

16. Zúñiga-Feest, A.; Bustos-Salazar, A.; Alves, F.; Martinez, V.; Smith-Ramírez, C. Physiological and morphological responses to permanent and intermittent waterlogging in seedlings of four evergreen trees of temperate swamp forests. *Tree Physiol.* **2017**, *37*, 779–789. [CrossRef]

17. Kreuzwieser, J.; Hauberg, J.; Howell, K.A.; Carroll, A.; Rennenberg, H.; Millar, A.H.; Whelan, J. Differential response of gray poplar leaves and roots underpins stress adaptation during hypoxia. *Plant Physiol.* **2009**, *149*, 461–473. [CrossRef]

18. Juntawong, P.; Sirikhachornkit, A.; Pimjan, R.; Sonthirod, C.; Sangsrakru, D.; Yoocha, T.; Tangphatsornruang, S.; Srinives, P. Elucidation of the molecular responses to waterlogging in Jatropha roots by transcriptome profiling. *Front. Plant Sci.* **2014**, *5*, 658. [CrossRef]

19. Ribeiro, I.M.; Vinson, C.C.; Coca, G.C.; Ferreira, C.D.S.; Franco, A.C.; Williams, T.C.R. Differences in the metabolic and functional mechanisms used to tolerate flooding in *Guazuma ulmifolia* (Lam.) from flood-prone Amazonian and dry *Cerrado savanna* populations. *Tree Physiol.* **2022**, *42*, 2116–2132. [CrossRef]

20. Zhang, Y.J.; Liu, G.Y.; Dong, H.Z.; Li, C.D. Waterlogging stress in cotton: Damage, adaptability, alleviation strategies, and mechanisms. *Crop J.* **2020**, *9*, 257–270. [CrossRef]

21. Kreuzwieser, J.; Rennenberg, H. Molecular and physiological responses of trees to waterlogging stress. *Plant Cell Environ.* **2014**, *37*, 2245–2259. [CrossRef]

22. Ferner, E.; Rennenberg, H.; Kreuzwieser, J. Effect of flooding on C metabolism of flood-tolerant (*Quercus robur*) and non-tolerant (*Fagus sylvatica*) tree species. *Tree Physiol.* **2012**, *32*, 135–145. [CrossRef]

23. Parent, C.; Crèvecoeur, M.; Capelli, N.; Dat, J.F. Contrasting growth and adaptive responses of two oak species to flooding stress: Role of non-symbiotic hemoglobin. *Plant Cell Environ.* **2011**, *34*, 1113–1126. [CrossRef]

24. Doupis, G.; Kavroulakis, N.; Psarras, G.; Papadakis, I.E. Growth, photosynthetic performance and antioxidative response of 'Hass' and 'Fuerte' avocado (*Persea americana* Mill.) plants grown under high soil moisture. *Photosynthetica* **2017**, *55*, 655–663. [CrossRef]

25. Pezeshki, S.R. Wetland plant responses to soil flooding. *Environ. Exp. Bot.* **2001**, *46*, 299–312. [CrossRef]

26. Ahmed, S.; Nawata, E.; Hosokawa, M.; Domae, Y.; Sakuratani, T. Alterations in photosynthesis and some antioxidant enzymatic activities of mungbean subjected to waterlogging. *Plant Sci.* **2002**, *163*, 117–123. [CrossRef]

27. Yan, K.; Zhao, S.; Cui, M.; Han, G.; Wen, P. Vulnerability of photosynthesis and photosystem I in Jerusalem artichoke (*Helianthus tuberosus* L.) exposed to waterlogging. *Plant Physiol. Biochem.* **2018**, *125*, 239–246. [CrossRef] [PubMed]

28. Christianson, J.A.; Llewellyn, D.J.; Dennis, E.S.; Wilson, I.W. Global gene expression responses to waterlogging in roots and leaves of cotton (*Gossypium hirsutum* L.). *Plant Cell Physiol.* **2010**, *51*, 21–37. [CrossRef]

29. Zhang, Y.; Kong, X.; Dai, J.; Luo, Z.; Li, Z.; Lu, H.; Xu, S.; Tang, W.; Zhang, D.; Li, W.; et al. Global gene expression in cotton (*Gossypium hirsutum* L.) leaves to waterlogging stress. *PLoS ONE* **2017**, *12*, e0185075. [CrossRef]

30. Wang, X.; Niu, Y.; Zheng, Y. Multiple functions of MYB transcription factors in abiotic stress responses. *Int. J. Mol. Sci.* **2021**, *22*, 6125. [CrossRef]

31. Guo, J.; Sun, B.; He, H.; Zhang, Y.; Tian, H.; Wang, B. Current understanding of bHLH transcription factors in plant abiotic stress tolerance. *Int. J. Mol. Sci.* **2021**, *22*, 4921. [CrossRef]

32. Pan, D.L.; Wang, G.; Wang, T.; Jia, Z.H.; Guo, Z.R.; Zhang, J.Y. *AdRAP2.3*, a novel ethylene response factor VII from *Actinidia deliciosa*, enhances waterlogging resistance in transgenic tobacco through improving expression levels of *PDC* and *ADH* Genes. *Int. J. Mol. Sci.* **2019**, *20*, 1189. [CrossRef]

33. Arbona, V.; Zandalinas, S.I.; Manzi, M.; González-Guzmán, M.; Rodriguez, P.L.; Gómez-Cadenas, A. Depletion of abscisic acid levels in roots of flooded Carrizo citrange (*Poncirus trifoliata* L. Raf. × *Citrus sinensis* L. Osb.) plants is a stress-specific response associated to the differential expression of PYR/PYL/RCAR receptors. *Plant Mol. Biol.* **2017**, *93*, 623–640. [CrossRef]

34. Bashar, K.K. Hormone dependent survival mechanisms of plants during post-waterlogging stress. *Plant Signal. Behav.* **2018**, *13*, e1529522. [CrossRef] [PubMed]

35. Pintó-Marijuan, M.; Munné-Bosch, S. Photo-oxidative stress markers as a measure of abiotic stress-induced leaf senescence: Advantages and limitations. *J. Exp. Bot.* **2014**, *65*, 3845–3857. [CrossRef]

36. Baxter, A.; Mittler, R.; Suzuki, N. ROS as key players in plant stress signalling. *J. Exp. Bot.* **2014**, *65*, 1229–1240. [CrossRef]

37. Pallavi, S.; Bhushan, J.A.; Shanker, D.R.; Mohammad, P. Reactive oxygen species, oxidative damage, and antioxidative defense mechanism in plants under stressful conditions. *J. Bot.* **2012**, *12*, 681.

38. Hasanuzzaman, M.; Bhuyan, M.; Zulfiqar, F.; Raza, A.; Mohsin, S.M.; Mahmud, J.A.; Fujita, M.; Fotopoulos, V. Reactive oxygen species and antioxidant defense in plants under abiotic stress: Revisiting the crucial role of a universal defense regulator. *Antioxidants* **2020**, *9*, 681. [CrossRef] [PubMed]

39. Hossain, Z.; López-Climent, M.; Arbona, V.; Pérez-Clemente, R.; Gómez-Cadenas, A. Modulation of the antioxidant system in Citrus under waterlogging and subsequent drainage. *J. Plant Physiol.* **2009**, *166*, 1391–1404. [CrossRef] [PubMed]

40. Han, X.; Zhao, Y.; Chen, Y.; Xu, J.; Jiang, C.; Wang, X. Lignin biosynthesis and accumulation in response to abiotic stresses in woody plants. *For. Res.* **2022**, *2*, 9. [CrossRef]

41. Gong, X.; Xu, Y.; Li, H.; Chen, X.; Song, Z. Antioxidant activation, cell wall reinforcement, and reactive oxygen species regulation promote resistance to waterlogging stress in hot pepper (*Capsicum annuum* L.). *BMC Plant Biol.* **2022**, *22*, 425. [CrossRef]

42. Nguyen, T.N.; Son, S.; Jordan, M.C.; Levin, D.B.; Ayele, B.T. Lignin biosynthesis in wheat (*Triticum aestivum* L.): Its response to waterlogging and association with hormonal levels. *BMC Plant Biol.* **2016**, *16*, 28. [CrossRef] [PubMed]

43. Zhao, X.; Huang, L.; Kang, L.; Jetter, R.; Yao, L.; Li, Y.; Xiao, Y.; Wang, D.; Xiao, Q.; Ni, Y.; et al. Comparative analyses of cuticular waxes on various organs of faba bean (*Vicia faba* L.). *Plant Physiol. Biochem.* **2019**, *139*, 102–112. [CrossRef] [PubMed]

44. Tellechea-Robles, L.E.; Cesea, M.S.; Bullock, S.H.; Cadena-Nava, R.D.; Méndez-Alonzo, R. Is leaf water-repellency and cuticle roughness linked to flooding regimes in plants of coastal wetlands? *Wetlands* **2019**, *40*, 515–525. [CrossRef]

45. Herrera, A. Responses to flooding of plant water relations and leaf gas exchange in tropical tolerant trees of a black-water wetland. *Front. Plant Sci.* **2013**, *4*, 106. [CrossRef]

46. Merchant, A.; Peuke, A.D.; Keitel, C.; Macfarlane, C.; Warren, C.R.; Adams, M.A. Phloem sap and leaf delta13C, carbohydrates, and amino acid concentrations in *Eucalyptus* globulus change systematically according to flooding and water deficit treatment. *J. Exp. Bot.* **2010**, *61*, 1785–1793. [CrossRef]

47. Grabherr, M.G.; Haas, B.J.; Yassour, M.; Levin, J.Z.; Thompson, D.A.; Amit, I. Full-length transcriptome assembly from RNA-Seq data without a reference genome. *Nat. Biotechnol.* **2011**, *29*, 644–652. [CrossRef]

48. Trapnell, C.; Williams, B.A.; Pertea, G.; Mortazavi, A.; Kwan, G.; van Baren, M.J.; Salzberg, S.L.; Wold, B.J.; Pachter, L. Transcript assembly and quantification by RNA-Seq reveals unannotated transcripts and isoform switching during cell differentiation. *Nat. Biotechnol.* **2010**, *28*, 511–515. [CrossRef]

49. Livak, K.J.; Schmittgen, T.D. Analysis of relative gene expression data using real-time quantitative PCR and the 2(-Delta Delta C(T)) Method. *Methods* **2001**, *25*, 402–408. [CrossRef]

50. Dunn, W.B.; Broadhurst, D.; Begley, P.; Zelena, E.; Francis-McIntyre, S.; Anderson, N. Procedures for large-scale metabolic profiling of serum and plasma using gas chromatography and liquid chromatography coupled to mass spectrometry. *Nat. Protoc.* **2011**, *6*, 1060–1083. [CrossRef]

51. Kind, T.; Wohlgemuth, G.; Lee, D.Y.; Lu, Y.; Palazoglu, M.; Shahbaz, S.; Fiehn, O. FiehnLib: Mass spectral and retention index libraries for metabolomics based on Quadrupole and Time-of-Flight Gas Chromatography/Mass Spectrometry. *Anal. Chem.* **2009**, *81*, 10038–10048. [CrossRef] [PubMed]

International Journal of
Molecular Sciences

MDPI

Review

Defining Mechanisms of C₃ to CAM Photosynthesis Transition toward Enhancing Crop Stress Resilience

Bowen Tan and Sixue Chen *

Department of Biology, University of Mississippi, Oxford, MS 38677, USA; btan@go.olemiss.edu
* Correspondence: schen8@olemiss.edu

Abstract: Global climate change and population growth are persistently posing threats to natural resources (e.g., freshwater) and agricultural production. Crassulacean acid metabolism (CAM) evolved from C_3 photosynthesis as an adaptive form of photosynthesis in hot and arid regions. It features the nocturnal opening of stomata for CO_2 assimilation, diurnal closure of stomata for water conservation, and high water-use efficiency. To cope with global climate challenges, the CAM mechanism has attracted renewed attention. Facultative CAM is a specialized form of CAM that normally employs C_3 or C_4 photosynthesis but can shift to CAM under stress conditions. It not only serves as a model for studying the molecular mechanisms underlying the CAM evolution, but also provides a plausible solution for creating stress-resilient crops with facultative CAM traits. This review mainly discusses the recent research effort in defining the C_3 to CAM transition of facultative CAM plants, and highlights challenges and future directions in this important research area with great application potential.

Keywords: Crassulacean acid metabolism; C_3 to CAM transition; facultative CAM; photosynthesis; climate change; water-use efficiency; CAM engineering

1. Introduction

Drastic climate change over the past decades can be reflected by the alternation in atmospheric CO_2 levels, tropospheric ozone concentrations, and other environmental indicators [1]. Climate change is not only affecting ecosystems, but also agriculture, food production, land, and water resources [2]. Arid or semi-arid land accounts for around 41% of the total surface on Earth, and it is expanding [3]. In 2035, global desertification is projected to be 65% of the total land surface in the subtropical regions [4]. With the rapid growth of the human population, the demand for food is increasing, and it is anticipated to surge by 70%. The current rate of global crop productivity only increases by ~2% per year, which cannot meet the demand for food [5]. To worsen the situation, the global decrease in freshwater from 1980 to 2015 has caused a 20.6% and 39.3% yield reduction in wheat and maize, respectively [6].

Photosynthesis, a pivotal biological process essential to all life, provides food and most of our energy resources [7]. There are three major modes of photosynthesis in vascular plants to assimilate atmospheric CO_2: C_3, C_4, and Crassulacean acid metabolism (CAM) [8,9] (Figure 1). CAM photosynthesis has evolved independently multiple times from C_3 as a photosynthetic adaptation to cope with the decreasing atmospheric CO_2 levels ~20 million years ago [10]. CAM plants are commonly found in harsh environments such as arid and semi-arid regions [11]. Other than those water-limited regions, CAM plants also inhabit the aquatic environment. With the release of the genome and transcriptome of an underwater CAM plant *Isoetes taiwanensis* [12], differences in the recruitment of phosphoenolpyruvate (PEP) carboxylase (PEPC) and core CAM pathway gene expression between aquatic and terrestrial plants demonstrate a different route of CAM evolution.

Citation: Tan, B.; Chen, S. Defining Mechanisms of C_3 to CAM Photosynthesis Transition toward Enhancing Crop Stress Resilience. *Int. J. Mol. Sci.* **2023**, *24*, 13072. https://doi.org/10.3390/ijms241713072

Academic Editor: Martin Bartas

Received: 3 July 2023
Revised: 12 August 2023
Accepted: 14 August 2023
Published: 22 August 2023

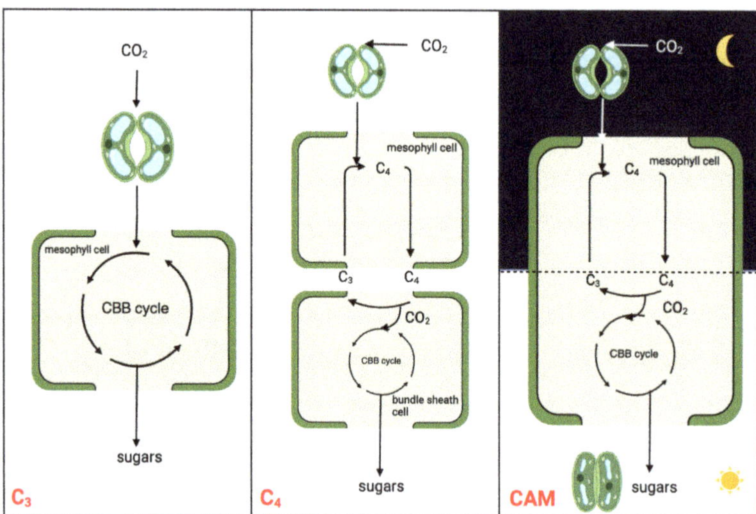

Figure 1. Simplified schematic to illustrate the molecular relationships and distinctions among C_3, C_4, and CAM photosynthesis mechanisms. The CBB cycle is the abbreviation of Calvin-Benson-Bassham cycle, which is also known as Calvin cycle.

CAM is a carbon concentrating mechanism, with the capability of assimilating CO_2 initially at night using PEPC in the cytosol, leading to the formation of a four-carbon malate, which is then stored in the vacuole [13,14]. The three-carbon acceptor in this reaction is PEP, which is replenished by the glycolytic breakdown of carbohydrate storage in the form of starch or other sugars. Unlike spatial decoupling of carboxylation and decarboxylation in C_4 photosynthesis, CAM photosynthesis separates these two processes in a temporal manner to shield ribulose-1,5-bisphosphate carboxylase-oxygenase (RuBisCO) from the oxygenase activity, minimizing photorespiration (Figure 1). CAM plants conduct gas exchange predominantly at night when the air temperature is low, thereby having a lower water loss by an order of magnitude than it would be during the day [15]. As such, CAM plants have water-use efficiency (WUE) several-fold higher than those of C_3 and C_4 plants under comparable conditions [16]. High WUE, together with enhanced heat and drought tolerance, drives the basic and applied research on CAM toward crop CAM engineering/bio-design. The typical diel cycle of CAM entails four phases: (I) nocturnal atmospheric CO_2 fixation by PEPC and malic acid storage in the vacuole; (II) RuBisCO activation just after dawn when, for a brief period, CO_2 is fixed by both PEPC and RuBisCO; (III) stored malate decarboxylated to CO_2, which is fixed by RuBisCO; (IV) the end of the light period when stomata reopen driven by the depletion of malate pool [13,17].

CAM plants normally exhibit the following features: the diurnal fluctuation of malic acids (accumulation during the night period and dissipation during the day); reciprocal diurnal fluctuation of storage carbohydrates such as starch, polyglucans, or soluble hexoses; a high level of PEPC and an active decarboxylase; large storage vacuoles that are in the same cells with chloroplasts; water-limitation related traits, such as dense trichomes, leaf succulence, and waxy cuticles [1]; and nocturnal net CO_2 uptake, which exhibits an inverse pattern of stomatal movement [18]. In the course of evolution, an intermediate CAM mechanism called facultative CAM arose. Plants conducting facultative CAM demonstrate the optional use of CAM under stress conditions, while remaining the use of C_3 or C_4 photosynthesis under normal conditions [19–23].

Given the characteristics of facultative CAM photosynthesis and climate change urgency, unraveling the molecular mechanisms underlying the C_3 to CAM transition has attracted growing interest. However, there is a lack of review on facultative CAM and especially C_3 to CAM transition. Herein we summarize recent advances in facultative C_3 to CAM transition, discuss current problems and challenges, and highlight future research directions (Figure 2).

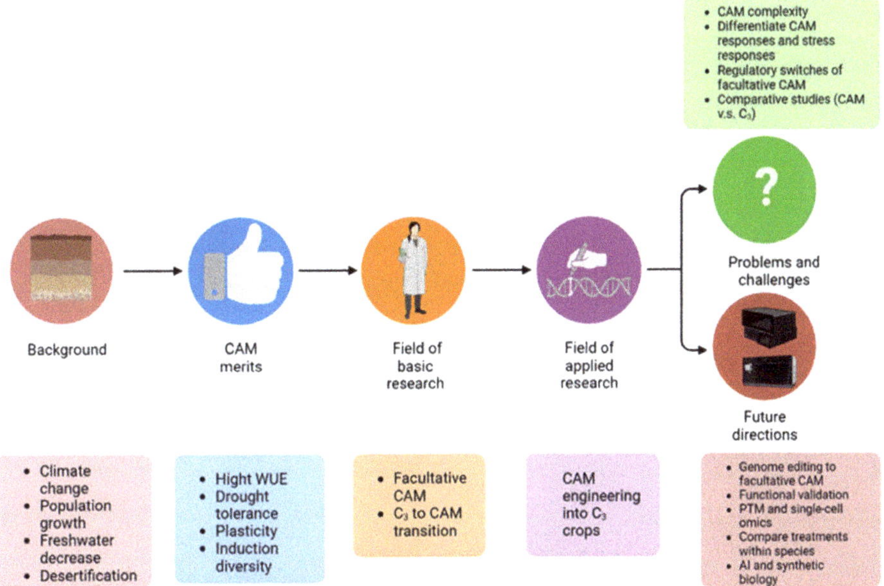

Figure 2. Graphical summary of the significance of studying C_3 to CAM transition of facultative CAM plants. The abbreviations used: C_3, C_3 photosynthesis; CAM, Crassulacean acid metabolism; WUE, water-use efficiency; and PTM, post-translational modification.

2. The Plasticity of CAM Is Best Represented in Facultative CAM

A remarkable hallmark of CAM plants is their considerable plasticity in expressing the four phases of CAM, while keeping the C_3 cycle fully functional [24]. Environmental factors such as light intensity, relative humidity, water availability [25], and developmental stages [16] affect the degree and duration of CAM expression [9]. CAM is highly plastic and can operate in different modes: (1) obligate/constitutive CAM or strong CAM, with high nocturnal acid accumulation (ΔH^+) and CO_2 fixation; (2) CAM-cycling, with daytime CO_2 fixation like C_3 and nocturnal fixation of CO_2 from respiration; (3) CAM idling, with stomata closed all the time and CAM fixation of CO_2 only from respiration; (4) facultative/inducible CAM, with C_3 mode of CO_2 fixation and zero ΔH^+ in the non-stressed state, and small nocturnal CO_2 fixation and ΔH^+ during C_3 to CAM transition in the stressed state [26]; (5) weak CAM, with similar CO_2 uptake pattern as strong CAM but less nocturnal acid accumulation.

Among the above five different modes of CAM, a preeminent model for elucidating the molecular underpinnings of CAM is facultative CAM [27,28]. In facultative CAM species, CAM may be induced by a variety of stimuli such as drought [29,30], salinity [19,31], high photosynthetic photon flux [6,32], abscisic acid (ABA) [33], photoperiod [34] and hydrogen peroxide [35]. Clearly, CAM plasticity is best represented by facultative CAM plants, which employ the C_3 photosynthesis under non-stress conditions to maximize growth, but are able to undergo a gradual C_3 to CAM transition to reduce water loss and maintain photosynthetic integrity under water-limited conditions. It ultimately translates into high WUE, survival, and reproductive success [36]. Facultative CAM plants have been identified in a wide range of plant families, such as *Bromeliaceae, Cactaceae, Aizoaceae, Montiaceae, Lamiaceae, Vitaceae, and Didiereaceae* [37], indicating multiple independent evolutionary events (Figure 3). Whether these independent events generated similar or different genetic and epigenetic changes that enable facultative CAM deserves immediate investigation, e.g., by identifying and utilizing the evolutionary pairs of C$_3$ and CAM species.

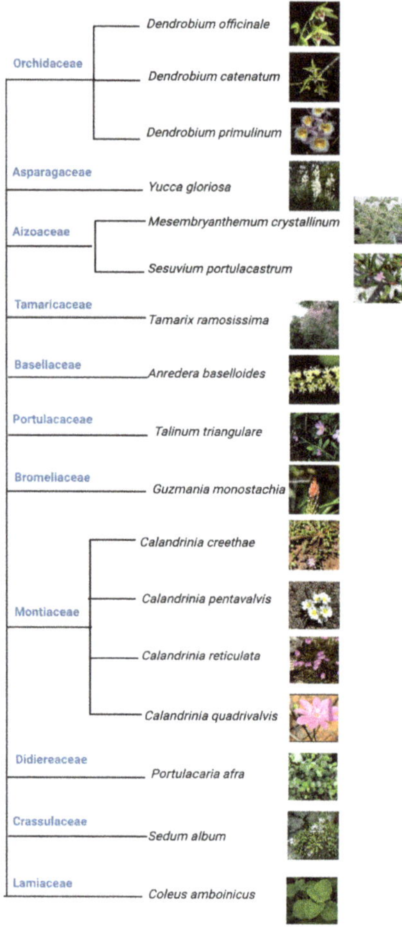

Figure 3. Tree view of all the facultative CAM plants investigated from 2017 to 2023. Source of each image was cited from top to bottom [38–53], respectively, except that the image of *M. crystallinum* was from the Chen lab.

3. Studies on C$_3$ to CAM Transition Revealed Important Molecular Players

Over the past decades, different model species were used for diverse aspects of research pertinent to CAM. Here are six main areas of CAM research: (1) CAM ecophysiology to study and discover new CAM species in different ecological environments (e.g., [27,54]); (2) CAM origin and evolution (e.g., [55,56]); (3) genomic features and molecular mechanisms regulating CAM (e.g., [57,58]); (4) C$_3$ to CAM transition (e.g., [19,20]); (5) CAM metabolic modeling (e.g., [59,60]); and (6) engineering CAM into C$_3$ plants (e.g., [5,61]). All these areas of basic research aim for the ultimate goal of exploiting the potential of CAM in crop improvement under climate change [56]. Based on recent publications, *Ananas comosus* (Pineapple), *Kalanchoë fedtschenkoi*, and *Mesembryanthemum crystallinum* (Common ice plant, Table 1) are the three most extensively studied CAM models. Their genomes have been fully sequenced [58,62,63]. For studying the C$_3$ to CAM transition, *M. crystallinum* has been the classic model, and *Talinum triangulare* is an emerging model [17]. *T. triangulare* is an herbaceous weed that shifts from C_3 to CAM photosynthesis on day 11 of drought treatment. The large evenly green leaves, rapid growth, relatively short life cycle, self-cross, and full reversibility of CAM make it a model system to study facultative CAM [28].

Table 1. Research progress on CAM in *M. crystallinum* from 2017 to 2022.

Research Focus	Key Findings	Reference
Identified C_3-CAM transition period and temporal physiological changes	- The shift in a 3-day period - CO_2 exchange reflects inversed stomatal behavior - A Python script was created for high throughput leaf area assay	[20]
Transcriptomics of guard cells during the C_3-CAM transition	- 18 transcription factors identified - Guard cell has its own transition	[19]
Nocturnal carboxylation is coordinated with starch degradation by the products of these pathways, such as carbohydrates	- Transitory starch is necessary for CAM operation - Carbohydrates coordinate the regulation of carboxylation and starch degradation	[64]
Functional CAM withdrawal in the de-salted plants	- Rapid downregulation of *PEPC1* and decrease in Δ malate was found - CAM-specific antioxidative enzyme activities exhibited transient & fully reversible responses to salt stress	[65]
Comparative proteomic changes in guard cells and mesophyll cells during the C_3-CAM transition	- Guard cells and mesophyll cells showed different changes in proteome during the transition - Proteins involved in osmotic adjustment, ion transport, energy metabolism and light response may play important roles in the transition	[66]
Phytohormones in the stomatal behavior during the C_3-CAM transition	- Diurnal balance of cytokinin and jasmonic acid regulates stomatal behavior	[67]
Genome sequencing, transcriptomics, and comparative genomics of leaves	- Chromosome rearrangements and gene loss in ice plant evolution - Several key CAM-related genes identified	[62]

Note: Δ malate represents the difference in malate concentration between the end of the night period and the beginning of the light phase.

The early research of C_3 to CAM transition mainly focused on several key CAM enzymes, such as PEPC [68], PEPC protein kinase (PPCK) and malic enzymes, as well as metabolite transport in the C_3 and CAM state [69,70]. Early studies suggest that ABA signaling, Ca^{2+} signaling, and protein phosphorylation/de-phosphorylation may play important roles in the C_3 to CAM transition, whereas no key players such as kinases/phosphatases were identified. Sequence analyses were also performed on *PPC1* and *PPC2*, which are CAM-specific genes that encode PEPC [71,72]. With the emergence of microarray technologies, large-scale mRNA profiling was carried out [73]. Instead of studying the transition, Cushman group compared the gene expression of ice plants between the non-stressed C_3 group and the induced-CAM group after 14 days of salt treatment. Gene expression of eight transporters was analyzed to study the inter-organellar metabolite transport between C_3 and the CAM group of ice plants [74]. However, there is a lack of understanding of the temporal metabolic and molecular control of the C_3 to CAM transition and a systems-level understanding was needed to reveal the regulatory changes underlying the transition.

With the advances in high-throughput omics technologies and computational biology, systems biology has become a prevalent approach for discovery (hypothesis generation) and functional studies (hypothesis testing). Beyond traditional physiological and biochemical methods, multi-omics (genomics, transcriptomics, proteomics and metabolomics) has generated a systems-level understanding of temporal molecular and metabolic controls underpinning CAM [75]. In *T. triangulare* leaves, targeted metabolite profiling and RNA sequencing were performed to reveal the rewiring of carbohydrate metabolism and candidate transcription factors (TFs) in the drought-induced CAM transition process [76]. Three years later, the same group identified seven candidate regulators of ABA-induced CAM including *HEAT SHOCK TF A2, NUCLEAR FACTOR Y, SUBUNITS A9*, and *JmjC DOMAIN-CONTAINING PROTEIN 27* [77]. In addition to the traditional drought and salt induction of CAM, hydrogen peroxide was shown to be able to induce CAM in *M. crystallinum* [35]. In the leaves of another facultative bromeliad *Guzmania monostachia*, increases in the expression of CAM-related genes (*PEPC1, PPCK, NAD-malate dehydrogenase*, aluminum-activated malate transporter 9 (*ALMT9*), PEP carboxykinase (*PEPCK*)) and *UREASE* transcripts were shown under drought. And the role of integrating N and C metabolism of urea was suggested [78]. The CAM gene expression, antioxidant activities, and chlorophyll fluorescence were compared between a C_3-CAM facultative species (*Sedum album*) and a C_4-CAM facultative species (*Portulaca oleracea*) [79]. The level of nitric oxide (NO) was found to be correlated with the CAM expression during CAM induction only in *S. album* but not *P. oleracea*. This suggests the different roles of NO in C_3 and C_4 species during CAM induction. All the aforementioned studies did not identify a critical transition period, which is key to capturing the molecular switches for CAM. Three years ago, the transition period of *M. crystallinum* was first defined during salt-induced C_3 to CAM shift, and further validated in independent studies through RNA-seq and physiological analyses [19,20,67]. Interestingly, three phytohormones, jasmonic acid (JA), cytokinin, and ABA were reported to play important roles in the inversed pattern of stomata opening/closing during the transition of *M. crystallinum* [67]. With the release of the ice plant genome [62], more studies can explore the genes and metabolites pertinent to the C_3 to CAM transition.

4. CAM Engineering toward Solving the Global Climate Challenges

The evolution of CAM is a natural innovation in response to the hotter and drier environment. Scientists are striving to gain a better understanding of CAM associated with high WUE, and to expand the CAM characteristics to agriculturally valuable C_3 crops. Although CAM mode compromises growth over survival, moving some of the CAM characteristics (e.g., inversed stomatal movement) into C_3 crops without compromising yield is highly attractive and promising. Here are the rationales: (1) The existence of facultative CAM plants, such as *M. crystallinum* (see discussion in the next paragraph); (2) Unlike C_4 photosynthesis, CAM is a single-cell phenomenon. All the genetic components, enzymes, and transporters of CAM are found in C_3 plants [17]; (3) CAM has emerged from ancestral C_3 photosynthesis independently in diverse plant lineages. The possible existence of multiple mechanisms presents opportunities for synthetic biology; (4) Recent omics/systems biology efforts have identified many molecular components important for the development of CAM [5,19,66,77,80]. They will certainly facilitate synthetic biology applications; (5) Chimeric/bifunctional promoters made inducible CAM a near reality. For example, guard cell-specific promoters that are also inducible by drought and switching off without drought [81,82]. The promoters will allow CAM to be turned on under adverse environmental conditions and turned off when conditions improve, so that crops become resilient and maintain productivity.

There is an ongoing debate on the C_3 to CAM continuum. Supported by the existence of facultative CAM plants, accumulation of malate at night in C_3 plants, and constraint-based modeling data [60], one side believes that the emergence of CAM could occur simply by increasing the pre-existing metabolic fluxes in diurnal decarboxylation from malate to CO_2, CO_2 re-carboxylation into the Calvin cycle, nocturnal CO_2 carboxylation to malate, and PEP replenishment [55,60]. However, the other side believes that CAM should be regarded as a discrete metabolic innovation and argues that the theory of C_3-CAM continuum underestimates the effort needed to activate CAM, which requires metabolic reprogramming [83]. After examining 30 CAM species and 40 C_3 species, the authors concluded that nocturnal acidification is the hallmark of CAM. Also, although CAM plants show plasticity and have a dispersed occurrence, they only account for a small proportion (~7%) of vascular plants [83]. The ecophysiological studies involving the carbon-isotope ratio showed that the facultative CAM state is not favored [28]. The evidence seems to point to the fact that CAM is not a facile trait to develop simply by increasing the metabolic fluxes. Disentangling how CAM occurs is essential to guide the CAM engineering direction toward success.

Functional analysis of CAM-related genes not only validates the omics discovery, but also lays a solid foundation for CAM engineering. Gene function analysis has been conducted in the reference plant *Arabidopsis thaliana* by assessing 13 key CAM enzymes and regulatory proteins from the ice plant [61]. Large cell size and succulence may be needed in terms of CAM engineering due to the storage of nocturnal organic acids and water. By overexpressing a TF *VvCEB1ₒₚₜ* from *Vitis vinifera* in *Arabidopsis*, the cell size and tissue succulence were enhanced [5,84]. PEPC enzymes and their coding genes have been mostly studied in the CAM field. A partial CAM pathway was assembled by expressing an engineered *Solanum tuberosum PEPC* in *A. thaliana* under the control of a dark-induced promoter from *A. thaliana* [85]. Overexpression of *PEPC* from *Agave americana* in tobacco showed improved biomass production under stress conditions [86]. TFs play vital roles in regulating various cellular processes, and the TF-based engineering approach has the potential to enhance abiotic stress tolerance in plants [11]. Several well-known TF families including homeobox (HB), NAM, ATAF1/2, and CUC2 (NAC), WRKY, and basic region/leucine zipper motif (bZIP) are linked to abiotic stress responses. *HB7* was highly upregulated in both facultative CAM plants, *T. triangulare* [76] and *M. crystallinum* [19] in stress-induced CAM. Later functional studies showed that overexpression of the TF *McHB7* in *M. crystallinum* [80] and *A. thaliana* [87] improved plant growth and salt tolerance. An ideal proposal for future CAM synthetic biology and engineering effort would be to enhance crop WUE and resilience (through stress-inducible promoters) without negatively affecting the yield so that crops can survive the drought and heat episodes and maintain productivity.

5. Pressing Problems and Challenges in CAM Research

Other than being a model to study the shift from C_3 to CAM, facultative CAM plants are also used to study the molecular mechanisms underlying stress tolerance [87,88]. As described in previous sections, *M. crystallinum* is a great model for elucidating molecular mechanisms of C_3 to CAM transition, it is challenging to distinguish stress-related responses from CAM-related responses. For example, transcripts involved in the ABA signaling pathway and sugar metabolism showed differential expressions. But it's challenging to know if these are stress or CAM responses. One way to overcome this dilemma is to focus on molecules/pathways that show shared changes during the CAM transition induced by different stresses, e.g., both drought and salinity [66,89,90]. These shared changes may represent evolutionary innovations for the C_3 to CAM transition, not just stress-specific responses. Additionally, there are a number of studies that employ comparative analyses between C_3 and CAM plants, such as comparative genomics [6], transcriptomics [91], and stomatal responses [92], which provide the static contrasts in genes, expression, regulation, and possible evolutionary mechanisms. A caveat of these studies is that phylogenetic closest pairs of C_3 and CAM species were not identified and used. These comparisons are more likely to lead to the original molecular changes/switches for CAM evolution. Besides, the dynamic changes during the C_3 to CAM transition are often overlooked.

Thanks to the release of genome sequencing data from obligate CAM plants, namely, *Phalaenopsis equestris* [93], pineapple [63], *K. fedtschenkoi* [58], *Carnegiea gigantea* [94], *I. taiwanensis* [12], and *Cissus rotundifolia* [95], the investigation of obligate CAM plants is considerably more comprehensive than facultative CAM plants. Paired with transcriptomics data, these works provide rich information not only on whole-genome duplication events during the CAM evolution and comparative genomics, but also on the mechanisms of how CAM operates and is modulated by studying the diel expression patterns of CAM-related genes. For example, the linkage between CAM and the circadian clock was first reported by showing that CAM genes were enriched with five circadian cis-regulatory elements (the Morning Element (CCACAC), the Evening Element (AAAATATCT), the CCA1-binding site (AAAAATCT), the G-box (CACGTG) and the TCP15-binding motif (NGGNCCCAC)) [63]. Four *K. fedtschenkoi* genes showed convergent changes in protein sequences and 60 genes showed convergent diel expression changes and convergent evolution in a variety of CAM species [58]. These results clearly documented specific components and requirements for CAM functionality. None of the model facultative CAM plants had available genome sequencing data until the first published assembly of the *M. crystallinum* genome in 2022 [62]. But the authors only sequenced the coding region and released some intermediate files which makes the genome information mostly inaccessible. Recently, the first available ice plant whole genome was released [96]. A total of 49,782 locus IDs were generated by next-generation sequencing. Further transcriptomics and proteogenomics experiments may be needed to study the expression and regulation of these identified genes.

CAM is an intricate trait that needs not only simply integrating different functional modules, but also the tight regulation of metabolic processes. The complexity of the C_3 to CAM transition also comes from the integration of diurnal and circadian rhythms, stomatal regulation, leaf anatomy, cell architecture, cell packing, and all the biochemical processes behind it. Different enzymes and pathways employed by different plant lineages (derived from the multiple independent evolution events) also complicate CAM engineering. Most research efforts have been committed to the carboxylation process, yet relatively little attention has been given to decarboxylation, regeneration of PEP, and energization processes. It's known that there are two carbon breakdown pathways, phosphorolytic and hydrolytic degradation. More complicated, there are two malate decarboxylation routes, via malate dehydrogenase (MDH) and PEP-CK, and various malic enzymes including cytosolic and chloroplastic NADP-malic enzyme (NADP-ME) and mitochondrial NAD-malic enzyme (NAD-ME) (Figure 4). Despite deeming ALMT9 as the influx transporter, the efflux transporter remains unknown and the control of these two steps still needs to be addressed. Importantly, the regulatory molecular switches, including epigenetic controls, alternative splicing, non-coding RNAs, small RNAs, TFs, kinases/phosphatases, and other posttranslational modification (PTM) regulators have not fully been investigated.

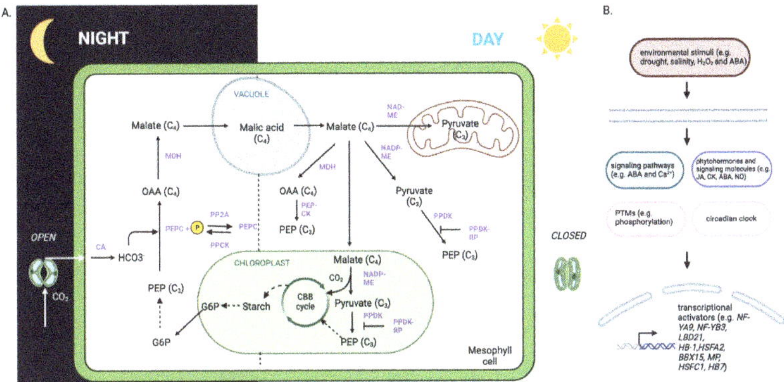

Figure 4. Summary of CAM diel cycle and potential CAM regulatory switches. (**A**) A simplified view of the CAM diel cycle including pathways, subcellular compartments, key intermediate metabolites, enzymes, and regulatory proteins. Solid arrow indicates a single-step process and dotted arrow indicates a multi-step process. (**B**) A diagram showing the potential regulatory switches of the C_3-CAM transition. Abbreviations: CA, carbonic anhydrase; PEP, phosphoenolpyruvate; PEPC, PEP carboxylase; PP2A, protein phosphatase 2A; PPCK, PEPC kinase; OAA, oxaloacetate; MDH, malate dehydrogenase; NAD-ME, NAD-dependent malic enzyme; NADP-ME, NADP-dependent malic enzyme; PEP-CK, PEP carboxykinase; PPDK, pyruvate phosphate dikinase; PPDK-RP, PPDK regulatory protein; CBB cycle, Calvin-Benson-Bassham cycle; G6P, glucose 6-phosphate; H_2O_2, hydrogen peroxide; ABA, abscisic acid; JA, jasmonic acid; NO, nitric oxide; CK, cytokinin; PTM, post-translational modification; NF-YA9, nuclear factor Y subunit A9; NF-YB3, nuclear factor Y subunit B3; LBD21, LOB domain-containing protein 21; HB, homeobox; HSFA2, heat-inducible transcription factor A2; BBX15, B-box type zinc finger protein 15; MP, MONOPTEROS; HSFC1, heat-inducible transcription factor C1.

6. Perspectives and Future Directions

Due to the limitations in genome availability, research on facultative CAM plants lags largely behind that on obligate CAM plants. Recently, genome editing tools like CRISPR/Cas9 and RNAi have been applied to two obligate CAM plants *K. fedtschenkoi* [97] and *K. laxiflora* [57]. With the availability of the genome of the classic facultative CAM model [62,96], *M. crystallinum*, more attention and research investment in facultative CAM plants and C_3 to CAM transition could be anticipated, and it will be exciting to see the expansion of single-cell omics approaches to facultative CAM plants.

Phosphorylation is one of the ubiquitous PTMs that regulate protein functions and plant physiological output. It's well known that PEPC activity is modulated through phosphorylation by a protein kinase PPCK. Some early papers showed that one of the decarboxylation enzymes, PEP-CK, is regulated by phosphorylation [98,99]. Acetylation, nitrosylation, and phosphorylation have been identified in NAD-ME and NADP-ME [17]. PTMs could be investigated to close the knowledge gap in facultative CAM, especially those important in triggering the transition from C_3 to CAM (Figure 4). Mass spectrometry is a powerful analytical tool for the discovery of proteins, metabolites, and PTMs. The sensitivity of mass spectrometers has been improved significantly toward single-cell analysis [100–102]. A large volume of data generated by omics experiments tend to have false positives. More functional validation studies through the community-wide effort are needed to validate the molecular components, changes, and regulations in the CAM transition process.

The computational/mathematical modeling of CAM dates back to 30 years ago [103]. The earlier type of model was the ordinary differential equations (ODE) model that simulated the CAM phenomenon by turning a simplified scheme of metabolic reactions of CAM into ODE to study diel rhythmicity [103]. With a different purpose, the flux balance analysis (FBA) model is used to model the metabolic fluxes in CAM plants under different conditions and predict the optimal flux distribution that maximizes or minimizes a specific purpose, such as biomass production, ATP synthesis, or malate accumulation, while satisfying the stoichiometric and thermodynamic constraints of the system [104]. Whereas the ODE model requires the input of quantitative data of the pathways to be modeled, such as enzyme activities, to accurately present the metabolite dynamics, the recent FBA models, on the other hand, focus on studying the metabolic steady-state flux distribution [59,60,105]. The FBA model doesn't require the input of quantitative data but also doesn't provide information that ODE can provide, e.g., metabolite concentrations and changes [104]. In a nutshell, ODE models offer temporal dynamics of CAM and provide a mechanistic understanding of the system but rely on high-quality data input and the models are more complicated. FBA models can be integrated with genome-scale data and excel in predicting metabolic flux distributions and metabolic phenotypes under varying environmental conditions or genetic perturbations. During these 30 years, a number of ODE and FBA models adapted from their previous generations have been improved. Better models to incorporate mesophyll conductance are to be expected [17]. With the advancement of single-cell omics [106–108], cell-type specific models will allow prediction of how CAM may function in response to different environmental factors at the single-cell resolution.

CAM can be induced in facultative CAM plants by different conditions. There are some comparative omics studies using different plant lineages. But there have not been any studies to compare the different treatments using the same species. For example, comparative studies may be performed on the signaling components under scrutiny to show if there are any shared pathways among the different treatments of inducing CAM, such as drought, salinity, or ABA. With the advancement of cutting-edge technologies, such as single-cell analysis, artificial intelligence in plant biology, and synthetic biology, the development of cell-type specific and stress-inducible CAM in major C_3 crops (e.g., soybean, cotton, and alfalfa) is within sight.

7. Concluding Remarks

As the global population is projected to reach 9 billion in 27 years, food production has become increasingly limited due to the demanding crop irrigation and the increasingly frequent drought episodes driven by climate change. CAM is a natural innovation for high WUE and stress resilience. In the era of system biology and synthetic biology, engineering the CAM characteristics into C_3 (or C_4) crops represents a potential breakthrough for meeting the global challenges of population growth and food security. The CAM research areas covered in this review are not mutually exclusive. Instead, they inform each other and deepen our comprehension of the evolutionary and molecular underpinnings of CAM. The C_3-CAM transition will provide advantages to plants in the following aspects: 1. water-use efficiency 2. drought tolerance 3. temperature resilience 4. plasticity in photosynthesis 5. carbon storage. Studying the C_3 to CAM transition using facultative CAM plants allows going back to evolution history and identifying the molecular switches (e.g., TFs and kinases) essential for the development of CAM characteristics. Once we know the "codes" for the CAM characteristics, we can use the synthetic biology "language" to build cell-type-specific circuitry for enabling important C_3 crops with enhanced WUE, stress resilience, and improved yield.

Author Contributions: Original draft preparation, B.T.; review, editing, and supervision, S.C. All authors have read and agreed to the published version of the manuscript.

Funding: This research was partly supported by a faculty startup fund from the University of Mississippi to S.C.

Institutional Review Board Statement: Not applicable.

Informed Consent Statement: Not applicable.

Data Availability Statement: Not applicable.

Acknowledgments: Qijie Guan is acknowledged for his input in assessing the two recently published ice plant genomes.

Conflicts of Interest: The authors declare no conflict of interest.

References

1. Leisner, C.P. Review: Climate change impacts on food security- focus on perennial cropping systems and nutritional value. *Plant Sci.* **2020**, *293*, 110412. [CrossRef] [PubMed]
2. Reidmiller, D.R.; Avery, C.W.; Easterling, D.R.; Kunkel, K.E.; Lewis, K.L.M.; Maycock, T.K.; Stewart, B.C. (Eds.) *USGCRP Impacts, Risks, and Adaptation in the United States: Fourth National Climate Assessment*; U.S. Global Change Research Program: Washington, DC, USA, 2018; Volume II, p. 1515. [CrossRef]
3. Yao, J.; Liu, H.; Huang, J.; Gao, Z.; Wang, G.; Li, D.; Yu, H.; Chen, X. Accelerated dryland expansion regulates future variability in dryland gross primary production. *Nat. Commun.* **2020**, *11*, 1665. [CrossRef] [PubMed]
4. Eshel, G.; Araus, V.; Undurraga, S.; Soto, D.C.; Moraga, C.; Montecinos, A.; Moyano, T.; Maldonado, J.; Díaz, F.P.; Varala, K.; et al. Plant ecological genomics at the limits of life in the Atacama Desert. *Proc. Natl. Acad. Sci. USA* **2021**, *118*, e2101177118. [CrossRef]
5. Lim, S.D.; Mayer, J.A.; Yim, W.C.; Cushman, J.C. Plant tissue succulence engineering improves water-use efficiency, water-deficit stress attenuation and salinity tolerance in Arabidopsis. *Plant J.* **2020**, *103*, 1049–1072. [CrossRef]
6. Hu, R.; Zhang, J.; Jawdy, S.; Sreedasyam, A.; Lipzen, A.; Wang, M.; Ng, V.; Daum, C.; Keymanesh, K.; Liu, D.; et al. Comparative genomics analysis of drought response between obligate CAM and C3 photosynthesis plants. *J. Plant Physiol.* **2022**, *277*, 153791. [CrossRef] [PubMed]
7. Blankenship, R.E. *Molecular Mechanisms of Photosynthesis*; Wiley-Blackwell: Hoboken, NJ, USA, 2008; pp. 1–10.
8. Crayn, D.M.; Winter, K.; Smith, J.A.C. Multiple origins of crassulacean acid metabolism and the epiphytic habit in the Neotropical family Bromeliaceae. *Proc. Natl. Acad. Sci. USA* **2004**, *101*, 3703–3708. [CrossRef]
9. Ceusters, J.; Van De Poel, B. Ethylene Exerts Species-Specific and Age-Dependent Control of Photosynthesis. *Plant Physiol.* **2018**, *176*, 2601–2612. [CrossRef]
10. Heyduk, K. Evolution of Crassulacean acid metabolism in response to the environment: Past, present, and future. *Plant Physiol.* **2022**, *190*, 19–30. [CrossRef]
11. Amin, A.B.; Rathnayake, K.N.; Yim, W.C.; Garcia, T.M.; Wone, B.; Cushman, J.C.; Wone, B.W.M. Crassulacean Acid Metabolism Abiotic Stress-Responsive Transcription Factors: A Potential Genetic Engineering Approach for Improving Crop Tolerance to Abiotic Stress. *Front. Plant Sci.* **2019**, *10*, 129. [CrossRef]
12. Wickell, D.; Kuo, L.-Y.; Yang, H.-P.; Ashok, A.D.; Irisarri, I.; Dadras, A.; de Vries, S.; de Vries, J.; Huang, Y.-M.; Li, Z.; et al. Underwater CAM photosynthesis elucidated by Isoetes genome. *Nat. Commun.* **2021**, *12*, 6348. [CrossRef]
13. Osmond, C.B. Crassulacean Acid Metabolism: A Curiosity in Context. *Annu. Rev. Plant Physiol.* **1978**, *29*, 379–414. [CrossRef]
14. Borland, A.M.; Guo, H.-B.; Yang, X.; Cushman, J.C. Orchestration of carbohydrate processing for crassulacean acid metabolism. *Curr. Opin. Plant Biol.* **2016**, *31*, 118–124. [CrossRef] [PubMed]
15. Bowyer, J.; Leegood, R. Photosynthesis. In *Plant Biochemistry*; Dey, P.M., Harborne, J.B., Eds.; Academic Press: London, UK, 1997; pp. 1–4, 49–110. [CrossRef]
16. Drennan, P.M.; Nobel, P.S. Responses of CAM species to increasing atmospheric CO_2 concentrations. *Plant Cell Environ.* **2000**, *23*, 767–781. [CrossRef]
17. Schiller, K.; Bräutigam, A. Engineering of Crassulacean Acid Metabolism. *Annu. Rev. Plant Biol.* **2021**, *72*, 77–103. [CrossRef] [PubMed]
18. Kluge, M.; Ting, I.P. *Crassulacean Acid Metabolism*; Springer: Berlin/Heidelberg, Germany, 1978; pp. 108–152. [CrossRef]
19. Kong, W.; Yoo, M.J.; Zhu, D.; Noble, J.D.; Kelley, T.M.; Li, J.; Kirst, M.; Assmann, S.M.; Chen, S. Molecular changes in *Mesembryanthemum crystallinum* guard cells underlying the C_3 to CAM transition. *Plant Mol. Biol.* **2020**, *103*, 653–667. [CrossRef]
20. Guan, Q.; Tan, B.; Kelley, T.M.; Tian, J.; Chen, S. Physiological changes in *Mesembryanthemum crystallinum* during the C_3 to CAM transition induced by salt stress. *Front. Plant Sci.* **2020**, *11*, 283. [CrossRef]
21. Ferrari, R.C.; Kawabata, A.B.; Ferreira, S.S.; Hartwell, J.; Freschi, L. A matter of time: Regulatory events behind the synchronization of C_4 and crassulacean acid metabolism in *Portulaca oleracea*. *J. Exp. Bot.* **2022**, *73*, 4867–4885. [CrossRef]
22. Gilman, I.S.; Moreno-Villena, J.J.; Lewis, Z.R.; Goolsby, E.W.; Edwards, E.J. Gene co-expression reveals the modularity and integration of C_4 and CAM in *Portulaca*. *Plant Physiol.* **2022**, *189*, 735–753. [CrossRef]
23. Ferrari, R.C.; Bittencourt, P.P.; Rodrigues, M.A.; Moreno-Villena, J.J.; Alves, F.R.R.; Gastaldi, V.D.; Boxall, S.F.; Dever, L.V.; Demarco, D.; Andrade, S.C.S.; et al. C_4 and crassulacean acid metabolism within a single leaf: Deciphering key components behind a rare photosynthetic adaptation. *New Phytol.* **2020**, *225*, 1699–1714. [CrossRef]
24. Gilman, I.S.; Edwards, E.J. Crassulacean acid metabolism. *Curr. Biol.* **2020**, *30*, R57–R62. [CrossRef]
25. Cushman, J.C. Crassulacean acid metabolism. A plastic photosynthetic adaptation to arid environments. *Plant Physiol.* **2001**, *127*, 1439–1448. [CrossRef] [PubMed]
26. Herrera, A.; Martin, C.E.; Tezara, W.; Ballestrini, C.; Medina, E. Induction by drought of Crassulacean acid metabolism in the terrestrial bromeliad, *Puya floccose*. *Photosynthetica* **2010**, *48*, 383–388. [CrossRef]

27. Winter, K. Ecophysiology of constitutive and facultative CAM photosynthesis. *J. Exp. Bot.* **2019**, *70*, 6495–6508. [CrossRef] [PubMed]
28. Winter, K.; Holtum, J.A.M. Facultative Crassulacean acid metabolism (CAM) plants: Powerful tools for unravelling the functional elements of CAM photosynthesis. *J. Exp. Bot.* **2014**, *65*, 3425–3441. [CrossRef] [PubMed]
29. Borland, A.M.; Griffiths, H. The regulation of CAM and respiratory recycling by water supply and light regime in the C$_3$ -CAM intermediate *Sedum telephium*. *Funct. Ecol.* **1990**, *4*, 33. [CrossRef]
30. Olivares, E.; Urich, R.; Montes, G.; Coronel, I.; Herrera, A. Occurrence of Crassulacean acid metabolism in *Cissus trifoliata* L. (Vitaceae). *Oecologia* **1984**, *61*, 358–362. [CrossRef] [PubMed]
31. Winter, K.; von Willert, D.J. NaCl-induzierter crassulaceensäurestoffwechsel bei *Mesembryanthemum crystallinum*. *Z. Pflanzenphysiol.* **1972**, *67*, 166–170. [CrossRef]
32. Maxwell, K. Resistance is useful: Diurnal patterns of photosynthesis in C$_3$ and Crassulacean acid metabolism epiphytic bromeliads. *Funct. Plant Biol.* **2002**, *29*, 679–687. [CrossRef]
33. Taybi, T.; Cushman, J.C. Abscisic acid signaling and protein synthesis requirements for phosphoenolpyruvate carboxylase transcript induction in the common ice plant. *J. Plant Physiol.* **2002**, *159*, 1235–1243. [CrossRef]
34. Brulfert, J.; Kluge, M.; Güçlü, S.; Queiroz, O. Interaction of photoperiod and drought as CAM inducing factors in *Kalanchoë* blossfeldiana Poelln., cv. Tom Thumb. *J. Plant Physiol.* **1988**, *133*, 222–227. [CrossRef]
35. Surówka, E.; Dziurka, M.; Kocurek, M.; Goraj, S.; Rapacz, M.; Miszalski, Z. Effects of exogenously applied hydrogen peroxide on antioxidant and osmoprotectant profiles and the C$_3$-CAM shift in the halophyte *Mesembryanthemum crystallinum* L. *J. Plant Physiol.* **2016**, *200*, 102–110. [CrossRef] [PubMed]
36. Winter, K.; Ziegler, H. Induction of Crassulacean acid metabolism in *Mesembryanthemum crystallinum* increases reproductive success under conditions of drought and salinity stress. *Oecologia* **1992**, *92*, 475–479. [CrossRef] [PubMed]
37. Nosek, M.; Gawrońska, K.; Rozpądek, P.; Sujkowska-rybkowska, M.; Miszalski, Z.; Kornaś, A. At the edges of photosynthetic metabolic plasticity—On the rapidity and extent of changes accompanying salinity stress-induced cam photosynthesis withdrawal. *Int. J. Mol. Sci.* **2021**, *22*, 8426. [CrossRef] [PubMed]
38. Snotch from Sapporo, Hokkaido, Japan. Available online: https://commons.wikimedia.org/wiki/File:Dendrobium_officinale_Kimura_%26_Migo_J._Shanghai_Sci._Inst._3-_122_(1936)_(33547665894)_-_cropped.jpg (accessed on 10 August 2023). This Image Is Licensed under the Creative Commons Attribution 2.0 Generic License.
39. Maria at Blogger. Available online: http://mariasorchids.blogspot.com/2018/02/a-display-of-dendrobiums.html (accessed on 10 August 2023). This Work Is Licensed under a Creative Commons Attribution-NonCommercial-ShareAlike 4.0 International License.
40. Ricardo. Available online: http://ricardogupi.blogspot.com/2010/10/dendrobium-primulinum-also-known-now-as.html (accessed on 10 August 2023). No License Found.
41. NC State Extension Gardener. Available online: https://plants.ces.ncsu.edu/plants/yucca-gloriosa/ (accessed on 10 August 2023). No License Found.
42. Dezmond Wells. Available online: https://www.inaturalist.org/taxa/130743-Sesuvium-portulacastrum (accessed on 10 August 2023). This Image Is under the Creative Commons Attribution-Share-Alike License 3.0.
43. Jerzy Opioła. Available online: https://commons.wikimedia.org/wiki/File:Tamarix_ramosissima_a2.jpg (accessed on 10 August 2023). This Work Is Licensed under the Creative Commons Attribution-Share Alike 3.0 Unported License.
44. Rojas-Sandoval, J.; Acevedo-Rodríguez, P. Anredera basselloides (Madeira vine). *CABI Compend.* **2014**. [CrossRef]
45. Alex Popovkin. Available online: https://ayurwiki.org/Ayurwiki/Talinum_triangulare (accessed on 10 August 2023). This Image Is Licensed under the Creative Commons Attribution 2.0 Generic License.
46. Cliff at Wikipedia. Available online: https://en.wikipedia.org/wiki/Guzmania_monostachia (accessed on 10 August 2023). This Work Is under the Creative Commons Attribution-ShareAlike License 4.0.
47. Roger Fryer; Jill Newland. Available online: http://www.northqueenslandplants.com/Australian%20Plant%20Families%20N-S/Portulacaceae/Calandrinia/Calandrinia%20creethae.html (accessed on 10 August 2023). This Work Is Licensed under a Creative Commons Attribution-Noncommercial 2.5 Australia License.
48. Dick Culbert. Available online: https://en.wikipedia.org/wiki/Calandrinia (accessed on 10 August 2023). This Image Is Licensed under the Creative Commons Attribution 2.0 Generic License.
49. Brian Freeman. Available online: https://davesgarden.com/guides/pf/go/248887/#b (accessed on 10 August 2023). No License Found.
50. Photography by C. Hortin & J. Hooper. Image Used with the Permission of the Western Australian Herbarium, Department of Biodiversity, Conservation and Attractions. Available online: https://florabase.dbca.wa.gov.au/help/copyright. (accessed on 10 August 2023).
51. Dinkum at Wikimedia Commons. Available online: https://commons.wikimedia.org/wiki/File:Portulacaria_afra_02.JPG (accessed on 10 August 2023). This Image Is Licensed under the Creative Commons CC0 License.
52. Frank Vincentz. Available online: https://commons.wikimedia.org/wiki/File:Sedum_album_03_ies.jpg (accessed on 10 August 2023).
53. Sajetpa at Malayalam Wikipedia. Available online: https://en.wikipedia.org/wiki/Coleus_amboinicus#/media/File:Leaf_-pani_koorkka.JPG (accessed on 10 August 2023). No License Found.
54. Winter, K.; Garcia, M.; Virgo, A.; Smith, J.A.C. Low-level CAM photosynthesis in a succulent-leaved member of the Urticaceae, *Pilea peperomioides*. *Funct. Plant Biol.* **2021**, *48*, 683–690. [CrossRef]

55. Bräutigam, A.; Schlüter, U.; Eisenhut, M.; Gowik, U. On the evolutionary origin of CAM photosynthesis. *Plant Physiol.* **2017**, *174*, 473–477. [CrossRef]

56. Yang, X.; Cushman, J.C.; Borland, A.M.; Edwards, E.J.; Wullschleger, S.D.; Tuskan, G.A.; Owen, N.A.; Griffiths, H.; Smith, J.A.C.; De Paoli, H.C.; et al. A roadmap for research on crassulacean acid metabolism (CAM) to enhance sustainable food and bioenergy production in a hotter, drier world. *New Phytol.* **2015**, *207*, 491–504. [CrossRef]

57. Boxall, S.F.; Kadu, N.; Dever, L.V.; Knerová, J.; Waller, J.L.; Gould, P.J.D.; Hartwell, J. *Kalanchoë* PPC1 is essential for Crassulacean acid metabolism and the regulation of core circadian clock and guard cell signaling genes. *Plant Cell* **2020**, *32*, 1136–1160. [CrossRef]

58. Yang, X.; Hu, R.; Yin, H.; Jenkins, J.; Shu, S.; Tang, H.; Liu, D.; Weighill, D.A.; Yim, W.C.; Ha, J.; et al. The *Kalanchoë* genome provides insights into convergent evolution and building blocks of crassulacean acid metabolism. *Nat. Commun.* **2017**, *8*, 1899. [CrossRef]

59. Shameer, S.; Baghalian, K.; Cheung, C.Y.M.; Ratcliffe, R.G.; Sweetlove, L.J. Computational analysis of the productivity potential of CAM. *Nat. Plants* **2018**, *4*, 165–171. [CrossRef]

60. Tay, I.Y.Y.; Odang, K.B.; Cheung, C.Y.M. Metabolic modeling of the C_3-CAM continuum revealed the establishment of a starch/sugar-malate cycle in CAM evolution. *Front. Plant Sci.* **2021**, *11*, 573197. [CrossRef] [PubMed]

61. Lim, S.D.; Lee, S.; Choi, W.G.; Yim, W.C.; Cushman, J.C. Laying the foundation for Crassulacean acid metabolism (CAM) biodesign: Expression of the C_4 metabolism cycle genes of CAM in *Arabidopsis*. *Front. Plant Sci.* **2019**, *10*, 101. [CrossRef] [PubMed]

62. Shen, S.; Li, N.; Wang, Y.; Zhou, R.; Sun, P.; Lin, H.; Chen, W.; Yu, T.; Liu, Z.; Wang, Z.; et al. High-quality ice plant reference genome analysis provides insights into genome evolution and allows exploration of genes involved in the transition from C_3 to CAM pathways. *Plant Biotechnol. J.* **2022**, *20*, 2107–2122. [CrossRef] [PubMed]

63. Ming, R.; VanBuren, R.; Wai, C.M.; Tang, H.; Schatz, M.C.; Bowers, J.E.; Lyons, E.; Wang, M.-L.; Chen, J.; Biggers, E.; et al. The pineapple genome and the evolution of CAM photosynthesis. *Nat. Genet.* **2015**, *47*, 1435–1442. [CrossRef]

64. Taybi, T.; Cushman, J.C.; Borland, A.M. Leaf carbohydrates influence transcriptional and post-transcriptional regulation of nocturnal carboxylation and starch degradation in the facultative CAM plant, *Mesembryanthemum crystallinum*. *J. Plant Physiol.* **2017**, *218*, 144–154. [CrossRef] [PubMed]

65. Nosek, M.; Gawrońska, K.; Rozpądek, P.; Szechyńska-Hebda, M.; Kornaś, A.; Miszalski, Z. Withdrawal from functional Crassulacean acid metabolism (CAM) is accompanied by changes in both gene expression and activity of antioxidative enzymes. *J. Plant Physiol.* **2018**, *229*, 151–157. [CrossRef]

66. Guan, Q.; Kong, W.; Zhu, D.; Zhu, W.; Dufresne, C.; Tian, J.; Chen, S. Comparative proteomics of *Mesembryanthemum crystallinum* guard cells and mesophyll cells in transition from C_3 to CAM. *J. Proteom.* **2021**, *231*, 104019. [CrossRef]

67. Wakamatsu, A.; Mori, I.C.; Matsuura, T.; Taniwaki, Y.; Ishii, R.; Yoshida, R. Possible roles for phytohormones in controlling the stomatal behavior of *Mesembryanthemum crystallinum* during the salt-induced transition from C_3 to Crassulacean acid metabolism. *J. Plant Physiol.* **2021**, *262*, 153448. [CrossRef]

68. Winter, K. Properties of phosphoenolpyruvate carboxylase in rapidly prepared, desalted leaf extracts of the Crassulacean acid metabolism plant *Mesembryanthemum crystallinum* L. *Planta* **1982**, *154*, 298–308. [CrossRef]

69. Häusler, R.E.; Baur, B.; Scharte, J.; Teichmann, T.; Eicks, M.; Fischer, K.L.; Flügge, U.I.; Schubert, S.; Weber, A.; Fischer, K. Plastidic metabolite transporters and their physiological functions in the inducible crassulacean acid metabolism plant *Mesembryanthemum crystallinum*. *Plant J.* **2000**, *24*, 285–296. [CrossRef]

70. Neuhaus, H.E.; Holtum, J.A.; Latzko, E. Transport of phosphoenolpyruvate by chloroplasts from *Mesembryanthemum crystallinum* L. exhibiting Crassulacean acid metabolism. *Plant Physiol.* **1988**, *87*, 64–68. [CrossRef] [PubMed]

71. Cushman, J.C.; Meyer, G.; Michalowski, C.B.; Schmitt, J.M.; Bohnert, H.J. Salt stress leads to differential expression of two isogenes of phosphoenolpyruvate carboxylase during Crassulacean acid metabolism induction in the common ice plant. *Plant Cell* **1989**, *1*, 715–725. [CrossRef]

72. Vaasen, A.; Begerow, D.; Hampp, R. Phosphoenolpyruvate carboxylase genes in C_3, Crassulacean acid metabolism (CAM) and C_3/CAM intermediate species of the genus *Clusia*: Rapid reversible C_3/CAM switches are based on the C_3 housekeeping gene. *Plant Cell Environ.* **2006**, *29*, 2113–2123. [CrossRef] [PubMed]

73. Cushman, J.C.; Tillett, R.L.; Wood, J.A.; Branco, J.M.; Schlauch, K.A. Large-scale mRNA expression profiling in the common ice plant, *Mesembryanthemum crystallinum*, performing C_3 photosynthesis and Crassulacean acid metabolism (CAM). *JXB* **2008**, *59*, 1875–1894. [CrossRef] [PubMed]

74. Kore-Eda, S.; Noake, C.; Ohishi, M.; Ohnishi, J.I.; Cushman, J.C. Transcriptional profiles of organellar metabolite transporters during induction of Crassulacean acid metabolism in *Mesembryanthemum crystallinum*. *Funct. Plant Biol.* **2005**, *32*, 451–466. [CrossRef]

75. Abraham, P.E.; Yin, H.; Borland, A.M.; Weighill, D.; Lim, S.D.; De Paoli, H.C.; Engle, N.; Jones, P.C.; Agh, R.; Weston, D.J.; et al. Transcript, protein and metabolite temporal dynamics in the CAM plant Agave. *Nat. Plants* **2016**, *2*, 16178. [CrossRef] [PubMed]

76. Brilhaus, D.; Bräutigam, A.; Mettler-Altmann, T.; Winter, K.; Weber, A.P. Reversible burst of transcriptional changes during induction of Crassulacean acid metabolism in *Talinum triangulare*. *Plant Physiol.* **2016**, *170*, 102–122. [CrossRef]

77. Maleckova, E.; Brilhaus, D.; Wrobel, T.J.; Weber, A.P.M. Transcript and metabolite changes during the early phase of abscisic acid-mediated induction of Crassulacean acid metabolism in *Talinum triangulare*. *JXB* **2019**, *70*, 6581–6596. [CrossRef]

78. Gonçalves, A.Z.; Mercier, H. Transcriptomic and biochemical analysis reveal integrative pathways between carbon and nitrogen metabolism in *Guzmania monostachia* (Bromeliaceae) under drought. *Front. Plant Sci.* **2021**, *12*, 715289. [CrossRef]

79. Habibi, G. Comparison of CAM expression, photochemistry and antioxidant responses in *Sedum album* and *Portulaca oleracea* under combined stress. *Physiol. Plant* **2020**, *170*, 550–568. [CrossRef]

80. Zhang, X.; Tan, B.; Zhu, D.; Dufresne, D.; Jiang, T.; Chen, S. Proteomics of homeobox7 enhanced salt tolerance in *Mesembryanthemum crystallinum. Int. J. Mol. Sci.* **2021**, *22*, 6390. [CrossRef] [PubMed]

81. Na, J.K.; Metzger, J.D. Chimeric promoter mediates guard cell-specific gene expression in tobacco under water deficit. *Biotechnol. Lett.* **2014**, *36*, 1893–1899. [CrossRef] [PubMed]

82. Rusconi, F.; Simeoni, F.; Francia, P.; Cominelli, E.; Conti, L.; Riboni, M.; Simoni, L.; Martin, C.R.; Tonelli, C.; Galbiati, M. The *Arabidopsis thaliana* MYB60 promoter provides a tool for the spatio-temporal control of gene expression in stomatal guard cells. *JXB* **2013**, *64*, 3361–3371. [CrossRef]

83. Winter, K.; Smith, J.A.C. CAM photosynthesis: The acid test. *New Phytol* **2022**, *233*, 599–609. [CrossRef] [PubMed]

84. Lim, S.D.; Yim, W.C.; Liu, D.; Hu, R.; Yang, X.; Cushman, J.C. A *Vitis vinifera* basic helix–loop–helix transcription factor enhances plant cell size, vegetative biomass and reproductive yield. *Plant Biotechnol. J.* **2018**, *16*, 1595–1615. [CrossRef]

85. Kebeish, R.; Niessen, M.; Oksaksin, M.; Blume, C.; Peterhaensel, C. Constitutive and dark-induced expression of *Solanum tuberosum* phosphoenolpyruvate carboxylase enhances stomatal opening and photosynthetic performance of *Arabidopsis thaliana*. *Biotechnol. Bioeng.* **2012**, *109*, 536–544. [CrossRef]

86. Liu, D.; Hu, R.; Zhang, J.; Guo, H.B.; Cheng, H.; Li, L.; Borland, A.M.; Qin, H.; Chen, J.G.; Muchero, W.; et al. Overexpression of an *Agave* phosphoenolpyruvate carboxylase improves plant growth and stress tolerance. *Cells* **2021**, *10*, 582. [CrossRef]

87. Zhang, X.; Tan, B.; Cheng, Z.; Zhu, D.; Jiang, T.; Chen, S. Overexpression of *McHB7* transcription factor from *Mesembryanthemum crystallinum* improves plant salt tolerance. *Int. J. Mol. Sci.* **2022**, *23*, 7879. [CrossRef]

88. Chiang, C.-P.; Yim, W.C.; Sun, Y.-H.; Ohnishi, M.; Mimura, T.; Cushman, J.C.; Yen, H.E. Identification of Ice Plant (*Mesembryanthemum crystallinum* L.) MicroRNAs Using RNA-Seq and Their Putative Roles in High Salinity Responses in Seedlings. *Front. Plant Sci.* **2016**, *7*, 1143. [CrossRef]

89. Chu, C.; Dai, Z.; Ku, M.S.; Edwards, G.E. Induction of Crassulacean acid metabolism in the facultative halophyte *Mesembryanthemum crystallinum* by abscisic acid. *Plant Physiol.* **1990**, *93*, 1253–1260. [CrossRef]

90. Cushman, J.C.; Borland, A.M. Induction of Crassulacean acid metabolism by water limitation. *Plant Cell Environ.* **2002**, *25*, 295–310. [CrossRef] [PubMed]

91. Heyduk, K.; Hwang, M.; Albert, V.; Silvera, K.; Lan, T.; Farr, K.; Chang, T.H.; Chan, M.T.; Winter, K.; Leebens-Mack, J. Altered gene regulatory networks are associated with the transition from C_3 to Crassulacean acid metabolism in *Erycina* (oncidiinae: Orchidaceae). *Front. Plant Sci.* **2019**, *9*, 2000. [CrossRef] [PubMed]

92. Santos, M.G.; Davey, P.A.; Hofmann, T.A.; Borland, A.; Hartwell, J.; Lawson, T. Stomatal responses to light, CO_2, and mesophyll tissue in *Vicia faba* and *Kalanchoë fedtschenkoi. Front. Plant Sci.* **2021**, *12*, 740534. [CrossRef]

93. Cai, J.; Liu, X.; Vanneste, K.; Proost, S.; Tsai, W.-C.; Liu, K.-W.; Chen, L.-J.; He, Y.; Xu, Q.; Bian, C.; et al. The genome sequence of the orchid *Phalaenopsis equestris. Nat. Genet.* **2014**, *47*, 65–72. [CrossRef] [PubMed]

94. Copetti, D.; Búrquez, A.; Bustamante, E.; Charboneau, J.L.M.; Childs, K.L.; Eguiarte, L.E.; Lee, S.; Liu, T.L.; McMahon, M.M.; Whiteman, N.K.; et al. Extensive gene tree discordance and hemiplasy shaped the genomes of North American columnar cacti. *Proc. Natl. Acad. Sci. USA* **2017**, *114*, 12003–12008. [CrossRef]

95. Xin, H.; Wang, Y.; Li, Q.; Wan, T.; Hou, Y.; Liu, Y.; Gichuki, D.K.; Zhou, H.; Zhu, Z.; Xu, C.; et al. A genome for *Cissus* illustrates features underlying the evolutionary success in dry savannas. *Hortic. Res.* **2022**, *9*, uhac208. [CrossRef] [PubMed]

96. Sato, R.; Kondo, Y.; Agarie, S. The first released available genome of the common ice plant (*Mesembryanthemum crystallinum* L.) extended the research region on salt tolerance, C_3-CAM photosynthetic conversion, and halophism [version 1; peer review: 1 approved with reservations]. *F1000Research* **2023**, *12*, 448. [CrossRef]

97. Liu, D.; Chen, M.; Mendoza, B.; Cheng, H.; Hu, R.; Li, L.; Trinh, C.T.; Tuskan, G.A.; Yang, X. CRISPR/Cas9-mediated targeted mutagenesis for functional genomics research of Crassulacean acid metabolism plants. *JXB* **2019**, *70*, 6621–6629. [CrossRef]

98. Walker, R.P.; Chen, Z.H.; Acheson, R.M.; Leegood, R.C. Effects of phosphorylation on phosphoenolpyruvate carboxykinase from the C_4 plant *Guinea* grass. *Plant Physiol.* **2002**, *128*, 165–172. [CrossRef]

99. Walker, R.P.; Leegood, R.C. Purification, and phosphorylation in vivo and in vitro, of phosphoenolpyruvate carboxykinase from cucumber cotyledons. *FEBS Lett.* **1995**, *362*, 70–74. [CrossRef]

100. Clark, N.M.; Elmore, J.M.; Walley, J.W. To the proteome and beyond: Advances in single-cell omics profiling for plant systems. *Plant Physiol.* **2022**, *188*, 726–737. [CrossRef] [PubMed]

101. Katam, R.; Lin, C.; Grant, K.; Katam, C.S.; Chen, S. Advances in plant metabolomics and its applications in stress and single-cell biology. *Int. J. Mol. Sci.* **2022**, *23*, 6985. [CrossRef] [PubMed]

102. Specht, H.; Emmott, E.; Petelski, A.A.; Huffman, R.G.; Perlman, D.H.; Serra, M.; Kharchenko, P.; Koller, A.; Slavov, N. Single-cell proteomic and transcriptomic analysis of macrophage heterogeneity using SCoPE2. *Genome Biol.* **2021**, *22*, 50. [CrossRef] [PubMed]

103. Nungesser, D.; Kluge, M.; Tolle, H.; Oppelt, W. A dynamic computer model of the metabolic and regulatory processes in Crassulacean acid metabolism. *Planta* **1984**, *162*, 204–214. [CrossRef]

104. Burgos, A.; Miranda, E.; Vilaprinyo, E.; Meza-Canales, I.D.; Alves, R. CAM Models: Lessons and Implications for CAM Evolution. *Front. Plant Sci.* **2022**, *13*, 893095. [CrossRef]
105. Töpfer, N.; Braam, T.; Shameer, S.; Ratcliffe, R.G.; Sweetlove, L.J. Alternative Crassulacean acid metabolism modes provide environment-specific water-saving benefits in a leaf metabolic model. *Plant Cell* **2020**, *32*, 3689–3705. [CrossRef]
106. Libault, M.; Chen, S. Editorial: Plant single cell type systems biology. *Front. Plant Sci.* **2016**, *7*, 35. [CrossRef]
107. Misra, B.B.; Assmann, S.M.; Chen, S. Plant single-cell and single-cell-type metabolomics. *Trends Plant Sci.* **2014**, *19*, 637–646. [CrossRef]
108. Su, L.; Xu, C.; Zeng, S.; Su, L.; Joshi, T.; Stacey, G.; Xu, D. Large-scale integrative analysis of soybean transcriptome using an unsupervised autoencoder model. *Front. Plant Sci.* **2022**, *13*, 831204. [CrossRef]

International Journal of
Molecular Sciences

Review

Abiotic Stress-Induced Leaf Senescence: Regulatory Mechanisms and Application

Shuya Tan [†], Yueqi Sha [†], Liwei Sun * and Zhonghai Li *

State Key Laboratory of Tree Genetics and Breeding, College of Biological Sciences and Technology, Beijing Forestry University, Beijing 100083, China
* Correspondence: lsun2013@bjfu.edu.cn (L.S.); lizhonghai@bjfu.edu.cn (Z.L.)
[†] These authors contributed equally to this work.

Abstract: Leaf senescence is a natural phenomenon that occurs during the aging process of plants and is influenced by various internal and external factors. These factors encompass plant hormones, as well as environmental pressures such as inadequate nutrients, drought, darkness, high salinity, and extreme temperatures. Abiotic stresses accelerate leaf senescence, resulting in reduced photosynthetic efficiency, yield, and quality. Gaining a comprehensive understanding of the molecular mechanisms underlying leaf senescence in response to abiotic stresses is imperative to enhance the resilience and productivity of crops in unfavorable environments. In recent years, substantial advancements have been made in the study of leaf senescence, particularly regarding the identification of pivotal genes and transcription factors involved in this process. Nevertheless, challenges remain, including the necessity for further exploration of the intricate regulatory network governing leaf senescence and the development of effective strategies for manipulating genes in crops. This manuscript provides an overview of the molecular mechanisms that trigger leaf senescence under abiotic stresses, along with strategies to enhance stress tolerance and improve crop yield and quality by delaying leaf senescence. Furthermore, this review also highlighted the challenges associated with leaf senescence research and proposes potential solutions.

Keywords: leaf senescence; abiotic stress; stress tolerance; transcription factor; *Arabidopsis*; crop

Citation: Tan, S.; Sha, Y.; Sun, L.; Li, Z. Abiotic Stress-Induced Leaf Senescence: Regulatory Mechanisms and Application. *Int. J. Mol. Sci.* **2023**, *24*, 11996. https://doi.org/10.3390/ijms241511996

Academic Editor: Martin Bartas

Received: 16 May 2023
Revised: 14 July 2023
Accepted: 19 July 2023
Published: 26 July 2023

1. Introduction

The leaves of plants serve as the primary sites for photosynthesis, where light energy is converted into chemical energy stored in carbohydrate molecules. These carbohydrates serve as the main energy source for all living organisms on Earth. Senescence, the final stage of leaf development, is a gradual and intricate biological process comprising initiation, progression, and terminal phases [1,2]. In this process, the leaves gradually turn yellow, shrivel, and fall off. During the later stages of leaf senescence, chlorophyll and chloroplasts deteriorate, accompanied by the breakdown of macro-molecules like proteins, lipids, and nucleic acids [1,2]. In annual plants, the nutrients released from senescent leaves are transferred to actively growing young leaves and seeds to enhance reproductive success. In the case of perennial plants, such as deciduous trees, nitrogen from leaf proteins is redirected to form bark storage proteins in phloem tissues. These proteins are stored throughout the winter and then mobilized and reused for spring shoot growth [3–5]. In agriculture, senescence is capable of remobilizing leaf nitrogen and micronutrients into the grain or fruit. The NAC transcription factor NAM-B1 plays an important role in the regulation of expressions of nitrogen transport-related genes during senescence [6]. A recent study revealed that OsDREB1C shortens lifespan but improves photosynthetic capacity and nitrogen utilization, and transgenic plants with overexpression of OsDREB1C have 41.3% to 68.3% higher yields than wild-type plants [7]. Consequently, the timing of leaf senescence plays a crucial role in facilitating nutrient cycling, environmental adaptation, and reproduction in plants [8].

The leaf senescence process is accompanied by changes in the expression of thousands of senescence-associated genes (*SAGs*) [9]. Studies have shown that several transcriptional regulators (TFs) regulate senescence by controlling *SAG* expression [10]. In one of them, a number of NAC TFs were identified as core regulators of senescence [11–13]. EIN3, a key TF that functions downstream of EIN2 in ethylene signaling pathway, increases the transcript levels of *ORE1/AtNAC092/AtNAC2* through the direct repression of *miR164* transcription [14]. WRKY53 positively regulates leaf senescence [15] via targeting various *SAGs* such as *SENRK1* [16].

The initiation and progression of leaf senescence are influenced by various internal and external factors [1,2,8]. Leaf senescence can be triggered as a defense mechanism in response to biotic stress factors such as pathogen infection or insect damage. Additionally, abiotic stress factors including drought, high salinity, high temperature, or nutrient deficiencies can accelerate leaf senescence [1,2,8,17–19]. These stressors can induce oxidative stress, leading to the accumulation of reactive oxygen species (ROS), which can cause DNA damage and activate SAGs [1,2,8,17,18,20].

Numerous studies conducted on crops like wheat and rice demonstrated that modifying leaf senescence processes can have a significant impact on crop yield and quality. For instance, in apple trees (*Malus domestica*), improving fruit quality was achieved by extending the lifespan of leaves through the modulation of senescence-associated transcription factors, *MbNAC25* and *MdbHLH3* [21,22]. Similarly, in tomato (*Solanum lycopersicon*), increasing fruit yield and sugar content was achieved by suppressing the expressions of *SlORESARA1* (*ORE1*) and *SlNAP*, which delayed leaf senescence [23,24]. Additionally, delaying leaf senescence in tobacco or cassava resulted in enhanced drought resistance [25–28]. Therefore, gaining a deeper understanding of the regulatory mechanisms underlying leaf senescence can aid researchers in developing more resilient plants that can withstand environmental stresses. This, in turn, would lead to improvements in crop yield, quality, and contribute to global food security and sustainability [8]. This manuscript provides a comprehensive review of the molecular mechanisms involved in leaf senescence induced by abiotic stresses such as nitrogen deficiency, drought, high salinity, and extreme temperature. It also discusses strategies to enhance stress tolerance, crop yield, and quality by delaying leaf senescence. Furthermore, the review highlights the challenges associated with leaf senescence research and explores potential solutions.

2. Abiotic Stress-Induced Leaf Senescence

2.1. Nitrogen Deficiency-Induced Leaf Senescence

Nitrogen, an essential macronutrient in plants, plays a crucial role in leaf senescence, and its deficiency triggers a rapid senescence process [29–31]. ORE1, a key regulator of leaf senescence, was identified as a major factor in nitrogen deficiency-induced leaf senescence [29,30]. In conditions of nitrogen deficiency, loss of ORE1 function results in delayed senescence, while overexpression of ORE1 accelerates leaf senescence, characterized by yellowing leaves, reduced chlorophyll content, and increased expression of SAG12 [29]. Interestingly, overexpression of nitrogen limitation adaptation (NLA) in ORE1 overexpressing plants mitigates the leaf senescence phenotype induced by nitrogen deficiency. NLA, which encodes a RING-type ubiquitin ligase [32], represses leaf senescence by promoting the ubiquitination and degradation of the nitrate transporter NRT1.7 [33]. In a similar mechanism, NLA interacts with ORE1 in the nucleus and regulates its stability through polyubiquitination, with the involvement of PHOSPHATE2 (PHO2). PHO2 encodes an E2 ubiquitin-conjugating enzyme (UBC) and is responsible for maintaining cellular phosphate homeostasis in Arabidopsis [34,35]. Consequently, nla and pho2 mutant plants exhibit accelerated leaf senescence under nitrogen-starvation conditions, whereas nla/ore1 and pho2/ore1 double mutant plants retain green leaves. These findings suggest that fine-tuning the levels of ORE1 through post-translational modifications by NLA/PHO2 ensures a regulated progression of senescence [29]. Interestingly, the deubiquitinases UBP12 and UBP13 were identified as regulators of ORE1 stability by deubiquitinating polyubiquitinated ORE1 and increasing its stability [30]. Plants overexpressing UBP12 or UBP13 display accelerated leaf senescence, which can be reversed by mutation of ORE1. Conversely, overexpression of ORE1 exacerbates the senescence phenotype when UBP12 or UBP13 is also overexpressed [30]. These studies provided a model that explains the molecular framework underlying the involvement of ORE1 in the regulation of nitrogen deficiency-induced leaf senescence [29,30]. Under normal conditions, ORE1 is polyubiquitinated by the E3/E2 enzyme complex, NLA/PHO2, and, subsequently, degraded by 26S proteasomes, leading to delayed leaf senescence. However, under nitrogen-deficient conditions, UBP12 and UBP13 counteract the effects of NLA/PHO2 by deubiquitinating polyubiquitinated ORE1, preventing its degradation. This elevated level of ORE1 activates the expression of downstream SAG genes, thereby accelerating leaf senescence.

Recently, a zinc finger transcription factor called growth, development, and splicing 1 (GDS1) [36] was discovered to have a role in repressing leaf senescence induced by nitrogen deficiency [31]. GDS1 functions as a crucial co-activator or co-protein in the early stages of pre-mRNA splicing and is essential for growth and development in Arabidopsis [36]. Mutants of gds1 exhibit early leaf senescence, reduced NO_3- content, and impaired nitrogen uptake under nitrogen-deficient conditions. Biochemical analysis revealed that GDS1 can bind to the G-box motifs present in the promoter regions of phytochrome-interacting factor 4 (PIF4) and PIF5, thereby repressing their expression [31]. PIF4 and PIF5 were identified as regulators of dark- and heat-induced as well as age-triggered leaf senescence in Arabidopsis [37–40]. Intriguingly, PIF4 and PIF5 also play a role in nitrogen deficiency-induced leaf senescence. Under nitrogen-deficient conditions, delayed leaf senescence was observed in pif4-2 and pif5-3 mutants compared to wild-type plants, while transgenic lines exhibited accelerated leaf senescence phenotypes. Expression levels of PIF4 and PIF5 in the leaves of wild-type plants were significantly higher under low nitrogen conditions compared to high nitrogen conditions [31]. This research presents a novel model to explain leaf senescence induced by low nitrogen levels [31]. Under nitrogen-sufficient conditions, GDS1 binds to the promoters of PIF4 and PIF5, inhibiting their expression and thereby suppressing the expression of downstream SAGs, resulting in delayed leaf senescence. However, under nitrogen-deficient conditions, the accumulation of anaphase-promoting complex or cyclosome proteins promotes the ubiquitination-mediated degradation of GDS1, leading to the release of PIF4 and PIF5 repression. Consequently, downstream SAGs are activated, promoting early leaf senescence.

Regarding both of these proposed models, which explain leaf senescence induced by low nitrogen levels [29–31]: are they independent or do they have any relationship? It was discovered that PIF4 and PIF5 directly bind to the promoter of *ORE1*, promoting its expression, thereby accelerating leaf senescence. Conversely, GDS1 directly binds to *PIF4* and *PIF5*, repressing their gene expression and mitigating low nitrogen-induced leaf senescence [31]. Future investigations will need to analyze whether GDS1 can directly regulate *ORE1* by binding to its promoters or indirectly influence its expression through PIF4/PIF5. Additionally, it would be interesting to explore if PUB12/14 and NLA/PHO2 can interact with GDS1. Furthermore, the relationship between these two regulatory pathways can be elucidated by generating multiple mutant combinations. These studies will contribute to a deeper understanding of leaf senescence induced by low nitrogen levels.

2.2. Drought Stress-Induced Leaf Senescence

Drought stress is a significant abiotic stress factor that has detrimental effects on plant growth and development [41], ultimately leading to leaf senescence [42]. The involvement of a NAC transcription factor, NTL4, in drought-induced leaf senescence has been identified [43]. Under normal conditions, there was no notable difference in the leaf senescence process between wild type plants, transgenic plants overexpressing *NTL4*, and *ntl4* mutants. However, under drought conditions, leaf senescence was accelerated in the transgenic plants while being significantly delayed in the *ntl4* mutant. NTL4 promotes the production of ROS by binding to the promoters of RBOHC and RBOHE under drought conditions. In turn, the elevated ROS production further stimulates *NTL4* gene expression, creating a feed-forward acceleration loop. Notably, *NTL4* is expressed at basal levels during vegetative growth stages and is rapidly induced in response to drought stress. The induction of *NTL4* expression under drought conditions is particularly evident in the distal leaf area, where leaf senescence initiates upon exposure to drought stress [2]. In response to drought, the distal regions of senescing leaves accumulate ROS and experience cell death [18]. This response facilitates the transfer of nutrients and metabolites from senescing leaves to absorptive organs and newly formed leaves, while minimizing water loss through transpiration [18]. Thus, NTL4-mediated leaf senescence enhances the chances of plant survival under drought conditions. Supporting this hypothesis, the overexpression of NAC transcription factors *ANAC019*, *ANAC055*, and *ANAC072* leads to early leaf senescence but increases drought tolerance [44]. Additionally, it has been found that the ABA receptor PYL9 promotes leaf senescence and enhances drought resistance [45]. By activating the signaling cascade of PP2Cs-SnRK2s-RAV1/ABF2-ORE1, the ABA receptor PYL9 promotes drought resistance by reducing transpirational water loss and triggering dormancy-like responses such as senescence in old leaves and growth inhibition in young tissues under severe drought conditions [45]. The accelerated leaf senescence observed in transgenic plants overexpressing *PYL9* (under the control of the *pRD29A* promoter) aids in generating a greater osmotic potential gradient, thereby allowing water to preferentially flow to developing tissues [45].

Nonetheless, when exposed to severe drought conditions, the expression of *NTL4* and the accumulation of ROS extend throughout the entire plant, resulting in necrosis of the entire plant body [28]. This observation suggests that delaying leaf senescence could potentially enhance drought tolerance. Under drought stress, maintaining a balance between growth and survival is crucial for the overall fitness of plants [46], yet the mechanisms underlying this balance remain poorly understood [8]. Gaining a deeper understanding of the molecular mechanisms involved in drought-induced leaf senescence holds promise for developing strategies to alleviate the detrimental effects of drought stress on plant growth and productivity [18]. In this regard, NTL4 emerges as a potential candidate gene for coordinating plant stress tolerance and growth by precisely regulating its gene expression to initiate leaf senescence at the appropriate time.

2.3. Salt Stress-Induced Leaf Senescence

Salinity, a significant environmental stressor, particularly in arid and semi-arid regions, poses a substantial threat to crop productivity, leading to significant crop losses [47,48]. Salt stress exerts its negative impact on crop growth through various mechanisms, including osmotic stress, toxicity from specific ions, nutrient imbalances, and disrupted hormonal regulation [49–51]. It is estimated that more than 6% of the Earth's land is affected by salinity, with approximately 20% of irrigated land being saline, resulting in substantial agricultural losses amounting to tens of billions of dollars annually [52–54]. The effect of salt stress on plant senescence varies depending on the salt concentration. Mild salt stress can induce early flowering in plants, while severe salt stress can trigger leaf senescence and cell death [17].

Several research studies focused on identifying transcription factors involved in the regulation of salt stress-induced leaf senescence [55,56]. One prominent family in this context is the NAC transcription factor family, which has been extensively studied for its role in salt stress-induced leaf senescence [55]. ANAC092/ORE1, a member of this family, was found to contribute to salt-promoted senescence by controlling gene expression in response to salt stress [57]. Overexpressing *ORE1* leads to salt-induced senescence, while *ANAC092* knockout plants exhibit delayed senescence [57]. Ethylene-insensitive 3 (EIN3), a key transcription factor in the ethylene signaling pathway, acts as an upstream regulator of ORE1, influencing both leaf senescence and the response to salt stress [14,58]. Consequently, the age-dependent trigeminal feed-forward pathway involving ANAC092/ORE1 potentially intersects with other developmental and environmental signals to govern leaf senescence and cell death processes [56].

ANAC016 and ANAC032 are additional transcription factors that contribute to the positive regulation of leaf senescence under salt stress by controlling the expression of *SAGs* [59–61]. Mutants of *nac016* were found to retain their green phenotype under salt stress conditions, while plants overexpressing *NAC016* exhibit rapid senescence [59]. Similarly, the expression of *ANAC032* was induced by salinity and promotes leaf senescence in response to salt stress [61]. Notably, the *ANAC032OX* line showed increased accumulation of hydrogen peroxide (H_2O_2), whereas the chimeric repressor line (*ANAC032-SRDX*) exhibited reduced H_2O_2 levels [61]. These findings suggest that the altered responses of ANAC032 transgenic lines to salt stress may involve differential accumulation of ROS [61]. ANAC047, another transcription factor induced by salinity, is also implicated in salt stress-induced senescence [62]. Transgenic plants expressing the chimeric inhibitor *ANAC047-SRDX* displayed enhanced salt tolerance, indicating that ANAC047 acts as a positive regulator of stress-induced senescence [62]. Conversely, ANAC083/VNI2 functions as a negative regulator of senescence in Arabidopsis [63]. Plants with high expression levels of *ANAC083* exhibited significant salt and drought tolerance, along with delayed senescence [63]. Moreover, increased *ANAC083* expression led to the upregulation of *COR/RD* genes [63]. ANAC042/JUNGBRUNNEN1 (JUB1), another negative regulator of senescence, promotes plant longevity and confers tolerance to abiotic stresses such as heat and salt in Arabidopsis [64]. *JUB1* expression is rapidly induced by the accumulation of H_2O_2, and its overexpression results in delayed natural senescence [64]. Recently, a transcription factor from the AP2/ERF family, ethylene-responsive factor 34 (ERF34), was identified as a negative regulator of salt stress-induced leaf senescence and a contributor to salt stress tolerance [56]. ERF34 directly binds to the promoters of early responsive to dehydration 10 (*ERD10*) and responsive to desiccation 29A (*RD29A*), activating their expression [56]. This study suggests that ERF34 may serve as a potential mediator that integrates salt stress signals with the leaf senescence program.

The pivotal role of stress response transcription factors as key regulators of leaf senescence was extensively demonstrated in crops and trees [65–67]. For instance, overexpression of the rice NAC gene *SNAC1* in transgenic cotton enhances drought and salt tolerance by promoting root development and reducing transpiration rate [68]. In rice, the salt stress response gene ONAC106 acts as a negative regulator of leaf senescence [69]. Gain-of-function mutants of *ONAC106*, such as *ONAC106-1D* transgenic plants with a 35S enhancer inserted into the *ONAC106* gene's promoter region, exhibited delayed senescence and improved salt stress tolerance [69]. Similarly, the overexpression of *ShNAC1* in *Solanum habrochaites* delays salt stress-induced leaf senescence [70]. In *Populus euphratica*, the overexpression of two NAC transcription factors, *PeNAC034* and *PeNAC036*, results in enhanced salt stress sensitivity and tolerance, respectively [71]. Notably, *PeNAC034* overexpression promotes leaf senescence, while *PeNAC036* overexpression inhibits it [72]. In addition to transcription factors, other regulatory genes also play a crucial role in salt-induced leaf senescence. In rice, the loss of function of the receptor-like kinase gene bilateral blade senescence 1 accelerates leaf senescence and reduces salt tolerance [73].

The overexpression of the salt-inducible protein *salT* in rice was shown to delay leaf senescence, potentially serving as a feedback regulation to suppress salt stress-induced senescence [74]. Furthermore, a comparative transcriptome analysis of Arabidopsis plants exposed to age-dependent and salt stress-induced leaf senescence revealed potential molecular mechanisms underlying the interplay between these two senescence scenarios, including the involvement of H_2O_2-mediated signaling [75]. Salt stress-induced leaf senescence is a complex process regulated by multiple genes and signaling pathways. However, the intricate mechanisms that integrate salt stress signaling with the leaf senescence program remain largely elusive [56]. Enhancing our understanding of the molecular mechanisms underlying salt-induced leaf senescence will contribute to the development of strategies aimed at improving plant stress tolerance and crop productivity [8].

2.4. Darkness-Induced Leaf Senescence

Light plays a crucial role in plant growth, morphology, and development [76]. However, when plants are exposed to shade or complete darkness for an extended period, it triggers leaf senescence [37,77–81]. Transcriptomic analysis has shown that gene expression changes induced by darkness closely resemble those observed during natural senescence [82–85]. In fact, more than 50% of the genes up-regulated during natural senescence are also up-regulated under dark treatment conditions [83]. As a result, dark treatments are widely employed as a rapid, convenient, and effective method to induce leaf senescence, making it easier to investigate the impact of additional regulators of senescence, such as phytohormones, sugars, and secondary metabolites [8,83].

Recent investigations unveiled several genes and signaling pathways associated with dark-induced leaf senescence. To identify mutants with delayed dark-induced senescence, an experiment utilizing an individually darkened leaf (IDL) setup was conducted on *Arabidopsis thaliana* Col-0 plants treated with ethyl methanesulfonate mutagenesis [80]. The study revealed that PIF5 loss-of-function mutants, specifically *pif5-621*, exhibited significantly delayed chlorophyll loss in the IDL [80]. Remarkably, the overall growth habit of *pif5-621* resembled that of wild-type plants, indicating a direct impact of the *pif5* mutation on senescence rather than an indirect effect through life cycle progression or overall growth [80]. One plausible hypothesis to explain the extended lifespan of *pif5-621* IDLs is that the cells decelerated their metabolism, particularly respiration, to minimize carbon consumption and prolong survival compared to wild-type IDLs. Supporting this notion, *pif* quadruple mutants (*pifQ*) *pif1 pif3 pif4 pif5*, which exhibit a constitutive photomorphogenic phenotype when grown in the dark, maintained green cotyledons even after 10 days of dark treatment, while cotyledons of the wild type turned completely yellow, indicating that PIFs promote senescence under light-deprived conditions [38,40]. PIF4 and PIF5 influence ABA signaling by modulating ABSCISIC ACID INSENSITIVE 5 (ABI5) and ENHANCED EM LEVELS (EEL), two sister genes encoding basic leucine zipper (bZIP) class A transcription factors, which exhibited significantly reduced induction after darkening in *pifQ* mutants compared to the wild type [40]. Correspondingly, the single mutants *abi5*, *eel*, and, particularly, the *abi5 eel* double mutant displayed delayed senescence under dark conditions. Furthermore, PIF4 or PIF5 stimulates ethylene signaling by directly regulating the transcription of *EIN3* [40]. Additionally, ethylene evolution is diminished in *pif4* mutants and elevated in *PIF4* and *PIF5* overexpressors [38,86]. Treatment of *pif4* mutants with ethylene partially restored the senescence phenotype, indicating that PIFs promote dark-induced senescence by inducing ethylene biosynthesis and signaling. Moreover, PIF4, PIF5, and their target transcription factors (ABI5, EEL, and EIN3) directly activate the transcription of *ORE1*, suggesting the establishment of multiple coherent feed-forward regulatory circuits involving these transcription factors to induce leaf senescence [37]. As expected, *ein3* and *ore1* mutants exhibited a significant delay in senescence compared to the wild type, as evidenced by higher chlorophyll content and Fv/Fm levels under dark conditions [14]. PIF4/PIF5 directly activates the expression of *ABI5* and *EIN3*, which, in turn, activate the transcription of *ORE1*. ORE1 collaborates with PIFs, ABI5, and EIN3 to up-regulate genes involved in chlorophyll degradation, including staygreen 1 (*SGR*) and non-yellow coloring 1 (*NYC1*) [38,40,87]. Conversely, ORE1 interacts with PIFs to suppress the chloroplast maintenance master regulators GOLDEN2-LIKE 1 (GLK1) and GLK2. This antagonistic action of ORE1 on GLKs shifts the balance from chloroplast maintenance to deterioration [88].

Apart from the ABA and ethylene signaling pathways, dark-induced leaf senescence also involves the participation of JA. The genes responsible for JA biosynthesis, namely lipoxygensase 2 (*LOX2*) and allene oxide synthase (*AOS*), are up-regulated during dark-induced leaf senescence, and the application of exogenous JA expedites the senescence process [89]. Overall, the induction of leaf senescence by dark treatment is governed by an intricate network of molecular mechanisms encompassing various genes and regulatory pathways. The identification of these pivotal genes and pathways offers valuable insights into the regulatory mechanisms underlying dark-induced leaf senescence and holds potential for the development of strategies aimed at delaying or preventing leaf senescence in crop plants.

2.5. Low Oxygen-Induced Leaf Senescence

Low oxygen, also referred to as hypoxia, represents an abiotic stress condition capable of triggering leaf senescence in plants [90–92]. In response to low oxygen levels, plants activate various adaptive mechanisms to maintain cellular homeostasis and minimize oxidative damage. However, prolonged exposure to hypoxia can accelerate leaf senescence, leading to reduced plant growth and yield. Notably, leaf senescence is a prominent visible symptom observed in plants subjected to extended submergence [90–92]. Chlorophyll degradation initiates during the hypoxic phase and becomes evident after prolonged submergence (typically lasting 5 to 7 days) in rice and Arabidopsis [90–92].

At the molecular level, the regulation of hypoxia-induced leaf senescence involves a complex interplay of genes and signaling pathways. Among the key contributors to this process are the transcription factors belonging to the group VII ethylene response factor (ERFVIIs), which stabilize under hypoxic conditions and activate downstream gene expression to facilitate plant adaptation to low oxygen levels [93–95]. In rice, the ERFVII transcription factor known as submergence 1a (SUB1A) functions as a regulator of submergence tolerance by attenuating leaf senescence during prolonged submergence. Through functional characterization, it was revealed that the induction of *SUB1A* expression during submergence restricts further ethylene production and reduces gibberellic acid responsiveness. As a result, shoot tissues experience a decrease in carbohydrate consumption, chlorophyll breakdown, amino acid accumulation, and elongation growth [90–92]. This quiescence response to submergence aids in preserving carbohydrate reserves and the capacity for photosynthesis. The prevention of carbohydrate depletion may contribute to the milder manifestation of leaf senescence observed during submergence [96].

Interestingly, ectopic overexpression of *SUB1A* not only delays darkness-induced leaf senescence but also limits ethylene production and responsiveness to JA and salicylic acid (SA). This suppression of ethylene, JA, and SA signaling pathways results in the preservation of chlorophyll and carbohydrates [97]. The delay in leaf senescence conferred by SUB1A contributes to enhanced tolerance to submergence, drought, and oxidative stress [96–98]. Collectively, the molecular mechanisms governing hypoxia-induced leaf senescence are intricate and multifaceted. Gaining a comprehensive understanding of these mechanisms is crucial for the development of strategies aimed at enhancing plant tolerance to hypoxia and mitigating its adverse effects on plant growth and yield.

2.6. Extreme Temperatures Stress-Induced Leaf Senescence

Heat stress is one of the major environmental factors that trigger precocious senescence in plans. Heat-stress-induced leaf senescence is associated with ethylene accumulation and chlorophyll loss [2,99]. High-temperature treatment increased ethylene production in soybean (*Glycine max*) leaves and pods, which may be due to higher ACC synthase activity [99]. Pheophytinase (PPH) could be one of enzymes that play key roles in regulating heat-accelerated chlorophyll degradation [100]. After heat stress, the survival rate of *pph* mutant plants was significantly higher than that of wild type plants. It also led to a significant decrease in chlorophyll content in wild type plants and *pph* mutants, but the decrease was greater in wild type plants. The previously mentioned PIF4 and PIF5 are key regulators of heat-induced senescence [37–40]. Under heat stress, leaf senescence was delayed in *pif4* and *pif5* mutants and accelerated in transgenic lines compared with the wild type. *NAC019*, *SAG113*, and *IAA29* were characterized as direct targets of PIF4 and PIF5. In addition, PIF4 and PIF5 proteins accumulate with the progression of heat stress-induced leaf senescence and are regulate at the transcriptional and posttranscriptional levels [101]. In addition, mutation of premature senescence leaf 50 (PSL50) led to higher heat sensitivity, reduced survival, excessive hydrogen peroxide (H_2O_2) content, and increased cell death under heat stress in rice. This result suggests that PSL50 improves heat tolerance by regulating H_2O_2 signaling under heat stress [102]. Low temperature and short day length could result in the decrease in cytokinin and the increase in abscisic acid in leaf tissue, which directly trigger/promote senescence [103], which was supported by another study [104]. So far, low-temperature-induced leaf senescence has not been well-studied, and the underlying molecular regulatory mechanisms remain to be explored.

2.7. Other Abiotic Stresses-Induced Leaf Senescence

Apart from the previously mentioned abiotic stresses, additional factors such as extreme temperatures, high sugar levels, and UV radiation can also trigger premature leaf senescence [2,8]. Elevated sugar levels within plant tissues lead to reduced photosynthesis and early onset of senescence. The loss of hexokinase-1 (HXK1) function results in a delayed senescence phenotype [105], whereas the overexpression of *Arabidopsis HXK1* (*AtHXK1*) in tomato plants accelerates senescence [106]. These findings indicate the involvement of the sugar sensor HXK1 in sugar signaling during senescence. Intriguingly, the *hxk/gin2* mutant does not accumulate hexose in senescing leaves [107]. Moreover, the *hxk/gin2* mutant exhibits a delay in senescence induction by externally supplied glucose [105], suggesting that HXK1 plays a role in sugar metabolism and response during senescence. Notably, growth on glucose in combination with low nitrogen supply induces leaf yellowing and alters gene expression patterns, characteristic of developmental senescence. Importantly, the senescence-specific gene *SAG12* is significantly upregulated by glucose. Additionally, two senescence-associated MYB transcription factor genes, production of anthocyanin pigment 1 (*PAP1*) and *PAP2*, are induced by glucose [108]. In Arabidopsis, glucose and fructose accumulate substantially during leaf developmental senescence, while the sucrose content remains relatively unchanged [107]. Generally, the sugar content in leaves gradually increases, reaching its peak during the mature green stage or early senescence stages. Although the mechanisms underlying the maintenance of carbon storage molecules, such as sugars and starch, during senescence are not fully understood, sugars undoubtedly play a crucial role in driving cellular processes in senescing leaves [109].

Moreover, the presence of heavy metal pollutants, such as cadmium, poses a significant environmental challenge, leading to detrimental effects on plant growth and development. Cadmium toxicity triggers the generation of ROS, disrupts the photosynthetic system, and disrupts nutrient balance, ultimately accelerating leaf senescence [110,111]. Intriguingly, the accumulation of cadmium in leaves increases exponentially during the senescence process [112], indicating a clear association between leaf senescence and cadmium accumulation. However, the exact mechanism of cadmium accumulation in senescing leaves and the causal relationship between cadmium accumulation and senescence remain unclear. In particular, senescing leaves of tall fescue (*Festuca arundinacea*) can serve as a means to remove cadmium from polluted soil through a sustainable approach known as phytoextraction [110,112].

3. Improvement of Stress Tolerance through Regulation of Leaf Senescence

Understanding leaf senescence holds great importance due to its potential for improving crop yield and quality. Manipulating the timing of leaf senescence enables plant breeders to enhance photosynthetic efficiency, nutrient absorption, and stress tolerance, leading to increased crop yield and improved quality. Additionally, leaf senescence plays a pivotal role in plant adaptation to environmental stress. Exploring leaf senescence provides valuable insights for developing stress-tolerant plants capable of withstanding adverse conditions like drought, heat, or cold, thereby minimizing the detrimental effects of these stressors on plant growth and productivity.

3.1. Utilization of Senescence-Specific or Stress-Associated Promoters

By utilizing the promoter of a senescence-specific gene *SAG12*, Gan and Amasino designed an ingenious and elegant auto-regulatory senescence-inhibition system, *pSAG12-IPT* [25] (Figure 1A). The promoter of *SAG12* was linked to the coding region of the isopentenyltransferase gene (IPT), which regulates the rate-limiting step in cytokinin biosynthesis, to form the chimeric gene *pSAG12-IPT* [25]. At the onset of senescence, this promoter activates IPT expression and increases cytokinin content to levels that prevent leaf senescence. Repression of senescence in turn attenuates promoter expression to prevent overproduction of cytokinin. The use of senescence promoters is essential to avoid premature IPT overexpression and CK hyper-production ahead of senescence. The auto-regulatory biosynthetic system using p*SAG12-IPT* was proven to be an effective strategy for developing transgenic plants to increase yield by delaying senescence and extending the shelf life of isolated organs such as leaves, flowers, and fruits [28]. The *pSAG12-IPT* system had been widely used in numerous plant species [28], including wheat (*Triticum aestivum* L.) [113], alfalfa (*Medicago sativa*) [114], lettuce (*Lactuca sativa* L. cv Evola) [115], cassava (*Manihot esculenta Crantz*) [27], and creeping bentgrass (*Agrostis stolonifera* L. 'Penncross') [116], etc. However, it should be noted that the pSAG12-IPT system possibly directly or indirectly affects plant development, including delayed flowering in transgenic lettuce [115], and reduced nitrogen accumulation in young leaves by altering sink-source relationships in tobacco [117]. To achieve maximum effectiveness, practical applications should carefully consider the advantages and disadvantages of this system.

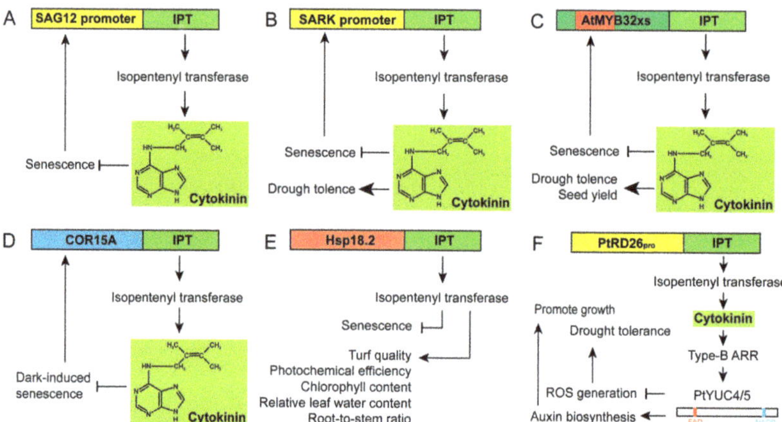

Figure 1. The diagrams show *pSAG12-IPT* and its various variants. (**A**) The auto-regulatory senescence-inhibition system of *pSAG12-IPT* [25]. (**B–F**) A range of variants have been developed based on the design concept of *pSAG12-IPT*, including (**B**) *pSARK-IPT* [26], (**C**) *AtMYB32xs-IPT* (Red rectangle represent the deleted root motif of 360bp in promoter) [118], (**D**) *COR15A-IPT* [119], (**E**) *HSP18.2-IPT* [116], and (**F**) *PtRD26pro-IPT* [120]. The yellow parts represent the senescence-specific promoters. The dark green part represents the developmental process-related promoter. The blue part represents the cold-induced promoter. The red part represents the heat shock promoter.

A range of variants were developed based on the design concept of *pSAG12-IPT*. One approach involves utilizing different promoters to control the expression of *IPT*. For instance, a modified version of this cytokinin (CK) auto-regulatory cycle strategy employed the promoter of *senescence-associated receptor kinase* (SARK) fused with the *IPT* gene (Figure 1B). Transgenic tobacco plants carrying *pSARK-IPT* exhibited enhanced survival under severe drought conditions, accompanied by improvements in photosynthetic rate and water use efficiency [26]. In these plants, the activation of the *SARK* promoter in response to drought-induced leaf senescence led to delayed senescence through cytokinin biosynthesis. However, it should be noted that premature activation of leaf senescence may occur during drought, and the benefits in terms of yield increase may not be realized under well-watered conditions when using stress-inducible promoters. To address the issues of stress inducibility and proper regulation of *IPT* genes, Spangenberg and colleagues ingeniously employed a modified promoter derived from the developmental process-related gene *AtMYB32* (*AtMYB32xs*) (Figure 1C), which removed the 360 bp root-specific motif [118]. Stable transgenic oilseed rape (*Brassica napus*) plants expressing *AtMYB32xs-IPT* exhibited delayed leaf senescence under controlled environment and field conditions. Remarkably, these *AtMYB32xs-IPT* plants achieved significantly higher seed yield during both rainy seasons and field irrigation conditions [118]. In petunia and chrysanthemum, transgenic plants known as *COR15A-IPT* were generated using the cold induction promoter from the cold-regulated15a (*COR15A*) gene of *Arabidopsis thaliana* (Figure 1D) [119]. Intriguingly, *COR15A-IPT* plants and their detached leaves remained green and healthy during extended dark storage (4 weeks at 25 °C) following an initial exposure to a brief period of cold induction (72 h at 4 °C). This study presented an approach to prolong the lifespan of transplants or excised leaves during storage under dark and cold conditions, which is particularly beneficial for long-distance transport. The heat shock promoter *HSP18.2* was fused with *IPT* to generate *HSP18.2-IPT* transgenic plants in creeping bentgrass (*Agrostis stolonifera*) (Figure 1E) [116]. The *HSP18.2-IPT* transgenic lines exhibited significantly improved turf quality, photochemical efficiency, chlorophyll content, relative leaf water content, and root-to-stem ratio. Furthermore, transgenic poplar lines expressing *IPT* under the control of the promoter of *PtRD26* (*PtRD26$_{pro}$-IPT*) (Figure 1F), a senescence and drought-inducible NAC transcription factor in poplar, displayed various phenotypic improvements, including enhanced growth and drought tolerance [120].

Another type of experimental design is to use promoters of *SAG12* to drive the expression of different genes. For example, tobacco plants overexpressing the maize homeobox gene *knotted1* (*kn1*) under the driver of *SAG12* promoter, designated as *pSAG12-kn1*, exhibited a significant delay in leaf senescence, with an increase in chlorophyll content and a decrease in the number of dead leaves. In the detached leaves of *pSAG12-kn1* plants, senescence was also postponed [121]. Collectively, these studies provided the possibility of regulating the onset of leaf senescence by cleverly using senescence and stress-related gene promoters to drive the expression of *IPT* and developmental genes, thereby improving crop resistance, yield, and quality.

3.2. Modulation of Expression of Senescence Associated Genes

An alternative approach to influencing the leaf senescence process involves manipulating the expression of crucial senescence genes, with the aim of enhancing crop resistance and yield. For instance, the knockout of *OsNAP*, a rice ortholog of *ANAC029/AtNAP* [122], resulted in prolonged grain-filling periods and increased grain yields compared to the wild type [123]. Therefore, precise regulation of *OsNAP* expression holds promise for improving stress resistance in rice. A noteworthy discovery is the potential use of naturally occurring *Stay-Green* (*OsSGR*) promoter and associated longevity variants in breeding programs to enhance rice yield [124]. Nam and colleagues conducted quantitative trait loci (QTL) mapping and identified genetic differences in life cycle and senescence patterns between two rice subspecies, indica, and japonica [124]. They found that promoter variations in the *OsSGR* gene, which encodes the chlorophyll-degrading Mg^{++}-dechelatase, triggered earlier and higher induction of *OsSGR* in indica, thereby accelerating senescence in indica cultivars. Introducing the japonica OsSGR allele into indica-type cultivars resulted in delayed senescence, increased grain yield, and improved photosynthetic capacity. This study highlighted the potential of modifying the senescence-related promoter region, in addition to gene coding regions, to achieve delayed leaf senescence and increased yield. The use of gene editing technologies like CRISPR/Cas9 offers a powerful tool for manipulating key genes involved in regulating leaf senescence, and further research is necessary to identify and manipulate additional genes involved in these processes [8].

To summarize, comprehending the molecular regulatory mechanisms underlying leaf senescence holds the potential to inform the development of molecular breeding strategies aimed at enhancing plant tolerance to abiotic stresses, increasing grain yield, and improving crop quality. A significant objective in leaf senescence research is the cultivation of plants with ideal leaf senescence phenotype (PILSP). PILSP exhibit the remarkable ability to effectively coordinate growth and stress tolerance, integrate both internal and external signals, and initiate leaf senescence at the optimal time. In the case of annual plants, leaves remain green in the initial stages of plant growth, resisting internal and external stresses, and only enter senescence when the leaves die, allowing for the complete transfer of photosynthetic products to the seeds. In contrast, perennial plants retain their leaves even under environmental stresses, and upon stress removal, leaf function is promptly restored. Consequently, when the leaves eventually die, the photosynthetic products can be fully channeled to the main stem or the growing organ.

4. Prospects

In recent years, considerable advancements have been made in leaf senescence research; however, several challenges remain in fully comprehending the molecular mechanisms of leaf senescence and effectively applying this knowledge to enhance crop improvement. One of the foremost challenges lies in the intricate nature of the regulatory network governing leaf senescence, posing difficulties in identifying pivotal genes and regulatory pathways. Additionally, the complexity is compounded by the fact that diverse environmental factors, such as drought, high temperature, and nutrient deficiency, can trigger leaf senescence through distinct pathways, further adding to the intricacy of the regulatory network.

To overcome these challenges, several strategies have been proposed. Firstly, a comprehensive analysis of the leaf senescence regulatory network and the identification of key genes and regulatory pathways can be achieved through the integration of multi-omics approaches, including genomics, transcriptomics, proteomics, and metabolomics [19,125–127]. These approaches enable a holistic understanding of the complex mechanisms involved. Secondly, advanced imaging techniques, such as live imaging and high-resolution microscopy, offer the opportunity to monitor the dynamic progression of leaf senescence and visualize the molecular events at play. For instance, confocal imaging fluorometer allows high spatio-temporal-resolution detection of chlorophyll fluorescence dynamics at the single chloroplast level [128]. Additionally, the combination of high-speed three-dimensional laser scanning confocal microscopy and high-sensitivity multiple-channel detection facilitates in-depth investigations of the spatial and temporal dynamics of chloroplast degradation during leaf senescence [129]. Thirdly, genetic engineering techniques, particularly CRISPR/Cas9-mediated genome editing, provide a means to manipulate the expression of key senescence-related genes and elucidate their roles in the process. A recent study successfully employed CRISPR/Cas9-mediated knockout to demonstrate the regulatory function of the peptide hormone CLE42 in leaf senescence [130].

In conclusion, leaf senescence research holds immense potential for enhancing crop yield and quality. However, addressing the existing challenges is crucial. By harnessing the power of multi-omics approaches, advanced imaging techniques, and genetic engineering, we can gain a deeper understanding of the molecular mechanisms underlying leaf senescence and effectively apply this knowledge to drive crop improvement.

Author Contributions: Conceptualization, Z.L.; writing—original draft preparation, S.T. and Y.S.; writing—reviewing and editing, S.T., Y.S., L.S. and Z.L.; supervision, Z.L. All authors have read and agreed to the published version of the manuscript.

Funding: This research was funded by Beijing Municipal Natural Science Foundation (5232015 to Z.L.), the Open Research Fund of the National Center for Protein Sciences at Peking University in Beijing (KF-202304), the National Natural Science Foundation of China (32170345, 31970196, and 32011540381 to Z.L.), and the High-level Talent Introduction Programme of Beijing Forestry University.

Institutional Review Board Statement: Not applicable.

Informed Consent Statement: Not applicable.

Data Availability Statement: No new data were created or analyzed in this study. Data sharing is not applicable to this article.

Acknowledgments: We sincerely apologize to all those authors whose work is not included in this review paper due to space limitations.

Conflicts of Interest: The authors declare no conflict of interest.

Abbreviations

ROS	Reactive Oxygen Species
SAG	Senescence-Associated Gene
NLA	Nitrogen Limitation Adaptation
PHO2	PHOSPHATE2
UBC	Ubiquitin-conjugating enzyme
UBP12	Ubiquitin-specific protease 12
GDS1	GROWTH, DEVELOPMENT AND SPLICING 1
ORE1	ORESARA1
JUB1	JUNGBRUNNEN1
ERF	ETHYLENE RESPONSE FACTOR
COR15A	COLD-REGULATED15A
ERD10	EARLY RESPONSIVE TO DEHYDRATION10
RD29A	RESPONSIVE TO DESICCATION29A
PIF	PHYTOCHROME-INTERACTING FACTOR
bZIP	basic LEUCINE ZIPPER
IDL	Individually Darkened Leaf
ABI5	ABSCISIC ACID INSENSITIVE 5
EEL	ENHANCED EM LEVELS
SGR1	STAYGREEN 1
NYC1	NON-YELLOW COLORING 1
GDL1	GOLDEN2-LIKE 1
LOX2	LIPOXYGENASE 2
AOS	ALLENE OXIDE SYNTHASE
SUB1A	SUBMERGENCE1A
HXK1	Hexokinase-1
PAP1	PRODUCTION OF ANTHOCYANIN PIGMENT 1
IPT	Isopentenyltransferase
SARK	Senescence-associated Receptor Kinase
Kn1	Knotted1
QTL	Quantitative Trait Loci
PILSP	Plants with Ideal Leaf Senescence Phenotype
ROS	Reactive Oxygen Species
SAG	Senescence-Associated Gene
NLA	Nitrogen Limitation Adaptation
PHO2	PHOSPHATE2
UBC	Ubiquitin-conjugating enzyme
UBP12	Ubiquitin-specific protease 12
GDS1	GROWTH, DEVELOPMENT AND SPLICING 1
ORE1	ORESARA1
JUB1	JUNGBRUNNEN1
ERF	ETHYLENE RESPONSE FACTOR
COR15A	COLD-REGULATED15A
ERD10	EARLY RESPONSIVE TO DEHYDRATION10

References

1. Guo, Y.; Gan, S. Leaf senescence: Signals, execution, and regulation. *Curr. Top. Dev. Biol.* **2005**, *71*, 83–112.
2. Lim, P.O.; Kim, H.J.; Nam, H.G. Leaf senescence. *Annu. Rev. Plant Biol.* **2007**, *58*, 115–136. [CrossRef]

3. Wang, H.L.; Zhang, Y.; Wang, T.; Yang, Q.; Yang, Y.; Li, Z.; Li, B.; Wen, X.; Li, W.; Yin, W.; et al. An alternative splicing variant of PtRD26 delays leaf senescence by regulating multiple NAC transcription factors in Populus. *Plant Cell* **2021**, *33*, 1594–1614. [CrossRef]

4. Keskitalo, J.; Bergquist, G.; Gardestrom, P.; Jansson, S. A cellular timetable of autumn senescence. *Plant Physiol.* **2005**, *139*, 1635–1648. [CrossRef] [PubMed]

5. Cooke, J.E.; Weih, M. Nitrogen storage and seasonal nitrogen cycling in Populus: Bridging molecular physiology and ecophysiology. *New Phytol.* **2005**, *167*, 19–30. [CrossRef] [PubMed]

6. Andleeb, T.; Knight, E.; Borrill, P. Wheat *NAM* genes regulate the majority of early monocarpic senescence transcriptional changes including nitrogen remobilization genes. *G3* **2023**, *13*, jkac275. [CrossRef]

7. Wei, S.; Li, X.; Lu, Z.; Zhang, H.; Ye, X.; Zhou, Y.; Li, J.; Yan, Y.; Pei, H.; Duan, F.; et al. A transcriptional regulator that boosts grain yields and shortens the growth duration of rice. *Science* **2022**, *377*, eabi8455. [CrossRef] [PubMed]

8. Guo, Y.; Ren, G.; Zhang, K.; Li, Z.; Miao, Y.; Guo, H. Leaf senescence: Progression, regulation, and application. *Mol. Hortic.* **2021**, *1*, 5. [CrossRef]

9. Cao, J.; Zhang, Y.; Tan, S.; Yang, Q.; Wang, H.-L.; Xia, X.; Luo, J.; Guo, H.; Zhang, Z.; Li, Z. LSD 4.0: An improved database for comparative studies of leaf senescence. *Mol. Hortic.* **2022**, *2*, 24. [CrossRef]

10. Cao, J.; Liu, H.; Tan, S.; Li, Z. Transcription Factors-Regulated Leaf Senescence: Current Knowledge, Challenges and Approaches. *Int. J. Mol. Sci.* **2023**, *24*, 9245. [CrossRef]

11. Moschen, S.; Di Rienzo, J.A.; Higgins, J.; Tohge, T.; Watanabe, M.; Gonzalez, S.; Rivarola, M.; Garcia-Garcia, F.; Dopazo, J.; Hopp, H.E.; et al. Integration of transcriptomic and metabolic data reveals hub transcription factors involved in drought stress response in sunflower (*Helianthus annuus* L.). *Plant Mol. Biol.* **2017**, *94*, 549–564. [CrossRef] [PubMed]

12. Moschen, S.; Bengoa Luoni, S.; Paniego, N.B.; Hopp, H.E.; Dosio, G.A.; Fernandez, P.; Heinz, R.A. Identification of candidate genes associated with leaf senescence in cultivated sunflower (*Helianthus annuus* L.). *PLoS ONE* **2014**, *9*, e104379. [CrossRef] [PubMed]

13. Trupkin, S.A.; Astigueta, F.H.; Baigorria, A.H.; Garcia, M.N.; Delfosse, V.C.; Gonzalez, S.A.; Perez de la Torre, M.C.; Moschen, S.; Lia, V.V.; Fernandez, P.; et al. Identification and expression analysis of NAC transcription factors potentially involved in leaf and petal senescence in *Petunia hybrida*. *Plant Sci.* **2019**, *287*, 110195. [CrossRef] [PubMed]

14. Li, Z.; Peng, J.; Wen, X.; Guo, H. Ethylene-insensitive3 is a senescence-associated gene that accelerates age-dependent leaf senescence by directly repressing miR164 transcription in Arabidopsis. *Plant Cell* **2013**, *25*, 3311–3328. [CrossRef] [PubMed]

15. Miao, Y.; Laun, T.; Zimmermann, P.; Zentgraf, U. Targets of the WRKY53 transcription factor and its role during leaf senescence in Arabidopsis. *Plant Mol. Biol.* **2004**, *55*, 853–867. [CrossRef] [PubMed]

16. Wang, Q.; Li, X.; Guo, C.; Wen, L.; Deng, Z.; Zhang, Z.; Li, W.; Liu, T.; Guo, Y. Senescence-Related Receptor Kinase 1 (SENRK1) functions downstream of WRKY53 in regulating leaf senescence in Arabidopsis. *J. Exp. Bot.* **2023**. [CrossRef] [PubMed]

17. Sakuraba, Y.; Kim, D.; Paek, N.C. Salt Treatments and Induction of Senescence. *Methods Mol. Biol.* **2018**, *1744*, 141–149.

18. Sade, N.; Del Mar Rubio-Wilhelmi, M.; Umnajkitikorn, K.; Blumwald, E. Stress-induced senescence and plant tolerance to abiotic stress. *J. Exp. Bot.* **2018**, *69*, 845–853. [CrossRef]

19. Woo, H.R.; Kim, H.J.; Lim, P.O.; Nam, H.G. Leaf Senescence: Systems and Dynamics Aspects. *Annu. Rev. Plant Biol.* **2019**, *70*, 347–376. [CrossRef]

20. Li, Z.; Kim, J.H.; Kim, J.; Lyu, J.I.; Zhang, Y.; Guo, H.; Nam, H.G.; Woo, H.R. ATM suppresses leaf senescence triggered by DNA double-strand break through epigenetic control of senescence-associated genes in Arabidopsis. *New Phytol.* **2020**, *227*, 473–484. [CrossRef]

21. Han, D.; Du, M.; Zhou, Z.; Wang, S.; Li, T.; Han, J.; Xu, T.; Yang, G. Overexpression of a Malus baccata NAC Transcription Factor Gene MbNAC25 Increases Cold and Salinity Tolerance in Arabidopsis. *Int. J. Mol. Sci.* **2020**, *21*, 1198. [CrossRef] [PubMed]

22. Hu, D.G.; Sun, C.H.; Zhang, Q.Y.; Gu, K.D.; Hao, Y.J. The basic helix-loop-helix transcription factor MdbHLH3 modulates leaf senescence in apple via the regulation of dehydratase-enolase-phosphatase complex 1. *Hortic. Res.* **2020**, *7*, 50. [CrossRef] [PubMed]

23. Lira, B.S.; Gramegna, G.; Trench, B.A.; Alves, F.R.R.; Silva, E.M.; Silva, G.F.F.; Thirumalaikumar, V.P.; Lupi, A.C.D.; Demarco, D.; Purgatto, E.; et al. Manipulation of a Senescence-Associated Gene Improves Fleshy Fruit Yield. *Plant Physiol.* **2017**, *175*, 77–91. [CrossRef]

24. Ma, X.; Zhang, Y.; Tureckova, V.; Xue, G.P.; Fernie, A.R.; Mueller-Roeber, B.; Balazadeh, S. The NAC Transcription Factor SlNAP2 Regulates Leaf Senescence and Fruit Yield in Tomato. *Plant Physiol.* **2018**, *177*, 1286–1302. [CrossRef]

25. Gan, S.; Amasino, R.M. Inhibition of leaf senescence by autoregulated production of cytokinin. *Science* **1995**, *270*, 1986–1988. [CrossRef]

26. Rivero, R.M.; Kojima, M.; Gepstein, A.; Sakakibara, H.; Mittler, R.; Gepstein, S.; Blumwald, E. Delayed leaf senescence induces extreme drought tolerance in a flowering plant. *Proc. Natl. Acad. Sci. USA* **2007**, *104*, 19631–19636. [CrossRef]

27. Zhang, P.; Wang, W.Q.; Zhang, G.L.; Kaminek, M.; Dobrev, P.; Xu, J.; Gruissem, W. Senescence-inducible expression of isopentenyl transferase extends leaf life, increases drought stress resistance and alters cytokinin metabolism in cassava. *J. Integr. Plant Biol.* **2010**, *52*, 653–669. [CrossRef]

28. Guo, Y.; Gan, S.S. Translational researches on leaf senescence for enhancing plant productivity and quality. *J. Exp. Bot.* **2014**, *65*, 3901–3913. [CrossRef]

29. Park, B.S.; Yao, T.; Seo, J.S.; Wong, E.C.C.; Mitsuda, N.; Huang, C.H.; Chua, N.H. Arabidopsis NITROGEN LIMITATION ADAPTATION regulates ORE1 homeostasis during senescence induced by nitrogen deficiency. *Nat. Plants* **2018**, *4*, 898–903. [CrossRef] [PubMed]

30. Park, S.H.; Jeong, J.S.; Seo, J.S.; Park, B.S.; Chua, N.H. Arabidopsis ubiquitin-specific proteases UBP12 and UBP13 shape ORE1 levels during leaf senescence induced by nitrogen deficiency. *New Phytol.* **2019**, *223*, 1447–1460. [CrossRef]

31. Fan, H.; Quan, S.; Ye, Q.; Zhang, L.; Liu, W.; Zhu, N.; Zhang, X.; Ruan, W.; Yi, K.; Crawford, N.M.; et al. A molecular framework underlying low-nitrogen-induced early leaf senescence in *Arabidopsis thaliana*. *Mol. Plant* **2023**, *16*, 756–774. [CrossRef] [PubMed]

32. Peng, M.; Hannam, C.; Gu, H.; Bi, Y.M.; Rothstein, S.J. A mutation in NLA, which encodes a RING-type ubiquitin ligase, disrupts the adaptability of Arabidopsis to nitrogen limitation. *Plant J.* **2007**, *50*, 320–337. [CrossRef]

33. Liu, W.; Sun, Q.; Wang, K.; Du, Q.; Li, W.X. Nitrogen Limitation Adaptation (NLA) is involved in source-to-sink remobilization of nitrate by mediating the degradation of NRT1.7 in Arabidopsis. *New Phytol.* **2017**, *214*, 734–744. [CrossRef] [PubMed]

34. Delhaize, E.; Randall, P.J. Characterization of a Phosphate-Accumulator Mutant of Arabidopsis thaliana. *Plant Physiol.* **1995**, *107*, 207–213. [CrossRef] [PubMed]

35. Dong, B.; Rengel, Z.; Delhaize, E. Uptake and translocation of phosphate by pho2 mutant and wild-type seedlings of Arabidopsis thaliana. *Planta* **1998**, *205*, 251–256. [CrossRef]

36. Kim, D.W.; Jeon, S.J.; Hwang, S.M.; Hong, J.C.; Bahk, J.D. The C3H-type zinc finger protein GDS1/C3H42 is a nuclear-speckle-localized protein that is essential for normal growth and development in Arabidopsis. *Plant Sci.* **2016**, *250*, 141–153. [CrossRef]

37. Liebsch, D.; Keech, O. Dark-induced leaf senescence: New insights into a complex light-dependent regulatory pathway. *New Phytol.* **2016**, *212*, 563–570. [CrossRef]

38. Song, Y.; Yang, C.; Gao, S.; Zhang, W.; Li, L.; Kuai, B. Age-triggered and dark-induced leaf senescence require the bHLH transcription factors PIF3, 4, and 5. *Mol. Plant* **2014**, *7*, 1776–1787. [CrossRef]

39. Li, N.; Bo, C.; Zhang, Y.; Wang, L. Phytochrome Interacting Factors PIF4 and PIF5 promote heat stress induced leaf senescence in Arabidopsis. *J. Exp. Bot.* **2021**, *72*, 4577–4589. [CrossRef] [PubMed]

40. Sakuraba, Y.; Jeong, J.; Kang, M.Y.; Kim, J.; Paek, N.C.; Choi, G. Phytochrome-interacting transcription factors PIF4 and PIF5 induce leaf senescence in Arabidopsis. *Nat. Commun.* **2014**, *5*, 4636. [CrossRef] [PubMed]

41. Zhang, H.; Zhu, J.; Gong, Z.; Zhu, J.K. Abiotic stress responses in plants. *Nat. Rev. Genet.* **2022**, *23*, 104–119. [CrossRef]

42. Munne-Bosch, S.; Alegre, L. Die and let live: Leaf senescence contributes to plant survival under drought stress. *Funct. Plant Biol.* **2004**, *31*, 203–216. [CrossRef] [PubMed]

43. Lee, S.; Seo, P.J.; Lee, H.J.; Park, C.M. A NAC transcription factor NTL4 promotes reactive oxygen species production during drought-induced leaf senescence in Arabidopsis. *Plant J.* **2012**, *70*, 831–844. [CrossRef] [PubMed]

44. Hickman, R.; Hill, C.; Penfold, C.A.; Breeze, E.; Bowden, L.; Moore, J.D.; Zhang, P.; Jackson, A.; Cooke, E.; Bewicke-Copley, F.; et al. A local regulatory network around three NAC transcription factors in stress responses and senescence in Arabidopsis leaves. *Plant J.* **2013**, *75*, 26–39. [CrossRef] [PubMed]

45. Zhao, Y.; Chan, Z.; Gao, J.; Xing, L.; Cao, M.; Yu, C.; Hu, Y.; You, J.; Shi, H.; Zhu, Y.; et al. ABA receptor PYL9 promotes drought resistance and leaf senescence. *Proc. Natl. Acad. Sci. USA* **2016**, *113*, 1949–1954. [CrossRef]

46. Claeys, H.; Inze, D. The agony of choice: How plants balance growth and survival under water-limiting conditions. *Plant Physiol.* **2013**, *162*, 1768–1779. [CrossRef]

47. Athar, H.U.; Zulfiqar, F.; Moosa, A.; Ashraf, M.; Zafar, Z.U.; Zhang, L.; Ahmed, N.; Kalaji, H.M.; Nafees, M.; Hossain, M.A.; et al. Salt stress proteins in plants: An overview. *Front. Plant Sci.* **2022**, *13*, 999058. [CrossRef]

48. Zulfiqar, F.; Ashraf, M. Nanoparticles potentially mediate salt stress tolerance in plants. *Plant Physiol. Biochem.* **2021**, *160*, 257–268. [CrossRef]

49. Munns, R.; Tester, M. Mechanisms of salinity tolerance. *Annu. Rev. Plant Biol.* **2008**, *59*, 651–681. [CrossRef] [PubMed]

50. van Zelm, E.; Zhang, Y.; Testerink, C. Salt Tolerance Mechanisms of Plants. *Annu. Rev. Plant Biol.* **2020**, *71*, 403–433. [CrossRef]

51. Deinlein, U.; Stephan, A.B.; Horie, T.; Luo, W.; Xu, G.; Schroeder, J.I. Plant salt-tolerance mechanisms. *Trends Plant Sci.* **2014**, *19*, 371–379. [CrossRef]

52. Fricke, W. Energy costs of salinity tolerance in crop plants: Night-time transpiration and growth. *New Phytol.* **2020**, *225*, 1152–1165. [CrossRef]

53. Munns, R.; Day, D.A.; Fricke, W.; Watt, M.; Arsova, B.; Barkla, B.J.; Bose, J.; Byrt, C.S.; Chen, Z.H.; Foster, K.J.; et al. Energy costs of salt tolerance in crop plants. *New Phytol.* **2020**, *225*, 1072–1090. [CrossRef]

54. Tyerman, S.D.; Munns, R.; Fricke, W.; Arsova, B.; Barkla, B.J.; Bose, J.; Bramley, H.; Byrt, C.; Chen, Z.; Colmer, T.D.; et al. Energy costs of salinity tolerance in crop plants. *New Phytol.* **2019**, *221*, 25–29. [CrossRef]

55. Kim, H.J.; Nam, H.G.; Lim, P.O. Regulatory network of NAC transcription factors in leaf senescence. *Curr. Opin. Plant Biol.* **2016**, *33*, 48–56. [CrossRef]

56. Park, S.J.; Park, S.; Kim, Y.; Hyeon, D.Y.; Park, H.; Jeong, J.; Jeong, U.; Yoon, Y.S.; You, D.; Kwak, J.; et al. Ethylene responsive factor34 mediates stress-induced leaf senescence by regulating salt stress-responsive genes. *Plant Cell Environ.* **2022**, *45*, 1719–1733. [CrossRef]

57. Balazadeh, S.; Siddiqui, H.; Allu, A.D.; Matallana-Ramirez, L.P.; Caldana, C.; Mehrnia, M.; Zanor, M.I.; Kohler, B.; Mueller-Roeber, B. A gene regulatory network controlled by the NAC transcription factor ANAC092/AtNAC2/ORE1 during salt-promoted senescence. *Plant J.* **2010**, *62*, 250–264. [CrossRef]

58. Peng, J.; Li, Z.; Wen, X.; Li, W.; Shi, H.; Yang, L.; Zhu, H.; Guo, H. Salt-induced stabilization of EIN3/EIL1 confers salinity tolerance by deterring ROS accumulation in Arabidopsis. *PLoS Genet.* **2014**, *10*, e1004664. [CrossRef]

59. Kim, Y.S.; Sakuraba, Y.; Han, S.H.; Yoo, S.C.; Paek, N.C. Mutation of the Arabidopsis NAC016 transcription factor delays leaf senescence. *Plant Cell Physiol.* **2013**, *54*, 1660–1672. [CrossRef]

60. Mahmood, K.; Xu, Z.; El-Kereamy, A.; Casaretto, J.A.; Rothstein, S.J. The Arabidopsis Transcription Factor ANAC032 Represses Anthocyanin Biosynthesis in Response to High Sucrose and Oxidative and Abiotic Stresses. *Front. Plant Sci.* **2016**, *7*, 1548. [CrossRef]

61. Mahmood, K.; El-Kereamy, A.; Kim, S.H.; Nambara, E.; Rothstein, S.J. ANAC032 Positively Regulates Age-Dependent and Stress-Induced Senescence in Arabidopsis thaliana. *Plant Cell Physiol.* **2016**, *57*, 2029–2046. [CrossRef]

62. Mito, T.; Seki, M.; Shinozaki, K.; Ohme-Takagi, M.; Matsui, K. Generation of chimeric repressors that confer salt tolerance in Arabidopsis and rice. *Plant Biotechnol. J.* **2011**, *9*, 736–746. [CrossRef]

63. Yang, S.D.; Seo, P.J.; Yoon, H.K.; Park, C.M. The Arabidopsis NAC transcription factor VNI2 integrates abscisic acid signals into leaf senescence via the COR/RD genes. *Plant Cell* **2011**, *23*, 2155–2168. [CrossRef]

64. Wu, A.; Allu, A.D.; Garapati, P.; Siddiqui, H.; Dortay, H.; Zanor, M.I.; Asensi-Fabado, M.A.; Munne-Bosch, S.; Antonio, C.; Tohge, T.; et al. JUNGBRUNNEN1, a reactive oxygen species-responsive NAC transcription factor, regulates longevity in Arabidopsis. *Plant Cell* **2012**, *24*, 482–506. [CrossRef]

65. Podzimska-Sroka, D.; O'Shea, C.; Gregersen, P.L.; Skriver, K. NAC Transcription Factors in Senescence: From Molecular Structure to Function in Crops. *Plants* **2015**, *4*, 412–448. [CrossRef]

66. Zhou, Y.; Huang, W.; Liu, L.; Chen, T.; Zhou, F.; Lin, Y. Identification and functional characterization of a rice NAC gene involved in the regulation of leaf senescence. *BMC Plant Biol.* **2013**, *13*, 132. [CrossRef]

67. Zhao, D.; Derkx, A.P.; Liu, D.C.; Buchner, P.; Hawkesford, M.J. Overexpression of a NAC transcription factor delays leaf senescence and increases grain nitrogen concentration in wheat. *Plant Biol.* **2015**, *17*, 904–913. [CrossRef]

68. Liu, G.; Li, X.; Jin, S.; Liu, X.; Zhu, L.; Nie, Y.; Zhang, X. Overexpression of rice NAC gene SNAC1 improves drought and salt tolerance by enhancing root development and reducing transpiration rate in transgenic cotton. *PLoS ONE* **2014**, *9*, e86895. [CrossRef]

69. Sakuraba, Y.; Piao, W.; Lim, J.H.; Han, S.H.; Kim, Y.S.; An, G.; Paek, N.C. Rice ONAC106 Inhibits Leaf Senescence and Increases Salt Tolerance and Tiller Angle. *Plant Cell Physiol.* **2015**, *56*, 2325–2339. [CrossRef]

70. Liu, H.; Zhou, Y.; Li, H.; Wang, T.; Zhang, J.; Ouyang, B.; Ye, Z. Molecular and functional characterization of ShNAC1, an NAC transcription factor from Solanum habrochaites. *Plant Sci.* **2018**, *271*, 9–19. [CrossRef]

71. Lu, X.; Zhang, X.; Duan, H.; Lian, C.; Liu, C.; Yin, W.; Xia, X. Three stress-responsive NAC transcription factors from Populus euphratica differentially regulate salt and drought tolerance in transgenic plants. *Physiol. Plant.* **2018**, *162*, 73–97. [CrossRef] [PubMed]

72. Li, Z.; Zhang, Y.; Zou, D.; Zhao, Y.; Wang, H.L.; Zhang, Y.; Xia, X.; Luo, J.; Guo, H.; Zhang, Z. LSD 3.0: A comprehensive resource for the leaf senescence research community. *Nucleic Acids Res.* **2020**, *48*, D1069–D1075. [CrossRef] [PubMed]

73. Zeng, D.D.; Yang, C.C.; Qin, R.; Alamin, M.; Yue, E.K.; Jin, X.L.; Shi, C.H. A guanine insert in OsBBS1 leads to early leaf senescence and salt stress sensitivity in rice (*Oryza sativa* L.). *Plant Cell Rep.* **2018**, *37*, 933–946. [CrossRef] [PubMed]

74. Zhu, K.; Tao, H.; Xu, S.; Li, K.; Zafar, S.; Cao, W.; Yang, Y. Overexpression of salt-induced protein (salT) delays leaf senescence in rice. *Genet. Mol. Biol.* **2019**, *42*, 80–86. [CrossRef]

75. Allu, A.D.; Soja, A.M.; Wu, A.; Szymanski, J.; Balazadeh, S. Salt stress and senescence: Identification of cross-talk regulatory components. *J. Exp. Bot.* **2014**, *65*, 3993–4008. [CrossRef]

76. de Wit, M.; Galvao, V.C.; Fankhauser, C. Light-Mediated Hormonal Regulation of Plant Growth and Development. *Annu. Rev. Plant Biol.* **2016**, *67*, 513–537. [CrossRef]

77. Weaver, L.M.; Amasino, R.M. Senescence is induced in individually darkened Arabidopsis leaves, but inhibited in whole darkened plants. *Plant Physiol.* **2001**, *127*, 876–886. [CrossRef]

78. Li, Z.; Zhao, T.; Liu, J.; Li, H.; Liu, B. Shade-Induced Leaf Senescence in Plants. *Plants* **2023**, *12*, 1550. [CrossRef]

79. Brouwer, B.; Ziolkowska, A.; Bagard, M.; Keech, O.; Gardestrom, P. The impact of light intensity on shade-induced leaf senescence. *Plant Cell Environ.* **2012**, *35*, 1084–1098. [CrossRef]

80. Liebsch, D.; Juvany, M.; Li, Z.; Wang, H.L.; Ziolkowska, A.; Chrobok, D.; Boussardon, C.; Wen, X.; Law, S.R.; Janeckova, H.; et al. Metabolic control of arginine and ornithine levels paces the progression of leaf senescence. *Plant Physiol.* **2022**, *189*, 1943–1960. [CrossRef]

81. Wu, H.Y.; Liu, L.A.; Shi, L.; Zhang, W.F.; Jiang, C.D. Photosynthetic acclimation during low-light-induced leaf senescence in post-anthesis maize plants. *Photosynth. Res.* **2021**, *150*, 313–326. [CrossRef]

82. Guo, Y.; Gan, S.S. Convergence and divergence in gene expression profiles induced by leaf senescence and 27 senescence-promoting hormonal, pathological and environmental stress treatments. *Plant Cell Environ.* **2012**, *35*, 644–655. [CrossRef]

83. Buchanan-Wollaston, V.; Page, T.; Harrison, E.; Breeze, E.; Lim, P.O.; Nam, H.G.; Lin, J.F.; Wu, S.H.; Swidzinski, J.; Ishizaki, K.; et al. Comparative transcriptome analysis reveals significant differences in gene expression and signalling pathways between developmental and dark/starvation-induced senescence in Arabidopsis. *Plant J.* **2005**, *42*, 567–585. [CrossRef] [PubMed]

84. van der Graaff, E.; Schwacke, R.; Schneider, A.; Desimone, M.; Flugge, U.I.; Kunze, R. Transcription analysis of arabidopsis membrane transporters and hormone pathways during developmental and induced leaf senescence. *Plant Physiol.* **2006**, *141*, 776–792. [CrossRef] [PubMed]

85. Lin, J.F.; Wu, S.H. Molecular events in senescing Arabidopsis leaves. *Plant J.* **2004**, *39*, 612–628. [CrossRef]

86. Khanna, R.; Shen, Y.; Marion, C.M.; Tsuchisaka, A.; Theologis, A.; Schafer, E.; Quail, P.H. The basic helix-loop-helix transcription factor PIF5 acts on ethylene biosynthesis and phytochrome signaling by distinct mechanisms. *Plant Cell* **2007**, *19*, 3915–3929. [CrossRef] [PubMed]

87. Qiu, K.; Li, Z.; Yang, Z.; Chen, J.; Wu, S.; Zhu, X.; Gao, S.; Gao, J.; Ren, G.; Kuai, B.; et al. EIN3 and ORE1 Accelerate Degreening during Ethylene-Mediated Leaf Senescence by Directly Activating Chlorophyll Catabolic Genes in Arabidopsis. *PLoS Genet.* **2015**, *11*, e1005399. [CrossRef] [PubMed]

88. Rauf, M.; Arif, M.; Dortay, H.; Matallana-Ramirez, L.P.; Waters, M.T.; Gil Nam, H.; Lim, P.O.; Mueller-Roeber, B.; Balazadeh, S. ORE1 balances leaf senescence against maintenance by antagonizing G2-like-mediated transcription. *EMBO Rep.* **2013**, *14*, 382–388. [CrossRef] [PubMed]

89. He, Y.; Fukushige, H.; Hildebrand, D.F.; Gan, S. Evidence supporting a role of jasmonic acid in Arabidopsis leaf senescence. *Plant Physiol.* **2002**, *128*, 876–884. [CrossRef]

90. Fukao, T.; Xu, K.; Ronald, P.C.; Bailey-Serres, J. A variable cluster of ethylene response factor-like genes regulates metabolic and developmental acclimation responses to submergence in rice. *Plant Cell* **2006**, *18*, 2021–2034. [CrossRef]

91. Lee, S.C.; Mustroph, A.; Sasidharan, R.; Vashisht, D.; Pedersen, O.; Oosumi, T.; Voesenek, L.A.; Bailey-Serres, J. Molecular characterization of the submergence response of the Arabidopsis thaliana ecotype Columbia. *New Phytol.* **2011**, *190*, 457–471. [CrossRef]

92. Vashisht, D.; Hesselink, A.; Pierik, R.; Ammerlaan, J.M.; Bailey-Serres, J.; Visser, E.J.; Pedersen, O.; van Zanten, M.; Vreugdenhil, D.; Jamar, D.C.; et al. Natural variation of submergence tolerance among Arabidopsis thaliana accessions. *New Phytol.* **2011**, *190*, 299–310. [CrossRef] [PubMed]

93. Gasch, P.; Fundinger, M.; Muller, J.T.; Lee, T.; Bailey-Serres, J.; Mustroph, A. Redundant ERF-VII Transcription Factors Bind to an Evolutionarily Conserved cis-Motif to Regulate Hypoxia-Responsive Gene Expression in Arabidopsis. *Plant Cell* **2016**, *28*, 160–180. [CrossRef]

94. Loreti, E.; Valeri, M.C.; Novi, G.; Perata, P. Gene Regulation and Survival under Hypoxia Requires Starch Availability and Metabolism. *Plant Physiol.* **2018**, *176*, 1286–1298. [CrossRef] [PubMed]

95. Giuntoli, B.; Perata, P. Group VII Ethylene Response Factors in Arabidopsis: Regulation and Physiological Roles. *Plant Physiol.* **2018**, *176*, 1143–1155. [CrossRef]

96. Fukao, T.; Yeung, E.; Bailey-Serres, J. The submergence tolerance regulator SUB1A mediates crosstalk between submergence and drought tolerance in rice. *Plant Cell* **2011**, *23*, 412–427. [CrossRef]

97. Fukao, T.; Yeung, E.; Bailey-Serres, J. The submergence tolerance gene SUB1A delays leaf senescence under prolonged darkness through hormonal regulation in rice. *Plant Physiol.* **2012**, *160*, 1795–1807. [CrossRef]

98. Alpuerto, J.B.; Hussain, R.M.; Fukao, T. The key regulator of submergence tolerance, SUB1A, promotes photosynthetic and metabolic recovery from submergence damage in rice leaves. *Plant Cell Environ.* **2016**, *39*, 672–684. [CrossRef]

99. Jespersen, D.; Yu, J.; Huang, B. Metabolite responses to exogenous application of nitrogen, cytokinin, and ethylene inhibitors in relation to heat-induced senescence in creeping bentgrass. *PLoS ONE* **2015**, *10*, e0123744. [CrossRef] [PubMed]

100. Jespersen, D.; Zhang, J.; Huang, B. Chlorophyll loss associated with heat-induced senescence in bentgrass. *Plant Sci.* **2016**, *249*, 1–12. [CrossRef]

101. Kim, C.; Kim, S.J.; Jeong, J.; Park, E.; Oh, E.; Park, Y.I.; Lim, P.O.; Choi, G. High Ambient Temperature Accelerates Leaf Senescence via Phytochrome-Interacting Factor 4 and 5 in Arabidopsis. *Mol. Cells* **2020**, *43*, 645–661. [PubMed]

102. He, Y.; Zhang, X.; Shi, Y.; Xu, X.; Li, L.; Wu, J.L. Premature Senescence Leaf 50 Promotes Heat Stress Tolerance in Rice (*Oryza sativa* L.). *Rice* **2021**, *14*, 53. [CrossRef] [PubMed]

103. Zhang, S.; Dai, J.; Ge, Q. Responses of Autumn Phenology to Climate Change and the Correlations of Plant Hormone Regulation. *Sci. Rep.* **2020**, *10*, 9039. [CrossRef] [PubMed]

104. Wang, H.; Gao, C.; Ge, Q. Low temperature and short daylength interact to affect the leaf senescence of two temperate tree species. *Tree Physiol.* **2022**, *42*, 2252–2265. [CrossRef]

105. Moore, B.; Zhou, L.; Rolland, F.; Hall, Q.; Cheng, W.H.; Liu, Y.X.; Hwang, I.; Jones, T.; Sheen, J. Role of the Arabidopsis glucose sensor HXK1 in nutrient, light, and hormonal signaling. *Science* **2003**, *300*, 332–336. [CrossRef]

106. Dai, N.; Schaffer, A.; Petreikov, M.; Shahak, Y.; Giller, Y.; Ratner, K.; Levine, A.; Granot, D. Overexpression of Arabidopsis hexokinase in tomato plants inhibits growth, reduces photosynthesis, and induces rapid senescence. *Plant Cell* **1999**, *11*, 1253–1266. [CrossRef]

107. Wingler, A.; Purdy, S.; MacLean, J.A.; Pourtau, N. The role of sugars in integrating environmental signals during the regulation of leaf senescence. *J. Exp. Bot.* **2006**, *57*, 391–399. [CrossRef]

108. Pourtau, N.; Jennings, R.; Pelzer, E.; Pallas, J.; Wingler, A. Effect of sugar-induced senescence on gene expression and implications for the regulation of senescence in Arabidopsis. *Planta* **2006**, *224*, 556–568. [CrossRef]

109. Kim, J. Sugar metabolism as input signals and fuel for leaf senescence. *Genes Genom.* **2019**, *41*, 737–746. [CrossRef]

110. Zhang, J.; Fei, L.; Dong, Q.; Zuo, S.; Li, Y.; Wang, Z. Cadmium binding during leaf senescence in *Festuca arundinacea*: Promotion phytoextraction efficiency by harvesting dead leaves. *Chemosphere* **2022**, *289*, 133253. [CrossRef]
111. Piacentini, D.; Corpas, F.J.; D'Angeli, S.; Altamura, M.M.; Falasca, G. Cadmium and arsenic-induced-stress differentially modulates Arabidopsis root architecture, peroxisome distribution, enzymatic activities and their nitric oxide content. *Plant Physiol. Biochem.* **2020**, *148*, 312–323. [CrossRef] [PubMed]
112. Fei, L.; Zuo, S.; Zhang, J.; Wang, Z. Phytoextraction by harvesting dead leaves: Cadmium accumulation associated with the leaf senescence in *Festuca arundinacea* Schreb. *Environ. Sci. Pollut. Res. Int.* **2022**, *29*, 79214–79223. [CrossRef]
113. Sykorova, B.; Kuresova, G.; Daskalova, S.; Trckova, M.; Hoyerova, K.; Raimanova, I.; Motyka, V.; Travnickova, A.; Elliott, M.C.; Kaminek, M. Senescence-induced ectopic expression of the A. tumefaciens ipt gene in wheat delays leaf senescence, increases cytokinin content, nitrate influx, and nitrate reductase activity, but does not affect grain yield. *J. Exp. Bot.* **2008**, *59*, 377–387. [CrossRef] [PubMed]
114. Calderini, O.; Bovone, T.; Scotti, C.; Pupilli, F.; Piano, E.; Arcioni, S. Delay of leaf senescence in Medicago sativa transformed with the ipt gene controlled by the senescence-specific promoter SAG12. *Plant Cell Rep.* **2007**, *26*, 611–615. [CrossRef] [PubMed]
115. McCabe, M.S.; Garratt, L.C.; Schepers, F.; Jordi, W.J.; Stoopen, G.M.; Davelaar, E.; van Rhijn, J.H.; Power, J.B.; Davey, M.R. Effects of P(SAG12)-IPT gene expression on development and senescence in transgenic lettuce. *Plant Physiol.* **2001**, *127*, 505–516. [CrossRef]
116. Xu, Y.; Burgess, P.; Zhang, X.; Huang, B. Enhancing cytokinin synthesis by overexpressing ipt alleviated drought inhibition of root growth through activating ROS-scavenging systems in *Agrostis stolonifera*. *J. Exp. Bot.* **2016**, *67*, 1979–1992. [CrossRef]
117. Cowan, A.K.; Freeman, M.; Bjorkman, P.O.; Nicander, B.; Sitbon, F.; Tillberg, E. Effects of senescence-induced alteration in cytokinin metabolism on source-sink relationships and ontogenic and stress-induced transitions in tobacco. *Planta* **2005**, *221*, 801–814. [CrossRef]
118. Kant, S.; Burch, D.; Badenhorst, P.; Palanisamy, R.; Mason, J.; Spangenberg, G. Regulated expression of a cytokinin biosynthesis gene IPT delays leaf senescence and improves yield under rainfed and irrigated conditions in canola (*Brassica napus* L.). *PLoS ONE* **2015**, *10*, e0116349. [CrossRef]
119. Khodakovskaya, M.; Li, Y.; Li, J.; Vankova, R.; Malbeck, J.; McAvoy, R. Effects of cor15a-IPT gene expression on leaf senescence in transgenic Petunia x hybrida and Dendranthema x grandiflorum. *J. Exp. Bot.* **2005**, *56*, 1165–1175. [CrossRef]
120. Wang, H.L.; Yang, Q.; Tan, S.; Wang, T.; Zhang, Y.; Yang, Y.; Yin, W.; Xia, X.; Guo, H.; Li, Z. Regulation of cytokinin biosynthesis using PtRD26(pro) -IPT module improves drought tolerance through PtARR10-PtYUC4/5-mediated reactive oxygen species removal in Populus. *J. Integr. Plant Biol.* **2022**, *64*, 771–786. [CrossRef]
121. Ori, N.; Juarez, M.T.; Jackson, D.; Yamaguchi, J.; Banowetz, G.M.; Hake, S. Leaf senescence is delayed in tobacco plants expressing the maize homeobox gene knotted1 under the control of a senescence-activated promoter. *Plant Cell* **1999**, *11*, 1073–1080. [CrossRef]
122. Guo, Y.; Gan, S. AtNAP, a NAC family transcription factor, has an important role in leaf senescence. *Plant J.* **2006**, *46*, 601–612. [CrossRef] [PubMed]
123. Liang, C.; Wang, Y.; Zhu, Y.; Tang, J.; Hu, B.; Liu, L.; Ou, S.; Wu, H.; Sun, X.; Chu, J.; et al. OsNAP connects abscisic acid and leaf senescence by fine-tuning abscisic acid biosynthesis and directly targeting senescence-associated genes in rice. *Proc. Natl. Acad. Sci. USA* **2014**, *111*, 10013–10018. [CrossRef]
124. Shin, D.; Lee, S.; Kim, T.H.; Lee, J.H.; Park, J.; Lee, J.; Lee, J.Y.; Cho, L.H.; Choi, J.Y.; Lee, W.; et al. Natural variations at the Stay-Green gene promoter control lifespan and yield in rice cultivars. *Nat. Commun.* **2020**, *11*, 2819. [CrossRef] [PubMed]
125. Kim, J.; Woo, H.R.; Nam, H.G. Toward Systems Understanding of Leaf Senescence: An Integrated Multi-Omics Perspective on Leaf Senescence Research. *Mol. Plant* **2016**, *9*, 813–825. [CrossRef] [PubMed]
126. Woo, H.R.; Kim, H.J.; Nam, H.G.; Lim, P.O. Plant leaf senescence and death-regulation by multiple layers of control and implications for aging in general. *J. Cell. Sci.* **2013**, *126 Pt 21*, 4823–4833. [CrossRef]
127. Zhang, Y.M.; Guo, P.; Xia, X.; Guo, H.; Li, Z. Multiple Layers of Regulation on Leaf Senescence: New Advances and Perspectives. *Front. Plant Sci.* **2021**, *12*, 788996. [CrossRef]
128. Tseng, Y.C.; Chu, S.W. High spatio-temporal-resolution detection of chlorophyll fluorescence dynamics from a single chloroplast with confocal imaging fluorometer. *Plant Methods* **2017**, *13*, 43. [CrossRef]
129. Iwai, M.; Yokono, M.; Kurokawa, K.; Ichihara, A.; Nakano, A. Live-cell visualization of excitation energy dynamics in chloroplast thylakoid structures. *Sci. Rep.* **2016**, *6*, 29940. [CrossRef]
130. Zhang, Y.; Tan, S.; Gao, Y.; Kan, C.; Wang, H.L.; Yang, Q.; Xia, X.; Ishida, T.; Sawa, S.; Guo, H.; et al. CLE42 delays leaf senescence by antagonizing ethylene pathway in Arabidopsis. *New Phytol.* **2022**, *235*, 550–562. [CrossRef]

International Journal of
Molecular Sciences

MDPI

Review

Roles of S-Adenosylmethionine and Its Derivatives in Salt Tolerance of Cotton

Li Yang [1,†], Xingxing Wang [2,3], Fuyong Zhao [1,†], Xianliang Zhang [2,3], Wei Li [2,3], Junsen Huang [2], Xiaoyu Pei [2], Xiang Ren [2], Yangai Liu [2], Kunlun He [2], Fei Zhang [2], Xiongfeng Ma [2,3,*] and Daigang Yang [2,*]

[1] College of Life Science, Yangtze University, Jingzhou 434025, China; yangli202203@126.com (L.Y.); fyzhao@yangtzeu.edu.cn (F.Z.)

[2] State Key Laboratory of Cotton Biology, Institute of Cotton Research, Chinese Academy of Agricultural Sciences, Anyang 455000, China; wangxingxing@caas.cn (X.W.); zhangxianliang@caas.cn (X.Z.); liwei@caas.cn (W.L.); huangs19980403@163.com (J.H.); peixiaoyu@caas.cn (X.P.); renxiang@caas.cn (X.R.); liuyangai@caas.cn (Y.L.); hekunlun@caas.cn (K.H.); zhangfei@caas.cn (F.Z.)

[3] Western Research Institute, Chinese Academy of Agricultural Sciences (CAAS), Changji 831100, China

*   Correspondence: maxiongfeng@caas.cn (X.M.); yangdaigang@caas.cn (D.Y.)

† These authors contributed equally to this work.

Abstract: Salinity is a major abiotic stress that restricts cotton growth and affects fiber yield and quality. Although studies on salt tolerance have achieved great progress in cotton since the completion of cotton genome sequencing, knowledge about how cotton copes with salt stress is still scant. S-adenosylmethionine (SAM) plays important roles in many organelles with the help of the SAM transporter, and it is also a synthetic precursor for substances such as ethylene (ET), polyamines (PAs), betaine, and lignin, which often accumulate in plants in response to stresses. This review focused on the biosynthesis and signal transduction pathways of ET and PAs. The current progress of ET and PAs in regulating plant growth and development under salt stress has been summarized. Moreover, we verified the function of a cotton SAM transporter and suggested that it can regulate salt stress response in cotton. At last, an improved regulatory pathway of ET and PAs under salt stress in cotton is proposed for the breeding of salt-tolerant varieties.

Keywords: salt stress; S-adenosylmethionine; ethylene; polyamines; SAM transporter; cotton

Citation: Yang, L.; Wang, X.; Zhao, F.; Zhang, X.; Li, W.; Huang, J.; Pei, X.; Ren, X.; Liu, Y.; He, K.; et al. Roles of S-Adenosylmethionine and Its Derivatives in Salt Tolerance of Cotton. *Int. J. Mol. Sci.* **2023**, *24*, 9517. https://doi.org/10.3390/ijms24119517

Academic Editor: Martin Bartas

Received: 13 April 2023
Revised: 19 May 2023
Accepted: 25 May 2023
Published: 30 May 2023

1. Introduction

Cotton is one of the most important cash crops in the world and the main source of natural fiber. With the continuous growth of the world's population, cultivated land decreases year by year, and cotton cultivation has gradually shifted to saline land. It is estimated that more than 50% of arable land will be salinized by 2050 [1]. Although cotton is a moderately salt-tolerant crop with a threshold of 7.7 dSm^{-1}, long-term and high salt stress affects its growth and yield, especially at the stages of germination and seedling [2,3]. Salt stress reduces the stomatal conductance in cotton, leading to a decrease in photosynthesis and reduced carbohydrate supply in developing boll, which ultimately affects the fiber quality and yield of cotton [4,5]. Currently, although a large number of studies have been conducted to analyze the genetic mechanism of salt tolerance in plants, the specific mechanism of salt tolerance in cotton is still unclear.

As genome sequences have increasingly become available for *Gossypium*, a large number of genes have been identified in response to abiotic stress in cotton [6,7]. For example, the NAC transcription factor gene *GhNAC072* has been overexpressed to improve salt and drought tolerance in Arabidopsis [8], and overexpression of *GhMPK3* alleviated the damage of salt stress in Arabidopsis [9]. S-adenosyl-L-methionine (SAM, also known as AdoMet) is a major methyl donor that not only participates in the methionine (Met) cycle, but also provides aminopropyl for the synthesis of PAs and ET [10–12]. A series of studies have revealed that ET is one of the important factors responding to salt stress in many species, as exemplified in Arabidopsis, rice, wheat, cucumber, tomato, and cotton [13–18]. PAs, a group of aliphatic amine compounds similar to phytohormones, exist mainly in three endogenous statuses (free, conjugated, bond) and function in the processes of plant development and stress response [19,20]. In soybean, PAs promote hypocotyl elongation by enhancing their own catabolism reactions and increasing the production of hydrogen peroxide (H_2O_2) under salt stress [21]. Put2 is a polyamine uptake protein, and overexpression of put2 in tomatoes can improve the antioxidant enzyme activity and salt tolerance of transgenic lines [22]. In addition, PAs respond to the saline–alkali stress in rapeseed through the dynamic adjustment of different PA statuses [23]. However, studies on the regulation of salt stress by PAs are still very few in cotton [24].

SAM, as the precursor of ET and PAs synthesis, needs to be transported to the corresponding organelles through SAM transporters to perform a biological function. SAM transporters were first identified in yeast and humans [25,26]. In Arabidopsis, only two SAM transporters (AtSAMC1 and AtSAMC2) have been identified. The impaired function of *AtSAMC1* will reduce the level of prenyl lipids, which mainly affect the chlorophyll pathway, thereby affecting photosynthesis and reducing tolerance to stress [11]. Thus far, studies on SAM transporters have mainly focused on humans and micro-organisms, with the exception of a few studies on Arabidopsis [27–29].

This review focuses on the recent findings on salt tolerance in cotton and systematically introduces the important role of SAM in the response to salt stress through its metabolic derivatives ET and PAs. Furthermore, we identified a putative SAM transporter (*GhSAMC*) from cotton through the transcriptome data of different cotton species and found that it can positively regulate the salt tolerance of cotton. The knowledge presented in this review may be used for breeding elite cotton varieties with high salt tolerance.

2. Research Status of Cotton on Salt Stress

Soil salinization affects about 20% of the world's arable land. This problem continues to exacerbate with global warming, the deterioration of the natural environment, and unreasonable irrigation methods [30,31]. Plants are sessile organisms and cannot escape extreme external environments as animals. Hence, plants have evolved a flexible system that can adjust their morphological, physiological, biochemical, and molecular mechanisms to respond to (a)biotic stresses. Salt stress causes plant cells to suffer from osmotic stress, ion stress, and other secondary stresses, especially oxidative stress [32]. Long-term salt stress will lead to soft and dark leaves in cotton, causing shortened functional periods, early shedding, weak stems, decreased fresh matter and plant height, increased salt concentration, and delayed flowering time [12,33]. In addition, increased relative shedding of flowers and bolls ultimately affects yield and fiber quality in cotton [34] (Figure 1).

2.1. Osmotic Regulation

In the early stage of salt stress, plants are subjected to osmotic stress and accumulate a large number of osmomodulating substances, such as inorganic ions (K^+, Cl^-, and inorganic salts, etc.), and organic solutes (proline, betaine, soluble sugars, polyols, and polyamines, etc.). Plants try to keep osmotic homeostasis depending on the dynamic adjustment of these osmomodulating substances [35].

2.2. Ionic Regulation

With the increase in salt concentration, plant cells accumulate a large amount of Na^+, causing a K^+/Na^+ homeostatic imbalance and Ca^{2+} dysfunction, resulting in ion stress. Plants mitigate ionic toxicity primarily by reducing Na^+ influx, separating and excreting Na^+ [36,37].

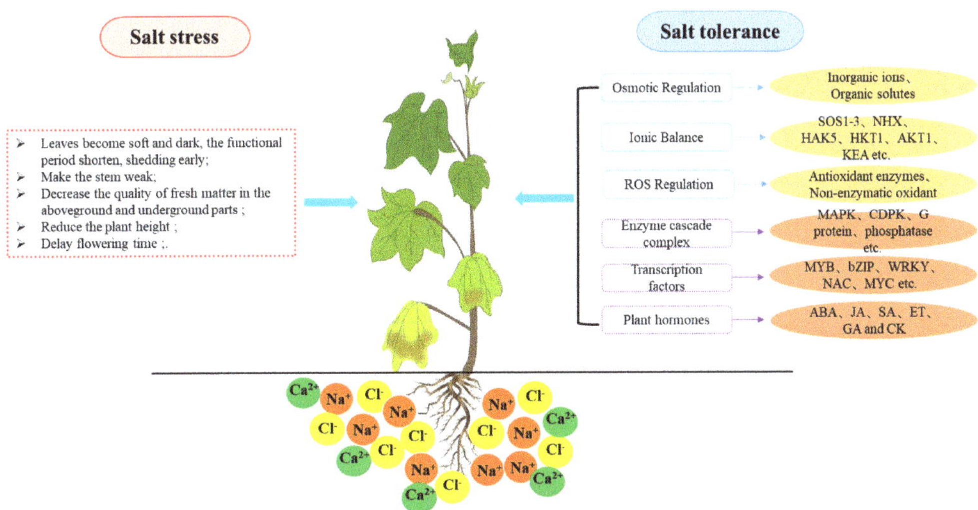

Figure 1. The response to salt stress in cotton. The left part describes the effects of salt stress on the growth and development of cotton; the right part illustrates the related mechanisms and factors of cotton salt tolerance.

At present, dozens of studies have been carried out on ion transporters in cotton. SOS1, a plasma membrane Na^+/H^+ reverse transporter, is responsible for the secretion of Na^+ in the cytoplasm [35]. Overexpression of the cotton SOS1 gene *GhSOS1* in Arabidopsis revealed that the Na^+/K^+ ratio and malondialdehyde (MDA) content decreased in the leaves of transgenic lines treated with salt stress, and the salt-tolerant ability was enhanced [38,39]. NHX, a vacuole Na^+/H^+ anti-transporter, is present in roots and leaves and enables excess Na^+ to be retained in the vacuole. The transport activity of NHX mainly depends on the H^+ gradient maintained by H^+-ATPase and H^+-PPase in the vacuole [40]. In a previous investigation, the overexpression of two cotton homologous genes *GhNHX1* and *GhNHX3D* enhanced the salt tolerance in cotton [41,42]. The K^+ transporter HAK5 and K^+ channel AKT1 are responsible for K^+ uptake in soil, and the Na^+/K^+ transporter HKT is responsible for transferring Na^+ from photosynthetic tissues to roots and extracting Na^+ from xylem [43–45]. While the homologous gene *SbHKT1* was overexpressed in cotton, it can improve the K^+ absorption ability in transgenic lines and maintain the homeostasis of reactive oxygen species (ROS) [46]. The K^+ efflux anti-transporter KEA enhances plant salt tolerance by regulating the homeostasis of K^+ and H^+ in cells. As an example, it was found that silencing of *GhKEA4* and *GhKEA12* reduced proline and soluble sugar content in cotton leaves and increased sensitivity to salt stress [47].

In addition to the above-mentioned genes that have been well studied in regulating Na^+/K^+ homeostasis, some new genes in the regulation of the Ca^{2+} and Na^+ flux have been identified in cotton in recent years, such as the cationic amino acid transporter gene (*GhCAT10D*), cyclic nucleotide-gated channel gene (*GhCNGC1/18/GhCNGC12/31/*), calcium-binding protein gene (*GhCLO6*), annexin gene (*GhANN1/GhANN8b*), and sodium bile acid transporter gene (*GhBASS2*) [48–52]. These observations indicate that cotton has a complex regulatory network to keep ionic homeostasis under salt stress as other plants. Therefore, exploring the key genes in the ionic regulation process is crucial for comprehensively dissecting the regulatory mechanism of salt tolerance in cotton.

2.3. Oxidation Regulation

With the extension time of salt stress, osmotic and ionic stresses will induce oxidative stress, causing the accumulation of a large number of ROS in plants. Additionally, chloroplasts, mitochondria, and peroxisomes are the three main sites for ROS accumulation. ROS mainly includes singlet oxygen ($^1O2$), superoxide ($O_2^{\bullet-}$), H_2O_2, and hydroxyl radicals (OH•) [53]. Under normal growth conditions, low concentrations of ROS in plants can be used as second messengers, involved in seed germination, growth and development, root growth and gravitropism, programmed cell death, and other processes [54,55]. Under salt stress conditions, plants accumulate a large amount of ROS. High concentrations of ROS can lead to protein denaturation, lipid peroxidation, and nucleic acid damage [56,57]. Although cotton is a relatively salt-tolerant crop, when cotton is in a high-salt environment, a large amount of ROS is produced in the plant. Excessive ROS can affect the growth and development of cotton plants and eventually lead to a decrease in cotton yield and fiber quality [58].

In order to avoid the damage caused by ROS, plants synthesize a series of antioxidant enzymes, such as superoxide dismutase (SOD), peroxidase (POD), catalase (CAT), ascorbyl peroxidase (APX), glutathione peroxidase (GPX), and non-enzymatic antioxidants such as carotenoids, phenolic compounds, flavonoids, ascorbic acid (AsA), and glutathione (GSH) [59,60]. In addition, the expression of related genes in plants can regulate the level of ROS and improve the tolerance of plants to salt stress. Recent studies have shown that *ghr-miR414c* can further regulate ROS metabolism by regulating the expression level of *GhFSD1* to affect the salt tolerance of cotton [61]. Cotton WRKY transcription factor *GhWRKY17* can improve the salt sensitivity of transgenic tobacco through ABA signal transduction and regulation of ROS production in plant cells [62]. Under salt stress conditions, the aldehyde dehydrogenase 21 gene (*ScALDH21*) of *S. canadensis* can scavenge excessive ROS in transgenic cotton, thereby improving the salt tolerance of cotton [63]. Silencing the Raf-like MAPKKK gene *GhRaf19* reduced the accumulation of ROS in cotton, thereby improving the tolerance of plants to salt stress [64].

2.4. Signal Transduction Regulation

Over the long evolution, plants have evolved various mechanisms to deal with adversity, such as the accumulation of permeable substances, the regulation of genes associated with Na^+/K^+ and Ca^{2+} homeostasis, and the production of antioxidant enzymes and non-enzymatic antioxidants. Each mechanism in above contains a complex signaling pathway, consisting of receptors, secondary messengers, phytohormones, and signal transductors [35]. Of these, the signal transductors include many enzymatic cascades such as mitogen-activated protein kinase (MAPK), calcium-dependent protein kinase (CDPK), G protein, phosphatase, and other components [65,66]. Various transcription factors (TFs), such as MYB [67,68], bZIP [69,70], WRKY [71–73], NAC [74–76], MYC [77], etc., act as a central regulator and a molecular switch in the signal transduction network under salt stress. These TFs regulate the expression pattern of downstream genes and affect the salt tolerance level by activating or inhibiting their interacting genes [78,79]. In addition, various phytohormones such as abscisic acid (ABA), jasmonic acid (JA), salicylic acid (SA), ethylene (ET), gibberellin (GA), and cytokinin (CK) also participate in the signal transduction of salt stress to regulate the response to salt stress [80,81].

At present, the salt tolerance of cotton can be improved to a certain extent through traditional breeding methods, molecular marker-assisted selection, transgenic breeding, and gene editing. However, salt tolerance in plants is a complex trait controlled by multiple quantitative trait loci (QTL), determined by the response of the whole plant [82]. It is very hard to significantly improve the salt-tolerant ability with changes in a single gene or protein. These require us to further explore the molecular mechanism of cotton salt tolerance to breed varieties with high salt-tolerant capability.

3. Role of SAM in Plant Salt Tolerance

As a precursor for the biosynthesis of ET and PAs, SAM is synthesized from the substrates ATP and Met by the catalysis of S-adenosylmethionine synthetase (SAMS) [83]. Related studies have shown that SAMS regulates plant responses to salt stress by promoting the synthesis of SAM. Heterologous expression of potato *SbSAMS* in Arabidopsis led to the accumulation of more SAM, S-adenosyl-L-homocysteine (SAHC), and ET in transgenic lines, showing higher salt and drought tolerance levels [83]. In another study, the overexpression of beet *BvM14-SAMS2* in Arabidopsis improved SAM content and the tolerant ability to salt in transgenic plants, in which the antioxidant system and polyamines' metabolism played an important role [84].. The advanced analysis found that the biosynthesis of PAs and 1-aminocyclopropane-1-carboxylic acid (ACC) was related to the *CsSAMS1* expression level in transgenic plants [85,86].

4. Role of Ethylene (ET) in Plant Growth and Development of Cotton under Salt Stress

When plants are exposed to salt stress, it leads to elevated ROS content in the body, especially including H_2O_2 and $O_2^{\bullet-}$; these increased ROS have a dual effect: when they are at low concentrations, they can act as signaling molecules that mediate salt tolerance, whereas they will produce oxidative damage to plant cells when at high concentrations, thereby disrupting the function of chloroplasts and even triggering plant cell death (PCD) [87–89]. As a volatile gas plant hormone, ET is usually considered as another stress hormone in addition to ABA, and its synthesis is induced by various (a)biotic environmental stresses [90,91]. ET also has a dual role in regulating salt stress: one is inhibiting the accumulation of ROS through ET signaling pathways, thereby maintaining the homeostasis of ROS in plants, and the other is promoting ROS production and signal transduction to activate Na^+/K^+ transport [92–94].

4.1. ET Biosynthesis

The biosynthesis of ET begins with SAMS catalyzing the conversion of Met to SAM. Next, the SAM is converted into ACC by 1-aminocyclopropane-1-carboxylic acid (ACC) synthetases (ACSs). Finally, using ACC as the substrate, ET is synthesized through ACC oxidases (ACOs) (Figure 2) [95–97].

ACSs and ACOs are commonly considered key biosynthetic enzymes in ET biosynthesis. Thus far, 12 *ACS* homologous genes have been identified in Arabidopsis [98]. Among these *ACS* genes, *AtACS1* encodes non-functional homodimers or inactivating catalytic enzymes, *AtACS3* is a pseudogene, *AtACS10* and *AtACS12* perform amino transfer functions, and the other 8 ACS genes (*AtACS2*, *AtACS4-9*, *AtACS11*) encode functional ACS [99]. Moreover, the presence of these eight functional active ACS enzymes and their ability to form active heterodimers will increase the diversity of ET biosynthetic reactions and improve the ability to regulate ET production at different developmental and environmental stages [100,101]. A total of 35 GhACS proteins with a conserved C-terminus were identified in upland cotton, whereas the N-terminus of ACS10 and ACS12 were divergent between cotton and Arabidopsis, indicating that they may have a different evolutionary trajectory. QRT-PCR analysis showed that the expression of *GhACS1* was significantly upregulated under salt stress, whereas *GhACS6.3*, *GhACS7.1*, and *GhACS10/12* could respond to cold stress [102]. ACO also plays an important role in ET biosynthesis. Arabidopsis encodes a total of five *ACO* genes, whose expression is regulated by growth and developmental and environmental stresses [103]. Through genome-wide identification, a total of 332 *GhACOs* were identified in upland cotton, and most of them co-expressed with other genes in response to salt and drought stresses. For example, the overexpression of *GhACO106-At* could improve salt tolerance in transgenic Arabidopsis [104]. At present, the ET biosynthetic pathway has been well elucidated in model species, but there is still little research on genes that promote ET biosynthesis and respond to salt stress in cotton, and further exploration is needed in the future.

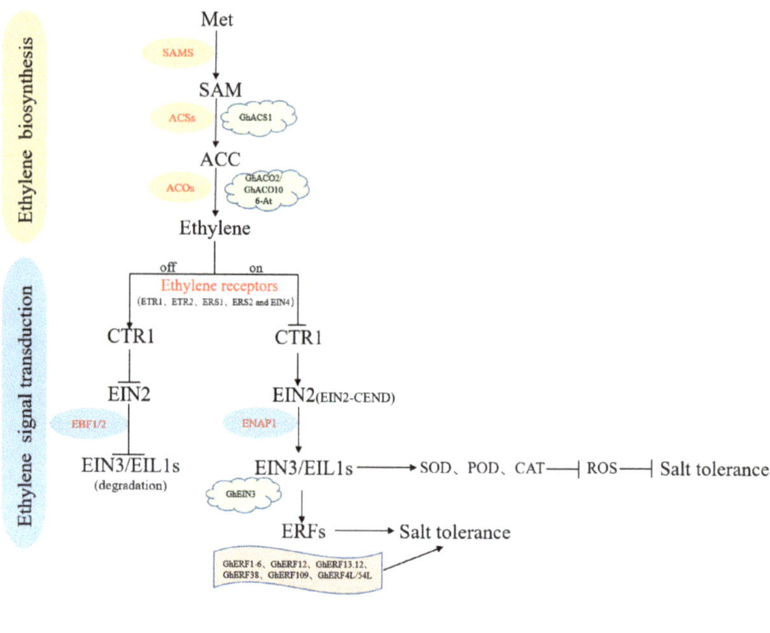

Figure 2. Ethylene biosynthesis and signal transduction. Met: methionine; SAM: S-adenosylmethionine; ACC: 1-Aminocyclopropanecarboxylic Acid; SAMS: SAM synthetase; ACSs: ACC synthases; ACOs: ACC oxidases; CTR1: Constitutive triple response1; EIN2: Ethylene-Insensitive2; EIN3/EIL1: Ethylene-Insensitive3/EIN3-like1; ENAP1: EIN2 nuclear-associated protein1; EBF1/2: EIN3 binding F-box protein1/2; ERFs: Ethylene-response factors.

4.2. ET Signaling Transduction

ET can be sensed and bound by ET receptors that are mainly located on the endoplasmic reticulum membrane to stimulate ET response [105,106]. In Arabidopsis, five ET receptors composed of Ethylene response 1 (ETR1), ETR2, Ethylene response sensor1 (ERS1), ERS2, and Ethylene insensitive 4 (EIN4) negatively regulate ET reactions [87,107,108]. ET activates downstream signaling pathways by inactivating these receptors. In the absence of ET, constitutive triple response factor 1 (CTR1) is activated by ethylene receptors and further phosphorylates the C-terminus of Ethylene-Insensitive 2 (EIN2-CEND), causing it to become inactive [109]. Ethylene-Insensitive 3 (EIN3) and EIL1 (EIN3-like1), downstream of EIN2, are degraded by two F-box proteins, EBF1/2 (EIN3 binds to F-box1/2) in the nucleus, thereby blocking the transcription of downstream target genes in the ethylene response [110–112]. Under stress conditions, ET increases in plants and is bound to ethylene receptors. As a result, CTR1 is inactivated. Subsequently, EIN2 is dephosphorylated and cleaved to release the EIN2-CEND, which can shuttle into the nucleus with the assistance of EIN2 nuclear-associated protein 1 (ENAP1) and enhance the activity of EIN3/EIL1 [113]. Due to the cascade reaction, downstream target genes that respond to ET, such as ethylene-response factors (ERFs), are regulated and then affect plant growth and development [114–117]. A summary of ethylene signal transduction pathways is shown in Figure 2.

CTR1 is a Raf-like Ser/Thr protein kinase that negatively regulates the ethylene signal transduction pathway, whereas *GhCTR1* expression was not induced by ET or salt but by SA, GA, and ABA [118]. EIN3/EIL1 are positive regulators of the ethylene signaling pathways and have been characterized in a lot of plants. Under the stimulation of exogenous ACC, the homologous gene *GhEIN3* of EIN3/EIL1 could be induced. A previous study showed that the transgenic Arabidopsis-containing *GhEIN3* had strong salt tolerance, indicating that *GhEIN3* may play a role in plant responses to salt stress by regulating the ROS and ABA pathways [119]. ERFs belong to the APETALA2(AP2)/ERFs family and play an important role in plant stress signaling pathways [120]. Thus far, a series of ERFs-related genes and their functions have been identified and analyzed in cotton, such as the overexpression of *GhERF38* and *GhERF13.12* in Arabidopsis, which improved salt tolerance and revealed dynamic changes in relevant biochemical parameters in transgenic lines [121,122]. In another study, virus-induced silencing of cotton *GhERF12* enhanced the sensitivity of transgenic plants to salt stress and reduced the activity of SOD and POD [123]. Similarly, the silencing of *GhERF4L* and *GhERF54L* in cotton led to reduced resistance to salt stress [124]. In addition, studies had found that each ERF transcription factor contained an AP2 domain, which was the DNA binding domain of ERFs and the key to the regulation of ERFs transcription [125]. A total of 220 genes containing a single AP2 domain sequence has been identified in upland cotton. Transcriptome and qRT-PCR analysis found that *Ghi-ERF2D.6, Ghi-ERF-12D.13, Ghi-ERF-6D.1, Ghi-ERF-7A.6,* and *Ghi-ERF11D.5* could respond to salt stress [126]. Transcriptome meta-analysis and network topology analysis showed that salt stress could increase the expression level of *GhERF109*, and further qRT-PCR analysis showed that *GhERF109* could indeed respond to salt stress [127]. Previous studies have found that *GhERF1, GhERF2, GhERF3, GhERF4, GhERF5,* and *GhERF6* mediate plant tolerance to salt and drought [128–131]. ET correlation factors in cotton are summarized in Table 1.

Table 1. The functions of ethylene-related factors in cotton.

Gene Name	Experimental Methods	Biological Function	Ref.
GhACS1	RNA-Seq data analysis and qRT-PCR analysis	Responsed to salt stress	(Li J et al., 2022 [102])
GhACO106-At	Overexpression in Arabidopsis	Promoted flowering and increased salt tolerance	(Wei H et al.,2021 [104])
GhEIN3	Overexpression in Arabidopsis VIGS in Cotton	Regulated ROS pathway and ABA signaling to response to salt	(Wang X et al., 2019 [119])
GhERF38	Overexpression in Arabidopsis	Responsed to salt/drought stress and ABA signaling	(Ma L et al., 2017 [121])
GhERF13.12	Overexpression in Arabidopsis VIGS in Cotton	Regulated ROS pathway and ABA signaling to response to salt stress	(Lu L et al., 2021 [122])
GhERF12	VIGS in Cotton	Regulated ROS pathway to response to salt stress	(Zhang J et al., 2021 [123])
GhERF4L/54L	VIGS in Cotton	Responsed to salt stress	(Long L et al., 2019 [124])
Ghi-ERF-2D.6/12D.13/6D.1/7A.6/11D.5	RNA-Seq data analysis and qRT-PCR analysis	Responsed to salt stress	(Zafar M et al., 2022 [126])
GhERF109	RNA-Seq data analysis and qRT-PCR analysis	Responsed to salt stress	(Bano N et al., 2022 [127])
GhERF1	Semi-qRT-PCR analysis	Responsed to ET, ABA, salt, cold, and drought stress	(Qiao Z et al., 2008 [128])
GhERF2/3/6	Semi-qRT-PCR analysis	Responsed to ET, ABA, salt, cold, and drought stress	(Jin L et al., 2010 [129])

Table 1. *Cont.*

Gene Name	Experimental Methods	Biological Function	Ref.
GhERF4	Semi-qRT-PCR analysis	Responded to ET, ABA, salt, cold, and drought stress	(Jin L and Liu J., 2008 [130])
GhERF5	Semi-qRT-PCR analysis	Responded to ET, ABA, salt, cold, and drought stress	(Jin L et al., 2009 [131])

5. Roles of Polyamines (PAs) in Plant Growth and Development under Salt Stress

PAs are a class of low-molecular aliphatic polycations with strong biological activity, widely present in various prokaryotes and eukaryotes [132]. PAs mainly consist of putrescine (Put, a diamine), spermidine (Spd, a triamine), spermine (Spm, a tetramine), and thermospermine (a structural isomer of spermine) [133,134]. In plants, PAs participate in a variety of physiological processes such as organogenesis, embryogenesis, flower development, leaf aging, fruit ripening, etc. [20,135,136]. Under abiotic stress, PAs can not only bind to negatively charged membrane phospholipids to maintain the permeability of cell membranes and the function of membrane proteins but also bind to anionic macromolecules (nucleic acids, proteins, etc.) to affect their physiological functions [137]. Moreover, PAs can partially replace the function of antioxidant enzymes to remove ROS and reduce oxidative damage [138,139]. Up to date, several reports have revealed that PAs regulate plant morphological growth parameters, photosynthetic pigment contents, stress-related indicators, antioxidant enzyme contents, and non-enzymatic antioxidant contents in rapeseed [140], cucumber [141], wheat [142], rice [143] and *Calendula officinalis* L. [144].

5.1. PAs Biosynthesis

5.1.1. Put Biosynthesis

Put is a central product in PAs' biosynthetic pathways [145]. Put mainly has three biosynthetic pathways. The first is the direct synthesis of Put from ornithine (ORN) catalyzed by ornithine decarboxylase (ODC) [146]. The second is to synthesize Put from arginine (Arg) under the ordered catalysis of arginine decarboxylase (ADC), argmatine iminohydrolase (AIH), and N-carbamoylputrescine amidohydrolase (CPA) [147]. In the last, Arg first degraded to citrulline (Cit) catalyzed by Arginine deaminase (ADI) and then catalyzed into Put by citrulline decarboxylase (CDC) decarboxylase (Figure 3) [148].

The Arabidopsis genome contains two ADCs genes (*AtADC1* and *AtADC2*), which have different expression patterns. *AtADC1* was almost not expressed in seeds, roots, and leaves, whereas *AtADC2* was highly expressed in these tissues [149]. Under salt stress, the overexpression of *AtADC2* can regulate the activity of SOD and CAT, thereby improving the salt tolerance of Arabidopsis [150]. The virus-induced silencing of cotton *GhADC2* enhanced plant tolerance to salt stress and reduced H_2O_2 content [24].

5.1.2. Spd and Spm Biosynthesis

Put, as a synthetic precursor for Spd and Spm, is catalyzed into Spd by Spd synthetase (SPDS), and then Spm synthase (SPMS) catalyzes Spd into Spm [151,152]. During this process, decarboxyl-SAM (dcSAM), a product catalyzed from SAM by SAM decarboxylase (SAMDC), provides aminopropyl groups for SPDS and SPMS-catalyzing reactions [153] (Figure 3).

Among these genes encoding key enzymes, SAMDC plays a crucial role in regulating the biosynthesis of Spd and Spm [136]. For example, the overexpression of *FvSAMDC* enhanced salt tolerance in tobacco [154]. In Arabidopsis, the overexpression of *CsSAMDC2* exhibited higher levels of salt and drought tolerance than its wild type [155]. Meanwhile, the ectopic expression of *GhSAMDC3* can improve salt tolerance in Arabidopsis by increasing Spd content and activating salt-tolerance-related genes [156]. In addition, studies have shown that *AhSAMDC* and *TrSAMDC1* can improve plant resistance to salt stress by increasing PAs content and antioxidant enzyme activities [157,158].

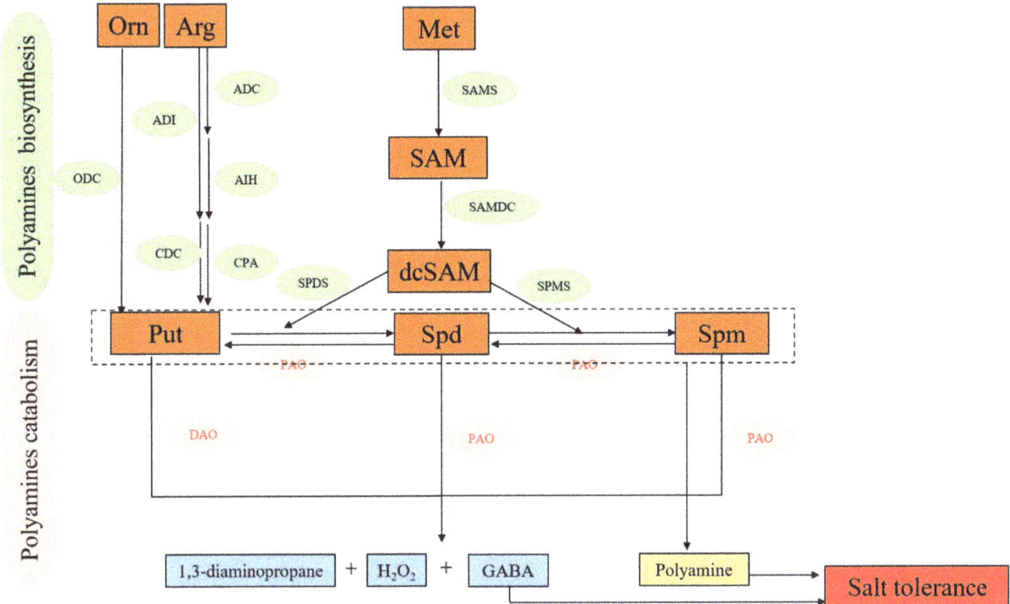

Figure 3. Polyamines biosynthesis and catabolism in plants. Met: methionine; SAM: S-adenosylmethionine; dcSAM: decarboxylated S-adenosylmethionine; GABA: γ-aminobutyric acid; ODC: ornithine decarboxylase; ADC: arginine decarboxylase; CDC: citrulline decarboxylase; AIH: argmatine iminohydrolase; CPA: N-carbamoylputrescine amidohydrolase; ADI: Agmatine deiminase; SAMDC: S-adenosylmethio-Nine decarboxylase; SPDS: spermidine synthase; SPMS: spermine synthase; DAO: Diamine oxidase; PAO: Polymine oxidase.

5.2. PAs Catabolism

The catabolism of PAs in plants relies on the oxidative deamination catalyzed by amino oxidases (AOs), mainly including cupramine oxidases (DAOs) and FAD-dependent polyamine oxidases (PAOs) [152,159]. DAOs convert Put into H_2O_2, amino, and Δ1-pyrroline. Δ1-pyrroline can be further converted to γ-aminobutyric acid (GABA) [160]. PAOs mainly decompose Spd and Spm to produce 1,3-diaminopropane and H_2O_2 (Figure 3) [161]. Five PAOs (*AtPAO1* to *AtPAO5*) have been identified in Arabidopsis, of which AtPAO1 and AtPAO5 are located in the cytoplasm, and AtPAO2 to AtPAO4 are located in peroxisomes [162–164].

PAOs play an important role in plant growth, development, and stress response. For example, *CsPAO2* can interact with *CsPSA3* to improve salt tolerance of cucumber by affecting photosynthesis and promoting PAs conversion [165]. Furthermore, the overexpression of *CsPAO3* in Arabidopsis can promote seed germination and root growth in NaCl-containing media and alleviate growth inhibition induced by salt stress [166]. Liu et al. [167] found that *OsPAO3* in rice was upregulated under salt stress during the germination stage, which increased PAO activity and led to an increase in PAs content of in salt-tolerant strains. Meanwhile, the increased PAs could improve the activity of ROS-scavenging enzymes to eliminate the excessive accumulation of H_2O_2 and ultimately elevate the salt tolerance of rice during germination. When *GhPAO3* was overexpressed in Arabidopsis, the salt tolerance was promoted [168]. In addition, PAs can also improve plant resistance to diseases and drought [169–172]. The above genes related to salt stress in P biosynthesis and catabolism in various crops are summarized in Table 2.

Table 2. Polyamines (PAs) biosynthesis and catabolism genes in mitigating salt stress in various crops.

PAs	Crops	Genes	Genes Response to Salt Stress	Ref.
Biosynthesis genes	Arabidopsis thaliana	*AtADC2*	Improved SOD and CAT activities	(Fu Y et al., 2017 [150])
	Cotton	*GhADC2*	Increased H_2O_2 content and oxidative stress	(Gu Q et al., 2021. [24])
	Fragaria vesca	*FvSAMDC*	Reduced H_2O_2 and $O_2^{\bullet-}$ content	(Kov'acs L et al., 2020. [154])
	Cleistogenes songorica	*CsSAMDC2*	Improved chlorophyll content and Photosynthetic capability	(Wu F et al., 2022. [155])
	Cotton	*GhSAMDC3*	Increased Spd content	(Tang X et al., 2021. [156])
	Peanut	*AhSAMDC*	Improved activities of antioxidant enzymes Increased Spd and Spm content	(Meng D et al., 2021. [157])
	White clover	*TrSAMDC1*	Improved SOD, POD, and CAT activities Reduced MDA and H_2O_2 content	(Jia T et al., 2021. [158])
Catabolic genes	Arabidopsis thaliana	*AtPAO1*	Increased ROS and H_2O_2 content	(Sagor G et al., 2016. [164])
	Cucumber	*CsPAO2*	Improved activities of antioxidant enzymes Reduced MDA content	(Wu J et al., 2022. [165])
	Cucumber	*CsPAO3*	Improved POD and CAT activities Reduced MDA and H_2O_2 content	(Wu J et al., 2022. [166])
	Rice	*OsPAO3*	Increased PAs content Improved Polyamine oxidase activities	(Liu G et al., 2022. [167])
	Cotton	*GhPAO3*	Increased PAs content	(Cheng X et al., 2017. [168])

6. Cross-Talk between ET and PAs under Salt Stress

ET and PAs share the same precursor SAM, indicating that changes in ET content may affect the homeostasis of PAs [173]. An increasing number of studies have found that salt treatment led to an increase in ET, ACC, Spd, and Spm in peppers, lettuce, spinach, and beetroot, indicating that the SAM pool is high enough to support ET and PAs biosynthesis under conditions, and ET and PAs can coexist without competition [174]. It can be seen that the interaction between ET and PAs depends more on the content of SAM in plants. In addition, co-regulation of endogenous ET and PAs in plants can enhance their tolerances to salt stress. For example, the interaction between *CsCDPK6* and *CsSAMS1* negatively regulates the biosynthesis of plant ET, inhibiting SAMDC, resulting in a decrease in Put conversion to Spd and Spm [86]. Wu et al. [155] overexpressed the PAs related gene *CsSAMDC2* in Arabidopsis to improve salt tolerance, and the expression levels of the ET and ABA responding genes also increased. In a recent report, Freitas et al. [175] found that ET regulated the H_2O_2 content derived from PA decomposition enzymes, inducing salt tolerance in maize. Moreover, ET participated in salt-induced oxidative stress in ripening tomato fruits and regulated the metabolic level of PAs [176]. In summary, a series of key genes in the biosynthesis pathways of ET and PAs have been studied to dissect the relationship between ET and PAs. However, the exact interaction mechanism between ET and PAs under salt stress still needs further research.

7. Roles of SAM Transporters in Plants

SAM is only synthesized in the cytoplasm and is necessary for the methylation of DNA, RNA, and protein in chloroplasts and mitochondria. Chloroplasts and mitochondria are semi-autonomous organelles and need the help of SAM transporters to transfer SAM into them [177,178]. Two genes, *AtSAMC1* (*At4g39460*) and *AtSAMC2* (*At1g34065*), homologous to yeast (*Sam5p*) and mammalian (*SAMC*) SAM transporters, are present in Arabidopsis. AtSAMC1 is located in the plastid, whereas AtSAMC2 is the dual localization of plastid and mitochondrial membranes [28]. The expression of *AtSAMC2* was almost undetectable in roots, stems, leaves, flowers, and seedlings, whereas *AtSAMC1* was expressed in the above organs and especially had a very high expression in seedlings [179]. Chloroplast SAM transporters are important bonds in many biological processes between cytoplasm and chloroplasts, such as the single-carbon metabolism for maintaining methylation reactions and other SAM-dependent functions in chloroplasts, as well as removing the byproduct SAHC derived from the methylation reaction [180]. Mitochondrial SAM transporters belong to the mitochondrial carrier (MC) family and are important for the methylation reaction in mitochondria [181]. SAM transporters are mainly responsible for the input of SAM in organelles and the output of SAHC to maintain the normal metabolism homeostasis of SAM in plant cells. Once the SAM transporter loses function, it will affect the synthesis of SAM, which is a precursor for ethylene, polyamines, glycine betaine, and lignin. Then, it may ultimately affect the survival of plants in adversity.

We screened out the salt-tolerant genes through comparative transcriptome analysis in *G. hirsutum*, *G. hirsutum* race *marie-galante*, *G. tomentosum*, and *G. barbadense* (PRJNA933089). Among them, *Gh_A05G2087* (named *GhSAMC*) was found to be a homologous gene of *AtSAMC1* and *AtSAMC2* in cotton after sequence alignment (Supplementary Figure S1). To investigate whether *GhSAMC* was affected by salt stress in cotton leaves, the TM-1 seedlings of upland cotton were first treated with 250 mM NaCl solution and then sampled at 0 h, 3 h, 6 h, 12 h, and 24 h for qRT-PCR analysis(The primers in the Supplementary Table S1). The results showed that *GhSAMC* was affected by salt stress, and its expression revealed a trend that it was rising first, reaching its peak at 6 h, and then falling with the extension of salt stress time (Figure 4E). We further performed virus induced gene silencing (VIGS) to test whether *GhSAMC* was necessary for cotton to tolerate salt stress (Materials and Methods in Supplementary). When newly emerged leaves were infiltrated with TRV:CLA1, they exhibited an albino phenotype (Figure 4A). The expression levels of *GhSAMC* were measured by qRT-PCR to confirm the silencing efficiency, and the results showed that *GhSAMC* had been silenced completely (Figure 4B). Treated *GhSAMC* silencing (TRV:SAMC) and control (TRV:00) plants with 250 mM NaCl solution at the three-leaf stage, it was found that TRV:SAMC seedlings exhibited more serious symptoms of wilting and lodging than TRV:00 plants (Figure 4C,D). Subsequently, we detected the malondialdehyde (MDA) content and total antioxidant capacity (T-AOC) of VIGS cotton plant leaves (Figure 4F,G), and the results showed that TRV:SAMC accumulated more ROS than TRV:00 seedlings. Therefore, we speculated that *GhSAMC* could regulate the accumulation of ROS under salt stress conditions to improve the salt tolerance of cotton. However, further research is needed to determine whether *GhSAMC* affects the synthesis of ET and PAs by regulating the transport of SAM, thereby regulating salt stress.

Furthermore, we carried out the transcriptomic analysis for leaves of VIGS plants (TRV:SAMC, TRV:00) treated with 250 mM of saline solution (PRJNA937453). Compared with CK (TRV:00), a total of 1169 differentially expressed genes (DEGs) were identified (Supplementary Table S2), including 470 upregulated genes and 699 downregulated genes (Figure 5C). GO enrichment analysis revealed that the DEGs were mainly divided into three categories, exemplified as biological processes, cell compositions, and molecular functions (Figure 5A). Moreover, GO categories were concentrated in the functional groups of oxidoreductase activity, oxidative stress response, cell membrane components, and cell walls. These observations indicated that the cell walls and membranes first played a barrier role in response to salt stress, and then the oxidoreductase genes were mobilized into cells to cope with the accumulation of ROS. Thus, the transcriptome data implied that *GhSAMC* may alleviate salt damage by regulating ROS. Plants can respond to salt stress through three ways: osmotic regulation, ion regulation, and oxidative regulation. Therefore, we mapped the differential genes related to these three regulatory methods into a heat map (Supplementary Figure S2). Among them, there were many DEGs related to oxidation–reduction, mainly including the POD and cytochrome P450 genes. The KEGG results enriched 22 metabolic pathways, including lipid and amino acid metabolism, glycolysis/gluconeogenesis, starch and sucrose metabolism, and carbon metabolism, suggesting that *GhSAMC* may be involved in regulating these metabolic pathways to help cotton cope with salt stress (Figure 5B).

Figure 4. *GhSAMC* silencing in cotton increases the sensitivity to salt stress. (**A**) Albino phenotype of TRV:CLA1 about two weeks after infiltration. (**B**) The silencing efficiency of *GhSAMC*. (**C**) Phenotype of silencing cotton seedlings before salt treatment. (**D**) Phenotype of silencing cotton seedlings after salt treatment. (**E**) Expression of *GhSAMC* over time under treatment with 250 mM saline solution. (**F**) MDA content of silencing cotton seedlings after salt treatment. (**G**) T-AOC content of silencing cotton seedlings after salt treatment (Values are means ± s.e.m (n = 3 biological replicates). Error bars represent the SD of three biological replicates. * $p < 0.05$ (Statistically significant), ** $p < 0.01$ (Statistically highly significant)).

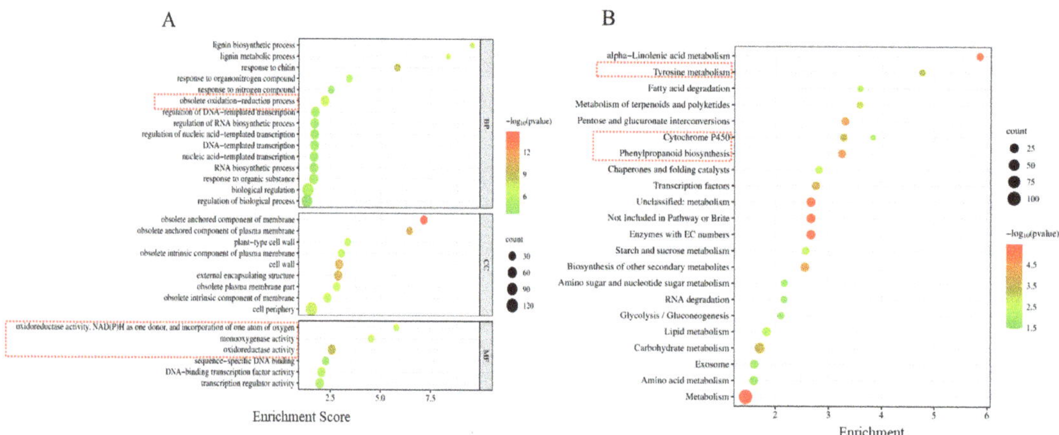

Figure 5. *Cont.*

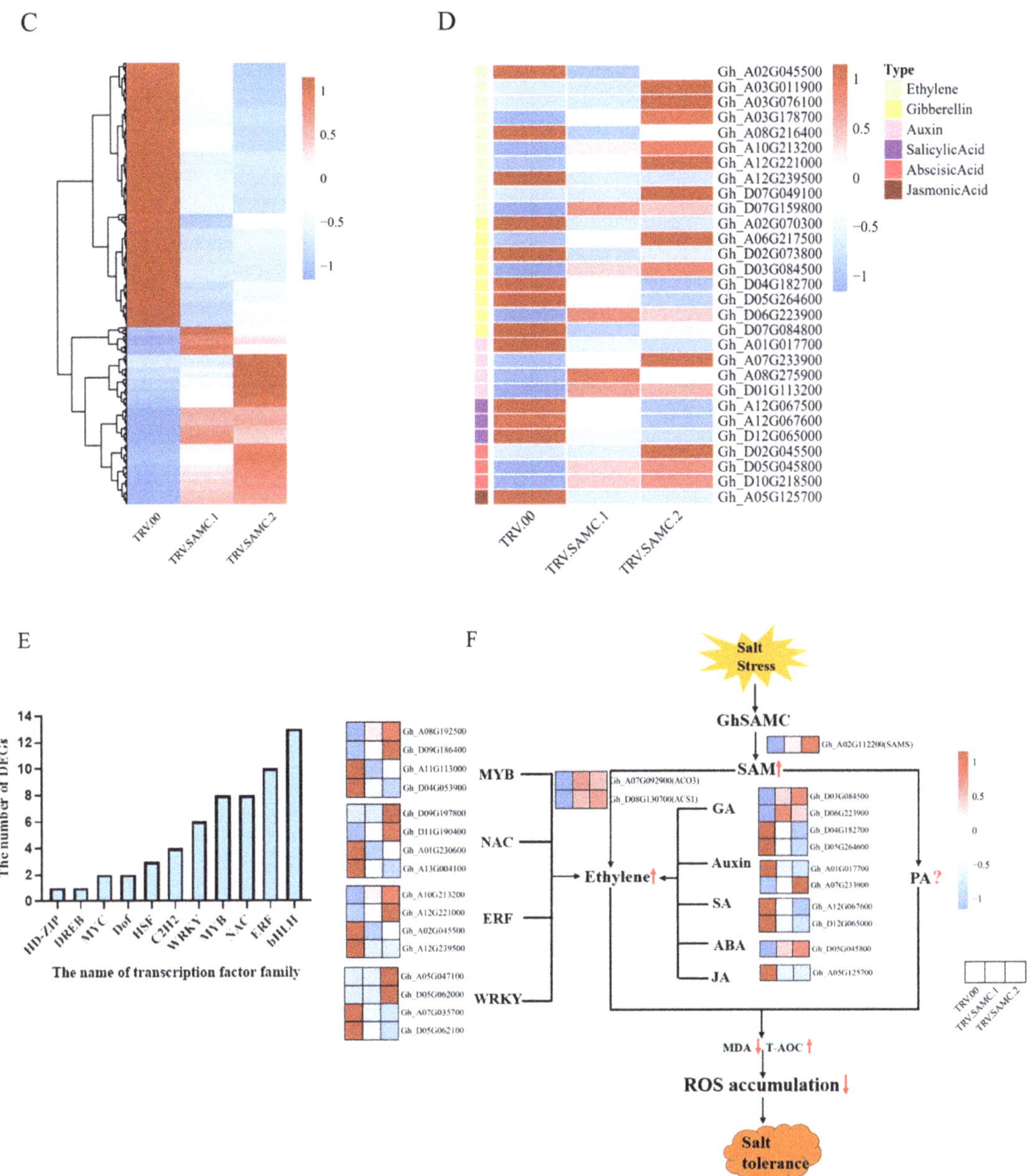

Figure 5. Transcriptome analysis of cotton leaves under salt stress. (**A**) GO enrichment analysis of differentially expressed genes. In the red boxes are some oxidation-reduction related pathways (**B**) KEGG enrichment analysis of differentially expressed genes. The three pathways in the red box, tyrosine metabolism, cytochrome P450, and phenylpropanoid metabolism, play an important role in salt stress. (**C**) The heat map of differentially expressed genes in cotton leaves under salt stress. (**D**) Heatmap of differentially expressed genes related to phytohormone. (**E**) Statistics of the number of differentially expressed transcription factors. (**F**) Model diagram of *GhSAMC* regulation in cotton under salt stress. The red upward arrow indicates an increase in the content of the substance. The red downward arrow indicates an decrease in the content of the substance. ? means that it is uncertain whether the content of the substance is increase or decrease.

The tolerance of plants to salt stress is regulated by a series of TFs. Through TFs analysis, 58 TFs related to salt stress were screened, including HD-ZIP, DREB, MYC, Dof, HSF, C2H2, WRKY, MYB, NAC, ERF, and bHLH (Figure 5E). Among them, the TFs families of bHLH, ERF, NAC, and MYB contained significantly more DEGs than other families, including 13, 10, 8, and 8 DEGs, respectively. Previous studies have shown that HD-ZIP [182,183], DREB [184,185], MYC [186,187], Dof [188,189], HSF [190,191], C2H2 [192,193], WRKY [194,195], MYB [196,197], NAC [198,199], and bHLH [200,201] could work alone or synergistically to regulate plant tolerance to salt stress. Additionally, the above TFs could crosslink with ET [202–208]. For example, the apple NAC transcription factor (*MdNAC047*) enhances plant resistance to salt stress by increasing ET synthesis in plants [209]. *MYB108A* can activate the expression of *ACS1* to promote the formation of ET and ultimately improve the salt tolerance of grapes [210]. The *WRKY29* transcription factor in Arabidopsis positively regulates the expression of the *ACS5, ACS6, ACS8, ACS11,* and *ACO5* genes, thereby promoting ET production [211].

In addition, phytohormones are also involved in the regulation of plant salt stress. Through the analysis of transcriptome data, a total of six phytohormones were screened, including ET, gibberellic acid (GA), auxin, salicylic acid (SA), ABA, and jasmonic acid (JA) (Figure 5D). Among them, ET and GA contained more differential genes, including 10 and 8 DEGs, respectively. The above six phytohormones play an important role in plant growth and development and can synergistically regulate plant tolerance to salt stress [212–214]. For example, under salt stress, the interaction between ET and JA inhibits the growth of rice seed roots [215]. Tomato *WRKY23* gene can regulate ET and auxin pathways to improve the tolerance of transgenic Arabidopsis to salt stress [216]. In addition, under normal conditions, GA and auxin can regulate plant growth together with ET, whereas SA, JA, and ABA have antagonistic effects on ET in various stress responses [212].

Finally, combining experimental data and transcriptome data, we summarized the working model for *GhSAMC* in cotton salt stress regulation: *GhSAMC* can regulate the rapid accumulation of SAM transporters when cotton is under salt stress. These SAM transporters accelerate the transport of SAM from cytoplasm to organelle and use SAM as a precursor to further increase the synthesis of ET and PAs. The increase in ET and PAs can inhibit the accumulation of ROS and maintain its homeostasis to alleviate salt stress in cotton (Figure 5F). In this process, related TFs (such as MYB, ERF, WRKY and NAC, etc.) and phytohormones (such as ABA, GA and SA, etc.) can also regulate ET biosynthesis and thus affect plant response to salt stress. The mechanism by which *GhSAMC* regulates plant salt stress needs further exploration.

8. Conclusions and Future Perspectives

Salt stress is considered to be one of the main factors limiting crop growth and yield. Therefore, plants have developed various strategies for survival under salt stress, such as osmotic regulation, ion, and ROS homeostasis [217]. Previous studies have shown that oxidative stress caused by salt stress is the most harmful to plants, and our analysis of transcriptome data also shows that a large number of oxidation–reduction-related factors are mobilized when plants are subjected to salt stress. In addition, plant tolerance to salt is regulated by multiple genes, not only a variety of transcription factors (such as EFR, WRKY, MYB, etc.) but also a variety of phytohormones (such as ET, ABA, SA, etc.).

As a general methyl donor for various methylation metabolites, SAM participates in the salt stress response through its derived metabolites (PAs and ET) [218,219]. This review focused on the biosynthesis and signal transduction of ET and PAs in salt stress, as well as the factors involved in these processes. In addition, the cotton SAM transporter gene *GhSAMC*, which is homologous to Arabidopsis *AtSAMC1* and *AtSAMC2*, was also identified. The virus-induced gene silencing experiment suggested that it may have positively regulated the salt tolerance of cotton.

Cotton cultivation has gradually shifted towards saline–alkali land because of the salinization and the need for more arable land to support the ever-increasing population.

To improve the tolerance of cotton to saline–alkali stress and increase yield and fiber quality, the following aspects can be considered in the future. First, a certain number of ion transporters have been discovered in cotton, which can alleviate salt stress damage by regulating ion homeostasis. Therefore, it is necessary to further dissect the exact regulatory mechanisms of these ion transporters, and then the knowledge learned from them can be used to improve the salt tolerance of cotton. Second, the impact of ET on cotton growth and development is multifaceted. To date, the regulatory mechanism of ET on cotton fiber development has been well studied, but the specific molecular mechanism of ET in salt stress regulation remains unclear [117,220,221]. Third, it has been proven that PAs can regulate the response to salt stress in many species; however, only a few reports have been found in cotton. Thus, it remains to be studied how PAs function in response to salt stress in cotton. Moreover, as a synthetic precursor for the synthesis of ET and PAs, SAM regulates plant salt stress by controlling the dynamic homeostasis of ET and PAs, thereby enabling SAM biosynthesis to affect responses to salt stress. Our study suggests that silencing *GhSAMC* enhances salt stress damage in cotton. Therefore, in the future, we can use transgenic technology to transfer *GhSAMC* into the cotton or use gene editing technology to knock out the *GhSAMC* gene in cotton, so as to further study the specific molecular mechanism of *GhSAMC* regulating cotton salt stress in cotton. Taken together, SAM and its derivates had very important roles in the response to salt stress, and the investigations summarized in this review revealed the complex biosynthesis and regulatory network of salt tolerance in plants and cotton.

Supplementary Materials: The supporting information can be downloaded at: https://www.mdpi.com/article/10.3390/ijms24119517/s1. References [222–227] are cited in the Supplementary Materials.

Author Contributions: X.W., X.M., F.Z. (Fuyong Zhao) and D.Y. conceived and designed the experiments. L.Y., X.Z., J.H., X.R., X.P., Y.L., K.H. and F.Z. (Fei Zhang) performed the experiments and analyzed the data. L.Y. drafted the manuscript. X.W., W.L., F.Z. (Fuyong Zhao), X.M. and D.Y. revised the manuscript. All authors have read and agreed to the published version of the manuscript.

Funding: This research was funded by the National Key R&D Program of China (2021YFF1000100), Science and Technology Program of Changji Hui Autonomous Prefecture (Grant No. 2021Z01-01), Xinjiang Tianshan Talents Program (2021), China Agriculture Research System (CARS-15-07), Innovation Project of the Chinese Academy of Agricultural Sciences (Grant No. CAAS-ASTIP-ICR-KP-2021-01), and Natural Science Foundation of Henan (Grant No. 202300410550).

Institutional Review Board Statement: Not applicable.

Informed Consent Statement: Not applicable.

Data Availability Statement: The sequencing data for this study have been deposited into the Sequence Read Archive under accession PRJNA933089 and PRJNA937453.

Acknowledgments: We sincerely thank Xiongfeng Ma (Cotton research institute) for his valuable advice and financial support of this research. To the entire research group, friends, and any other person who contributed, we are grateful for your help so much.

Conflicts of Interest: The authors declare no conflict of interest.

References

1. Ashraf, M. Breeding for Salinity Tolerance in Plants. *Crit. Rev. Plant Sci.* **1994**, *13*, 17–42. [CrossRef]
2. Maas, E.V.; Hoffman, G.J. Crop salt tolerance–current assessment. *J. Irrig. Drain. Div.* **1977**, *103*, 115–134. [CrossRef]
3. Ahmad, S.; Khan, N.I.; Iqbal, M.Z.; Hussain, A.; Hassan, M. Salt Tolerance of Cotton (*Gossypium hirsutum* L.). *Asian J. Plant Sci.* **2002**, *1*, 715–719. [CrossRef]
4. Abdelraheem, A.; Esmaeili, N.; O'Connell, M.; Zhang, J.F. Progress and perspective on drought and salt stress tolerance in cotton. *Ind. Crops Prod.* **2019**, *130*, 118–129. [CrossRef]
5. Brugnoli, E.; Lauteri, M. Effects of Salinity on Stomatal Conductance, Photosynthetic Capacity, and Carbon Isotope Discrimination of Salt-Tolerant (*Gossypium hirsutum* L.) and Salt-Sensitive (*Phaseolus vulgaris* L.) C(3) Non-Halophytes. *Plant Physiol.* **1991**, *95*, 628–635. [CrossRef] [PubMed]

6. Huang, G.; Wu, Z.; Percy, R.G.; Bai, M.; Li, Y.; Frelichowski, J.E.; Hu, J.; Wang, K.; Yu, J.Z.; Zhu, Y. Genome sequence of *Gossypium herbaceum* and genome updates of *Gossypium herbaceum* and *Gossypium hirsutum* provide insights into cotton A-genome evolution. *Nat. Genet.* **2020**, *52*, 516–524. [CrossRef]

7. Wang, K.; Wang, Z.; Li, F.; Ye, W.; Wang, J.; Song, G.; Yue, Z.; Cong, L.; Shang, H.; Zhu, S.; et al. The draft genome of a diploid cotton *Gossypium raimondii*. *Nat. Genet.* **2012**, *44*, 1098–1103. [CrossRef]

8. Mehari, T.G.; Hou, Y.; Xu, Y.; Umer, M.J.; Shiraku, M.L.; Wang, Y.; Wang, H.; Peng, R.; Wei, Y.; Cai, X.; et al. Overexpression of cotton *GhNAC072* gene enhances drought and salt stress tolerance in transgenic *Arabidopsis*. *BMC Genom.* **2022**, *23*, 648. [CrossRef]

9. Sadau, S.B.; Ahmad, A.; Tajo, S.M.; Ibrahim, S.; Kazeem, B.B.; Wei, H.; Yu, S. Overexpression of *GhMPK3* from Cotton Enhances Cold, Drought, and Salt Stress in *Arabidopsis*. *Agronomy* **2021**, *11*, 1049. [CrossRef]

10. Fontecave, M.; Atta, M.; Mulliez, E. S-adenosylmethionine: Nothing goes to waste. *Trends Biochem. Sci.* **2004**, *29*, 243–249. [CrossRef]

11. Bouvier, F.; Linka, N.; Isner, J.C.; Mutterer, J.; Weber, A.P.; Camara, B. *Arabidopsis* SAMT1 defines a plastid transporter regulating plastid biogenesis and plant development. *Plant Cell* **2006**, *18*, 3088–3105. [CrossRef]

12. Munawar, W.; Hameed, A.; Khan, M. Differential Morphophysiological and Biochemical Responses of Cotton Genotypes under Various Salinity Stress Levels during Early Growth Stage. *Front. Plant Sci.* **2021**, *12*, 622309. [CrossRef]

13. Vaseva, I.I.; Simova-Stoilova, L.; Kirova, E.; Mishev, K.; Depaepe, T.; Van Der Straeten, D.; Vassileva, V. Ethylene signaling in salt-stressed *Arabidopsis thaliana* ein2-1 and ctr1-1 mutants A dissection of molecular mechanisms involved in acclimation. *Plant Physiol. Biochem.* **2021**, *167*, 999–1010. [CrossRef] [PubMed]

14. Choudhury, A.R.; Roy, S.K.; Trivedi, P.; Choi, J.; Cho, K.; Yun, S.H.; Walitang, D.I.; Park, J.; Kim, K.; Sa, T. Label-free proteomics approach reveals candidate proteins in rice (*Oryza sativa* L.) important for ACC deaminase producing bacteria-mediated tolerance against salt stress. *Environ. Microbiol.* **2022**, *8*, 3612–3624. [CrossRef] [PubMed]

15. Khan, S.; Sehar, Z.; Fatma, M.; Mir, I.R.; Iqbal, N.; Tarighat, M.A.; Abdi, G.; Khan, N.A. Involvement of ethylene in melatonin-modified photosynthetic-N use efficiency and antioxidant activity to improve photosynthesis of salt grown wheat. *Plant Physiol.* **2022**, *174*, e13832. [CrossRef] [PubMed]

16. Shakar, M.; Yaseena, M.; Mahmoodb, R.; Ahmad, I. Calcium carbide induced ethylene modulate biochemical profile of Cucumis sativus at seed germination stage to alleviate salt stress. *Sci. Hortic.* **2016**, *213*, 179–185. [CrossRef]

17. Gharbi, E.; Martínez, J.P.; Benahmed, H.; Lepoint, G.; Vanpee, B.; Quinet, M.; Lutts, S. Inhibition of ethylene synthesis reduces salt-tolerance in tomato wild relative species *Solanum chilense*. *J. Plant Physiol.* **2017**, *210*, 24–37. [CrossRef] [PubMed]

18. Ahmad, Z.; Tahir, S.; Rehman, A.; Niazi, N.K.; Abid, M.; Amanullah, M. Effect of Substrate Dependent Ethylene on Cotton (*Gossypium hirsutum* L.) at Physiological and Molecular Levels under Salinity Stress. *J. Plant Nutr.* **2015**, *38*, 1913–1928. [CrossRef]

19. Gill, S.S.; Tuteja, N. Polyamines and abiotic stress tolerance in plants. *Plant Signal. Behav.* **2010**, *5*, 26–33. [CrossRef]

20. Gholami, M.; Fakhari, A.R.; Ghanati, F. Selective Regulation of Nicotine and Polyamines Biosynthesis in Tobacco Cells by Enantiomers of Ornithine. *Chirality* **2013**, *25*, 22–27. [CrossRef] [PubMed]

21. Campestre, M.P.; Bordenave, C.D.; Origone, A.C.; Menéndez, A.B.; Ruiz, O.A.; Rodríguez, A.A.; Maiale, S.J. Polyamine catabolism is involved in response to salt stress in soybean hypocotyls. *J. Plant Physiol.* **2011**, *168*, 1234–1240. [CrossRef] [PubMed]

22. Zhong, M.; Yue, L.; Liu, W.; Qin, H.; Lei, B.; Huang, R.; Yang, X.; Kang, Y. Genome-Wide Identification and Characterization of the Polyamine Uptake Transporter (Put) Gene Family in Tomatoes and the Role of Put2 in Response to Salt Stress. *Antioxidants* **2023**, *12*, 228. [CrossRef] [PubMed]

23. Lechowska, K.; Wojtyla, A.; Quinet, M.; Kubala, S.; Lutts, S.; Garnczarska, M. Endogenous Polyamines and Ethylene Biosynthesis in Relation to Germination of Osmoprimed *Brassica napus* Seeds under Salt Stress. *Int. J. Mol. Sci.* **2022**, *23*, 349. [CrossRef] [PubMed]

24. Gu, Q.; Ke, H.; Liu, C.; Lv, X.; Sun, Z.; Liu, Z.; Rong, W.; Yang, J.; Zhang, Y.; Wu, L.; et al. A stable QTL *qSalt-A04-1* contributes to salt tolerance in the cotton seed germination stage. *Theor. Appl. Genet.* **2021**, *134*, 2399–2410. [CrossRef] [PubMed]

25. Agrimi, G.; Di Noia, M.A.; Marobbio, C.M.; Fiermonte, G.; Lasorsa, F.M.; Palmieri, F. Identification of the human mitochondrial S-adenosylmethionine transporter: Bacterial expression, reconstitution, functional characterization and tissue distribution. *Biochem. J.* **2004**, *379*, 183–190. [CrossRef] [PubMed]

26. Petrotta-Simpson, T.F.; Talmadge, J.E.; Spence, K.D. Specificity and genetics of S-adenosylmethionine transport in Saccharomyces cerevisiae. *J. Bacteriol.* **1975**, *123*, 516–522. [CrossRef] [PubMed]

27. Dridi, L.; Ahmed Ouameur, A.; Ouellette, M. High Affinity S-Adenosylmethionine Plasma Membrane Transporter of Leishmania Is a Member of the Folate Biopterin Transporter (FBT) Family. *J. Biol. Chem.* **2010**, *285*, 19767–19775. [CrossRef] [PubMed]

28. Palmieri, L.; Arrigoni, R.; Blanco, E.; Carrari, F.; Zanor, M.I.; Studart-Guimaraes, C.; Fernie, A.R.; Palmieri, F. Molecular Identification of an *Arabidopsis* S-Adenosylmethionine Transporter. Analysis of Organ Distribution, Bacterial Expression, Reconstitution into Liposomes, and Functional Characterization. *Plant Physiol.* **2006**, *142*, 855–865. [CrossRef]

29. Kraidlova, L.; Schrevens, S.; Tournu, H.; Van Zeebroeck, G.; Sychrova, H.; Van Dijck, P. Characterization of the Candida albicans Amino Acid Permease Family: Gap2 Is the Only General Amino Acid Permease and Gap4 Is an S-Adenosylmethionine (SAM) Transporter Required for SAM-Induced Morphogenesis. *Msphere* **2016**, *1*, e00284-16. [CrossRef]

30. Abiala, M.A.; Abdelrahman, M.; Burritt, D.J.; Tran, L. Salt stress tolerance mechanisms and potential applications of legumes for sustainable reclamation of salt-degraded soils. *Land Degrad. Dev.* **2018**, *29*, 3812–3822. [CrossRef]

31. Park, H.J.; Kim, W.Y.; Yun, D.J. A New Insight of Salt Stress Signalling in Plant. *Mol. Cells.* **2016**, *39*, 447–459. [CrossRef]

32. Muchate, N.S.; Nikalje, G.C.; Rajurkar, N.S.; Suprasanna, P.; Nikam, T.D. Plant Salt Stress: Adaptive Responses, Tolerance Mechanism and Bioengineering for Salt Tolerance. *Bot. Rev.* **2016**, *82*, 371–406. [CrossRef]
33. Sharif, I.; Aleem, S.; Farooq, J.; Rizwan, M.; Younas, A.; Sarwar, G.; Chohan, S.M. Salinity stress in cotton: Effects, mechanism of tolerance and its management strategies. *Physiol. Mol. Biol. Plants* **2019**, *25*, 807–820. [CrossRef] [PubMed]
34. Maryum, Z.; Luqman, T.; Nadeem, S.; Khan, S.; Wang, B.; Ditta, A.; Khan, M. An overview of salinity stress, mechanism of salinity tolerance and strategies for its management in cotton. *Front. Plant Sci.* **2022**, *13*, 907937. [CrossRef]
35. Sanchez, D.H.; Siahpoosh, M.R.; Roessner, U.; Udvardi, M.; Kopka, J. Plant metabolomics reveals conserved and divergent metabolic responses to salinity. *Physiol. Plant.* **2008**, *132*, 209–219. [CrossRef] [PubMed]
36. Flowers, T.J.; Colmer, T.D. Plant salt tolerance: Adaptations in halophytes. *Ann. Bot.* **2015**, *115*, 327–331. [CrossRef]
37. Mishra, A.; Tanna, B. Halophytes: Potential Resources for Salt Stress Tolerance Genes and Promoters. *Front. Plant Sci.* **2017**, *8*, 829. [CrossRef]
38. Feki, K.; Quintero, F.J.; Khoudi, H.; Leidi, E.O.; Masmoudi, K.; Pardo, J.M.; Brini, F. A constitutively active form of a durum wheat Na^+/H^+ antiporter SOS1 confers high salt tolerance to transgenic *Arabidopsis*. *Plant Cell Rep.* **2014**, *33*, 277–288. [CrossRef] [PubMed]
39. Che, B.N.; Cheng, C.; Fang, J.J.; Liu, Y.M.; Jiang, L.; Yu, B.J. The Recretohalophyte Tamarix *TrSOS1* Gene Confers Enhanced Salt Tolerance to Transgenic Hairy Root Composite Cotton Seedlings Exhibiting Virus-Induced Gene Silencing of *GhSOS1*. *Int. J. Mol. Sci.* **2019**, *20*, 2930. [CrossRef] [PubMed]
40. Fahmideh, L.; Fooladvand, Z. Isolation and Semi Quantitative PCR of Na+/H+ Antiporter (SOS1 and NHX) Genes under Salinity Stress in Kochia scoparia. *Biol. Proced. Online* **2018**, *20*, 11. [CrossRef] [PubMed]
41. Cushman, K.R.; Pabuayon, I.C.M.; Hinze, L.L.; Sweeney, M.E.; de Los Reyes, B.G. Networks of Physiological Adjustments and Defenses, and Their Synergy with Sodium (Na+) Homeostasis Explain the Hidden Variation for Salinity Tolerance across the Cultivated *Gossypium hirsutum* Germplasm. *Front. Plant Sci.* **2020**, *11*, 588854. [CrossRef] [PubMed]
42. Feng, J.; Ma, W.; Ma, Z.; Ren, Z.; Zhou, Y.; Zhao, J.; Li, W.; Liu, W. GhNHX3D, a Vacuolar-Localized Na(+)/H(+) Antiporter, Positively Regulates Salt Response in Upland Cotton. *Int. J. Mol. Sci.* **2021**, *22*, 4047. [CrossRef] [PubMed]
43. Nieves-Cordones, M.; Miller, A.J.; Aleman, F.; Martinez, V.; Rubio, F. A putative role for the plasma membrane potential in the control of the expression of the gene encoding the tomato high-affinity potassium transporter HAK5. *Plant Mol. Biol.* **2008**, *68*, 521–532. [CrossRef] [PubMed]
44. Pilot, G.; Gaymard, F.; Mouline, K.; Chérel, I.; Sentenac, H. Regulated expression of *Arabidopsis* Shaker K+ channel genes involved in K+ uptake and distribution in the plant. *Plant Mol. Biol.* **2003**, *51*, 773–787. [CrossRef] [PubMed]
45. Ali, A.; Raddatz, N.; Pardo, J.M.; Yun, D.J. HKT sodium and potassium transporters in *Arabidopsis thaliana* and related halophyte species. *Physiol. Plant.* **2021**, *171*, 546–558. [CrossRef] [PubMed]
46. Guo, Q.; Meng, S.; Tao, S.; Feng, J.; Fan, X.; Xu, P.; Xu, Z.; Shen, X. Overexpression of a samphire high-affinity potassium transporter gene SbHKT1 enhances salt tolerance in transgenic cotton. *Acta Physiol. Plant.* **2020**, *42*, 36. [CrossRef]
47. Li, Y.; Feng, Z.; Wei, H.; Cheng, S.; Hao, P.; Yu, S.; Wang, H. Silencing of *GhKEA4* and *GhKEA12* Revealed Their Potential Functions under Salt and Potassium Stresses in Upland Cotton. *Front. Plant Sci.* **2021**, *12*, 789775. [CrossRef]
48. Chen, X.G.; Wu, Z.; Yin, Z.J.; Zhang, Y.X.; Rui, C.; Wang, J.; Malik, W.A.; Lu, X.K.; Wang, D.L.; Wang, J.J.; et al. Comprehensive genomic characterization of cotton cationic amino acid transporter genes reveals that *GhCAT10D* regulates salt tolerance. *BMC Plant Biol.* **2022**, *22*, 441. [CrossRef]
49. Zhao, J.; Peng, S.; Cui, H.; Li, P.; Li, T.; Liu, L.; Zhang, H.; Tian, Z.; Shang, H.; Xu, R. Dynamic Expression, Differential Regulation and Functional Diversity of the *CNGC* Family Genes in Cotton. *Int. J. Mol. Sci.* **2022**, *23*, 2041. [CrossRef]
50. Fu, X.; Yang, Y.; Kang, M.; Wei, H.; Lian, B.; Wang, B.; Ma, L.; Hao, P.; Lu, J.; Yu, S.; et al. Evolution and Stress Responses of *CLO* Genes and Potential Function of the *GhCLO06* Gene in Salt Resistance of Cotton. *Front. Plant Sci.* **2021**, *12*, 801239. [CrossRef]
51. Myo, T.; Wei, F.; Zhang, H.; Hao, J.; Zhang, B.; Liu, Z.; Cao, G.; Tian, B.; Shi, G. Genome-wide identification of the *BASS* gene family in four *Gossypium* species and functional characterization of *GhBASSs* against salt stress. *Sci. Rep.* **2021**, *11*, 11342. [CrossRef] [PubMed]
52. Zhang, D.; Li, J.; Niu, X.; Deng, C.; Song, X.; Li, W.; Cheng, Z.; Xu, Q.A.; Zhang, B.; Guo, W. GhANN1 modulates the salinity tolerance by regulating ABA biosynthesis, ion homeostasis and phenylpropanoid pathway in cotton. *Environ. Exp. Bot.* **2021**, *185*, 104427. [CrossRef]
53. Hasanuzzaman, M.; Raihan, M.R.H.; Masud, A.A.C.; Rahman, K.; Nowroz, F.; Rahman, M.; Nahar, K.; Fujita, M. Regulation of Reactive Oxygen Species and Antioxidant Defense in Plants under Salinity. *Int. J. Mol. Sci.* **2021**, *22*, 9326. [CrossRef] [PubMed]
54. Singh, R.; Singh, S.; Parihar, P.; Mishra, R.K.; Tripathi, D.K.; Singh, V.P.; Chauhan, D.K.; Prasad, S.M. Reactive Oxygen Species (ROS): Beneficial Companions of Plants' Developmental Processes. *Front. Plant Sci.* **2016**, *7*, 1299. [CrossRef] [PubMed]
55. Choudhury, S.; Panda, P.; Sahoo, L.; Panda, S.K. Reactive oxygen species signaling in plants under abiotic stress. *Plant Signal. Behav.* **2013**, *8*, e23681. [CrossRef] [PubMed]
56. Mohsin, S.; Hasanuzzaman, M.; Bhuyan, M.; Parvin, K.; Fujita, M. Exogenous Tebuconazole and Trifloxystrobin Regulates Reactive Oxygen Species Metabolism toward Mitigating Salt-Induced Damages in Cucumber Seedling. *Plants* **2019**, *8*, 428. [CrossRef]

57. Mangal, V.; Lal, M.K.; Tiwari, R.K.; Altaf, M.A.; Sood, S.; Kumar, D.; Bharadwaj, V.; Singh, B.; Singh, R.K.; Aftab, T. Molecular Insights into the Role of Reactive Oxygen, Nitrogen and Sulphur Species in Conferring Salinity Stress Tolerance in Plants. *J. Plant Growth Regul.* **2023**, *42*, 554–574. [CrossRef]

58. Xie, F.; Wang, Q.; Sun, R.; Zhang, B. Deep sequencing reveals important roles of microRNAs in response to drought and salinity stress in cotton. *J. Exp. Bot.* **2015**, *66*, 789–804. [CrossRef]

59. Rahman, M.M.; Mostofa, M.G.; Keya, S.S.; Siddiqui, M.N.; Ansary, M.M.U.; Das, A.K.; Rahman, M.A.; Tran, L.S. Adaptive Mechanisms of Halophytes and Their Potential in Improving Salinity Tolerance in Plants. *Int. J. Mol. Sci.* **2021**, *22*, 10733. [CrossRef]

60. Khalid, M.F.; Huda, S.; Yong, M.; Li, L.; Li, L.; Chen, Z.; Ahmed, T. Alleviation of drought and salt stress in vegetables: Crop responses and mitigation strategies. *Plant Growth Regul.* **2023**, *99*, 177–194. [CrossRef]

61. Wang, W.; Liu, D.; Chen, D.; Cheng, Y.; Zhang, X.; Song, L.; Hu, M.; Dong, J.; Shen, F. MicroRNA414c affects salt tolerance of cotton by regulating reactive oxygen species metabolism under salinity stress. *RNA Biol.* **2019**, *16*, 362–375. [CrossRef]

62. Yan, H.; Jia, H.; Chen, X.; Hao, L.; An, H.; Guo, X. The Cotton WRKY Transcription Factor *GhWRKY17* Functions in Drought and Salt Stress in Transgenic *Nicotiana benthamiana* through ABA Signaling and the Modulation of Reactive Oxygen Species Production. *Plant Cell Physiol.* **2014**, *55*, 2060–2076. [CrossRef] [PubMed]

63. Yang, H.; Yang, Q.; Zhang, D.; Wang, J.; Cao, T.; Bozorov, T.A.; Cheng, L.; Zhang, D. Transcriptome Reveals the Molecular Mechanism of the ScALDH21 Gene from the Desert Moss *Syntrichia caninervis* Conferring Resistance to Salt Stress in Cotton. *Int. J. Mol. Sci.* **2023**, *24*, 5822. [CrossRef] [PubMed]

64. Jia, H.; Hao, L.; Guo, X.; Liu, S.; Yan, Y.; Guo, X. A Raf-like MAPKKK gene, *GhRaf19*, negatively regulates tolerance to drought and salt and positively regulates resistance to cold stress by modulating reactive oxygen species in cotton. *Plant Sci.* **2016**, *252*, 267–281. [CrossRef] [PubMed]

65. Atkinson, N.J.; Urwin, P.E. The interaction of plant biotic and abiotic stresses: From genes to the field. *J. Exp. Bot.* **2012**, *63*, 695–709. [CrossRef] [PubMed]

66. Mishra, S.; Kumar, S.; Saha, B.; Awasthi, J.; Dey, M.; Panda, S.K.; Sahoo, L. *Crosstalk between Salt, Drought, and Cold Stress in Plants: Toward Genetic Engineering for Stress Tolerance*; Tuteja, N., Gill, S.S., Eds.; Wiley-VCH Verlag GmbH & Co. KGaA: Weinheim, Germany, 2016; pp. 57–88. [CrossRef]

67. Kim, D.; Jeon, S.J.; Yanders, S.; Park, S.C.; Kim, H.S.; Kim, S. MYB3 plays an important role in lignin and anthocyanin biosynthesis under salt stress condition in *Arabidopsis*. *Plant Cell Rep.* **2022**, *41*, 1549–1560. [CrossRef]

68. Ullah, A.; Ul, Q.M.; Nisar, M.; Hazrat, A.; Rahim, G.; Khan, A.H.; Hayat, K.; Ahmed, S.; Ali, W.; Khan, A.; et al. Characterization of a novel cotton MYB gene, *GhMYB108-like* responsive to abiotic stresses. *Mol. Biol. Rep.* **2020**, *47*, 1573–1581. [CrossRef] [PubMed]

69. Khanale, V.; Bhattacharya, A.; Satpute, R.; Char, B. Brief bioinformatics identification of cotton bZIP transcription factors family from *Gossypium hirsutum*, *Gossypium arboreum* and *Gossypium raimondii*. *Plant Biotechnol. Rep.* **2021**, *15*, 493–511. [CrossRef]

70. Azeem, F.; Tahir, H.; Usman, I.; Tayyaba, S. A genome-wide comparative analysis of bZIP transcription factors in *G. arboreum* and *G. raimondii* (Diploid ancestors of present-day cotton). *Physiol. Mol. Biol. Plants* **2020**, *3*, 433–444. [CrossRef] [PubMed]

71. Ullah, A.; Sun, H.; Hakim; Yang, X.; Zhang, X. A novel cotton WRKY gene, *GhWRKY6-like*, improves salt tolerance by activating the ABA signaling pathway and scavenging of reactive oxygen species. *Physiol. Plant.* **2018**, *162*, 439–454. [CrossRef]

72. Çelik, Ö.; Meriç, S.; Ayan, A.; Atak, Ç. Epigenetic analysis of WRKY transcription factor genes in salt stressed rice (*Oryza sativa* L.) plants. *Environ. Exp. Bot.* **2019**, *159*, 121–131. [CrossRef]

73. Guo, X.; Ullah, A.; Siuta, D.; Kukfisz, B.; Iqbal, S. Role of WRKY Transcription Factors in Regulation of Abiotic Stress Responses in Cotton. *Life* **2022**, *12*, 1410. [CrossRef] [PubMed]

74. Shah, S.T.; Pang, C.; Fan, S.; Song, M.; Arain, S.; Yu, S. Isolation and expression profiling of *GhNAC* transcription factor genes in cotton (*Gossypium hirsutum* L.) during leaf senescence and in response to stresses. *Gene* **2013**, *531*, 220–234. [CrossRef] [PubMed]

75. Latif, A.; Azam, S.; Shahid, N.; Javed, M.R.; Haider, Z.; Yasmeen, A.; Sadaqat, S.; Shad, M.; Husnain, T.; Rao, A.Q. Overexpression of the *AGL42* gene in cotton delayed leaf senescence through downregulation of NAC transcription factors. *Sci. Rep.* **2022**, *12*, 21093. [CrossRef] [PubMed]

76. Trishla, V.S.; Kirti, P.B. Structure-function relationship of *Gossypium hirsutum* NAC transcription factor, *GhNAC4* with regard to ABA and abiotic stress responses. *Plant Sci.* **2021**, *302*, 110718. [CrossRef] [PubMed]

77. Sun, H.; Chen, L.; Li, J.; Hu, M.; Ullah, A.; He, X.; Yang, X.; Zhang, X. The JASMONATE ZIM-Domain Gene Family Mediates JA Signaling and Stress Response in Cotton. *Plant Cell Physiol.* **2017**, *58*, 2139–2154. [CrossRef]

78. Abid, M.A.; Liang, C.; Malik, W.; Meng, Z.; Tao, Z.; Meng, Z.; Ashraf, J.; Guo, S.; Zhang, R. Cascades of Ionic and Molecular Networks Involved in Expression of Genes Underpin Salinity Tolerance in Cotton. *J. Plant Growth Regul.* **2018**, *37*, 668–679. [CrossRef]

79. Shah, W.H.; Rasool, A.; Saleem, S.; Mushtaq, N.U.; Tahir, I.; Hakeem, K.R.; Rehman, R.U. Understanding the Integrated Pathways and Mechanisms of Transporters, Protein Kinases, and Transcription Factors in Plants under Salt Stress. *Int. J. Genom.* **2021**, *2021*, 5578727. [CrossRef]

80. Choudhary, P.; Pramitha, L.; Rana, S.; Verma, S.; Aggarwal, P.R.; Muthamilarasan, M. Hormonal crosstalk in regulating salinity stress tolerance in graminaceous crops. *Physiol. Plant.* **2021**, *173*, 1587–1596. [CrossRef]

81. Singh, P.; Choudhary, K.K.; Chaudhary, N.; Gupta, S.; Sahu, M.; Tejaswini, B.; Sarkar, S. Salt stress resilience in plants mediated through osmolyte accumulation and its crosstalk mechanism with phytohormones. *Front. Plant Sci.* **2022**, *13*, 1006617. [CrossRef]

82. Hanin, M.; Ebel, C.; Ngom, M.; Laplaze, L.; Masmoudi, K. New Insights on Plant Salt Tolerance Mechanisms and Their Potential Use for Breeding. *Front. Plant Sci.* **2016**, *7*, 1787. [CrossRef] [PubMed]

83. Kim, S.H.; Kim, S.H.; Palaniyandi, S.A.; Yang, S.H.; Suh, J.W. Expression of potato S-adenosyl-L-methionine synthase (*SbSAMS*) gene altered developmental characteristics and stress responses in transgenic *Arabidopsis* plants. *Plant Physiol. Biochem.* **2015**, *87*, 84–91. [CrossRef]

84. Ma, C.; Wang, Y.; Gu, D.; Nan, J.; Chen, S.; Li, H. Overexpression of S-Adenosyl-l-Methionine Synthetase 2 from Sugar *Beet M14* Increased *Arabidopsis* Tolerance to Salt and Oxidative Stress. *Int. J. Mol. Sci.* **2017**, *18*, 847. [CrossRef] [PubMed]

85. He, M.W.; Wang, Y.; Wu, J.Q.; Shu, S.; Sun, J.; Guo, S.R. Isolation and characterization of S-Adenosylmethionine synthase gene from cucumber and responsive to abiotic stress. *Plant Physiol. Biochem.* **2019**, *141*, 431–445. [CrossRef]

86. Zhu, H.; He, M.; Jahan, M.S.; Wu, J.; Gu, Q.; Shu, S.; Sun, J.; Guo, S. CsCDPK6, a CsSAMS1-Interacting Protein, Affects Polyamine/Ethylene Biosynthesis in Cucumber and Enhances Salt Tolerance by Overexpression in Tobacco. *Int. J. Mol. Sci.* **2021**, *22*, 11133. [CrossRef]

87. Riyazuddin, R.; Verma, R.; Singh, K.; Nisha, N.; Keisham, M.; Bhati, K.K.; Kim, S.T.; Gupta, R. Ethylene: A Master Regulator of Salinity Stress Tolerance in Plants. *Biomolecules* **2020**, *10*, 959. [CrossRef] [PubMed]

88. Rossi, F.R.; Krapp, A.R.; Bisaro, F.; Maiale, S.J.; Pieckenstain, F.L.; Carrillo, N. Reactive oxygen species generated in chloroplasts contribute to tobacco leaf infection by the necrotrophic *fungus Botrytis cinerea*. *Plant J.* **2017**, *92*, 761–773. [CrossRef]

89. Petrov, V.; Hille, J.; Mueller-Roeber, B.; Gechev, T.S. ROS-mediated abiotic stress-induced programmed cell death in plants. *Front. Plant Sci.* **2015**, *6*, 69. [CrossRef] [PubMed]

90. Ecker, J.R.U.O. The ethylene signal transduction pathway in plants. *Science* **1995**, *268*, 667–675. [CrossRef] [PubMed]

91. Mehrotra, R.; Bhalothia, P.; Bansal, P.; Basantani, M.K.; Bharti, V.; Mehrotra, S. Abscisic acid and abiotic stress tolerance different tiers of regulation. *J. Plant Physiol.* **2014**, *171*, 486–496. [CrossRef] [PubMed]

92. Naing, A.H.; Campol, J.R.; Kang, H.; Xu, J.; Chung, M.Y.; Kim, C.K. Role of Ethylene Biosynthesis Genes in the Regulation of Salt Stress and Drought Stress Tolerance in Petunia. *Front. Plant Sci.* **2022**, *13*, 844449. [CrossRef] [PubMed]

93. Binder, B.M. Ethylene signaling in plants. *J. Biol. Chem.* **2020**, *295*, 7710–7725. [CrossRef]

94. Lockhart, J. Salt of the Earth: Ethylene Promotes Salt Tolerance by Enhancing Na^+/K^+ Homeostasis. *Plant Cell* **2013**, *25*, 3150. [CrossRef] [PubMed]

95. Pattyn, J.; Vaughan-Hirsch, J.; Van de Poel, B. The regulation of ethylene biosynthesis: A complex multilevel control circuitry. *New Phytol.* **2021**, *229*, 770–782. [CrossRef] [PubMed]

96. Boller, T.; Herner, R.C.; Kende, H. Assay for and enzymatic formation of an ethylene precursor, 1-aminocyclopropane-1-carboxylic acid. *Planta* **1979**, *145*, 293–303. [CrossRef]

97. Druege, U. Ethylene and Plant Responses to Abiotic Stress. In *Ethylene Action in Plants*; Springer: Berlin/Heidelberg, Germany, 2006; pp. 81–118. [CrossRef]

98. Jakubowicz, M.; Sadowski, J. 1-Aminocyclopropane-l-carboxylate synthase-genes and expression. *Acta Physiol. Plant.* **2002**, *24*, 459–478. [CrossRef]

99. Yamagami, T.; Tsuchisaka, A.; Yamada, K.; Haddon, W.F.; Harden, L.A.; Theologis, A. Biochemical Diversity among the 1-Aminocyclopropane-1-Carboxylate Synthase Isozymes Encoded by the *Arabidopsis* Gene Family. *J. Biol. Chem.* **2003**, *278*, 49102–49112. [CrossRef] [PubMed]

100. Lee, H.Y.; Chen, Y.C.; Kieber, J.J.; Yoon, G.M. Regulation of the turnover of ACC synthases by phytohormones and heterodimerization in *Arabidopsis*. *Plant J.* **2017**, *91*, 491–504. [CrossRef]

101. Tsuchisaka, A.; Theologis, A. Heterodimeric interactions among the 1-amino-cyclopropane-1-carboxylate synthase polypeptides encoded by the *Arabidopsis* gene family. *Proc. Natl. Acad. Sci. USA* **2004**, *101*, 2275–2280. [CrossRef]

102. Li, J.; Zou, X.; Chen, G.; Meng, Y.; Ma, Q.; Chen, Q.; Wang, Z.; Li, F. Potential Roles of 1-Aminocyclopropane-1-carboxylic Acid Synthase Genes in the Response of *Gossypium* Species to Abiotic Stress by Genome-Wide Identification and Expression Analysis. *Plants* **2022**, *11*, 1524. [CrossRef]

103. Gómez-Lim, M.A.; Valdés-López, V.; Cruz-Hernandez, A.; Saucedo-Arias, L.J. Isolation and characterization of a gene involved in ethylene biosynthesis from *Arabidopsis thaliana*. *Gene* **1993**, *134*, 217–221. [CrossRef] [PubMed]

104. Wei, H.; Xue, Y.; Chen, P.; Hao, P.; Wei, F.; Sun, L.; Yang, Y. Genome-Wide Identification and Functional Investigation of 1-Aminocyclopropane-1-carboxylic Acid Oxidase (ACO) Genes in Cotton. *Plants* **2021**, *10*, 1699. [CrossRef] [PubMed]

105. Schott-Verdugo, S.; Müller, L.; Classen, E.; Gohlke, H.; Groth, G. Structural Model of the ETR1 Ethylene Receptor Transmembrane Sensor Domain. *Sci. Rep.* **2019**, *9*, 8869. [CrossRef] [PubMed]

106. Bleecker, A.B. The ethylene-receptor family from *Arabidopsis*: Structure and function. *Philos. Trans. R. Soc. B Biol. Sci.* **1998**, *353*, 1405–1412. [CrossRef] [PubMed]

107. Sakai, H.; Hua, J.; Chen, Q.G.; Chang, C.; Medrano, L.J.; Bleecker, A.B.; Meyerowitz, E.M. ETR2 is an ETR1-like gene involved in ethylene signaling in *Arabidopsis*. *Proc. Natl. Acad. Sci. USA* **1998**, *95*, 5812–5817. [CrossRef] [PubMed]

108. Gallie, D.R. Appearance and elaboration of the ethylene receptor family during land plant evolution. *Plant Mol. Biol.* **2015**, *87*, 521–539. [CrossRef]

109. Park, H.L.; Seo, D.H.; Lee, H.Y.; Bakshi, A.; Park, C.; Chien, Y.; Kieber, J.J.; Binder, B.M.; Yoon, G.M. Ethylene-triggered subcellular trafficking of CTR1 enhances the response to ethylene gas. *Nat. Commun.* **2023**, *14*, 365. [CrossRef]

110. Bisson, M.M.A.; Groth, G. New Insight in Ethylene Signaling: Autokinase Activity of ETR1 Modulates the Interaction of Receptors and EIN2. *Mol. Plant* **2010**, *3*, 882–889. [CrossRef]

111. Merchante, C.; Brumos, J.; Yun, J.; Hu, Q.; Spencer, K.R.; Enríquez, P.; Binder, B.M.; Heber, S.; Stepanova, A.N.; Alonso, J.M. Gene-Specific Translation Regulation Mediated by the Hormone-Signaling Molecule EIN2. *Cell* **2015**, *163*, 684–697. [CrossRef]

112. Iqbal, N.; Trivellini, A.; Masood, A.; Ferrante, A.; Khan, N.A. Current understanding on ethylene signaling in plants: The influence of nutrient availability. *Plant Physiol. Biochem.* **2013**, *73*, 128–138. [CrossRef]

113. Dolgikh, V.A.; Pukhovaya, E.M.; Zemlyanskaya, E.V. Shaping Ethylene Response: The Role of EIN3/EIL1 Transcription Factors. *Front. Plant Sci.* **2019**, *10*, 1030. [CrossRef] [PubMed]

114. Debbarma, J.; Sarki, Y.N.; Saikia, B.; Boruah, H.; Singha, D.L.; Chikkaputtaiah, C. Ethylene Response Factor (ERF) Family Proteins in Abiotic Stresses and CRISPR-Cas9 Genome Editing of ERFs for Multiple Abiotic Stress Tolerance in Crop Plants: A Review. *Mol. Biotechnol.* **2019**, *61*, 153–172. [CrossRef] [PubMed]

115. Klay, I.; Pirrello, J.; Riahi, L.; Bernadac, A.; Cherif, A.; Bouzayen, M.; Bouzid, S. Ethylene Response Factor Sl-ERF.B.3 Is Responsive to Abiotic Stresses and Mediates Salt and Cold Stress Response Regulation in Tomato. *Sci. World J.* **2014**, *2014*, 167681. [CrossRef] [PubMed]

116. Fatma, M.; Asgher, M.; Iqbal, N.; Rasheed, F.; Sehar, Z.; Sofo, A.; Khan, N.A. Ethylene Signaling under Stressful Environments: Analyzing Collaborative Knowledge. *Plants* **2022**, *11*, 2211. [CrossRef] [PubMed]

117. Yu, D.; Li, X.; Li, Y.; Ali, F.; Li, F.; Wang, Z. Dynamic roles and intricate mechanisms of ethylene in epidermal hair development in *Arabidopsis* and cotton. *New Phytol.* **2022**, *234*, 375–391. [CrossRef] [PubMed]

118. Yu, F.; Guo, R.; Wu, C.; Li, H.; Guo, X. Molecular cloning and expression characteristics of a novel MAPKKK gene, *GhCTR1*, from cotton (*Gossypium hirsutum* L.). *S. Afr. J. Bot.* **2012**, *78*, 211–219. [CrossRef]

119. Wang, X.Q.; Han, L.H.; Zhou, W.; Tao, M.; Hu, Q.Q.; Zhou, Y.N.; Li, X.B.; Li, D.D.; Huang, G.Q. GhEIN3, a cotton (*Gossypium hirsutum*) homologue of *AtEIN3*, is involved in regulation of plant salinity tolerance. *Plant Physiol. Biochem.* **2019**, *143*, 83–93. [CrossRef]

120. Phukan, U.J.; Jeena, G.S.; Tripathi, V.; Shukla, R.K. Regulation of Apetala2/Ethylene Response Factors in Plants. *Front. Plant Sci.* **2017**, *8*, 150. [CrossRef]

121. Ma, L.; Hu, L.; Fan, J.; Amombo, E.; Khaldun, A.; Zheng, Y.; Chen, L. Cotton *GhERF38* gene is involved in plant response to salt/drought and ABA. *Ecotoxicology* **2017**, *26*, 841–854. [CrossRef]

122. Lu, L.; Qanmber, G.; Li, J.; Pu, M.; Chen, G.; Li, S.; Liu, L.; Qin, W.; Ma, S.; Wang, Y.; et al. Identification and Characterization of the ERF Subfamily B3 Group Revealed *GhERF13.12* Improves Salt Tolerance in Upland Cotton. *Front. Plant Sci.* **2021**, *12*, 705883. [CrossRef]

123. Zhang, J.; Zhang, P.; Huo, X.; Gao, Y.; Chen, Y.; Song, Z.; Wang, F.; Zhang, J. Comparative Phenotypic and Transcriptomic Analysis Reveals Key Responses of Upland Cotton to Salinity Stress during Postgermination. *Front. Plant Sci.* **2021**, *12*, 639104. [CrossRef] [PubMed]

124. Long, L.; Yang, W.; Liao, P.; Guo, Y.; Kumar, A.; Gao, W. Transcriptome analysis reveals differentially expressed ERF transcription factors associated with salt response in cotton. *Plant Sci.* **2019**, *281*, 72–81. [CrossRef] [PubMed]

125. Wessler, S.R. Homing into the origin of the AP2 DNA binding domain. *Trends Plant Sci.* **2005**, *10*, 54–56. [CrossRef] [PubMed]

126. Zafar, M.M.; Rehman, A.; Razzaq, A.; Parvaiz, A.; Mustafa, G.; Sharif, F.; Mo, H.; Youlu, Y.; Shakeel, A.; Ren, M. Genome-wide characterization and expression analysis of Erf gene family in cotton. *BMC Plant Biol.* **2022**, *22*, 134. [CrossRef]

127. Bano, N.; Fakhrah, S.; Mohanty, C.S.; Bag, S.K. Transcriptome Meta-Analysis Associated Targeting Hub Genes and Pathways of Drought and Salt Stress Responses in Cotton (*Gossypium hirsutum*): A Network Biology Approach. *Front. Plant Sci.* **2022**, *13*, 818472. [CrossRef]

128. Qiao, Z.X.; Huang, B.; Liu, J.Y. Molecular cloning and functional analysis of an ERF gene from cotton (*Gossypium hirsutum*). *Biochim Biophys Acta* **2008**, *1779*, 122–127. [CrossRef] [PubMed]

129. Jin, L.; Li, H.; Liu, J. Molecular Characterization of Three Ethylene Responsive Element Binding Factor Genes from Cotton. *J. Integr. Plant Biol.* **2010**, *52*, 485–495. [CrossRef]

130. Jin, L.G.; Liu, J.Y. Molecular cloning, expression profile and promoter analysis of a novel ethylene responsive transcription factor gene GhERF4 from cotton (*Gossypium hirstum*). *Plant Physiol. Biochem.* **2008**, *46*, 46–53. [CrossRef] [PubMed]

131. Jin, L.; Huang, B.; Li, H.; Liu, J. Expression profiles and transactivation analysis of a novel ethylene-responsive transcription factor gene GhERF5 from cotton. *Prog. Nat. Sci.* **2009**, *19*, 563–572. [CrossRef]

132. Kusano, T.; Yamaguchi, K.; Berberich, T.; Takahashi, Y. Advances in polyamine research in 2007. *J. Plant Res.* **2007**, *120*, 345–350. [CrossRef]

133. Alcazar, R.; Altabella, T.; Marco, F.; Bortolotti, C.; Reymond, M.; Koncz, C.; Carrasco, P.; Tiburcio, A.F. Polyamines: Molecules with regulatory functions in plant abiotic stress tolerance. *Planta* **2010**, *231*, 1237–1249. [CrossRef]

134. Takano, A.; Kakehi, J.; Takahashi, T. Thermospermine is not a minor polyamine in the plant kingdom. *Plant Cell Physiol.* **2012**, *53*, 606–616. [CrossRef] [PubMed]

135. Vuosku, J.; Suorsa, M.; Ruottinen, M.; Sutela, S.; Muilu-Makela, R.; Julkunen-Tiitto, R.; Sarjala, T.; Neubauer, P.; Haggman, H. Polyamine metabolism during exponential growth transition in Scots pine embryogenic cell culture. *Tree Physiol.* **2012**, *32*, 1274–1287. [CrossRef] [PubMed]

136. Tiburcio, A.F.; Altabella, T.; Bitrian, M.; Alcazar, R. The roles of polyamines during the lifespan of plants: From development to stress. *Planta* **2014**, *240*, 1–18. [CrossRef] [PubMed]
137. Mansour, M. Plasma membrane permeability as an indicator of salt tolerance in plants. *Biol. Plant* **2013**, *57*, 1–10. [CrossRef]
138. Lightfoot, H.L.; Hall, J. Endogenous polyamine function-the RNA perspective. *Nucleic Acids Res.* **2014**, *42*, 11275–11290. [CrossRef] [PubMed]
139. Bueno, M.; Cordovilla, M.P. Polyamines in Halophytes. *Front. Plant Sci.* **2019**, *10*, 439. [CrossRef] [PubMed]
140. ElSayed, A.I.; Mohamed, A.H.; Rafudeen, M.S.; Omar, A.A.; Awad, M.F.; Mansour, E. Polyamines mitigate the destructive impacts of salinity stress by enhancing photosynthetic capacity, antioxidant defense system and upregulation of calvin cycle-related genes in rapeseed (*Brassica napus* L.). *Saudi J. Biol. Sci.* **2022**, *29*, 3675–3686. [CrossRef] [PubMed]
141. Korbas, A.; Kubiś, J.; Rybus-Zając, M.; Chadzinikolau, T. Spermidine Modify Antioxidant Activity in Cucumber Exposed to Salinity Stress. *Agronomy* **2022**, *12*, 1554. [CrossRef]
142. Pál, M.; Ivanovska, B.; Oláh, T.; Tajti, J.; Hamow, K.Á.; Szalai, G.; Khalil, R.; Vanková, R.; Dobrev, P.; Misheva, S.P.; et al. Janda Role of polyamines in plant growth regulation of Rht wheat mutants. *Plant Physiol. Biochem.* **2019**, *137*, 189–202. [CrossRef]
143. Sagor, G.H.M.; Inoue, M.; Kusano, T.; Berberich, T. Expression profile of seven polyamine oxidase genes in rice (*Oryza sativa*) in response to abiotic stresses, phytohormones and polyamines. *Physiol. Mol. Biol. Plants* **2021**, *27*, 1353–1359. [CrossRef] [PubMed]
144. Baniasadi, F.; Saffari, V.R.; Moud, A. Physiological and growth responses of *Calendula officinalis* L. plants to the interaction effects of polyamines and salt stress. *Sci. Hortic.* **2018**, *234*, 312–317. [CrossRef]
145. Michael, A.J. Biosynthesis of polyamines and polyamine-containing molecules. *Biochem. J.* **2016**, *473*, 2315–2329. [CrossRef] [PubMed]
146. Fuell, C.; Elliott, K.A.; Hanfrey, C.C.; Franceschetti, M.; Michael, A.J. Polyamine biosynthetic diversity in plants and algae. *Plant Physiol. Biochem.* **2010**, *48*, 513–520. [CrossRef] [PubMed]
147. Hanfrey, C.; Sommer, S.; Mayer, M.J.; Burtin, D.; Michael, A.J. *Arabidopsis* polyamine biosynthesis: Absence of ornithine decarboxylase and the mechanism of arginine decarboxylase activity. *Plant J.* **2001**, *27*, 551–560. [CrossRef] [PubMed]
148. Pegg, A.E. Functions of Polyamines in Mammals. *J. Biol. Chem.* **2016**, *291*, 14904–14912. [CrossRef]
149. Hummel, I.; Bourdais, G.; Gouesbet, G.; Couee, I.; Malmberg, R.L.; El, A.A. Differential gene expression of ARGININE DECAR-BOXYLASE ADC1 and ADC2 in *Arabidopsis thaliana*: Characterization of transcriptional regulation during seed germination and seedling development. *New Phytol.* **2004**, *163*, 519–531. [CrossRef]
150. Fu, Y.; Guo, C.; Wu, H.; Chen, C. Arginine decarboxylase ADC2 enhances salt tolerance through increasing ROS scavenging enzyme activity in *Arabidopsis thaliana*. *Plant Growth Regul.* **2017**, *83*, 253–263. [CrossRef]
151. Marco, F.; Alcazar, R.; Tiburcio, A.F.; Carrasco, P. Interactions between polyamines and abiotic stress pathway responses unraveled by transcriptome analysis of polyamine overproducers. *OMICS* **2011**, *15*, 775–781. [CrossRef]
152. Saha, J.; Giri, K. Molecular phylogenomic study and the role of exogenous spermidine in the metabolic adjustment of endogenous polyamine in two rice cultivars under salt stress. *Genes* **2017**, *609*, 88–103. [CrossRef]
153. Napieraj, N.; Reda, M.G.; Janicka, M.G. The role of NO in plant response to salt stress: Interactions with polyamines. *Funct. Plant Biol.* **2020**, *47*, 865–879. [CrossRef] [PubMed]
154. Kovacs, L.; Mendel, A.; Szentgyorgyi, A.; Fekete, S.; Sore, F.; Posta, K.; Kiss, E. Comparative analysis of overexpressed Fragaria vesca S-adenosyl-l-methionine synthase (*FvSAMS*) and decarboxylase (*FvSAMDC*) during salt stress in transgenic *Nicotiana benthamiana*. *Plant Growth Regul.* **2020**, *91*, 53–73. [CrossRef]
155. Wu, F.; Muvunyi, B.P.; Yan, Q.; Kanzana, G.; Ma, T.; Zhang, Z.; Wang, Y.; Zhang, J. Comprehensive genome-wide analysis of polyamine and ethylene pathway genes in *Cleistogenes songorica* and CsSAMDC2 function in response to abiotic stress. *Environ. Exp. Bot.* **2022**, *202*, 105029. [CrossRef]
156. Tang, X.; Wu, L.; Wang, F.; Tian, W.; Hu, X.; Jin, S.; Zhu, H. Ectopic Expression of GhSAMDC3 Enhanced Salt Tolerance Due to Accumulated Spd Content and Activation of Salt Tolerance-Related Genes in *Arabidopsis thaliana*. *DNA Cell Biol.* **2021**, *40*, 1144–1157. [CrossRef] [PubMed]
157. Meng, D.Y.; Yang, S.; Xing, J.Y.; Ma, N.N.; Wang, B.Z.; Qiu, F.T.; Guo, F.; Meng, J.; Zhang, J.L.; Wan, S.B.; et al. Peanut (*Arachis hypogaea* L.) S-adenosylmethionine decarboxylase confers transgenic tobacco with elevated tolerance to salt stress. *Plant Biol.* **2021**, *23*, 341–350. [CrossRef] [PubMed]
158. Jia, T.; Hou, J.; Iqbal, M.Z.; Zhang, Y.; Cheng, B.; Feng, H.; Li, Z.; Liu, L.; Zhou, J.; Feng, G.; et al. Overexpression of the white clover TrSAMDC1 gene enhanced salt and drought resistance in *Arabidopsis thaliana*. *Plant Physiol. Biochem.* **2021**, *165*, 147–160. [CrossRef] [PubMed]
159. Cona, A.; Rea, G.; Angelini, R.; Federico, R.; Tavladoraki, P. Functions of amine oxidases in plant development and defence. *Trends Plant Sci.* **2006**, *11*, 80–88. [CrossRef]
160. Moschou, P.N.; Wu, J.; Cona, A.; Tavladoraki, P.; Angelini, R.; Roubelakis-Angelakis, K.A. The polyamines and their catabolic products are significant players in the turnover of nitrogenous molecules in plants. *J. Exp. Bot.* **2012**, *63*, 5003–5015. [CrossRef]
161. Khajuria, A.; Sharma, N.; Bhardwaj, R.; Ohri, P. Emerging Role of Polyamines in Plant Stress Tolerance. *Curr. Protein Pept. Sci.* **2018**, *19*, 1114–1123. [CrossRef]
162. Kim, D.W.; Watanabe, K.; Murayama, C.; Izawa, S.; Niitsu, M.; Michael, A.J.; Berberich, T.; Kusano, T. Polyamine Oxidase5 Regulates *Arabidopsis* Growth through Thermospermine Oxidase Activity. *Plant Physiol.* **2014**, *165*, 1575–1590. [CrossRef]

163. Fincato, P.; Moschou, P.N.; Ahou, A.; Angelini, R.; Roubelakis-Angelakis, K.A.; Federico, R.; Tavladoraki, P. The members of *Arabidopsis thaliana PAO* gene family exhibit distinct tissue and organ-specific expression pattern during seedling growth and flower development. *Amino Acids* **2012**, *42*, 831–841. [CrossRef] [PubMed]

164. Sagor, G.H.; Zhang, S.; Kojima, S.; Simm, S.; Berberich, T.; Kusano, T. Reducing Cytoplasmic Polyamine Oxidase Activity in *Arabidopsis* Increases Salt and Drought Tolerance by Reducing Reactive Oxygen Species Production and Increasing Defense Gene Expression. *Front. Plant Sci.* **2016**, *7*, 214. [CrossRef] [PubMed]

165. Wu, J.; Zhu, M.; Liu, W.; Jahan, M.S.; Gu, Q.; Shu, S.; Sun, J.; Guo, S. *CsPAO2* Improves Salt Tolerance of Cucumber through the Interaction with *CsPSA3* by Affecting Photosynthesis and Polyamine Conversion. *Int. J. Mol. Sci.* **2022**, *23*, 12413. [CrossRef] [PubMed]

166. Wu, J.; Liu, W.; Jahan, M.S.; Shu, S.; Sun, J.; Guo, S. Characterization of polyamine oxidase genes in cucumber and roles of *CsPAO3* in response to salt stress. *Environ. Exp. Bot.* **2022**, *194*, 104696. [CrossRef]

167. Liu, G.; Jiang, W.; Tian, L.; Fu, Y.; Tan, L.; Zhu, Z.; Sun, C.; Liu, F. Polyamine oxidase 3 is involved in salt tolerance at the germination stage in rice. *J. Genet. Genom.* **2022**, *49*, 458–468. [CrossRef]

168. Cheng, X.Q.; Zhu, X.F.; Tian, W.G.; Cheng, W.H.; Hakim; Sun, J.; Jin, S.X.; Zhu, H.G. Genome-wide identification and expression analysis of polyamine oxidase genes in upland cotton (*Gossypium hirsutum* L.). *Plant Cell Tissue Organ Cult.* **2017**, *129*, 237–249. [CrossRef]

169. Gerlin, L.; Baroukh, C.; Genin, S. Polyamines: Double agents in disease and plant immunity. *Trends Plant Sci.* **2021**, *26*, 1061–1071. [CrossRef]

170. Majumdar, R.; Minocha, R.; Lebar, M.D.; Rajasekaran, K.; Long, S.; Carter-Wientjes, C.; Minocha, S.; Cary, J.W. Contribution of Maize Polyamine and Amino Acid Metabolism toward Resistance against Aspergillus flavus Infection and Aflatoxin Production. *Front. Plant Sci.* **2019**, *10*, 692. [CrossRef] [PubMed]

171. Nandy, S.; Mandal, S.; Gupta, S.K.; Anand, U.; Ghorai, M.; Mundhra, A.; Rahman, M.H.; Ray, P.; Mitra, S.; Ray, D.; et al. Role of Polyamines in Molecular Regulation and Cross-Talks against Drought Tolerance in Plants. *J. Plant Growth Regul.* **2022**. [CrossRef]

172. Momtaz, O.A.; Hussein, E.M.; Fahmy, E.M.; Ahmed, S.E. Expression of S-adenosyl methionine decarboxylase gene for polyamine accumulation in Egyptian cotton Giza 88 and Giza 90. *GM Crops* **2010**, *1*, 257–266. [CrossRef] [PubMed]

173. Grzesiak, M.; Filek, M.; Barbasz, A.; Kreczmer, B.; Hartikainen, H. Relationships between polyamines, ethylene, osmoprotectants and antioxidant enzymes activities in wheat seedlings after short-term PEG- and NaCl-induced stresses. *Plant Growth Regul.* **2013**, *69*, 177–189. [CrossRef]

174. Zapata, P.J.; Serrano, M.; Garcia-Legaz, M.F.; Pretel, M.T.; Botella, M.A. Short Term Effect of Salt Shock on Ethylene and Polyamines Depends on Plant Salt Sensitivity. *Front. Plant Sci.* **2017**, *8*, 855. [CrossRef] [PubMed]

175. Freitas, V.S.; Miranda, R.D.S.; Costa, J.H.; Oliveira, D.F.D.; Paula, S.D.O.; Miguel, E.D.C.; Freire, R.S.; Prisco, J.T.; Gomes-Filho, E. Ethylene triggers salt tolerance in maize genotypes by modulating polyamine catabolism enzymes associated with H_2O_2 production. *Environ. Exp. Bot.* **2018**, *145*, 75–86. [CrossRef]

176. Takacs, Z.; Czekus, Z.; Tari, I.; Poor, P. The role of ethylene signalling in the regulation of salt stress response in mature tomato fruits: Metabolism of antioxidants and polyamines. *J. Plant Physiol.* **2022**, *277*, 153793. [CrossRef] [PubMed]

177. Cheng, Z.; Sattler, S.; Maeda, H.; Sakuragi, Y.; Bryant, D.A.; DellaPenna, D. Highly divergent methyltransferases catalyze a conserved reaction in tocopherol and plastoquinone synthesis in cyanobacteria and photosynthetic eukaryotes. *Plant Cell* **2003**, *15*, 2343–2356. [CrossRef] [PubMed]

178. Yasuno, R.; Wada, H. The biosynthetic pathway for lipoic acid is present in plastids and mitochondria in *Arabidopsis thaliana*. *FEBS Lett.* **2002**, *517*, 110–114. [CrossRef]

179. Monné, M.; Marobbio, C.M.T.; Agrimi, G.; Palmieri, L.; Palmieri, F. Mitochondrial transport and metabolism of the major methyl donor and versatile cofactor S-adenosylmethionine, and related diseases: A review. *IUBMB Life* **2022**, *74*, 573–591. [CrossRef]

180. Ravanel, S.; Block, M.A.; Rippert, P.; Jabrin, S.; Curien, G.; Rébeillé, F.; Douce, R. Methionine metabolism in plants: Chloroplasts are autonomous for de novo methionine synthesis and can import S-adenosylmethionine from the cytosol. *J. Biol. Chem.* **2004**, *279*, 22548–22557. [CrossRef]

181. Nunes-Nesi, A.; Cavalcanti, J.; Fernie, A.R. Characterization of In Vivo Function(s) of Members of the Plant Mitochondrial Carrier Family. *Biomolecules* **2020**, *10*, 1226. [CrossRef]

182. Li, Y.; Yang, Z.; Zhang, Y.; Guo, J.; Liu, L.; Wang, C.; Wang, B.; Han, G. The roles of HD-ZIP proteins in plant abiotic stress tolerance. *Front. Plant Sci.* **2022**, *13*, 1027071. [CrossRef]

183. Sharif, R.; Raza, A.; Chen, P.; Li, Y.; El-Ballat, E.M.; Rauf, A.; Hano, C.; El-Esawi, M.A. HD-ZIP Gene Family: Potential Roles in Improving Plant Growth and Regulating Stress-Responsive Mechanisms in Plants. *Genes* **2021**, *12*, 1256. [CrossRef] [PubMed]

184. Rai, K.K.; Rai, N.; Rai, S.P. Prediction and validation of DREB transcription factors for salt tolerance in Solanum lycopersicum L.: An integrated experimental and computational approach. *Environ. Exp. Bot.* **2019**, *165*, 1–18. [CrossRef]

185. Hassan, S.; Berk, K.; Aronsson, H. Evolution and identification of DREB transcription factors in the wheat genome: Modeling, docking and simulation of DREB proteins associated with salt stress. *J. Biomol. Struct. Dyn.* **2022**, *40*, 7191–7204. [CrossRef] [PubMed]

186. Wang, S.; Wang, Y.; Yang, R.; Cai, W.; Liu, Y.; Zhou, D.; Meng, L.; Wang, P.; Huang, B. Genome-Wide Identification and Analysis Uncovers the Potential Role of JAZ and MYC Families in Potato under Abiotic Stress. *Int. J. Mol. Sci.* **2023**, *24*, 6706. [CrossRef] [PubMed]

187. Valenzuela, C.E.; Acevedo-Acevedo, O.; Miranda, G.S.; Vergara-Barros, P.; Holuigue, L.; Figueroa, C.R.; Figueroa, P.M. Salt stress response triggers activation of the jasmonate signaling pathway leading to inhibition of cell elongation in *Arabidopsis* primary root. *J. Exp. Bot.* **2016**, *67*, 4209–4220. [CrossRef] [PubMed]
188. Su, Y.; Liang, W.; Liu, Z.; Wang, Y.; Zhao, Y.; Ijaz, B.; Hua, J. Overexpression of *GhDof1* improved salt and cold tolerance and seed oil content in *Gossypium hirsutum*. *J. Plant Physiol.* **2017**, *218*, 222–234. [CrossRef]
189. Zou, X.; Sun, H. DOF transcription factors: Specific regulators of plant biological processes. *Front. Plant Sci.* **2023**, *14*, 1044918. [CrossRef]
190. Iqbal, M.Z.; Jia, T.; Tang, T.; Anwar, M.; Ali, A.; Hassan, M.J.; Zhang, Y.; Tang, Q.; Peng, Y. A Heat Shock Transcription Factor TrHSFB2a of White Clover Negatively Regulates Drought, Heat and Salt Stress Tolerance in Transgenic *Arabidopsis*. *Int. J. Mol. Sci.* **2022**, *23*, 12769. [CrossRef]
191. Guo, M.; Liu, J.; Ma, X.; Luo, D.; Gong, Z.; Lu, M. The Plant Heat Stress Transcription Factors (HSFs): Structure, Regulation, and Function in Response to Abiotic Stresses. *Front. Plant Sci.* **2016**, *7*, 114. [CrossRef]
192. Han, G.; Li, Y.; Qiao, Z.; Wang, C.; Zhao, Y.; Guo, J.; Chen, M.; Wang, B. Advances in the Regulation of Epidermal Cell Development by C2H2 Zinc Finger Proteins in Plants. *Front. Plant Sci.* **2021**, *12*, 754512. [CrossRef]
193. Liu, Y.; Khan, A.R.; Gan, Y. C2H2 Zinc Finger Proteins Response to Abiotic Stress in Plants. *Int. J. Mol. Sci.* **2022**, *23*, 2730. [CrossRef] [PubMed]
194. Bankaji, I.; Sleimi, N.; Vives-Peris, V.; Gómez-Cadenas, A.; Pérez-Clemente, R.M. Identification and expression of the Cucurbita WRKY transcription factors in response to water deficit and salt stress. *Sci. Hortic.* **2019**, *256*, 108562. [CrossRef]
195. Hichri, I.; Muhovski, Y.; Žižková, E.; Dobrev, P.I.; Gharbi, E.; Franco-Zorrilla, J.M.; Lopez-Vidriero, I.; Solano, R.; Clippe, A.; Errachid, A.; et al. The *Solanum lycopersicum* WRKY3 Transcription Factor SlWRKY3 Is Involved in Salt Stress Tolerance in Tomato. *Front. Plant Sci.* **2017**, *8*, 1343. [CrossRef] [PubMed]
196. Beathard, C.; Mooney, S.; Al-Saharin, R.; Goyer, A.; Hellmann, H. Characterization of *Arabidopsis thaliana* R2R3 S23 MYB Transcription Factors as Novel Targets of the Ubiquitin Proteasome-Pathway and Regulators of Salt Stress and Abscisic Acid Response. *Front. Plant Sci.* **2021**, *12*, 629208. [CrossRef] [PubMed]
197. Dossa, K.; Mmadi, M.A.; Zhou, R.; Liu, A.; Yang, Y.; Diouf, D.; You, J.; Zhang, X. Ectopic expression of the sesame MYB transcription factor SiMYB305 promotes root growth and modulates ABA-mediated tolerance to drought and salt stresses in *Arabidopsis*. *Aob Plants* **2020**, *12*, z81. [CrossRef]
198. Punia, H.; Tokas, J.; Malik, A.; Sangwan, S.; Rani, A.; Yashveer, S.; Alansi, S.; Hashim, M.J.; El-Sheikh, M.A. Genome-Wide Transcriptome Profiling, Characterization, and Functional Identification of NAC Transcription Factors in Sorghum under Salt Stress. *Antioxidants* **2021**, *10*, 1605. [CrossRef]
199. Alshareef, N.O.; Wang, J.Y.; Ali, S.; Al-Babili, S.; Tester, M.; Schmöckel, S.M. Overexpression of the NAC transcription factor JUNGBRUNNEN1 (JUB1) increases salinity tolerance in tomato. *Plant Physiol. Biochem.* **2019**, *140*, 113–121. [CrossRef]
200. Ahmad, A.; Niwa, Y.; Goto, S.; Ogawa, T.; Shimizu, M.; Suzuki, A.; Kobayashi, K.; Kobayashi, H. bHLH106 Integrates Functions of Multiple Genes through Their G-Box to Confer Salt Tolerance on *Arabidopsis*. *PLoS ONE* **2015**, *10*, e126872. [CrossRef]
201. Krishnamurthy, P.; Vishal, B.; Khoo, K.; Rajappa, S.; Loh, C.; Kumar, P.P. Expression of AoNHX1 increases salt tolerance of rice and *Arabidopsis*, and bHLH transcription factors regulate AtNHX1 and AtNHX6 in *Arabidopsis*. *Plant Cell Rep.* **2019**, *38*, 1299–1315. [CrossRef]
202. Manavella, P.A.; Dezar, C.A.; Bonaventure, G.; Baldwin, I.T.; Chan, R.L. HAHB4, a sunflower HD-Zip protein, integrates signals from the jasmonic acid and ethylene pathways during wounding and biotic stress responses. *Plant J.* **2008**, *56*, 376–388. [CrossRef]
203. Rehman, S.; Mahmood, T. Functional role of DREB and ERF transcription factors: Regulating stress-responsive network in plants. *Acta Physiol. Plant.* **2015**, *37*, 178. [CrossRef]
204. Hu, Y.; Sun, H.; Han, Z.; Wang, S.; Wang, T.; Li, Q.; Tian, J.; Wang, Y.; Zhang, X.; Xu, X.; et al. ERF4 affects fruit ripening by acting as a JAZ interactor between ethylene and jasmonic acid hormone signaling pathways. *Hortic. Plant J.* **2022**, *8*, 689–699. [CrossRef]
205. Feng, B.; Han, Y.; Xiao, Y.; Kuang, J.; Fan, Z.; Chen, J.; Lu, W. The banana fruit Dof transcription factor MaDof23 acts as a repressor and interacts with MaERF9 in regulating ripening-related genes. *J. Exp. Bot.* **2016**, *67*, 2263–2275. [CrossRef] [PubMed]
206. Wang, Y.; Zhou, Y.; Wang, R.; Xu, F.; Tong, S.; Song, C.; Shao, Y.; Yi, M.; He, J. Ethylene Response Factor LlERF110 Mediates Heat Stress Response via Regulation of LlHsfA3A Expression and Interaction with LlHsfA2 in Lilies (*Lilium longiflorum*). *Int. J. Mol. Sci.* **2022**, *23*, 16135. [CrossRef] [PubMed]
207. Han, Y.; Fu, C.; Kuang, J.; Chen, J.; Lu, W. Two banana fruit ripening-related C2H2 zinc finger proteins are transcriptional repressors of ethylene biosynthetic genes. *Postharvest Biol. Technol.* **2016**, *116*, 8–15. [CrossRef]
208. Alessio, V.M.; Cavaçana, N.; Dantas, L.L.D.B.; Lee, N.; Hotta, C.T.; Imaizumi, T.; Menossi, M. The FBH family of bHLH transcription factors controls ACC synthase expression in sugarcane. *J. Exp. Bot.* **2018**, *69*, 2511–2525. [CrossRef] [PubMed]
209. An, J.P.; Yao, J.F.; Xu, R.R.; You, C.X.; Wang, X.F.; Hao, Y.J. An apple NAC transcription factor enhances salt stress tolerance by modulating the ethylene response. *Physiol. Plant.* **2018**, *164*, 279–289. [CrossRef]
210. Xu, L.; Xiang, G.; Sun, Q.; Ni, Y.; Jin, Z.; Gao, S.; Yao, Y. Melatonin enhances salt tolerance by promoting MYB108A-mediated ethylene biosynthesis in grapevines. *Hortic. Res.-Engl.* **2019**, *6*, 114. [CrossRef]
211. Wang, Z.; Wei, X.; Wang, Y.; Sun, M.; Zhao, P.; Wang, Q.; Yang, B.; Li, J.; Jiang, Y. WRKY29 transcription factor regulates ethylene biosynthesis and response in *Arabidopsis*. *Plant Physiol. Biochem.* **2023**, *194*, 134–145. [CrossRef]

212. Zhou, Y.; Xiong, Q.; Yin, C.C.; Ma, B.; Chen, S.Y.; Zhang, J.S. Ethylene Biosynthesis, Signaling, and Crosstalk with Other Hormones in Rice. *Small Methods* **2020**, *4*, 1900278. [CrossRef]
213. Yu, Z.; Duan, X.; Luo, L.; Dai, S.; Ding, Z.; Xia, G. How Plant Hormones Mediate Salt Stress Responses. *Trends Plant Sci.* **2020**, *25*, 1117–1130. [CrossRef] [PubMed]
214. Ryu, H.; Cho, Y. Plant hormones in salt stress tolerance. *J. Plant Biol.* **2015**, *58*, 147–155. [CrossRef]
215. Zou, X.; Liu, L.; Hu, Z.; Wang, X.; Zhu, Y.; Zhang, J.; Li, X.; Kang, Z.; Lin, Y.; Yin, C. Salt-induced inhibition of rice seminal root growth is mediated by ethylene-jasmonate interaction. *J. Exp. Bot.* **2021**, *72*, 5656–5672. [CrossRef] [PubMed]
216. Singh, D.; Debnath, P.; Sane, A.P.; Sane, V.A. Tomato (*Solanumlycopersicum*) WRKY23 enhances salt and osmoticstress tolerance by modulating the ethylene and auxin pathways in transgenic *Arabidopsis*. *Plant Physiol. Biochem.* **2023**, *195*, 11. [CrossRef] [PubMed]
217. Afzal, M.; Hindawi, S.E.S.; Alghamdi, S.S.; Migdadi, H.H.; Khan, M.A.; Hasnain, M.U.; Arslan, M.; Habib Ur Rahman, M.; Sohaib, M. Potential Breeding Strategies for Improving Salt Tolerance in Crop Plants. *J. Plant Growth Regul.* **2022**, *42*, 3365–3387. [CrossRef]
218. Ogawa, S.; Mitsuya, S. S-methylmethionine is involved in the salinity tolerance of *Arabidopsis thaliana* plants at germination and early growth stages. *Physiol. Plant.* **2012**, *144*, 13–19. [CrossRef]
219. Watanabe, M.; Chiba, Y.; Hirai, M.Y. Metabolism and Regulatory Functions of O-Acetylserine, S-Adenosylmethionine, Homocysteine, and Serine in Plant Development and Environmental Responses. *Front. Plant Sci.* **2021**, *12*, 643403. [CrossRef]
220. Verma, P.; Venugopalan, M.V.; Blaise, D.; Waghmare, V.N. Ethylene mediated regulation of fiber development in Asiatic cotton (*Gossypium arboreum* L.). *S. Afr. J. Bot.* **2020**, *135*, 349–354. [CrossRef]
221. Yousaf, S.; Rehman, T.; Tabassum, B.; Aftab, F.; Qaisar, U. Genome scale analysis of *1-aminocyclopropane-1-carboxylate oxidase* gene family in *G. barbadense* and its functions in cotton fiber development. *Sci. Rep.* **2023**, *13*, 4004. [CrossRef]
222. Liu, Z.; Ge, X.; Yang, Z.; Zhang, C.; Zhao, G.; Chen, E.; Liu, J.; Zhang, X.; Li, F. Genome-wide identification and characteriza tion of SnRK2 gene family in cotton (*Gossypium hirsutum* L.). *BMC Genet.* **2017**, *18*, 54. [CrossRef]
223. Livak, K.J.; Schmittgen, T.D. Analysis of Relative Gene Expression Data Using Real-Time Quantitative PCR and the $2^{-\Delta\Delta CT}$ Method. *Methods* **2001**, *25*, 402–408. [CrossRef] [PubMed]
224. Tuttle, J.R.; Haigler, C.H.; Robertson, D.N. Virus-induced gene silencing of fiber-related genes in cotton. *Methods Mol. Biol.* **2015**, *1287*, 219–234. [CrossRef] [PubMed]
225. Pertea, M.; Kim, D.; Pertea, G.M.; Leek, J.T.; Salzberg, S.L. Transcript-level expression analysis of RNA-seq experiments with HISAT, StringTie and Ballgown. *Nat. Protoc.* **2016**, *11*, 1650–1667. [CrossRef] [PubMed]
226. Anders, S.; Pyl, P.T.; Huber, W. HTSeq-a Python framework to work with high-throughput sequencing data. *Bioinformatics* **2015**, *31*, 166–169. [CrossRef] [PubMed]
227. Chen, C.; Chen, H.; Zhang, Y.; Thomas, H.R.; Frank, M.H.; He, Y.; Xia, R. TBtools: An Integrative Toolkit Developed for Interactive Analyses of Big Biological Data. *Mol. Plant* **2020**, *13*, 1194–1202. [CrossRef]

International Journal of
Molecular Sciences

MDPI

Review

Bridging the Gap: From Photoperception to the Transcription Control of Genes Related to the Production of Phenolic Compounds

Adriana Volná [1], Jiří Červeň [2], Jakub Nezval [1], Radomír Pech [1] and Vladimír Špunda [1,3,*]

[1] Department of Physics, University of Ostrava, 710 00 Ostrava, Czech Republic; adriana.volna@osu.cz (A.V.); jakub.nezval@osu.cz (J.N.); radomir.pech@osu.cz (R.P.)
[2] Department of Biology and Ecology, University of Ostrava, 710 00 Ostrava, Czech Republic; jiri.cerven@osu.cz
[3] Global Change Research Institute, Czech Academy of Sciences, 603 00 Brno, Czech Republic
* Correspondence: vladimir.spunda@osu.cz

Abstract: Phenolic compounds are a group of secondary metabolites responsible for several processes in plants—these compounds are involved in plant–environment interactions (attraction of pollinators, repelling of herbivores, or chemotaxis of microbiota in soil), but also have antioxidative properties and are capable of binding heavy metals or screening ultraviolet radiation. Therefore, the accumulation of these compounds has to be precisely driven, which is ensured on several levels, but the most important aspect seems to be the control of the gene expression. Such transcriptional control requires the presence and activity of transcription factors (TFs) that are driven based on the current requirements of the plant. Two environmental factors mainly affect the accumulation of phenolic compounds—light and temperature. Because it is known that light perception occurs via the specialized sensors (photoreceptors) we decided to combine the biophysical knowledge about light perception in plants with the molecular biology-based knowledge about the transcription control of specific genes to bridge the gap between them. Our review offers insights into the regulation of genes related to phenolic compound production, strengthens understanding of plant responses to environmental cues, and opens avenues for manipulation of the total content and profile of phenolic compounds with potential applications in horticulture and food production.

Keywords: photoreceptors; radiation; temperature; transcription factors

Citation: Volná, A.; Červeň, J.; Nezval, J.; Pech, R.; Špunda, V. Bridging the Gap: From Photoperception to the Transcription Control of Genes Related to the Production of Phenolic Compounds. *Int. J. Mol. Sci.* **2024**, *25*, 7066. https://doi.org/10.3390/ijms25137066

Academic Editor: Abir U. Igamberdiev

Received: 31 May 2024
Revised: 21 June 2024
Accepted: 25 June 2024
Published: 27 June 2024

1. Introduction

Phenolic compounds (PheCs) are a group of secondary metabolites with antioxidative properties and UV screening function [1]. It was documented that several abiotic environmental factors lead to their increased accumulation—for example drought, decreased temperature, increased soil salinity, and exposure to heavy metals or UV radiation [2–6]. From these reasons, their involvement in antioxidative defense is assumed as well as their contribution to the increased stress tolerance. It is also assumed that their accumulation in the upper layers of the epidermis can greatly affect the total amount of UV radiation penetrating the deeper layers of mesophyll and therefore prevent excess radiation from reaching the lower cell layers of the leaf [7]. These facts suggest the irreplaceable role of PheCs in plant protection against abiotic environmental cues, although to date, direct proof of their contribution to antioxidative protection or data supporting their involvement in the processes leading to increased stress tolerance are scarce. Therefore, we decided to summarize the current knowledge about the transcription control of genes related to phenolic compound production and to explain how the transcription factors (TFs) driving the expression are related to light perception via the photoreceptors.

In addition to UV radiation, another crucial environmental factor driving (among others) the accumulation of PheCs is the irradiance and spectral composition of photosynthetically active radiation (PAR). In general, incident radiation is perceived in two possible ways—by the direct and indirect light perceptions. Indirect light sensing does not require any specialized molecules and is perceived via the multiple sensory mechanisms related to photosynthetic performance, and thus the availability of sugars [8–12], or increased accumulation of reactive oxygen species [13–15] and related alterations in the redox state of the cells. On the contrary, direct light perception is ensured via specialized photoreceptors (more details can be found in Sections 2.1–2.5). These proteins have a light-absorbing molecule (this can be pigment, flavin nucleotide, or a cluster of aromatic amino acids) that is responsible for the perception itself. In general, after the light absorption, the redox state of the chromophore is changed [16] which leads to alterations in the higher-order protein structures (depending on the specific photoreceptor it can be either monomerization, dimerization, or changes in the protein folding) which in turn affects the capability to interact with specific proteins or protein complexes (more details about direct light perception in plants can be found in Section 2).

In general, after the light-induced activation of the photoreceptors (Figure 1), the interaction with the COP1/SPA (CONSTITUTIVE PHOTOMORPHOGENIC 1 and SUPPRESSOR OF PHYTOCHROME A-105) complex (more details can be found in Section 4) follows (with the exception of zeitlupes and phototropins). This large protein complex can induce the ubiquitination and proteolysis of several dozens of substrates including the transcription factors [17] affecting the expression of genes related to the production of phenolic compounds. Therefore, the COP1/SPA protein complex integrates the vast majority of the light sensing from the various photoreceptors [17] and coordinates the expression of target genes via balancing the availability of transcription factors (TF, especially of HY5 TF, which is involved in the induction of the transcription in genes related to the production of PheCs). More details about the structure, function, and interactome of the COP1/SPA complex can be found in Section 4.

General scheme of light perception

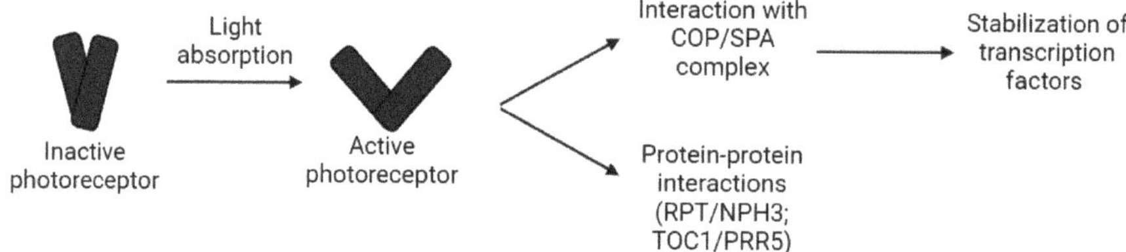

Figure 1. General scheme of direct light perception in plants (simplified). The absorption of light turns inactive photoreceptors into active states and causes conformational changes, which differ between the different types of photoreceptors (in some cases monomerization occurs, while in others dimerization, and other changes in higher-order structures are documented). Such changes significantly affect the protein–protein interactions that follow and are different for different photoreceptors. The vast majority of activated photoreceptors interact with the COP/SPA/E3 ubiquitin ligase complex and therefore lead to the stabilization of transcription factors driving the gene expression of target genes. The remaining photoreceptors (zeitlupes and phototropins) interact with other proteins and do not contribute to the COP/SPA signaling.

The genes responsible for producing phenolic compounds can undergo regulation through either a single relatively specific transcription factor [18] which directly interacts with a particular region in the gene's promoter, or they can form a specialized complex composed of several such proteins that collectively regulate the same gene [19]. Depending on the position in the biosynthetic pathway, we can distinguish the early and late genes related to the phenolic compounds production depending on the position in the biosynthetic pathway. Early genes have a rather simple regulatory mechanism via the single transcription factor (no complex is needed), which requires the presence of a specific sequence in the promoter region. On the contrary, the late genes related to the production of PheCs have a more complex regulatory control, which is ensured via the MBW complex consisting of three main components (M—MYB proteins, B—bHLH proteins, and W—WD40 repeat proteins). Therefore, for control of these "late genes" of PheCs biosynthesis, all three components of the MBW complex are necessary [19]. Details about the MBW's complex structure, function, and regulation can be found in Section 5.6.

The final part of the review (Section 7. Future Perspectives) is dedicated to the summarization of the presented knowledge and highlighting possible applications in the horticulture and indoor cultivation of important agronomical plants and crops, which might lead to the production of more resilient plants with adjusted content of protective metabolites without the adverse effects of the imbalanced light conditions on plant production [20,21] and morphology. To reach these aims we have to elucidate the mechanisms and regulation of light signaling and their crosstalks with sensing of other environmental cues.

2. Direct Light Sensing

As mentioned above (Section 1), direct light sensing is ensured by specialized proteins called photoreceptors. These proteins vary in the domain architecture, structure, bound chromophore, and absorption properties (as depicted in Figure 2) and some of them even in the interactions after the light-induced activation (as shown in Figure 3).

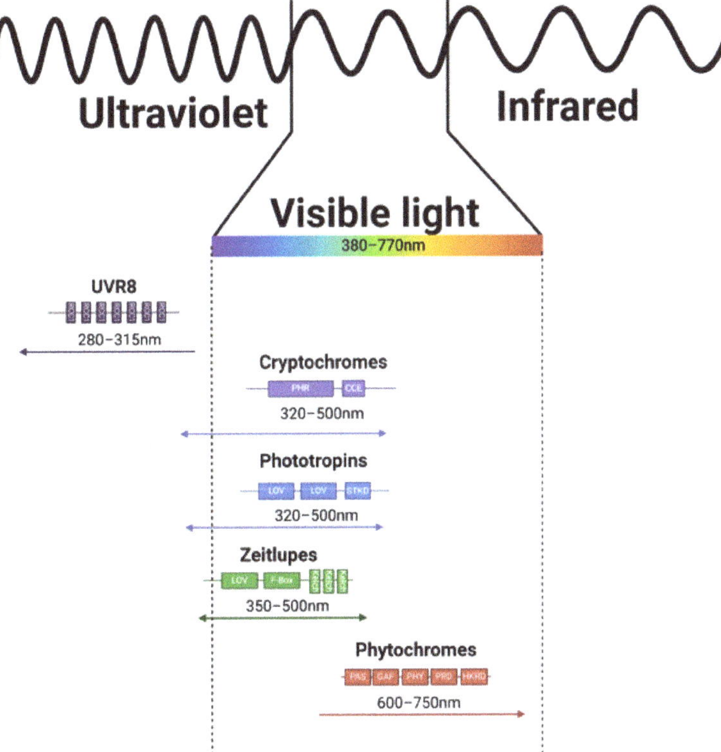

Figure 2. Domain architecture and absorption ranges of the photoreceptors. This scheme represents each group of known plant photoreceptors with highlighted absorption ranges, domain architecture, and the corresponding part of the spectra these photoreceptors play a role in (adopted and modified based on publications [22–25]).

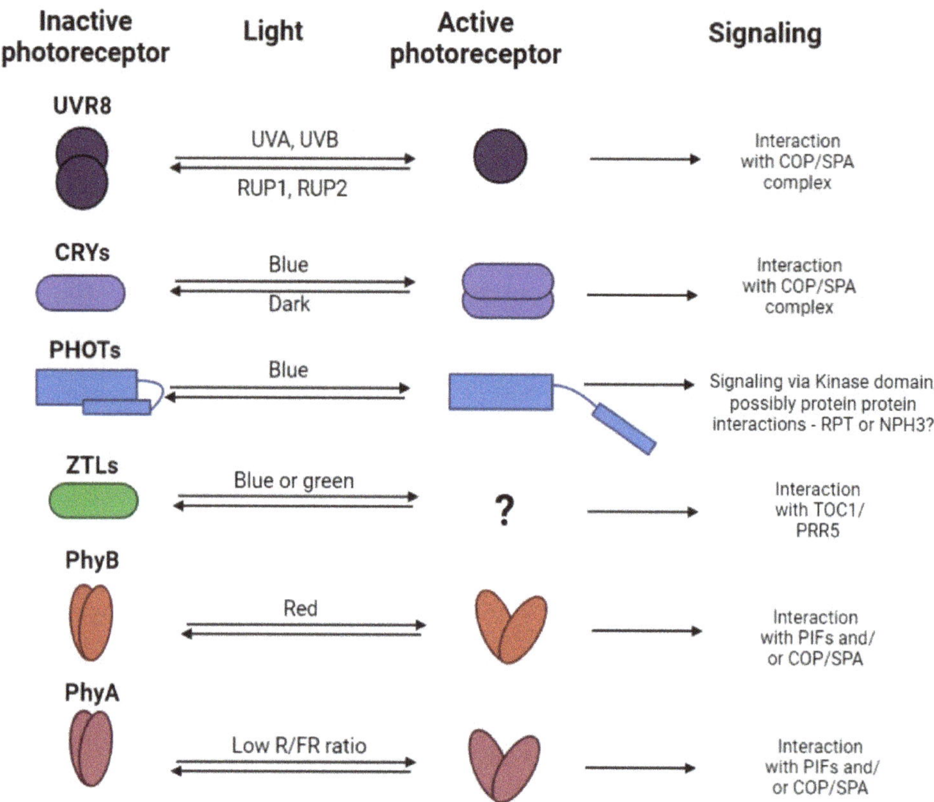

Figure 3. The mechanism of the photoreceptors sensing, signaling, and regeneration. For each plant photoreceptor, the basic mechanism of the photo sensing is indicated (it can be either monomerization, dimerization, or conformational changes), together with the light region that induced the active state of photoreceptors, and the signaling pathway where the light signal is further propagated (the related literature is cited in the text accompanying this scheme).

2.1. Ultraviolet-B Receptor

The specific group of photoreceptors are UVR8s. The absorption range is from 280 to 315 nm [26] and differs for the monomeric and dimeric state [27], therefore these photoreceptors are able to absorb UV-A and also UV-B radiation.

The specificity of UVR8 lies in the fact that it does not contain a nonprotein chromophore, but instead, it has several RCC1 domains resulting in a beta-propeller structure with aromatic amino acids (tryptophans) situated at the propeller's core. Tryptophans, being aromatic amino acids, possess the capability to directly absorb light and transmit excitation energy. Within the UVR8 protein, these amino acids aggregate into three tryptophan clusters, with six tryptophans forming the distal cluster, three in the proximal cluster, and four in the central cluster, effectively fulfilling the role typically played by nonprotein ligands (such as FAD or FMN) in other photoreceptors. In Arabidopsis, the transport of excitation energy within UVR8 is facilitated through resonance energy transfer [28], starting from the distal tryptophan cluster (W39, W92, W144, W196, W300, W352), then passing through the proximal cluster (W198, W250, W302), and finally reaching the central cluster (W233, W285, W337, W94). Although the ultimate destiny of the excitation energy remains unknown, it is established that conformational change occurs within the entire UVR8 structure, leading to UVR8 monomerization and consequent alterations in its potential interactions with other proteins.

Inactive UVR8 forms dimers, but once the UV radiation photon is absorbed, the dissociation occurs and UVR8 forms an active monomer. Active UVR8 monomer binds the COP1/SPA/E3 ubiquitin ligase complex and induces COP1/SPA dissociation from the enzymatic complex (CUL4/DDB1/E3 ubiquitin ligase; Figure 3). Active complex UVR8/COP1/SPA then migrates to the nucleus. Here, the UVR8/COP1/SPA complex directly interacts with the HY5 transcription factor and therefore increases its stability. In addition, HY5s binding to its own promoter [29] can be induced.

The equilibrium between the active and inactive UVR8 photoreceptor [30] is maintained via the interaction with the RUP proteins (specifically the RUP1 and RUP 2 sharing 63% of their sequence [31]), which physically interact with the monomeric UVR8 protein and facilitate its return to the ground dimeric state [32].

2.2. Cryptochromes

Direct light sensing occurs through photoreceptors that operate within specific spectral intervals of solar radiation (Figure 1). Blue light is detected by a diverse group of photoreceptors, including cryptochromes (CRYs), phototropins (PHOTs), and zeitlupes (ZTLs). CRYs, which are relatively short proteins (over 480 amino acids in *Hordeum vulgare*), bind flavin adenine dinucleotide, enabling light absorption [33] in the range of 320–500 nm [34]. CRY1 and CRY2 are distinguished within the CRYs, and their signaling can modulate the expression of 10–20% of *Arabidopsis thaliana* coding genes [35]. In monocot plants, CRY1 is encoded by two different genes, designated CRY1a and CRY1b, while CRY2 is encoded independently, differing from CRY1 in gene sequence length, exon–intron structure, presence or absence of 5′UTR, and amino acid composition of the resulting protein [36]. Structurally, they contain PHR (Photolyase Homologous Region) and CCE (CRY C-terminal Extension) domains [36]. In addition to CRY1 and CRY2, *Arabidopsis thaliana's* genome also contains CRY3 [37]. Unlike CRY1 and CRY2, CRY3 exhibits a distinct structure characterized by the absence of a C-terminal extension necessary for COP/SPA complex interaction [37]. This structural variance translates into functional differences as well. It appears that CRY3 diverges from the typical photoreceptor role in plants and instead functions as a DNA photolyase [37]. Homologs of CRY3 were also identified in other plant species—for example in tomato and rice [37].

The general model of CRYs sensing involves blue light absorption by incorporated FAD, leading to conformational changes that form CRY homodimers. The active dimer is phosphorylated and interacts with the COP1/SPA (Figure 3) E3 ubiquitin ligase complex, which leads to the initiation of gene expression [38]. Additionally, the transition of CRYs from monomers to dimers can be repressed by interaction with BIC1 and BIC2 proteins (BLUE-LIGHT INHIBITOR OF CRYPTOCHROMES 1, and BLUE-LIGHT INHIBITOR OF CRYPTOCHROMES 2), providing additional signaling control. In addition, recent knowledge suggests the potential for CRYs to form tetramers [39–41].

2.3. Phototropines

The second group of blue light photoreceptors, PHOTs, are approximately 900 amino acids long in *Arabidopsis thaliana* and possess two LOV (Light-oxygen-voltage-sensing domain) domains, a serine/threonine kinase domain, and a Jα helix [42]. Light absorption by their chromophores, FMNs [43], induces conformational changes, activating the kinase domain [43–46]. Although the PHOT signaling pathway is not fully elucidated, proteins such as NPH3 [47] and RPT2 [48,49] have been identified to interact with PHOT1(Figure 3), playing roles in photoreceptor complex constitution, phototropism, and stomata opening.

2.4. Zeitlupes

Another set of photoreceptors capable of light perception is represented by adagio proteins. These proteins are known for sensing in the blue/green light regions, with an absorption range from 350 to 500 nm [50], covering UVA, blue, cyan, and green spectral components. ZTLs feature LOV, F-box, and kelch domains, with the LOV domain, containing flavin (FAD, FMN, or riboflavin [51]) bound to the LOV protein via cysteine residues. Together with FKF1 (Flavin-kelch-Fbox-1) and LKP (LOV kelch protein 1), ZTLs primarily function in sensing photoperiod duration and flowering time [50]. ZTLs and FKF1 can form heterodimers with the GI protein, induced by blue light. Decreasing light dose [50], particularly its blue component, during light/dark transitions releases ZTL from the GI (GIGANTEA) complex, enabling interaction with TOC1 (Timing Of Cab expression 1; transcription repressor) for degradation during the dark period, thereby influencing the transcription of target genes. Similarly, decreasing light dose leads to the release of FKF1 from the GI heterodimer, allowing interaction with CDF (Cycling Dof Factor) to trigger the transcription of related genes [50,52].

2.5. Phytochromes

The exclusive photoreceptors responsible for perceiving red and far-red light are phytochromes (PHYs). Phytochromes are approximately 124 kDa proteins that form homodimers and contain covalently bound pigments, specifically phytochromobilin [53]. Their tertiary structure comprises N-terminal extension, PAS (per-ARNT-sim), GAF (cGMP phosphodiesterase/adenylyl cyclase/FhlA domain), PHY (phytochrome specific domain), PRD (PAS-related domain), and HKRD (histidine kinase-related domain) [54]. The phylogenetic analysis allows the distinction of phytochrome sub-family members, which appear to be genus specific, resulting in varying numbers of members among plant species. For instance, *Arabidopsis thaliana* exhibits PhyA-PhyE, while spring barley displays PhyA-PhyC [42]. However, studies indicate that PhyA [55] and PhyB [56,57] are the most crucial for environmental sensing, while the roles of other PHY family members remain to be fully elucidated. The precise mechanism of Phy-related signaling will be detailed below.

Phytochromes exist in two distinct states: inactive (Pr) and active (Pfr), whose transitions are triggered by red and far-red radiation, respectively, as depicted in Figure 4 and their respective ratios. While red light activates PhyB, far-red light deactivates it (a process that can be induced by increased temperature). These transitions involve conformational changes that result in shifts of absorption spectra and alter the potential for interaction with other proteins. When in the active state, phytochrome can interact with phytochrome-interacting factors (PIFs—PIF4 and 7), which are responsible for positively controlling gene expression as transcriptional activators. This interaction leads to ubiquitination and subsequent degradation by the 26S proteasome, thereby reducing gene expression. In contrast, when phytochromes are inactivated (due to increased temperature or exposure to far-red light), they are unable to interact with PIFs, allowing for the initiation of transcription of target genes [56,58].

PIFs target genes that contain G-boxes in their promoters, such as genes related to PheCs biosynthesis which feature this sequence motif, or G/PBE boxes. PIFs can form homodimers or heterodimers with different PIFs or completely different proteins [59]. An example of this is their interaction with HY5 ([59] Protein LONG HYPOCOTYL 5), a participant in light-related signaling.

In addition to their direct interaction with activated phytochromes, the transcription of PIFs (such as PIF4) and the translation of PIF7 are finely regulated by alternative mechanisms. For instance, under lower temperatures, the expression of PIF4 is inhibited by the ELF3 (EARLY FLOWERING 3) protein. However, as temperatures rise, ELF3 undergoes phase separation, leading to the initiation of transcription of target genes. Similarly, a hairpin structure localized in the 5'-UTR of PIF7 mRNA undergoes a temperature-induced change in structure, directly correlating with increased synthesis of PIF7 proteins [61].

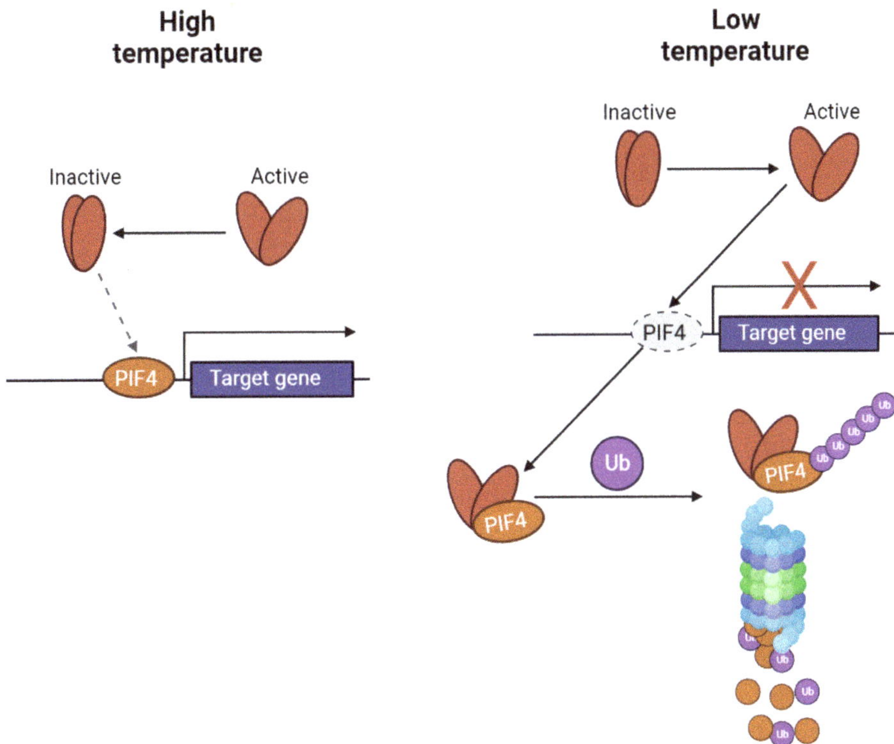

Figure 4. Mechanism of the transcription control of target genes by PhyB interaction with PIFs (scheme was adopted from [60] and modified).

3. Photoreceptors and Temperature

Furthermore, photoreceptors involved in UV and blue/green light sensing, including PHOTs, CRYs, ZTLs, and UVR8 [62], may also participate in thermosensing processes. For example, in *Marchantia polymorpha*, PHOTs exhibited prolonged activity under decreased temperatures [63], while in *Arabidopsis thaliana*, a greater population of active CRYs was observed at lower temperatures compared to moderate ones [64]. The activity of UVR8 photoreceptors also appears to be influenced by temperature, with a reduced proportion of inactive UVR8 dimers observed at lower temperatures [30], indicating increased monomerization. Additionally, the number of active UVR8 monomers can be modulated by RUP (REPRESSOR OF UV-B PHOTOMORPHOGENESIS1 and REPRESSOR OF UV-B PHOTOMORPHOGENESIS2) proteins [30]. ZTL inactivation progresses with increasing temperatures, yet ZTL proteins are believed to be involved in heat stress tolerance, as evidenced by studies with ZTL-deficient and ZTL-overexpressing mutant lines of *Arabidopsis thaliana* [65]. Thus, plant photoreceptors not only play a primary role in photosensory functions but also share a role in thermosensing, regulated by temperature-induced changes in their active state duration. Lower temperatures prolong the duration of the active state, whereas higher temperatures decrease it. While an increasing number of publications support this assumption, direct experiments are required to confirm or refute this hypothesis. Additionally, plants possess numerous other thermosensing mechanisms [66] beyond the scope of this review.

4. Constitutive Photomorphogenesis Protein 1/Protein SPA1-Related Protein Complex

The majority of activated photoreceptors, excluding zeitlupes and phototropins, interact with the COP1/SPA protein complex (see Figure 3). This complex comprises the COP1 protein, which exists as a single copy gene in *Arabidopsis thaliana*, and one of four SPA proteins. Research has shown that COP1/SPA is part of a larger complex alongside CUL4 (Cullin-4), DDB1 (DNA Damage-Binding Protein 1), and RBX (RING-Box Protein) [17].

In *Arabidopsis thaliana*, this enzymatic complex plays a role in degrading numerous proteins, including transcription factors (TFs) such as BIT1, HY5, HYH, PIF1, PIF5, PIF8, and PIL1, as well as photoreceptors (CRYs, PHYs) [17] involved in various processes like photomorphogenesis, light perception, anthocyanin biosynthesis, and phytohormone signaling. Typically, in darkness, different substrates, primarily TFs, undergo ubiquitination, leading to the suppression of target genes. However, upon light activation of photoreceptors, the activity of the COP1/SPA complex is inhibited, resulting in reduced substrate degradation and increased TF population, thereby activating downstream genes. Although the COP1/SPA complex generally plays an inhibitory role by inducing substrate degradation, it exhibits unique interactions. For instance, when interacting with activated UVR8, it stabilizes the HY5 transcription factor. Additionally, in association with PHYs, it facilitates the degradation of PIFs, transcriptional repressors.

5. Transcription Factors Involved in PheCs Biosynthesis Control

PheCs biosynthetic pathway forms an important branch of higher plant secondary metabolism. PheCs content is effectively regulated by several environmental cues and their co-action (such as PAR, UV, LT, low N, high CO_2, etc.) and they are often studied for their possible contribution to stress-related responses and enhancement of plant tolerance. Below, we intend to describe in detail how the transcriptional control of genes is directly involved in the production of such universal/multipurpose substances. TFs responsible for the coordination of the gene expression contain one of the following motifs: bHLH, bZIP, HTH, or WRKY, and are listed in Table 1.

5.1. Proteins Containing Basic Helix–Loop–Helix Motifs (So-Called bHLH)

This diverse group of proteins contains (as suggested by the acronym bHLH) a basic helix–loop–helix motif and comprises 26 protein subfamilies [67], and all of them are responsible for direct interaction with the DNA and transcription control. These proteins are present in almost all plant species and in some of them, more than 100 different bHLH proteins were documented. The estimated origin of this group of proteins dates back 440 million years ago to the first land plants [67]. They are involved in the various processes under both optimum [68] and stress conditions [69].

An important protein containing a bHLH structural motif is TT8 (TRANSPARENT TESTA 8; also known as AtbHLH42, bHLH42, or Transcription factor EN32) [42]. According to the literature, this protein is a 518 amino acids long transcription activator, when associated with MYB75 (PAP1) or MYB90 (PAP2) [42]. In addition, this protein controls the expression of the DFR (DIHYDROFLAVONOL 4-REDUCTASE; Table 1) and BAN (BANYULS) genes [70] in *Arabidopsis thaliana* siliques. DFR gene codes for the DFR protein which catalyzes the formation of leucopelargonidin from dihydrokaempferol or the formation of leucocyanidin from dihydroquercetin [70]. BANYULS gene codes for a DFR-like protein [71] which is a negative regulator of anthocyanin biosynthesis. Plants with knocked-out BAN displayed a higher accumulation of anthocyanins in the seed coat. In addition, it was even proposed that this gene codes for leucoanthocyanin reductase (LAR), but no direct evidence was reported. On the contrary, it was documented that BAN genes of *Medicago truncatula* and *Arabidopsis thaliana* encode for anthocyanidin reductase catalyzing the formation of 2,3-cis-flavan-3-ols from anthocyanins and therefore the reduction in coloration [72].

5.2. Basic ZIPper-Containing Proteins (bZIP)

The second group of the TFs related to the PheCs production are basic (leucine) zipper (bZIP)-containing proteins (Table 1). Their common ancestor (the divergence point varies depending on the plant species; monocots and eudicots are estimated 140–150 million years ago) probably had four bZIP genes, and those proteins were probably involved in the oxidative stress responses and also in the responses to light [73]. This assumption is supported also by current literature—where bZIP family members are referred to as key players in the transcription control of the genes triggered in the response to light and photomorphogenesis. Moreover, this diverse group of TFs is involved also in responses to other abiotic stresses [74], as well as in the responses to pathogens [75].

The most important member of this protein family is AtZIP56 (*Arabidopsis thaliana* basic leucine zipper 56) also known as HY5 (ELONGATED HYPOCOTYL 5). This TF interacts directly with the G-box motif (-GCCACGTGC/GC-; but interactions with E-box, GATA-box, Z-box, C-box, and other hybrid boxes of the DNA, including promoters related to the PheCs production, are documented as well [76]). Due to the interaction with the COP/SPA protein complex, which is involved in the signaling of the environmental cues (both light intensity and spectral quality, as well as temperature), HY5 is designated as an "integrator of light and temperature signals" [77]. In *Arabidopsis thaliana*, HY5 is a 168 amino acids long protein containing disordered bZIP, leucine zipper, basic motif, and an interaction site for COP1 (-ESDEEIRRVPEF-; [42]). In the dark, HY5 is ubiquitinated after the interaction with the COP1 and degraded at the 26S proteasome complex [78]. On the contrary, in the light COP1 is inhibited, and therefore the proteins HY5/HYH/CAM7 (HY5—ELONGATED HYPOCOTYL 5, HYH—HY5 HOMOLOG, CAM7—Calmodulin-7) can interact with the HY5 promoter and induce HY5s transcription. This results in the temporary accumulation of HY5 proteins which can trigger the expression of the genes with the G box [78]-containing promoters, including those related to the PheCs production.

The second and equally important member of the bZIP family is HYH (HY5 HOMOLOG). In *Arabidopsis thaliana* it is 149 amino acids long protein with the disordered, bZIP, basic motif, leucine zipper regions, and interaction site for COP1 [42]. HY5 and HYH share an identity of 49% (73 out of 149 amino acid residues), and in the DNA-binding region, the homology in the amino acid sequence steeply increases to 87.5% (21 out of 24 amino acid residues) [79]. Similarly to HY5, also HYH is involved in the regulation of the expression of genes related to the biosynthesis of PheCs. It was documented that HYH directly interacts with the COP1's WD-40 domain and the main consequence is the degradation of the HYH protein. In addition, both proteins (HY5 and HYH) are involved in the regulation of the transcription in the same pathway—biosynthesis of anthocyanins under the low temperature in the *Arabidopsis thaliana* [80].

5.3. Transcription Factors Containing Helix–Turn–Helix Motifs (HTH TFs)

The third group of the TFs is helix–turn–helix motif-containing proteins (HTHs). MYB proteins (MYBs; the acronym is derived from Avian myeloblastosis virus, where the first protein containing such structure was reported), for example, belong to this diverse group of proteins. MYBs have a conserved DNA-binding domain that consists of amino acid repeats (R meaning one repeat in the structure descriptions below) which can be imperfect and are approximately 50 amino acids long. Each of the repeats forms α-helices, but the most important are the second and third helices of each repeat—those directly recognize (the third one), interact (the second one) with the DNA, and form HLH structure [81]. In *Arabidopsis thaliana*, 197 genes coding for MYB TFs were identified in all 5 chromosomes which vary in the number of the repeats and therefore form 4 subfamilies [82]: MYB-R1R2R3, MYB-R2R3, MYB-related genes, and atypical MYB genes.

TT2 (Protein TRANSPARENT TESTA2, also known as AtMYB123) is a 258 amino acids long transcription factor. When associated with bHLH2/EGL3/MYC146 (basic Helix Loop Helix protein 2, also known as Protein ENHANCER OF GLABRA 3, or Transcription factor EN 30), bHLH12/EN52/MYC1 (basic Helix Loop Helix protein 12, also known as

Transcription factor EN 58, or Myelocytomatosis protein 1), or bHLH42/tt8 (basic Helix Loop Helix protein 42, also known as TRANSPARENT TESTA8) work as transcription activator involved in the control of late anthocyanin genes (Table 1; [83]).

In *Arabidopsis thaliana*, PAP1 (Probable plastid-lipid-associated protein 1, also known as FBN1a, FIB1a, or PGL35) is 318 amino acids long protein involved in the transcription control (genes controlled by the PAP1 are mentioned below). Expression of this protein stimulates the jasmonic acid biosynthesis and it was documented that increased accumulation of PAP1 is linked with the increased tolerance of the PSII against photoinhibition in Arabidopsis [84]. On top of that, the PAP1 protein is a key player in the PheCs regulation for example in *Arabidopsis thaliana* [85], *Salvia miltiorrhiza* [86], *Brassica napus* [87], and *Nicotiana tabacum* [88].

PAP2 (also known as FBN1b or FIB1b) is 310 amino acid residues long transcription factor in *Arabidopsis thaliana*. Similarly to PAP1, PAP2 is involved in stress responses and its expression also stimulates jasmonic acid biosynthesis [89]. In addition, PAP2 in *Arabidopsis thaliana* is involved in the protection of PS II against photooxidative stress (induced by high light in combination with cold). According to the Plant Transcription Factor Database (Table 1), PAP2 stimulates the expression of TTG2 in *Arabidopsis thaliana*. Therefore, both proteins (PAP1 and PAP2) are connected with growth regulation and PSII protection, but also with anthocyanin accumulation under stress conditions [89].

According to the current knowledge, other members of the MYB family are equally important. An example of these members is the trio MYB11, MYB12, and MYB111, which are involved in the control of genes involved in PheCs biosynthesis. All three proteins have the same type of the MYB domain—R2R3, have a similar length (MYB11—343 amino acids, MYB12—371 amino acids, MYB111—342 amino acids), and are designated as flavonol-specific TFs because all of them positively regulate the transcription of the *CHS, CHI, F3H,* and *FLS1* genes in *Arabidopsis thaliana*.

5.4. WRKY Proteins

This group of TFs is also involved in very diverse processes related to various types of abiotic stresses [90]. The characteristic of those TFs is that all members of the WRKY family have WRKYGQ repetitions near the N-terminus and can directly interact with W-box ((C/T) TGAC (T/C))-containing promoters [90]. In *Arabidopsis thaliana*, there were identified 74 genes coding for the WRKY TFs [90]. W-box motif in the promoter region is present mainly in genes involved in the biosynthesis of lignin [91], flavonols, or tannins [92]. WRKY TFs therefore contribute to the control of the secondary metabolism. For example, in the *Arabidopsis thaliana*, WRKY23 probably controls the expression of *TT4, TT5, TT6,* and *TT7* genes [93] (TRANSPARENT TESTA proteins 4–7; the genes coding them are also known as *CHS*, Chalcone-flavanone isomerase, Naringenin,2-oxoglutarate 3-dioxygenase, and Flavonoid 3′-monooxygenase) [42]. The WRKYGQ repeat tract is responsible for the interaction with the DNA, which directly interacts with the W-box sequence in the major groove as was documented in *Solanum lycopersicum* (WRKY3 and WRKY4) [94].

Several subclasses of the WRKY proteins were described based on their structure—I, II, and III. Group I WRKY proteins have two WRKY domains, group II proteins have one WRKY domain and one zinc finger (Cys2-His2) motif, and group three have one WRKY domain and one zinc finger (but different from the group II—Cys2-His/Cys Cys2-His2 zinc-finger motif) [95]. In addition to the numeric description of the domain architecture, other higher-order structure features can be indicated by the lowercase letters a–e. The wide structural diversity of the WRKY proteins therefore enables them to interact with each other and further increase their regulatory potential. For example, in the *Arabidopsis thaliana*, members of the IIb group can interact with each other (AtWRKY6/AtWRKY42), but also with the IIa group (AtWRKY36/AtWRKY40). Similarly, the WRKY proteins belonging to the IIa group can both interact with each other or with the IIb group members (AtWRKY36/AtWRKY60). Lastly, the members of group III can interact with each other (AtWRKY30/AtWRKY53), but also with the IIa group members [95] (AtWRKY40/AtWRKY38). The limited regulatory potential of the IId members is quite interesting, whose exclusive interaction with the

IIa group (OsWRKY71/OsWRKY51) was documented in *Oryza sativa*. In addition to this (WRKY-WRKY protein interactions), the interactions with other proteins were documented. For example with the VQ motif-containing proteins (HAIKU1, MAPK4, MKS1, or even with the chromatin remodeling proteins (HDA19) [96].

An interesting member of the WRKY TFs family involved in light-related signaling control is WRKY36, which is the transcriptional inhibitor of the *HY5* gene coding for the HY5 protein. It was documented that UVR8 can directly interact with the WRKY36 in *Arabidopsis thaliana* and also that UVB exposure increases the amount of UVR8/WRKY36 dimers [97]. In combination with the fact that UVB leads to increased UVR8 accumulation in the nucleus, authors proposed [97] that those active UVR8 monomers (formed after the UVB exposure) migrate to the nucleus and pull the WRKY36 transcription inhibitor from the promoter of the *HY5* gene, and thus lead to the increased transcription of the central light and temperature integrator (HY5) which controls both other TFs (e.g., MYB12) as well as PheCs-related genes.

5.5. WD-40

WD-40 proteins are characterized by a WD-40 domain containing multiple repeats [98], where each repeat contains 44–60 amino acids [99]. Each unit contains dipeptides (either glycine-histidine (GH; in the proximity of the N-termini) or tryptophan-asparagine (WD; in the proximity of the C-termini)) and can be folded into the four-stranded antiparallel beta-sheet, which is one segment of the beta-propeller structure (a similar structure to the UVR8 photoreceptor in the monomeric state; [99]). Each WD-40 protein has at least four such segments [98].

In *Arabidopsis thaliana*, 269 WD-40 proteins were identified and can be distinguished in approximately 113 subfamilies which have significant homology with their human or fruit fly counterparts suggesting their importance [99].

For example, the COP1 (CONSTITUTIVE PHOTOMORPHOGENIC 1) protein is a member of the WD-40 family having seven WD repeats in Arabidopsis [98]. This particular protein is involved in the light signaling and control of photomorphogenesis via the direct repression of the transcription activators [98]. Recently, it was also documented that it stabilizes the transcription repressors (PIFs, EIN3, or EIL1; [100]).

Another example of the WD-40 protein can be SPA proteins [101]. In *Arabidopsis thaliana*, four SPA (SPA1, SPA2, SPA3, SPA4) proteins were identified and each one of them can interact with the COP1 [101].

5.6. MYB-bHLH-WD-40 Complex

For a late group of PheCs-related genes, transcription is controlled at multiple levels. In addition to the direct interaction of the promoters with the various TFs (for example TT2), the MBW complex (consisting of the MYB-bHLH-WD-40 proteins) also contributes to the coordination of the gene expression. In addition, *Arabidopsis thaliana* has two different MBW complexes—the first one for the control of the anthocyanin (PAPs-EGL3/GL3/TT8-TTG1) (Lloyd et al., 2017) and the second one for the control of proanthocyanidin (TT2/TT8/TTG1) biosynthetic genes [102] in the plant body (anthocyanins) and seeds (proanthocyanidins)—which further highlight the importance of MBW complex. In this complex (MBW) the PAP1 and TT8 are responsible for the DNA binding, and consequent regulation of transcriptional activity, while the TTG1 is necessary for the MBW complex activity; and it is assumed that it also prevents the promoter interaction with the transcriptional inhibitors [77].

The MBW complex itself positively regulates transcription of the PheCs-related genes, but interactions with other proteins are documented and result in the alteration (usually the decrease) of the late genes expression. For example, MYBL2 (MYB-like 2 protein) probably competes with the R2R3 MYBs at the MBW complex and interacts with the bHLH proteins which leads to decreased anthocyanin production [103]. Similarly, the SPL9 (SQUAMOSA PROMOTER BINDING PROTEIN-LIKE 9) protein directly competes with the TT8 protein for interaction in the anthocyanin-related MBW complex. If the SPL9 is incorporated, the reduced transcription of the target genes related to the anthocyanin biosynthesis is observed (DFR, BAN, etc.) in *Arabidopsis thaliana* [103], and therefore (FLS and DFR compete for substrates) the flavonol biosynthesis [103] is increased. In addition, expression of the SPL9 is controlled at the posttranscriptional level by the miR156 and miR157 via PTGS [103].

6. Main Transcription Regulators Coordinating Expression of Genes Related to the Production of the Phenolic Compounds

To further summarize available knowledge focused on the TFs driving the expression of genes related to the production of phenolic compounds, we selected 13 crucial TFs based on different structural motifs mentioned above in the text. We browsed the corresponding information in the PlantTFDB database [104] for *Arabidopsis thaliana*. These selected TFs are summarized in Table 1, including the information about their structure and target genes these regulate (positively and if any, negatively as well).

Firstly, the TT8 TF (having the bHLH structural motif) according to the database positively affects the expression of the genes related to the phenolic compounds production (BAN, and DFR), putative methylesterases (MES4, and MES6), WRKY44 TF (TTG2), bZIP TF (GL2), MYB-like 2 and its own expression. From the data in the database, so far, no gene was documented to be repressed by the TT8.

Although the vast majority of the reported studies documented the stimulative effects of the HY5 on the target genes, according to the PlantTFDB database (Table 1), three target genes are repressed, by the HY5 TF—NPF6.3 (gene coding for NRT1 Protein—that is also known as Nitrate transporter 1.1), FHY1 (gene coding for FAR-RED ELONGATED HYPOCOTYL 1 Protein, involved in the FR/R responses), and FHL (gene coding for the FAR-RED-ELONGATED HYPOCOTYL 1-LIKE Protein). On the other hand, HY5 positively regulates the expression of 19 genes related to various processes—from those related to the production of phenolic compounds (MYB12, CHS, DFR, UGT84A1, LDOX), via those related to the production of chlorophyll and photosynthesis (HEMA1, ELIP1, CAB2, CAB1, PSBD, RBCS1A), those related to the phytohormones (IAA7, IAA14, ABI5, HB-8), and those related to the nitrogen and phosphorus metabolism (NIA2, PHR1), or light responses (ELF4, LZF1).

HY-5like TFs are similar in function, namely HYH1, HYH2, HYH3, and HYH4 targeting genes that activate (PHR1, NIA2, ELIP1, PEX11B, CHS) and repress (NPF6.3) as well as TT2 TF, similar to TT8 regulating the expression of genes related to phenolic compound production (ANS, DFR, BAN) and other TFs (TT8, TTG2, GL2), and its own as well (TT2).

Table 1. Summary of selected plant transcription factors of *Arabidopsis thaliana* related to the control of the genes involved in PheCs biosynthesis (data were extracted from the PlantTFDB database [104]; question marks represent no available data in the source database dealing with the activated or repressed target genes). Dash represents no detected repressed genes, while question marks indicate no available data in the database.

TF	TT8	HY5	HYH_1	HYH_2	HYH_3	HYH_4	TT2	PAP1	MYB11	MYB12	MYB111	WRKY23	WRKY36
Motif	bHLH	bZIP	bZIP	bZIP	bZIP	bZIP	HTH	HTH	HTH	HTH	HTH	WRKY	WRKY
ID	AT4G09820	AT5G11260	AT3G17609.1	AT3G17609.2	AT3G17609.3	AT3G17609.4	AT5G35550.1	AT1G56650.1	AT3G62610.1	AT5G49330.1	AT5G49330.1	AT2G47260.1	AT1G69810.1
Target genes activated	BAN, TT8, MYBL2, DFR, GL2, MES6, MES4, TTG2	RBCS1A, ELIP1, PHR1, LZF1, IAA7, CHS, CAB2, ABI5, IAA14, DFR, CAB1, UGT84A1, PSBD, NIA2, ELF4, LDOX, HEMA1, MYB12, HB-8	PHR1, NIA2, ELIP1, PEX11B, CHS	PHR1, NIA2, ELIP1, PEX11B, CHS	PHR1, NIA2, ELIP1, PEX11B, CHS	PHR1, NIA2, ELIP1, PEX11B, CHS	ANS, TT8, TT2, TTG2, GL2, DFR, BAN	CHS, CHI, DFR, MYB3, TT8, UF3GT, PAP2, A5GT, UGT78D2, 5MAT, MBD2.2, GST	CHS, CHI, F3H, FLS1	CHS, CHI, F3H, FLS1	CHS, CHI, F3H, FLS1	???	???
Target genes repressed	-	NPF6.3, FHY1, FHL	NPF6.3	NPF6.3	NPF6.3	NPF6.3	-	scpl10	-	-	-	???	???

Although the PAP1 TF has been repeatedly documented to have a positive effect on the transcription of the target genes, according to the PlantTFDB database it inhibits transcription of the scpl10 gene coding for the Serine carboxypeptidase-like 10 protein that is involved in the sinapoylation of anthocyanins (adding of the acyl group to the anthocyanins consequently affecting the antioxidative properties, charge, and therefore probably the localization within the cell as well). On the other hand, PAP1 positively regulates the transcription of the genes related to the production of the phenolic compounds (CHS, CHI, DFR, UF3GT, A5GT, UGT78D2, 5MAT), but also other proteins and transcription regulators (GST, MBD2.2, MYB3, TT8).

The other three selected proteins with a HTH structural motif are MYB11, MYB12, and MYB112. All these three proteins do not (according to the database) repress any target gene, but positively affect the transcription of genes related to the production of flavonols (CHS, CHI, F3H, FLS). The last two selected proteins containing the WRKY structural motif are WRKY23 and WRKY36—although these TFs are reported in the literature as proteins affecting the expression of genes related to the production of phenolic compounds [95,97], so far, the PlantTFDB database does not have any records about their target genes—neither activated, nor repressed.

To conclude from above, TFs first regulate the target genes related to the phenolic compounds (but some of them are also other key enzymes—one example can be the HY5 positively regulating the expression of the genes related to the photosynthesis, biosynthesis of chlorophyll, and phytohormones), secondly, they can regulate each other's expression (for example the TT2 TF positively regulates the expression of TT8 TF) and lastly, some of them have an autoregulatory function (for example, the TT8 TF positively regulates its own gene expression). In addition to this, some of these TFs also have an inhibitory role in the expression of some target genes (for example, the HY5 homologs with negative effects on the expression of NPF6.3 gene coding for the nitrate transporter). On the other hand, transcription control of these genes is not ensured exclusively by the TFs—so far, it has been shown, that alterations in the epigenetic modifications [105] or formation of local structures [106] have a considerable role as well.

An example can be the transcription regulation via the micro RNAs (miRNAs). These are relatively short (22–35 nucleotides long) RNA molecules complementary to "common" mRNAs (coding the final protein) able to either pause, or stop the expression of its target (via the post-transcriptional gene silencing, or RNA-induced methylation of the gene) [107]. Although it seems unlikely, miRNAs also contribute to the control of the production of phenolic compounds (e.g., miR156, miR858, or miR828). For example, miR858 [108] targets the MYB11, MYB12, and MYB111 TFs transcripts (with stimulative effects on the expression of genes coding for the enzymes of the phenylpropanoid pathway), but other members of the MYB R2R3 family are also putative targets of mir858. Similarly, the miR828 [109,110] targets the MYB TFs driving the expression of genes related to anthocyanin production via the degradation of MYB113, MYB82, and TAS4 (trans-acting small interfering RNA). This results in the negative regulation of MYB75 (PAP1) and MYB90 (PAP2)—here, a regulatory loop between PAP1 and TAS4 was also proposed (while PAP1 stimulates the accumulation of TAS4, the TAS4 has an inhibitory effect on PAP1, PAP2 and MYB113) [111].

To better explain the transcriptional control of the genes related to the production of the phenolic compounds, we summarized the text from this section into a simplified scheme below (Figure 5).

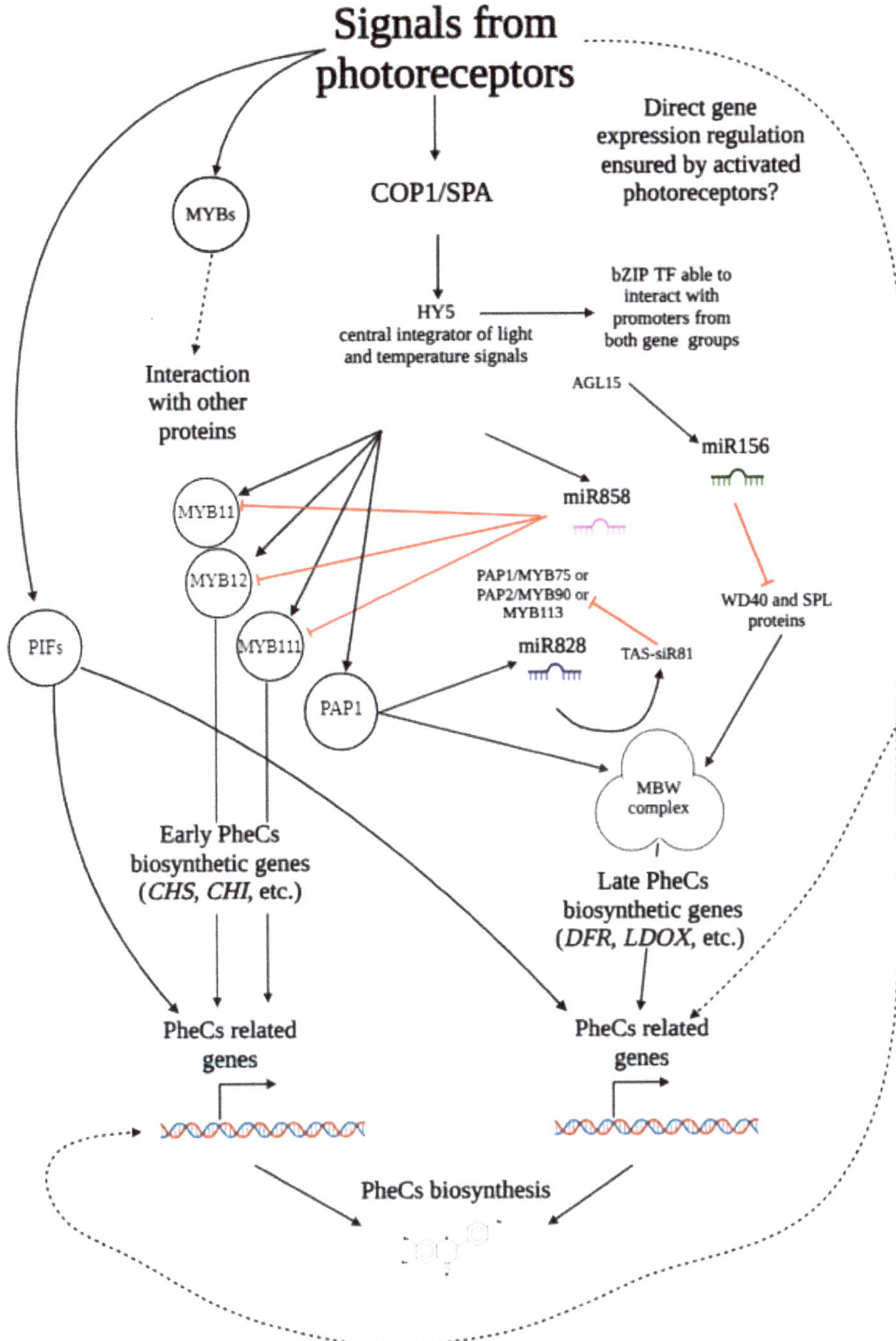

Figure 5. The simplified scheme of the transcriptional control of the genes related to the production of phenolic compounds (scheme was adopted from [112] and modified).

7. Future Perspectives

In future studies, it would be beneficial to investigate the extent to which different photoreceptors drive the genes of interest. For instance, genes related to the production of phenolic compounds are known to be influenced by photoreceptors and their associated signaling cascades. While it is well documented that blue light and UV radiation can stimulate these genes, the specific contributions of various photoreceptors remain unclear. For example, cryptochromes are known to play a major role in blue light stimulation, but it is uncertain whether ZTLs or phototropins also contribute to this effect, and if so, to what extent. Additionally, how signals from multiple photoreceptors are integrated and prioritized, particularly in natural sunlight that activates more than one photoreceptor, needs further exploration. Understanding how the COP/SPA/Cul/DDB complex differentiates between signals from different photoreceptors and whether subsequent steps in the COP/SPA signaling cascade vary based on the originating photoreceptor is crucial.

Regarding nomenclature, establishing unified and widely accepted acronyms for key transcription factors would be highly beneficial. Currently, transcription factors involved in regulating phenolic compounds often have multiple aliases, complicating literature reviews and our understanding of light-induced responses. For example, PAP2 (Probable plastid-lipid-associated protein 2, chloroplastic) is also known as FBN1b or FIB1b (derived from Fibrillin-1b) and MYB90. Standardizing these names would facilitate research and communication.

Another research gap is the lack of a comprehensive summary of known regulatory proteins (transcription factors) that directly control specific genes and their transcription. A detailed list of proteins driving the expression of Chalcone synthase, the initial enzyme in the phenylpropanoid biosynthetic pathway, along with the promoter motifs ensuring interaction with other proteins, would greatly aid in understanding the complex regulation of these genes. Such a database, also including other genes, could clarify observed effects and discrepancies in gene expression and inspire new hypotheses regarding the regulation of protective metabolite production.

Furthermore, summarizing current knowledge on epigenetic modifications, such as DNA modifications, histone modifications, and non-B-DNA structures from multiple sources (sequencing and in silico predictions) would be valuable. Testing whether these modifications can be introduced into a given gene and assessing their impact on gene expression in the lab could provide significant insights.

Despite substantial progress in understanding transcriptional control of gene expression, our knowledge remains limited. Future studies should aim to unify nomenclature and test more complex hypotheses about gene expression coordination.

8. Conclusions

In this review, we combined current knowledge focused on the plant photoreceptors and related signaling pathways with the knowledge about the plant transcription factors driving the expression of genes related to the production of phenolic compounds. We are convinced that this thorough summarization of the current literature on this topic is very useful and can be used as inspiration for the design of new studies aimed at a comprehensive understanding of the signal mediated by different intensities and spectral composition of PAR and UV (and temperature) on PheCS production and profile. In addition to this, we pointed in the previous section (Section 7) to the current challenges and gaps in this field, together with the possible applications of this knowledge.

Author Contributions: Conceptualization, A.V.; resources, A.V.; writing—original draft preparation, A.V.; writing—review and editing, J.Č., R.P., J.N. and V.Š.; visualization, A.V.; supervision, J.Č. and V.Š.; funding acquisition, V.Š. and R.P. All authors have read and agreed to the published version of the manuscript.

Funding: This research was funded by the Czech Science Foundation (GACR 21-18532S to V.Š., A.V., R.P., and J.N.), the European Union under the LERCO project (number CZ.10.03.01./00/22_003/0000003) via the Operational Programme Just Transition, the University of Ostrava (SGS08/PřF/2024 to A.V., R.P.,

J.N.), and the Moravian-Silesian Region (RRC/10/2021 to R.P.). In addition, V.Š. was partially financially supported by the project of the Ministry of Education, Youth and Sports of the Czech Republic (AdAgriF; CZ.02.01.01/00/22_008/0004635).

Institutional Review Board Statement: Not applicable.

Informed Consent Statement: Not applicable.

Data Availability Statement: No new data were created or analyzed in this study. Data sharing is not applicable to this article.

Acknowledgments: We would like to thank ChatGPT (https://www.openai.com/; 15 May 2024) for assistance in rephrasing problematic sections of this article. All schemes accompanying this publication were created using BioRender software (https://biorender.com/). Martin Bartas, also deserves our gratitude, who performed a thorough review of this text.

Conflicts of Interest: The authors declare no conflicts of interest.

References

1. Csepregi, K.; Hideg, É. Phenolic Compound Diversity Explored in the Context of Photo-Oxidative Stress Protection. *Phytochem. Anal.* **2018**, *29*, 129–136. [CrossRef] [PubMed]
2. Sarker, U.; Oba, S. Drought stress enhances nutritional and bioactive compounds, phenolic acids and antioxidant capacity of Amaranthus leafy vegetable. *BMC Plant Biol.* **2018**, *18*, 258. [CrossRef] [PubMed]
3. Younis, M.E.-B.; Hasaneen, M.N.A.-G.; Abdel-Aziz, H.M.M. An enhancing effect of visible light and UV radiation on phenolic compounds and various antioxidants in broad bean seedlings. *Plant Signal Behav.* **2010**, *5*, 1197–1203. [CrossRef] [PubMed]
4. Christie, P.J.; Alfenito, M.R.; Walbot, V. Impact of low-temperature stress on general phenylpropanoid and anthocyanin pathways: Enhancement of transcript abundance and anthocyanin pigmentation in maize seedlings. *Planta* **1994**, *194*, 541–549. [CrossRef]
5. Catalá, R.; Medina, J.; Salinas, J. Integration of low temperature and light signaling during cold acclimation response in Arabidopsis. *Proc. Natl. Acad. Sci. USA* **2011**, *108*, 16475–16480. [CrossRef] [PubMed]
6. Michalak, A. Phenolic Compounds and Their Antioxidant Activity in Plants Growing under Heavy Metal Stress. *Pol. J. Environ. Stud.* **2006**, *15*, 523–530.
7. Hunt, L.; Klem, K.; Lhotáková, Z.; Vosolsobě, S.; Oravec, M.; Urban, O.; Špunda, V.; Albrechtová, J. Light and CO_2 Modulate the Accumulation and Localization of Phenolic Compounds in Barley Leaves. *Antioxidants* **2021**, *10*, 385. [CrossRef] [PubMed]
8. Saile, J.; Wießner-Kroh, T.; Erbstein, K.; Obermüller, D.M.; Pfeiffer, A.; Janocha, D.; Lohmann, J.; Wachter, A. SNF1-RELATED KINASE 1 and TARGET OF RAPAMYCIN control light-responsive splicing events and developmental characteristics in etiolated Arabidopsis seedlings. *Plant Cell* **2023**, *35*, 3413–3428. [CrossRef] [PubMed]
9. Wu, R.; Lin, X.; He, J.; Min, A.; Pang, L.; Wang, Y.; Lin, Y.; Zhang, Y.; He, W.; Li, M.; et al. Hexokinase1: A glucose sensor involved in drought stress response and sugar metabolism depending on its kinase activity in strawberry. *Front. Plant Sci.* **2023**, *14*, 1069830. [CrossRef] [PubMed]
10. Moore, B.; Zhou, L.; Rolland, F.; Hall, Q.; Cheng, W.-H.; Liu, Y.-X.; Hwang, I.; Jones, T.; Sheen, J. Role of the *Arabidopsis* Glucose Sensor HXK1 in Nutrient, Light, and Hormonal Signaling. *Science* **2003**, *300*, 332–336. [CrossRef]
11. Avidan, O.; Moraes, T.A.; Mengin, V.; Feil, R.; Rolland, F.; Stitt, M.; Lunn, J.E. In vivo protein kinase activity of SnRK1 fluctuates in Arabidopsis rosettes during light-dark cycles. *Plant Physiol.* **2023**, *192*, 387–408. [CrossRef]
12. Riegler, S.; Servi, L.; Scarpin, M.R.; Godoy Herz, M.A.; Kubaczka, M.G.; Venhuizen, P.; Meyer, C.; Brunkard, J.O.; Kalyna, M.; Barta, A.; et al. Light regulates alternative splicing outcomes via the TOR kinase pathway. *Cell Rep.* **2021**, *36*, 109676. [CrossRef] [PubMed]
13. Apel, K.; Hirt, H. Reactive oxygen species: Metabolism, oxidative stress, and signal transduction. *Annu. Rev. Plant Biol.* **2004**, *55*, 373–399. [CrossRef] [PubMed]
14. El-Esawi, M.; Arthaut, L.-D.; Jourdan, N.; d'Harlingue, A.; Link, J.; Martino, C.F.; Ahmad, M. Blue-light induced biosynthesis of ROS contributes to the signaling mechanism of Arabidopsis cryptochrome. *Sci. Rep.* **2017**, *7*, 13875. [CrossRef] [PubMed]
15. Zandalinas, S.I.; Sengupta, S.; Burks, D.; Azad, R.K.; Mittler, R. Identification and characterization of a core set of ROS wave-associated transcripts involved in the systemic acquired acclimation response of Arabidopsis to excess light. *Plant J.* **2019**, *98*, 126–141. [CrossRef] [PubMed]
16. Möglich, A.; Yang, X.; Ayers, R.A.; Moffat, K. Structure and Function of Plant Photoreceptors. *Annu. Rev. Plant Biol.* **2010**, *61*, 21–47. [CrossRef] [PubMed]
17. Ponnu, J.; Hoecker, U. Illuminating the COP1/SPA Ubiquitin Ligase: Fresh Insights Into Its Structure and Functions during Plant Photomorphogenesis. *Front. Plant Sci.* **2021**, *12*, 662793. [CrossRef]
18. Zvi, M.M.B.; Shklarman, E.; Masci, T.; Kalev, H.; Debener, T.; Shafir, S.; Ovadis, M.; Vainstein, A. PAP1 transcription factor enhances production of phenylpropanoid and terpenoid scent compounds in rose flowers. *New Phytol.* **2012**, *195*, 335–345. [CrossRef] [PubMed]

19. Xu, W.; Dubos, C.; Lepiniec, L. Transcriptional control of flavonoid biosynthesis by MYB–bHLH–WDR complexes. *Trends Plant Sci.* **2015**, *20*, 176–185. [CrossRef]
20. Pennisi, G.; Sanyé-Mengual, E.; Orsini, F.; Crepaldi, A.; Nicola, S.; Ochoa, J.; Fernandez, J.A.; Gianquinto, G. Modelling Environmental Burdens of Indoor-Grown Vegetables and Herbs as Affected by Red and Blue LED Lighting. *Sustainability* **2019**, *11*, 4063. [CrossRef]
21. Zhang, S.; Ma, J.; Zou, H.; Zhang, L.; Li, S.; Wang, Y. The combination of blue and red LED light improves growth and phenolic acid contents in *Salvia miltiorrhiza* Bunge. *Ind. Crops Prod.* **2020**, *158*, 112959. [CrossRef]
22. Gupta, S.K.; Sharma, M.; Deeba, F.; Pandey, V. Plant Response. In *UV-B Radiation*; John Wiley & Sons, Ltd.: Hoboken, NJ, USA, 2017; pp. 217–258. ISBN 978-1-119-14361-1.
23. Kong, S.-G.; Okajima, K. Diverse photoreceptors and light responses in plants. *J. Plant Res.* **2016**, *129*, 111–114. [CrossRef] [PubMed]
24. Legris, M.; Ince, Y.Ç.; Fankhauser, C. Molecular mechanisms underlying phytochrome-controlled morphogenesis in plants. *Nat. Commun.* **2019**, *10*, 5219. [CrossRef] [PubMed]
25. Paik, I.; Huq, E. Plant photoreceptors: Multi-functional sensory proteins and their signaling networks. *Semin. Cell Dev. Biol.* **2019**, *92*, 114–121. [CrossRef] [PubMed]
26. Christie, J.M.; Arvai, A.S.; Baxter, K.J.; Heilmann, M.; Pratt, A.J.; O'Hara, A.; Kelly, S.M.; Hothorn, M.; Smith, B.O.; Hitomi, K.; et al. Plant UVR8 Photoreceptor Senses UV-B by Tryptophan-Mediated Disruption of Cross-Dimer Salt Bridges. *Science* **2012**, *335*, 1492–1496. [CrossRef] [PubMed]
27. Mathes, T.; Heilmann, M.; Pandit, A.; Zhu, J.; Ravensbergen, J.; Kloz, M.; Fu, Y.; Smith, B.O.; Christie, J.M.; Jenkins, G.I.; et al. Proton-Coupled Electron Transfer Constitutes the Photoactivation Mechanism of the Plant Photoreceptor UVR8. *J. Am. Chem. Soc.* **2015**, *137*, 8113–8120. [CrossRef] [PubMed]
28. Li, X.; Ren, H.; Kundu, M.; Liu, Z.; Zhong, F.W.; Wang, L.; Gao, J.; Zhong, D. A leap in quantum efficiency through light harvesting in photoreceptor UVR8. *Nat. Commun.* **2020**, *11*, 4316. [CrossRef] [PubMed]
29. Tossi, V.E.; Regalado, J.J.; Iannicelli, J.; Laino, L.E.; Burrieza, H.P.; Escandón, A.S.; Pitta-Álvarez, S.I. Beyond Arabidopsis: Differential UV-B Response Mediated by UVR8 in Diverse Species. *Front. Plant Sci.* **2019**, *10*, 780. [CrossRef] [PubMed]
30. Findlay, K.M.W.; Jenkins, G.I. Regulation of UVR8 photoreceptor dimer/monomer photo-equilibrium in Arabidopsis plants grown under photoperiodic conditions. *Plant Cell Environ.* **2016**, *39*, 1706–1714. [CrossRef] [PubMed]
31. Gruber, H.; Heijde, M.; Heller, W.; Albert, A.; Seidlitz, H.K.; Ulm, R. Negative feedback regulation of UV-B–induced photomorphogenesis and stress acclimation in Arabidopsis. *Proc. Natl. Acad. Sci. USA* **2010**, *107*, 20132–20137. [CrossRef]
32. Wang, L.; Wang, Y.; Chang, H.; Ren, H.; Wu, X.; Wen, J.; Guan, Z.; Ma, L.; Qiu, L.; Yan, J.; et al. RUP2 facilitates UVR8 redimerization via two interfaces. *Plant Commun.* **2022**, *4*, 100428. [CrossRef]
33. Orth, C.; Niemann, N.; Hennig, L.; Essen, L.-O.; Batschauer, A. Hyperactivity of the Arabidopsis cryptochrome (cry1) L407F mutant is caused by a structural alteration close to the cry1 ATP-binding site. *J. Biol. Chem.* **2017**, *292*, 12906–12920. [CrossRef]
34. Lopez, L.; Fasano, C.; Perrella, G.; Facella, P. Cryptochromes and the Circadian Clock: The Story of a Very Complex Relationship in a Spinning World. *Genes* **2021**, *12*, 672. [CrossRef] [PubMed]
35. Lin, C.; Todo, T. The cryptochromes. *Genome Biol.* **2005**, *6*, 220. [CrossRef]
36. Barrero, J.M.; Downie, A.B.; Xu, Q.; Gubler, F. A Role for Barley CRYPTOCHROME1 in Light Regulation of Grain Dormancy and Germination. *Plant Cell* **2014**, *26*, 1094–1104. [CrossRef] [PubMed]
37. Klar, T.; Pokorny, R.; Moldt, J.; Batschauer, A.; Essen, L.-O. Cryptochrome 3 from *Arabidopsis thaliana*: Structural and Functional Analysis of its Complex with a Folate Light Antenna. *J. Mol. Biol.* **2007**, *366*, 954–964. [CrossRef]
38. Tissot, N.; Ulm, R. Cryptochrome-mediated blue-light signalling modulates UVR8 photoreceptor activity and contributes to UV-B tolerance in Arabidopsis. *Nat. Commun.* **2020**, *11*, 1323. [CrossRef] [PubMed]
39. Ma, L.; Guan, Z.; Wang, Q.; Yan, X.; Wang, J.; Wang, Z.; Cao, J.; Zhang, D.; Gong, X.; Yin, P. Structural insights into the photoactivation of Arabidopsis CRY2. *Nat. Plants* **2020**, *6*, 1432–1438. [CrossRef] [PubMed]
40. Palayam, M.; Ganapathy, J.; Guercio, A.M.; Tal, L.; Deck, S.L.; Shabek, N. Structural insights into photoactivation of plant Cryptochrome-2. *Commun. Biol.* **2021**, *4*, 1–11. [CrossRef] [PubMed]
41. Shao, K.; Zhang, X.; Li, X.; Hao, Y.; Huang, X.; Ma, M.; Zhang, M.; Yu, F.; Liu, H.; Zhang, P. The oligomeric structures of plant cryptochromes. *Nat. Struct. Mol. Biol.* **2020**, *27*, 480–488. [CrossRef]
42. The UniProt Consortium UniProt: A worldwide hub of protein knowledge. *Nucleic Acids Res.* **2019**, *47*, D506–D515. [CrossRef]
43. Christie, J.M. Phototropin Blue-Light Receptors. *Annu. Rev. Plant Biol.* **2007**, *58*, 21–45. [CrossRef] [PubMed]
44. Christie, J.M.; Swartz, T.E.; Bogomolni, R.A.; Briggs, W.R. Phototropin LOV domains exhibit distinct roles in regulating photoreceptor function. *Plant J.* **2002**, *32*, 205–219. [CrossRef] [PubMed]
45. Inoue, S.; Takemiya, A.; Shimazaki, K. Phototropin signaling and stomatal opening as a model case. *Curr. Opin. Plant Biol.* **2010**, *13*, 587–593. [CrossRef] [PubMed]
46. Nakasone, Y.; Ohshima, M.; Okajima, K.; Tokutomi, S.; Terazima, M. Photoreaction Dynamics of Full-Length Phototropin from *Chlamydomonas reinhardtii*. *J. Phys. Chem. B* **2019**, *123*, 10939–10950. [CrossRef] [PubMed]
47. Motchoulski, A.; Liscum, E. Arabidopsis NPH3: A NPH1 Photoreceptor-Interacting Protein Essential for Phototropism. *Science* **1999**, *286*, 961–964. [CrossRef] [PubMed]

48. Inada, S.; Ohgishi, M.; Mayama, T.; Okada, K.; Sakai, T. RPT2 Is a Signal Transducer Involved in Phototropic Response and Stomatal Opening by Association with Phototropin 1 in Arabidopsis thaliana. *Plant Cell* **2004**, *16*, 887–896. [CrossRef]
49. Sakai, T.; Wada, T.; Ishiguro, S.; Okada, K. RPT2: A Signal Transducer of the Phototropic Response in Arabidopsis. *Plant Cell* **2000**, *12*, 225–236. [CrossRef] [PubMed]
50. Pudasaini, A.; Zoltowski, B.D. Zeitlupe Senses Blue-Light Fluence to Mediate Circadian Timing in Arabidopsis thaliana. *Biochemistry* **2013**, *52*, 7150–7158. [CrossRef]
51. Pudasaini, A.; Shim, J.S.; Song, Y.H.; Shi, H.; Kiba, T.; Somers, D.E.; Imaizumi, T.; Zoltowski, B.D. Kinetics of the LOV domain of ZEITLUPE determine its circadian function in Arabidopsis. *eLife* **2017**, *6*, e21646. [CrossRef]
52. Ito, S.; Song, Y.H.; Imaizumi, T. LOV Domain-Containing F-Box Proteins: Light-Dependent Protein Degradation Modules in Arabidopsis. *Mol. Plant* **2012**, *5*, 573–582. [CrossRef] [PubMed]
53. Rockwell, N.C.; Su, Y.-S.; Lagarias, J.C. Phytochrome Structure and Signaling Mechanisms. *Annu. Rev. Plant Biol.* **2006**, *57*, 837–858. [CrossRef] [PubMed]
54. Li, J.; Li, G.; Wang, H.; Deng, X.W. Phytochrome Signaling Mechanisms. *Arab. Book* **2011**, *9*, e0148. [CrossRef] [PubMed]
55. Franklin, K.A.; Allen, T.; Whitelam, G.C. Phytochrome A is an irradiance-dependent red light sensor. *Plant J.* **2007**, *50*, 108–117. [CrossRef]
56. Legris, M.; Klose, C.; Burgie, E.S.; Rojas, C.C.R.; Neme, M.; Hiltbrunner, A.; Wigge, P.A.; Schäfer, E.; Vierstra, R.D.; Casal, J.J. Phytochrome B integrates light and temperature signals in *Arabidopsis*. *Science* **2016**, *354*, 897–900. [CrossRef] [PubMed]
57. Qiu, Y.; Li, M.; Kim, R.J.-A.; Moore, C.M.; Chen, M. Daytime temperature is sensed by phytochrome B in Arabidopsis through a transcriptional activator HEMERA. *Nat. Commun.* **2019**, *10*, 140. [CrossRef] [PubMed]
58. Bianchetti, R.; De Luca, B.; de Haro, L.A.; Rosado, D.; Demarco, D.; Conte, M.; Bermudez, L.; Freschi, L.; Fernie, A.R.; Michaelson, L.V.; et al. Phytochrome-Dependent Temperature Perception Modulates Isoprenoid Metabolism. *Plant Physiol.* **2020**, *183*, 869–882. [CrossRef] [PubMed]
59. Pham, V.N.; Kathare, P.K.; Huq, E. Phytochromes and Phytochrome Interacting Factors. *Plant Physiol.* **2018**, *176*, 1025–1038. [CrossRef]
60. Sakamoto, T.; Kimura, S. Plant Temperature Sensors. *Sensors* **2018**, *18*, 4365. [CrossRef]
61. Chung, B.Y.W.; Balcerowicz, M.; Di Antonio, M.; Jaeger, K.E.; Geng, F.; Franaszek, K.; Marriott, P.; Brierley, I.; Firth, A.E.; Wigge, P.A. An RNA thermoswitch regulates daytime growth in Arabidopsis. *Nat. Plants* **2020**, *6*, 522–532. [CrossRef]
62. Hayes, S.; Schachtschabel, J.; Mishkind, M.; Munnik, T.; Arisz, S.A. Hot topic: Thermosensing in plants. *Plant Cell Environ.* **2021**, *44*, 2018–2033. [CrossRef] [PubMed]
63. Fujii, Y.; Tanaka, H.; Konno, N.; Ogasawara, Y.; Hamashima, N.; Tamura, S.; Hasegawa, S.; Hayasaki, Y.; Okajima, K.; Kodama, Y. Phototropin perceives temperature based on the lifetime of its photoactivated state. *Proc. Natl. Acad. Sci. USA* **2017**, *114*, 9206–9211. [CrossRef] [PubMed]
64. Pooam, M.; Dixon, N.; Hilvert, M.; Misko, P.; Waters, K.; Jourdan, N.; Drahy, S.; Mills, S.; Engle, D.; Link, J.; et al. Effect of temperature on the Arabidopsis cryptochrome photocycle. *Physiol. Plant.* **2021**, *172*, 1653–1661. [CrossRef] [PubMed]
65. Salomé, P.A. In the Heat of the Moment: ZTL-Mediated Protein Quality Control at High Temperatures. *Plant Cell* **2017**, *29*, 2685–2686. [CrossRef] [PubMed]
66. Noguchi, M.; Kodama, Y. Temperature Sensing in Plants: On the Dawn of Molecular Thermosensor Research. *Plant Cell Physiol.* **2022**, *63*, 737–743. [CrossRef] [PubMed]
67. Pires, N.; Dolan, L. Origin and Diversification of Basic-Helix-Loop-Helix Proteins in Plants. *Mol. Biol. Evol.* **2010**, *27*, 862–874. [CrossRef]
68. Hao, Y.; Zong, X.; Ren, P.; Qian, Y.; Fu, A. Basic Helix-Loop-Helix (bHLH) Transcription Factors Regulate a Wide Range of Functions in Arabidopsis. *Int. J. Mol. Sci.* **2021**, *22*, 7152. [CrossRef] [PubMed]
69. Qian, Y.; Zhang, T.; Yu, Y.; Gou, L.; Yang, J.; Xu, J.; Pi, E. Regulatory Mechanisms of bHLH Transcription Factors in Plant Adaptive Responses to Various Abiotic Stresses. *Front. Plant Sci.* **2021**, *12*, 677611. [CrossRef] [PubMed]
70. Nesi, N.; Debeaujon, I.; Jond, C.; Pelletier, G.; Caboche, M.; Lepiniec, L. The TT8 Gene Encodes a Basic Helix-Loop-Helix Domain Protein Required for Expression of DFR and BAN Genes in Arabidopsis Siliques. *Plant Cell* **2000**, *12*, 1863–1878. [CrossRef] [PubMed]
71. Devic, M.; Guilleminot, J.; Debeaujon, I.; Bechtold, N.; Bensaude, E.; Koornneef, M.; Pelletier, G.; Delseny, M. The BANYULS gene encodes a DFR-like protein and is a marker of early seed coat development. *Plant J.* **1999**, *19*, 387–398. [CrossRef]
72. Xie, D.-Y.; Sharma, S.B.; Paiva, N.L.; Ferreira, D.; Dixon, R.A. Role of Anthocyanidin Reductase, Encoded by BANYULS in Plant Flavonoid Biosynthesis. *Science* **2003**, *299*, 396–399. [CrossRef]
73. Corrêa, L.G.G.; Riaño-Pachón, D.M.; Schrago, C.G.; dos Santos, R.V.; Mueller-Roeber, B.; Vincentz, M. The Role of bZIP Transcription Factors in Green Plant Evolution: Adaptive Features Emerging from Four Founder Genes. *PLoS ONE* **2008**, *3*, e2944. [CrossRef]
74. Yu, Y.; Qian, Y.; Jiang, M.; Xu, J.; Yang, J.; Zhang, T.; Gou, L.; Pi, E. Regulation Mechanisms of Plant Basic Leucine Zippers to Various Abiotic Stresses. *Front. Plant Sci.* **2020**, *11*, 1258. [CrossRef] [PubMed]
75. Alves, M.S.; Dadalto, S.P.; Gonçalves, A.B.; De Souza, G.B.; Barros, V.A.; Fietto, L.G. Plant bZIP Transcription Factors Responsive to Pathogens: A Review. *Int. J. Mol. Sci.* **2013**, *14*, 7815–7828. [CrossRef] [PubMed]

76. Gangappa, S.N.; Botto, J.F. The Multifaceted Roles of HY5 in Plant Growth and Development. *Mol. Plant* **2016**, *9*, 1353–1365. [CrossRef] [PubMed]
77. Nguyen, N.H. HY5, an integrator of light and temperature signals in the regulation of anthocyanins biosynthesis in Arabidopsis. *AIMS Mol. Sci.* **2020**, *7*, 70–81. [CrossRef]
78. Xu, D.; Jiang, Y.; Li, J.; Lin, F.; Holm, M.; Deng, X.W. BBX21, an Arabidopsis B-box protein, directly activates HY5 and is targeted by COP1 for 26S proteasome-mediated degradation. *Proc. Natl. Acad. Sci. USA* **2016**, *113*, 7655–7660. [CrossRef] [PubMed]
79. Holm, M.; Ma, L.-G.; Qu, L.-J.; Deng, X.-W. Two interacting bZIP proteins are direct targets of COP1-mediated control of light-dependent gene expression in Arabidopsis. *Genes Dev.* **2002**, *16*, 1247–1259. [CrossRef] [PubMed]
80. Zhang, Y.; Zheng, S.; Liu, Z.; Wang, L.; Bi, Y. Both HY5 and HYH are necessary regulators for low temperature-induced anthocyanin accumulation in Arabidopsis seedlings. *J. Plant Physiol.* **2011**, *168*, 367–374. [CrossRef]
81. Dubos, C.; Stracke, R.; Grotewold, E.; Weisshaar, B.; Martin, C.; Lepiniec, L. MYB transcription factors in Arabidopsis. *Trends Plant Sci.* **2010**, *15*, 573–581. [CrossRef]
82. Katiyar, A.; Smita, S.; Lenka, S.K.; Rajwanshi, R.; Chinnusamy, V.; Bansal, K.C. Genome-wide classification and expression analysis of MYB transcription factor families in rice and Arabidopsis. *BMC Genom.* **2012**, *13*, 544. [CrossRef] [PubMed]
83. Nesi, N.; Jond, C.; Debeaujon, I.; Caboche, M.; Lepiniec, L. The Arabidopsis TT2 Gene Encodes an R2R3 MYB Domain Protein That Acts as a Key Determinant for Proanthocyanidin Accumulation in Developing Seed. *Plant Cell* **2001**, *13*, 2099–2114. [CrossRef] [PubMed]
84. Yang, Y.; Sulpice, R.; Himmelbach, A.; Meinhard, M.; Christmann, A.; Grill, E. Fibrillin expression is regulated by abscisic acid response regulators and is involved in abscisic acid-mediated photoprotection. *Proc. Natl. Acad. Sci. USA* **2006**, *103*, 6061–6066. [CrossRef] [PubMed]
85. Shi, M.-Z.; Xie, D.-Y. Features of anthocyanin biosynthesis in pap1-D and wild-type Arabidopsis thaliana plants grown in different light intensity and culture media conditions. *Planta* **2010**, *231*, 1385–1400. [CrossRef] [PubMed]
86. Zhang, Y.; Yan, Y.-P.; Wang, Z.-Z. The Arabidopsis PAP1 Transcription Factor Plays an Important Role in the Enrichment of Phenolic Acids in Salvia miltiorrhiza. *J. Agric. Food Chem.* **2010**, *58*, 12168–12175. [CrossRef] [PubMed]
87. Li, X.; Gao, M.-J.; Pan, H.-Y.; Cui, D.-J.; Gruber, M.Y. Purple Canola: Arabidopsis PAP1 Increases Antioxidants and Phenolics in Brassica napus Leaves. *J. Agric. Food Chem.* **2010**, *58*, 1639–1645. [CrossRef] [PubMed]
88. Mitsunami, T.; Nishihara, M.; Galis, I.; Alamgir, K.M.; Hojo, Y.; Fujita, K.; Sasaki, N.; Nemoto, K.; Sawasaki, T.; Arimura, G. Overexpression of the PAP1 Transcription Factor Reveals a Complex Regulation of Flavonoid and Phenylpropanoid Metabolism in Nicotiana tabacum Plants Attacked by Spodoptera litura. *PLoS ONE* **2014**, *9*, e108849. [CrossRef] [PubMed]
89. Youssef, A.; Laizet, Y.; Block, M.A.; Maréchal, E.; Alcaraz, J.-P.; Larson, T.R.; Pontier, D.; Gaffé, J.; Kuntz, M. Plant lipid-associated fibrillin proteins condition jasmonate production under photosynthetic stress. *Plant J.* **2010**, *61*, 436–445. [CrossRef]
90. Jiang, J.; Ma, S.; Ye, N.; Jiang, M.; Cao, J.; Zhang, J. WRKY transcription factors in plant responses to stresses. *J. Integr. Plant Biol.* **2017**, *59*, 86–101. [CrossRef]
91. Guillaumie, S.; Mzid, R.; Méchin, V.; Léon, C.; Hichri, I.; Destrac-Irvine, A.; Trossat-Magnin, C.; Delrot, S.; Lauvergeat, V. The grapevine transcription factor WRKY2 influences the lignin pathway and xylem development in tobacco. *Plant Mol. Biol.* **2010**, *72*, 215–234. [CrossRef]
92. Phukan, U.J.; Jeena, G.S.; Shukla, R.K. WRKY Transcription Factors: Molecular Regulation and Stress Responses in Plants. *Front. Plant Sci.* **2016**, *7*, 760. [CrossRef]
93. Grunewald, W.; De Smet, I.; Lewis, D.R.; Löfke, C.; Jansen, L.; Goeminne, G.; Vanden Bossche, R.; Karimi, M.; De Rybel, B.; Vanholme, B.; et al. Transcription factor WRKY23 assists auxin distribution patterns during Arabidopsis root development through local control on flavonol biosynthesis. *Proc. Natl. Acad. Sci. USA* **2012**, *109*, 1554–1559. [CrossRef]
94. Aamir, M.; Singh, V.K.; Meena, M.; Upadhyay, R.S.; Gupta, V.K.; Singh, S. Structural and Functional Insights into WRKY3 and WRKY4 Transcription Factors to Unravel the WRKY–DNA (W-Box) Complex Interaction in Tomato (*Solanum lycopersicum* L.). A Computational Approach. *Front. Plant Sci.* **2017**, *8*, 819. [CrossRef]
95. Chen, L.; Song, Y.; Li, S.; Zhang, L.; Zou, C.; Yu, D. The role of WRKY transcription factors in plant abiotic stresses. *Biochim. Biophys. Acta (BBA)-Gene Regul. Mech.* **2012**, *1819*, 120–128. [CrossRef]
96. Chi, Y.; Yang, Y.; Zhou, Y.; Zhou, J.; Fan, B.; Yu, J.-Q.; Chen, Z. Protein–Protein Interactions in the Regulation of WRKY Transcription Factors. *Mol. Plant* **2013**, *6*, 287–300. [CrossRef]
97. Yang, Y.; Liang, T.; Zhang, L.; Shao, K.; Gu, X.; Shang, R.; Shi, N.; Li, X.; Zhang, P.; Liu, H. UVR8 interacts with WRKY36 to regulate HY5 transcription and hypocotyl elongation in Arabidopsis. *Nat. Plants* **2018**, *4*, 98–107. [CrossRef]
98. van Nocker, S.; Ludwig, P. The WD-repeat protein superfamily in Arabidopsis: Conservation and divergence in structure and function. *BMC Genom.* **2003**, *4*, 50. [CrossRef]
99. Mishra, A.K.; Puranik, S.; Prasad, M. Structure and regulatory networks of WD40 protein in plants. *J. Plant Biochem. Biotechnol.* **2012**, *21*, 32–39. [CrossRef]
100. Pan, Y.; Shi, H. Stabilizing the Transcription Factors by E3 Ligase COP1. *Trends Plant Sci.* **2017**, *22*, 999–1001. [CrossRef] [PubMed]
101. Laubinger, S.; Fittinghoff, K.; Hoecker, U. The SPA Quartet: A Family of WD-Repeat Proteins with a Central Role in Suppression of Photomorphogenesis in Arabidopsis. *Plant Cell* **2004**, *16*, 2293–2306. [CrossRef] [PubMed]

102. Lloyd, A.; Brockman, A.; Aguirre, L.; Campbell, A.; Bean, A.; Cantero, A.; Gonzalez, A. Advances in the MYB–bHLH–WD Repeat (MBW) Pigment Regulatory Model: Addition of a WRKY Factor and Co-option of an Anthocyanin MYB for Betalain Regulation. *Plant Cell Physiol.* **2017**, *58*, 1431–1441. [CrossRef]

103. Li, S. Transcriptional control of flavonoid biosynthesis. *Plant Signal. Behav.* **2014**, *9*, e27522. [CrossRef] [PubMed]

104. Jin, J.; Tian, F.; Yang, D.-C.; Meng, Y.-Q.; Kong, L.; Luo, J.; Gao, G. PlantTFDB 4.0: Toward a central hub for transcription factors and regulatory interactions in plants. *Nucleic Acids Res.* **2017**, *45*, D1040–D1045. [CrossRef] [PubMed]

105. Yu, L.; Sun, Y.; Zhang, X.; Chen, M.; Wu, T.; Zhang, J.; Xing, Y.; Tian, J.; Yao, Y. ROS1 promotes low temperature-induced anthocyanin accumulation in apple by demethylating the promoter of anthocyanin-associated genes. *Hortic. Res.* **2022**, *9*, uhac007. [CrossRef] [PubMed]

106. Pečinka, P.; Bohálová, N.; Volná, A.; Kundrátová, K.; Brázda, V.; Bartas, M. Analysis of G-Quadruplex-Forming Sequences in Drought Stress-Responsive Genes, and Synthesis Genes of Phenolic Compounds in Arabidopsis thaliana. *Life* **2023**, *13*, 199. [CrossRef] [PubMed]

107. Volná, A.; Bartas, M.; Pečinka, P.; Špunda, V.; Červeň, J. What Do We Know about Barley miRNAs? *Int. J. Mol. Sci.* **2022**, *23*, 14755. [CrossRef] [PubMed]

108. Sharma, D.; Tiwari, M.; Pandey, A.; Bhatia, C.; Sharma, A.; Trivedi, P.K. MicroRNA858 Is a Potential Regulator of Phenylpropanoid Pathway and Plant Development. *Plant Physiol.* **2016**, *171*, 944–959. [CrossRef] [PubMed]

109. Yang, F.; Cai, J.; Yang, Y.; Liu, Z. Overexpression of microRNA828 reduces anthocyanin accumulation in Arabidopsis. *Plant Cell Tiss. Organ Cult.* **2013**, *115*, 159–167. [CrossRef]

110. Jia, X.; Shen, J.; Liu, H.; Li, F.; Ding, N.; Gao, C.; Pattanaik, S.; Patra, B.; Li, R.; Yuan, L. Small tandem target mimic-mediated blockage of microRNA858 induces anthocyanin accumulation in tomato. *Planta* **2015**, *242*, 283–293. [CrossRef]

111. Luo, Q.-J.; Mittal, A.; Jia, F.; Rock, C.D. An autoregulatory feedback loop involving PAP1 and TAS4 in response to sugars in Arabidopsis. *Plant Mol. Biol.* **2012**, *80*, 117–129. [CrossRef]

112. Pech, R.; Volná, A.; Hunt, L.; Bartas, M.; Červeň, J.; Pečinka, P.; Špunda, V.; Nezval, J. Regulation of Phenolic Compound Production by Light Varying in Spectral Quality and Total Irradiance. *Int. J. Mol. Sci.* **2022**, *23*, 6533. [CrossRef]

MDPI AG
Grosspeteranlage 5
4052 Basel
Switzerland
Tel.: +41 61 683 77 34

International Journal of Molecular Sciences Editorial Office
E-mail: ijms@mdpi.com
www.mdpi.com/journal/ijms